The propagation of radio waves

The propagation of radio waves

THE THEORY OF RADIO WAVES OF LOW POWER IN THE IONOSPHERE AND MAGNETOSPHERE

K.G. BUDDEN, F.R.S.

Fellow of St John's College and Emeritus Reader in Physics in the University of Cambridge

CAMBRIDGE UNIVERSITY PRESS

Cambridge

New York New Rochelle

Melbourne Sydney

CAMBRIDGE UNIVERSITY PRESS
Cambridge, New York, Melbourne, Madrid, Cape Town, Singapore,
São Paulo, Delhi, Dubai, Tokyo, Mexico City

Cambridge University Press
The Edinburgh Building, Cambridge CB2 8RU, UK

Published in the United States of America by
Cambridge University Press, New York

www.cambridge.org
Information on this title: www.cambridge.org/9780521369527

First published 1985
First paperback edition (with corrections) 1988

A catalogue record for this publication is available from the British Library

Library of Congress catalogue card number: 84–28498

ISBN 978-0-521-25461-2 Hardback
ISBN 978-0-521-36952-7 Paperback

Contents

	Preface	xiii
1	***The ionosphere and magnetosphere***	**1**
1.1	The earth's atmosphere	1
1.2	Plane and spherical radio waves	3
1.3	Waves in ion plasmas	3
1.4	Relation to other kinds of wave propagation	5
1.5	Height dependence of electron concentration: the Chapman layer	7
1.6	Collision frequencies	10
1.7	Observations of the ionosphere	12
1.8	The structure of the ionosphere	14
1.9	The magnetosphere	17
1.10	Disturbances of the ionosphere and magnetosphere	19
	Problems 1	20
2	***The basic equations***	**22**
2.1	Units and symbols	22
2.2	Definitions of electric intensity ***e*** and magnetic intensity ***h***	23
2.3	The current density ***j*** and the electric polarisation ***p***	24
2.4	The electric displacement ***d*** and magnetic induction ***b***	25
2.5	Harmonic waves and complex vectors	25
2.6	Maxwell's equations	27
2.7	Cartesian coordinate system	28
2.8	Progressive plane waves	29
2.9	Plane waves in free space	31
2.10	The notation $\boldsymbol{\mathscr{H}}$ and ***H***	31
2.11	The power input to the plasma from a radio wave	32
2.12	The flow of energy. The Poynting vector	33
2.13	Complex refractive index	34
2.14	Evanescent waves	35
2.15	Inhomogeneous plane waves	35
	Problems 2	37

3	***The constitutive relations***	**38**
3.1	Introduction	38
3.2	Free, undamped electrons	38
3.3	The Lorentz polarisation term	40
3.4	Electron collisions. Damping of the motion	42
3.5	The Debye length	44
3.6	Effect of the earth's magnetic field on the motion of electrons	45
3.7	Effect of the magnetic field of the wave on the motion of electrons	46
3.8	Electric neutrality of the plasma. Plasma oscillations	48
3.9	The susceptibility matrix	49
3.10	Complex principal axes	50
3.11	Properties of principal axis elements of the permittivity. Effect of ions	53
3.12	Collisions. The Sen–Wyller formulae	57
3.13	Electron–electron collisions. Electron–ion collisions	61
	Problems 3	62
4	***Magnetoionic theory 1. Polarisation and refractive index***	**66**
4.1	Plane wave and homogeneous plasma	66
4.2	Isotropic plasma	67
4.3	Anisotropic plasma. The wave polarisation	68
4.4	Properties of the polarisation equation	71
4.5	Alternative measure of the polarisation. Axis ratio and tilt angle	73
4.6	Refractive index 1. The dispersion relation	74
4.7	Longitudinal component of electric polarisation and electric field	75
4.8	The flow of energy for a progressive wave in a magnetoplasma	76
4.9	Refractive index 2. Alternative derivations and formulae	77
4.10	Zeros and infinity of refractive index. Equal refractive indices	79
4.11	Dependence of refractive index on electron concentration 1. $Y < 1$	80
4.12	Dependence of refractive index on electron concentration 2. $Y > 1$	84
4.13	Effect of collisions included	86
4.14	The transition collision frequency	86
4.15	The terms 'ordinary' and 'extraordinary'	88
4.16	Dependence of refractive index on electron concentration 3. Collisions allowed for	89
4.17	Approximations for refractive indices and wave polarisations	94
	Problems 4	99
5	***Magnetoionic theory 2. Rays and group velocity***	**103**
5.1	Introduction	103
5.2	Refractive index surfaces	104
5.3	The ray. Ray surfaces	108
5.4	Properties of ray surfaces	111
5.5	Crystal optics	113
5.6	Classification of refractive index and ray surfaces. C.M.A. type diagrams	116
5.7	Dependence of refractive index on frequency	124
5.8	Group velocity	128

5.9 Properties of the group velocity 131
5.10 Effect of electron collisions on the group refractive index 137
Problems 5 139

6 *Stratified media. The Booker quartic* **141**
6.1 Introduction 141
6.2 The variable q 142
6.3 The Booker quartic. Derivation 144
6.4 Some properties of the Booker quartic 146
6.5 Some special cases of the Booker quartic 151
6.6 The discriminant of the Booker quartic 152
6.7 The Booker quartic for east–west and west–east propagation 153
6.8 The Booker quartic for north–south and south–north propagation 155
6.9 Effect of electron collisions on solutions of the Booker quartic 160
6.10 The electromagnetic fields 162
Problems 6 163

7 *Slowly varying medium. The W.K.B. solutions* **165**
7.1 Introduction 165
7.2 The differential equations for an isotropic ionosphere 166
7.3 The phase memory concept 167
7.4 Loss-free medium. Constancy of energy flow 168
7.5 W.K.B. solutions 169
7.6 The W.K.B. method 170
7.7 Discrete strata 172
7.8 Coupling between upgoing and downgoing waves 174
7.9 Liouville method and Schwarzian derivative 175
7.10 Conditions for the validity of the W.K.B. solutions 177
7.11 Properties of the W.K.B. solutions 178
7.12 W.K.B. solutions for oblique incidence and vertical polarisation 180
7.13 Differential equations for anisotropic ionosphere 181
7.14 Matrix theory 183
7.15 W.K.B. solutions for anisotropic ionosphere 187
7.16 The matrices $\mathbf{S}$ and $\mathbf{S}^{-1}$ 190
7.17 W.K.B. solutions for vertical incidence 192
7.18 Ray theory and 'full wave' theory 193
7.19 The reflection coefficient 194
Problems 7 195

8 *The Airy integral function and the Stokes phenomenon* **197**
8.1 Introduction 197
8.2 Linear height distribution of electron concentration and isolated zero of q 197
8.3 The differential equation for horizontal polarisation and oblique incidence 199
8.4 The Stokes differential equation 200
8.5 Qualitative discussion of the solutions of the Stokes equation 201
8.6 Solutions of the Stokes equation expressed as contour integrals 202

8.7 Solutions of the Stokes equation expressed as Bessel functions 204
8.8 Tables of the Airy integral functions. Computing 205
8.9 Zeros and turning points of Ai(ζ) and Bi(ζ) 205
8.10 The W.K.B. solutions of the Stokes equation 206
8.11 Asymptotic expansions 206
8.12 The Stokes phenomenon of the 'discontinuity of the constants' 209
8.13 Stokes lines and anti-Stokes lines 209
8.14 The Stokes diagram 211
8.15 Definition of the Stokes multiplier 212
8.16 Furry's derivation of the Stokes multipliers for the Stokes equation 212
8.17 The range of validity of asymptotic approximations 213
8.18 The choice of a fundamental system of solutions of the Stokes equation 214
8.19 Connection formulae, or circuit relations 215
8.20 Stratified ionosphere. Uniform approximation 216
8.21 The phase integral method for reflection 218
8.22 The intensity of light near a caustic 223
Problems 8 227

9 ***Integration by steepest descents*** **229**
9.1 Introduction 229
9.2 Some properties of complex variables and complex functions 229
9.3 Saddle points 231
9.4 Error integrals and Fresnel integrals 233
9.5 Contour maps 237
9.6 Integration by the method of steepest descents 238
9.7 Application to solutions of the Stokes equation 241
9.8 The method of stationary phase 247
9.9 Higher order approximation in steepest descents 249
9.10 Double steepest descents 251
Problems 9 252

10 ***Ray tracing in a loss-free stratified medium*** **254**
10.1 Introduction 254
10.2 The ray path 255
10.3 Wave packets 257
10.4 Equations of the ray path 259
10.5 The reversibility of the path 261
10.6 The reflection of a wave packet 261
10.7 An example of a ray path at oblique incidence 263
10.8 Poeverlein's construction 264
10.9 Propagation in magnetic meridian plane. The 'Spitze' 266
10.10 Ray paths for the extraordinary ray when $Y < 1$ 268
10.11 Extraordinary ray when $Y > 1$ 270
10.12 Lateral deviation at vertical incidence 273
10.13 Lateral deviation for propagation from (magnetic) east to west or west to east 275
10.14 Lateral deviation in the general case 276
10.15 Calculation of attenuation, using the Booker quartic 277

10.16 Phase path. Group or equivalent path 278
10.17 Ray pencils 279
10.18 Caustics 281
10.19 The field where the rays are horizontal 285
10.20 The field near a caustic surface 286
10.21 Cusps. Catastrophes 287
10.22 The skip distance 288
10.23 Edge focusing 289
Problems 10 292

11 *Reflection and transmission coefficients* 295
11.1 Introduction 295
11.2 The reference level for reflection coefficients 295
11.3 The reference level for transmission coefficients 297
11.4 The four reflection coefficients and the four transmission coefficients 298
11.5 Reflection and transmission coefficient matrices 299
11.6 Alternative forms of the reflection coefficient matrix 300
11.7 Wave impedance and admittance 301
11.8 Reflection at a sharp boundary 1. Isotropic plasma 304
11.9 Properties of the Fresnel formulae 306
11.10 Reflection at a sharp boundary 2. Anisotropic plasma 307
11.11 Normal incidence. Anisotropic plasma with free space below it 308
11.12 Normal incidence. Two anisotropic plasmas 309
11.13 Probing the ionosphere by the method of partial reflection 311
11.14 Spherical waves. Choice of reference level 313
11.15 Goos–Hänchen shifts for radio waves 315
11.16 The shape of a pulse of radio waves 320
Problems 11 325

12 *Ray theory results for isotropic ionosphere* 328
12.1 Introduction 328
12.2 Vertically incident pulses 329
12.3 Effect of collisions on phase height $h(f)$ and equivalent height $h'(f)$ 330
12.4 Equivalent height for a parabolic height distribution of electron concentration 332
12.5 Effect of a 'ledge' in the electron height distribution 336
12.6 The calculation of electron concentration $N(z)$, from $h'(f)$ 337
12.7 Ray paths at oblique incidence 342
12.8 Equivalent path P' at oblique incidence 345
12.9 Maximum usable frequency, MUF 348
12.10 The forecasting of MUF 349
12.11 Martyn's theorem for attenuation of radio waves 352
Problems 12 353

13 *Ray theory results for anisotropic plasmas* 356
13.1 Introduction 356
13.2 Reflection levels and penetration frequencies 357

13.3	The calculation of equivalent height, $h'(f)$	359
13.4	Ionograms	362
13.5	Topside sounding	365
13.6	The calculation of electron concentration $N(z)$ from $h'(f)$	368
13.7	Faraday rotation	372
13.8	Whistlers	376
13.9	Ion cyclotron whistlers	380
13.10	Absorption, non-deviative and deviative	391
13.11	Wave interaction 1. General description	393
13.12	Wave interaction 2. Outline of theory	395
13.13	Wave interaction 3. Kinetic theory	397
	Problems 13	398
14	***General ray tracing***	**400**
14.1	Introduction	400
14.2	The eikonal function	402
14.3	The canonical equations for a ray path	403
14.4	Properties of the canonical equations	405
14.5	The Haselgrove form of the equations	407
14.6	Fermat's principle	409
14.7	Equivalent path and absorption	412
14.8	Signal intensity in ray pencils	414
14.9	Complex rays. A simple example	417
14.10	Real pseudo rays	422
14.11	Complex rays in stratified isotropic media	424
14.12	Complex rays in anisotropic absorbing media	425
14.13	Reciprocity and nonreciprocity with rays 1. The aerial systems	428
14.14	Reciprocity and nonreciprocity with rays 2. The electric and magnetic fields	431
14.15	Reciprocity and nonreciprocity with rays 3. Applications	433
	Problems 14	436
15	***Full wave solutions for isotropic ionosphere***	**438**
15.1	Introduction	438
15.2	Linear electron height distribution	439
15.3	Reflection at a discontinuity of gradient	441
15.4	Piecewise linear models	443
15.5	Vertical polarisation at oblique incidence 1. Introductory theory	446
15.6	Vertical polarisation 2. Fields near zero of refractive index	448
15.7	Vertical polarisation 3. Reflection coefficient	450
15.8	Exponential electron height distribution	453
15.9	Parabolic electron height distribution 1. Phase integrals	456
15.10	Parabolic electron height distribution 2. Full wave solutions	460
15.11	Parabolic electron height distribution 3. Equivalent height of reflection	464
15.12	The differential equations of theoretical physics	466
15.13	The hypergeometric equation and its circuit relations	467
15.14	Epstein distributions	470

15.15	Reflection and transmission coefficients for Epstein layers	472
15.16	Ionosphere with gradual boundary	473
15.17	The 'sech^2' distribution	475
15.18	Other electron height distributions	476
15.19	Collisions. Booker's theorem	477
	Problems 15	479
16	***Coupled wave equations***	**480**
16.1	Introduction	480
16.2	First order coupled equations	482
16.3	Coupled equations near a coupling point	485
16.4	Application to vertical incidence	489
16.5	Coupling and reflection points in the ionosphere	492
16.6	Critical coupling	495
16.7	Phase integral method for coupling	499
16.8	The Z-trace	502
16.9	Additional memory	505
16.10	Second order coupled equations	507
16.11	Försterling's coupled equations for vertical incidence	509
16.12	Properties of the coupling parameter ψ	510
16.13	The method of 'variation of parameters'	514
16.14	The coupling echo	517
	Problems 16	518
17	***Coalescence of coupling points***	**520**
17.1	Introduction	520
17.2	Further matrix theory	521
17.3	Coalescence of the first kind, C1	523
17.4	Coalescence of the second kind, C2	525
17.5	Ion cyclotron whistlers	530
17.6	Radio windows 1. Coalescence	532
17.7	Radio windows 2. Formulae for the transparency	534
17.8	Radio windows 3. Complex rays	540
17.9	Radio windows 4. The second window	542
17.10	Limiting polarisation 1. Statement of the problem	543
17.11	Limiting polarisation 2. Theory	546
18	***Full wave methods for anisotropic stratified media***	**550**
18.1	Introduction	550
18.2	Integration methods	552
18.3	Alternative methods 1. Discrete strata	553
18.4	Alternative methods 2. Vacuum modes	556
18.5	Alternative methods 3. The matrizant	558
18.6	Starting solutions at a great height	560
18.7	Finding the reflection coefficient	562
18.8	Allowance for the earth's curvature	563

18.9 Admittance matrix as dependent variable 566
18.10 Other forms, and extensions of the differential equations 569
18.11 Numerical swamping 574
18.12 Reciprocity 576
18.13 Resonance 579
Problems 18 581

19 ***Applications of full wave methods*** **583**
19.1 Introduction 583
19.2 Vertical incidence and vertical magnetic field 584
19.3 Oblique incidence and vertical magnetic field 587
19.4 Resonance and barriers 591
19.5 Isolated resonance 593
19.6 Resonance tunnelling 596
19.7 Inversion of ionospheric reflection measurements 602
19.8 Full wave solutions at higher frequencies 606
Answers to problems 609
Bibliography 612
Index of definitions of the more important symbols 643
Subject and name index 652

Preface

Radio waves in the ionosphere was first published in 1961. Since then there have been many major advances that affect the study of radio propagation, and three in particular. The first is the use of space vehicles and rockets, which have enabled the top side of the ionosphere to be studied and have revealed the earth's magnetosphere and magnetotail. The second is the advance of the subject 'plasma physics' which has transformed our knowledge of the physical processes in ion plasmas. The third is the use of computers, large and small, which, by removing the need for laborious calculations, have changed our attitude to theoretical results, even though they are sometimes wrongly used as a substitute for clear physical thinking. Moreover some important textbooks have appeared in the intervening 20 years. They are too numerous all to be mentioned, but four have a special bearing on the subjects of this book. These are: T.H. Stix (1962), *Theory of plasma waves*; K. Rawer and K. Suchy (1967), *Radio observations of the ionosphere*; K. Davies (1969), *Ionospheric radio waves*; and V.L. Ginzburg (1970), *Propagation of electromagnetic waves in plasma.*

Some of the topics mentioned in *Radio waves in the ionosphere*, here abbreviated to RWI, are of importance in other branches of physics. Two in particular have formed the subject of numerous mathematical papers. These are (a) 'W.K.B. solutions', and (b) the study of ordinary linear homogeneous differential equations of the second order.

W.K.B. solutions have a special physical significance connected with ray theory. They are approximations that can be extended in many different ways to give higher accuracy. This is of the greatest mathematical interest, though sometimes complicated. The practical radio propagation engineer can use computed full wave solutions and rarely or never needs extensions of the W.K.B. solutions.

The differential equations governing radio propagation in a stratified ionosphere can often be reduced, over a small range of height, to equations of the second order, and a few such equations must be studied to reveal some important physical ideas.

But attempts are often made to use the same second order equation for the whole ionosphere. Many authors have tried to find all possible electron height distribution functions that can be used in this way to give solutions expressible in terms of known mathematical functions. This has led to some new results of mathematical interest, but they are only rarely of use to the propagation engineer.

For these reasons, unfortunately, the word 'mathematical' has come to mean something that the practical engineer can often disregard. The present book is concerned with ideas of theoretical physics that are useful in practical problems. The word 'mathematical' is avoided as far as possible.

This book covers roughly the same topics as RWI, but most of them have had to be treated rather differently, and some account has been added of phenomena at low frequencies where the effect of heavy ions is important. This is therefore a new book with a new title. It is not just a new edition of RWI. It is hoped that it will serve both as a textbook, for those comparatively new to the subject, and as a reference book for more experienced readers. The stress throughout is on understanding the basic physical principles, since the radio engineer who really understands the physics is well equipped to tackle practical problems.

The reader is assumed to be familiar with calculus, the theory of complex variables, vectors including the differential vector operators, matrices, and electromagnetic theory as far as Maxwell's equations and Poynting's theorem.

It is inevitable that many important topics are omitted. In most of the book it is assumed that the ionosphere is horizontally stratified, but this is only approximately true, since horizontal variations and irregularities play an important part in radio propagation. The extensive work on waves in media with irregularities is not covered here. Nor is there any discussion of reflections or scattering of radio waves from cylindrical structures such as meteor trails. The statistical mechanics of an ion plasma, and phenomena such as wave interaction, which depend on electron temperature, are only briefly referred to. In the theory of the propagation of radio waves between widely separated points near the earth's surface, the space between the earth and the ionosphere is often treated as a wave guide. This is a very large topic, beyond the scope of this book. A most powerful recent radio method of studying the ionosphere from the ground is the technique of incoherent scatter or Thomson scatter. Any account of this needs some detailed treatment of the physics of warm plasmas, and it is too large a subject to include here.

At the end of this book is a long list of references both to text books and to published papers. But no attempt has been made to provide a complete list of all important papers on the subject, since they are far too numerous. References have two main purposes. First they show the reader where to obtain more information on some topic. Second they are used to indicate who was the originator of some particular concept. This second purpose is sometimes difficult to achieve, especially when a new idea is worked out by several authors at about the same time. It must often happen that a reference cites an author who gives the clearest treatment even though he may not have been the first to publish the idea. An outstanding example of this is the dispersion relation for a cold electron magnetoplasma, often called the 'Appleton–Hartree' formula. Readers interested in the general history of ionospheric physics might find it useful to study the papers by Professor C.S. Gillmor (1976, 1978, 1981, 1982).

In the preface to the English edition of his book on electromagnetic waves in plasmas, already mentioned, Professor V.L. Ginzburg wrote: '[In the book] ··· by Dr. K.G. Budden, *Radio Waves in the Ionosphere,* ··· out of 243 references to the literature, there is only one to Soviet work, and the date is 1948'. My reply to this just criticism is that, regrettably, I do not read the Russian language and this applies also to most readers outside the USSR. In 1961 when RWI was published, translations of Russian works were not readily available. The first purpose of the references was therefore best achieved by papers in western European languages. The position now is very different and there are many excellent English translations of textbooks and papers by Soviet authors. The second historical purpose of the references entails a rather tedious search of the journals, which is the task of the historian rather than the physicist. This book contains more references to Soviet work than RWI but I cannot claim that all the important ones are included. The interested reader is referred to Professor Ginzburg's own book.

I am most grateful to those readers of RWI who took the time and trouble to send in lists of errors. Particular thanks are accorded to Dr J.R. Wilts of the US Naval Electronics Laboratory Center, San Diego, California, who supplied the longest list.

I continue to be deeply indebted to all those colleagues who were mentioned in the preface of RWI. Since then others, including my research students, have contributed many useful ideas. Particular mention should be made of the group at Leicester University under Professor T.B. Jones, of workers at the Appleton laboratories (formerly at Slough, now part of the Rutherford–Appleton Laboratory at Chilton), and of workers at the British Antarctic Survey, Cambridge, especially Dr D. Jones. Special thanks are accorded to Professor J. Heading, University College of Wales, Aberystwyth, who, in correspondence over several years, has helped me to avoid rushing too far from the straight and narrow path into those complex regions where the angels fear to tread. Professor K. Suchy, Department of Theoretical Physics, Düsseldorf, West Germany, has, in his published work, contributed many important ideas and techniques to the subject. It was a pleasure to welcome him to Cambridge for the year 1980–1 as an Overseas Visiting Fellow at St John's College.

Since 1979 I have been in regular correspondence with Professor Ya.L. Al'pert, working in Moscow and we have collaborated in a research project (see Al'pert, Budden, Moiseyev and Stott, 1983). The work has contributed many valuable ideas to this book. Professor Al'pert has been refused permission to leave Russia and has been working in comparative isolation. His own recent book (Al'pert, 1983) is a very full treatise and amplifies many of the topics discussed here.

But I am indebted most of all to the two colleagues who were mainly instrumental in getting me interested in radio waves. Professor Henry G. Booker's contributions to the theory are perhaps the most important and far-reaching of all. As an undergraduate I had the good fortune to attend his lectures in Cambridge. He has been kind enough to allow me to make free use of his extensive published work. Mr J.A. Ratcliffe was my *de facto* supervisor when I was a research student, and he was head of the Radio Section of the Cavendish Laboratory until 1960. Without his help and guidance since 1936, this book could not have been written. His own books *Magnetoionic theory* (1959) and *An introduction to the ionosphere and magnetosphere* (1972) are models of clear exposition and should be read by all readers of the present book.

Finally I must express my thanks to my wife who typed the whole book including the formulae, and to Mrs I. Tabecka for her skilful work on the diagrams.

Cavendish Laboratory,
Cambridge, 1984 K.G. Budden

1

The ionosphere and magnetosphere

1.1. The earth's atmosphere

This book is mainly concerned with the effect on radio wave propagation of the ionised regions of the earth's atmosphere. Near the ground the air is almost unionised and its electrical conductivity is negligibly small, because the ionising radiations have all been absorbed at greater heights. When any part of it is in equilibrium, its state is controlled by the earth's gravitational field so that it is a horizontally stratified system. Although it is never in complete equilibrium, gravity has a powerful controlling effect up to about 1000 km from the ground.

The molecules of the neutral atmosphere have an electric polarisability which means that the refractive index for radio waves is very slightly greater than unity, about 1.000 26, near the ground. The water vapour also affects the refractive index. Thus the neutral air can very slightly refract radio waves. This can lead to important effects in radio propagation. For example in stable meteorological conditions a duct can form near the surface of the sea, acting as a wave guide in which high frequency radio waves can propagate to great distances (see Booker and Walkinshaw, 1946; Brekhovskikh, 1960; Budden, 1961b; Wait, 1962). Spatial irregularities of the refractive index of the air can cause scintillation of radio signals, and also scattering which can be used to achieve radio propagation beyond the horizon (Booker and Gordon, 1950). These effects are beyond the scope of this book. It is here assumed that the non-conducting atmosphere below the ionosphere is the same as free space, with refractive index unity.

Incoming ultra-violet radiation and X-rays from the sun ionise the atmosphere. At great heights of the order 1000 km or more the medium is almost fully ionised, but so tenuous that the ion and electron concentrations are small. At lower levels there is more gas to be ionised so that the radiation is more strongly absorbed and the ion concentrations are greater. At still lower levels the radiation has been used up so that the degree of ionisation is again small. The region where the ion and electron concentrations are greatest is the ionosphere. The height where maximum ionisation occurs depends on the absorption coefficient of the air for the ionising constituent of the sun's radiation, and on the mechanism by which ions and electrons are removed. Different ionising constituents give maxima at different heights. Thus there are several different ionospheric layers of ionisation. The principal ones are the E-layer at about 100 km and the F-layer at about 300 km. Their formation and structure is discussed in more detail in §§ 1.5–1.8. In the F-layer the ion and electron concentrations are only about 10^{-5} of the concentration of neutral particles, and in the E-layer the figure is about 10^{-11}. Thus gravity, acting mainly on the neutral particles, exerts a dominating control on the hydrostatic configuration of the medium. This means that the ionised layers tend to be horizontally stratified.

Above about 1000 km the atmosphere is a fully ionised ion plasma and therefore a relatively good electrical conductor. In these conditions the earth's magnetic field has a strong effect on the structure of the medium. For a perfectly conducting fluid in a magnetic field of induction $\boldsymbol{B}$ there is an effective stress like a pressure $\frac{1}{2}\boldsymbol{B}^2/\mu_0$ in directions perpendicular to $\boldsymbol{B}$ and a tension of the same magnitude in the direction of $\boldsymbol{B}$; see, for example, Ferraro and Plumpton (1966, § 1.7). The earth's magnetic field $\boldsymbol{B}$ is approximately that of a magnetic dipole near the earth's centre. Thus, at a fixed magnetic latitude, its magnitude B is roughly proportional to the inverse cube of the distance from the centre. If the plasma were perfectly conducting, the magnetic force per unit volume would be the spatial gradient of $\frac{1}{2}B^2/\mu_0$. The gravitational force per unit volume is ρg where ρ is the density and g is the gravitational acceleration. It can be shown that these two forces are about equal at a height of 300–400 km. Here the conductivity, though large, is not large enough to allow the magnetic forces to have their full effect. Gravity still exerts a strong control up to heights of the order of 1000 km. Above this the atmosphere is not horizontally stratified but its structure is controlled by the magnetic field. This region is the magnetosphere and it extends to about 14 R_e on the sunlit side of the earth. The unit R_e is one earth radius, about 6400 km. Distances in R_e are usually measured from the centre of the earth.

When the magnetic field within a conducting medium tends to change, the changing flux density induces electric fields which cause currents to flow, and these currents themselves give a magnetic field. Lenz's law shows that this new field must have a direction which opposes the change that originated it. Thus the effect of the conductivity is to oppose any change of the magnetic field. The rate at which a magnetic field can change depends on the conductivity and on the size of the conductor. (See problem 1.1.) Within a perfect conductor the magnetic induction $\boldsymbol{B}$ cannot change at all. This effect can be observed with metals which become superconducting when cooled to very low temperatures, and it led to the phrase 'freezing in of the magnetic field'. The magnetosphere is like a blob of conducting fluid surrounding the earth, with the earth's magnetic field frozen into it.

The shape of the magnetosphere is greatly influenced by the gas that streams out from the sun, known as the solar wind. This gas is itself a fully ionised plasma and therefore a good conductor. Thus the earth's magnetic field cannot quickly penetrate into it, because it is 'frozen out'. But the solar wind distorts the magnetosphere so that it is drawn out into a long tail, the 'magneto-tail' on the side remote from the sun. This extends out to a distance greater than the radius of the moon's orbit.

Further details of the structure of the ionosphere and magnetosphere are given in §§ 1.5–1.8 and 1.9 respectively.

1.2. Plane and spherical radio waves

Radio waves travelling from a transmitter to a receiver near the earth's surface may take one of several possible paths. A wave may travel over the earth's surface, and it is then known as the ground wave. The earth has an imperfectly conducting curved surface. The problem for a vertical electric dipole transmitter near a plane surface was first propounded and studied by Sommerfeld (1909), and it has intrigued many mathematical writers ever since. One of the best treatments of it is that of Baños (1966). This topic is within the field of surface waves and guided waves and is beyond the scope of this book.

Another wave may travel up to the ionosphere, be reflected there, and return to the receiver. It is with a single reflection of this kind that the present book is largely concerned. The wave comes from a source of small dimensions so that the wave front is approximately spherical, but by the time it reaches the ionosphere the radius of curvature is so large that the wave can often be treated as plane. The approximation involved is examined in §§ 11.4, 11.5 and the error is shown to be negligible except in a few special cases. Similarly the ionospheric layers are curved because of the earth's curvature, but in most problems this curvature can be neglected.

Radio receivers are often used in rockets and satellites within the ionosphere or magnetosphere. It is thus important to be able to study the spatial distribution of the electromagnetic field within these media. At the higher frequencies this can be done by ray tracing (chs. 10, 14). At lower frequencies it is necessary to use the computed solutions of the governing differential equations (chs. 18, 19).

1.3. Waves in ion plasmas

The ionosphere and magnetosphere are composed of ionised gases known as ion plasmas. The negative ions are almost entirely electrons. The plasma must, on the average, be electrically neutral. If any space charge is present it can be shown (§ 3.8) that it must oscillate with the plasma frequency. This is the phenomenon of plasma oscillations. These are damped out with a time constant of a few milliseconds or less. But a radio wave in the plasma modifies the ion and electron concentrations, so that there is, in general, a space charge oscillating with the frequency of the wave.

An ion is about 2000 to 60 000 times more massive than an electron. Thus at the frequencies used for radio communication, that is, greater than a few kilohertz, the range of movement of an ion caused by the electric field of a radio wave is smaller than that of an electron, by about the same factor. This means that the ions can for most purposes be ignored. The effective particles in the plasma are the electrons. The positive ions simply provide a background to keep the plasma electrically neutral on the average. In the lowest parts of the ionosphere there can also be massive negative ions formed by the attachment of electrons to air molecules, and these play some part in the very complicated chemistry of the ionisation process (Rishbeth and Garriott, 1969). But electrons are easily detached from these ions so that their concentration cannot be large. Radio propagation effects attributable to massive negative ions have never been observed as far as the author knows.

The earth's magnetic field has a dominating effect on radio wave propagation in the atmospheric plasmas, and they are therefore sometimes called magnetoplasmas. Because of this they are doubly refracting for radio waves, so that in many problems it is necessary to study a differential equation of the fourth order. The two refractive indices depend strongly on the direction of the wave normal (chs. 4, 5) and the medium is said to be anisotropic. For some purposes, however, it is convenient to neglect the earth's magnetic field, so that the differential equation is only of the second order. There is then only one refractive index, and it is independent of the direction of the wave normal. The medium is said to be isotropic. This is useful for establishing some physical principles.

The neglect of the massive ions is permissible only for frequencies which greatly exceed the ion gyro-frequencies. The gyro- or cyclotron frequency for an ion with charge e and mass m_i in a magnetic field B is $eB/2\pi m_i$. For protons in the upper ionosphere it is about 400 Hz. In the lower ionosphere there are few protons and the gyro-frequencies of the heavier ions are smaller still. Thus the ions need only be considered for frequencies low in the audio range. This is well below the range used for radio communication, but the naturally occurring radio signals such as whistlers and chorus extend down to this range. They are important in studies of the ionosphere and magnetosphere. Some account of the effect of massive ions at these frequencies in the upper ionosphere and magnetosphere is therefore included (§§ 3.11, 13.8, 13.9, 17.5).

The ionosphere is quite hot, 800 K in the E-region and 1000–2000 K in the F-region. Thus an electron has a random thermal velocity of the order of $10^5\ \mathrm{m\,s^{-1}}$, whereas a typical radio wave imparts to it an additional velocity of order only $10^3\ \mathrm{m\,s^{-1}}$. In spite of this it is found, rather surprisingly, that the thermal motions can be neglected. This is called the 'cold plasma' treatment. It can be shown to be justified by using a full kinetic treatment of the problem with the Boltzmann–Vlasov equations. This is done in the numerous books on plasma physics (see, for example:

Stix, 1962; Clemmow and Dougherty, 1969; Shkarofsky, Johnston and Bachynski, 1966).

When the temperature of the plasma is allowed for it is found that, besides the electromagnetic (radio) waves, other types of wave, known collectively as plasma waves, can propagate in the plasma. The main ones are the electron plasma wave also known as the Langmuir wave, and the ion-acoustic wave. These waves are, in general, much more heavily attenuated than radio waves and do not travel over great distances, so they cannot be used for communication. A transition from radio waves to plasma waves sometimes occurs (§ 19.5), but in most radio problems the plasma waves can be ignored. They are however of dominating importance in the theory of the technique of incoherent scatter (§ 1.7).

For most purposes in this book, therefore, the temperature of the electrons can be neglected. But it does have to be considered in the study of electron collisions, § 3.12, and in wave interaction, §§ 13.11–13.13.

Lorentz (1909) studied the theory of the effect of electron collisions on an electromagnetic wave in a plasma and concluded that they have the same effect as a retarding force proportional to the velocity imparted to an electron by the field of the wave. In most of this book the simple Lorentz form of the retarding force is used. It has been realised, however, that the average collision frequency ν of an electron may depend on its total velocity v, including the random thermal velocity which is often very large. If ν were independent of velocity, the Lorentz result would still be true but as a result of laboratory measurements it is now thought that ν is proportional to v^2. This has led to a more complicated treatment of collisions (Sen and Wyller, 1960) described in § 3.12. Collisions can sometimes be neglected, for example in the magnetosphere where ν is very small, or at high frequencies for which 'ray theory' methods are used. Collision damping is important in the lower part of the ionosphere (below about 300 km) and at low frequencies where, usually, the wavelength is so great that ray theory methods are inapplicable and a 'full wave' treatment must be used. Chs. 15–19 are devoted to this. It is often convenient to ignore collisions when describing and classifying the properties of the waves.

The refractive indices n of the two waves in a cold magnetoplasma are strongly frequency dependent and we say that the medium is highly dispersive for these waves. If collisions are ignored the squares of the refractive indices are real. If n^2 is negative the wave is said to be evanescent (§ 2.14) and the frequencies where n^2 is zero are called 'cut-off' frequencies. One n^2 can be infinite at certain frequencies and this is called 'resonance'.

1.4. Relation to other kinds of wave propagation

The theory of radio wave propagation in the terrestrial plasmas is closely related to other branches of physics that deal with wave propagation in media whose properties vary from place to place. For example in wave mechanics a study is made of the propagation of electron waves in a potential field. The variation of potential is analogous to the variation of the square of the refractive index for radio waves. But the potential is real in nearly all problems of wave mechanics, so that there is nothing analogous to the damping forces which, for radio waves, lead to a complex value of the squared refractive index.

For electron waves in a crystal the wave velocity in general depends on the direction of the wave normal so that the 'medium' is anisotropic. There is an important analogy between the direction and magnitude of the particle velocity, and the ray direction and group velocity of a radio wave in a magnetoplasma. The waves of wave mechanics are highly dispersive and can show cut-off and resonance. Some of the material of chs. 5, 7, 8 and 15 is of great importance when applied to wave mechanics.

In oceanography a study is made of the propagation of sound waves in the ocean and its bed. The system is often horizontally stratified because the temperature, salinity and other properties depend on depth. For sound waves the ocean bed can behave like an inverted ionosphere, and many of the results of chs. 7–19 can be applied to it. But it can nearly always be assumed that sound waves are not dispersive. A plane sound wave does not show cut-off or resonance.

In a solid three kinds of elastic wave can propagate, two transverse and one longitudinal, so that the medium might be considered to be triply refracting. These waves are studied in seismology (Bullen and Bolt, 1985; Jeffreys, 1976), and in crystal acoustics (Musgrave, 1970). The interest has been mainly in propagation through homogeneous solids and in reflection and transmission at boundaries. Seismic waves in continuously varying stratified media have been treated by Kennett (1983). Matrix equations very similar to the set (7.80) that forms the basic equations for much of the theory in this book were used for elastic waves by Gilbert and Backus (1966), and for acoustic waves in the ocean by Abramovici (1968). Seismic waves in anisotropic solids have been discussed by Stoneley (1955, 1963) and Crampin (1970) who were mainly interested in surface (guided) waves. Plane seismic waves, like sound waves, are not dispersive and do not show cut-off or resonance.

Finally mention must be made of sound waves of very long period, hours or days, in the earth's atmosphere. The wavelength is of the order of thousands of kilometres so that the waves extend through a large part of the atmosphere. They are strongly influenced by gravity and are known as acoustic-gravity waves or planetary waves. Most of their energy is well below the ionosphere where the air density is large. They cause movements of the atmosphere, including movements of the ionospheric layers, that can be observed from their effect on reflected radio waves. Thus radio probing of the ionosphere is an important method of studying these waves. One type of wave of this kind is known as a travelling ionospheric disturbance, abbreviated TID; see Munro (1953), Munro and Heisler (1956a, b), Murata (1974).

1.5. Height dependence of electron concentration: the Chapman layer

Before the reflecting properties of the ionosphere for radio waves can be calculated, it is necessary to know how the electron concentration $N(z)$ depends on height z above the ground. To study this, some assumption must be made about how the ionospheric layers are formed. A most important contribution to this problem was made by Chapman (1931a, b, 1939) who derived an expression for $N(z)$ now known as the Chapman law. The full theory of the formation of ionospheric layers has been refined and extended by Chapman and others, and is beyond the scope of this book (see Chapman, 1931a, b, 1939; Rishbeth and Garriott, 1969; Hargreaves, 1979). Only the simplest version of the Chapman theory is given here.

Assume that the air is constant in composition and at a constant temperature T. Then the air density ρ at height z is

$$\rho = \rho_0 \exp(-z/H) \tag{1.1}$$

where ρ_0 is the density at the ground, $H = KT/Mg$, K is Boltzmann's constant, M is the mean mass of a molecule and g is the gravitational acceleration. The curvature of the earth is neglected and g is assumed constant. H is called the 'scale height' of the atmosphere and is approximately 10 km at the ground. The sun's radiation enters the atmosphere at an angle χ from the zenith. Let the mass absorption coefficient of the air for the radiation be σ, and assume that the rate of production of electrons, q, is proportional to the rate of absorption of radiation per unit volume. Let the flux of energy in the incident radiation be I_0 outside the earth's atmosphere, and I at a height z. The energy flux decreases as the radiation is absorbed on passing down through the atmosphere, so that

$$\mathrm{d}I = I\sigma\rho \sec\chi \,\mathrm{d}z. \tag{1.2}$$

This is combined with (1.1) and integrated to give

$$I = I_0 \exp(-\sigma\rho_0 H \sec\chi\, \mathrm{e}^{-z/H}). \tag{1.3}$$

Now let $z_0 = H\ln(\sigma\rho_0 H)$. Then (1.3) gives

$$I = I_0 \exp[-\sec\chi \exp\{-(z-z_0)/H\}]. \tag{1.4}$$

The rate of absorption of energy per unit volume at height z is $\cos\chi\,(\mathrm{d}I/\mathrm{d}z)$ and since this is proportional to the rate of production of electrons q, we have from (1.4)

$$q = q_0 \exp[1-(z-z_0)/H - \sec\chi \exp\{-(z-z_0)/H\}] \tag{1.5}$$

where $q_0 = I_0/\mathrm{e}H$ is a constant (here e is the exponential). The maximum value of q is $q_0 \cos\chi$ and it occurs where $(z-z_0)/H = \ln(\cos\chi)$, so that q_0 is the maximum rate of electron production when $\chi = 0$.

Next it is necessary to consider how electrons are removed. The process is very complicated (Rishbeth and Garriott, 1969) but in the lower part of the ionosphere below about 250 km, the effect is the same as if the electrons simply recombined with positive ions. Assume that the only ions present are electrons and positive ions. Then they both have concentration N, and the rate of removal of electrons is αN^2 per unit volume, where α is a constant called the recombination coefficient. The variation of N with time t is then given by

$$\mathrm{d}N/\mathrm{d}t = q - \alpha N^2. \tag{1.6}$$

It was at one time believed that in the F-region the electrons were removed by attachment to neutral particles so that $\mathrm{d}N/\mathrm{d}t = q - \beta N$ and the constant β was called the attachment coefficient. But it is now known that at these greater heights the processes are much more complicated and electron diffusion plays an important part (see Rishbeth and Garriott, 1969). Here we shall study (1.6) which applies fairly

Fig. 1.1. Dependence of electron concentration N on height z according to the simple Chapman theory for a flat earth, for various values of the sun's zenith angle χ.

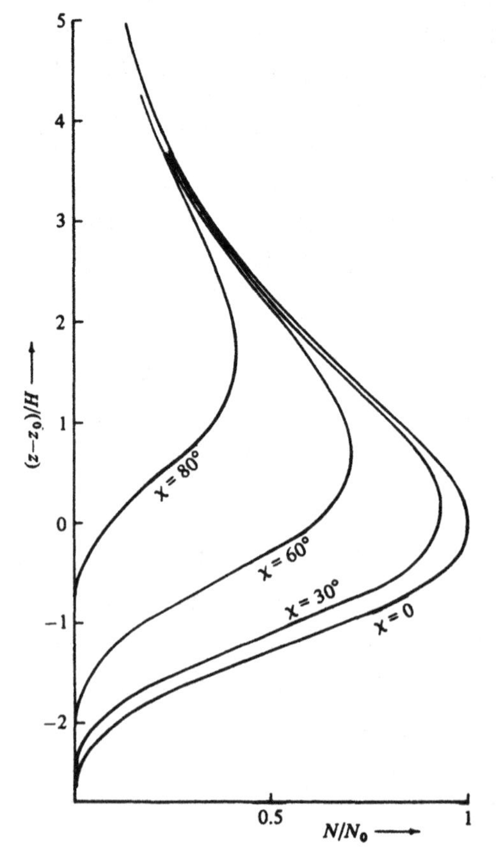

well in the E-region and below. Now α is of the order $10^{-13}\text{m}^3\text{s}^{-1}$ and the time constant $(\alpha N)^{-1}$ is about 10 min. Thus the processes of formation and removal of electrons come into equilibrium in times of this order. As a first approximation, therefore, $\mathrm{d}N/\mathrm{d}t$ can be neglected compared with the other terms in (1.6) and

$$N = (q/\alpha)^{\frac{1}{2}}. \tag{1.7}$$

This, with (1.5), gives

$$N = N_0 \exp [\![\tfrac{1}{2}\{1 - (z - z_0)/H - \sec\chi \exp[-(z - z_0)/H]\}]\!]. \tag{1.8}$$

This expression will be called the 'Chapman law'. It is based on a very much oversimplified picture of the actual mechanism of production and removal of electrons. For example it assumes that α is independent of height. But it does give a guide as to how the electron concentration might vary with height. In fig. 1.1 the expression (1.8) is plotted against height z for various values of χ. It is seen that N has a maximum value $N_{\mathrm{m}} = N_0(\cos\chi)^{\frac{1}{2}} = (I_0 \cos\chi/\mathrm{e}\alpha H)^{\frac{1}{2}}$ at the height $z = z_{\mathrm{m}} = z_0 + H \ln(\sec\chi) = H \ln(\sigma\rho_0 H \sec\chi)$. It falls off quite steeply below this, and less steeply above it. The maximum concentration N_m depends on H, I_0 and χ but not on the absorption coefficient σ. The height of the maximum z_m depends on H, σ and χ but not on the intensity I_0 of the radiation.

This property suggests that several different ionospheric layers occur because the absorption coefficient σ has different values for the various components of the ionising radiation from the sun. Thus the F-layer is produced largely by ultra-violet radiation in the wavelength range very roughly 17 to 80 nm. The E and D layers are produced largely by X-rays in the wavelength range very roughly 0.1 to 17 nm. (1 nm = 10 ångström units). The solar radiation contains some narrow spectral lines including Lyman-α and Lyman-β which come from atomic hydrogen, and are in the ultra violet range. It happens that Lyman-α is not strongly absorbed at F region heights but makes an important contribution to the ionisation in the D region.

Many radio observations have been used to test the Chapman law (1.8) and have shown good agreement when the ionosphere is not disturbed. For example the penetration frequency (§ 13.2) of a Chapman layer is proportional to $N_{\mathrm{m}}^{\frac{1}{2}}$, that is to $(\cos\chi)^{\frac{1}{4}}$, and this has been tested (Appleton, 1937, § 2). In the lowest part of the D-region, the dependence on χ of the height where a fixed small N occurs was studied by Budden, Ratcliffe and Wilkes (1939) and found to agree with (1.8).

An alternative form of (1.8) is got by taking

$$\zeta = (z - z_{\mathrm{m}})/H = (z - z_0)/H - \ln(\sec\chi) \tag{1.9}$$

so that ζH is height measured from the level of maximum N. Then

$$N = N_{\mathrm{m}} \exp\{\tfrac{1}{2}(1 - \zeta - \mathrm{e}^{-\zeta})\}, \tag{1.10}$$

which shows that the 'shape' of a Chapman layer is independent of the sun's zenith angle χ. The curvature of the curve of $N(z)$ at its maximum is $\frac{1}{2}N_{\mathrm{m}}/H^2$. This is the same as the curvature at the apex of a parabolic layer (see § 12.4) whose 'half thickness' is $2H$.

The electrons' height distribution function $N(z)$ appears in the differential equation which has to be solved to find the reflecting properties of the ionosphere for radio waves. This is usually done by numerical computing (chs. 18, 19). But it is also useful to select small ranges of z and use approximate simple expressions for $N(z)$, which permit the differential equation to be reduced to a simple standard form whose solutions have well known properties. For example, it is often permissible to assume, for a small range of z, that $N(z)$ is a linear function. This case is of the greatest importance and is the subject of ch. 8 and §§15.2, 15.4. Near a maximum of $N(z)$ the linear law is not satisfactory but it is useful to approximate $N(z)$ by a parabola (§§ 15.9–15.11). In the lower part of a Chapman layer it is sometimes useful to treat $N(z)$ as an exponential function (§ 15.8).

1.6. Collision frequencies

Radio waves in the ionosphere and magnetosphere are subject to some attenuation because the motions of the electrons and ions are damped through collisions with other particles. In the simplest treatment, originally used by Lorentz (1909), it is assumed that the damping of the electron motions occurs because of a retarding force $-m\nu V$, (3.11) below, where m is the electron's mass and V is the part of its velocity associated with the ordered motion imposed on it by the electromagnetic field. This is discussed in more detail in §§ 3.4, 3.12. Here ν is an effective collision frequency for electrons. A similar treatment can be used for the collisions of ions. In the formulae for the refractive indices, ν appears in the ratio $Z = \nu/\omega$, where $\omega = 2\pi f$ is the angular frequency. At high frequencies Z is small and can often be neglected.

One way of finding ν is to make measurements of the attenuation of radio waves. The attenuation occurs over a range of height z and care is needed in finding the particular height to which the measured ν applies. There are techniques for dealing with this and careful measurements have given useful values. See, for example, Davies (1969, ch. 6), *Journal of Atmospheric and Terrestrial Physics* (1962, special volume).

Measurements of this type include the use of rockets going through the lower parts of the ionosphere (Kane, 1962), and studies of the effect of electron collisions on the ionospheric reflection coefficients for radio waves of very low frequency (Deeks, 1966a, b, § 4).

There are many other methods of measuring ν. Two radio methods of great importance are measurements of partial reflections, § 11.3, and measurements of wave interaction, §§ 13.11–13.13. Measurements have also been made on plasmas in the laboratory under conditions that are intended to simulate the ionosphere. See, for example, Huxley and Crompton (1974), Phelps and Pack (1959). For a full list of references see Shkarofsky, Johnston and Bachynski (1966, pp. 177–87, 199–201).

The subject of collisions between particles in a plasma is complicated and is studied in books on plasma physics, for example Clemmow and Dougherty (1969, esp. § 9.5), Shkarofsky, Johnston and Bachynski (1966, chs. 4, 5). Any collision frequency is a statistical average with respect to the velocities of the colliding particles and the directions of their paths before and after the collision. In a full theory it is found that in different applications the averages must be taken in different ways, so that many different collision frequencies have to be defined (Suchy and Rawer, 1971; Suchy, 1974b). The one that is needed for studying the refractive indices for radio waves is the one that is used in the theory of transport phenomena, such as thermal conductivity and electrical conductivity. It is called the 'collision frequency for momentum transfer'. For a summary of the main points of the theory see Budden (1965).

In the Lorentz formula (3.11), V is the ordered part of the electron's velocity. It is superimposed on the random thermal velocity v which is much larger (see problem 3.8). The collision frequency $\nu(v)$ for electrons must depend on v because, if it did not do so, the phenomenon of wave interaction, §§13.11–13.13, would not occur. It was at one time supposed that all electrons have the same mean free path λ_e that is independent of v. Then $\nu(v) = v/\lambda_e$ is proportional to v. As a result of laboratory measurements, however, it is often assumed that $\nu(v)$ is proportional to v^2. This subject is dealt with in § 3.12. The simple Lorentz formula (3.11) uses an effective value of ν, often written ν_{eff}, that is independent of v, and it must be related to the average value ν_{av} of $\nu(v)$. It is usually assumed that, in the ionosphere, the electron velocities have a Maxwellian distribution and this can be used to find ν_{av}. It is thus shown in § 3.12 that if $\nu(v) \propto v^2$ then $\nu_{eff} = \frac{5}{3}\nu_{av}$ (3.87), and if $\nu(v) \propto v$ then $\nu_{eff} = \frac{4}{3}\nu_{av}$ (3.93).

The propagation properties of radio waves are not very sensitive to the exact value of ν so that it is not necessary to know ν with high precision. Throughout this book the Lorentz formula (3.11) is used and the symbol ν is used for the effective collision frequency, ν_{eff} above. It is useful to have some idea of how this ν depends on height z in the ionosphere. Below about 100 km electron collisions are predominantly with neutral molecules and ν is proportional to their concentration. In an isothermal atmosphere of constant composition it would be expected that ν is proportional to the pressure, thus

$$\nu = \nu_0 \exp(-z/H) \tag{1.11}$$

where H is the scale height defined in § 1.5 and ν_0 is a constant. In practice H takes different values at different heights because of height variations of temperature and composition, and this law can only be expected to hold over ranges of z so small that H may be taken as constant. Throughout the ionosphere the concentration of positive ions is much less than that of neutral particles but the forces between electrons and ions are of longer range. Consequently at 100 km and above, the contribution to ν from ions must be allowed for; see §§ 3.4, 3.13.

A useful summary of the factors that affect the value of ν was made by Nicolet (1953, 1959). Based on this and other data, a curve was given by Budden (1961a, fig. 1.2) to show how ν depends on height z according to the best estimates then available. A new version of this curve is given in fig. 1.2. It has been revised slightly in the light of data from measurements of partial reflection, § 11.3, and of wave interaction §§ 13.11–13.13, but it differs very little from the 1961 version.

The value of ν does change slightly with time of day and with season, and the height distribution of ν depends on latitude. These changes are, however, very small and for many purposes they can be ignored. In this respect ν is very different from the electron concentration $N(z)$; see § 1.8. Fig. 1.2 applies for temperate latitudes but gives a useful first estimate of ν for any hour and season and latitude.

Because changes of ν affect the propagation of radio waves far less than changes of electron concentration $N(z)$; see §1.8. Fig. 1.2 applies for temperate latitudes but ranges of height z. This is done for many of the topics discussed in this book.

1.7. Observations of the ionosphere

Up to about 1960 our knowledge of the structure of the ionosphere was obtained almost entirely by radio methods, using signals from transmitters near the ground. The principal technique is known as ionosonde. Short pulses of radio waves are sent vertically upwards and return to a receiver on the ground after reflection in the ionosphere. The time of travel t is measured. This is the same as the technique used in

Fig. 1.2. Dependence of the effective electron collision frequency ν upon height z.

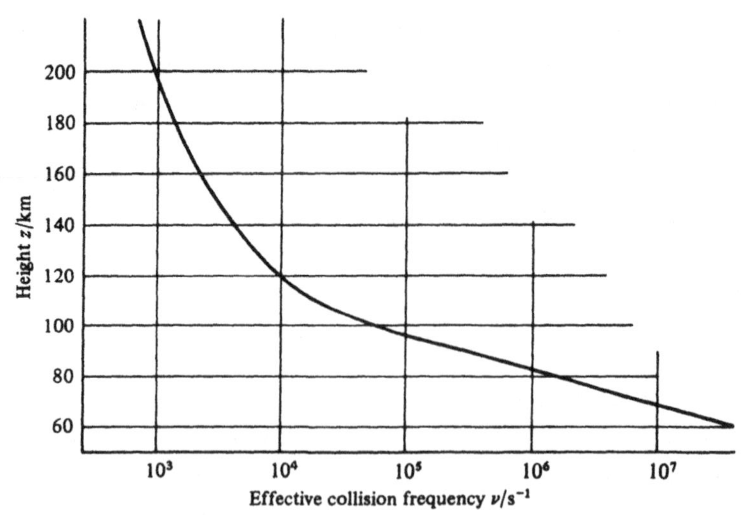

radar. It was first applied to the ionosphere by Breit and Tuve (1926). If the pulse travelled entirely in free space it would have to go to a height $h' = \frac{1}{2}ct$ called the equivalent height of reflection. This is strongly frequency dependent. The reflected pulse usually has two components, the ordinary and extraordinary waves with different values of h'.

Ionosonde equipments are commercially available and are used at observing stations all over the world. They produce curves, or records on magnetic tape, of equivalent height versus frequency $h'(f)$, which can be processed by a computer to calculate $N(z)$ (§ 13.6). Some ionosonde equipments now supply the data in digital form (see Bibl and Reinisch, 1978; Downing, 1979), and some include a computer for calculating $N(z)$, (see Reinisch and Huang, 1983). There are more elaborate equipments in which the ionosonde is linked to a computer that controls its frequency and aerial systems. For example an equipment known as the Dynasonde (Wright and Pitteway, 1979) gives information about polarisation, angle of arrival, phase, and other features of the reflected pulses.

Other ground based radio techniques include the following: reflection of waves of very low frequency (VLF), chs. 18, 19; partial reflection, § 11.13; wave interaction or cross modulation, §§ 13.11–13.13; absorption, §§ 1.6, 13.10. The device known as a riometer measures the absorption of the radio signal that comes in from outside the earth, mainly from the galaxy. This method was first suggested by Little and Leinbach (1959). For details see Rawer and Suchy (1967, § 46), Davies (1969, § 6.6.2). There are other methods, not described in this book, including Doppler shift of reflected radio waves (Davies, Watts and Zacharisen, 1962; Jones, T.B., 1964; Jones and Wand, 1965); refraction of radio signals received at the ground from satellites (Titheridge, 1964).

From about 1960 onwards rockets and satellites began to be used and it became possible for the first time to study the ionosphere at heights above the maximum of $N(z)$ in the F-layer, that is the region called the top side of the ionosphere. Two main methods were used. The first was the use of a measuring device or probe carried on the vehicle and designed to measure the electron concentration $N(z)$ and the electron temperature T_e. Such probes are called Langmuir probes. One difficulty is that the plasma being studied is disturbed by the vehicle moving through it (see Boyd, 1968; Willmore, 1970). The second method is the use of the ionosonde technique with a transmitter and receiver in a satellite above the F-region. This is called topside sounding. It is similar to the use of an ionosonde on the ground, but it shows many new features (§ 13.5). A third technique of some interest is the measurement at the ground of the Faraday rotation of a linearly polarised radio signal emitted by a satellite and travelling right down through the ionosphere; see § 13.7.

In recent years a new probing technique has been coming into use. This is called incoherent scatter or Thomson scatter. It was suggested by Gordon (1958) and first successfully used by Bowles (1958, 1961). A radio beam of high frequency (typically 40–400 MHz) and high power is sent into the ionosphere from a transmitter on the ground. The plasma scatters some of the radiation and this is detected by a receiver also on the ground. The observed scattering region can be selected by adjustment of the highly directive transmitting and receiving aerials, or by time gating of the receiver.

The received scattered signal is extremely weak so that sophisticated signal integration methods have to be used in the receiver. But the method works well and gives the electron and ion concentrations and temperatures. For an account of it see Dougherty and Farley (1960), Fejer (1960, 1961), Evans (1969), Beynon (1974), Bauer (1975). The theory of the method involves a detailed study of the physics of warm plasmas and is beyond the scope of this book.

For a survey and assessment of various methods of measuring $N(z)$ and other properties of the ionosphere, see Booker and Smith (1970).

1.8. The structure of the ionosphere

A great deal of knowledge has now accrued about the structure of the ionosphere and how it depends on time of day, on season, on latitude and longitude and on solar activity. This has been incorporated into a computer program known as the International Reference Ionosphere (abbreviation: IRI), which predicts the height distribution of electron concentration $N(z)$ and of other features such as ion concentration and electron, ion and neutral temperatures. It is published jointly by the International Union of Radio Science (URSI) and the Committee on Space Research (COSPAR). For details see Rawer (1981, 1984). It is available in the computer libraries of some of the larger computers. It does not give concentrations of neutral particles or pressure (these can be obtained from the International Reference Atmosphere, CIRA 1972), and it does not give electron collision frequencies (see § 1.6). The version of IRI used here for fig. 1.3 is dated 1978. The program is revised from time to time and the user is advised to check that he has the latest version. A new version was announced by URSI in March 1983.

The ionosphere consists of two main layers known as E and F. For the E-layer, the electron concentration $N(z)$ has a maximum at a height of 100 to 110 km. Above this $N(z)$ is only slightly less for a height range extending right up to the base of the F-layer. Thus the E- and F-layers are not really distinct. For many purposes it is useful to assume that the part of the E-layer below its maximum is a Chapman layer having $N(z)$ given by (1.8), with $H = 10\,\text{km}$, $z_0 = 115\,\text{km}$ and $N_0 = 2.8 \times 10^{11}\,\text{m}^{-3}$. This means that the penetration frequency (§ 13.2) is $4.7 \times (\cos \chi)^{\frac{1}{4}}$ MHz. The actual behaviour of the E-layer is very much more complicated than this. At night the E-layer has a penetration frequency of the order of 0.5 MHz, corresponding to an electron concentration of $3 \times 10^9\,\text{m}^{-3}$.

A thin but intensely ionised layer sometimes appears at a height near 115 km. Its occurrence is sporadic and it is therefore known as 'sporadic E', or E_S. It nearly always gives partial reflection and penetration of radio waves over a wide frequency range. There are several different types of E_S. One form is believed to have horizontal variations of electron concentration N so that radio waves can penetrate through where N is small, but are reflected from where N is large. Another form is believed to be a uniform thin layer so that partial penetration and reflection occurs because of the thinness. For the theory of this see § 19.8. An E_S layer can be so strong that its effective penetration frequency is comparable with or greater than the penetration frequency of the F-layer, and then radio observations of the F-layer from the ground are made difficult or impossible.

Several mechanisms may contribute to the formation of E_S. It is believed that wind shears, that is a large vertical gradient of horizontal wind speed, play a major part. Ionisation caused by meteorites may have some effect. In auroral latitudes precipitation of charged particles may contribute. For a full account see Smith, E. K. and Matsushita (1962), Rawer and Suchy (1967, § 40), Rishbeth and Garriott (1969, § 6.3).

The F-layer is more heavily ionised, with a maximum of $N(z)$ in the range 200 to 400 km. Its diurnal and seasonal variations are more complicated than those of the E-layer. During daylight the function $N(z)$ often shows a bulge below the maximum (see fig. 1.3). This is known as the F1-layer, and it may occasionally attain an actual maximum. The main maximum above it is known as the F2-layer. It has been suggested that the formation of the whole F-layer is caused mainly by a single ionising agency which would form a simple layer like a Chapman layer if the attachment coefficient β (§ 1.5) were constant at all heights. Bradbury (1938) suggested that β decreases as the height z increases, so that the maximum of $N(z)$ in the F2-layer arises, not from a fast rate of production but because of a slow rate of removal of electrons. But this alone would not be adequate to explain the F2 maximum. For a detailed discussion see Rishbeth and Garriott (1969, § 3.62 and Ch. IV).

In chs. 10 and 12 some account is given of methods of finding the function $N(z)$ in the F-layer from radio observations using the ionosonde technique. This work shows that, at the maximum, the curvature of the $N(z)$ curve is about the same as that at the apex of a parabola of half thickness 100 km, or that of a Chapman layer with scale height about 50 km. The F-layer is therefore much thicker than the E-layer, as well as more heavily ionised. In temperate latitudes the penetration frequency of the F-layer ranges from about 2 MHz at night to 8 MHz in a summer day. These correspond to electron concentrations of 5×10^{10} and $8 \times 10^{11} \mathrm{m}^{-3}$ respectively.

In the daytime there is another layer below the E-layer, known as the D-layer with a maximum of $N(z)$ near $z = 80$ km. This overlaps the lower part of the E-layer so that although $N(z)$ is enhanced, it is still a monotonically increasing function as shown in the example of fig. 1.3. The first experimental study of the D-layer came from radio observations at very low frequencies (VLF, of the order of 16 kHz). Here the wavelength is so great that interpretation of the observations is less direct than at high frequencies. One of the most important features of the full wave theory given in this book is that it has helped to disentangle the numerous radio observations at VLF. The procedure most often used is to adopt a trial function $N(z)$ and work out its reflecting properties. If these do not agree with observations, some other $N(z)$ is

Fig. 1.3. Examples of the dependence of electron concentration on height in the ionosphere as computed from the International Reference Ionosphere. The continuous curves are for June and the broken curves for January. R is the sunspot number. $R = 100$ is typical of a year near sunspot maximum and $R = 10$ of a year near sunspot minimum.

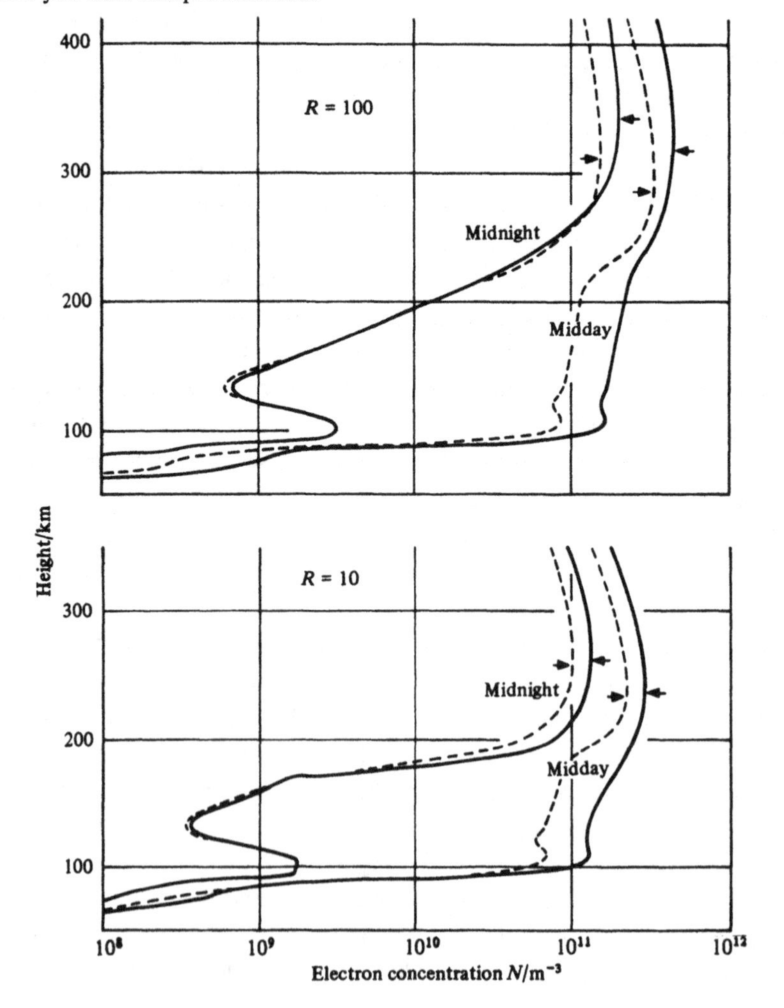

tried until a satisfactory result is obtained. The process is laborious because of the variability of the radio observations with time of day and season, and with ionospheric disturbances, but it has been successfully used by Deeks (1966a, b). The problem of finding $N(z)$ from the observations, directly without a trial and error process, is known as 'inversion', and techniques for doing it have been given by Backus and Gilbert (1967, 1968, 1970). Several workers have used this kind of technique for radio observations. For details see § 19.7.

1.9. The magnetosphere

For many purposes the earth's magnetic field can be regarded as that of a magnetic dipole at the centre of the earth. For greater accuracy the dipole is displaced from the centre and quadrupole and higher multipole terms must be included but these refinements are not needed in this book. If the earth were in a vacuum the strength of its magnetic field would decrease as the inverse cube of the distance from the centre. This is approximately true in the ionosphere but does not hold beyond about three or four earth's radii from the centre because outside the earth there is a fully ionised plasma. This is a stream of gas coming out from the sun, called the 'solar wind'. Near the earth's orbit its outward speed is about $200\,\mathrm{km\,s^{-1}}$ and it consists mainly of electrons and protons each with a concentration of about $10^7\,\mathrm{m^{-3}}$. This plasma has a strong effect on the outer parts of the earth's magnetic field.

When a magnetic field is present within an electric conductor, it cannot be changed quickly because any change induces e.m.f.'s that cause currents to flow, whose magnetic fields oppose the attempted change. This effect is well known for metals made superconducting by cooling them to the temperature of liquid helium. Any change of magnetic field in a superconductor is impossible and the field is said to be 'frozen in'. In a conducting body that is not superconducting, the magnetic field can change but only within a time period of order $\tau \approx \mu_0 \sigma l^2$ that depends on the electric conductivity σ and the linear dimension l of the body. Here μ_0 is the permeability of a vacuum; § 2.1. (See problem 1.1.) For the same reasons, a magnetic field cannot quickly penetrate into a conducting body. For the solar wind σ is of order $0.03\,\mathrm{S\,m^{-1}}$. If l is of order two or three times the earth's radius $\mathrm{R_e}$, this gives τ of order several months.

The earth's magnetic field therefore cannot penetrate into this advancing plasma stream. On the sunward side of the earth there is a surface at about 10 to 12 $\mathrm{R_e}$ from the centre where the earth's field ends abruptly. It is called the 'magnetopause' and surrounds the earth as sketched in fig. 1.4. Beyond it is a region of turbulence called the 'magnetosheath' and outside this on the sunward side is a shock front on which the solar wind impinges. The region inside the magnetopause is the magnetosphere, and the plasma here moves with the earth and is regarded as part of the earth's upper atmosphere. The solar wind plasma flows round the outside of the magnetosphere.

On the side of the earth remote from the sun the earth's magnetic field and the magnetosphere extend back into a region called the 'magnetotail', that goes out to a distance of 100 R_e or more.

At heights greater than about 1000 km above the earth's surface the plasma in the magnetosphere is a good conductor and its distribution is strongly influenced by the magnetic field, as explained in § 1.1. The effect of gravity here is small. Observations have shown that the magnetospheric plasma is in two main parts. The inner part is continuous with the upper ionosphere and is called the 'plasmasphere'. It extends out to a surface which, at the equator on the sunward side, is at about 4 to 5 R_e and it follows the magnetic lines of force; see fig. 1.4. It is called the 'plasmapause'. Just inside it the electron concentration is very roughly $10^8 m^{-3}$ and just outside it about $10^7 m^{-3}$. Thus there is a fairly sudden decrease when the plasmapause is crossed. This was first discovered in observations of whistlers, § 13.8, Carpenter (1963, 1968, 1983), Carpenter and Park (1973), and has since been confirmed and studied by measurements in space vehicles.

The figures given in this section and in fig. 1.4 for the dimensions of the magnetosphere and for the electron and ion concentrations are very approximate. In

Fig. 1.4. Sketch showing cross section of the magnetosphere by a plane containing the magnetic axis MM and the sun–earth line, at a time when this plane is at right angles to the ecliptic. The continuous lines with arrows are magnetic lines of force. The short arrows show the direction of flow of the solar wind. The letters CC denote polar cusp regions. For explanation of these and of the neutral sheet see references in text.

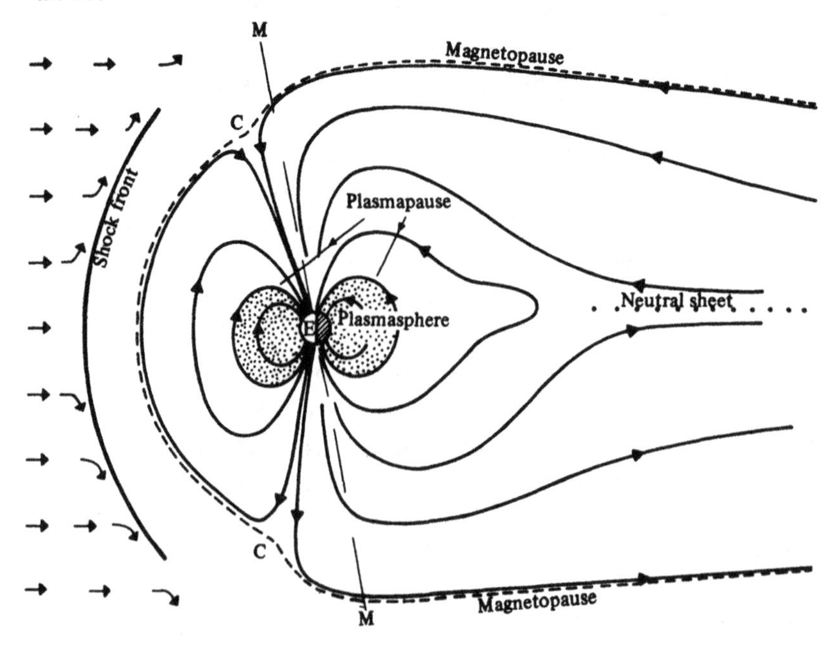

practice they vary with time and are strongly influenced by radiation from the sun, by fluctuations in the solar wind and by the magnetic fields frozen into the solar wind plasma.

For a most illuminating discussion of the processes that result from the solar wind impinging on the earth's magnetic field see Ratcliffe (1972). For a description of the magnetosphere and further references see Al'pert (1983, especially tables 2.1, 2.2), Ratcliffe (1970, 1972), Kennel, Lanzerotti and Parker (1979).

1.10. Disturbances of the ionosphere and magnetosphere

The earth's atmosphere is subjected to many disturbing influences that affect the ionosphere and magnetosphere. These do not play any essential part in the theoretical topics discussed in this book, but some of them should be mentioned briefly. Most of them originate in events on the sun and are more frequent and intense near times of maximum sunspot number. A good general account is given by Ratcliffe (1970, 1972).

An eruption on the sun's surface, known as a solar flare, can occur very suddenly. It is seen as a brightening of a small area, usually near a sunspot, and it results in intense emission of ultraviolet radiation that causes ionisation in the D-region of the ionosphere. This is known as a 'sudden ionospheric disturbance' (SID). It occurs sometimes within a few seconds and results in strong absorption of radio waves, so that communication circuits that rely on ionospheric reflection are interrupted. This is referred to as a 'radio fadeout', or 'short wave fadeout'. There are accompanying disturbances in the earth's magnetic field, and marked changes in the propagation of radio waves of very low frequency (10 to 100 kHz) which are strongly influenced by the D-region. The ionisation decreases again and returns to its normal value in a time of order one hour.

Energetic charged particles are emitted from the sun, particularly from solar flares. Those that enter the magnetosphere are deviated by the earth's magnetic field. They move on helical paths in the direction of the lines of force but cannot easily move across the field. They can therefore reach low altitudes only at high latitudes, that is in polar regions. Here they produce increased ionisation in the D-region. This results in increased absorption of radio waves, known as 'polar cap absorption' (PCA). The particles that cause this are mainly protons.

A prolonged and complicated disturbance can be produced by changes in the solar wind. There can sometimes be an increase in the concentration and the velocity of the particles in the wind, accompanied by a magnetic field frozen in to the moving plasma. When this strikes the earth it causes changes in the earth's magnetic field, known as a 'magnetic storm', changes in the ionosphere, known as an 'ionosphere storm', and changes in the magnetosphere. The current flows in the ionosphere, magnetosphere and magnetotail are altered and the size of the plasmasphere becomes much reduced. The modified solar wind can reach the earth suddenly and this gives a 'sudden commencement'. The subsequent disturbance evolves in a complicated way and may last several days. The main effect in the ionosphere is an increase of ionisation. This may be in the D-region and then leads to increased absorption of radio waves. There may also be an increase at higher levels including the production of sporadic E-layers (§§ 1.8, 19.8).

Meteors and meteorites can produce enhanced ionisation, mainly in the D- and E-regions. The ionisation in a meteor trail has a cylindrical structure and its effect on radio waves is an interesting theoretical problem but beyond the scope of this book.

Eclipses of the sun disturb the production process of ionisation in the upper atmosphere. Radio observations during eclipses have given valuable information about the complicated chemical processes involved in the formation and decay of ionised layers. See Beynon and Brown (1956), Rishbeth and Garriott (1969, § 6.2).

The ionosphere can be influenced by large scale motions of the earth's atmosphere. Both solar and lunar tidal motions have been observed. Acoustic gravity waves or planetary waves have already been mentioned; see end of § 1.4. Waves of this type were generated in the atmosphere by a large chemical explosion at Flixborough, England, in 1974 and their effect on the ionosphere was studied by radio methods; see Jones and Spracklen (1974).

Many of the results of radio observations can be explained by assuming that the ionosphere is horizontally stratified, that is that the electron concentration and collision frequency are functions only of the height z. This assumption is adopted throughout this book, except in § 11.13. There are, however, many other observations that show that there must also be horizontal variations of electron concentration. These are often irregular and are subject both to steady movements and to random change. Scattering from ionospheric irregularities can be used for some forms of communication link (Booker, 1959). One of their effects is to cause fading or scintillation of radio signals. These permit measurements to be made of steady motions, usually called winds, in the ionosphere. For reviews see Briggs and Spencer (1954), Ratcliffe (1956), Yeh and Liu (1982).

This subject has become part of the much wider study of waves in random media, with applications in radio astronomy and in ocean acoustics. See Uscinski (1977).

PROBLEMS 1

1.1. Derive a partial differential equation relating the variation with space and time coordinates of a magnetic field in a homogeneous isotropic solid of high electrical conductivity, when the displacement current is negligible compared with the conduction current, and obtain a solution for it in one case of physical interest. In what circumstances is the neglect of the displacement current permissible?

Estimate a rough order of magnitude for the decay time of a magnetic field in (a) a copper sphere of diameter 1 m and resistivity $1.7 \times 10^{-8}\,\Omega$m, (b) a cloud of interplanetary plasma of diameter 6×10^4 km containing electrons with concentration $10^8 \mathrm{m}^{-3}$ if the average number of collisions made by each electron is 10 per second.

[Natural Sciences Tripos. Part II Physics, 1968.]

2
The basic equations

2.1. Units and symbols

Electromagnetic theory is often based on the results, verifiable directly or indirectly by experiments, that the force F_q between two electric charges q_1, q_2, and the force F_m between two magnetic poles m_1, m_2, are given respectively by

$$F_q = \frac{q_1 q_2}{4\pi\varepsilon_0 r^2}, \quad F_m = \frac{4\pi m_1 m_2}{\mu_0 r^2} \tag{2.1}$$

where r is the separation in each case. Some authors argue that free magnetic poles do not exist, and base their definitions of magnetic quantities on the forces between currents, or between moving charges. Others use a different version of the second equation (2.1) in which μ_0 is in the numerator, thereby implying a different meaning for the term 'magnetic pole'. These considerations are unimportant in the ionosphere and magnetosphere whose magnetic permeabilities are always taken to be the same as in a vacuum.

The constants ε_0 and μ_0 are known, respectively, as the permittivity and the permeability of a vacuum. Their numerical values depend on which system of units is being used. The factor 4π appears in (2.1) for rationalised systems of units. When ε_0 and μ_0 are left in a formula as symbols, that formula is valid in any self-consistent system of units. It is often implied that such formulae are restricted to the International System of Units (SI), but this is incorrect. Equations (2.1), and all other formulae in this book that do not use specific numerical values, are valid in any self-consistent rationalised units, including SI units. The recommendations in *Quantities, Units and Symbols* (The Royal Society, 1975) are followed as far as possible. The symbol c is used for the speed of electromagnetic waves in a vacuum. It is occasionally used with other meanings when there should be no risk of confusion. In the International System $\mu_0 = 4\pi \times 10^{-7}$ and $\varepsilon_0 = 1/(\mu_0 c^2)$.

Some mathematical works still use Gaussian units, which are often defined by saying that electrical quantities, including the electric intensity $\boldsymbol{E}$ or $\boldsymbol{e}$ are in (unrationalised) electrostatic units, and magnetic quantities, including the magnetic intensity $\boldsymbol{H}$ or $\boldsymbol{h}$ are in (unrationalised) electromagnetic units. But such a system of mixed units is, in a sense, not self-consistent, and many physicists regard it as bad practice to use symbols with different systems of units in the same formula. Gaussian units have some advantages in mathematical work, but may not be familiar to radio engineers, and their use is declining. The advantages of the Gaussian system are here achieved in a different way by using for magnetic intensity a measure $\mathscr{H}$ defined in §2.10.

In most of this book the electromagnetic fields are those of waves which are harmonic in time with angular frequency ω, and the symbols used for the field vectors are capital letters, for example $\boldsymbol{E}$, $\boldsymbol{H}$, $\boldsymbol{D}$, $\boldsymbol{B}$. These are complex vectors as explained in § 2.5. In the present chapter, however, the discussion is not restricted to harmonically varying fields. The actual field vectors are, in general, functions of time t and of the space coordinates, and small letters are used for them, for example, $\boldsymbol{e}$, $\boldsymbol{h}$, $\boldsymbol{d}$, $\boldsymbol{b}$ which are real vectors. In most of this book the complex $\boldsymbol{B}$ in a wave is not needed. The symbol $\boldsymbol{B}$ is used to denote the real value of the magnetic induction of the earth's magnetic field, assumed to be independent of time.

2.2. Definitions of electric intensity $\boldsymbol{e}$ and magnetic intensity $\boldsymbol{h}$

The electric intensity $\boldsymbol{e}$ at a point in free space is defined as follows. A small charge δq is placed at the point and the force $\delta \boldsymbol{f}$ acting on it is measured. Then

$$\boldsymbol{e} = \lim_{\delta q \to 0} (\delta \boldsymbol{f}/\delta q).$$

In the ionosphere or magnetosphere there are on the average N free electrons and ions per unit volume. The electric intensity defined as above must vary markedly in the free space between them, that is over a distance of the order $N^{-\frac{1}{3}}$. But when $\boldsymbol{e}$ and the other field variables are used in Maxwell's equations, they are assumed to represent vector fields continuously distributed in space and approximately constant over distances which are large compared with $N^{-\frac{1}{3}}$ but small compared to a wavelength. The plasma is thus treated as a continuous medium, and the use of Maxwell's equations implies a 'smoothing out' process over a distance large compared with $N^{-\frac{1}{3}}$. Within the plasma, therefore, definitions of $\boldsymbol{e}$ and $\boldsymbol{h}$ must be found which effect this smoothing out.

For a medium which is not free space, the electric intensity $\boldsymbol{e}$ is usually defined in terms of a long thin cavity. The component of $\boldsymbol{e}$ in a given direction is defined by the following imagined experiment. A long thin cavity is cut in the medium parallel to the given direction. Its length must be so small that the electric state of the medium does not change appreciably within it. For waves in a homogeneous plasma, this means that the length is small compared to a wavelength. An infinitesimal test charge δq is placed at the centre of the cavity, and the component δF of the force acting on it in the direction of the cavity is measured. The component of $\boldsymbol{e}$ in this direction is $\lim_{\delta q \to 0} \delta F/\delta q$. For a medium which contains discrete ions and electrons. this definition can still be used provided that the area of cross-section of the cavity is very large compared with $N^{-\frac{2}{3}}$. It is the electric intensity so defined that is to be used in Maxwell's equations. The cavity definition thus effects the 'smoothing out' process mentioned above.

The magnetic intensity $\boldsymbol{h}$ might be defined in a similar way by using a small magnetic pole in the cavity. But since magnetic poles do not exist, it is better first to define the magnetic induction $\boldsymbol{b}$, as in § 2.4, and then to use

$$\boldsymbol{b} = \mu_0 \boldsymbol{h} \tag{2.2}$$

as the definition of $\boldsymbol{h}$.

2.3. The current density $\boldsymbol{j}$ and the electric polarisation $\boldsymbol{p}$

In the field of a radio wave in the ionosphere or magnetosphere there is a current density $\boldsymbol{j}$ which arises from the motion of charges. In many problems it will be sufficiently accurate to consider $\boldsymbol{j}$ as arising from the movement of electrons only, but in some cases the movement of heavy ions may also contribute (see §§ 3.11, 3.13, 4.9, 13.8, 13.9, 17.5).

Let $\boldsymbol{r}$ be the average vector displacement of an electron from the position it would have occupied if there were no field. Then the average electron velocity is $\partial \boldsymbol{r}/\partial t$ and the current density is

$$\boldsymbol{j} = Ne\partial \boldsymbol{r}/\partial t \tag{2.3}$$

where e is the charge on one electron, and is a negative number. The average concentration of electrons, N, can only be defined for a volume large enough to contain many electrons, and (2.3) therefore implies a 'smoothing out' over a distance large compared with $N^{-\frac{1}{3}}$. The volume used to define the concentration N should be greater than l_D^3 where l_D is the Debye length, discussed in § 3.5.

Now it is convenient to define the electric polarisation $\boldsymbol{p}$ thus:

$$\boldsymbol{p} = Ne\boldsymbol{r}, \tag{2.4}$$

so that

$$\boldsymbol{j} = \partial \boldsymbol{p}/\partial t. \tag{2.5}$$

The electric polarisation must not be confused with the wave polarisation described in §§ 4.3–4.5.

For dielectrics the electric polarisation is usually defined to be the electric dipole moment per unit volume, but this definition is not suitable for a medium containing free electrons. For if the electrons are randomly distributed and are then all given small equal displacements in the same direction, they remain randomly distributed and it is not obvious that the medium has become polarised. The definition (2.4) of $\boldsymbol{p}$

is therefore somewhat ambiguous because the origin for $\boldsymbol{r}$ can never be specified and it would be impossible to devise an experiment to measure $\boldsymbol{p}$ at a given point. But the velocity $\partial\boldsymbol{r}/\partial t$ has a definite meaning and an experiment to measure $\boldsymbol{j}$, in (2.3) or (2.5), could easily be suggested. Strictly speaking, therefore, the whole theory should be formulated in terms of $\boldsymbol{j}$. But $\boldsymbol{p}$ appears in the last Maxwell equation (2.23) only through its time derivative, so the ambiguity is removed. Hence we use $\boldsymbol{p}$ because it makes the formulae simpler.

2.4. The electric displacement *d* and magnetic induction *b*

The electric displacement $\boldsymbol{d}$ is defined thus

$$\boldsymbol{d} = \varepsilon_0 \boldsymbol{e} + \boldsymbol{p} \tag{2.6}$$

where $\boldsymbol{e}$ is defined as in § 2.2. The time derivative $\partial\boldsymbol{d}/\partial t$ is called the total current density. It includes $\partial\boldsymbol{p}/\partial t$ which is the contribution (2.5) from the moving electrons and ions, and $\varepsilon_0\partial\boldsymbol{e}/\partial t$ which is the displacement current density contributed by the electric field.

An alternative definition of $\boldsymbol{d}$ equivalent to (2.6) uses a flat plate-like cavity. The component of $\boldsymbol{d}$ in any given direction is defined by the following imagined experiment. In the undisturbed plasma, before the fields are applied, a thin flat cavity is cut with its plane perpendicular to the given direction. Its thickness must be large compared with $N^{-\frac{1}{3}}$ but small compared with a wavelength. An infinitesimal test charge δq is placed in the centre of the cavity and the electromagnetic fields are then applied to the plasma. The component δF of the force on the charge in the direction normal to the cavity is measured. The component of $\boldsymbol{d}$ in this direction is then $\varepsilon_0 \lim_{\delta q\to 0}\delta F/\delta q$. It is the electric displacement $\boldsymbol{d}$ so defined that is used in Maxwell's equations. When this definition is used for a plasma it is important to remember that the cavity is imagined to be cut before the electrons and ions are displaced by the fields; no electrons or ions must cross the central plane of the cavity.

The magnetic induction $\boldsymbol{b}$ may also be defined in terms of a plate-like cavity. The component of $\boldsymbol{b}$ in a given direction is found as follows. Before the electrons and ions are displaced by the fields, a thin flat plate-like cavity is imagined to be cut in the medium perpendicular to the given direction. A straight wire carrying an infinitesimal test current δI is placed in the central plane of the cavity and the component ∂f of the force on unit length of it, in a direction perpendicular to the wire and in the central plane, is measured. The component of $\boldsymbol{b}$ normal to the cavity is $\lim_{\delta I\to 0}\delta f/\delta I$. These definitions effect the 'smoothing out' that is implied when Maxwell's equations are used.

2.5. Harmonic waves and complex vectors

In nearly all the problems discussed in this book, the field is assumed to arise from a wave that varies harmonically with time t, which means that all field variables, that is, components of $\boldsymbol{e}, \boldsymbol{h}, \boldsymbol{p}, \boldsymbol{j}, \boldsymbol{d}$ and $\boldsymbol{b}$, oscillate with the same angular frequency ω. Let $F_0(t)$ be some field component. It is a real variable. Then

$$F_0 = A\cos(\omega t + \phi) = \mathrm{Re}(F\mathrm{e}^{\mathrm{i}\omega t}) \tag{2.7}$$

where

$$F = A\mathrm{e}^{\mathrm{i}\phi}. \tag{2.8}$$

and A and ϕ are real. The complex number F determines the field component completely. It is multiplied by $\mathrm{e}^{\mathrm{i}\omega t}$ and the real part is then taken. Its modulus $A = |F|$ is called the amplitude of F_0 and $\phi = \arg F$ is called the phase.

The same argument applies to each component of every field vector so that, for example,

$$\boldsymbol{e}(t) = \mathrm{Re}(\boldsymbol{E}\mathrm{e}^{\mathrm{i}\omega t}) \tag{2.9}$$

where $\boldsymbol{E}$ is a complex vector. In a similar way capital letters are used to denote the complex vectors representing the other field components. But $\boldsymbol{B}$ and $\boldsymbol{Y}$ are exceptions; they are used for real vectors independent of time. In this book vectors are printed in bold type, their magnitudes in italic type and their components in italic type with subscripts. For example the electric intensity $\boldsymbol{E}$ has Cartesian components E_x, E_y, E_z which in general are complex, and magnitude $E = (E_x^2 + E_y^2 + E_z^2)^{\frac{1}{2}}$ which also, in general, is complex.

The differential equations (2.20) to (2.23) satisfied by the field vectors, are linear and homogeneous, and it is shown in ch. 3 that the relation between $\boldsymbol{D}$ or $\boldsymbol{P}$ or $\boldsymbol{J}$ and $\boldsymbol{E}$ is also linear and homogeneous. Hence if the complex representations, $\boldsymbol{E}\mathrm{e}^{\mathrm{i}\omega t}$, $\boldsymbol{H}\mathrm{e}^{\mathrm{i}\omega t}$ etc., for the field vectors are substituted, the equations may be separated into their real and imaginary parts which must be satisfied separately. Further, the factor $\mathrm{e}^{\mathrm{i}\omega t}$ appears in every term and so it may be cancelled. It follows that in all these equations, the actual real vectors $\boldsymbol{e}$, $\boldsymbol{h}$ etc. may be replaced by $\boldsymbol{E}$, $\boldsymbol{H}$ etc. It is implied that, to interpret the equations, the factor $\mathrm{e}^{\mathrm{i}\omega t}$ is restored in each term, and the real part of the term is then used.

This device of the omitted complex time factor $\mathrm{e}^{\mathrm{i}\omega t}$ is widely used in electrical engineering and in many branches of physics. It cannot be used without modification in problems that involve non-linear terms. For example, problems of energy flow involve products of field quantities, and the complex representations cannot be used immediately because the real part of the product of two complex numbers is not the same as the product of their real parts. It is then possible, however, to extend the use of complex numbers so that products of two complex vectors can be handled. This is done in the theory of energy input to the plasma, § 2.11, and of the Poynting vector, § 2.12.

To illustrate further the physical meaning of a complex vector, a description will now be given for the vector $\boldsymbol{E}\mathrm{e}^{\mathrm{i}\omega t}$. The same results apply for any of the other field vectors. The complex scalar product of $\boldsymbol{E}$ with itself is

$$\boldsymbol{E}\cdot\boldsymbol{E} = |\boldsymbol{E}\cdot\boldsymbol{E}|e^{2i\alpha} \tag{2.10}$$

where $\alpha = \frac{1}{2}\arg(\boldsymbol{E}\cdot\boldsymbol{E})$ is a real angle. Let

$$\boldsymbol{E}_0 = \boldsymbol{E}e^{-i\alpha} = \boldsymbol{u} + i\boldsymbol{v} \tag{2.11}$$

where $\boldsymbol{u}$ and $\boldsymbol{v}$ are real vectors. Then

$$\boldsymbol{E}_0\cdot\boldsymbol{E}_0 = |\boldsymbol{E}\cdot\boldsymbol{E}| = u^2 - v^2 + 2i\boldsymbol{u}\cdot\boldsymbol{v} \tag{2.12}$$

is real so that $\boldsymbol{u}\cdot\boldsymbol{v} = 0$. Thus $\boldsymbol{u}$ and $\boldsymbol{v}$ are at right angles. Now the electric intensity vector is

$$\mathrm{Re}(\boldsymbol{E}e^{i\omega t}) = \mathrm{Re}\{\boldsymbol{E}_0 e^{i(\omega t + \alpha)}\} = \boldsymbol{u}\cos(\omega t + \alpha) - \boldsymbol{v}\sin(\omega t + \alpha). \tag{2.13}$$

This has two components at right angles and oscillating in quadrature. If the vector (2.13) is drawn from a fixed origin, its other end traces out an ellipse with $\boldsymbol{u}$, $\boldsymbol{v}$ as principal semi-axes. They are, from (2.11), (2.10) the real and imaginary parts of

$$\boldsymbol{E}\cdot e^{-i\alpha} = \boldsymbol{E}\{\boldsymbol{E}\cdot\boldsymbol{E}/|\boldsymbol{E}\cdot\boldsymbol{E}|\}^{-\frac{1}{2}}. \tag{2.14}$$

Now (2.11) gives (a star * denotes a complex conjugate)

$$\boldsymbol{E}\cdot\boldsymbol{E}^* = \boldsymbol{E}_0\cdot\boldsymbol{E}_0^* = u^2 + v^2. \tag{2.15}$$

Hence, from (2.12), the squares of the lengths of the principal semi-axes are

$$u^2, v^2 = \tfrac{1}{2}\{\boldsymbol{E}\cdot\boldsymbol{E}^* \pm |\boldsymbol{E}\cdot\boldsymbol{E}|\}. \tag{2.16}$$

The ellipse is a circle if $u = v$ that is, from (2.12), (2.10)

$$\boldsymbol{E}\cdot\boldsymbol{E} = 0. \tag{2.17}$$

It is a straight line if either $u = 0$ or $v = 0$, that is, from (2.16)

$$\boldsymbol{E}\cdot\boldsymbol{E}^* = |\boldsymbol{E}\cdot\boldsymbol{E}|. \tag{2.18}$$

Finally, from (2.11)

$$i\boldsymbol{E} \wedge \boldsymbol{E}^* = i(\boldsymbol{u} + i\boldsymbol{v}) \wedge (\boldsymbol{u} - i\boldsymbol{v}) = 2\boldsymbol{u} \wedge \boldsymbol{v}. \tag{2.19}$$

This is a real vector of length $2uv$ at right angles to $\boldsymbol{u}$ and $\boldsymbol{v}$. It is therefore normal to the plane of the ellipse and its length is $2/\pi$ times the area of the ellipse.

2.6. Maxwell's equations

The electromagnetic field in a plasma is governed by the four Maxwell equations:

$$\mathrm{div}\,\boldsymbol{d} = 0, \tag{2.20}$$

$$\mathrm{div}\,\boldsymbol{b} = 0, \tag{2.21}$$

$$\mathrm{curl}\,\boldsymbol{e} = -\frac{\partial \boldsymbol{b}}{\partial t} = -\mu_0\frac{\partial \boldsymbol{h}}{\partial t}, \tag{2.22}$$

$$\mathrm{curl}\,\boldsymbol{h} = \frac{\partial \boldsymbol{d}}{\partial t}. \tag{2.23}$$

The first, (2.20), is a differential form of Gauss's theorem and results from the inverse square law of force in electrostatics. It is assumed that there is no permanent space charge, but that the contribution to the electric polarisation $\boldsymbol{p}$ of every electron and ion is included. The second equation, (2.21) is similarly a result of the inverse square law of force in magnetism.

The third equation, (2.22), is a differential form of Faraday's law of electromagnetic induction, which states that $\int \boldsymbol{e}\cdot\mathrm{d}\boldsymbol{l}$ for a closed circuit is minus the rate of change of the magnetic flux linking the circuit. The integral is proportional to the work done in taking a small charge round the circuit. This could be found by imagining a long thin cavity to be cut along the line of the circuit. The cross section must be large compared with $N^{-\frac{1}{3}}$ but may still be so small that a negligible amount of material is removed, so that the disturbance of the fields is inappreciable. The test-charge is then moved in the cavity right round the circuit and the work measured. This argument shows that the electric intensity $\boldsymbol{e}$ used in (2.22) must be as defined by a long thin cavity, as in § 2.2.

The fourth equation, (2.23), is a differential form of Ampère's circuital theorem using the total current density $\partial\boldsymbol{d}/\partial t$. This is made up of a part $\partial\boldsymbol{p}/\partial t$ arising from the movement of electrons (and possibly ions) and a part $\varepsilon_0(\partial\boldsymbol{e}/\partial t)$ which is the contribution to the displacement current from the changing electric intensity.

2.7. Cartesian coordinate system

Let x, y, z be right-handed Cartesian coordinates, and let $\boldsymbol{i}, \boldsymbol{j}, \boldsymbol{k}$, be unit vectors in the directions of the x, y, z axes. Subscripts x, y, z are used to denote the x, y, z components respectively of a vector. For example, the components of $\boldsymbol{F}$ are written F_x, F_y, F_z. There is a very useful expression for the operator curl, which may be written in the form of a determinant

$$\operatorname{curl}\boldsymbol{F} = \begin{vmatrix} \boldsymbol{i} & \boldsymbol{j} & \boldsymbol{k} \\ \partial/\partial x & \partial/\partial y & \partial/\partial z \\ F_x & F_y & F_z \end{vmatrix}. \tag{2.24}$$

When this is evaluated the operators in the second row must always precede the field components in the third row.

For harmonic waves, the complex representation of § 2.5 is used, so that capital letters are now used for the field vectors and their components, and all field variables contain the time t only through the factor $\mathrm{e}^{\mathrm{i}\omega t}$, (2.2). Hence the operator $\partial/\partial t$ is equivalent to multiplication by $\mathrm{i}\omega$. If this and (2.2), (2.24) are used, the last two Maxwell equations (2.22), (2.23) become:

$$\left.\begin{aligned} \frac{\partial E_z}{\partial y} - \frac{\partial E_y}{\partial z} &= -\mathrm{i}\omega\mu_0 H_x, & \frac{\partial H_z}{\partial y} - \frac{\partial H_y}{\partial z} &= \mathrm{i}\omega D_x \\ \frac{\partial E_x}{\partial z} - \frac{\partial E_z}{\partial x} &= -\mathrm{i}\omega\mu_0 H_y, & \frac{\partial H_x}{\partial z} - \frac{\partial H_z}{\partial x} &= \mathrm{i}\omega D_y \end{aligned}\right\} \tag{2.25}$$

$$\frac{\partial E_y}{\partial x}-\frac{\partial E_x}{\partial y}=-\mathrm{i}\omega\mu_0 H_z,\quad \frac{\partial H_y}{\partial x}-\frac{\partial H_x}{\partial y}=\mathrm{i}\omega D_z$$

In most of chs. 2–5 we consider only a single plane wave in a homogeneous medium, and the z axis is chosen to be the direction of the wave normal. For ch. 6 onwards there may be several plane waves with wave normals in different directions and the z axis is then usually chosen to be vertically upwards.

2.8. Progressive plane waves

A plane wave is defined to be a disturbance in which there is no variation of any field component in any plane parallel to a fixed plane. The z axis may be chosen to be normal to this fixed plane, and is called the 'wave normal'. The derivatives $\partial/\partial x$, $\partial/\partial y$ are then zero for all field components, so that (2.25) become

$$\frac{\partial E_y}{\partial z}=\mathrm{i}\omega\mu_0 H_x,\quad \frac{\partial H_x}{\partial z}=\mathrm{i}\omega D_y \tag{2.26}$$

$$\frac{\partial E_x}{\partial z}=-\mathrm{i}\omega\mu_0 H_y,\quad \frac{\partial H_y}{\partial z}=-\mathrm{i}\omega D_x \tag{2.27}$$

$$H_z=0,\quad D_z=0. \tag{2.28}$$

The equations (2.28) show that, for plane waves, $\boldsymbol{D}$ and $\boldsymbol{H}$ are perpendicular to the wave normal, and they are therefore said to be 'transverse'. In an isotropic medium $\boldsymbol{D}$ is proportional to $\boldsymbol{E}$, so that $\boldsymbol{E}$ also is transverse. But a magnetoplasma is not isotropic and it is shown later, §§ 4.3, 4.7, that $\boldsymbol{E}$ often has a longitudinal component, that is, a component in the direction of the wave normal.

For a homogeneous isotropic medium we may write

$$\boldsymbol{D}=\varepsilon_0 n^2\boldsymbol{E}=\varepsilon_0\varepsilon\boldsymbol{E} \tag{2.29}$$

where n is the refractive index, derived later, chs. 4 and 5, and is a constant (in general complex) at a given frequency. This may be substituted in (2.26), (2.27). The equations (2.26) then involve E_y, H_x only, and (2.27) involve E_x, H_y only. These pairs are therefore independent, and the variables in one pair can vanish without affecting the other. If this happens, the wave is said to be linearly polarised. Elimination of H_y from (2.27) gives

$$\frac{\partial^2 E_x}{\partial z^2}+k^2n^2E_x=0 \tag{2.30}$$

where

$$k=\omega/c,\quad c=(\varepsilon_0\mu_0)^{-\frac{1}{2}}. \tag{2.31}$$

This meaning for k appears in many texts and papers on radio propagation, and it is used throughout this book. It is different from the usage in some other branches of physics, where k means $\omega n/c$.

Two independent solutions of (2.30) are

$$E_x = E_x^{(1)} e^{-iknz} \tag{2.32}$$

$$E_x = E_x^{(2)} e^{+iknz} \tag{2.33}$$

where $E_x^{(1)}$, $E_x^{(2)}$ are independent of z and t. These two solutions represent waves travelling in the direction of positive and negative z, respectively. Any other solution of (2.30) can be expressed as the sum of multiples of the two independent solutions (2.32), (2.33).

Substitution of (2.32) into (2.27) gives

$$n\varepsilon_0^{\frac{1}{2}} E_x = \mu_0^{\frac{1}{2}} H_y. \tag{2.34}$$

Thus the ratio of E_x to H_y is a constant so that E_x and H_y both depend on z only through the factor e^{-iknz}. A plane wave of this kind, in which the dependence of all field quantities upon z is the same, is called a 'progressive plane wave'. The operator $\partial/\partial z$ must then be equivalent to multiplication by a constant (in this case $\partial/\partial z \equiv -ikn$), The ratio E_x/H_y is called the 'wave impedance', so that a progressive wave is one for which the wave impedance is independent of z.

Similarly, substitution of (2.33) into (2.27) gives

$$n\varepsilon_0^{\frac{1}{2}} E_x = -\mu_0^{\frac{1}{2}} H_y, \tag{2.35}$$

so that E_x and H_y both vary with z only through the factor e^{iknz}, and $\partial/\partial z = ikn$ for both field components. Thus (2.33) is also a progressive plane wave. Consider the expression

$$E_x = E_0 \cos(knz) \tag{2.36}$$

which is a solution of (2.30). It is a plane wave, and can be expressed as the sum of terms like (2.32) and (2.33) with equal modulus. Substitution of (2.36) in (2.27) gives

$$\mu_0^{\frac{1}{2}} H_y = -i\varepsilon_0^{\frac{1}{2}} n E_0 \sin(knz). \tag{2.37}$$

It is at once clear that this does not have the properties of a progressive wave. In this case the two component progressive waves have equal modulus and the wave is called a 'standing wave'. If the moduli of the component progressive wave were unequal, the wave would be called a 'partial standing wave'.

It can be shown in a similar way that (2.26) leads to two linearly polarised progressive plane waves, with the electric vector parallel to the y axis. For the one travelling in the positive z direction, (2.32) and (2.34) are replaced by

$$E_y = E_y^{(1)} e^{-iknz} \tag{2.38}$$

and

$$n\varepsilon_0^{\frac{1}{2}} E_y = -\mu_0^{\frac{1}{2}} H_x. \tag{2.39}$$

Similarly (2.33) and (2.35) give the same expressions with the sign of n reversed. It has thus been shown that for progressive plane waves travelling in, say, the positive z direction, there are two independent solutions, each linearly polarised with their electric vectors at right angles. They may be combined with any moduli and relative phase to give a resultant wave which is, in general, elliptically polarised. Any polarisation ellipse can be produced by a suitable combination of the component plane waves and, once established, the ellipse does not change as long as the wave remains in the same homogeneous isotropic medium. In contrast to this, it will be shown later, § 4.3, that in a homogeneous cold magnetoplasma, progressive waves can only have one of two possible polarisations. A wave of any other polarisation must in general be made up of two component waves travelling with different velocities.

2.9. Plane waves in free space

In free space there are no electrons and $\boldsymbol{D} = \varepsilon_0 \boldsymbol{E}$ so that $n = 1$. Then the progressive plane wave solution (2.32) gives

$$E_x = E_x^{(1)} \mathrm{e}^{-\mathrm{i}kz} \tag{2.40}$$

where $c = (\varepsilon_0 \mu_0)^{-\frac{1}{2}}$ is the speed of the waves, and (2.34) gives

$$\varepsilon_0^{\frac{1}{2}} E_x = \mu_0^{\frac{1}{2}} H_y. \tag{2.41}$$

Thus the ratio E_x/H_y is real which shows that E_x and H_y are in phase, and the vectors $\boldsymbol{E}$, $\boldsymbol{H}$ and the wave normal, in that order, form a right-handed system. The wave impedance is

$$E_x/H_y = (\mu_0/\varepsilon_0)^{\frac{1}{2}} = Z_0, \tag{2.42}$$

which is called the 'characteristic impedance of free space'.

2.10. The notation $\mathscr{H}$ and $\boldsymbol{H}$

It is now convenient to adopt for the magnetic intensity $\boldsymbol{H}$ a different measure which simplifies the equations and will be used throughout this book. Take

$$\mathscr{H} = Z_0 \boldsymbol{H}. \tag{2.43}$$

Thus $\mathscr{H}$ measures the magnetic field in terms of the electric field that would be associated with it in a progressive plane wave in free space. It has the same effect as if $\boldsymbol{E}$ and $\boldsymbol{H}$ were measured in Gaussian units (see § 2.1). The vector $\mathscr{H}$ has the same physical dimensions as the electric intensity $\boldsymbol{E}$ (which is not true for $\boldsymbol{E}$ and $\boldsymbol{H}$ in Gaussian units). The last two Maxwell equations (2.22), (2.23) now become, for harmonic waves

$$\operatorname{curl} \boldsymbol{E} = -\mathrm{i}k\mathscr{H}, \quad \operatorname{curl} \mathscr{H} = \mathrm{i}k\varepsilon_0^{-1} \boldsymbol{D} \tag{2.44}$$

or written in full, in Cartesian coordinates:

$$\left.\begin{aligned}
\frac{\partial E_z}{\partial y} - \frac{\partial E_y}{\partial z} &= -\mathrm{i}k\mathscr{H}_x, & \frac{\partial \mathscr{H}_z}{\partial y} - \frac{\partial \mathscr{H}_y}{\partial z} &= \mathrm{i}k\varepsilon_0^{-1} D_x \\
\frac{\partial E_x}{\partial z} - \frac{\partial E_z}{\partial x} &= -\mathrm{i}k\mathscr{H}_y, & \frac{\partial \mathscr{H}_x}{\partial z} - \frac{\partial \mathscr{H}_z}{\partial x} &= \mathrm{i}k\varepsilon_0^{-1} D_y \\
\frac{\partial E_y}{\partial x} - \frac{\partial E_x}{\partial y} &= -\mathrm{i}k\mathscr{H}_z, & \frac{\partial \mathscr{H}_y}{\partial x} - \frac{\partial \mathscr{H}_x}{\partial y} &= \mathrm{i}k\varepsilon_0^{-1} D_z
\end{aligned}\right\}. \tag{2.45}$$

These equations are the starting-point for much of the later work in this book.

2.11. The power input to the plasma from a radio wave

It is shown in books on electromagnetic theory that when a small change $\delta\boldsymbol{d}$ is made in the electric displacement, the electric forces must supply energy to unit volume of the medium, equal to

$$\delta\mathscr{S}_{\mathrm{E}} = \boldsymbol{e}\cdot\delta\boldsymbol{d}. \tag{2.46}$$

A simple way of showing this is to imagine that a homogeneous sample of the plasma is enclosed in a condenser with parallel plates of area A separated by a distance s. Let the charge per unit area on the plates be σ. Then $\sigma = d$ where d is the component of $\boldsymbol{d}$ perpendicular to the plates. The electric intensity $\boldsymbol{e}$ cannot have a component parallel to the conducting plates. It is therefore normal to the plates. Let its magnitude be e. Then the potential difference between the plates is es. If σ is now increased by a small amount $\delta\sigma$, the electric energy supplied is $esA\delta\sigma$ that is energy $e\delta\sigma$ per unit volume. This is equal to $\boldsymbol{e}\cdot\delta\boldsymbol{d}$ as in (2.46). A more general proof is given by Stratton (1941, § 2.8). Thus the power input to unit volume of the plasma is

$$\partial\mathscr{S}_{\mathrm{E}}/\partial t = \boldsymbol{e}\cdot\partial\boldsymbol{d}/\partial t. \tag{2.47}$$

The part of this that arises from the conduction current is dissipated as heat. The remainder gives stored energy in the plasma. The discussion of this topic is continued in § 2.12. See also problems 3.2, 3.3.

It can be shown that energy is also supplied by the magnetic field, given by

$$\delta\mathscr{S}_{\mathrm{M}} = \boldsymbol{h}\cdot\delta\boldsymbol{b} \tag{2.48}$$

per unit volume. In a vacuum, or in any medium for which $\boldsymbol{b} = \mu_0\boldsymbol{h}$, including the plasmas of the earth's upper atmosphere, (2.48) gives $\delta\mathscr{S}_{\mathrm{M}} = \frac{1}{2}\mu_0\delta(\boldsymbol{h}^2)$. There is no mechanism by which magnetic energy is dissipated as heat, so that $\mathscr{S}_{\mathrm{M}}$ is stored energy per unit volume, given by

$$\mathscr{S}_{\mathrm{M}} = \tfrac{1}{2}\mu_0\boldsymbol{h}^2 = \tfrac{1}{2}\boldsymbol{h}\cdot\boldsymbol{b}. \tag{2.49}$$

For a radio wave in which the fields vary harmonically in time, it is of interest to use (2.47) to find the average value $(\partial\mathscr{S}_{\mathrm{E}}/\partial t)_{\mathrm{av}}$ of the power input over a time very long compared to the period $2\pi/\omega$. The complex representations $\boldsymbol{E}$, $\boldsymbol{D}$ of $\boldsymbol{e}$, $\boldsymbol{d}$ cannot immediately be used because (2.47) involves a product. But

$$\boldsymbol{e} = \tfrac{1}{2}(\boldsymbol{E}\mathrm{e}^{\mathrm{i}\omega t} + \boldsymbol{E}^*\mathrm{e}^{-\mathrm{i}\omega t}) \tag{2.50}$$

and there is a similar expression for $\boldsymbol{d}$. Hence (2.47) gives

$$\boldsymbol{e}\cdot\partial\boldsymbol{d}/\partial t = \tfrac{1}{4}\mathrm{i}\omega(\boldsymbol{E}\mathrm{e}^{\mathrm{i}\omega t} + \boldsymbol{E}^*\mathrm{e}^{-\mathrm{i}\omega t})\cdot(\boldsymbol{D}\mathrm{e}^{\mathrm{i}\omega t} - \boldsymbol{D}^*\mathrm{e}^{-\mathrm{i}\omega t}). \tag{2.51}$$

Here there is a term containing $\mathrm{e}^{2\mathrm{i}\omega t}$, and another containing $\mathrm{e}^{-2\mathrm{i}\omega t}$. When the average over a period is taken, the contribution from these oscillatory terms is zero. The remaining terms give

$$(\partial\mathscr{S}_{\mathrm{E}}/\partial t)_{\mathrm{av}} = \tfrac{1}{4}\mathrm{i}\omega(\boldsymbol{E}^*\cdot\boldsymbol{D} - \boldsymbol{E}\cdot\boldsymbol{D}^*) = \tfrac{1}{2}\omega\,\mathrm{Im}(\boldsymbol{E}\cdot\boldsymbol{D}^*). \tag{2.52}$$

It would be expected that the electric energy stored in the medium is proportional to the square of the amplitudes of the fields, that is to $|E|^2, |D|^2$ (compare (2.49) for the magnetic energy). These amplitudes are constant. The stored energy may oscillate in a time comparable to the period $2\pi/\omega$, but it cannot continue to increase in a much longer time period. The power (2.52) must therefore be the average rate at which energy is dissipated as heat in the medium.

For most radio waves the fields are small enough to permit the assumption that $\boldsymbol{D}$ depends linearly on $\boldsymbol{E}$. The relation is given by

$$\boldsymbol{D} = \varepsilon_0 \boldsymbol{\varepsilon} \boldsymbol{E} \tag{2.53}$$

where $\boldsymbol{\varepsilon}$ is called the relative electric permittivity. Expressions for it are derived in the following chapter. For an isotropic plasma $\boldsymbol{\varepsilon}$ is a scalar ε, in general complex, and (2.52) gives $-\frac{1}{2}\omega \boldsymbol{E}\cdot\boldsymbol{E}^* \operatorname{Im}(\varepsilon)$. This is zero if ε is real. The medium cannot then dissipate energy as heat, and is said to be 'loss free'. For an anisotropic plasma $\boldsymbol{\varepsilon}$ is a tensor of rank 3, represented by a 3×3 matrix. It is then convenient to write (2.53) with suffix notation

$$D_i = \varepsilon_0 \varepsilon_{ij} E_j, \quad i,j = x, y, z \tag{2.54}$$

where it is implied that any term containing the same suffix twice is equal to sum of terms with all possible values of that suffix. This is called the 'summation convention for repeated suffixes'. (See, for example, Jeffreys, 1931). Then (2.52) gives

$$(\partial \mathscr{S}_{\mathrm{E}}/\partial t)_{\mathrm{av}} = \tfrac{1}{4}\mathrm{i}\omega E_i^* E_j(\varepsilon_{ij} - \varepsilon_{ji}^*). \tag{2.55}$$

If the medium is loss free, this must be zero for all possible fields E_i, which requires that

$$\varepsilon_{ij} = \varepsilon_{ji}^*. \tag{2.56}$$

Then $\boldsymbol{\varepsilon}$ is said to be Hermitian, that is the matrix representing $\boldsymbol{\varepsilon}$ is the complex conjugate of its transpose.

2.12. The flow of energy. The Poynting vector

Equations (2.46), (2.48) show that the power being supplied per unit volume of a plasma is

$$\partial \mathscr{S}/\partial t = \partial(\mathscr{S}_{\mathrm{E}} + \mathscr{S}_{\mathrm{M}})/\partial t = \boldsymbol{e}\cdot\partial \boldsymbol{d}/\partial t + \boldsymbol{h}\cdot\partial \boldsymbol{b}/\partial t. \tag{2.57}$$

This may be integrated over some volume V of the medium. Then

$$\frac{\partial}{\partial t}\int_V \mathscr{S}\,\mathrm{d}V = \int_V \{\boldsymbol{e}\cdot\partial \boldsymbol{d}/\partial t + \boldsymbol{h}\cdot\partial \boldsymbol{b}/\partial t\}\,\mathrm{d}V \tag{2.58}$$

which is the rate at which the electric and magnetic forces are supplying energy to the volume. If, now, the Maxwell equations (2.22), (2.23) are used, (2.58) becomes

$$\int_V (\boldsymbol{e}\cdot\operatorname{curl}\boldsymbol{h} - \boldsymbol{h}\cdot\operatorname{curl}\boldsymbol{e})\,\mathrm{d}V = -\int_V \operatorname{div}(\boldsymbol{e}\wedge\boldsymbol{h})\,\mathrm{d}V \tag{2.59}$$

from a well known identity of vector analysis. The last integral can be expressed as a surface integral by using the 'divergence theorem'. The result is

$$\frac{\partial}{\partial t}\int_V \mathscr{S}\,\mathrm{d}V = -\int_S (\boldsymbol{e}\wedge\boldsymbol{h})\cdot \mathrm{d}\boldsymbol{S} \tag{2.60}$$

where the second integral is evaluated over the whole of the surface S enclosing V, and $\mathrm{d}\boldsymbol{S}$ is a vector in the direction of the outward normal, of magnitude equal to an element of surface area.

Equation (2.60) is Poynting's theorem. It suggests that the Poynting vector

$$\boldsymbol{\Pi} = \boldsymbol{e}\wedge\boldsymbol{h} \tag{2.61}$$

gives the flux of energy in the electromagnetic field. This result cannot be regarded as proved, and the difficulties are discussed in books on electromagnetic theory; see for example Stratton (1941, § 2.19). In this book it is assumed that $\boldsymbol{\Pi}$ gives the flux of energy. There are no known cases where this assumption leads to incorrect results.

The Poynting vector (2.61) is the product of two field quantities. For a harmonic wave, these cannot simply be replaced by their complex representations. Instead we have

$$\boldsymbol{\Pi} = \tfrac{1}{4}(\boldsymbol{E}\mathrm{e}^{\mathrm{i}\omega t} + \boldsymbol{E}^*\mathrm{e}^{-\mathrm{i}\omega t}) \wedge (\boldsymbol{H}\mathrm{e}^{\mathrm{i}\omega t} + \boldsymbol{H}^*\mathrm{e}^{-\mathrm{i}\omega t}). \tag{2.62}$$

Now take the average of this over one period. The terms containing factors $\mathrm{e}^{2\mathrm{i}\omega t}$ and $\mathrm{e}^{-2\mathrm{i}\omega t}$ are oscillatory and give zero. The remaining terms give

$$\boldsymbol{\Pi}_{\mathrm{av}} = \tfrac{1}{4}(\boldsymbol{E}\wedge\boldsymbol{H}^* + \boldsymbol{E}^*\wedge\boldsymbol{H}) = \tfrac{1}{2}\mathrm{Re}(\boldsymbol{E}\wedge\boldsymbol{H}^*). \tag{2.63}$$

The product $\tfrac{1}{2}\boldsymbol{E}\wedge\boldsymbol{H}^*$ for a harmonic wave is called the complex Poynting vector. Its real part is assumed to be the time average of the flux of energy. It is extensively used in later chapters for waves in stratified media. Its component $\boldsymbol{\Pi}_z$ perpendicular to the strata is given by

$$4Z_0\Pi_z = E_x\mathscr{H}_y^* - E_y\mathscr{H}_x^* - E_y^*\mathscr{H}_x + E_x^*\mathscr{H}_y. \tag{2.64}$$

2.13. Complex refractive index

In § 2.8 a progressive plane wave in a homogeneous medium was defined to be one in which all field quantities vary in space only through a factor $\mathrm{e}^{-\mathrm{i}knz}$ where z is measured in the direction of the wave normal. The refractive index n need not be real and it is shown in ch. 4 that in general n is complex. Hence let

$$n = \mu - \mathrm{i}\chi \tag{2.65}$$

where μ and χ are real. Then for the progressive wave (2.38)

$$E_y = E_y^{(1)}\mathrm{e}^{-\mathrm{i}k\mu z}\mathrm{e}^{-k\chi z}. \tag{2.66}$$

This represents a linearly polarised wave travelling with wave velocity c/μ and attenuated at a rate depending on χ. The wave must be losing energy which is absorbed by the medium (see problem 2.1). If χ were negative, the wave would grow in amplitude as it travelled. This can happen in some plasmas where there is a source of energy that can be supplied to the wave. It may occur, for example, when there are charged particles streaming through the plasma or when other waves are present. This is part of the subject of 'plasma instabilities', which is discussed in books on plasma physics. In this book, however, it is assumed that there is no source of wave energy in the medium and the plasmas are said to be 'passive'. Then it is expected that μ and χ always have the same sign, and this is shown to be true for the refractive indices derived later, ch. 4. When μ and χ are both negative, the wave (2.66) travels in the direction of negative z and is attenuated as it travels.

In a variable medium the refractive index n is a function of the space coordinates. For some purposes, §§7.19, 8.21, 14.9–14.12, 16.5–16.7, the process of analytic continuation is used to find n at complex values of these coordinates. The restriction that μ and χ have the same sign does not apply when the coordinates are complex, even though the medium is passive.

2.14. Evanescent waves

If n^2 is negative and real, n is purely imaginary, so that $\mu = 0$, and (2.66) becomes

$$E_y = E_y^{(1)} e^{-k\chi z}. \tag{2.67}$$

This appears to represent a wave travelling with infinite wave-velocity. Every field component varies harmonically in time, but there is no harmonic variation in space. The phase is the same for all values of z. A disturbance of this kind is called an 'evanescent' wave. It is still a 'progressive' wave according to the definition adopted here. Other examples of such waves are given in §7.11.

The expressions (2.34), (2.39) are true whether n is real or complex. If it is real, then the electric and magnetic fields of a linearly polarised wave are in phase or antiphase. In an evanescent wave n is purely imaginary, so that the electric and magnetic fields are in quadrature. This means that the time average value of the Poynting vector (2.63) is zero. Hence there is no net energy flow for such a wave in an isotropic medium. For an evanescent wave in an anisotropic medium there can be some flow of energy perpendicular to the wave normal (see §4.8). If n has the general complex value (2.65), the phase difference between the electric and magnetic fields is given by $\arctan(\chi/\mu)$.

If n^2 is real and positive, a progressive plane wave is unattenuated, and there is no absorption of energy. If n^2 is real and negative, the only possible progressive wave is evanescent, and there is no energy flow and again no absorption of energy. A medium for which n^2 is real, whether positive or negative, is said to be 'loss-free'.

2.15. Inhomogeneous plane waves

For the progressive waves considered in previous sections it was convenient to choose the z axis to be in the direction of the wave normal. When the wave normal has some other direction, with direction cosines $\boldsymbol{l} = l_x, l_y, l_z$, the exponential in (2.32) or (2.38) is replaced by

$$\exp\{-\mathrm{i}kn(l_x x + l_y y + l_z z)\}. \tag{2.68}$$

The vector $\boldsymbol{l}$ need not be real. Let

$$\boldsymbol{l} = \boldsymbol{a} + \mathrm{i}\boldsymbol{b} \tag{2.69}$$

where $\boldsymbol{a}$ and $\boldsymbol{b}$ are real vectors of lengths a and b respectively. Since $\boldsymbol{l}$ is a unit vector it follows that

$$a^2 - b^2 = 1, \quad \boldsymbol{a}\cdot\boldsymbol{b} = 0 \tag{2.70}$$

Thus $\boldsymbol{a}$ and $\boldsymbol{b}$ are at right angles. Now choose new real Cartesian axes with the x, y, z axes parallel to $\boldsymbol{a}$, $\boldsymbol{a} \wedge \boldsymbol{b}$, $\boldsymbol{b}$ respectively. In these axes the direction cosines of the wave normal are $(a, 0, \mathrm{i}b)$ and the exponential (2.68) becomes

$$\exp\{-\mathrm{i}kn(ax + \mathrm{i}bz)\} = \exp(-\mathrm{i}knax)\exp(knbz). \tag{2.71}$$

A wave of this kind is called an 'inhomogeneous plane wave'. Nearly all the component waves considered in chs. 6, 15–19, are of this kind.

Now suppose that n is real. Then in (2.71) the second exponential is real. The first exponential alone represents a plane wave with wave fronts x = constant. These real planes are not true wave fronts, except when $b = 0$, and it is better to call them 'planes of constant phase'. The second exponential in (2.71) shows that the waves change in amplitude when z varies, and the planes z = constant are 'planes of constant amplitude'. Next suppose that n is purely imaginary. Then the planes x = constant are planes of constant amplitude, and the planes z = constant are planes of constant phase. It has thus been shown that for a loss free medium, in which n^2 must be real and either positive or negative, the planes of constant phase and the planes of constant amplitude are at right angles.

Consider the special case where the wave is linearly polarised with $\boldsymbol{E}$ parallel to the y axis. Then (2.45) shows that

$$\mathscr{H}_x = \mathrm{i}nbE_y, \mathscr{H}_z = naE_y. \tag{2.72}$$

This result can be used to study the time averaged energy flux $\boldsymbol{\Pi}_{\mathrm{av}}$ (2.63) (see problem 2.1). It shows that $\mathscr{H}_x$ and $\mathscr{H}_z$ are in quadrature. Thus if $\boldsymbol{\mathscr{H}}$ is drawn from a fixed origin its other end traces out a real ellipse, that is in a plane containing both the real and imaginary parts $\boldsymbol{a}$, $\boldsymbol{b}$ of the complex wave normal vector $\boldsymbol{l}$. It has a very different meaning from the polarisation ellipse of an elliptically polarised wave. A similar result can be derived for a linearly polarised wave with $\boldsymbol{\mathscr{H}}$ parallel to the y axis. Then (2.72) is replaced by

$$E_x = -\frac{\mathrm{i}b}{n}\mathscr{H}_y, \quad E_z = \frac{a}{n}\mathscr{H}_y \tag{2.73}$$

For a wave of any other polarisation, a combination of (2.72) and (2.73) can be used.

If n is complex and given by (2.65), the exponential in (2.71) becomes

$$\exp\{-ik(\mu ax + \chi bz) + k(\mu bz - \chi ax)\}. \tag{2.74}$$

Here the planes of constant phase and of constant amplitude are no longer at right angles. The angle between them is, from (2.74),

$$\arctan\{-ab(\mu^2 + \chi^2)/\mu\chi\}. \tag{2.75}$$

PROBLEMS 2

2.1. For a progressive wave in an isotropic homogeneous medium whose refractive index is complex and whose wave normal is in a real direction, find the time average $\boldsymbol{\Pi}_{av}$ of the flux of energy, and find the rate P at which energy is being dissipated as heat in the medium. Verify that $P = -\operatorname{div}\boldsymbol{\Pi}_{av}$. Do the same for an inhomogeneous plane wave.

2.2. In a certain radio wave, the electric field at a given fixed point varies harmonically with time as given by a factor $e^{i\omega t}$, and its components have the values given below. In each case find its state of polarisation, and, where appropriate, find the area of the polarisation ellipse, the axis ratio, and the direction cosines of the normal to its plane. (a) 1, i, 0. (b) i, $-$i, 2. (c) $2 - i\sqrt{3}$, $2 + i\sqrt{3}$, 4. (d) 2, 2, $\frac{1}{2}$. (e) $1 - i$, $1 + i$, 0. (f) ib, 0, a.

2.3. An inhomogeneous plane wave of angular frequency ω travels in a vacuum. Its wave normal has direction cosines $a, 0, i(a^2 - 1)^{\frac{1}{2}}$ where a and the square root are real and positive. It is elliptically polarised with $E_y = \mathscr{H}_y e^{i\alpha}$ where α is real. Thus $\mathrm{Re}(\boldsymbol{E}e^{i\omega t})$ and $\mathrm{Re}(\boldsymbol{\mathscr{H}}e^{i\omega t})$, when drawn from a fixed origin, both trace out ellipses. Prove that these two ellipses have the same size and shape and that their planes make equal and opposite angles β with the plane $y = 0$, where $\tan^2\beta = a^2(a^2 - 1)/(a^2 - \cos^2\alpha)$. Prove that the ratio of their principal axes is $(a^2 - \cos\alpha)/(a^2 + \cos\alpha)$.

2.4. For an inhomogeneous plane wave in a loss free isotropic medium, the electric field is in a real direction perpendicular to both the real and imaginary parts of the complex wave normal vector and is given by $E_0 e^{i\omega t}$. Prove that when the magnetic intensity vector $\mathrm{Re}(\boldsymbol{H}e^{i\omega t})$ is drawn from a fixed origin, it traces out an ellipse whose principal axes are parallel and perpendicular to the planes of constant phase. Show that the component of $\boldsymbol{H}$ parallel to the planes of constant amplitude is in quadrature with E_0. Hence show that the time averaged Poynting vector $\frac{1}{2}\mathrm{Re}(\boldsymbol{E} \wedge \boldsymbol{H}^*)$ is parallel to the planes of constant amplitude. Investigate also the behaviour of the instantaneous Poynting vector $\mathrm{Re}(\boldsymbol{E}e^{i\omega t}) \wedge \mathrm{Re}(\boldsymbol{H}e^{i\omega t})$. Consider the two cases where the refractive index n is (i) real, and (ii) purely imaginary.

3

The constitutive relations

3.1. Introduction

Before (2.45) can be applied to the theory of wave propagation in a plasma, it is necessary to express the electric displacement $\boldsymbol{D}$, and therefore the electric polarisation $\boldsymbol{P}$, in terms of the electric intensity $\boldsymbol{E}$. The resulting expressions are called the constitutive relations of the plasma and are derived in this chapter. The subject of wave propagation is resumed in ch. 4.

From now on, except in § 3.8, all fields are assumed to vary harmonically in time, and are designated by capital letters representing complex vectors, as explained in § 2.5. Time derivatives of some of the fields appear in the constitutive relations, and since $\partial/\partial t \equiv \mathrm{i}\omega$, it follows that the angular frequency ω appears in the expression for the electric permittivity. When this happens the medium is said to be time dispersive.

In this and the following chapter the plasma is assumed to be homogeneous. In later chapters the results are applied to an inhomogeneous plasma. This is justified provided that the plasma is sufficiently slowly varying. The meaning of 'slowly varying' is discussed in § 7.10. It is further assumed, in the present chapter, that the spatial variation of the fields can be ignored. This means that, over a distance large compared with $N^{-\frac{1}{3}}$, the fields can be treated as uniform. Thus spatial derivatives of the fields do not appear in the expression for the permittivity. When this assumption is not true, the expression for the permittivity contains the refractive index vector $\boldsymbol{n}$; § 5.1. Then the medium is said to have spatial dispersion. This is important in phenomena where $|\boldsymbol{n}| \gg 1$. For example it plays an essential part in the dispersion relations for plasma waves, and ion acoustic waves, which depend on the plasma temperature, and whose wave velocities are small, so that $|\boldsymbol{n}|$ is large. See also problem 13.2. But for radio waves the effect of spatial dispersion is negligible.

3.2. Free, undamped electrons

Consider a volume of plasma as large as possible consistent with the assumption that within it the electric intensity $\boldsymbol{E}$ is uniform (see § 3.5). Then every electron experiences the same electric force $\boldsymbol{E}e$. The electrons have random velocities because of their thermal motions, but these are in all directions and the average over many electrons is zero. Let $\boldsymbol{r}$ be the vector displacement of an electron from the position it would occupy if there were no field $\boldsymbol{E}$. It is a harmonically varying quantity and has the complex representation

$$\boldsymbol{r} = \boldsymbol{R}e^{i\omega t} \tag{3.1}$$

where it is implied that the real part of the right-hand side is used as explained in §2.5. It is now assumed, as a first approximation, that all other forces on the electrons are negligible. Thus collisions and the effect of the earth's magnetic field are ignored. The force exerted by the magnetic field $\boldsymbol{\mathcal{H}}$ of the wave is neglected; the justification for this is discussed in § 3.7. Then the equation of motion of one electron is

$$\boldsymbol{E}e = m\partial^2\boldsymbol{r}/\partial t^2 \tag{3.2}$$

where m is the mass of an electron. Hence from (2.4) and (3.1) on multiplying (3.2) by Ne:

$$\boldsymbol{E}Ne^2 = -\omega^2 m\boldsymbol{P}. \tag{3.3}$$

This gives the contribution to $\boldsymbol{P}$ from the electrons. An exactly similar argument can be used to find the contribution from the ions. Since an ion is of the order of 2000 to 60 000 times more massive than an electron, its contribution to $\boldsymbol{P}$ in (3.3) is smaller in the same proportion. Thus, for radio waves, it is nearly always permissible to neglect the ions' contributions. The discussions in this book are therefore given for electrons, and the ions are ignored, except where specifically stated. The same results can be used for the ions provided that m and e are replaced by the appropriate values. Some examples where the effect of ions is studied are given in §§3.11, 3.13, 4.9, 13.9. If the ions are ignored, (3.3) gives for the constitutive relation

$$\boldsymbol{P} = -\varepsilon_0 X\boldsymbol{E}, \tag{3.4}$$

$$X = Ne^2/(\varepsilon_0 m\omega^2). \tag{3.5}$$

The quantity X is very important and appears throughout the theory in this book. In Appleton's original paper (1932), and in many important early papers on magnetoionic theory, the symbol x was used, but for many years this has now been replaced by X to avoid confusion with the coordinate x. Sometimes X is written

$$X = \omega_N^2/\omega^2 = f_N^2/f^2, \quad \omega_N^2 = Ne^2/\varepsilon_0 m, \quad \omega_N = 2\pi f_N. \tag{3.6}$$

Here f_N is called the 'plasma frequency', and ω_N is the angular plasma frequency. Its square is proportional to the electron concentration N. X also is proportional to N and inversely proportional to the square of the wave frequency f. Note that its value is independent of the sign of the electronic charge.

The following are useful numerical values. The plasma frequency f_N is in hertz and the electron concentration N is in m^{-3}:

$$f_N^2 = 80.61\,N, \quad f_N = 8.98\,N^{\frac{1}{2}}, \quad N = 1.240 \times 10^{-2} f_N^2. \tag{3.7}$$

3.3. The Lorentz polarisation term

In the preceding section it was assumed that the same electric field $\boldsymbol{E}$ acts on all the electrons in the volume of plasma considered, and in later sections it is assumed that this $\boldsymbol{E}$ is the same as the average or 'smoothed out' $\boldsymbol{E}$ as defined, in § 2.2, in terms of a long thin cavity, and as used in Maxwell's equations (2.45). In some types of dielectric this assumption is not true. Lorentz (1909) considered dielectrics in which all the electrons are bound within molecules. When an electric field $\boldsymbol{E}$ is applied, each molecule is distorted so that it acquires an electric dipole moment $\boldsymbol{M}_E$, and the electric polarisation is $\boldsymbol{P} = N_M \boldsymbol{M}_E$ where N_M is the concentration of the molecules. Thus $\boldsymbol{P}$ is the electric dipole moment per unit volume, and the difficulty mentioned in § 2.3, connected with the origin of $\boldsymbol{r}$, does not arise. Lorentz showed that if the molecules of the dielectric are arranged on a cubic lattice, or if they are arranged randomly in space, as in a fluid or in glass, for example, then the average electric field acting on a molecule is $\boldsymbol{E} + \frac{1}{3}\boldsymbol{P}/\varepsilon_0$. Here the second term is called the Lorentz polarisation term. Some authors assumed that this expression applied also to the random arrangement of free electrons in a plasma. If this were so the constitutive relation (3.4) would become

$$P = -\varepsilon_0 E X/(1 + \tfrac{1}{3}X) \tag{3.8}$$

and there would be corresponding changes in the forms (3.14), (3.23), (3.35), given later. It is now known that for a plasma the Lorentz term must be omitted.

The reasons for this may be summarised as follows. In this discussion the word 'average' implies an average over very many particles, at a fixed time. The electric force acting on any one electron is made up of the electrostatic forces from the other electrons and from the ions, as well as the force $\boldsymbol{E}e$. The particles are randomly distributed in space so that the total electrostatic force is different for different electrons, and it is required to find the average value. To do this consider a very large number N_c of 'complexions' of the plasma, and for each one choose the origin of coordinates to be at the position of one test electron. To find the average force on it, we add together the forces in the different complexions and then divide by N_c. This is achieved by imagining that the complexions are all superimposed. In any one complexion, two electrons are not likely to be extremely close together because of the electrostatic repulsion. Let s_e be the distance where the electrostatic energy is about equal to the average thermal energy so that

$$e^2/4\pi\varepsilon_0 s_e \approx KT \tag{3.9}$$

where K is Boltzmann's constant and T is the temperature. Then for all the complexions there will be very few electrons at distances less than s_e from the test electron. Apart from this the electrons are randomly distributed in space, so that when the complexions are superimposed the distribution of electronic charge at distances greater than s_e approaches uniformity as $N_c \to \infty$. At smaller distances the spatial charge density gets less but the charge distribution is spherically symmetrical, with the test charge at the centre. A spherical distribution of charge gives no electric intensity at its centre.

Similarly the centre of a positive ion cannot be at a distance less than the ion radius s_i from the test electron. At this distance there are repulsive forces much stronger than the electrostatic attraction. Apart from this the positive ions are randomly distributed in space. When the complexions are superimposed the positive charge of the ions is uniformly distributed but with a reduced charge density at distances less than s_i. Again the positive charge has a spherical distribution and gives no electric intensity at its centre. Thus the average force acting on the test electron is simply $\boldsymbol{E}e$.

The result of applying the field $\boldsymbol{E}$ is that the electrons have displacements that are the same for all, and similarly the ions are displaced all by the same amount, and the ions and electrons move relative to each other. But after these displacements, the spatial distributions of the particles are still random, and so the foregoing argument still applies. Thus, for a plasma, the Lorentz term is absent.

It is useful to see how this kind of argument can give a Lorentz term for other kinds of dielectric. Consider, therefore, a fluid containing molecules which become electrically polarised when the electric field $\boldsymbol{E}$ is applied. For simplicity suppose that the molecules are spherical. Each polarised molecule behaves as though it has an electric dipole of moment $\boldsymbol{M}_E$ at its centre. The molecules are randomly distributed in space, except that no two molecule centres can be closer together than the molecular diameter d_M. It is required to find the average electric field acting on one test molecule, and this is done by imagining a large number N_c of complexions to be superimposed, with the test molecule at the origin. There is then a spatial distribution of electric dipoles, that is a distribution of electric polarisation which, outside the radius d_M, tends to uniformity as $N_c \to \infty$, and has the value $N_c N_M \bar{\boldsymbol{M}}_E$ where $\bar{\boldsymbol{M}}_E$ is the average value of $\boldsymbol{M}_E$. The test molecule is in a spherical cavity of radius d_M within this distribution. Standard electrostatic theory now shows that the electric field within this cavity is uniform and equal to $\frac{1}{3}N_c N_M \bar{\boldsymbol{M}}_E/\varepsilon_0$. Thus its average value is $\frac{1}{3}N_M \bar{\boldsymbol{M}}_E/\varepsilon_0 = \frac{1}{3}\boldsymbol{P}/\varepsilon_0$. This is the Lorentz term.

The difference between the plasma and the molecular dielectric is that for the plasma there are two distributions of charges that are displaced relative to each other and each behaves, on average, as a spherical distribution of charge, with zero electric intensity at the centre. For the dielectric there is only one distribution of polarised molecules which, on average, behaves as a spherical distribution of electric polarisation, with the Lorentz field $\frac{1}{3}\boldsymbol{P}/\varepsilon_0$ at its centre.

Many arguments for and against the inclusion of the Lorentz term have been given. Some are theoretical (Darwin, 1934, 1943) and others are based on observations (Farmer and Ratcliffe, 1936; Ratcliffe, 1939; N. Smith, 1941;

Beynon, 1947). The author is indebted to Professor C.O. Hines for allowing him to see a draft of an unpublished paper in which arguments similar to those given here were very thoroughly reviewed. The strongest observational evidence that the Lorentz term should not be included is probably from the theory of whistlers; see § 13.8 and Eckersley (1935).

3.4. Electron collisions. Damping of the motion

An electron in a stationary volume of plasma is subjected to a force exerted on it by other particles. Its magnitude and direction vary rapidly in time because all particles have thermal motions. For an electron whose average velocity is zero, the time average of this force must be zero because the plasma as a whole is not accelerated. But if the electron is caused to move, by the field of a radio wave, it does on average experience a force caused by encounters with other particles. These encounters are known as collisions.

In the ionosphere the concentration of neutral particles is about $2 \times 10^{22} \mathrm{m}^{-3}$ at height 70 km, and $2 \times 10^{18} \mathrm{m}^{-3}$ at height 300 km. The concentration of electrons is about 10^{11} to $3 \times 10^{11} \mathrm{m}^{-3}$ at the maximum of the F2-layer, that is about 300 km, and is always less than this at other heights. Thus in the ionosphere the concentration of neutrals exceeds the concentration of charged particles by a factor 10^7 or more, so most of the encounters made by an electron are with neutral particles. The force exerted by a neutral particle on an electron is of very short range and therefore the encounter is of short duration, of order 10^{-14}s (see problem 3.8). This is a minute fraction of the period of a typical radio wave. Most encounters in the ionosphere can therefore be treated as instantaneous.

For an encounter of an electron with a charged particle, the force is a modified form of the inverse square law of electrostatic force, called the Coulomb force (see (3.18) below). It is of longer range, and these encounters cannot be considered to be instantaneous. An electron experiences electric forces from the other charges at all times. This type of encounter can be dealt with by plasma kinetic theory using a Boltzmann–Vlasov equation of Fokker–Planck type. (See, for example, Shkarofsky, Johnston and Bachynski, 1966). It is found, however, that the final result can still be expressed in the form (3.11) below, provided that a suitable value is used for the effective collision frequency ν. In the magnetosphere the plasma is almost fully ionised so that the concentration of neutral particles is very small, and electron encounters are mainly with charged particles. But the concentrations are small so that the effective collision frequency is small, of order $100 \mathrm{s}^{-1}$, and for many purposes it can be neglected. In the theory that follows we examine the effect of electron collisions that are assumed to be instantaneous.

Suppose that each electron makes, on average, ν instantaneous collisions per unit time with other particles. If τ is the time between two such successive collisions of one electron then the average value of τ is $1/\nu$. It is shown in books on statistical mechanics that the probability that τ lies in the range τ to $\tau + \mathrm{d}\tau$ is $\nu \mathrm{e}^{-\nu\tau}\mathrm{d}\tau$. Suppose that each electron is subjected to a steady force $\boldsymbol{f}$. In time τ an electron then moves a distance $\boldsymbol{s} = \frac{1}{2}(\boldsymbol{f}/m)\tau^2$. This movement is superimposed on the random thermal movements. The average value of $\boldsymbol{s}$ for many electrons is therefore

$$\boldsymbol{s} = \frac{\boldsymbol{f}}{2m}\int_0^\infty \tau^2 \nu \mathrm{e}^{-\nu\tau}\mathrm{d}\tau = \boldsymbol{f}/m\nu^2 \tag{3.10}$$

which is the average displacement between two collisions. Since there are, on average, ν collisions per unit time, the average velocity is given by

$$\boldsymbol{V} = \boldsymbol{f}/m\nu. \tag{3.11}$$

This is a steady velocity so that each electron behaves as though $\boldsymbol{f}$ is balanced by a retarding force $-m\nu\boldsymbol{V}$ proportional to its velocity. This concept was first suggested by Lorentz (1909).

In a radio wave $\boldsymbol{f}$ is a sinusoidally varying force. If ν is large compared with the angular wave frequency ω, the above treatment might be expected to be approximately true, but when ν is comparable with, or much less than ω, the analysis is more complicated. It can be shown, however, that the relation can still be expected to hold. This is proved in numerous books and papers that deal with plasma kinetic theory. See, for example, Clemmow and Dougherty (1969, § 9.5), Rawer and Suchy (1967, §§ 2–4), Allis (1956). A version of the theory that uses the notation and methods of this book was given by Budden (1965). A very illuminating discussion of electron collisions was given by Ratcliffe (1959).

Equation (3.11) shows that an electron is subjected to a retarding or damping force, proportional to the velocity $\boldsymbol{V} = \partial\boldsymbol{r}/\partial t$. This is not the total velocity of the electron but only that part of it that is imparted to it by the electric field of a wave. The equation will be assumed to hold for all values of ν. When the damping force is included the equation of motion (3.2) becomes

$$\boldsymbol{E}e = m\partial^2\boldsymbol{r}/\partial t^2 + m\nu\partial\boldsymbol{r}/\partial t. \tag{3.12}$$

Now it is often assumed that the collision frequency ν is the same for all electrons, regardless of their total velocity. If this is done, (3.12) may be multiplied by Ne, and from (2.4), (3.1)

$$\boldsymbol{E}Ne^2 = -\omega^2 m(1 - \mathrm{i}\nu/\omega)\boldsymbol{P}. \tag{3.13}$$

The assumption that ν is independent of electron velocity is not true, and a modification of the theory is needed to deal with this, given in § 3.12. It is shown there, however, that (3.13) is still useful provided that ν is replaced by a suitable 'effective' value, sometimes written ν_{eff}. Here we shall omit the subscript $_{\mathrm{eff}}$ and assume that ν has the appropriate effective value.

Equation (3.13) would be the same as (3.3) if the electron mass m were replaced by the complex value $m(1 - \mathrm{i}\nu/\omega)$. Thus the whole of magnetoionic theory, ch. 4, can be treated by deriving the theory of a collisionless plasma and then replacing the electron and ion masses by their appropriate complex values. This idea was first suggested by Smith, R.L. and Brice (1964).

In (3.13) $\boldsymbol{P}$ is the contribution to the electric polarisation from the electrons. If the contribution from the ions is neglected, as in § 3.2, (3.13) gives the required constitutive relation when the earth's magnetic field is also neglected. It may be written

$$\boldsymbol{P} = -\varepsilon_0 \frac{X}{1 - \mathrm{i}Z} \boldsymbol{E} \tag{3.14}$$

where X is defined by (3.5) and

$$Z = \nu/\omega. \tag{3.15}$$

This Z is a useful measure of the collision frequency. In Appleton's original theory (1932) the symbol z was used, but this is now replaced by Z, to avoid confusion with the coordinate z. In much of this book the notation

$$U = 1 - \mathrm{i}Z \tag{3.16}$$

is used. If $\nu = 0$, then $U = 1$.

From (3.14), (3.15) it is clear that the effect of collisions is small when the frequency ω is large. For many purposes it is useful and simpler to ignore collisions altogether. The plasma is then described as a 'collisionless plasma'.

Since $\boldsymbol{D} = \varepsilon_0 \varepsilon \boldsymbol{E}$ it follows from (2.29), (3.14), (3.16) that

$$\varepsilon = 1 - X/U = n^2 \tag{3.17}$$

3.5. The Debye length

In the forms (3.4), (3.14) of the constitutive relation, and in the more complicated forms to be derived later, the plasma is treated as though it is a continuum, and the fields $\boldsymbol{E}$ and $\boldsymbol{P}$ are 'smoothed out' or average values, as discussed in §§ 2.2–2.4. The discrete nature of the charges in the plasma is ignored. This treatment is used for radio waves throughout this book and the justification for it must be examined. The relations (3.4), (3.14) could not be applied to a small volume of plasma with linear dimensions of order $N^{-\frac{1}{3}}$, which would contain only a few charges. They can only be applied to phenomena whose length scale, usually the wavelength of a radio wave, greatly exceeds some length l_D. An estimate of this length is now required.

In free space the force exerted on one charge by another is given by the inverse square law, that is the Coulomb law. In a plasma this is not true if the two charges are well separated, because the fields are modified by the intervening charges. If one charge is an electron, it repels other electrons and attracts positive charges, so that it is surrounded by a spherical distribution of particles with positive charges predominating. Thus its field is partially screened. Its unscreened field would have the Coulomb potential $\phi = e/(4\pi\varepsilon_0 r)$ at radius r. With screening present the potential is approximately

$$\phi = \frac{e}{4\pi\varepsilon_0 r}\exp(-2^{\frac{1}{2}}r/l_D) \tag{3.18}$$

where

$$l_D = (\varepsilon_0 KT/Ne^2)^{\frac{1}{2}} \tag{3.19}$$

is the Debye length. Here K is Boltzmann's constant and it is assumed that the ions and electrons have the same temperature T. If their temperatures are different, a slightly more complicated expression must be used. The result (3.18) was first derived by Debye and Hückel (1923) for mobile ions in electrolytes. The theory is given in many books on plasma physics; see, for example, Clemmow and Dougherty (1969, pp. 233 ff).

If an electron undergoes a sudden change in its motion, if, for example, it makes a collision, the other charges near it would at once experience a change in the electric force arising from it. For more remote charges this change would be smaller, and for charges at distances exceeding l_D the change would be negligibly small.

In this book the Debye length is important for two reasons. First, it establishes the length scale for the smoothing out of the fields, $\boldsymbol{E}$, $\boldsymbol{P}$, etc. as described in ch. 2. Second, it affects the theory of collisions between charged particles. To calculate the effective collision frequency ν_{eff}, an integral must be evaluated that gives the collision cross section. If the Coulomb law of force is used, this integral diverges and no useful value exists. But with the potential (3.18), the integral exists and a formula can be derived for ν_{eff}. The theory and the formula are given by Ginzburg (1970).

For the ionosphere typical approximate values of l_D are 0.1 m at height 70 km, 3×10^{-3} m at height 100 km, 7×10^{-3} m at height 300 km. For the magnetosphere the value may be 1 or 2 m. The radio propagation problems in this book involve length scales much greater than these values, so that it will only rarely be necessary to make any reference to the Debye length.

3.6. Effect of the earth's magnetic field on the motion of electrons

Let $\boldsymbol{B}$ be the constant magnetic induction of the earth's field. A charge e moving with velocity $\partial\boldsymbol{r}/\partial t$ through it is subjected to a force $e(\partial\boldsymbol{r}/\partial t) \wedge \boldsymbol{B}$. The equation of motion of an electron (3.12) has an additional term thus

$$\boldsymbol{E}e + e(\partial\boldsymbol{r}/\partial t) \wedge \boldsymbol{B} = m\partial^2\boldsymbol{r}/\partial t^2 + m\nu\partial\boldsymbol{r}/\partial t. \tag{3.20}$$

The operator $\partial/\partial t$ is now replaced by $\mathrm{i}\omega$. If it is assumed that ν is the same for all electrons, (3.20) may be multiplied by $Ne/m\omega^2$ to give

$$\frac{Ne^2}{m\omega^2}\boldsymbol{E} + \frac{\mathrm{i}e}{m\omega}\boldsymbol{P} \wedge \boldsymbol{B} = -\boldsymbol{P}(1 - \mathrm{i}Z) \tag{3.21}$$

where (2.4), (3.1), (3.15) have been used. Let

$$\boldsymbol{Y} = e\boldsymbol{B}/m\omega. \tag{3.22}$$

For electrons e is a negative number, so $\boldsymbol{Y}$ is antiparallel to $\boldsymbol{B}$. Now (3.21) is rearranged and (3.5) is used, to give

$$-\varepsilon_0 X\boldsymbol{E} = U\boldsymbol{P} + \mathrm{i}\boldsymbol{P} \wedge \boldsymbol{Y} \tag{3.23}$$

where U is given by (3.16). If the contributions of the ions to $\boldsymbol{P}$ are negligible, (3.23) is the required constitutive relation. It is used for most of the theory in this book, and in particular it is used in ch. 4 to derive the important formulae of magnetoionic theory. A plasma for which (3.23) can be used is called an 'electrons only' plasma, or simply an 'electron plasma'. Positive ions must be present to ensure that the plasma is electrically neutral, but their movements are so small that they can otherwise be ignored.

The length of the vector $\boldsymbol{Y}$ is denoted by

$$Y = |\boldsymbol{Y}| = \omega_H/\omega, \quad \omega_H = |e\boldsymbol{B}/m| = 2\pi f_H. \tag{3.24}$$

In Appleton's original theory (1932) the symbol y was used instead of Y. An electron moving in the magnetic field $\boldsymbol{B}$ traverses a helical path, and makes one turn of the helix in a time $1/f_H$. This time is independent of the electron's speed provided that this is small enough for relativistic effects to be neglected. The frequency f_{H} is called the 'gyro-frequency' for electrons. Some authors call it the 'cyclotron frequency'. ω_H is the angular gyro-frequency.

The equation (3.23) is a vector equation. When written out in Cartesian components it gives three equations, but these may be written as a single matrix equation. Let l_x, l_y, l_z be the direction cosines of $\boldsymbol{Y}$. Then (3.23) gives

$$-\varepsilon_0 X\begin{pmatrix} E_x \\ E_y \\ E_z \end{pmatrix} = \begin{pmatrix} U & \mathrm{i}Yl_z & -\mathrm{i}Yl_y \\ -\mathrm{i}Yl_z & U & \mathrm{i}Yl_x \\ \mathrm{i}Yl_y & -\mathrm{i}Yl_x & U \end{pmatrix}\begin{pmatrix} P_x \\ P_y \\ P_z \end{pmatrix}. \tag{3.25}$$

3.7. Effect of the magnetic field of the wave on the motion of electrons

In the preceding sections the force exerted on an electron by the magnetic field $\mathscr{H}$ of the wave has been neglected. To justify this, it is interesting to estimate the order of magnitude of this force by considering the motion of one electron. Suppose a wave of angular frequency ω is travelling in the positive z direction, and is linearly polarised with its electric vector in the x direction. Let x denote the coordinate of one electron, and neglect damping forces. Let the electric field have amplitude E_0. The following argument involves products of harmonically varying quatities, so that complex numbers should be avoided. Hence we write

$$m\frac{\partial^2 x}{\partial t^2} = eE_0 \cos \omega t, \quad \frac{\partial x}{\partial t} = \frac{eE_0}{m\omega}\sin \omega t. \tag{3.26}$$

Since damping is neglected, the refractive index n is real (see ch. 4). Hence the magnetic field is $H_y = (\varepsilon_0/\mu_0)^{\frac{1}{2}} nE_0 \cos \omega t$, from (2.34). Then the instantaneous force exerted on the electron by this field is

$$\mu_0 e H_y \frac{\partial x}{\partial t} = \frac{ne^2 E_0^2}{m\omega c} \sin \omega t \cos \omega t. \tag{3.27}$$

This is in the direction of the z axis, that is, of the wave normal. Notice that it varies with twice the wave frequency, and that its time average value is zero.

If damping forces are included, the magnetic field H_y and the velocity $\partial x/\partial t$ are no longer in quadrature, and the force then has an average value different from zero. There is thus an average force in the direction of the wave normal, which gives rise to radiation pressure. It is usually neglected in the theory of radio waves in the ionosphere, but it is discussed in books on electromagnetic theory (see, for example Clemmow, 1973).

The maximum value of the force (3.27) is $ne^2E_0^2/(2m\omega c) = F_{\mathrm{M}}$. The ratio of this to the maximum electric force F_{E} is

$$F_{\mathrm{M}}/F_{\mathrm{E}} = neE_0/(2m\omega c). \tag{3.28}$$

The value of E_0 at 100 km from an omnidirectional radio transmitter radiating 1 MW is of the order 0.08 V/m. The electric field encountered in the ionosphere from man-made radio transmitters will rarely reach this value, and the frequency is always greater than 10 kHz. The refractive index n is of order unity. Hence the ratio (3.28) will not exceed about 3.7×10^{-4}, and will usually be very much less than this, so that the effect of the magnetic field of the wave can be neglected.

Cases have arisen where very much larger electric fields are encountered in the ionosphere and magnetosphere. For example, when a radio transmitter is carried on a space vehicle, the fields near to it are large. In the early 1970s experiments on heating the ionosphere artificially were made at Platteville, Colorado and shortly afterwards at Arecibo, Puerto Rico. A narrow beam of radiation was sent vertically upwards from a directive aerial on the ground. The powers used were of the order of 2 MW (Platteville) and 160 kW (Arecibo). This arrangement gives wave fields in the ionosphere very much greater than those from a commercial radio transmitter, and the forces arising from the magnetic field of the wave cannot then be neglected. One of the most important results observed in these experiments was the appearance of parametric instabilities which arose because of non-linear effects, including this magnetic force. A special number of *Radio Science* (November 1974) was devoted to this subject, and it includes several excellent summaries. Reviews have been given by Fejer (1975, 1979). The general theory of non-linear effects in plasmas is discussed by Gurevich (1978).

Electromagnetic energy is radiated from a lightning flash, which may give electric fields in the ionosphere much greater than 0.1 V m^{-1}, and magnetic fields which are comparable with the earth's magnetic field. In this case the received signals are known as 'atmospherics', and they are studied with receivers which accept frequencies down to 100 Hz or less. The magnetic field of the wave, and other non-linear effects, may play a part in the mechanism of reflection of these signals from the ionosphere, at points near to the flash.

In commercial radio communication the field intensities rarely reach levels where non-linear effects are important, and in this book they are almost entirely disregarded.

3.8. Electric neutrality of the plasma. Plasma oscillations

It is assumed that, before the plasma is disturbed by a radio wave, it is electrically neutral. The justification for this must be examined. In this section the field variables may have a general time dependence and they are therefore denoted by small letters. If the electrons in a plasma are given small displacements $\boldsymbol{r}$ from their normal positions, and if $\boldsymbol{r}$ depends on the space coordinates of the electrons, there will be an increase of electron concentration at some places and a depletion at others. Thus there is a charge density ρ given approximately by

$$\rho = -\operatorname{div}\boldsymbol{p} = -Ne\operatorname{div}\boldsymbol{r}. \tag{3.29}$$

It is here assumed that $\boldsymbol{p}$, $\boldsymbol{r}$ and ρ are 'smoothed out' as though the plasma is a continuum (see §§ 2.2, 2.3, 3.5). It is also assumed that the resulting change of N is so small that a term $-e\boldsymbol{r}\cdot\operatorname{grad} N$ may be neglected in (3.29).

The first Maxwell equation (2.20) gives

$$\varepsilon_0 \operatorname{div}\boldsymbol{e} = -\operatorname{div}\boldsymbol{p} = \rho. \tag{3.30}$$

Conservation of the charges requires that

$$\operatorname{div}\boldsymbol{j} = -\partial\rho/\partial t = Ne\operatorname{div}(\partial\boldsymbol{r}/\partial t) \tag{3.31}$$

where $\boldsymbol{j}$ is the current density, § 2.3. If the earth's magnetic field is neglected, the equation of motion of any one electron is (3.12). Multiply this equation by Ne/m and take its divergence. With (3.29)–(3.31) and (3.6) this gives

$$\frac{\partial^2\rho}{\partial t^2} + \nu\frac{\partial\rho}{\partial t} + \omega_N^2\rho = 0. \tag{3.32}$$

This applies at every point in the plasma. If ν is small an approximate solution is

$$\rho = \rho_0\cos(\omega_N t + \alpha)\exp(-\tfrac{1}{2}\nu t) \tag{3.33}$$

where ρ_0 and α are independent of the time t but may depend on the space coordinates. The charge density oscillates with the plasma frequency $\omega_N/2\pi$. This is known as a 'plasma oscillation'. It decays with a time constant $2/\nu$. In the ionosphere this is about 1 ms or less. Thus an appreciable charge density cannot be sustained for a time longer than this.

The charge density ρ could be expressed as a three-dimensional Fourier integral representation in space. For any one Fourier component ρ_F we can use the complex representation and include the time dependence (3.33) thus

$$\rho_F \propto \exp\{i(\omega_N t - \boldsymbol{K}\cdot\boldsymbol{x})\}\exp(-\tfrac{1}{2}\nu t) \tag{3.34}$$

where $\boldsymbol{x}$ denotes the space coordinates x, y, z. This is a wave whose wave vector is $\boldsymbol{K}$. Its components may have any real values so any wavelength is possible and any wave normal direction is possible but the angular frequency is always ω_N. The charge density (3.34) is in strata parallel to the wave fronts, and gives an electric field parallel to the wave normal. Maxwell's equations (2.22) show that there is no magnetic field in this wave. In a plasma where the electron temperature is allowed for, a wave of this kind is known as an electron plasma wave or Langmuir wave. Then the frequency is not restricted to ω_N but can take values in a small range near ω_N.

If the earth's magnetic field $\boldsymbol{B}$ is included, it is convenient to use Cartesian coordinates x, y, z with the z axis parallel to $\boldsymbol{B}$. The electron's equation of motion is (3.20) and has the additional term $e(\partial\boldsymbol{r}/\partial t)\wedge\boldsymbol{B}$. If the electron velocities $\partial\boldsymbol{r}/\partial t$ are all parallel to $\boldsymbol{B}$, this term is zero and the equation for ρ is still (3.32). But in general the constituent waves (3.34) now have electric fields which give to $\partial\boldsymbol{r}/\partial t$ components in the x and y directions, so that it does not remain parallel to $\boldsymbol{B}$. In one special case this does not happen. It is when $\boldsymbol{r}$ and $\partial\boldsymbol{r}/\partial t$ are independent of x and y, so that ρ is in strata perpendicular to $\boldsymbol{B}$. The electron motions are then parallel to $\boldsymbol{B}$ and are unaffected by it. The direction of $\boldsymbol{K}$ is now restricted to be parallel to $\boldsymbol{B}$ but the wavelength may have any value. The time dependence is given by (3.33).

In the general case when the term $e(\partial\boldsymbol{r}/\partial t)\wedge\boldsymbol{B}$ is not zero, the Fourier analysis in space may still be done but the frequency depends on $\boldsymbol{K}$ and is found from the dispersion relation. This idea is the basis of some studies of the oscillations of a magnetoplasma. See, for example, Dougherty and Monaghan (1966). A more detailed theory of plasma oscillations is given by Clemmow and Dougherty (1969).

3.9. The susceptibility matrix

In the matrix form (3.25) of the constitutive relations the components of $\boldsymbol{E}$ are expressed in terms of the components of $\boldsymbol{P}$. In later chapters it will be necessary to have the components of $\boldsymbol{P}$ expressed in terms of those of $\boldsymbol{E}$. Equation (3.25) is equivalent to three simultaneous equations for P_x, P_y, P_z. These can be solved, and the process is equivalent to inversion of the 3×3 matrix. The result is

$$\frac{1}{\varepsilon_0}\begin{pmatrix}P_x\\P_y\\P_z\end{pmatrix} = -\frac{X}{U(U^2-Y^2)} \times\begin{pmatrix} U^2-l_x^2Y^2 & -il_zYU-l_xl_yY^2 & il_yYU-l_xl_zY^2\\ il_zYU-l_xl_yY^2 & U^2-l_y^2Y^2 & -il_xYU-l_yl_zY^2\\ -il_yYU-l_xl_zY^2 & il_xYU-l_yl_zY^2 & U^2-l_z^2Y^2\end{pmatrix}\begin{pmatrix}E_x\\E_y\\E_z\end{pmatrix}. \tag{3.35}$$

The expression multiplying (E_x, E_y, E_z) is called the susceptibility matrix of the ionosphere, and is denoted by $\boldsymbol{M}$. It was given in this form by Banerjea (1947). Its components are denoted by $M_{ij}(i,j = x, y, z)$.

It can be shown that the elements of $\boldsymbol{M}$ have the following properties:

$$M_{xx}(M_{yz} + M_{zy}) = M_{xy}M_{zx} + M_{yx}M_{xz}, \tag{3.36}$$

$$M_{xx}M_{yy} - M_{yz}M_{zy} = \frac{X^2}{U^2 - Y^2} \tag{3.37}$$

and four other relations obtained by permuting the suffixes in (3.36) and (3.37).

$$M_{xx}M_{yy}M_{zz} + M_{xy}M_{yz}M_{zx} + M_{yx}M_{zy}M_{xz} - M_{xx}M_{yz}M_{zy}$$
$$- M_{yy}M_{xz}M_{zx} - M_{zz}M_{yx}M_{xy} = -\frac{X^3}{U(U^2 - Y^2)}. \tag{3.38}$$

These relations are needed later.

If the medium is loss-free, $Z = 0$ and $U = 1$. Then $M_{ij} = M_{ji}^*$ (where a star denotes a complex conjugate) so that the matrix $\boldsymbol{M}$ is Hermitian. The electric relative permittivity is $\boldsymbol{\varepsilon} = \mathbf{1} + \boldsymbol{M}$ where $\mathbf{1}$ is the unit matrix, so $\boldsymbol{\varepsilon}$ is also Hermitian. It was shown in § 2.11 that the time average rate at which energy goes into the plasma from an electromagnetic field is then zero.

3.10. Complex principal axes

To derive the refractive indices of a homogeneous electron plasma, the constitutive relation can be used in the form (3.25), and this is done in §§ 4.3, 4.6. It is not necessary to invert the matrix and get explicit expressions for the components of $\boldsymbol{P}$. But for a plasma in which more than one species of particle contributes to $\boldsymbol{P}$, the constitutive relation cannot be written in the form (3.25). It is necessary first to derive an expression of the form (3.25) for each species. The matrices are then inverted to give the contribution to $\boldsymbol{P}$ from each species. The total $\boldsymbol{P}$ is the sum of these contributions, expressed as a 3×3 matrix multiplying $\boldsymbol{E}$, as in (3.35). This sum is the constitutive relation. Inversion of the matrices in the form (3.35) is algebraically very complicated. Fortunately there is a simpler method which proves to be very powerful and useful, both in magnetoionic theory, ch. 4, and in the more general theory of radio wave propagation.

The values of X, Y, U are different for each species of charged particle and later we shall attach subscripts to indicate the values for different species. For the present, however, (3.25) applies to any one species and the subscripts are omitted.

The relation (3.25) can be simplified by a special choice of the real Cartesian axis system x, y, z. We therefore choose the z axis to be parallel to $\boldsymbol{Y}$, so that $l_x = l_y = 0$, $l_z = 1$, and (3.25) becomes:

$$-\varepsilon_0 X \begin{pmatrix} E_x \\ E_y \\ E_z \end{pmatrix} = \begin{pmatrix} U & \mathrm{i}Y & 0 \\ -\mathrm{i}Y & U & 0 \\ 0 & 0 & U \end{pmatrix} \begin{pmatrix} P_x \\ P_y \\ P_z \end{pmatrix}. \tag{3.39}$$

For many crystalline dielectrics the matrix that expresses $\boldsymbol{E}$ in terms of $\boldsymbol{P}$ is symmetric. It is then possible to choose real orthogonal Cartesian axes so that the matrix is diagonal. These axes are called principal axes. The inverted matrix is then found simply by taking the reciprocals of the elements of the original diagonal matrix. The matrix in (3.39) is not symmetric and so in general a plasma has no real principal axes. It is still possible, however, to transform the axis system so that the matrix is diagonal. The resulting new axis system is complex.

In the old axis system of (3.39) the new complex axes are eigen column vectors of the matrix. Standard matrix theory (see, for example, Graham, 1979) shows that they are

$$\boldsymbol{u}_1 = 2^{-\frac{1}{2}}\begin{pmatrix} 1 \\ -\mathrm{i} \\ 0 \end{pmatrix}, \quad \boldsymbol{u}_2 = 2^{-\frac{1}{2}}\begin{pmatrix} 1 \\ \mathrm{i} \\ 0 \end{pmatrix}, \quad \boldsymbol{u}_3 = \begin{pmatrix} 0 \\ 0 \\ 1 \end{pmatrix} \tag{3.40}$$

with eigen values $U + Y$, $U - Y$, U respectively. Each vector in (3.40) might be multiplied by any constant. The constants have been chosen so that (3.40) satisfies the normalisation condition used in (3.44) below.

Note that $\boldsymbol{u}_1 \cdot \boldsymbol{u}_2$ is not zero so that the new axes are not all mutually orthogonal. In a system of oblique axes there are two ways of specifying the components of a vector, namely the contravariant and covariant forms (see, for example, Mathews and Walker 1970, § 15-3). We here use the contravariant form.

The contravariant components E_1, E_2, E_3 of the vector $\boldsymbol{E}$ in the new axis system are a linear combination of E_x, E_y, E_z and are found by multiplication by a transforming matrix $\mathcal{U}$ thus

$$\begin{pmatrix} E_1 \\ E_2 \\ E_3 \end{pmatrix} = \mathcal{U} \begin{pmatrix} E_x \\ E_y \\ E_z \end{pmatrix} \tag{3.41}$$

and the same applies for any other vector, including $\boldsymbol{P}$. Now (3.39) may be multiplied on the left by $\mathcal{U}$ to give

$$-\varepsilon_0 X \mathcal{U} \begin{pmatrix} E_x \\ E_y \\ E_z \end{pmatrix} = \mathcal{U} \begin{pmatrix} U & \mathrm{i}Y & 0 \\ -\mathrm{i}Y & U & 0 \\ 0 & 0 & U \end{pmatrix} \mathcal{U}^{-1} \mathcal{U} \begin{pmatrix} P_x \\ P_y \\ P_z \end{pmatrix}. \tag{3.42}$$

For the matrix $\mathcal{U}(\ldots.)\mathcal{U}^{-1}$ to be diagonal, the three columns of $\mathcal{U}^{-1}$ must be those in (3.40). Hence

$$\mathcal{U}^{-1} = 2^{-\frac{1}{2}} \begin{pmatrix} 1 & 1 & 0 \\ -\mathrm{i} & \mathrm{i} & 0 \\ 0 & 0 & 2^{\frac{1}{2}} \end{pmatrix}, \quad \mathcal{U} = 2^{-\frac{1}{2}} \begin{pmatrix} 1 & \mathrm{i} & 0 \\ 1 & -\mathrm{i} & 0 \\ 0 & 0 & 2^{\frac{1}{2}} \end{pmatrix}. \tag{3.43}$$

The inverse of $\mathcal{U}$ is the transpose of its complex conjugate, and $\mathcal{U}$ is therefore called a unitary matrix. This condition may be written in subscript notation with the summation convention, thus

$$\mathcal{U}_{ij}\mathcal{U}^*_{kj} = \delta_{ik} \tag{3.44}$$

where δ_{ik} is the Kronecker delta.

Now (3.41) with (3.43) shows that

$$E_1 = 2^{-\frac{1}{2}}(E_x + \mathrm{i}E_y), \quad E_2 = 2^{-\frac{1}{2}}(E_x - \mathrm{i}E_y), \quad E_3 = E_z$$
$$E_x = 2^{-\frac{1}{2}}(E_1 + E_2), \quad E_y = -\mathrm{i}2^{-\frac{1}{2}}(E_1 - E_2), \tag{3.45}$$

and there are similar expressions for the components of any other field vector.

The result (3.42) may now be written

$$-\varepsilon_0 X \begin{pmatrix} E_1 \\ E_2 \\ E_3 \end{pmatrix} = \begin{pmatrix} U+Y & 0 & 0 \\ 0 & U-Y & 0 \\ 0 & 0 & U \end{pmatrix} \begin{pmatrix} P_1 \\ P_2 \\ P_3 \end{pmatrix}. \tag{3.46}$$

The matrix can be inverted at once by taking the reciprocals of its elements. Hence

$$P_1 = -\frac{\varepsilon_0 X}{U+Y} E_1, \quad P_2 = -\frac{\varepsilon_0 X}{U-Y} E_2, \quad P_3 = -\frac{\varepsilon_0 X}{U} E_3. \tag{3.47}$$

For an 'electrons only' plasma, this is the constitutive relation in complex principal axes. The factors

$$\frac{-X}{U+Y}, \quad \frac{-X}{U-Y}, \quad \frac{-X}{U} \tag{3.48}$$

are the elements of the diagonalised susceptibility tensor. We may also use $\boldsymbol{D} = \varepsilon_0 \boldsymbol{E} + \boldsymbol{P} = \varepsilon_0 \boldsymbol{\varepsilon} \boldsymbol{E}$ whence

$$D_1 = \varepsilon_0 \varepsilon_1 E_1, \quad D_2 = \varepsilon_0 \varepsilon_2 E_2, \quad D_3 = \varepsilon_0 \varepsilon_3 E_3 \tag{3.49}$$

where

$$\varepsilon_1 = 1 - X/(U+Y), \quad \varepsilon_2 = 1 - X/(U-Y), \quad \varepsilon_3 = 1 - X/U. \tag{3.50}$$

These are the principal axis values of the elements of the diagonalised electric permittivity tensor $\boldsymbol{\varepsilon}$ for an 'electrons only' plasma. They were first used for the ionospheric plasma by Westfold (1949). See also Lange-Hesse (1952), Davids (1953). Later in this book the following combinations of ε_1, ε_2, ε_3 appear often and it is useful to introduce the notation

$$G = \tfrac{1}{2}(\varepsilon_1 + \varepsilon_2) - \varepsilon_3, \quad J = \tfrac{1}{2}\varepsilon_3(\varepsilon_1 + \varepsilon_2) - \varepsilon_1 \varepsilon_2. \tag{3.51}$$

For an 'electrons only' plasma in which ν is assumed to be independent of velocity, it can be shown from (3.50) that

$$G = -J = -XY^2/U(U^2 - Y^2). \tag{3.52}$$

But this is not true for a plasma with more than one contributing species of ion, nor when ν is velocity dependent; see § 3.12.

When $\varepsilon_1, \varepsilon_2, \varepsilon_3$ have been found, the equations may be transformed back to give the permittivity tensor $\boldsymbol{\varepsilon}$ in any other axis system. For example in the real axis system used for (3.39), it becomes

$$\boldsymbol{\varepsilon} = \mathcal{U}^{-1} \begin{pmatrix} \varepsilon_1 & 0 & 0 \\ 0 & \varepsilon_2 & 0 \\ 0 & 0 & \varepsilon_3 \end{pmatrix} \mathcal{U} = \begin{pmatrix} \frac{1}{2}(\varepsilon_1 + \varepsilon_2) & \frac{1}{2}\mathrm{i}(\varepsilon_1 - \varepsilon_2) & 0 \\ -\frac{1}{2}\mathrm{i}(\varepsilon_1 - \varepsilon_2) & \frac{1}{2}(\varepsilon_1 + \varepsilon_2) & 0 \\ 0 & 0 & \varepsilon_3 \end{pmatrix}. \tag{3.53}$$

In magnetoionic theory, ch. 4, we use a real axis system in which the vector Y is in the x-z plane at an angle Θ to the z axis, where tan Θ is positive. In this system it can be shown that

$$\boldsymbol{\varepsilon} = \begin{pmatrix} \frac{1}{2}(\varepsilon_1+\varepsilon_2)\cos^2\Theta+\varepsilon_3\sin^2\Theta & \frac{1}{2}\mathrm{i}(\varepsilon_1-\varepsilon_2)\cos\Theta & G\sin\Theta\cos\Theta \\ -\frac{1}{2}\mathrm{i}(\varepsilon_1-\varepsilon_2)\cos\Theta & \frac{1}{2}(\varepsilon_1+\varepsilon_2) & -\frac{1}{2}\mathrm{i}(\varepsilon_1-\varepsilon_2)\sin\Theta \\ G\sin\Theta\cos\Theta & \frac{1}{2}\mathrm{i}(\varepsilon_1-\varepsilon_2)\sin\Theta & \frac{1}{2}(\varepsilon_1+\varepsilon_2)\sin^2\Theta+\varepsilon_3\cos^2\Theta \end{pmatrix}. \tag{3.54}$$

When studying waves in a horizontally stratified ionosphere, the main subject of this book, we use a coordinate system in which the vector Y has direction cosines l_x, l_y, l_z. In this system it can be shown that

$$\boldsymbol{\varepsilon} = \begin{pmatrix} \frac{1}{2}(\varepsilon_1+\varepsilon_2)-l_x^2G & -l_xl_yG+\frac{1}{2}\mathrm{i}l_z(\varepsilon_1-\varepsilon_2) & -l_xl_zG-\frac{1}{2}\mathrm{i}l_y(\varepsilon_1-\varepsilon_2) \\ -l_xl_yG-\frac{1}{2}\mathrm{i}l_z(\varepsilon_1-\varepsilon_2) & \frac{1}{2}(\varepsilon_1+\varepsilon_2)-l_y^2G & -l_yl_zG+\frac{1}{2}\mathrm{i}l_x(\varepsilon_1-\varepsilon_2) \\ -l_xl_zG+\frac{1}{2}\mathrm{i}l_y(\varepsilon_1-\varepsilon_2) & -l_yl_zG-\frac{1}{2}\mathrm{i}l_x(\varepsilon_1-\varepsilon_2) & \frac{1}{2}(\varepsilon_1+\varepsilon_2)-l_z^2G \end{pmatrix}. \tag{3.55}$$

Suppose now that the superimposed magnetic field is everywhere reversed in direction, so that Y is reversed and l_x, l_y, l_z all change sign. Then (3.55) shows that $\boldsymbol{\varepsilon}$ is replaced by its transpose.

3.11. Properties of principal axis elements of the permittivity. Effect of ions

All the dispersive and refracting properties of a cold magnetoplasma can be expressed in terms of the three elements ε_1, ε_2, ε_3 of the diagonalised permittivity tensor. It is therefore useful to set out their main properties. This is done below in a numbered sequence (1) to (16). Items (1) to (6) apply to an 'electrons only' plasma. The remaining items allow for ions of non-infinite mass.

In an 'electrons only' plasma, positive ions must be present to ensure that the plasma is electrically neutral, § 3.8, but it is implied that they are infinitely massive. Then (3.5) shows that X is zero for the ions. Consequently, for the ions, the terms (3.48) are zero and the ions make no contribution to $\boldsymbol{P}$. Then, if the electron collision frequency ν is assumed to be independent of velocity, $\varepsilon_1, \varepsilon_2, \varepsilon_3$ are given by (3.50), and this is their meaning in nearly the whole of this book. For many purposes it is useful to neglect collisions and this is done in the rest of this section. The discussion of collisions is resumed in § 3.12.

(1) From (3.50):

$$\varepsilon_1 = 1 - X/(1+Y) = 1 - f_N^2/\{f(f+f_H)\} \tag{3.56}$$

$$\varepsilon_2 = 1 - X/(1-Y) = 1 - f_N^2/\{f(f-f_H)\} \tag{3.57}$$

$$\varepsilon_3 = 1 - X = 1 - f_N^2/f^2. \tag{3.58}$$

With collisions neglected, all three εs are real when X is real. But note that, in some later sections, complex values of X are used. Curves showing how these elements depend on frequency are given in fig. 3.1. The behaviour is slightly different for $f_N < f_H$ and $f_N > f_H$, and examples of both are given.

(2) The zeros of the εs are given by

$$\varepsilon_1: X = 1 + Y \quad f = \pm(\tfrac{1}{4}f_H^2 + f_N^2)^{\frac{1}{2}} - \tfrac{1}{2}f_H \tag{3.59}$$

$$\varepsilon_2: X = 1 - Y \quad f = \pm(\tfrac{1}{4}f_H^2 + f_N^2)^{\frac{1}{2}} + \tfrac{1}{2}f_H \tag{3.60}$$

$$\varepsilon_3: X = 1 \qquad f = f_N. \tag{3.61}$$

The frequencies (3.59) (3.60) are called 'cut-off' frequencies, and (3.61) is called the 'window' frequency (see §§ 17.6–17.9), although some authors call this also a cut-off frequency. In each case there is only one real positive value of f that gives a zero. The three zeros are marked in fig. 3.1.

(3) ε_2 is infinite when $f = f_H$, the electron gyro-frequency. The other two are never infinite when $f > 0$.

(4) All three εs tend to infinity when $f \to 0$. For very small f, ε_1 and ε_3 are negative and ε_2 is positive.

(5) ε_1 and ε_2 are never equal (see item (15) below).

(6) $\varepsilon_1 + \varepsilon_2$ is zero when

$$X = 1 - Y^2; \quad f = \pm f_U; \quad f_U = (f_N^2 + f_H^2)^{\frac{1}{2}} \quad \text{(positive value)} \tag{3.62}$$

f_U is called the 'upper hybrid resonance frequency', and is marked U in fig. 3.1. Its significance is explained in § 5.7 item (7).

When ions of non-infinite mass are allowed for, the contributions to $\boldsymbol{P}$ in (3.47) from electrons and from each species of ion must be added. It is now assumed that all ions have a single positive charge. A subscript is added to X, Y etc. to distinguish the

Fig. 3.1. Frequency dependence of ε_1, ε_2, ε_3 for a cold collisionless electron plasma. In the left-hand figure $f_N/f_H = 2$, and in the right-hand figure $f_N/f_H = 0.5$. The ordinate is the modulus on a logarithmic scale. Continuous curves are used for positive values and broken curves for negative values of the εs. U denotes the upper hybrid resonance. Abscissa scale is linear.

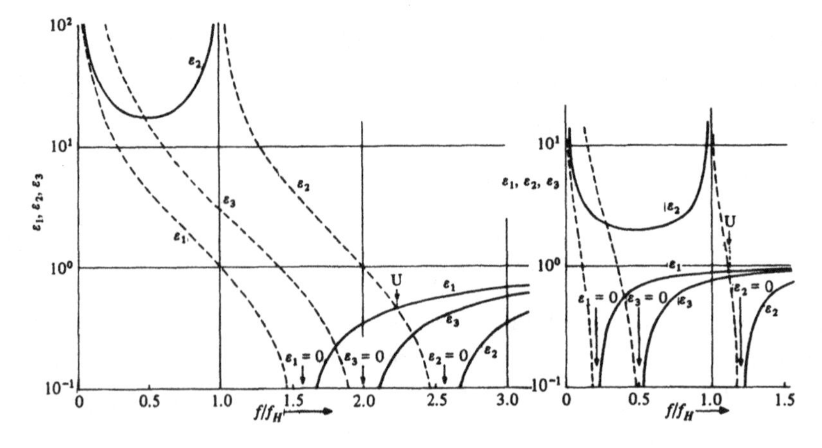

values for electrons and ions. Thus we use the integer $i = 1, 2, 3 \ldots\ldots$ to designate the ion species in order of increasing mass.

The frequencies

$$f_{Ni} = \frac{1}{2\pi}(N_i e^2/\varepsilon_0 m_i)^{\frac{1}{2}}, \quad f_{Hi} = \frac{1}{2\pi}|eB/m_i| \tag{3.63}$$

are called the ion plasma frequency and the ion gyro-frequency respectively. For electrons the subscript i is replaced by e. In most of this section $\nu_i = Z_i = 0$; $U_i = 1$ for all i including $i = \mathrm{e}$.

(7) Let $C_i = N_i/N_e$. Since the plasma is electrically neutral it is necessary that

$$N_e = \sum_i N_i, \quad \sum_i C_i = 1. \tag{3.64}$$

In cases where $\sum_i C_i < 1$ it is implied that positive ions of infinite mass are present with concentration $N_e - \sum_i N_i$.

(8) The expressions for the εs in (1) are replaced by

$$\varepsilon_1 = 1 - X_e/(U_e + Y_e) - \sum_i X_i(U_i - Y_i) \tag{3.65}$$

$$\varepsilon_2 = 1 - X_e/(U_e - Y_e) - \sum_i X_i(U_i + Y_i) \tag{3.66}$$

$$\varepsilon_3 = 1 - X_e/U_e - \sum_i X_i/U_i. \tag{3.67}$$

The U_is and U_e are retained here for future reference. As in (1) all three εs are real when collisions are neglected. Curves giving examples of how they depend on frequency are shown in fig. 3.2 for a plasma with two species of positive ion.

(9) Each of (3.65)–(3.67) contains a 'sum over species'. Some authors use a different notation for these. A fairly widely used system is that of Stix (1962), namely

$$\varepsilon_1 = L, \quad \varepsilon_2 = R, \quad \varepsilon_3 = P \tag{3.68}$$

(10)

$$\varepsilon_1(f) = \varepsilon_2(-f) \tag{3.69}$$

This is useful, for example, when calculating ε_1 and ε_2 with a programmable pocket calculator, when collisions are neglected. The same program will do both. It is true also for an 'electrons only' plasma.

(11) (3.5) and (3.24) show that

$$X_i = N_i e^2/(\varepsilon_0 m_i \omega^2) = C_i X_e(m_e/m_i) \tag{3.70}$$

$$Y_i = eB/\omega m_i \quad = Y_e(m_e/m_i). \tag{3.71}$$

Now m_e/m_i is very small, about 5.4×10^{-4} if i refers to protons and smaller still for heavier ions. Hence ε_3 in (3.67) is only very slightly different from (3.58) at all frequencies. For (3.65), (3.66), if f is greater than four or five times the greatest ion gyro-frequency f_{H1}, the Σ terms are small and ε_1, ε_2 are very close to their values (3.56) (3.57) for electrons only.

(12) ε_3 has one zero where $f=f_0$ and this is slightly greater than, but very close to f_{Ne}. In practical cases the difference is negligible. f_0 is called the 'window' frequency; Budden and Stott (1980). See §§ 17.6–17.9. ε_1 and ε_2 have zeros very close to (3.59), (3.60) respectively. It can be shown that ε_2 has no other zeros, but the total number of zeros of ε_1, for positive f, is equal to the total number of positive ion species, including ions of infinite mass.

(13) ε_2 is infinite when $f=f_{He}$, the electron gyro-frequency, as in (3). ε_1 is infinite when $f=f_{Hi}$, that is at each of the ion gyro-frequencies.

(14) When $f\to 0$, $\varepsilon_3\to\infty$ as in (4). If f is very small and there are no ions of infinite mass, it can be shown, by using (3.64), that

$$\left.\begin{matrix}\varepsilon_1\\ \varepsilon_2\end{matrix}\right\} = 1 + \rho/\varepsilon_0 B^2 \pm 2\pi N_e M f/(\varepsilon_0 e B^3) + \mathrm{O}(f^2) \tag{3.72}$$

where

$$\rho = N_e m_e + \sum_i N_i m_i, \quad M = \sum_i C_i m_i^2 - m_e^2. \tag{3.73}$$

Fig. 3.2. Similar to fig. 3.1 but for a plasma in which the positive ions are protons and singly charged helium ions in equal concentrations. $f_{Ne}/f_H = 0.5$. Both scales are logarithmic.

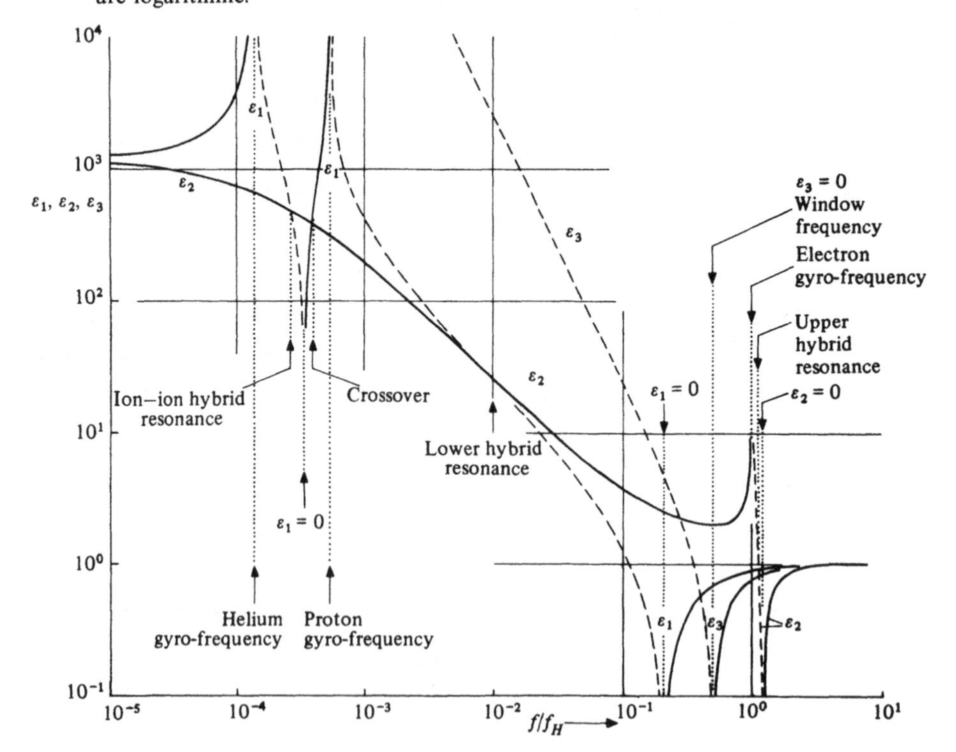

and the unit charge e is here a positive number. Thus as $f \to 0$, ε_1 and ε_2 approach the same non-infinite value. See fig. 3.2.

(15) Frequencies $f \neq 0$ where $\varepsilon_1 = \varepsilon_2$ are called 'crossover' frequencies (Smith and Brice, 1964). It can be shown from (3.65) (3.66) that there are none if there is only one positive ion species. A crossover frequency occurs between each consecutive pair of ion gyro-frequencies. Fig. 3.2 shows an example. When $\varepsilon_1 = \varepsilon_2$, (3.53) shows that $\boldsymbol{\varepsilon}$ is diagonal for the real axes used there, and two of its three elements are equal. A dielectric with this property is said to be 'uniaxial'. For some examples of its properties see §§ 5.5, 19.3, problem 4.7. Such a medium is about the simplest possible kind of anisotropic medium and is most useful for illustrating the theory of wave phenomena. See, for example, Clemmow (1966, ch. 8), Felsen (1964).

(16) Frequencies for which $\varepsilon_1 + \varepsilon_2 = 0$ are called hybrid frequencies. The greatest is the upper hybrid frequency f_{U} and is still given approximately by (3.62). This result is very little affected by the positive ions. The next greatest is denoted by f_{L} and is called the lower hybrid frequency. It exceeds the greatest ion gyro-frequency f_{H1}, and it is given approximately by

$$f_{\mathrm{L}} = \frac{f_{H\mathrm{e}} f_{N\mathrm{e}} m_{\mathrm{e}}^{\frac{1}{2}}}{f_{\mathrm{U}}} \left(\sum_i \frac{C_i}{m_i} \right)^{\frac{1}{2}}. \tag{3.74}$$

There is also a hybrid frequency, called an ion–ion hybrid frequency, between each pair of ion gyro-frequencies. There is an example in fig. 3.2. The total number of hybrid frequencies is equal to the number of ion species of non-infinite mass, including electrons.

For a more detailed description of the effect of positive ions on the properties of a plasma see Al'pert (1983), Al'pert, Budden *et al.* (1983), Budden and Stott (1980), Booker (1984).

3.12. Collisions. The Sen–Wyller formulae

In this section we again consider an 'electrons only' plasma. If the collision frequency ν is independent of electron velocity, the εs are given by (3.50). But in practice ν is a function of the velocity v and allowance must be made for this. In most of this book ν means the effective collision frequency used in (3.15), but in this section it denotes the velocity dependent collision frequency $\nu(v)$. The effective collision frequency is here temporarily written ν_{eff}. The contribution to $\boldsymbol{P}$ from one species of charged particle is given by (3.47), and the total $\boldsymbol{P}$ is found by adding the contributions for all the species. Consider now, those electrons whose thermal velocity v is in the range v to $v + \delta v$, and let their concentration be δN. It might at first be thought that these various groups of δN electrons can be treated as though they are different species, and that their contributions to $\boldsymbol{P}$ in (3.47) should be added. This, however, is oversimplified. A full kinetic theory shows that the groups δN cannot be treated in this way, as though they are independent of each other.

Collisions are of greatest importance in the lower part of the ionosphere, below about 120 km, where the velocity distribution function of the electrons may with fair accuracy be taken as Maxwellian and isotropic in velocity space. Thus

$$\delta N = 4\pi^{-\frac{1}{2}} N(\tfrac{1}{2}m/KT)^{\frac{3}{2}} v^2 \exp(-\tfrac{1}{2}mv^2/KT)\delta v \tag{3.75}$$

where T is the electron temperature. Instead of v it is convenient to use the variable

$$u = \tfrac{1}{2}mv^2/KT. \tag{3.76}$$

When $\nu(v)$ is expressed as a function of u, let it be denoted by $\check{\nu}(u)$. For the distribution (3.75) the full theory shows that

$$P_1 = -\tfrac{4}{3}\pi^{-\frac{1}{2}}\varepsilon_0 X E_1 \int_0^\infty \frac{u^{\frac{3}{2}} e^{-u}\, du}{U(u) + Y} \tag{3.77}$$

with similar expressions for P_2 and P_3. If the contributions (3.47) were simply added, the factor $u^{\frac{3}{2}}$ in (3.77) would be replaced by $\frac{3}{2}u^{\frac{1}{2}}$. The full theory leading to (3.77) is given by Sen and Wyller (1960), Sen (1967). See also Phelps (1960). A version of it that uses the notation and methods of this book was given by Budden (1965).

Now (3.47) is used to give the three elements of the permittivity tensor. In the expression relating P_2 to E_2, the Y in (3.77) is replaced by $-Y$. For P_3 and E_3 the Y is omitted. Then:

$$\varepsilon_1 = 1 - \tfrac{4}{3}\pi^{-\frac{1}{2}} X \int_0^\infty \frac{u^{\frac{3}{2}} e^{-u} du}{U(u) + Y}, \quad \varepsilon_2 = 1 - \tfrac{4}{3}\pi^{-\frac{1}{2}} X \int_0^\infty \frac{u^{\frac{3}{2}} e^{-u} du}{U(u) - Y},$$

$$\varepsilon_3 = 1 - \tfrac{4}{3}\pi^{-\frac{1}{2}} X \int_0^\infty \frac{u^{\frac{3}{2}} e^{-u} du}{U(u)}. \tag{3.78}$$

If ν is independent of v, so that U is a constant, the integrals in (3.78) can be evaluated at once, and the result (3.50) is obtained.

The collision frequency ν cannot be independent of v, because, if it were, the phenomenon of wave interaction, also called the Luxembourg effect, could not occur; see §§ 13.11–13.13. It has often been assumed, especially in studies of the theory of wave interaction, that the electrons have a constant mean free path λ_e. This would make ν proportional to v thus

$$\nu = v/\lambda_e, \quad \langle \nu \rangle = \left(\frac{8KT}{\pi m}\right)^{\frac{1}{2}} \frac{1}{\lambda_e} \tag{3.79}$$

where the Maxwellian distribution (3.75) has been used to find the average value $\langle \nu \rangle$. Some results for this case have been given by Margenau (1946), Ginzburg (1970). But laboratory experiments (Crompton, Huxley and Sutton, 1953; Phelps and Pack, 1959; Huxley and Crompton, 1974) have shown that for electrons of small energy, down to about 0.1 eV, the collision frequency is more nearly proportional to v^2. It will therefore now be assumed that

$$\nu(v) = \beta v^2, \quad \nu_{av} = 3\beta KT/m, \quad \check{\nu}(u) = \tfrac{2}{3}\nu_{av} u \tag{3.80}$$

where (3.75) was used to find ν_{av}. Then the three integrals in (3.78) can all be written in the form

$$\mathscr{I} = w\int_0^\infty \frac{u^{\frac{3}{2}}e^{-u}du}{s - iu} \tag{3.81}$$

where

$$w = \tfrac{3}{2}\omega/\nu_{av} \tag{3.82}$$

and s is $w(1 + Y)$, $w(1 - Y)$, w for ε_1, ε_2, ε_3 respectively.

The integrals can be separated into their real and imaginary parts, and all six resulting integrals can be expressed in terms of the standard functions

$$\mathscr{C}_p(s) = \frac{1}{p!}\int_0^\infty \frac{u^p e^{-u}du}{u^2 + s^2} \tag{3.83}$$

thus

$$\left.\begin{aligned}
\varepsilon_1 &= 1 - X[w^2(1 + Y)\mathscr{C}_{\frac{3}{2}}\{w(1 + Y)\} + iw\tfrac{5}{2}\mathscr{C}_{\frac{5}{2}}\{w(1 + Y)\}]\\
\varepsilon_2 &= 1 - X[w^2(1 - Y)\mathscr{C}_{\frac{3}{2}}\{w(1 - Y)\} + iw\tfrac{5}{2}\mathscr{C}_{\frac{5}{2}}\{w(1 - Y)\}]\\
\varepsilon_3 &= 1 - X[w^2\mathscr{C}_{\frac{3}{2}}(w) + iw\tfrac{5}{2}\mathscr{C}_{\frac{5}{2}}(w)].
\end{aligned}\right\} \tag{3.84}$$

The functions (3.83) appear in various problems of kinetic theory. They have been tabulated by Dingle, Arndt and Roy (1956 b). A way of calculating them that is very useful in computing was given by Hara (1963).

Sen and Wyller (1960), Sen (1967) and others have used (3.84) in magnetoionic theory to calculate the refractive indices and polarisations for radio waves in a cold magnetoplasma. The resulting formulae are known as the Sen–Wyller formulae. Some authors call this theory 'generalised magnetoionic theory' but this is incorrect. The Appleton–Lassen version of the theory is a special case where ν is assumed to be independent of v. The Sen–Wyller version is another special case where ν is assumed to be proportional to v^2.

Instead of using the average value ν_{av} (3.80), some authors use a value $\nu_m = \frac{2}{3}\nu_{av}$ which they call the 'monoenergetic collision frequency'. This term is misleading and should be abandoned. For a physical variable, the natural and widely understood parameter to use in describing it is the average value.

If ν_{av}/ω is small enough, so that s is large, the factor $(1 - iu/s)^{-1}$ in (3.81) may be expanded by the binomial theorem, and the series may then be integrated term by term. This gives an asymptotic (divergent) expansion for the function $\mathscr{I}$; see § 8.11. Thus

$$\mathscr{I} = \tfrac{3}{4}\pi^{\frac{1}{2}}\frac{w}{s}\left[1 + \frac{5i}{2s} - \frac{35}{4s^2}\cdots\right] = \tfrac{3}{4}\pi^{\frac{1}{2}}\frac{w}{s}\left[1 - \frac{5i}{2s} + \frac{10}{4s^2}\right]^{-1}. \tag{3.85}$$

If s is so large that only the first two terms need to be used, (3.78), with the last expression in (3.85), gives

$$\varepsilon_1 = 1 - X/(1 + Y - iZ_{eff}),\ \varepsilon_2 = 1 - X/(1 - Y - iZ_{eff}),\ \varepsilon_3 = 1 - X/(1 - iZ_{eff}) \tag{3.86}$$

where

$$Z_{\text{eff}} = 5/2w, \quad \nu_{\text{eff}} = \omega Z_{\text{eff}} = \tfrac{5}{3}\nu_{\text{av}}. \tag{3.87}$$

Thus the standard result (3.50) may still be used provided that the correct 'effective' collision frequency ν_{eff} (3.87) is used. This conclusion is satisfactory for nearly all radio propagation problems in the ionosphere and magnetosphere.

Results of calculations that compared the Appleton–Lassen formulae and the Sen–Wyller formulae were presented by several authors. Thus Budden (1965) included a study of very low frequencies where $Z_{\text{eff}} = \nu_{\text{eff}}/\omega$ is large so that failure of (3.86) might be expected for ε_3. Smith (1975) studied the case $f \approx f_H$ so that $Y \approx 1$ and failure of (3.86) must occur for ε_2. Johler and Harper (1962), Deeks (1966b) and others used both formulae in studies of the ionospheric reflection of radio waves of very low frequency. Numerous other cases were given by Sen and Wyller (1960), Sen (1967). The results show that in nearly all cases there is very little difference. The electron collision frequency is not known with sufficient accuracy to be able to distinguish between the two in any actual radio measurements. The author does not know of any radio observations that cannot be explained by the Appleton–Lassen formulae.

The electron collision frequency can be shown to be given by $\nu = N_{\text{m}} A v$ where N_{m} is the concentration of the particles collided with, and A is the 'transport area of cross section'. If ν is proportional to v^2, this shows that A must be very small for slowly moving electrons, and that their mean free path $1/N_{\text{m}}A$ must be very large. But this seems most unlikely on general physical grounds. For an electron temperature of 800 K, typical for a height of about 150 km, three quarters of the electrons have energies less than about 0.1 eV. The assumption (3.80) is based on laboratory measurements, in most of which electron energies of 0.1 to 0.2 eV or more were used. Measurements for lower energies are more difficult but some have been reported. There may be some uncertainty about how well a confined laboratory plasma can simulate the ionospheric plasma. There seems to be no strong evidence that the result $\nu \propto v^2$ can be used for the electrons of lowest energy in the ionosphere. For these reasons it seems preferable at present to retain the simpler formulae (3.47), as is done in this book.

It is possible that, for the small electron thermal velocities of interest in the ionosphere, the dependence of ν on v does not follow any power law. But for the reasons given above it is the author's opinion that the assumption $\nu \propto v$ is likely to be better than $\nu \propto v^2$. It may therefore be useful to give the relevant formulae. Results equivalent to (3.91) below were derived by Margenau (1946). See also Allis (1956, §42). Instead of (3.80) we now use (3.79) and the three integrals in (3.78) can then all be written in the form

$$\mathscr{J} = W^{\frac{1}{2}} \int_0^\infty \frac{u^{\frac{3}{2}} e^{-u} \, du}{t^{\frac{1}{2}} - iu^{\frac{1}{2}}} \tag{3.88}$$

where

$$W = \frac{4}{\pi}\left(\frac{\omega}{\langle v\rangle}\right)^2 = \tfrac{1}{2}\omega^2\lambda_e^2 m/KT \tag{3.89}$$

and t is $(1+Y)^2W, (1-Y)^2W, W$ for $\varepsilon_1, \varepsilon_2, \varepsilon_3$ respectively. The integrals (3.88) can be separated into their real and imaginary parts, and all six resulting integrals can be expressed in terms of the standard functions

$$\mathscr{A}_p(t) = \frac{1}{p!}\int_0^\infty \frac{u^p e^{-u} du}{t+u}. \tag{3.90}$$

These, like the functions (3.83), have been tabulated (Dingle, Arndt and Roy, 1956a). When used in (3.78) they give finally

$$\begin{aligned}
\varepsilon_1 &= 1 - X\left[W(1+Y)\mathscr{A}_{\frac{3}{2}}\{(1+Y)^2W\} + \mathrm{i}W^{\frac{1}{2}}\frac{8}{3\sqrt{\pi}}\mathscr{A}_2\{(1+Y)^2W\}\right]\\
\varepsilon_2 &= 1 - X\left[W(1-Y)\mathscr{A}_{\frac{3}{2}}\{(1-Y)^2W\} + \mathrm{i}W^{\frac{1}{2}}\frac{8}{3\sqrt{\pi}}\mathscr{A}_2\{(1-Y)^2W\}\right]\\
\varepsilon_3 &= 1 - X\left[W\mathscr{A}_{\frac{3}{2}}(W) + \mathrm{i}W^{\frac{1}{2}}\frac{8}{3\sqrt{\pi}}\mathscr{A}_2(W)\right].
\end{aligned} \tag{3.91}$$

If the average collision frequency $\langle v\rangle$ is small so that W and t are large, the factor $\{1 - \mathrm{i}(u/t)^{\frac{1}{2}}\}^{-1}$ in (3.88) may be expanded by the binomial theorem, and the series integrated term by term. This gives the asymptotic series

$$\mathscr{I} = \frac{3\sqrt{\pi}}{4}\left(\frac{W}{t}\right)^{\frac{1}{2}}\left\{1 - \frac{8}{3\sqrt{\pi}t^{\frac{1}{2}}}\mathrm{i} + \frac{0.237}{t} + \ldots\right\}^{-1}. \tag{3.92}$$

If t is so large that only two terms need to be used, the result is the same as (3.86) where now

$$Z_{\mathrm{eff}} = \tfrac{4}{3}\langle v\rangle/\omega, \quad v_{\mathrm{eff}} = \tfrac{4}{3}\langle v\rangle = \frac{4}{3\lambda_e}\left(\frac{8KT}{\pi m}\right)^{\frac{1}{2}}, \tag{3.93}$$

compare (3.87).

3.13. Electron–electron collisions. Electron–ion collisions

Two other aspects of collisions in a plasma have received some study and must be briefly mentioned. The first is collisions between like particles, particularly between electrons. In the small volume of plasma considered in this chapter (see § 3.5) the electrons all have, on average, the same ordered velocity. Thus for any two electrons the momentum associated with the ordered motion is the same. If these two electrons collide, the total ordered momentum that they share cannot change. This suggests that electron–electron collisions should not, on average, affect the ordered part of the motion, so that the constitutive relation is also unaffected by them. There is, however, a way in which these collisions can have an effect. In a wave the ordered velocities vary in space over a distance of about a wavelength. The part of the ordered velocity perpendicular to the wave normal thus has a spatial gradient of the same kind as is studied in the theory of viscosity in a gas. Transfer of ordered electron momentum between regions of different transverse velocity gives rise to a viscous force. This can lead to an additional damping effect which has been called 'electron viscosity'. It depends on the spatial variation of the wave fields and is therefore an example of 'spatial dispersion' mentioned in § 3.1. It is unlikely to be important for radio waves, since their wavelength is long, but can be important for other kinds of wave in a plasma. For a discussion of electron–electron collisions see Al'pert (1980b, 1983), Akhiezer, Lapshin and Stepanov (1976), Ginzburg (1970).

The other effect to be mentioned concerns collisions between unlike particles. In § 3.4 it was implied that the particles that an electron collides with have an average velocity of zero. This is not strictly true. If these particles are ions, they have an ordered velocity component imparted to them by the wave. In (3.12) the collision damping force is $m\nu\partial \boldsymbol{r}_e/\partial t$ where $\partial \boldsymbol{r}_e/\partial t$ is the ordered velocity of the electrons, as now indicated by the subscript e. Consider the contribution to this term from the various ion species $i = 1, 2, \ldots$, and let ν_{ei} be the appropriate effective collision frequencies. Then $\partial \boldsymbol{r}_e/\partial t$ must be replaced by the relative velocity of the electrons and ions, so that the correct damping force is

$$m\sum_i \nu_{ei}\partial(\boldsymbol{r}_e - \boldsymbol{r}_i)/\partial t. \tag{3.94}$$

For many frequencies, including those of the radio waves studied in this book, the ion velocities $\partial \boldsymbol{r}_i/\partial t$ are small enough to be neglected, but this is not true for frequencies near to or less than the ion gyro-frequency. This effect has been studied by Al'pert (1980b, 1983) who gave examples showing that it can be important. Ions and electrons can, through collisions, impart to the neutral particles an ordered velocity $\partial \boldsymbol{r}_n/\partial t$ and the summation in (3.94) should include a term to allow for this. A full theory that does this was given by Al'pert, Budden, Moiseyev and Stott (1983, appendix A).

PROBLEMS 3

3.1. A radio transmitter on the ground, with a power of 100 kW, emits continuous waves of frequency 2 MHz uniformly in all directions above the earth's surface. Its radiated field acts on an electron in the lower ionosphere at 100 km from the transmitter. Find the ordered velocity imparted to the electron if the electric field is the same as it would be in free space. Compare this with the average thermal velocity of the electron if the temperature is 500 K.

3.2. The energy supplied to unit volume of a plasma by an electric field $\boldsymbol{e}$, that is applied starting at time $t = 0$, is

$$\mathscr{S}_E = \int_0^t \boldsymbol{e}\frac{\partial \boldsymbol{d}}{\partial t}\mathrm{d}t.$$

Apply this to a cold electron plasma in which there is no superimposed magnetic field, and collisions are negligible. Since the energy cannot be dissipated as heat it must be stored in the plasma. Show that, when $\boldsymbol{e}$ is the field $E_0 \sin \omega t$ of a linearly polarised radio wave, this stored energy is

$$\tfrac{1}{2}E_0^2\left(\varepsilon_0 \sin^2 \omega t + \frac{Ne^2}{m\omega^2}\cos^2 \omega t\right).$$

Show that the first term is the same as the energy that would be stored in the electric field if it were in a vacuum, and the second term is the kinetic energy of the electrons. Note that the energy $\frac{1}{2}\mu_0 H^2$ supplied by the magnetic field H of the wave is not included.

(Two methods of solving this problem were given by Budden, 1961a, pp. 33–5.)

3.3. A cold plasma contains free electrons. There is no superimposed magnetic field and collisions are negligible. A plane linearly polarised radio wave is present whose electric field at a given point is $E_0 \sin \omega t$. The wave is progressive if $\omega > \omega_N$, or evanescent if $\omega < \omega_N$. Prove that the total stored energy per unit volume, including contributions from both electric and magnetic fields, is

$$\tfrac{1}{2}\varepsilon_0 E_0^2\{1 + (X - 1)\cos 2\omega t\} \quad \text{when} \quad \omega > \omega_N,$$
$$\tfrac{1}{2}\varepsilon_0 E_0^2\{X + (X - 1)\cos 2\omega t\} \quad \text{when} \quad \omega < \omega_N,$$

where $X = \omega_N^2/\omega^2$.

3.4. A uniform cold electron plasma occupies the space between two parallel planes and there is a vacuum outside. There is no superimposed magnetic field and collisions are negligible. The electrons are all displaced by the same small distance perpendicular to the planes, and are then released from rest. Show that they oscillate with angular frequency ω_N. (Assume that *all* the electrons, including those near the boundary, experience the same electric field as an electron near the centre.)

Repeat this problem for a plasma enclosed in a sphere and show that the oscillation frequency is now $\omega_N/\sqrt{3}$. Try it also for a plasma in a cylinder if the initial electron displacements are perpendicular to the axis. (Oscillation frequency $\omega_N/\sqrt{2}$.)

(Note. These oscillations are sometimes called 'plasma oscillations' but they are very different from those described in § 3.8. For an application to radio propagation, see: Herlofson, 1951; Pitteway, 1958, 1960.)

3.5. In a plasma formed of ionised helium gas, half the ions are singly charged and the other half are doubly charged. Show that this plasma has one crossover frequency, $f = f_{H1}\sqrt{2}$ approximately, where f_{H1} is the gyro-frequency of the singly charged ions.

3.6. A system of principal axis coordinates x_1, x_2, x_3 is given in terms of the Cartesian coordinates x, y, z by $x_1 = 2^{-\frac{1}{2}}(x + \mathrm{i}y)$, $x_2 = 2^{-\frac{1}{2}}(x - \mathrm{i}y)$, $x_3 = z$. Similar relations, (3.41), hold for the contravariant principal axis components of any vector.

V is a scalar function of position. Prove the following results:

(a) $\boldsymbol{a} \cdot \boldsymbol{b} = a_1 b_2 + a_2 b_1 + a_3 b_3$

(b) $\operatorname{div} \boldsymbol{a} = \partial a_1/\partial x_1 + \partial a_2/\partial x_2 + \partial a_3/\partial x_3$

The contravariant principal axis components of the following vectors are:

(c) $\boldsymbol{a} \wedge \boldsymbol{b}$: $\mathrm{i}\{a_3 b_1 - a_1 b_3, a_2 b_3 - a_3 b_2, a_1 b_2 - a_2 b_1\}$

(d) $\boldsymbol{a} \wedge \boldsymbol{a}^*$: $\mathrm{i}\{a_3 a_2^* - a_1 a_3^*, a_2 a_3^* - a_3 a_1^*, a_1 a_1^* - a_2 a_2^*\}$

(e) $\operatorname{curl} \boldsymbol{a}$: $\mathrm{i}\{\partial a_1/\partial x_3 - \partial a_3/\partial x_2, \partial a_3/\partial x_1 - \partial a_2/\partial x_3, \partial a_2/\partial x_2 - \partial a_1/\partial x_1\}$

(f) grad V: $\partial V/\partial x_2, \partial V/\partial x_1, \partial V/\partial x_3$

(Notes. (i) Some of these results are useful in one method of deriving the dispersion relation for a cold magnetoplasma. It is the method that uses Maxwell's equations (2.20)–(2.23) expressed in the principal axis system. See problem 4.12. (ii) The covariant principal axis components of grad V are $\partial V/\partial x_1$, $\partial V/\partial x_2$, $\partial V/\partial x_3$.)

3.7. An electron of mass m and speed v collides with a molecule of mass $M \gg m$ initially at rest. The angle between the electron's paths before and after the collision is χ. Show that the fraction of kinetic energy lost by the electron is $2m/M(1 - \cos\chi) + O(m/M)^2$. In many such collisions suppose that the final paths can have any possible direction with equal probability. Show that the average fractional loss of kinetic energy is $G \approx 2m/M$.

(Experimental measurements show that in practice G can be considerably larger than this. A full treatment of this problem must make allowance for the thermal velocities of the molecules. See §§ 13.12, 13.13 and the references given there.)

3.8. Make a rough estimate of the duration τ of the encounter of an electron with (a) a massive ion of charge e, (b) a massive neutral molecule such as N_2 or O_2. (Hint: the electron's motion will be appreciably affected by the ion if it is within a range equal to the Debye length l_D. The duration τ will be very roughly the time it takes for an electron with the average thermal velocity to travel a distance l_D. For a diatomic molecule let d be the separation of the atoms. The range of the force that the molecule exerts on an electron may be of the order $2d$. The duration τ will be very roughly the time it takes the electron to travel a distance $4d$. The values of d for N_2 and O_2 are 0.11, 0.12 nm respectively. The mean molecular diameters are about three times this).

3.9. An electron approaches a very massive neutral spherical molecule along a line which, when produced, passes at shortest distance b from the molecule's centre, and b is called the 'impact parameter'. After the encounter, the electron has been deflected through an angle χ, and $A = 2\pi \int_0^\infty (1 - \cos\chi) b \, db$ is called the 'transport area of cross section' for the encounter. As a simple model, the molecule can be thought of as an infinitely massive, perfectly conducting small sphere, at rest.

(a) Show that an electron at distance r experiences an attractive force proportional to r^{-5}.

(b) If the electron approaches with velocity v, show that A is proportional to $1/v$. (Hint: the differential equation of the orbit is

$$\frac{d^2u}{d\theta^2} + u = \frac{Ku^3}{b^2v^2}$$

where $u = 1/r$, K is a constant and θ is the polar coordinate angle. Show that χ is given by

$$\pi + \chi = 2\int_0^{x_0} \{1 - x^2 + Kx^4/(2v^2b^4)\}^{-\frac{1}{2}} dx$$

where x_0 is the greatest value of b/r. Hence show that $\cos\chi$ depends on b and v only through the combination b^2v. Use this in the integral for A. It is not necessary to evaluate the integrals).

(c) Hence show that the effective collision frequency is independent of v.

4

Magnetoionic theory 1. Polarisation and refractive index

4.1. Plane wave and homogeneous plasma

Magnetoionic theory is that branch of our subject that studies the properties of a plane electromagnetic wave in a homogeneous cold magnetoplasma, by using Maxwell's equations together with the electric permittivity tensor as given by the constitutive relation, ch.3. A plane wave was defined in § 2.8 and the z axis was there chosen to be parallel to the wave normal. This choice is used here. A progressive wave was defined as one in which the z dependence of all field components is only through the same factor $\exp(-iknz)$ where $k = \omega/c$ and n is the refractive index. Thus for operations on any field component

$$\partial/\partial x \equiv 0, \quad \partial/\partial y \equiv 0, \quad \partial/\partial z \equiv -ikn, \tag{4.1}$$

and hence Maxwell's equations in the form (2.45) give

$$nE_y = -\mathscr{H}_x \tag{4.2}$$

$$n\mathscr{H}_y = \varepsilon_0^{-1} D_x \tag{4.3}$$

$$nE_x = \mathscr{H}_y \tag{4.4}$$

$$n\mathscr{H}_x = -\varepsilon_0^{-1} D_y \tag{4.5}$$

$$\mathscr{H}_z = 0 \tag{4.6}$$

$$D_z = 0 \tag{4.7}$$

In later chapters a different choice of coordinates x, y, z is used so that the field components in (4.2)–(4.7) there have different meanings.

Equations (4.6)–(4.7) show that $\boldsymbol{\mathscr{H}}$ and $\boldsymbol{D}$ are perpendicular to the wave normal. This must always apply to any plane wave. The constitutive relation gives $\boldsymbol{D}$ in terms of $\boldsymbol{E}$ thus

$$\boldsymbol{D} = \varepsilon_0 \boldsymbol{E} + \boldsymbol{P} = \varepsilon_0 \boldsymbol{\varepsilon} \boldsymbol{E}. \tag{4.8}$$

This and equations (4.2)–(4.7) are the starting-point for the study of magnetoionic theory in this and the following chapter.

4.2. Isotropic plasma

For an isotropic plasma $\boldsymbol{\varepsilon}$ is a scalar ε and $\boldsymbol{D}$ is parallel to $\boldsymbol{E}$. Then the ratio $\mathscr{H}_x/E_y$ may be eliminated from (4.2), (4.5) to give

$$n^2 = \varepsilon = 1 - X/U \tag{4.9}$$

where (3.17) has been used. The same result is obtained by eliminating $\mathscr{H}_y/E_x$ from (4.3), (4.4). These two pairs of equations are independent of each other, which shows that the two linearly polarised plane waves, with their electric fields parallel to the y and x axes respectively, are propagated independently and (4.9) applies to both. They therefore travel with the same wave velocity $v = c/n$. Two such linearly polarised waves with any amplitudes and relative phase could be combined to give a wave which would in general be elliptically polarised.

For a collisionless plasma (4.9) gives

$$n^2 = 1 - X = 1 - f_N^2/f^2. \tag{4.10}$$

Fig. 4.1. Shows how the real part μ and the imaginary part $-\chi$ of the refractive index $n = \mu - i\chi$ depend on X for an isotropic plasma. The numbers by the curves are the values of $Z = \nu/\omega$.

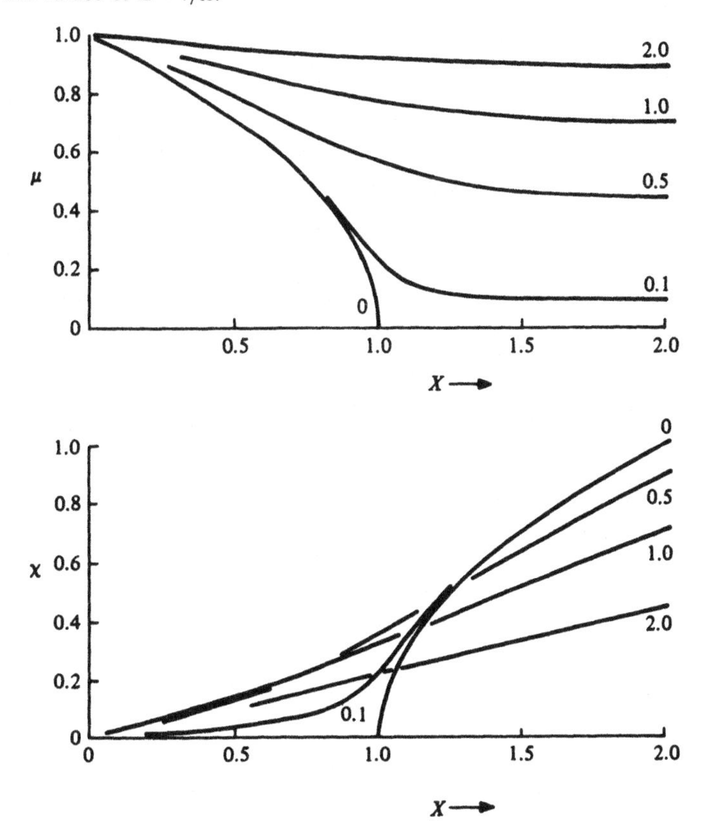

This is the same as ε_3, (3.58), for a collisionless electron magnetoplasma. Its frequency dependence is shown in fig. 3.1. Thus if $f > f_N$, n is real and the wave is propagated without attenuation. If $f < f_N$, n is purely imaginary and the wave is evanescent; see § 2.14. The plasma frequency f_N is the cut-off frequency. The dispersion law (4.10) is the same as for one mode in a loss-free wave guide.

If collisions are included, (4.9) shows that n is complex. We choose the value with positive real part, and the imaginary part must then be negative. The wave travels in the direction of z increasing and is attenuated as it travels; see § 2.13. Fig. 4.1 shows how the real and imaginary parts of n depend on X for several different values of Z. More extensive curves of this kind and a full discussion are given by Ratcliffe (1959). It is clear that if Z is non-zero and real, there is no real value of X that makes $n^2 = 0$. In later chapters, however, it will prove useful to assume that X and Z are analytic functions of the height h (in later chapters denoted by z.) They must be real and positive when h is real, but when h is complex, X and Z can take complex values. Then n^2 can be zero for one or more complex values of h. These zeros of n^2 in the complex h plane play a very important part in the theory of the reflection of radio waves.

4.3. Anisotropic plasma. The wave polarisation

In an anisotropic plasma D_y cannot be expressed in terms of E_y alone, and D_x cannot be expressed in terms of E_x alone. Hence the pairs of equations (4.2), (4.5) and (4.3), (4.4) are no longer independent. This leads to a restriction on the possible states of polarisation of the wave, as will now be shown.

The z axis is chosen parallel to the wave normal but the x and y axes may be rotated about the z axis, so that we are free to impose a further condition. The axes are therefore chosen so that the vector $\boldsymbol{Y}$ is in the x-z plane at an angle Θ to the z axis, and

$$l_x = \sin\Theta, \quad l_z = \cos\Theta \tag{4.11}$$

have the same sign, so that $\tan\Theta$ is positive. Then $l_y = 0$ and the constitutive relation (3.25) gives

$$-\varepsilon_0 X \begin{pmatrix} E_x \\ E_y \\ E_z \end{pmatrix} = \begin{pmatrix} U & \mathrm{i}Y\cos\Theta & 0 \\ -\mathrm{i}Y\cos\Theta & U & \mathrm{i}Y\sin\Theta \\ 0 & -\mathrm{i}Y\sin\Theta & U \end{pmatrix} \begin{pmatrix} P_x \\ P_y \\ P_z \end{pmatrix}. \tag{4.12}$$

For the more general plasma, these same axes were used to give the form (3.54) of the permittivity tensor.

With this choice of axes, the wave polarisation ρ of a progressive wave is defined thus

$$\rho = E_y / E_x. \tag{4.13}$$

Equations (4.2), (4.4) then show that

$$\rho = -\mathscr{H}_x/\mathscr{H}_y. \tag{4.14}$$

The component E_z may not be zero, but it does not enter into the definition of ρ. The component $\mathscr{H}_z$ is always zero.

The wave polarisation ρ is a complex number which shows how the transverse component of $\boldsymbol{E}$ varies with time. For example if ρ is real, E_x and E_y have the same phase and the transverse component of $\boldsymbol{E}$ is always parallel to a fixed line; this is called linear polarisation. If $\rho = \mathrm{i}$, E_x and E_y are in quadrature and have equal amplitudes, and the polarisation is circular. To an observer looking in the direction of the wave normal, that is in the direction of positive z, the electric vector observed at a fixed point would appear to rotate anticlockwise. This is called left-handed circular polarisation. If $\rho = -\mathrm{i}$ the rotation is clockwise, and the polarisation is called right-handed circular polarisation. These conventions are used in some recent physics text books, and are followed in this book. They accord with the recommendations of the Institute of Electrical and Electronics Engineers (1969), and of the International Astronomical Union (1973). Some authors use different conventions, so it is important to state which system is being used. If ρ is complex, the polarisation is elliptical, and (4.13), (4.14) show that the magnetic vector $\mathscr{H}$ and the transverse component of $\boldsymbol{E}$ traverse similar ellipses in the same sense. The ellipses for $\boldsymbol{E}$ and for $\mathscr{H}$ have their major axes at right angles. There are alternative ways of describing the wave polarisation; see § 4.5. For a full discussion see Rawer and Suchy (1967, § 7).

We shall now apply (4.13), (4.14) to an electron plasma and use the form (4.12) of the constitutive relation. Equations (4.7), (4.8) show that $\varepsilon_0 E_z + P_z = 0$. If this is used in the z component of (4.12) it gives

$$(U - X)P_z = \mathrm{i}Y\sin\Theta\, P_y. \tag{4.15}$$

Now divide (4.5) by (4.3) and use (4.13), (4.14) and (4.8). This gives

$$\frac{D_y}{D_x} = -\frac{\mathscr{H}_x}{\mathscr{H}_y} = \frac{E_y}{E_x} = \frac{P_y}{P_x} = \rho. \tag{4.16}$$

The x and y components of (4.12) give

$$-\varepsilon_0 X E_x = UP_x + \mathrm{i}Y\cos\Theta\, P_y \tag{4.17}$$

$$-\varepsilon_0 X E_y = -\mathrm{i}Y\cos\Theta\, P_x + \left(U - \frac{Y^2\sin^2\Theta}{U - X}\right)P_y \tag{4.18}$$

where (4.15) has been used in (4.18). Now divide (4.18) by (4.17) and use (4.16). Then

$$\rho = \frac{-\mathrm{i}Y\cos\Theta + \rho\{U - Y^2\sin^2\Theta/(U - X)\}}{U + \mathrm{i}Y\rho\cos\Theta} \tag{4.19}$$

whence

$$\rho^2 - i\rho\frac{Y\sin^2\Theta}{(U-X)\cos\Theta} + 1 = 0. \tag{4.20}$$

This is a quadratic equation for ρ and shows that, in a homogeneous electron magnetoplasma, with ν independent of electron velocity, a progressive wave must have a polarisation given by one of the two solutions

$$\rho = \frac{\frac{1}{2}iY\sin^2\Theta \pm i\{\frac{1}{4}Y^2\sin^4\Theta + \cos^2\Theta(U-X)^2\}^{\frac{1}{2}}}{(U-X)\cos\Theta}. \tag{4.21}$$

Equation (4.20) is the polarisation equation of magnetoionic theory and plays a most important part in the theory of later chapters.

If a plane wave is present with its wave normal parallel to the z axis and with any given polarisation, it must be resolved into two component waves with the two polarisations (4.21). These waves in general have different refractive indices (see (4.46) below) so that the composite wave is not a progressive wave because its fields do not have a unique z dependence satisfying (4.1).

If $X = 0$, the medium is free space. Yet (4.21) shows that apparently ρ can have one of two values only. But now (4.46) shows that both refractive indices are unity. A wave with any desired polarisation can be resolved into components with the polarisations (4.21), but these both travel with wave velocity c and the resultant polarisation does not change. The composite wave remains a progressive wave. This property is important in the theory of the limiting polarisation for a wave emerging from the ionosphere; see §§ 17.10, 17.11.

For a more general type of plasma, with ions allowed for or with a velocity dependent collision frequency, or both, we use the form (3.54) for the permittivity tensor with double subscripts x, y, z to indicate its components. Then (4.7) gives

$$E_z = -(\varepsilon_{zx}E_x + \varepsilon_{zy}E_y)/\varepsilon_{zz}. \tag{4.22}$$

This replaces (4.15). Division of D_y by D_x gives, from (4.16) with (4.22) after some reduction

$$\rho = \frac{\varepsilon_{zz}\varepsilon_{yx} - \varepsilon_{yz}\varepsilon_{zx} + (\varepsilon_{yy}\varepsilon_{zz} - \varepsilon_{yz}\varepsilon_{zy})\rho}{\varepsilon_{xx}\varepsilon_{zz} - \varepsilon_{xz}\varepsilon_{zx} + (\varepsilon_{xy}\varepsilon_{zz} - \varepsilon_{xz}\varepsilon_{zy})\rho}. \tag{4.23}$$

The elements of $\boldsymbol{\varepsilon}$ from (3.54) are now used, and J from (3.51). Then (4.23) is the quadratic equation

$$\rho^2 - i\rho\frac{2J\sin^2\Theta}{\varepsilon_3(\varepsilon_1 - \varepsilon_2)\cos\Theta} + 1 = 0. \tag{4.24}$$

This applies for the general cold magnetoplasma, but if the special values (3.50) are used, it is the same as (4.20).

4.4. Properties of the polarisation equation

From (4.21) it is clear that when collisions are neglected, so that $U = 1$, both values of ρ are purely imaginary. The two polarisation ellipses then have their major axes parallel to either the x or the y axis.

The condition that two solutions of the quadratic (4.20) shall be equal is

$$\tfrac{1}{4}Y^2 \sin^4\Theta + \cos^2\Theta(U - X)^2 = 0. \tag{4.25}$$

When X and Z are real, this is only possible if

$$X = 1 \text{ and } Z = Z_t \tag{4.26}$$

where

$$Z_t = \tfrac{1}{2}Y \sin^2\Theta/|\cos\Theta| \tag{4.27}$$

or equivalently

$$\nu_t = \omega_t \quad \text{where} \quad \omega_t = \tfrac{1}{2}\omega_H \sin^2\Theta/|\cos\Theta|. \tag{4.28}$$

Thus ω_t is a 'transition' value of the collision frequency, and is independent of the wave frequency. Many authors, including Budden (1961a), have called it the critical frequency and used for it the symbol ω_c, but the word 'critical' is now used for too many different things in physics. The transition frequency ω_t depends only on the strength of the earth's magnetic field and its inclination $\pi - \Theta$ to the wave normal. Suppose that $\cos\Theta$ is positive, as it is in the northern hemisphere for a radio wave with its wave normal vertically upwards. Then, if the condition (4.26) holds, both values of ρ are -1, so that, for both polarisations, the electric field is in the same plane at $-45°$ to the x axis. In these conditions it is possible, however, for another wave to exist whose electric field has a component at right angles to this plane; see §4.14. In the southern hemisphere, when (4.26) is true, both values of ρ are $+1$; see fig. 4.2.

Let the two values (4.21) of ρ be ρ_O and ρ_E. The subscripts O and E mean 'ordinary' and 'extraordinary', respectively. The meanings of these terms are discussed in the following section and in §§4.11, 4.15, 5.6. Now (4.20) shows that

$$\rho_O \rho_E = 1. \tag{4.29}$$

Choose new axes x', y', z', formed from x, y, z by rotation through an angle ψ about the z axis and let E'_x, E'_y, ρ'_O, ρ'_E be the new values of E_x, E_y, ρ_O, ρ_E. Then for a wave with $E_y = \rho_O E_x$

$$E'_x = E_x(\cos\psi + \rho_O \sin\psi), \quad E'_y = E_x(-\sin\psi + \rho_O \cos\psi) \tag{4.30}$$

whence

$$\rho'_O = (\rho_O - \tan\psi)/(1 + \rho_O \tan\psi). \tag{4.31}$$

Similarly it can be shown that

$$\rho'_E = (\rho_E - \tan\psi)/(1 + \rho_E \tan\psi). \tag{4.32}$$

Now (4.29), (4.20) show that if $\psi = \frac{1}{4}\pi$

$$-\rho'_{\mathrm{O}} = \rho'_{\mathrm{E}} = \frac{\mathrm{i}\{\frac{1}{4}Y^2 \sin^4\Theta + (U-X)^2\cos^2\Theta\}^{\frac{1}{2}}}{\frac{1}{2}\mathrm{i}Y\sin^2\Theta + (U-X)\cos\Theta} \tag{4.33}$$

where the square root has a positive real part when $Z < Z_t$, (4.27). Hence the two polarisation ellipses are mirror images of each other in a plane at 45° to the original axes, that is to the magnetic meridian plane. The ellipses are identical in shape, and their field vectors have opposite directions of rotation. The two major axes are in the same quadrant; see fig. 4.2.

The polarisation ρ is defined by (4.13) in terms of a right-handed system of axes in which the z axis is the wave normal. For a wave travelling in the opposite direction, the z axis and one of the other axes must be reversed. To ensure that $\tan\Theta$ remains positive (§ 4.3 and (4.11)), this other axis must be the x axis. The middle term of (4.20) then changes sign so that both values of ρ change sign. There can be two waves with anti-parallel wave normals and ρ values differing only in sign. The polarisation ellipses for these waves are exactly the same. The ρs have opposite signs only because of the change of axes. This convention is used in this and the following chapter. In some problems discussed in later chapters, waves in both directions are present, but it is convenient to define ρ in terms of the *same* set of axes for both waves. Then, for ordinary waves travelling in either direction, ρ has the same value ρ_{O}, and similarly for extraordinary waves.

Fig. 4.2. The two polarisation ellipses of the horizontal component of the electric field for waves with vertical wave normals, as seen looking downwards in the ionosphere. (a) is for the northern hemisphere and (b) for the southern, and N E S W denote the magnetic cardinal points. The arrows show the sense of rotation when $X < 1$, and should be reversed for $X > 1$. The ellipses for the magnetic field are obtained by rotating these figures through 90°.

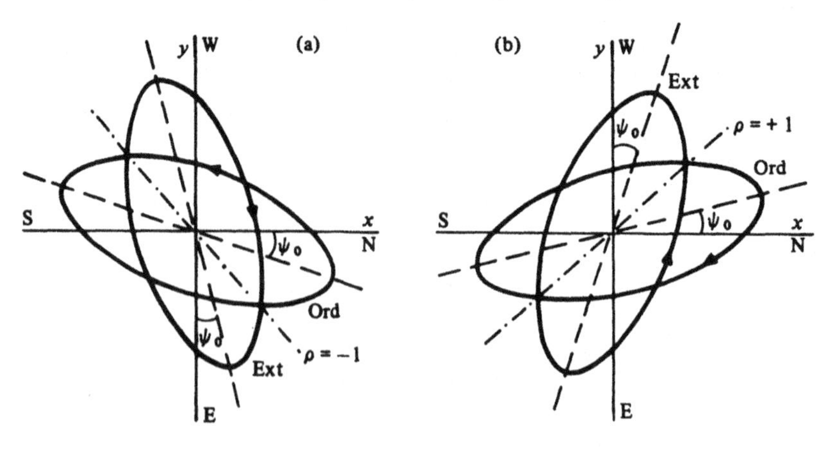

4.5. Alternative measure of the polarisation. Axis ratio and tilt angle

Let
$$\rho_{\mathrm{O}} = \tan\gamma = u + \mathrm{i}v \tag{4.34}$$
and
$$\gamma = \psi_{\mathrm{O}} + \mathrm{i}\phi_{\mathrm{O}} \tag{4.35}$$
where u, v, ψ_{O}, ϕ_{O} are real. The value ρ'_{O} (4.31), referred to new axes as in the preceding section is then
$$\rho'_{\mathrm{O}} = \tan(\gamma - \psi) \tag{4.36}$$
which shows that ρ'_{O} is purely imaginary when $\psi = \psi_{\mathrm{O}}$ or $\psi = \psi_{\mathrm{O}} + \frac{1}{2}\pi$. These are the inclinations to the x axis of the major and minor axes respectively of the polarisation ellipse for the ordinary wave. Similarly the polarisation ellipse of the extraordinary wave has its major and minor axes at angles $\pm\frac{1}{2}\pi - \psi_{\mathrm{O}}$ and $-\psi_{\mathrm{O}}$ to the x axis. Substitution of (4.34) in the quadratic (4.20) gives
$$\sin 2\gamma = -2\mathrm{i}\,Y^{-1}(1 - X - \mathrm{i}Z)\cos\Theta/\sin^2\Theta \tag{4.37}$$
whence
$$\sin 2\psi_{\mathrm{O}}\cosh 2\phi_{\mathrm{O}} = -2Y^{-1}Z\cos\Theta/\sin^2\Theta, \tag{4.38}$$
$$\cos 2\psi_{\mathrm{O}}\sinh 2\phi_{\mathrm{O}} = -2Y^{-1}(1 - X)\cos\Theta/\sin^2\Theta. \tag{4.39}$$
This gives two values of ψ_{O} and we may choose the value for which $|\psi_{\mathrm{O}}| \leqslant \frac{1}{2}\pi$. If $\cos\Theta$ is changed but does not become zero or change sign, (4.38) shows that ψ_{O} does not change sign, and (4.39) shows that, if X is kept constant, $|\psi_{\mathrm{O}}|$ never attains the value $\frac{1}{4}\pi$. It is shown in §4.11 that for transverse propagation, $\cos\Theta = 0$, the ordinary wave is linearly polarised with $\boldsymbol{E}$ parallel to the x axis. The polarisation ellipse for $\boldsymbol{E}$ has degenerated to a line, and its major axis is parallel to the x axis. But (4.38) shows that now $\psi_{\mathrm{O}} = 0$. Thus ψ_{O} is the direction of the major axis of the $\boldsymbol{E}$ ellipse for the ordinary wave. When $\cos\Theta$ is small, $|\psi_{\mathrm{O}}| < \frac{1}{4}\pi$ and this remains true for all Θ. The angle ψ_{O} is called the 'tilt angle' for the ordinary wave. It is the angle between the magnetic meridian and the major axis of the polarisation ellipse for the component of $\boldsymbol{E}$ perpendicular to the wave normal.

Let ψ in (4.36) take the value ψ_{O}. Then ρ'_{O} is purely imaginary and its modulus is the ratio: minor axis/major axis. This is called the 'axis ratio'. Its value, from (4.34), (4.35) is $|\tan \mathrm{i}\phi_{\mathrm{O}}| = |\tanh\phi_{\mathrm{O}}|$, and it can be shown from (4.34) that
$$\tanh 2\phi_{\mathrm{O}} = 2v/(1 + u^2 + v^2). \tag{4.40}$$
The concepts of 'axis ratio' and 'tilt angle' were used for radio waves by Kelso *et al.* (1951).

Consider a radio wave travelling vertically upwards in the ionosphere in the northern hemisphere. Here the vertical component of the earth's magnetic field is downwards, so that the vertical component of $\boldsymbol{Y}$ is upwards, and $\cos\Theta$ is positive. Then (4.38) shows that ψ_{O} is negative. Thus, where it is permissible to treat the ionosphere as a homogeneous electron plasma, the major axes of the polarisation ellipses for electric field of both waves must be in the quadrant between magnetic north and magnetic east; see fig. 4.2(a). No progressive wave is possible with these major axes in the north-west and south-east quadrants. This applies also for vertically downgoing waves because the polarisation ellipses are the same as for upgoing waves; see end of §4.4. In the southern hemisphere $\cos\Theta$ is negative for upgoing waves. The major axes for the $\boldsymbol{E}$ ellipses are here always in the north-west and south-east quadrants. This is illustrated in fig. 4.2(b).

A more detailed discussion of the wave polarisation of a downcoming radio wave is given in §§ 17.10, 17.11.

For the more general type of plasma, when the quadratic (4.24) is used, the result (4.37) becomes

$$\sin 2\gamma = -2\mathrm{i}\frac{\varepsilon_3(\varepsilon_1 - \varepsilon_2)\cos\Theta}{J\sin^2\Theta} \tag{4.41}$$

whence the axis ratio and tilt angle can be found exactly as described above.

4.6. Refractive index 1. The dispersion relation

If $\mathscr{H}_y$ is eliminated from (4.3), (4.4), they give

$$D_x = \varepsilon_0 n^2 E_x \tag{4.42}$$

and similarly (4.2), (4.5) give

$$D_y = \varepsilon_0 n^2 E_y. \tag{4.43}$$

These equations are the same as would be obtained for an isotropic dielectric, in which case n^2, the square of the refractive index, is equal to the scalar electric permittivity. This result may not, however, have any particular physical significance. It would not be true if the magnetic permeability differed from unity, and it is not true for the z components because $D_z = 0$ but in general $E_z \neq 0$.

From (4.8) with (4.42), (4.43)

$$P_x = \varepsilon_0(n^2 - 1)E_x, \quad P_y = \varepsilon_0(n^2 - 1)E_y. \tag{4.44}$$

From this substitute E_x in (4.17) for the electron plasma, and divide by P_x. Since $P_y/P_x = \rho$ from (4.16), the result is

$$X/(n^2 - 1) = -U - \mathrm{i}\rho Y\cos\Theta \tag{4.45}$$

whence

$$n^2 = 1 - \frac{X}{U + \mathrm{i}\rho Y\cos\Theta}. \tag{4.46}$$

Now substitute for ρ from (4.21). This gives

$$n^2 = 1 - \frac{X(U-X)}{U(U-X) - \frac{1}{2}Y^2\sin^2\Theta \pm \{\frac{1}{4}Y^4\sin^4\Theta + Y^2\cos^2\Theta(U-X)^2\}^{\frac{1}{2}}}. \tag{4.47}$$

This formula, for the square of the two refractive indices of a cold electron magnetoplasma, with the electron collision frequency independent of velocity, has for many years been known as the Appleton–Hartree formula. This form, apart from the notation, is essentially the same as Appleton's (1932) form and the method of deriving it given above is almost the same as Appleton's method. Both Appleton (1932) and Hartree (1931) included the Lorentz polarisation term (§ 3.3), but Appleton also gave a version without it. It is now known that the Lorentz term should be omitted; see § 3.3. A version of (4.47) without it was given by Goldstein (1928). An equivalent formula to (4.47), also without the Lorentz term, was derived by Lassen (1927) who used a somewhat different method. Some authors now call (4.47), or its equivalent, the Appleton–Lassen formula, and this name seems to be coming into increasing use; see Rawer and Suchy (1976). The formula (4.47) is often needed for a collisionless plasma so that $U = 1$. The result is given here for future reference:

$$n^2 = 1 - \frac{X(1-X)}{1 - X - \frac{1}{2}Y^2\sin^2\Theta \pm \{\frac{1}{4}Y^4\sin^4\Theta + Y^2\cos^2\Theta(1-X)^2\}^{\frac{1}{2}}}. \qquad (4.48)$$

For the more general plasma, (4.17) cannot be used, and instead we substitute E_z, from (4.22), into the D_x element of (4.8). With (4.42) and $E_y/E_x = \rho$, this gives

$$n^2 = [(\varepsilon_{zz}\varepsilon_{xx} - \varepsilon_{xz}\varepsilon_{zx}) + \rho(\varepsilon_{zz}\varepsilon_{xy} - \varepsilon_{xz}\varepsilon_{zy})]/\varepsilon_{zz} \qquad (4.49)$$

and on using the form (3.54) for the elements of $\boldsymbol{\varepsilon}$, it becomes

$$n^2 = \frac{\varepsilon_3(\varepsilon_1 + \varepsilon_2) + i\rho\cos\Theta\,\varepsilon_3(\varepsilon_1 - \varepsilon_2)}{(\varepsilon_1 + \varepsilon_2)\sin^2\Theta + 2\varepsilon_3\cos^2\Theta}. \qquad (4.50)$$

Then on substituting the solutions of (4.24) for ρ we obtain finally

$$n^2 = \frac{\varepsilon_1\varepsilon_2\sin^2\Theta + \frac{1}{2}\varepsilon_3(\varepsilon_1 + \varepsilon_2)(1 + \cos^2\Theta) \pm \{J^2\sin^4\Theta + \varepsilon_3^2(\varepsilon_1 - \varepsilon_2)^2\cos^2\Theta\}^{\frac{1}{2}}}{(\varepsilon_1 + \varepsilon_2)\sin^2\Theta + 2\varepsilon_3\cos^2\Theta}. \qquad (4.51)$$

4.7. Longitudinal component of electric polarisation and electric field

Equation (4.7) shows that $D_z = 0$, but E_z and P_z in general are not zero. The components E_z, P_z are parallel to the wave normal and are called longitudinal components. For the electron plasma, from (4.15), (4.16) and (4.20)

$$P_z/P_x = \rho P_z/P_y = \rho i y \sin\Theta/(U - X) = (\rho^2 + 1)\cot\Theta. \qquad (4.52)$$

Thus the components of the complex vector $\boldsymbol{P}$ are in the ratio

$$P_x : P_y : P_z \propto U - X : \rho(U - X) : iY\rho\sin\Theta \propto 1 : \rho : (\rho^2 + 1)\cot\Theta. \qquad (4.53)$$

If collisions are neglected, ρ is purely imaginary. Then P_x and P_z are in phase or antiphase and P_y is in quadrature with them. The normal to the polarisation ellipse for $\boldsymbol{P}$ is coplanar with $\boldsymbol{Y}$ and the wave normal. The vector $\boldsymbol{P}$ is of interest because it is proportional to the electron displacements caused by the wave.

From (4.52), with (4.7) and (4.44)

$$E_z/E_x = \rho E_z/E_y = -\rho \mathrm{i} Y \sin\Theta\,(n^2-1)/(U-X) = -\cot\Theta(\rho^2+1)(n^2-1). \tag{4.54}$$

If collisions are neglected so that n^2 is real, E_z and E_x are in phase or antiphase and E_y is in quadrature with them. The normal to the polarisation ellipse for $\boldsymbol{E}$ is coplanar with $\boldsymbol{Y}$ and the wave normal.

Equation (4.54) shows that the longitudinal component E_z of the electric field is zero only if either $\sin\Theta = 0$ (equivalently $\rho = \pm \mathrm{i}$), or $\rho = 0$, or $n^2 = 1$. The first of these conditions holds when the wave normal is parallel or antiparallel to the earth's magnetic field, and this is called longitudinal propagation. The second, $\rho = 0$, requires that $\cot\Theta = 0$ from (4.24), so that the earth's magnetic field is perpendicular to the wave normal, and this is called transverse propagation. The earth's field is also parallel to the electric field $\boldsymbol{E}$ of the wave. Then the only motion imparted to the electrons is parallel to the earth's magnetic field, and the electron motion is unaffected by the field. The third condition $n^2 = 1$ applies only when $X = 0$, that is for free space. It can be shown that one value of n^2 is unity when $X = 1, Z = 0$, but in this case ρ is infinite and E_z in (4.54) is not zero.

4.8. The flow of energy for a progressive wave in a magnetoplasma

It is interesting to evaluate the time average value $\boldsymbol{\Pi}_{\mathrm{av}}$ of the Poynting vector, given by (2.63), for a progressive wave in an electron magnetoplasma, so as to find the flux of energy. Since $\mathscr{H}_z = 0$, the coponents of $\boldsymbol{\Pi}_{\mathrm{av}}$ are

$$\Pi_x = \tfrac{1}{4}Z_0^{-1}(E_z\mathscr{H}_y^* + E_z^*\mathscr{H}_y),\quad \Pi_y = \tfrac{1}{4}Z_0^{-1}(E_z\mathscr{H}_x^* + E_z^*\mathscr{H}_x),$$
$$\Pi_z = \tfrac{1}{4}Z_0^{-1}(E_x\mathscr{H}_y^* + E_x^*\mathscr{H}_y - E_y\mathscr{H}_x^* - E_y^*\mathscr{H}_x). \tag{4.55}$$

In an isotropic medium E_z also is zero so that Π_x, Π_y are zero and the energy flow is in the direction of the wave normal, but this is not in general true for an anisotropic medium. It is convenient to express the components of $\boldsymbol{E}$ and $\boldsymbol{\mathscr{H}}$ in terms of E_x by using (4.2), (4.4), (4.13), (4.14) and (4.54). Hence

$$\left.\begin{aligned}
\Pi_x &= -\frac{\mathrm{i}Y\sin\Theta}{4Z_0}\left\{\frac{n\rho^*(n^{*2}-1)}{U^*-X} - \frac{n^*\rho(n^2-1)}{U-X}\right\}|E_x|^2,\\
\Pi_y &= \frac{\mathrm{i}Y\sin\Theta\,\rho\rho^*}{4Z_0}\left\{\frac{(n^2-1)n^*}{U-X} - \frac{(n^{*2}-1)n}{U^*-X}\right\}|E_x|^2,\\
\Pi_z &= \frac{1}{4Z_0}(n+n^*)(1+\rho\rho^*)|E_x|^2.
\end{aligned}\right\} \tag{4.56}$$

These give the magnitude and direction of $\boldsymbol{\Pi}_{\mathrm{av}}$ when collisions are allowed for, and the values of n and ρ for either of the two characteristic waves may be inserted. The expressions are of interest for the special case when collisions are neglected so that n^2 is real, ρ is purely imaginary, and $U = U^* = 1$. Then $\rho^* = -\rho$. There are two important cases. The first is when n is real so that $n^* = n$. Then

$$\Pi_x = \frac{\mathrm{i}\rho n(n^2-1)Y\sin\Theta}{2Z_0(1-X)}|E_x|^2, \quad \Pi_y = 0, \quad \Pi_z = \frac{n}{2Z_0}(1-\rho^2)|E_x|^2, \tag{4.57}$$

which show that $\mathbf{\Pi}_{\mathrm{av}}$ is in the plane containing the earth's magnetic field and the wave normal, and makes an angle

$$\arctan\left\{\frac{\mathrm{i}Y\rho\sin\Theta(n^2-1)}{(1-X)(1-\rho^2)}\right\} = \arctan\left[\pm\frac{\frac{1}{2}Y\sin\Theta\cos\Theta\,(n^2-1)}{\{\frac{1}{4}Y^2\sin^4\Theta+(1-X)^2\cos^2\Theta\}^{\frac{1}{2}}}\right] \tag{4.58}$$

with the wave normal. Here ρ and $1/\rho$ from (4.21) have been used. The + sign applies for the ordinary wave and the − sign for the extraordinary wave. This result is derived by a different method in § 5.3 and its significance is discussed there.

The second important case is when n is purely imaginary, so that the wave is evanescent, and $n^* = -n$. Then (4.56) gives

$$\Pi_x = \Pi_z = 0, \quad \Pi_y = \frac{\mathrm{i}n\rho^2 Y\sin\Theta(n^2-1)}{2Z_0(1-X)}|E_x|^2 \tag{4.59}$$

which shows that $\mathbf{\Pi}_{\mathrm{av}}$ is perpendicular both to the wave normal and to the earth's magnetic field.

4.9. Refractive index 2. Alternative derivations and formulae

The refractive index n can be derived directly without going through the intermediate step of finding the polarisation ρ. We use (4.42), (4.43), (4.7) which come from Maxwell's equations. They are substituted in the constitutive relation (4.8) to give

$$\left.\begin{aligned}(\varepsilon_{xx}-n^2)E_x+\varepsilon_{xy}E_y+\varepsilon_{xz}E_z&=0,\\ \varepsilon_{yx}E_x+(\varepsilon_{yy}-n^2)E_y+\varepsilon_{yz}E_z&=0,\\ \varepsilon_{zx}E_x+\varepsilon_{zy}E_y+\varepsilon_{zz}E_z&=0.\end{aligned}\right\} \tag{4.60}$$

For these three equations to have a solution, it is necessary that the determinant of the coefficients of E_x, E_y, E_z shall be zero. The form (3.54) for the elements of $\boldsymbol{\varepsilon}$ is now used. This gives

$$\begin{vmatrix}\frac{1}{2}(\varepsilon_1+\varepsilon_2)\cos^2\Theta+\varepsilon_3\sin^2\Theta-n^2 & \frac{1}{2}\mathrm{i}(\varepsilon_1-\varepsilon_2)\cos\Theta & G\sin\Theta\cos\Theta\\ -\frac{1}{2}\mathrm{i}(\varepsilon_1-\varepsilon_2)\cos\Theta & \frac{1}{2}(\varepsilon_1+\varepsilon_2)-n^2 & -\frac{1}{2}\mathrm{i}(\varepsilon_1-\varepsilon_2)\sin\Theta\\ G\sin\Theta\cos\Theta & \frac{1}{2}\mathrm{i}(\varepsilon_1-\varepsilon_2)\sin\Theta & \frac{1}{2}(\varepsilon_1+\varepsilon_2)\sin^2\Theta+\varepsilon_3\cos^2\Theta\end{vmatrix}=0 \tag{4.61}$$

and on multiplying out, it gives a quadratic equation for n^2 which can be written in the various forms:

$$n^4\{\tfrac{1}{2}(\varepsilon_1+\varepsilon_2)\sin^2\Theta+\varepsilon_3\cos^2\Theta\}-n^2\{\varepsilon_1\varepsilon_2\sin^2\Theta+\tfrac{1}{2}\varepsilon_3(\varepsilon_1+\varepsilon_2)(1+\cos^2\Theta)\}$$
$$+\varepsilon_1\varepsilon_2\varepsilon_3=0, \tag{4.62}$$

$$\varepsilon_3(n^2-\varepsilon_1)(n^2-\varepsilon_2)+n^2\sin^2\Theta(Gn^2+J)=0, \tag{4.63}$$

$$\tfrac{1}{2}(\varepsilon_1+\varepsilon_2)(n^2-\varepsilon_3)\left(n^2-\frac{2\varepsilon_1\varepsilon_2}{\varepsilon_1+\varepsilon_2}\right)-n^2\cos^2\Theta(Gn^2+J)=0. \tag{4.64}$$

If the values (3.50) for the electron plasma are used this becomes

$$An^4-2Bn^2+C=0 \tag{4.65}$$

where

$$\left.\begin{aligned} A&=U^2(U-X)-Y^2(U-X)-XY^2\sin^2\Theta,\\ B&=U(U-X)^2-Y^2(U-X)-\tfrac{1}{2}XY^2\sin^2\Theta,\\ C&=(U-X)^3-Y^2(U-X). \end{aligned}\right\} \tag{4.66}$$

The solution of (4.62) has already been given at (4.51). The solution of (4.65) is

$$n^2=\frac{(U-X)(U^2-UX-Y^2)-\tfrac{1}{2}XY^2\sin^2\Theta\pm XY\{\tfrac{1}{4}Y^2\sin^4\Theta+\cos^2\Theta(U-X)^2\}^{\frac{1}{2}}}{U^2(U-X)-Y^2(U-X\cos^2\Theta)} \tag{4.67}$$

This can be shown to be the same as (4.47).

The form (4.47) of the solution is rather unusual first because of the 1 on the right-hand side and second because the square root is in the denominator. There is no simple corresponding form of the solution of (4.62) for the more general plasma (but see problem 4.13). This form seems to have appeared because it was derived from (4.12) without the need to invert the matrix, so that ρ appears in the denominator of (4.46).

Equation (4.65) can be treated as a quadratic for $1/n^2$ and solved to give

$$\begin{aligned} n^2&=\frac{2\varepsilon_1\varepsilon_2\varepsilon_3}{\varepsilon_1\varepsilon_2\sin^2\Theta+\tfrac{1}{2}\varepsilon_3(\varepsilon_1+\varepsilon_2)(1+\cos^2\Theta)\pm\{J^2\sin^4\Theta+\varepsilon_3^2(\varepsilon_1-\varepsilon_2)^2\cos^2\Theta\}^{\frac{1}{2}}}\\ &=\frac{(U-X)(U-X-Y)(U-X+Y)}{(U-X)(U^2-UX-Y^2)-\tfrac{1}{2}XY^2\sin^2\Theta\pm XY\{\tfrac{1}{4}Y^2\sin^4\Theta+(U-X)^2\cos^2\Theta\}^{\frac{1}{2}}} \end{aligned}$$

(electron plasma) (4.68)

Both (4.67) and (4.68) are useful. But (4.47) is much more widely used and is simpler for some forms of computation.

Another form of the dispersion relation is sometimes needed. To derive it we abandon the assumption that the wave normal is parallel to the z axis, and assume that it is in some general direction so that the factor $\exp(-iknz)$ used in (4.1)–(4.7) is replaced by

$$\exp\{-ik(xn_x+yn_y+zn_z)\}. \tag{4.69}$$

The refractive index is here treated as a vector $\boldsymbol{n}$ with components n_x, n_y, n_z; see § 5.1. Then if (4.69) is used in Maxwell's equations (2.44), with the constitutive relation (4.8), it gives

$$\boldsymbol{\Gamma}\boldsymbol{E}_0=\boldsymbol{\mathscr{H}}_0,\quad \boldsymbol{\Gamma}\boldsymbol{\mathscr{H}}_0=-\boldsymbol{\varepsilon}\boldsymbol{E}_0 \tag{4.70}$$

where $\mathbf{\Gamma}$ is the matrix

$$\mathbf{\Gamma} = \begin{pmatrix} 0 & -n_z & n_y \\ n_z & 0 & -n_x \\ -n_y & n_x & 0 \end{pmatrix}. \tag{4.71}$$

By multiplying (4.70) by $\mathbf{\Gamma}$ on the left, $\mathscr{H}_0$ may be eliminated to give

$$(\mathbf{\Gamma}^2 + \boldsymbol{\varepsilon})\boldsymbol{E}_0 = 0. \tag{4.72}$$

This can only have a solution if the determinant of the matrix on the left is zero; compare (4.60), (4.61). Hence

$$\begin{vmatrix} \varepsilon_{xx} - n_y^2 - n_z^2 & \varepsilon_{xy} + n_x n_y & \varepsilon_{xz} + n_x n_z \\ \varepsilon_{yx} + n_x n_y & \varepsilon_{yy} - n_x^2 - n_z^2 & \varepsilon_{yz} + n_y n_z \\ \varepsilon_{zx} + n_x n_z & \varepsilon_{zy} + n_y n_z & \varepsilon_{zz} - n_x^2 - n_y^2 \end{vmatrix} = 0. \tag{4.73}$$

This is the dispersion relation. The axes used here are different from those in (4.61) and so the elements of $\boldsymbol{\varepsilon}$ are different. (4.73) is a generalisation of (4.61) for the new axis system. It is used in § 14.2.

4.10. Zeros and infinity of refractive index. Equal refractive indices

The quadratic (4.65) with (4.66) or the solutions (4.68) show that one value of n^2, but only one, is zero when

$$X = U + Y, \quad \text{or} \quad X = U - Y, \quad \text{or} \quad X = U. \tag{4.74}$$

These are called 'cut-off' conditions, and the frequencies where they are satisfied are called cut-off frequencies. The corresponding conditions for the general plasma are, from (4.62)

$$\varepsilon_1 = 0 \quad \text{or} \quad \varepsilon_2 = 0 \quad \text{or} \quad \varepsilon_3 = 0 \tag{4.75}$$

When the third condition (4.74) is satisfied, the other non-zero value of of n^2 is $n^2 = 1$. For the general plasma when $\varepsilon_3 = 0$, the other non-zero n^2 is $2\varepsilon_1\varepsilon_2/(\varepsilon_1 + \varepsilon_2)$. An important feature of these conditions is that they are independent of the angle Θ between the wave normal and $\boldsymbol{Y}$.

If collisions are included, these conditions cannot be satisfied for any real values of X, but they are important when $X(h)$ and $Z(h)$ are treated as complex analytic functions of the height h regarded as a complex variable (in later chapters denoted by z). Points in the complex h plane where $n = 0$ are called reflection points.

The quadratic (4.65) with (4.66) or the solutions (4.67) show that one value of n^2, but only one, is infinite when

$$X = X_\infty \equiv U(U^2 - Y^2)/(U^2 - Y^2 \cos^2\Theta) \tag{4.76}$$

and for the general plasma

$$\tfrac{1}{2}(\varepsilon_1 + \varepsilon_2)\sin^2\Theta + \varepsilon_3 \cos^2\Theta = 0\,. \tag{4.77}$$

This condition is called 'resonance'. It does depend on the angle Θ. Its effect on wave propagation is discussed in §§ 18.13, 19.4–19.6.

The dispersive properties for radio waves, that is, the formulae for n^2 and ρ, are practically unaffected in most conditions when allowance is made for the electron and ion temperatures. But this is no longer true when the resonance condition is approached. Then (4.54) shows that for large n, E_z/E_x, E_z/E_y are of order n^2 and (4.2), (4.4) then show that $E_z/\mathscr{H}_x$, $E_z/\mathscr{H}_y$ are of order n. Thus the predominant field component is E_z and the wave resembles a longitudinal plasma wave. The full theory for a warm plasma (Ginzburg, 1970; Goland and Piliya, 1972) shows that the value of n does not actually become infinite when (4.76) or (4.77) is satisfied. It becomes very large and goes over continuously to the value for a plasma wave. See § 19.5 for further discussion.

Equation (4.46) shows that the two values of n^2 can be equal only when the two ρs are equal and the condition for this for an electron plasma was given at (4.26), (4.27). If collisions are neglected and X is real, this condition becomes

$$X = 1, \quad \sin\Theta = 0. \tag{4.78}$$

If $X(h)$ and $Z(h)$ are analytic functions of the height h (in later chapters denoted by z), the condition becomes

$$X(h) = 1 - \mathrm{i}\{Z(h) \pm Z_\mathrm{t}\} \tag{4.79}$$

where Z_t is given by (4.27). Points in the complex h plane where this is satisfied are called coupling points. They are studied in §§ 4.14, 16.5. There can be a coupling point on the real h axis only if

$$Z(h) = Z_\mathrm{t} \quad \text{where} \quad X(h) = 1. \tag{4.80}$$

The condition for equal roots with (4.51), (4.62) for the more general plasma is discussed in § 13.9.

4.11. Dependence of refractive index on electron concentration 1. $Y < 1$

We shall now examine the dependence of the refractive index on electron concentration N for an electron plasma when collisions are neglected. This subject is particularly important in the study of the probing of the ionosphere by vertically incident radio waves, as in the ionosonde technique, § 1.7.

Since n^2 is real, it is useful to plot curves of n^2 versus X, which is proportional to N. Frequencies $f > f_H$ are considered first so that $Y < 1$. The form of the curves is then as shown in fig. 4.3. When $X = 0$ the medium is just a vacuum and $n^2 = 1$. Thus all the curves go through the point $X = 0$, $n^2 = 1$.

Consider now the special case $\Theta = \pm\pi/2$. Then (4.48) shows that the two solutions are

$$n_\mathrm{O}^2 = 1 - X, \quad n_\mathrm{E}^2 = 1 - \frac{X(1-X)}{1 - X - Y^2}. \tag{4.81}$$

The subscripts are explained below. These two curves are shown as chain lines in fig. 4.3. The wave normal is perpendicular to Y and this case is therefore called 'transverse propagation'. It can be shown from (4.46), (4.29) that the two solutions (4.81) give

$$\rho_{\mathrm{O}} = 0, \quad \rho_{\mathrm{E}} = \infty \tag{4.82}$$

respectively. The first solution (4.81) is the same as for an isotropic plasma, and (4.82) with (4.53) (4.54) shows that $\boldsymbol{E}$ and $\boldsymbol{P}$ are parallel or antiparallel to $\boldsymbol{Y}$, that is to the earth's magnetic field. Thus the ordered part of the electron motions is unaffected by this field, and the wave is called the 'ordinary' wave. This is indicated by the subscript O.

The other wave is the 'extraordinary' wave and is indicated by the subscript E. It shows two zeros where $X = 1 \pm Y$ and a resonance where $X = 1 - Y^2$. This is known as the upper hybrid resonance; see § 3.11, item (6). The second equation (4.82) shows that $E_x = 0$, and (4.54) shows that

$$E_z/E_y = \pm \mathrm{i}XY/(1 - X - Y^2) \tag{4.83}$$

Fig. 4.3. Dependence of n^2 on X for a cold collisionless electron plasma when $Y<1$. In this example $Y \approx 0.6$. Broken lines are for $\sin\Theta = 0$, and chain lines for $\cos\Theta = 0$. The continuous lines are for an intermediate value of Θ, in this example 30°. Curves within the region shaded with dots are for the ordinary wave and curves in regions shaded with lines are for the extraordinary wave. RES is the value of X where $n^2 \to \infty$ for the extraordinary wave. UHR means 'upper hybrid resonance'.

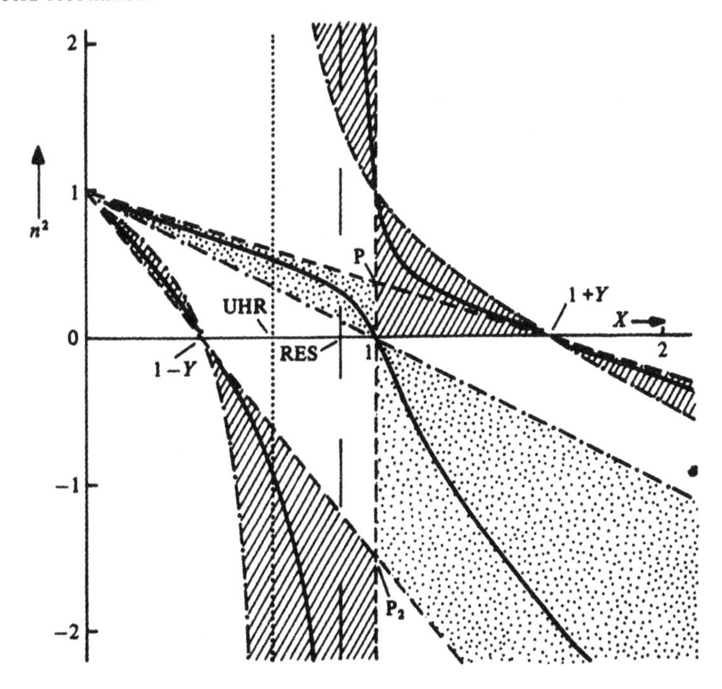

(minus sign if $\sin\Theta$ is negative). Thus E_z and E_y are in quadrature.

When Θ differs slightly from $\pi/2$, the curve in fig. 4.3 for the ordinary wave moves away from the straight line $n^2 = 1 - X$, but it still goes through the zero at $X = 1$. Similarly the curves for the extraordinary waves go through the same zeros as before, but the resonance moves to a greater value of X, namely $(1 - Y^2)/(1 - Y^2\cos^2\Theta)$.

Next consider the special case $\sin\Theta = 0$. Then (4.48) shows that the two solutions are

$$n^2 = 1 - \frac{X}{1+Y}, \quad n^2 = 1 - \frac{X}{1-Y}. \tag{4.84}$$

These represent two straight lines through the zeros at $X = 1 \pm Y$, shown as broken lines in fig. 4.3. These solutions were discussed by Lorentz (1909, ch. IV). The wave normal is parallel or antiparallel to Y and this case is therefore called 'longitudinal propagation'. For upgoing waves near the north magnetic pole $\cos\Theta = +1$, and (4.46) shows that for the first of (4.84) $\rho = -\mathrm{i}$ and for the second $\rho = +\mathrm{i}$. These correspond to circularly polarised waves with a right-handed and a left-handed sense respectively. The formulae (4.84) do not show the zero at $X = 1$ nor the infinity, which (4.76) shows is also at $X = 1$. Neither do the two values of n^2 become equal where (4.79) is satisfied. This is because (4.84) and (4.78) were obtained from (4.48) and (4.20) respectively by setting $\sin^2\Theta = 0$, but this process is not valid when $X = 1$.

If $\sin\Theta$ is very small but not quite zero, the curves of n^2 vs X are as shown in fig. 4.4. The infinity for the extraordinary wave is close to $X = 1$, and both curves have sharp bends near $X = 1$. The curve for the ordinary wave may be followed from left to right. For $X < 1$, the corresponding value of ρ is close to $-\mathrm{i}$ but when the almost

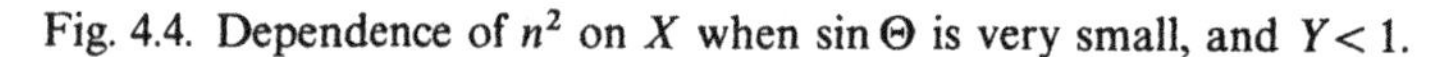

Fig. 4.4. Dependence of n^2 on X when $\sin\Theta$ is very small, and $Y < 1$.

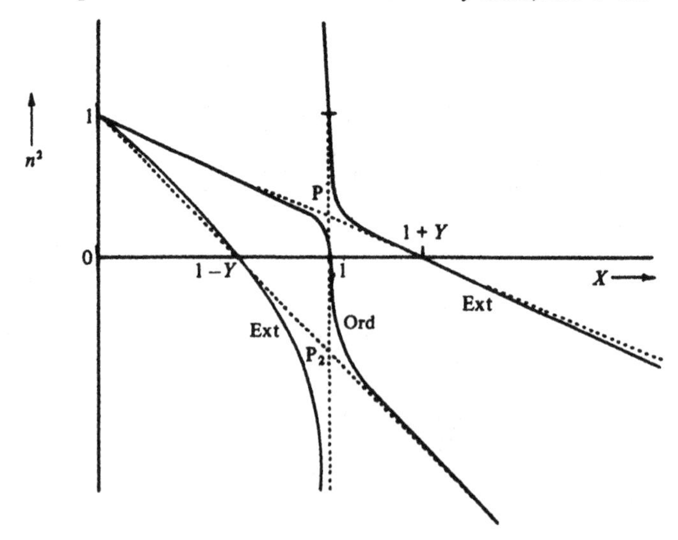

vertical part of the curve near $X = 1$ is traversed, ρ changes rapidly and goes over to a value close to $+\mathrm{i}$, which is retained for $X > 1$. It will be shown later (ch. 16) that this rapid change of polarisation is associated with strong coupling between the ordinary and extraordinary waves. If an ordinary wave enters a medium where it encounters an electron concentration that increases as it travels (increasing X) under the conditions of fig. 4.4, it would cause some extraordinary wave to be generated near $X = 1$. In the limit when $\sin\Theta = 0$, the extraordinary wave takes over completely where $X = 1$, and for $X > 1$ the only wave present would be the extraordinary wave.

This explains how the curves of fig. 4.4 go over, in the limit $\sin\Theta \to 0$, into the broken lines of fig. 4.3. The nearly vertical parts of the curves in fig. 4.4 become, in the limit, the vertical line $X = 1$ which is shown as a broken line in fig. 4.3 since it is really part of the curves for $\sin\Theta = 0$. On this line the refractive index n can take any value, and (4.52), (4.54) show that $E_x = E_y = P_x = P_y = 0$. The wave is a purely longitudinal wave and has all the properties of the plasma oscillation described near the end of § 3.8.

These results also show, by taking the curves of fig. 4.4 to the limit $\sin\Theta \to 0$, that in (4.84) when $X < 1$ the subscripts on n^2 should be $_{\mathrm{O}}$, $_{\mathrm{E}}$ respectively, and when $X > 1$ they should be $_{\mathrm{E}}$, $_{\mathrm{O}}$ respectively. The part of the vertical line $X = 1$ between $n^2 = -Y/(1 - Y)$ at P_2 and $n^2 = Y/(1 + Y)$ at P applies to the ordinary wave and the rest applies to the extraordinary wave. The curves for ordinary waves and extraordinary waves touch at P and P_2 where they have sharp bends and these two points are known as 'window points'. The point P is important in the theory of radio windows, §§ 17.6–17.9. The point P_2 is less important because there n^2 is negative and the waves are evanescent.

It can be shown, see § 5.2 item (12), that when $|\sin\Theta|$ lies between 0 and 1, the curves of n^2 vs X can never cross those for $\sin\Theta = 0$, or for $|\sin\Theta| = 1$. Thus they lie entirely within the regions that are shown shaded in fig. 4.3. For purely transverse propagation, $\cos\Theta = 0$, comparison of (4.81) and (4.47) shows that if the term $\pm\{\cdots\}^{\frac{1}{2}}$ in (4.48) is positive it gives the first equation (4.81), that is the value of n^2 for the ordinary wave. When Θ is now changed so that $\cos\Theta \neq 0$, the square root cannot change sign for any real $\cos\Theta$. Therefore, by continuity with the transverse case, we define n^2 for the ordinary wave as that value of (4.48) that has a positive value of the term $\pm\{\cdots\}^{\frac{1}{2}}$. Thus the curves for the ordinary wave in fig. 4.3 must lie in the region shaded with dots. Similarly n^2 for the extraordinary wave uses a negative value of the term $\pm\{\cdots\}^{\frac{1}{2}}$, and the curves must lie in the region shaded with lines. Further discussion of the terms 'ordinary' and 'extraordinary' is given in §§ 4.15, 4.16, 5.6.

The following further properties of the two refractive indices should be noted.
(a) When $X > 1$ the ordinary wave is evanescent for all Θ, since n^2 is negative.

(b) The curve for the extraordinary wave has two branches. The wave for the right-hand branch in fig. 4.3 is called the Z-mode. The curve goes through the point $X = 1$, $n^2 = 1$ for all Θ. The wave is evanescent when $X > 1 + Y$.

(c) When $(1 - Y^2)/(1 - Y^2 \cos^2 \Theta) < X < 1$ the refractive index for the Z-mode is real and > 1 so that the wave velocity of the wave is less than the speed of light. Some authors call this wave the 'extraordinary slow wave' (Al'pert, 1983).

4.12. Dependence of refractive index on electron concentration 2. $Y > 1$

When the frequency is less than the electron gyro-frequency, about 1 MHz in the ionosphere at temperate latitudes, the form of the curves of n^2 vs X is somewhat different. At these low frequencies the effect of collisions becomes important but it is still of interest to consider first the consequences of neglecting collisions. The zeros and infinity of n^2 are still given by (4.74), (4.76) with $U = 1$ but the value of X at one of the zeros, $1 - Y$, is now negative. At the infinity X may be either positive in the range $X > 1$, or negative in the range $X < 1 - Y^2$. To illustrate these properties it is useful to draw the curves for negative as well as positive values of X, although negative values cannot occur at real heights. See figs. 4.5, 4.6.

The curves for purely transverse propagation are still given by (4.81) but now with

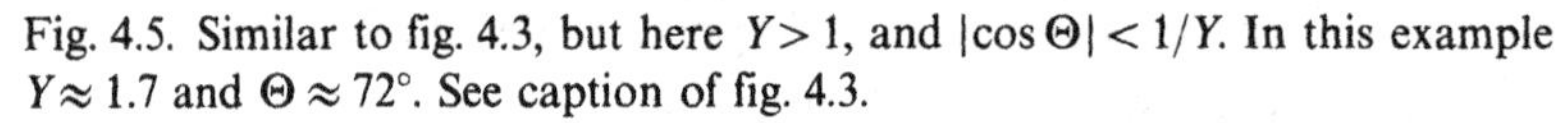
Fig. 4.5. Similar to fig. 4.3, but here $Y > 1$, and $|\cos \Theta| < 1/Y$. In this example $Y \approx 1.7$ and $\Theta \approx 72°$. See caption of fig. 4.3.

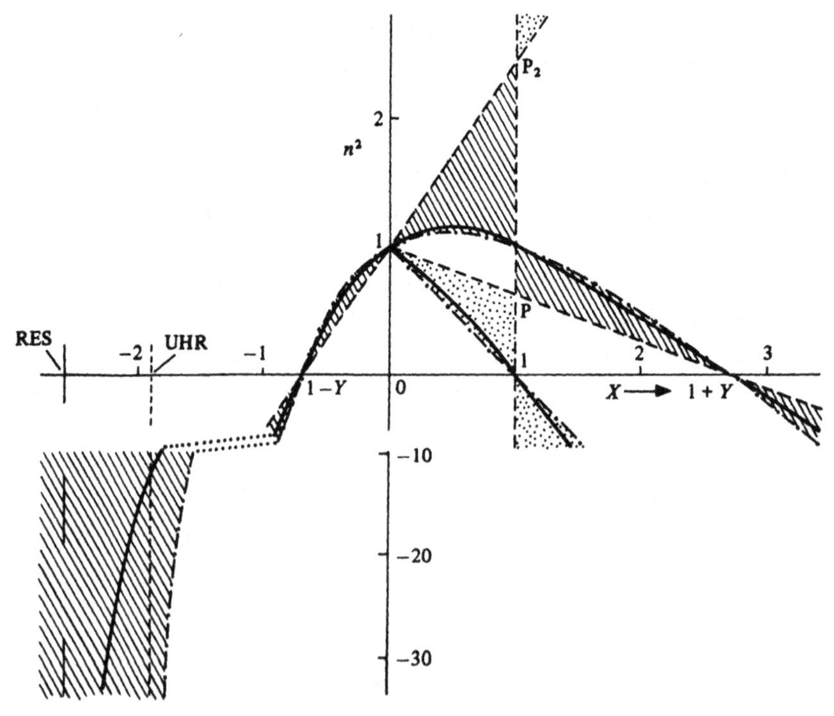

$Y > 1$. They are shown as chain lines in figs. 4.5, 4.6. Similarly the curves for purely longitudinal propagation are given by (4.84) together with the line $X = 1$; these are shown as broken curves in figs. 4.5, 4.6. On the line $X = 1$ there are two window points where $n^2 = Y/(Y-1)$ at P_2 and where $n^2 = Y/(Y+1)$ at P. For both of them n is real and they are associated with radio windows of two different types. See §§ 17.6–17.9.

The meaning of the terms 'ordinary' and 'extraordinary' is the same as for $Y < 1$ in § 4.11, so that n^2 for the ordinary wave uses a positive sign for the term $\pm\{\ldots\}^{\frac{1}{2}}$ in (4.47). Thus the curves for the ordinary wave lie in the region shaded with dots in figs. 4.5, 4.6, and the curves for the extraordinary wave in the region shaded with lines. The part of the line $X = 1$ between P and P_2 applies to the extraordinary wave, and the rest applies to the ordinary wave.

Fig. 4.5 illustrates the behaviour when $Y^2\cos^2\Theta < 1$, which includes the transverse and near transverse cases. The infinity, (4.76) with $U = 1$, is where X is negative and occurs for the extraordinary wave. For positive X each curve has only one branch. The ordinary wave is evanescent for all $X > 1$, and the extraordinary wave is evanescent for all $X > 1 + Y$.

Fig. 4.6 is for $Y^2\cos^2\Theta > 1$ which includes the longitudinal and near longitudinal

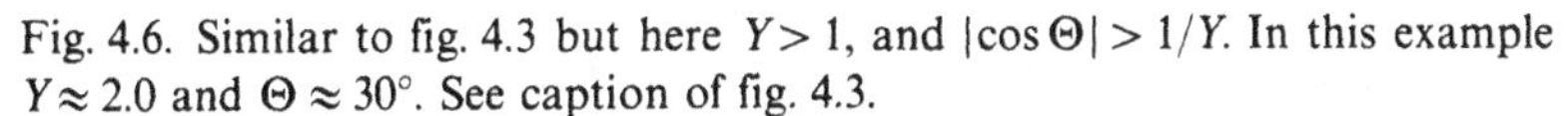

Fig. 4.6. Similar to fig. 4.3 but here $Y > 1$, and $|\cos\Theta| > 1/Y$. In this example $Y \approx 2.0$ and $\Theta \approx 30°$. See caption of fig. 4.3.

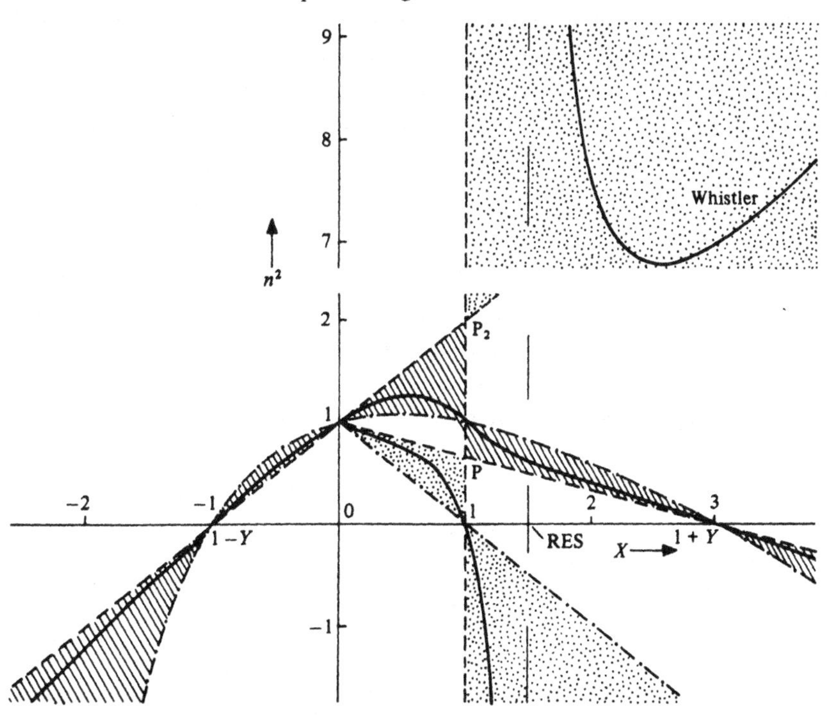

cases. The infinity is now where X is positive and > 1, and it occurs for the ordinary wave whose curve therefore has two branches. On the branch where $X > (Y^2 - 1)/(Y^2 \cos^2\Theta - 1)$ the value of n^2 is positive and large, and tends to $+\infty$ when $X \to +\infty$. This is called the whistler branch and the wave is called the whistler mode. The phenomenon of whistlers is described in §13.8.

4.13. Effect of collisions included

When Z is not zero the two values of n^2 given by (4.47) are complex and we take $n = \mu - i\chi$, (2.65), where μ and χ are real. It is now convenient to draw curves showing how μ and χ depend on X. These are slightly different in form from the curves of n^2 vs X used previously.

It was mentioned in §§2.13, and 4.2 that for a passive isotropic medium μ and χ must always be both positive (or both negative). This is still true when the earth's magnetic field is allowed for. It is necessary on physical grounds because the medium must always absorb energy from the wave, but it can be proved formally as follows. It is required to prove that $\mathrm{Im}(n^2)$ is negative, so that $\mathrm{Im}(1/n^2)$ is positive. The dispersion relation (4.62) may be written

$$\cos^2\Theta\left(\frac{1}{\varepsilon_1}-\frac{1}{n^2}\right)\left(\frac{1}{\varepsilon_2}-\frac{1}{n^2}\right)+\sin^2\Theta\left(\frac{1}{\varepsilon_3}-\frac{1}{n^2}\right)\left\{\frac{1}{2}\left(\frac{1}{\varepsilon_1}+\frac{1}{\varepsilon_2}\right)-\frac{1}{n^2}\right\}=0. \quad (4.85)$$

From the definitions (3.65)–(3.67) it can be shown that for real X_e, X_i, Z_e, Z_i each of $\varepsilon_1, \varepsilon_2, \varepsilon_3$ has a negative imaginary part and their reciprocals in (4.85) therefore have positive imaginary parts. If $1/n^2$ had a negative imaginary part, the imaginary part of the left side of (4.85) would be non-zero and positive, and the equation would not be satisfied. Hence $\mathrm{Im}(1/n^2)$ must be positive and non-zero if any collisions are included. An alternative proof that applies only to an electron plasma was given by Budden (1961a, §6.11).

4.14. The transition collision frequency

It was shown in §4.4 that for the electron plasma the two possible values of the polarisation ρ are both equal to ± 1 when $X = 1$, $Z = Z_t$ where $Z_t = \frac{1}{2}Y\sin^2\Theta/|\cos\Theta|$. This is simply the condition that the square root in the Appleton–Lassen formula (4.47) shall be zero, so that the two refractive indices are equal. Now it follows from (4.34), (4.35) and (4.38) that $\mathrm{Re}(\rho)$ and $\cos\Theta$ have opposite signs. Hence (4.46) shows that the one refractive index is then given by

$$n^2 = 1 - X/\{1 - i(Z + Y|\cos\Theta|)\}. \quad (4.86)$$

For upgoing waves in the northern hemisphere $\cos\Theta$ is positive. For the transition condition, therefore, $\rho = -1$. This means that for both waves the electric vector is parallel to the plane $x = -y$.

It is shown later (§7.13) that the differential equations which govern the propagation of waves in a given direction (here taken to be parallel to the z axis) in a homogeneous medium are equivalent to a single differential equation of the fourth order. This should have four independent solutions, and it has been shown that in general this is so, for there are two solutions which represent progressive waves travelling in the direction of positive z, and two more for waves travelling in the opposite direction. For the transition condition described above, however, it might appear that there are only two independent solutions representing waves travelling in opposite directions, both with $\rho = -1$ and with n given by (4.86). But in this case two more solutions can be found, in which the electric vector has a component at right angles to the plane $x = -y$. This may be shown as follows.

Suppose that the medium is very nearly, but not quite, at the transition condition, so that $\rho = -1 \pm \varepsilon$ where ε is so small that its square and higher powers may be neglected, and let the associated values of the refractive index be n_+ and n_-. Then (4.46) gives

$$n_+^2 - n_-^2 = 2\mathrm{i}XY\varepsilon\cos\Theta\,(U - \mathrm{i}Y\cos\Theta)^{-2} \tag{4.87}$$

$$n_+ - n_- = \mathrm{i}XY\varepsilon\cos\Theta\, n^{-1}(U - \mathrm{i}Y\cos\Theta)^{-2} \tag{4.88}$$

where n (without subscript) is written for the average value of n_+ and n_-. Choose new axes x', y', z formed from the original axes (of §4.3) by rotation through $-45°$ about the z axis, and let E'_x E'_y be the components of $\boldsymbol{E}$ parallel to the x' and y' axes, and let ρ' denote the new value of the polarisation ρ. Then (4.31) and (4.32) show that $\rho' = \pm\frac{1}{2}\varepsilon$. Let the two waves with these polarisations have equal amplitudes, so that their components E'_x annul each other when $z = 0$. Then

$$E'_x = A\{\exp(-\mathrm{i}kn_+z) - \exp(-\mathrm{i}kn_-z)\}, \tag{4.89}$$

where A is a constant. The other field components may be found from (4.16) and (4.2)–(4.5). To the first order in ε they are given by:

$$\left.\begin{aligned} E'_x &= -\mathrm{i}kz(n_+ - n_-)A\mathrm{e}^{-\mathrm{i}knz}, \\ E'_y &= A\varepsilon\mathrm{e}^{-\mathrm{i}knz}, \\ \mathscr{H}'_x &= -nA\varepsilon\mathrm{e}^{-\mathrm{i}knz}, \\ \mathscr{H}'_y &= A(n_+ - n_-)(1 - \mathrm{i}knz)\mathrm{e}^{-\mathrm{i}knz} \end{aligned}\right\} \tag{4.90}$$

Now let A tend to ∞, and ε tend to zero in such a way that $A\varepsilon$ remains constant and equal to B, say. Then by using (4.88) the field components may be written:

$$\left.\begin{aligned} E'_x &= BkXY\cos\Theta\, n^{-1}(U - \mathrm{i}Y\cos\Theta)^{-2}z\mathrm{e}^{-\mathrm{i}knz}, \\ E'_y &= B\mathrm{e}^{-\mathrm{i}knz}, \\ \mathscr{H}'_x &= -nB\mathrm{e}^{-\mathrm{i}knz}, \\ \mathscr{H}'_y &= \mathrm{i}BXY\cos\Theta\, n^{-1}(U - \mathrm{i}Y\cos\Theta)^{-2}(1 - \mathrm{i}knz)\mathrm{e}^{-\mathrm{i}knz} \end{aligned}\right\} \tag{4.91}$$

The wave field described by (4.91) has been derived by superimposing two progressive waves, travelling in the direction of positive z. The result is not a progressive wave since E'_x and $\mathscr{H}'_y$ have components with a factor z, but it is a solution of Maxwell's equations in this critical case. A fourth solution could be found that represented a similar wave travelling in the direction of negative z.

The wave (4.91) does not play any essential part in the theory of later chapters. The following is an example of a problem where it would be needed. Consider a homogeneous medium with a plane sharp boundary, and suppose that the transition condition holds for a direction normal to the boundary. A plane wave is incident normally on the boundary, and it is required to find the amplitude and polarisation of the reflected wave. The boundary conditions cannot in general be satisfied unless there are two waves in the medium, namely a linearly polarised progressive wave with refractive index given by (4.86), and a wave of the type (4.91). The detailed solution of this problem is complicated, however, and of no special practical interest.

4.15. The terms 'ordinary' and 'extraordinary'

When electron collisions are allowed for, it is desirable to have a formal definition of the terms 'ordinary' and 'extraordinary'. The following definition applies for a plasma in which the only effective charged particles are electrons, and only when the angle Θ between the wave normal and the earth's magnetic field is real. It still applies, however, when X and Z take complex values. This occurs when, in later chapters, the functions $X(h)$ and $Z(h)$ are continued analytically so as to use complex values of the height h, in later chapters denoted by z.

The Appleton–Lassen formula (4.47) and the formula (4.21) for ρ may be written

$$n^2 = 1 - \frac{X(U-X)}{U(U-X) - \frac{1}{2}Y^2 \sin^2\Theta + S_{\mathrm{R}}}, \quad \rho = \frac{\mathrm{i}(\frac{1}{2}Y^2 \sin^2\Theta - S_{\mathrm{R}})}{Y(U-X)\cos\Theta} \tag{4.92}$$

where

$$S_{\mathrm{R}}^2 = \tfrac{1}{4}Y^4 \sin^4\Theta + Y^2(U-X)^2\cos^2\Theta. \tag{4.93}$$

Definition: When $\mathrm{Re}(S_{\mathrm{R}})$ is positive, the values of n and ρ from (4.92) are written n_{O}, ρ_{O} and are the refractive index and polarisation of the ordinary wave. When $\mathrm{Re}(S_R)$ is negative, the values are written n_{E}, ρ_{E} and are the refractive index and polarisation of the extraordinary wave.

When $\cos\Theta \to 0$, that is transverse propagation, this definition gives $n_{\mathrm{O}}^2 = 1 - X/U$, $\rho_{\mathrm{O}} = 0$, so it accords with the definition given in §4.11.

When Θ, X, Z are all real, and $Z < Z_t$, (4.26)–(4.28), and $\sin\Theta \neq 0$, the real part of S_{R}^2 is positive and never zero. Then S_{R} necessarily has a non-zero real part, and there is no ambiguity in the terms ordinary and extraordinary. When $Z > Z_t$, or when $\sin\Theta = 0$ and $Z = 0$, there is ambiguity when $X \equiv 1$. This is discussed in the following section. When X and Z are complex, $\mathrm{Re}(S_{\mathrm{R}})$ can sometimes be zero and there is then ambiguity in the definition. When $X(z)$ and $Z(z)$ are functions of the height z as in later chapters, the condition $\text{Re}(S_R) = 0$ is the equation of a line in the complex z plane. It is sometimes convenient to use this line as a branch cut; see §§ 16.5–16.7.

The above definition is useful for radio waves vertically incident on a horizontally stratified ionosphere, for then Θ has a fixed real value. For oblique incidence the problem is more complicated. Instead of the refractive indices $\pm n_O$, $\pm n_E$, the four roots q of the Booker quartic equation must be used; ch. 6. The angle Θ between the wave normal and the earth's magnetic field now has different values for the four waves. It is in general complex and varies with height. The above definition could in principle still be used but is rarely or never invoked, because the need for a precise definition does not arise.

At extremely low frequencies where the effect of heavy ions is important a definition similar to that given above can be formulated; see, for example, Al'pert, Budden *et al.* (1983, § 8(a)). But at these low frequencies the waves are usually described by other names (Booker, 1984), and the terms ordinary and extraordinary are not often used.

4.16. Dependence of refractive index on electron concentration 3. Collisions allowed for

To extend the results shown in figs. 4.3–4.6 so as to allow for electron collisions, it would be necessary to plot curves of both μ and χ (2.65) versus X. To cover all cases of interest a very large number of curves would be needed and no attempt has been made to supply them in this book. Some curves of this kind have been given by Booker (1934), Ratcliffe (1959), Budden (1961a), Ginzburg (1970), Kelso (1964), Rawer and Suchy (1967). When a computer is available, μ and χ can be calculated very quickly and it is easy to plot the curves as needed.

For purely transverse propagation the first formula (4.81) is for the ordinary wave and is now replaced by (4.9). This is the same as for any wave in an isotropic electron plasma, and typical curves of μ and χ for this case are shown in fig. 4.1. The second formula (4.81) is for the extraordinary wave and is now replaced by

$$n_E^2 = 1 - \frac{X(U-X)}{U(U-X) - Y^2}. \tag{4.94}$$

Curves of μ and χ were given by Budden (1961a) for the two typical cases $Y < 1$ and $Y > 1$.

For purely longitudinal propagation the formulae (4.84) are replaced by

$$n^2 = 1 - \frac{X}{U+Y}, \quad n^2 = 1 - \frac{X}{U-Y} \tag{4.95}$$

The dependence on X of μ and χ from the first formula (4.95) is given by fig. 4.1 provided that the abscissa X is replaced by $X/(1+Y)$ and the numbers by the curves are the values of $Z/(1+Y)$. The same applies to the second formula (4.95) if $Y<1$; the abscissa is now $X/(1-Y)$ and the numbers by the curves are the values of $Z/(1-Y)$. If $Y>1$ a new set of curves is needed, as given by Budden (1961a).

It was shown in § 4.11 that when $Z=0$ and $\sin\Theta$ is zero, the curve of n^2 versus X makes a steep traverse near $X=1$ from the curve given by the first formula (4.84) to that given by the second, and the two sections of the resulting curve are joined by a segment of the vertical line $X=1$. When $\sin\Theta$ is small but not quite zero this steep traverse still occurs as shown in fig. 4.4 where the curve is nearly vertical near $X=1$. When $Z\neq 0$ and when $\sin\Theta$ is zero or small, this steep traverse does not occur. It appears only when Θ exceeds a transition value Θ_t given by

$$\sin^2\Theta_t/|\cos\Theta_t| = 2Z/Y. \tag{4.96}$$

This is the same as (4.27) except that the subscript t appears on Θ, and not on Z. In deriving (4.27), Θ was regarded as held constant and a transition value of Z was sought. Here we suppose that Z is held constant and seek a transition value Θ_t of Θ. When (4.27) or (4.96) is satisfied the two values of the polarisation ρ are both equal to -1 (for upgoing waves in the northern hemisphere) and the two refractive indices are equal.

Fig. 4.7. Shows how μ and χ depend on X for a cold electron plasma with $Y=0.5$, $Z=0.007\,655$, $\Theta_t=10°$. These curves should be studied as part of the sequence, figs. 4.7–4.11. In (a), (b) $\Theta=0°$. In (c) (d) $\Theta=6°$.

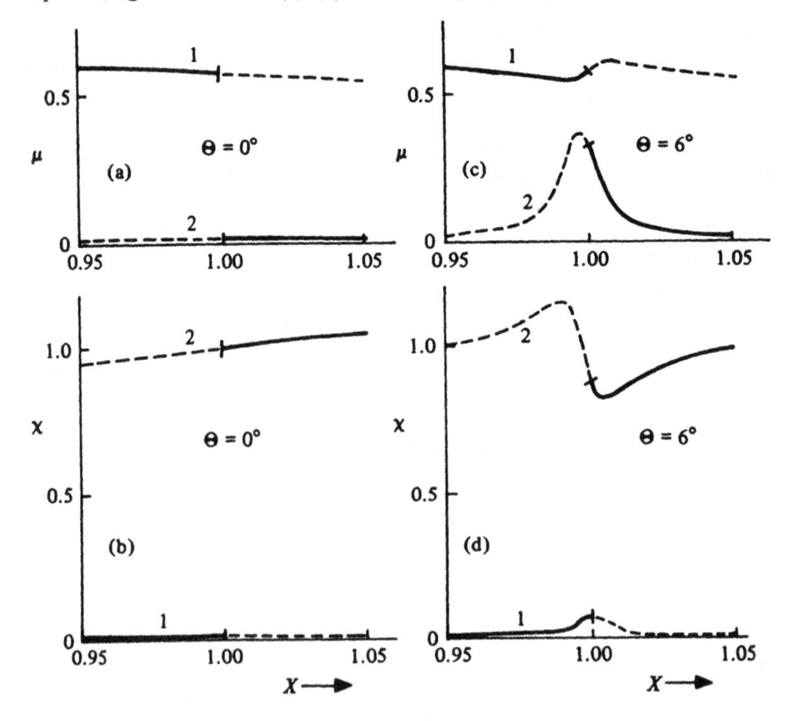

Figs. 4.7–4.11 show curves of μ and χ versus X for a fixed value of Z and for a sequence of values of Θ. They cover a small range of X near $X = 1$ where the transition occurs. In this example the transition value of Θ is $\Theta_t = 10°$. Figs. 4.7 (a, b) are for $\Theta = 0$. The values of μ and χ are very close to those that would be obtained from the straight lines (4.95) in fig. 4.3, where U was equal to 1. Thus for the first expression (4.95) with $+\ Y$, upper straight broken line in fig. 4.3, n^2 is almost real and positive but now the non-zero Z gives it a small imaginary part, so χ is non-zero but small. The resulting curves are labelled 1 in figs. 4.7(a, b). For the second expression (4.95) with $-\ Y$, lower straight broken line in fig. 4.3, n^2 is almost real and negative, but again it now has a small imaginary part. Thus μ is non-zero but small. The curves in figs. 4.7(a, b) are labelled 2. There is nothing in these two figures to correspond to

Fig. 4.8. Dependence of μ and χ on X for the same plasma as in fig. 4.7. The right-hand figure shows the complex ρ plane with the values of ρ for the two waves. The arrows show the direction of increasing X and the bars are where $X = 1$. Here $\Theta = 9°$.

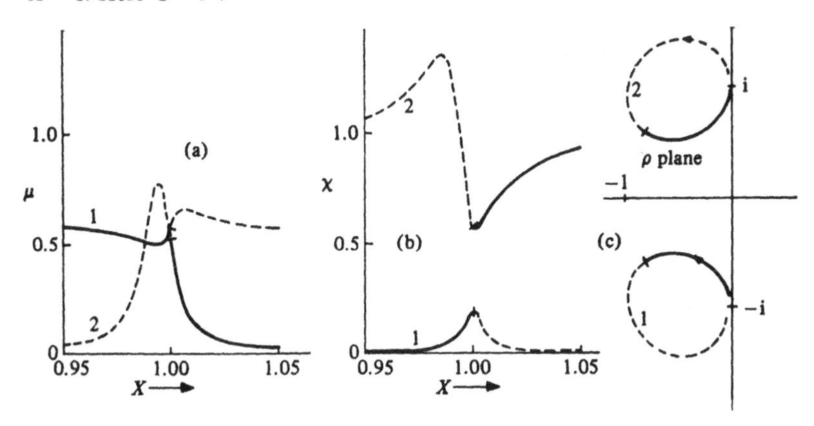

Fig. 4.9. Similar to fig. 4.8 but for the transition case $\Theta = \Theta_t = 10°$.

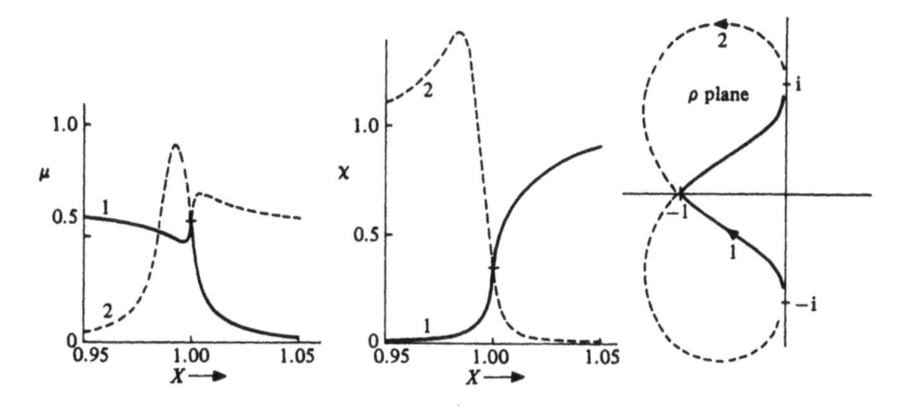

the vertical line $X = 1$ in fig. 4.3. For curves 1 the polarisation ρ is $-\mathrm{i}$ for all X, and for curves 2 it is $+\mathrm{i}$.

The wave corresponding to the first expression (4.95), described by curves 1, was, in § 4.11, called the 'ordinary wave' for $X < 1$, and the 'extraordinary wave' for $X > 1$. A wave with $\rho = -\mathrm{i}$ which enters a plasma where $X < 1$, with $\Theta = 0$, is an ordinary wave. Suppose now that the value of X increases with distance as the wave travels. The values of μ and χ are given by curves 1 of figs. 4.7(a, b), and they are smoothly varying functions of distance. If the medium varies very slowly with distance, the wave travels on without reflection, and it continues to do so when it enters a region where $X > 1$. After it has passed the point where $X = 1$, however, it is called an 'extraordinary wave'. There is no physical process associated with this change, and in particular no 'mode conversion', and no change of polarisation. All that has changed is the adjective used to describe the wave. See § 16.6 for an example.

Figs. 4.7(c, d) show similar curves of μ and χ versus X for $\Theta = 6°$. This again is less than Θ_t. The curves are continuous and the same remarks apply. The curves for the

Fig. 4.10. Similar to fig. 4.8 but here $\Theta = 10.5°$, slightly greater than the transition value $\Theta_t = 10°$.

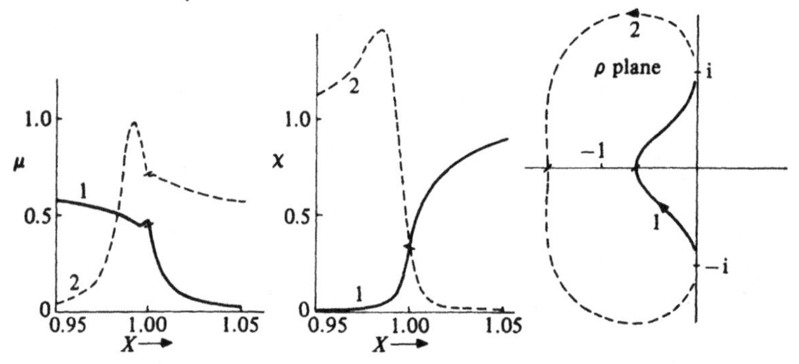

Fig. 4.11. Similar to fig. 4.8, but here $\Theta = 11°$.

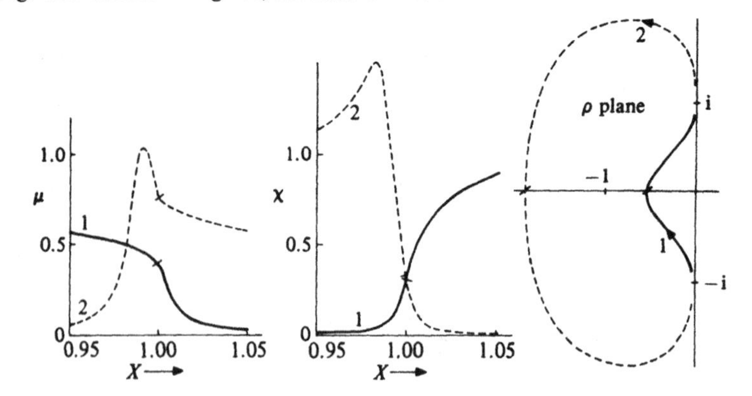

ordinary wave are shown as continuous lines and those for the extraordinary wave as broken lines, and this convention applies for the whole sequence, figs. 4.7–4.11. There is a change of description in figs. 4.7(c, d) when X passes the value 1, as before. The polarisations ρ remain close to $-\mathrm{i}$ and $+\mathrm{i}$ for curves 1 and 2 respectively. There are now small departures from these values when X is near to 1.

Fig. 4.8 shows curves for $\Theta = 9°$ which is very slightly less than Θ_t. The general behaviour is still similar to that of the cases in fig. 4.7. The two curves for μ in fig. 4.8(a) now cross each other twice. It is the crossing nearest to $X = 1$ that is important in the present discussion. The curves for χ do not cross. Fig. 4.8(c) shows how the two polarisations ρ change when X changes. When X is near 1 they now depart appreciably from the values $\pm\mathrm{i}$ and have both moved nearer to the value $\rho = -1$. For figs. 4.7, 4.8 the conditions are sometimes said to be 'less than critical'. This term is explained in §§ 16.5, 16.6.

Fig. 4.9 shows curves for the transition case $\Theta = \Theta_t = 10°$. Now the curves for μ have a crossing where $X = 1$, and those for χ have now touched each other where $X = 1$. The curves for ρ both go through the value $\rho = -1$ when $X = 1$. Fig. 4.10 shows what happens when $\Theta = 10.5°$, slightly greater than Θ_t. The curves have separated again but are now linked in a different way. The curves for χ now cross near $X = 1$, and those for μ do not cross there. Both curves are continuous at $X = 1$. The wave that is ordinary for $X < 1$ now remains an ordinary wave for $X > 1$, and similarly for the extraordinary wave. There is now no change of the descriptive adjectives. Fig. 4.10 shows that the polarisation of the ordinary wave now goes over from near $-\mathrm{i}$ when $X < 1$, to near $+\mathrm{i}$ when $X > 1$, and the opposite transition occurs for the extraordinary wave. The functions μ and χ change rapidly as X varies, near $X = 1$, and this is the counterpart of the steep traverse near $X = 1$ in fig. 4.4. Finally fig. 4.11 shows the behaviour for $\Theta = 11°$ where the condition $\Theta > \Theta_t$ is well established. The behaviour of μ, χ and ρ is now just what might be expected if the continuous curves in fig. 4.3 near $X = 1$ were used to plot μ and χ, but modified to include a non-zero value of Z. For figs. 4.10, 4.11 the conditions are sometimes said to be 'greater than critical', and this is explained in §§ 16.5, 16.6.

The transition phenomena just described, and illustrated by figs. 4.7–4.11 were for the special case $Y = \frac{1}{2}$ and are typical for $Y < 1$. When $Y > 1$ the phenomena are very similar but there are minor differences. For example when Θ is slightly less than Θ_t, the two curves $\chi(X)$ cross near $X = 1$ and the two curves $\mu(X)$ do not cross. When $\Theta > \Theta_t$ the $\mu(X)$ curves cross and the $\chi(X)$ curves do not cross.

For a plasma with the effect of ions allowed for, this type of transition is important in the theory of ion-cyclotron whistlers § 13.9.

The interchange of the adjectives 'ordinary' and 'extraordinary' at $X = 1$ for $\Theta < \Theta_t$ could be avoided by adopting different definitions of these adjectives when $X > 1$. But this does not remove the need for an interchange. It would simply mean that if X is held constant at some value > 1, and if Θ is then varied continuously from $< \Theta_t$ to $> \Theta_t$, the two adjectives would have to be interchanged when Θ passes the value Θ_t. There can be no choice of definitions that will altogether avoid the need for aninterchange. The choice made here is the one adopted in figs. 4.7–4.11. It is used throughout this book and by most authors, but not all.

The transition just described occurs when two solutions of the quadratic (4.20) for ρ are equal, and the condition is given by (4.25) which with (4.27) leads to

$$X = 1 - \mathrm{i}(Z \pm Z_t). \tag{4.97}$$

This represents two points in the complex X plane. The one with the $+$ sign can never be real, but the other is real when $Z = Z_t$, which is equivalent to (4.96), and this is the transition condition. The point $X = 1 - \mathrm{i}(Z - Z_t)$ is a branch point of the function $n(X)$. It is in the negative imaginary half of the complex X plane when $Z > Z_t$ or equivalently $\Theta < \Theta_t$. As Z is decreased, or Θ increased, so that the transition condition is approached, this branch point moves up to the real X axis and lies on it at $X = 1$ at the exact transition. When $Z < Z_t$, or equivalently $\Theta > \Theta_t$, it is in the positive imaginary half of the complex X plane. This branch point is called a 'coupling point'. In a medium which is not very slowly varying near $X = 1$, one of the two waves can give rise to some of the other wave by a mode conversion process called coupling. It is studied in §§ 16.5–16.7.

In this section we have studied the behaviour of $\mu(X)$ and $\chi(X)$ when the point X moves along the line $\mathrm{Im}(X) = 0$. The two values of $n = \mu - \mathrm{i}\chi$ are those of the double valued complex function $n(X)$. If the branch point is where $\mathrm{Im}(X)$ is negative, the two curves for $\mu(X)$ cross each other, near $X = 1$, and the curves for $\chi(X)$ do not cross. When the branch point is on the line $\mathrm{Im}(X) = 0$, both curves cross at $X = 1$ where the two complex values of n are equal. When the branch point is where $\mathrm{Im}(X)$ is positive, the curves for $\chi(X)$ cross and those for $\mu(X)$ do not cross near $X = 1$. It can be shown that this behaviour always occurs for any two-valued complex function with a single branch point. It was studied for magnetoionic theory by Booker (1934).

For any X and Z, including real values, the two refractive indices are equal when Θ has some complex value, Θ_g say. If curves of μ and χ are plotted against $\mathrm{Im}(\Theta)$, with $\mathrm{Re}(\Theta)$ held constant, their behaviour when $\mathrm{Im}(\Theta)$ is near to $\mathrm{Im}(\Theta_g)$ would be similar to that of the curves $\mu(X)$ and $\chi(X)$ described above. The curves, for μ or χ, that cross when $\mathrm{Re}(\Theta) < \mathrm{Re}(\Theta_g)$ would not cross when $\mathrm{Re}(\Theta) > \mathrm{Re}(\Theta_g)$ and vice versa. There could again be a need to interchange the adjectives 'ordinary' and 'extraordinary' when $\mathrm{Im}(\Theta) > \mathrm{Im}(\Theta_g)$.

4.17. Approximations for refractive indices and wave polarisations

Various approximations have been used for n^2 and for ρ. These were useful in the days before computers and calculators were available, when calculations with the exact formulae were laborious. In theoretical physics many of the approximations of a function, $f(z)$ say, for values of z near to some specified value z_0, use expansions in powers of $z - z_0$. One type is an asymptotic approximation. This is discussed in §§ 8.11, 8.17 and is not considered further in the present section. Another type applies when $f(z)$ is analytic for z equal to or near to z_0. Then $f(z)$ is expanded in a Taylor series whose successive terms are $f(z_0)$, $(z - z_0)(\mathrm{d}f/\mathrm{d}z)_0$, $\frac{1}{2}(z - z_0)^2(\mathrm{d}^2 f/\mathrm{d}z^2)_0$, etc., where the subscript 0 indicates that the value at $z = z_0$ is used. The approximation consists in retaining only two or three terms of the series. If r terms are retained, the approximation gives $f(z_0)$ and its derivatives up to $(\mathrm{d}^{r-1}f/\mathrm{d}z^{r-1})_0$ exactly. When $z \neq z_0$ the error in the approximation is of order $(z - z_0)^r$ and therefore small when $|z - z_0|$ is small. Approximations that are expansions with this analytic property are widely used in theoretical physics and there are many examples in this book.

Any approximation, whether based on an expansion or on some other method, cannot be correct under all conditions. The most important thing to know about it is where it fails, so that errors are avoided.

In magnetoionic theory one group of approximate formulae are called quasi-longitudinal (QL), and quasi-transverse (QT) approximations. They have been known for some time and are here called the 'conventional' QL and QT approximations. They are used only for an electron plasma, and are as follows:

QL:

$$n^2 \approx 1 - X/(U \pm Y\cos\Theta), \quad \rho \approx \mp \mathrm{i}. \tag{4.98}$$

If $(1 - X)(1 - Y)\cos\Theta$ is positive, the upper signs are for the ordinary wave and the lower for the extraordinary wave. If it is negative, the upper signs are for the extraordinary and the lower for the ordinary wave.

QT, ordinary wave:

$$n^2 \approx 1 - X/\{U + (U - X)\cot^2\Theta\}, \quad \rho \approx 0 \tag{4.99}$$

QT, extraordinary wave:

$$n^2 \approx 1 - X(U - X)/\{U(U - X) - Y^2\sin^2\Theta\}, \quad 1/\rho \approx 0. \tag{4.100}$$

The QL approximations (4.98) are exactly correct when $\sin\Theta = 0$, and the QT approximations (4.99), (4.100) are exactly correct when $\cos\Theta = 0$. But they fail in the following conditions:

(a) Except for (4.99), they do not make n^2 zero at the correct values of X, and the values of X where $n^2 = 0$ depend on Θ.

(b) They do not make n^2 infinite at the correct value of X.

(c) They do not give the correct derivatives $\mathrm{d}(n^2)/\mathrm{d}(\sin^2\Theta) = n\mathrm{d}^2 n/\mathrm{d}\Theta^2$ at $\sin\Theta = 0$ for QL, nor $\mathrm{d}(n^2)/\mathrm{d}(\cos^2\Theta) = n\mathrm{d}^2 n/\mathrm{d}\Theta^2$ at $\cos\Theta = 0$ for QT. Thus they do not give the correct values of the curvature of the refractive index surface in these two cases.

It might be thought that the term 'quasi-longitudinal' implies that the approximation is good in conditions at and near to longitudinal, that is near $\sin^2\Theta = 0$, and that it gives correct values of one or two of the derivatives when $\Theta \to 0$. The term 'quasi-transverse' could be taken to have similar implications. But the failure (c) shows that this is not so. In the analytic sense mentioned above, the conventional QL and QT approximations are not correct. In spite of their failures they have been used successfully but it seems that no clear rule has ever been given to show when they are useful and when they fail.

The form (4.63) of the dispersion relation shows that near $\sin^2\Theta = 0$, n^2 is an analytic function of $\sin^2\Theta$ so that it can be expanded in a series of powers of $\sin^2\Theta$. Similarly (4.64) shows that near $\cos^2\Theta = 0$, n^2 is an analytic function of $\cos^2\Theta$ and a similar expansion is possible. These expansions were given by Budden (1983), both for an electron plasma and for the more general plasma. They will be referred to here by the abbreviations QL2 and QT2. They are as follows:

QL2:

$$n^2 \approx \left(1 - \frac{X}{1+Y}\right)\left\{1 - \frac{XY\sin^2\Theta}{2(U-X)(U+Y)} + \dots\right\} \tag{4.101}$$

$$n^2 \approx \varepsilon_1\left\{1 - \frac{\varepsilon_1 - \varepsilon_3}{2\varepsilon_3}\sin^2\Theta + \dots\right\}. \tag{4.102}$$

These are for one of the two characteristic waves. For the other, the sign of Y is reversed in (4.101), and ε_1 is replaced by ε_2 in (4.102).

QT2 ordinary wave:

$$n^2 \approx \left(1 - \frac{X}{U}\right)\left\{1 + \frac{X}{U}\cos^2\Theta + \dots\right\} \tag{4.103}$$

$$n^2 \approx \varepsilon_3\left\{1 - \frac{(\varepsilon_1 - \varepsilon_3)(\varepsilon_2 - \varepsilon_3)}{J}\cos^2\Theta + \dots\right\}. \tag{4.104}$$

QT2 extraordinary wave:

$$n^2 \approx \frac{(U-X)^2 - Y^2}{U(U-X) - Y^2}\left\{1 - \frac{X(U-X)}{U(U-X) - Y^2}\cos^2\Theta + \dots\right\} \tag{4.105}$$

$$n^2 \approx \frac{2\varepsilon_1\varepsilon_2}{\varepsilon_1 + \varepsilon_2}\left\{1 - \frac{\varepsilon_3(\varepsilon_1 - \varepsilon_2)^2}{2(\varepsilon_1 + \varepsilon_2)J}\cos^2\Theta + \dots\right\}. \tag{4.106}$$

The coefficients of the terms in $\sin^4\Theta$ and $\cos^4\Theta$ in these expansions were given by Budden (1983), who also gave the corresponding expansions for the wave polarisations ρ, but these details are not needed here.

The formulae (4.101)–(4.106) give the zeros of n^2 at the correct values of X and they give the correct values of $n\, \mathrm{d}^2n/\mathrm{d}\Theta^2$ for $\sin\Theta = 0$ (QL2) and $\cos\Theta = 0$ (QT2) respectively. Thus they do not fail under conditions (a), (c) above. But they do not make n^2 infinite at the correct value of X and so they fail under condition (b).

The radius of convergence of the QL2 expansion is the smallest value of $|\sin^2\Theta|$ for which n^2 has a singularity, and there are two of these. First there is a pole where

the denominator of (4.67) or of (4.51) is zero, but this appears in only one of the two solutions, namely the extraordinary wave if $Y < 1$, and the ordinary wave if $Y > 1$. It gives as a convergence criterion:
For QL2:

$$|\sin^2\Theta| < \left|\frac{(U-X)(U^2-Y^2)}{XY^2}\right|, \quad |\sin^2\Theta| < \left|\frac{\varepsilon_3}{G}\right| \tag{4.107}$$

The second singularity is the branch point where the square root in (4.47) or (4.51) is zero. This gives as a convergence criterion:
For QL2:

$$|\sin^2\Theta| < 2\left|\frac{U-X}{Y}[U-X \pm \{(U-X)^2 - Y^2\}^{\frac{1}{2}}]\right|$$

$$|\sin^2\Theta| < \left|\frac{\varepsilon_3(\varepsilon_1-\varepsilon_2)^2}{2J}[\varepsilon_3 \pm \{\varepsilon_3^2 - 4J^2(\varepsilon_1-\varepsilon_2)^{-2}\}^{\frac{1}{2}}]\right| \tag{4.108}$$

where it is implied that the smaller values are used.

In a similar way we obtain convergence criteria for $|\cos^2\Theta|$ in the QT2 approximations. The radius of convergence set by the pole, for one wave only, is:
For QT2:

$$|\cos^2\Theta| < \left|\frac{U(U^2-UX-XY^2)}{XY^2}\right|$$

$$|\cos^2\Theta| < \left|1 + \frac{\varepsilon_3}{G}\right|. \tag{4.109}$$

The range given by the branch point is a rather complicated expression not needed here but given by Budden (1983).

The convergence criteria give some indication of where the QL2 and QT2 approximations may and may not be used. Budden (1983) gave examples to illustrate that for frequencies in the radio range, that is not much less than half the electron gyro-frequency, the QL2 and QT2 versions are considerably better than the conventional QL and QT. Johler and Walters (1960) studied the conventional QL approximation for very low frequencies in the range 10 to 80 kHz. They describe it as crude and warn that it should be used only with considerable caution.

Consider now the application of these formulae to extremely low frequencies, in the range studied in magnetohydrodynamics. For frequencies less than the smallest ion gyro-frequency ε_3 is very large and negative, and ε_1, ε_2 are positive and approximately equal (see fig. 3.2). Now (4.109) second equation gives an extremely small convergence range for QT2, for the wave that has an infinity of n^2 (Booker and Vats, 1985). This is a clear example where QT2 is not useful. But for the other wave there is no infinity of n^2 and (4.109) does not apply to it. For this wave QT2 is still useful.

Other forms of approximation are possible that avoid failure at an infinity of n^2. Instead of finding an approximation for n^2, the method is to find an approximation for $n^2\{\frac{1}{2}(\varepsilon_1+\varepsilon_2)\sin^2\Theta+\varepsilon_3\cos^2\Theta\}$. Here it is the second factor, in brackets, that is zero when one n^2 is infinite. This method was used by Booker and Vats (1985). The author is greatly indebted to Professor H.G. Booker for supplying a copy of this paper in advance of publication, and for a stimulating correspondence. Booker and Vats showed that, for the wave that has an infinity, high accuracy is attained by using either the QL or QT form of this version, over a wide range of frequency and angle Θ. But for the other wave their method gives infinite n^2 where in fact there is no infinity.

Another form of approximation should be mentioned. For an electron plasma the dispersion relation is (4.65), (4.66), and it is easy to show that as $\omega\to 0$, $X\to\infty$, both values of n^2 tend to infinity. But (4.65) can be written as an equation for n^2/X, and its solutions expanded in powers of $1/X$. For a collisionless plasma, $U=1$, the first terms of this expansion for the two waves are

$$n^2=-X/(1\pm Y\cos\Theta). \tag{4.110}$$

The radius of convergence of the expansion imposes the restriction that

$$1/X<|(Y^2\cos^2\Theta-1)/(Y^2-1)|. \tag{4.111}$$

The formula (4.110), with the lower sign, has been found useful for studying the propagation of whistlers (§ 13.8) in the magnetosphere. Here Y is of order 10 to 100 and X is of order 10^2 to 10^4. The condition (4.111) is amply satisfied except when $\cos\Theta$ is very close to $1/Y$. It can be shown that the second term of the expansion is small. The value $\cos\Theta=1/Y$ makes n^2 infinite in (4.110). The correct value is $(1/Y)\{1+(Y^2-1)/X\}^{\frac{1}{2}}$ so the error is small when X is large.

The formula (4.110) happens to be the same as the conventional QL approximation, (4.98) without the 1, and some authors have regarded it as an angular approximation rather than as depending on a large value of X. Its simplicity makes it especially useful. For example the formulae (5.75) for the group refractive index, and (5.83) for the angle α between the ray and the wave normal are simple, and Gendrin (1960, 1961) has used them in a study of whistler propagation. Booker (1984) has shown that the formula $\mathscr{U}=\partial\omega/\partial\boldsymbol{\kappa}$, (5.72), in this case can be expressed very simply in Cartesian coordinates, whereas it is not so easy with the exact formula; see § 5.8.

The formula (4.110) may be unreliable if the frequency is too great, so that X is not large or if the frequency is so small that the effect of heavy ions cannot be ignored. The useful frequency range is roughly from the lower hybrid resonance frequency (3.74) to the electron gyro-frequency.

Approximations for n^2 are sometimes needed for very high frequencies so that X, Y and Z are very small. For small X both values of n^2-1 are of order X. Suppose that X^2 and higher powers can be neglected. Then the dispersion relation (4.65)

(4.66) is expressed as a quadratic for $(n^2 - 1)/X$ and in the coefficients we set $X = 0$. The solution is

$$n^2 - 1 \approx X \frac{\frac{1}{2}Y^2 \sin^2\Theta - U^2 \pm Y\{\frac{1}{4}Y^2 \sin^4\Theta + U^2 \cos^2\Theta\}^{\frac{1}{2}}}{U(U^2 - Y^2)}. \tag{4.112}$$

This result is used in the study of limiting polarisation, § 17.11. At high frequencies, Z and Y are small. If $|\cos\Theta/\sin^2\Theta| \gg \frac{1}{2}Y$, and if Z and Y^2 and higher powers of Y can be neglected, this gives

$$n^2 \approx 1 - X(1 \mp Y\cos\Theta); \quad n \approx 1 - \tfrac{1}{2}X(1 \mp Y\cos\Theta). \tag{4.113}$$

The upper sign is for the ordinary wave. This result is used in the study of Faraday rotation, § 13.7.

This discussion has shown that there are many ways of finding approximations for n^2 and ρ, and some of them are useful for specified applications. The subject has been fully discussed by Booker (1984). In nearly all cases it is advisable, before using them, to make some check of accuracy and this usually means comparing them with the exact formula. But then the exact formula might just as well be used from the start and there is no need to go to the trouble of making a check. Even the most complicated forms of the cold plasma dispersion relation can be solved very quickly with modern computers. In the author's opinion approximations of the kind described in this section are rarely needed for frequencies in the range used for radio communication, that is when X and Y are neither very large nor very small, and are likely to go out of use. For warm and hot plasmas the situation is quite different. The full form of the dispersion relation is extremely complicated and approximations are invariably used.

PROBLEMS 4

4.1. An elliptically polarised plane electromagnetic wave has its wave normal parallel to the z axis. Its wave polarisation is $E_y/E_x = R$ (complex). What are the axis ratio and the square of the length of the semi-major-axis of the polarisation ellipse, in terms of $|E_x|$ and R?

4.2. The wave of problem 4.1 is incident normally on a cold magnetoplasma in which the characteristic polarisations are ρ and $1/\rho$. Find the ratio of the major axes of the polarisation ellipses of the two resulting waves in the plasma, and check that the answer is right for the special cases $R = \rho$ or $1/\rho$.

4.3. For a plane progressive radio wave in a cold collisionless electron magnetoplasma, show, by using (2.19) or otherwise, that the angle between the wave normal and the normal to the polarisation ellipse for the electric polarisation $\boldsymbol{P}$ is $\arctan\{-iY\rho\sin\Theta/(1 - X)\}$.

4.4. Two elliptically polarised plane electromagnetic waves of the same frequency, with their wave normals parallel to the z axis, are combined. Their wave polarisations E_y/E_x are respectively $\mathrm{i}a$ and $-\mathrm{i}/a$ where a is real, positive and less than unity. The major axes of their polarisation ellipses are equal. What are the smallest and greatest possible axis ratio (minor axis/major axis) of the polarisation ellipse of the resultant wave and in each case what is the angle between the major axis and the x axis?

4.5. An electromagnetic wave travels through a cold magnetoplasma with its wave normal always parallel to the superimposed magnetic field. The medium attenuates the waves and its thickness is such that the amplitude of the left-handed circularly polarised characteristic wave is reduced by a factor F and that of the right-handed wave is reduced by a factor $\frac{1}{2}F$. What is the state of polarisation of the emergent wave, if the incident wave is (a) linearly polarised (b) unpolarised?

4.6. In an isotropic electron plasma with collisions the refractive index is $\mu - \mathrm{i}\chi$ where μ and χ are real and non-negative. X is held constant and is less than 4/3. Show that if Z is increased from zero, χ at first increases and then attains a maximum value given by $\chi^2 = X^2/\{8(2-X)\}$ when $Z^2 = (4-3X)/(4-X)$. What happens if X exceeds 4/3?

4.7. A homogeneous plasma contains free electrons, and collision damping is negligible. There is a superimposed steady magnetic field which is so strong that it prevents the electrons from moving at right angles to it. The electrons can therefore move only in the direction of the field. A wave of angular frequency ω travels with its wave normal at an angle Θ to the field. It is linearly polarised with its electric vector in the plane containing the wave normal and the direction of the field. Write down the equation of motion of an electron. Hence find the relation between the electric polarisation $\boldsymbol{P}$ and the electric intensity $\boldsymbol{E}$. Show that the refractive index μ is given by:

$$\mu^2 = \frac{1-X}{1-X\cos^2\Theta}$$

where $X = (\omega_N/\omega)^2$ and $\omega_N/2\pi$ is the plasma frequency.

4.8. In a cold electron plasma there is no superimposed magnetic field and electron collisions are negligible. The angular plasma frequency is ω_N and $k = \omega_N/c$. Show that the following inhomogeneous plane waves of angular frequency ω_N are solutions of Maxwell's equations and satisfy the constitutive relations, where A is a constant:

(a) $E_x = \mathrm{i}E_0 \exp\{-A(\mathrm{i}x + z)\}, \quad E_z = E_0 \exp\{-A(\mathrm{i}x + z)\},$
$E_y = \mathscr{H}_x = \mathscr{H}_y = \mathscr{H}_z = 0.$

(b) $E_y = kE_0 \exp\{-A(\mathrm{i}x + z)\}, \quad \mathscr{H}_x = \mathrm{i}AE_0 \exp\{-A(\mathrm{i}x + z)\},$
$\mathscr{H}_z = AE_0 \exp\{-A(\mathrm{i}x + z)\}, \quad E_x = E_z = \mathscr{H}_y = 0.$

In case (b) find the Poynting vector and its time average. (See Budden, 1961a, § 4.9.)

4.9. A homogeneous medium contains N free electrons per unit volume with negligible collision damping, and there is a constant superimposed magnetic field. A linearly polarised plane electromagnetic wave of angular frequency ω travels with its wave normal and its electric vector both perpendicular to the magnetic field. Show that the refractive index for this wave is given by

$$\mu^2 = 1 - X(1-X)/(1-X-Y^2).$$

Show that the electric field has a longitudinal component in quadrature with the transverse component and that the ratio of the longitudinal to the transverse amplitudes is $XY/(1-X-Y^2)$.

Use the Poynting vector to find the instantaneous direction of the energy flow. Hence find the direction of average energy flow when

(a) $X < 1 - Y$ or $X > 1 + Y$, (b) $1 - Y < X < 1 + Y$.
(Maths Tripos 1958, Part III.)

4.10. Discuss the propagation of electromagnetic waves in an isotropic material medium, treating the electric field at each point as the resultant of the electric fields of the incident wave and of the wavelets scattered by elements of volume of the medium.

What extensions of the theory would you expect to have to make to deal with (a) anisotropic media, (b) magnetic rotation of the plane of polarisation?
(Natural Sciences Tripos, 1952, Part II, Physics.)
(See §18.4; Darwin, 1924; Hartree, 1929, 1931b).

4.11. For the two plane progressive waves with the same wave normal direction, and with refractive indices $n = \mu - i\chi$, in a cold collisionless magnetoplasma, prove that if $Z = Z_t$, (4.27), or equivalently $\Theta = \Theta_t$(4.96), the four curves of μ and χ versus X, (fig. 4.9) have infinite slope where $X = 1$, and the two curves that show how ρ varies in the complex ρ plane are at $\pm 45°$ to the real ρ axis.

4.12. Use the result (e) of problem 3.6 to express Maxwell's equations in principal axis coordinates. Apply them to a plane progressive wave in a homogeneous cold magnetoplasma, with refractive index n and wave normal at an angle Θ to the superimposed magnetic field. Hence derive the dispersion relation in the form

$$\begin{vmatrix} 1+\cos^2\Theta - 2\varepsilon_1/n^2 & -\sin^2\Theta & -\sin\Theta\cos\Theta \\ -\sin^2\Theta & 1+\cos^2\Theta - 2\varepsilon_2/n^2 & -\sin\Theta\cos\Theta \\ -\sin\Theta\cos\Theta & -\sin\Theta\cos\Theta & \sin^2\Theta - \varepsilon_3/n^2 \end{vmatrix} = 0.$$

Show that this is the same as (4.63).

4.13. Show that the solutions of (4.63), giving the two refractive indices for a cold magnetoplasma, may be written

$$n^2 = -\frac{J}{G} - \frac{\frac{1}{2}\varepsilon_3(\varepsilon_1-\varepsilon_2)^2(\varepsilon_1-\varepsilon_3)(\varepsilon_2-\varepsilon_3)}{G\{\frac{1}{2}\varepsilon_3(\varepsilon_1-\varepsilon_2)^2 + GJ\sin^2\Theta\} \pm G^2\{J^2\sin^4\Theta + \varepsilon_3^2(\varepsilon_1-\varepsilon_2)^2\cos^2\Theta\}^{\frac{1}{2}}}.$$

Show that, for an electron plasma, on using (3.50), (3.52) this arrangement of the terms gives the Appleton–Lassen formula (4.47).

4.14. In a collisionless electron plasma there is an extraordinary wave with its wave normal at an angle Θ to the earth's magnetic field. The z axis is parallel to the wave normal and the x axis is coplanar with the wave normal and the earth's field. Show that, when $X = 1$, the fields of this wave satisfy $E_x = \mathscr{H}_y = \mathscr{H}_z = 0$, $\mathscr{H}_x = -E_y$, $E_y/E_z = \mathrm{i}Y \sin\Theta$. Discuss the behaviour of the fields of this wave in the limits (a) $\Theta \rightarrow 0$, (b) $Y \rightarrow 0$.

4.15. Unpolarised radiation at radio frequencies comes into the earth's ionosphere from the galaxy (at great heights it is reasonable to ignore collisions). Show that within the ionosphere half the energy flux is in the ordinary and half in the extraordinary wave.
(See Budden and Hugill, 1964, especially § 4.)

5

Magnetoionic theory 2. Rays and group velocity

5.1. Introduction

In ch. 4 we studied the propagation of a plane progressive radio wave in a cold homogeneous plasma and derived expressions for the two refractive indices. We then examined how they depend on the electron concentration N for various fixed values of the frequency ω and the angle Θ between the wave normal and the vector $\boldsymbol{Y}$. The present chapter deals with the same problem, and is concerned with the dependence of the refractive indices on Θ and on ω. It is necessary to examine situations where more than one progressive wave is present with different wave normal directions. It is therefore convenient now to introduce a new Cartesian coordinate system ξ, η, ζ with fixed axes, and with the ζ axis parallel to the vector $\boldsymbol{Y}$. The wave normal has polar angles Θ, Φ, with the ζ axis as the polar axis. It is convenient to think of the refractive index as a vector $\boldsymbol{n}$ with magnitude n and in the direction of the wave normal. Thus $\boldsymbol{n}$ has the Cartesian components

$$n_\xi = n\sin\Theta\cos\Phi, \quad n_\eta = n\sin\Theta\sin\Phi, \quad n_\zeta = n\cos\Theta. \tag{5.1}$$

For any one of the plane waves studied here all field components depend on ξ, η, ζ only through an exponential factor which may be written in any of the forms

$$\begin{aligned}\exp\{-ik(\xi n_\xi + \eta n_\eta + \zeta n_\zeta)\} &= \exp(-ik\boldsymbol{n}\cdot\boldsymbol{g}) \\ &= \exp\{-ikn(\xi\sin\Theta\cos\Phi + \eta\sin\Theta\sin\Phi + \zeta\cos\Theta)\}.\end{aligned} \tag{5.2}$$

where $\boldsymbol{g}$ is the vector (ξ, η, ζ). The dispersion relation (4.47) or (4.51) or (4.63) or (4.64) shows that n depends only on Θ and not on Φ. Thus let

$$n_\gamma = n\sin\Theta = (n_\xi^2 + n_\eta^2)^{\frac{1}{2}} \tag{5.3}$$

so that

$$n^2 = n_\gamma^2 + n_\zeta^2. \tag{5.4}$$

If these are inserted in (4.63) they give for the dispersion relation

$$\begin{aligned}\tfrac{1}{2}(\varepsilon_1 + \varepsilon_2)n_\gamma^4 + \varepsilon_3 n_\zeta^4 + n_\gamma^2 n_\zeta^2\{\tfrac{1}{2}(\varepsilon_1 + \varepsilon_2) + \varepsilon_3\} \\ - n_\gamma^2\{\tfrac{1}{2}(\varepsilon_1 + \varepsilon_2)\varepsilon_3 + \varepsilon_1\varepsilon_2\} - n_\zeta^2(\varepsilon_1 + \varepsilon_2)\varepsilon_3 + \varepsilon_1\varepsilon_2\varepsilon_3 = 0.\end{aligned} \tag{5.5}$$

This is a quadratic either for n_γ^2 or for n_ζ^2, and may be solved to give one of these quantities in terms of the other. These solutions are useful in some problems; see, for example, Al'pert, Budden, Moiseyev and Stott (1983). The form (5.5) can also be written

$$\tfrac{1}{2}n_\gamma^2\left(\frac{1}{n^2-\varepsilon_1}+\frac{1}{n^2-\varepsilon_2}\right)+n_\zeta^2\frac{1}{n^2-\varepsilon_3}=1. \tag{5.6}$$

5.2. Refractive index surfaces

Consider now a system of Cartesian coordinates n_ξ, n_η, n_ζ with axes parallel to the ξ, η, ζ axes of ordinary space. It defines a three-dimensional space called 'refractive index space', and in it we use the same polar angles Θ, Φ, as in ordinary space. Let a surface be constructed such that the radius from the origin to any point on it has the direction of the wave normal and length equal to the refractive index n. For any cold plasma there are two such surfaces, one for the ordinary and one for the extraordinary wave. These surfaces are called 'refractive index surfaces'. For a magnetoplasma the shape is shown by a cross section in any plane Φ = constant, and it is simply the curve $n(\Theta)$ in polar coordinates with radius n and angle Θ. Some examples are given figs. 5.4–5.13.

The main properties of these surfaces will now be listed. They are given for a collisionless plasma and mainly for the electron plasma in which X and Y are held constant, but some results for the more general plasma, with ions, are included too. Because of the symmetry property (2) it is sufficient to consider only the range $0<\Theta<\frac{1}{2}\pi$.

(1) Since $n(\Theta)$ is independent of Φ, each surface is a surface of revolution about the n_ζ axis.
(2) Since $n(\pi-\Theta)=n(\Theta)$, the plane $\Theta=\pi/2$, that is $n_\zeta=0$, is a plane of symmetry.
(3) When $X\to 0$ the plasma goes over to free space and both values of n tend to unity. The two refractive index surfaces tend to unit spheres.
(4) For an isotropic plasma there is only one value of n and it is independent of Θ. The refractive index surface is a sphere of radius n.
(5) The square root in (4.48), or (4.67) with $U=1$, is never zero when Θ is not 0 or $\frac{1}{2}\pi$, so the two refractive index surfaces never cross. If collisions are allowed for the refractive index surface is a complex surface. The two complex ns are equal where $\Theta=\Theta_t$ given by (4.96). This defines a real 'transition cone' in refractive index space. It is analogous to the optic axes of a crystal; see § 5.5. (Davids, 1953.)
(6) The equation (5.5) or (5.6) of the system of refractive index surfaces is of fourth degree in n_ξ, n_η, n_ζ. Thus no straight line in refractive index space can cut the two surfaces in more than four points. This applies in particular to a straight line through the origin. Hence for any given Θ there can at most be two real positive

values of n. Each polar curve $n(\Theta)$ is single valued, continuous except where $n \to \infty$, and cannot have loops or cusps; see (11) below.

(7) When $X \to 1$ parts of the refractive index surfaces shrink down towards segments of the line $\Theta = 0$. The behaviour is shown in figs. 5.4, 5.5, 5.9–5.11. In the limit $X = 1$ the two surfaces touch at the points $n = \pm\{Y/(Y+1)\}^{\frac{1}{2}}$, $\Theta = 0$, and also at the points $n = \pm\{Y/(Y-1)\}^{\frac{1}{2}}$, $\Theta = 0$. The last two are real only if $Y > 1$. This defines four points in refractive index space, known as window points. See §§ 4.11, 4.12.

(8) One refractive index surface is a sphere in the three following cases:

(a) Window frequency, $\varepsilon_3 = 0$, $n^2 = 2\varepsilon_1\varepsilon_2/(\varepsilon_1 + \varepsilon_2)$ (in electron plasma $X = 1$, $n^2 = 1$),

(b) Crossover $\varepsilon_1 = \varepsilon_2 = n^2$,

(c) $\varepsilon_1 = \varepsilon_3 = n^2$.

Only the first of these cases can occur in an electron plasma. In the terrestrial plasmas the third case only occurs where ε_3 is negative so n is imaginary, and a fourth case $\varepsilon_2 = \varepsilon_3$ cannot occur at all.

(9) A refractive index surface does not go in to zero radius for any Θ, unless $X = 1 \pm Y$, (ε_1 or ε_2 zero), and then one value of n is zero for all Θ.

(10) If ε_1 is small and positive, the equation for one refractive index surface is, from (4.68)

$$n^2 \approx 2\varepsilon_1/(1 + \cos^2\Theta) \approx 2(1 - X + Y)/\{(1 + Y)(1 + \cos^2\Theta)\} \tag{5.7}$$

where the second form is for the collisionless electron plasma. The surface is approximately a small ellipsoid. A similar result applies when ε_2 is small and positive; ε_1 is replaced by ε_2 and the sign of Y is changed in (5.7). Thus, near cut-off, if n is to be real, the element ε_1 or ε_2 that is zero at the cut-off must now be positive, which means that the frequency must slightly exceed the cut-off frequency (compare § 5.7 item (5)), or equivalently, X must be slightly less than its value at cut-off.

(11) One value of n is infinite where

$$\Theta = \Theta_r, \quad \cos^2\Theta_r = (Y^2 + X - 1)/XY^2. \tag{5.8}$$

This defines a cone in refractive index space known as a resonance cone. There cannot be more than one. It can occur for real Θ only if

$$Y < 1 \quad \text{and} \quad 1 - Y^2 \leqslant X \leqslant 1 \tag{5.9}$$

or

$$Y > 1 \quad \text{and} \quad X \geqslant 1 (\text{or}\, X \leqslant 1 - Y^2). \tag{5.10}$$

The condition in brackets is for negative X and of no great practical importance, but see figs. 4.5, 4.6. When (5.10) holds, the n that is infinite for $\Theta = \Theta_r$ is real for $\Theta < \Theta_r$ and the cone $\Theta = \Theta_r$ is called a forward resonance cone. This occurs for the whistler mode, §§4.12, 13.8. When (5.9) holds, the n that is infinite for $\Theta = \Theta_r$ is real for $\Theta > \Theta_r$ and the cone $\Theta = \Theta_r$ is called a reversed resonance cone. This occurs for the Z-mode, §§4.11, 16.8. It can be shown, from the dispersion relation (4.63) or (4.64) or from its solution (4.51) or (4.68) respectively that near resonance

$$n \approx K \sin^{-\frac{1}{2}}(\Theta - \Theta_r) \tag{5.11}$$

where K is a constant.

(12) By differentiating (4.65) with respect to Θ and using (4.66) it can be shown that

$$\frac{1}{n}\frac{dn}{d\Theta} = \pm\tfrac{1}{2}\sin\Theta\cos\Theta\, Y(n^2-1)\{\tfrac{1}{4}Y^2\sin^4\Theta + (1-X)^2\cos^2\Theta\}^{-\frac{1}{2}} \tag{5.12}$$

where the + sign applies for the ordinary wave and the − sign for the extraordinary wave. This shows that $dn/d\Theta$ is zero when $\Theta = 0$ or $\frac{1}{2}\pi$. The Appleton–Lassen formula (4.48) shows that when $X \neq 0$, n can never be 1 unless $X = 1$, and then one value of n is 1 for all Θ, so the refractive index surface is the unit sphere. Apart from this case and, for the more general plasma, cases (a), (b), (c) of item (8) above, $dn/d\Theta$ is never zero when $\Theta \neq 0$, and $\neq \frac{1}{2}\pi$. Thus in the range of Θ where it is real and positive, n is a monotonic function of Θ. This proves the statement in §4.11 that the continuous curves of figs. 4.3, 4.5, 4.6 must lie within the shaded regions as shown. This monotonic property also holds for the general plasma with ions allowed for; see problem 5.4.

In some problems it is necessary to know the curvature of the refractive index surface. It is not needed for the topics discussed later in this book, but some formulae and references are given here for completeness. A full discussion has been given by Stott (1983).

Budden and Hugill (1964) used the Gaussian curvature, that is, the product of the two principal curvatures, and gave formulae for it for a collisionless electron plasma. The curvature of the curve $n(\Theta)$ is proportional to d^2n_ζ/dn_y^2 and formulae for this were given by Al'pert, Budden, Moiseyev and Stott (1983).

A point of inflection of the curve $n(\Theta)$ may occur at an angle $\Theta = \Theta_s$ which defines a cone called a Storey cone because Storey (1953) drew attention to one of the most important occurrences of it, in the theory of whistlers. The condition is $d^2n_\zeta/dn_y^2 = 0$ from which Θ_s can be found by solving an equation of degree six. This was given by Clemmow and Mullaly (1955) for an electron plasma, and by Budden and Stott (1980) and Stott (1983) for the more general plasma. If the curve $n(\Theta)$ is concave towards the origin when $\Theta < \Theta_s$, the cone $\Theta = \Theta_s$ is called a (forward) Storey cone, and if it is convex towards the origin for $\Theta < \Theta_s$ the cone is called a reversed Storey cone.

Stott (1983) has shown that for any one quadrant $0 < \Theta < \frac{1}{2}\pi$ there cannot be more than two values of Θ_s, and when there are two they must occur in the same refractive index surface, that is both ordinary or both extraordinary.

The curvature when $\Theta = 0$ is most easily found from the series expansion (4.102) or (4.101) and is given, for one wave, by

$$(\mathrm{d}^2 n_\zeta/\mathrm{d}n_y^2)_0 = -\frac{1}{2}\frac{\varepsilon_1 + \varepsilon_3}{\varepsilon_3 \varepsilon_1^{\frac{1}{2}}} = \frac{Y(2-X) - 2(X-1)}{2(X-1)(1+Y)}\left(\frac{1+Y}{1+Y-X}\right)^{\frac{1}{2}} \tag{5.13}$$

where subscript 0 denotes the value at $\Theta = 0$. Thus Θ_s is zero when

$$\varepsilon_1 + \varepsilon_3 = 0 \tag{5.14}$$

(Walker, 1977b) or, for the electron plasma, when

$$Y = 2(X-1)/(2-X), \quad X = 2(1+Y)/(2+Y). \tag{5.15}$$

(Clemmow and Mullaly, 1955). This condition gives a transition in the C.M.A. diagram, described later, §5.4. For the other wave, ε_1 is replaced by ε_2, or equivalently the sign of Y is reversed; see (5.50).

For $\Theta = \frac{1}{2}\pi$ the curvature of the refractive index surface for the ordinary wave is given, in a similar way, from the expansion (4.104) or (4.103), by

$$(\mathrm{d}^2 n_y/\mathrm{d}n_\zeta^2)_A = \varepsilon_3^{\frac{1}{2}} G/J = -(1-X)^{\frac{1}{2}} \tag{5.16}$$

and for the extraordinary wave, from the expansion (4.106) or (4.105) by

$$\begin{aligned}(\mathrm{d}^2 n_y/\mathrm{d}n_\zeta^2)_B &= \frac{\varepsilon_1\varepsilon_2(\varepsilon_1 + \varepsilon_2) - \varepsilon_3(\varepsilon_1^2 + \varepsilon_2^2)}{J\{2\varepsilon_1\varepsilon_2(\varepsilon_1 + \varepsilon_2)\}^{\frac{1}{2}}} \\ &= \frac{X^2 + Y^2 - 1}{[(1 - X - Y^2)\{(1-X)^2 - Y^2\}]^{\frac{1}{2}}}\end{aligned} \tag{5.17}$$

where subscripts A, B denote the values at $\Theta = \frac{1}{2}\pi$ for the ordinary and extraordinary waves respectively. The curvature (5.16) is independent of Y but is not the same as the curvature $-n^{-1} = -(1-X)^{-\frac{1}{2}}$ for an isotropic plasma. There is a discontinuity in the curvature of both refractive index surfaces at $\Theta = \frac{1}{2}\pi$ when Y changes from zero to a small non-zero value.

The curvature (5.17) is zero when

$$\varepsilon_1\varepsilon_2(\varepsilon_1 + \varepsilon_2) - \varepsilon_3(\varepsilon_1^2 + \varepsilon_2^2) = 0 \tag{5.18}$$

(Walker, 1977b), or for the electron plasma

$$X^2 + Y^2 = 1 \tag{5.19}$$

(Clemmow and Mullaly, 1955). This condition gives a transition in the C.M.A. diagram; see §5.6. Again (5.17), when $Y = 0$, is not the same as for an isotropic plasma. There is a discontinuity of behaviour when Y changes from zero to a non-zero value.

When collisions are allowed for, all three components of $\boldsymbol{n}$ are in general complex and refractive index space is a three-dimensional complex space, that requires six real dimensions to represent it. The refractive index surfaces are complex surfaces in this space, though they are still surfaces of revolution about the n_ζ axis. Thus they can be studied by using a cross section plane Φ = constant. If only real values of Θ are of interest it is then useful to plot the polar diagrams of $\mu(\Theta)$ and $\chi(\Theta)$ where $n = \mu - i\chi$. This has been done, for example, by Jones (1972). In many problems, however, we need to consider complex values of Θ. There is then no simple graphical way of representing the refractive index surfaces and we have to rely on algebraic methods. The topic is discussed again later, § 14.12.

5.3. The ray. Ray surfaces

We now ask: what is the direction of the time averaged Poynting vector $\mathbf{\Pi}_{av}$ for a progressive plane wave in a homogeneous loss free plasma? The answer is that it is normal to the refractive index surface (Al'pert, 1948).

The proof is as follows. For the plane wave with real refractive index vector $\boldsymbol{n}$, Maxwell's third and fourth equations in the form (2.44) give, on cancellation of a factor $-ik$;

$$\boldsymbol{n} \wedge \boldsymbol{E} = \boldsymbol{\mathcal{H}} \tag{5.20}$$

$$\boldsymbol{n} \wedge \boldsymbol{\mathcal{H}} = -\varepsilon_0^{-1}\boldsymbol{D}. \tag{5.21}$$

Consider a second plane wave whose refractive index vector is $\boldsymbol{n} + \delta\boldsymbol{n}$ where $\delta\boldsymbol{n}$ is infinitesimally small. Let the fields be $\boldsymbol{E} + \delta\boldsymbol{E}$, $\boldsymbol{\mathcal{H}} + \delta\boldsymbol{\mathcal{H}}$. Then (5.20), (5.21) may be differentiated to give

$$\boldsymbol{n} \wedge \delta\boldsymbol{E} + \delta\boldsymbol{n} \wedge \boldsymbol{E} = \delta\boldsymbol{\mathcal{H}} \tag{5.22}$$

$$\boldsymbol{n} \wedge \delta\boldsymbol{\mathcal{H}} + \delta\boldsymbol{n} \wedge \boldsymbol{\mathcal{H}} = -\varepsilon_0^{-1}\delta\boldsymbol{D}. \tag{5.23}$$

Now take the scalar product of (5.22) with $\boldsymbol{\mathcal{H}}^*$ and on the left hand side use the permutation property for the product of three vectors $\boldsymbol{a}\cdot(\boldsymbol{b} \wedge \boldsymbol{c}) = \boldsymbol{b}\cdot(\boldsymbol{c} \wedge \boldsymbol{a}) = \boldsymbol{c}\cdot(\boldsymbol{a} \wedge \boldsymbol{b})$. Then

$$\delta\boldsymbol{E}\cdot(\boldsymbol{\mathcal{H}}^* \wedge \boldsymbol{n}) + \delta\boldsymbol{n}\cdot(\boldsymbol{E} \wedge \boldsymbol{\mathcal{H}}^*) = \boldsymbol{\mathcal{H}}^*\cdot\delta\boldsymbol{\mathcal{H}} \tag{5.24}$$

Take the scalar product of (5.23) with $\boldsymbol{E}^*$ and permute the terms on the left in a similar way. Then

$$\delta\boldsymbol{\mathcal{H}}\cdot(\boldsymbol{E}^* \wedge \boldsymbol{n}) - \delta\boldsymbol{n}\cdot(\boldsymbol{E}^* \wedge \boldsymbol{\mathcal{H}}) = -\varepsilon_0^{-1}\boldsymbol{E}^*\cdot\delta\boldsymbol{D}. \tag{5.25}$$

For the vector products in the first terms of (5.24), (5.25) use the complex conjugates of (5.21), (5.20) respectively. Subtract the two equations and rearrange. This gives:

$$4\delta\boldsymbol{n}\cdot\mathbf{\Pi}_{av} - \varepsilon_0^{-1}(\boldsymbol{E}^*\cdot\delta\boldsymbol{D} - \boldsymbol{D}^*\cdot\delta\boldsymbol{E}) = 0 \tag{5.26}$$

where (2.63) for $\mathbf{\Pi}_{av}$ has been used. Now use the subscript notation form (2.54) for $\boldsymbol{D}$. Since the plasma is loss-free, $\boldsymbol{\varepsilon}$ is Hermitian, (2.56). Then

$$\boldsymbol{D}^*\cdot\delta\boldsymbol{E} = \varepsilon_0\varepsilon_{ij}^*E_j^*\delta E_i = \varepsilon_0 E_j^*\varepsilon_{ji}\delta E_i = \boldsymbol{E}^*\cdot\delta\boldsymbol{D}. \tag{5.27}$$

This shows that the second term of (5.26) is zero, so that

$$\delta \boldsymbol{n} \cdot \boldsymbol{\Pi}_{\mathrm{av}} = 0. \tag{5.28}$$

Now $\delta \boldsymbol{n}$ must lie in the refractive index surface, and $\boldsymbol{\Pi}_{\mathrm{av}}$ is perpendicular to all possible $\delta \boldsymbol{n}$s with this property. Hence $\boldsymbol{\Pi}_{\mathrm{av}}$ is normal to the surface.

A result equivalent to this will now be derived in a different way. Suppose that there is a point source at the origin of coordinates. If the medium were homogeneous the source would emit a wave whose wavefronts are outward travelling spheres. We need to find the shape of the wavefronts in an anisotropic medium. In any plane $\zeta = \text{constant}$, each field component is a function of ξ, η which can be expressed as a two-dimensional Fourier integral, whose integrand contains a factor $\exp\{-\mathrm{i}(\xi\kappa_\xi + \eta\kappa_\eta)\}$, where $\kappa_\xi = kn_\xi$, $\kappa_\eta = kn_\eta$. But this is the same as the ξ and η dependence of the field (5.2) of a plane wave. Thus the fields in the chosen plane $\zeta = \text{constant}$ are a doubly infinite spectrum of plane waves. The fields in any other plane can now be found by using the known properties of the plane waves as derived in the preceding chapter. The dispersion relation (5.3)–(5.5) shows that there are two different possible values of n_ζ. Thus the plane wave spectrum can be resolved into two separate systems, for the ordinary and extraordinary waves. We confine attention to one of these systems. Then any field component in the wave of that system emitted by the point source is given by

$$F(\xi, \eta, \zeta) = \int_{-\infty}^{\infty} \int_{-\infty}^{\infty} A(n_\xi, n_\eta) \exp\{-\mathrm{i}k(\xi n_\xi + \eta n_\eta + \zeta n_\zeta)\} \, \mathrm{d}n_\xi \, \mathrm{d}n_\eta. \tag{5.29}$$

The exponential represents a plane wave, and (5.29) is said to express the field as an 'angular spectrum of plane waves'. For a full description see Clemmow (1966) and for an example of its use see Al'pert, Budden, Moiseyev and Stott (1983). The function A is a 'slowly varying' function of n_ξ, n_η (see § 9.6). The exponential is rapidly varying when ξ, η and ζ are large. In the present discussion it is adequate to treat A as constant for small ranges of n_ξ, n_η, n_ζ. The quantity

$$\varphi = k(\xi n_\xi + \eta n_\eta + \zeta n_\zeta) \tag{5.30}$$

is called the phase of the integrand. The integral (5.29) is the sum of numerous complex numbers whose phases in general are different so that there is partial cancellation and the sum is not large. But there can be regions of the n_ξ–n_η plane where the phase does not change much for small changes $\delta n_\xi, \delta n_\eta$ and the resulting small change δn_ζ derived from the dispersion relation. From these regions the contributions to the integral are large. Thus the main contributions to (5.29) come from regions where the phase is stationary for small variations δn_ξ, δn_η. We seek a refractive index vector $\boldsymbol{n}$ such that

$$\delta\varphi = 0: \quad \xi \delta n_\xi + \eta \delta n_\eta + \zeta \delta n_\zeta = 0\,. \tag{5.31}$$

Now any $\delta \boldsymbol{n}$ lies in the refractive index surface, so that the vector $\boldsymbol{g} = \xi, \eta, \zeta$ must be normal to the refractive index surface and $\boldsymbol{n}$ must be found to achieve this. The vector $\boldsymbol{g}$ is represented by a line drawn out from the origin in a fixed direction, and known as the 'ray'. The more formal definition of the term 'ray' is discussed later, §§ 10.1, 14.1–14.3.

The particular vector $\boldsymbol{n}$ where (5.31) is exactly satisfied is called the predominant $\boldsymbol{n}$ of the spectrum in (5.29). This integral can be evaluated by the method of steepest descents (§ 9.6) to give an expression for $F(\boldsymbol{g})$ containing a factor $\exp(-ik\boldsymbol{n}\cdot\boldsymbol{g})$. Thus at points on a given ray $\boldsymbol{g}$, the direction of $\boldsymbol{n}$ is the predominant wave normal on that ray. It is usually called simply the 'wave normal'. It is different for different rays. Because the refractive index surface is a surface of revolution about the ζ axis, it follows that the ray, the wave normal and the ζ axis are coplanar. Let β be the angle between the ray and the ζ axis so that

$$\alpha = \Theta - \beta \tag{5.32}$$

is the angle between the ray and the wave normal. Then it follows by a geometrical construction shown in fig. 5.1 that

$$\tan \alpha = \frac{1}{n} \frac{\partial n}{\partial \Theta}. \tag{5.33}$$

An expression for this, for an electron plasma without collisions, has already been given at (5.12).

Fig. 5.1. Cross section of refractive index surface by a plane containing the direction of the earth's magnetic field. CX and CA are the normal and tangent, respectively, at the point C. The line OC, of length n, is parallel to the wave normal. CB is perpendicular to OA and has length $\approx n\delta\Theta$.

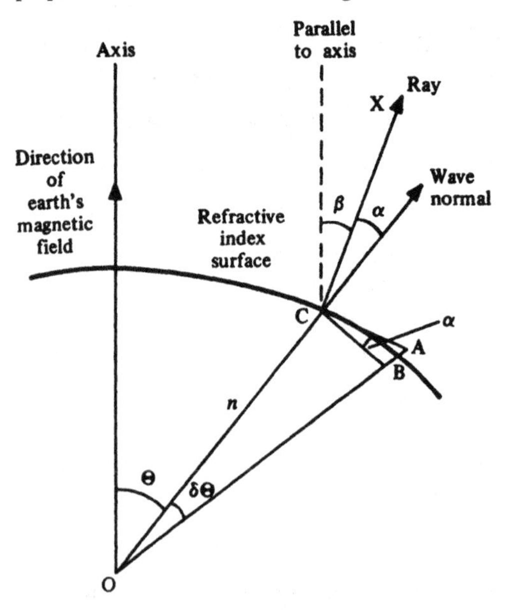

The contributing plane waves in (5.29) whose refractive index vectors are close to $\boldsymbol{n}$ have nearly equal phases and add together constructively at points on the ray, but, except when $\alpha = 0$, not at points on a wave normal through the origin. The velocity of a wave front in the direction of the wave normal is

$$v = c/n \tag{5.34}$$

and is called the wave velocity. The velocity of a point where the wave front intersects a ray is

$$v \sec \alpha = c/(n \cos \alpha) = V = c/\mathscr{M} \tag{5.35}$$

and is called the ray velocity. It is the magnitude of a vector $\boldsymbol{V}$ in the direction of the ray and we have seen, at (5.28), that it is also the direction of the time averaged Poynting vector, $\boldsymbol{\Pi}_{av}$. The product $n \cos \alpha = \mathscr{M}$ is called the 'ray refractive index', and is used in §§13.7, 14.6.

Now a surface is constructed whose points are given by the vector $\boldsymbol{V}$ drawn from the origin. It is called the 'ray surface'. A point on a ray travelling outwards with velocity $\boldsymbol{V}$ is a point where the phase stays constant as the wave travels. This applies for all rays and the locus of these points is the surface where the phase is constant. In other words it is the wave front of the wave field going out from the point source at the origin. Thus the ray surface gives the shape of the wave front going out from a point source (Budden, 1961a; Walker, 1977a). The normal to the local wave front is the wave normal and has the direction of $\boldsymbol{n}$. Thus the predominant refractive index vector $\boldsymbol{n}$ is normal to the ray surface.

5.4. Properties of ray surfaces

The ray surface can be plotted by first choosing a value of the wave normal direction Θ and computing $n(\Theta)$ from the dispersion relation. Then α is found from (5.33) or equivalently (5.12), β from (5.32) and V from (5.35). This is a possible method for computing, but β is not at first known. A succession of values of Θ must be used until all the required values of β have been covered. For a plasma with collisions allowed for, Θ is complex and the process may be laborious.

A radio signal does not, in general, travel in the direction of the wave normal, but along the ray, and in many practical problems it is the ray direction β that is given, but the wave normal direction Θ is at first unknown so that the dispersion relation cannot immediately be used to find n or α or V. A formula is required that gives these quantities when β, but not Θ, is given. This formula has been given, for an electron plasma, by Budden and Daniell (1965), and for the more general plasma by Budden and Stott (1980) and Stott (1983). It is an equation for finding n^2 and is of degree six. It is not used in this book, except in the present section, and it is too long to be worth quoting here. It is given in full in the references cited above. Stott (1983) has shown that, for a loss-free plasma, at most four of the solutions n^2 can be real and positive, and that, for all β, one of the two ray surfaces cannot have more than one real positive value of n^2. The equation is useful for finding some of the properties of ray surfaces. One of its main uses is for finding the saddle points for integrals like (5.29) representing an angular spectrum of plane waves; see Al'pert, Budden, Moiseyev and Stott (1983).

The main properties of ray surfaces will now be listed. As for refractive index surfaces, § 5.2, they are given for a collisionless plasma and mainly for the electron plasma, but some results for the more general plasma are included. The ray surfaces are surfaces in a space with coordinates V_ξ, V_η, V_ζ and axes parallel to the ξ, η, ζ axes of ordinary space. The ray vector V makes an angle β with the V_ζ axis and has azimuth angle Φ, the same Φ as for the wave normal. Because of the symmetry property (2) it is sufficient to consider only the range $0 \leqslant \beta \leqslant \frac{1}{2}\pi$.

(1) $V(\beta)$ is independent of Φ. Each ray surface is a surface of revolution about the V_ζ axis.
(2) $V(\pi - \beta) = V(\beta)$. The plane $\beta = \frac{1}{2}\pi$ is a plane of symmetry.
(3) In free space both ray surfaces are spheres of radius c.
(4) For an isotropic plasma both ray surfaces are spheres of radius c/n.
(5) A radius drawn out from the origin can cross a ray surface in one, two or three points. For examples see figs. 5.6, 5.9–5.11.
(6) To each point $\boldsymbol{V}$ on a ray surface there is a corresponding point $\boldsymbol{n}$ on the refractive index surface. From (5.34), (5.35) it follows that

$$\boldsymbol{V}\cdot\boldsymbol{n} = Vn\cos\alpha = V_\xi n_\xi + V_\eta n_\eta + V_\zeta n_\zeta = c. \tag{5.36}$$

Points with this property are said to be reciprocal with respect to the sphere of radius c.
(7) The normal to the ray surface at $\boldsymbol{V}$ is parallel to the corresponding $\boldsymbol{n}$. The normal to the refractive index surface at $\boldsymbol{n}$ is parallel to the corresponding $\boldsymbol{V}$.
(8) Cut-off occurs when $\varepsilon_1 \to 0$ or $\varepsilon_2 \to 0$. Consider the first of these. When ε_1 is very small the refractive index surface is approximately the small ellipsoid (5.7) whence it can be shown from (5.33), (5.32), (5.35) that

$$\cos^2\beta \approx 4\cos^2\Theta/(1 + 3\cos^2\Theta), \quad V^2 \approx \tfrac{1}{2}c^2/\{\varepsilon_1(1 - \tfrac{3}{4}\cos^2\beta)\}. \tag{5.37}$$

Thus the ray surface is a large ellipsoid and, at exact cut-off, $V \to \infty$ for all β. The behaviour when $\varepsilon_2 \to 0$ is exactly similar.
(9) Near a resonance, $\Theta \to \Theta_r$, the refractive index is given by (5.11) which shows, from (5.33) that

$$\tan\alpha = -\tfrac{1}{2}\cot(\Theta - \Theta_r) \tag{5.38}$$

so that $\alpha \to \pm\pi/2$ and $\beta \to \beta_r = \Theta_r - \alpha$. Thus $\cos\alpha \to 0$. Although $n \to \infty$, the product

$$n\cos\alpha \approx 2K\sec(\Theta - \Theta_r)\sin^{\frac{1}{2}}(\Theta - \Theta_r) \tag{5.39}$$

tends to zero. Hence $V \to \infty$. The angle $\beta = \beta_r$ defines a resonance cone in the ray surface. Suppose that β_r is in the quadrant $0 < \beta_r < \frac{1}{2}\pi$. Then two cases can occur. First Θ_r may be in the range $-\frac{1}{2}\pi < \Theta_r < 0$. Then V is real when $\beta < \beta_r$ and n is real when $\Theta < |\Theta_r|$. The resonance cones are forward resonance cones; see item (11) of §5.2. See figs. 5.5, 5.6 for examples. Second Θ_r may be in the range $\frac{1}{2}\pi < \Theta_r < \pi$. Then V is real for $\beta > \beta_r$ and n is real for $\pi - \Theta_r < \Theta < \Theta_r$. The resonance cones are reversed resonance cones. See figs. 5.8, 5.9 for examples.

(10) When $\beta = 0$ and V is real, one value of Θ is 0, and $n = c/V$. There may be other real values of Θ, n and V; see for example figs. 5.6, 5.11. Similarly when $\beta = \frac{1}{2}\pi$ and V is real, one value of Θ is $\frac{1}{2}\pi$ and $n = c/V$. Again there may be other real values of Θ, n and V; see for example fig. 5.9.

(11) The ray and wave normal are parallel, that is $\alpha = 0$, when $\Theta = \beta = 0$ and $\Theta = \beta = \frac{1}{2}\pi$ as in (10). They are parallel for all $\beta = \Theta$ when the refractive index surface is a sphere as in (3), (4) above or in item (8) of § 5.2. There are no other cases where they are parallel.

(12) The ray direction β, normal to the refractive index surface, attains a maximum or minimum value when the point $\boldsymbol{n}$ on the refractive index surface moves through a point of inflection, $\Theta = \Theta_s$. Here let $\beta = \beta_s$. The value of Θ gives the direction of the normal to the ray surface and here changes smoothly and continuously. It follows that the ray surface has a cusp where $\beta = \beta_s$. This defines a Storey cone in the ray surface if V is real where $\beta < \beta_s$ and a reversed Storey cone if V is real where $\beta > \beta_s$. The angles β_s and Θ_s, may have opposite signs (see, for example, fig. 5.11) or the same sign (see, for example, figs. 5.6, 5.9, 5.10).

5.5. Crystal optics

The properties of a plane progressive radio wave in a homogeneous magnetoplasma are similar in some respects to those of an electromagnetic wave of optical frequency in a crystal, and it is useful to make a brief comparison.

For many purposes a crystalline medium can be treated as a non-magnetic dielectric. Spatial dispersion effects can sometimes be important and they give rise to 'optical activity' (see problem 13.2) but here they are neglected. For most crystals energy losses are negligible and the dielectric tensor $\boldsymbol{\varepsilon}$ is real and symmetric. It is then possible to choose a Cartesian coordinate system ξ, η, ζ with real axes, so that $\boldsymbol{\varepsilon}$ is diagonal with real positive elements $\varepsilon_a, \varepsilon_b, \varepsilon_c$ and the axes are chosen so that $\varepsilon_a < \varepsilon_b < \varepsilon_c$.

If $\mathscr{H}$ is eliminated from Maxwell's equations (2.44) they give

$$\operatorname{curl}\operatorname{curl}\boldsymbol{E} = k^2\varepsilon_0^{-1}\boldsymbol{D} = k^2\boldsymbol{\varepsilon}\boldsymbol{E}. \tag{5.40}$$

For a plane progressive wave with refractive index vector $\boldsymbol{n} = (n_\xi, n_\eta, n_\zeta)$. We may use curl $\boldsymbol{E} = -ik\boldsymbol{n} \wedge \boldsymbol{E}$ whence (5.40) gives

$$\begin{pmatrix} n_\xi^2 - n^2 + \varepsilon_a & n_\xi n_\eta & n_\xi n_\zeta \\ n_\xi n_\eta & n_\eta^2 - n^2 + \varepsilon_b & n_\eta n_\zeta \\ n_\xi n_\zeta & n_\eta n_\zeta & n_\zeta^2 - n^2 + \varepsilon_c \end{pmatrix} \begin{pmatrix} E_\xi \\ E_\eta \\ E_\zeta \end{pmatrix} = 0. \tag{5.41}$$

For solutions to exist, the determinant of the matrix must be zero and this gives the dispersion relation. It can be shown that this leads to

$$\frac{n_\xi^2}{n^2 - \varepsilon_a} + \frac{n_\eta^2}{n^2 - \varepsilon_b} + \frac{n_\zeta^2}{n^2 - \varepsilon_c} = 1. \tag{5.42}$$

This is the equation of the refractive index surfaces. Its properties are well known and are described in textbooks of optics; see for example Born and Wolf (1970, ch. 14) or Lipson and Lipson (1969, § 5.4). It should be compared with (5.6) which is the analogous form for a magnetoplasma.

If $\boldsymbol{n}$ has a given real direction, (5.42) is a quadratic for n^2 which always has two real positive solutions. There is no cut-off and no resonance (compare § 5.2 items (9), (10), (11)). Any radius drawn outwards from the origin cuts the refractive index surfaces in two real points. It can be shown that the wave polarisations of the two waves are linear. As in a magnetoplasma, $\boldsymbol{E}$ may have a component parallel to the wave normal.

The origin is a centre of symmetry of the refractive index surfaces, and the planes $n_\xi = 0, n_\eta = 0, n_\zeta = 0$ are planes of symmetry, called principal planes. But the surfaces are not surfaces of revolution about any axis. In each principal plane the cross section of the surfaces is a circle and an ellipse. For $n_\zeta = 0$ the circle is entirely outside the ellipse, and for $n_\xi = 0$ it is entirely inside the ellipse. For the plane $n_\eta = 0$, equation (5.42) gives

$$(n^2 - \varepsilon_b)\left(\frac{n_\xi^2}{\varepsilon_c} + \frac{n_\zeta^2}{\varepsilon_a} - 1\right) = 0. \tag{5.43}$$

The first factor gives the circle of radius $\varepsilon_b^{\frac{1}{2}}$ and the second gives the ellipse with major axis $\varepsilon_c^{\frac{1}{2}}$ and minor axis $\varepsilon_a^{\frac{1}{2}}$. These curves intersect at four points satisfying

$$\frac{n_\xi^2}{n_\zeta^2} = \tan^2 \Theta_w = \frac{\varepsilon_c(\varepsilon_b - \varepsilon_a)}{\varepsilon_a(\varepsilon_c - \varepsilon_b)}. \tag{5.44}$$

This defines four directions, one in each quadrant of the plane $n_\eta = 0$, that is $\Phi = 0$. They are called 'optic axes'. It can be shown that these are the only directions where the two values of n are equal. Near an optic axis the refractive index surfaces have the shape of two cones with their apices touching. The ray direction is normal to these cones and an infinite number of directions is possible, all lying in the surface of another cone. This gives the phenomenon of internal conical refraction; Born and Wolf (1970, p. 688). For a collisionless magnetoplasma there are no real optic axes.

The closest analogue to them is the transition cone $\Theta = \Theta_t$, § 5.2 item (5), in a plasma with collisions allowed for.

The ray velocity vector $\boldsymbol{V}$ is defined just as for a magnetoplasma. It is normal to the refractive index surface and satisfies (5.36). It can be shown that the equation of the ray surfaces is

$$\frac{V_\xi^2}{V^2 - c^2/\varepsilon_a} + \frac{V_\eta^2}{V^2 - c^2/\varepsilon_b} + \frac{V_\zeta^2}{V^2 - c^2/\varepsilon_c} = 1. \tag{5.45}$$

Thus their general geometrical properties are very similar to those for the refractive index, (5.42). In particular, in the plane $V_\eta = 0$ there are four directions, one in each quadrant, where the two values of V are equal, given by

$$\frac{V_\xi^2}{V_\zeta^2} = \tan^2\Theta_x = \frac{\varepsilon_b - \varepsilon_a}{\varepsilon_c - \varepsilon_b}. \tag{5.46}$$

These directions are called optic ray axes and are not the same as the optic axes. For an optic ray axis an infinite number of wave normal directions, that is normals to the ray surface, are possible all lying within the surface of a cone. This gives the phenomenon of external conical refraction; Born and Wolf (1970, p. 689).

Optic ray axes can occur in a magnetoplasma, and they are then always parallel to the earth's magnetic field. There are examples in figs. 5.6, 5.11, which show cross sections of the ray surface including the region where it has the form of two cones with touching apices. No radio experiment has been devised to show external conical refraction, though it would be possible in principle. But the field from a point source immersed in a plasma is enhanced in the direction of the optic ray axis, that is of the earth's magnetic field; Al'pert (1983); Al'pert, Budden, Moiseyev and Stott (1983).

A crystal in general has optic axes in two non-parallel directions, and is said to be biaxial. But if two of $\varepsilon_a, \varepsilon_b, \varepsilon_c$ are equal there is only one such direction, and the crystal is said to be uniaxial. We here suppose that $\varepsilon_b = \varepsilon_a$. Then (5.44) for the optic axes shows that $n_\xi = 0$, $\Theta_w = 0$ or π. Similarly (5.46) for the optic ray axes shows that $V_\xi = 0$, $\Theta_x = 0$ or π. The optic axes and the optic ray axes are both parallel to the n_ζ or V_ζ axis. On this axis the ray and the wave normal each have only one possible direction, parallel to the axis, so there is now no phenomenon of conical refraction. Equation (5.42) becomes:

$$(n^2 - \varepsilon_b)\{n_\xi^2 + n_\eta^2)/\varepsilon_c + n_\zeta^2/\varepsilon_b - 1\} = 0. \tag{5.47}$$

One refractive index surface is a sphere and the other is an ellipsoid outside it but touching it on the n_ζ axis. For the case $\varepsilon_b = \varepsilon_c > \varepsilon_a$ the optic axis would be the n_ξ axis and the ellipsoid would be inside the sphere, but otherwise the treatment is very similar. Equation (5.45) becomes

$$(V^2 - c^2/\varepsilon_b)\{(V_\xi^2 + V_\eta^2)\varepsilon_c + V_\zeta^2\varepsilon_b - c^2\} = 0. \tag{5.48}$$

One ray surface is a sphere and the other is an ellipsoid, inside it but touching it on the V_ζ axis.

A magnetoplasma is uniaxial when $\varepsilon_1 = \varepsilon_2$. This cannot happen exactly for an electron plasma, but it can happen for the more general collisionless magnetoplasma, always when ε_3 is negative. Then equation (5.6) for the refractive index surfaces is

$$(n^2 - \varepsilon_1)(n_y^2/\varepsilon_3 + n_\zeta^2/\varepsilon_1 - 1) = 0. \tag{5.49}$$

One surface is a sphere of radius $\varepsilon_1^{\frac{1}{2}}$ and the other is a hyperboloid with resonances given by $\tan^2\Theta_r = -\varepsilon_3/\varepsilon_1$. Similarly one ray surface is a sphere and the other is a hyperboloid.

For the electron plasma it is sometimes useful to assume that the earth's magnetic field is large enough, or the frequency is small enough to make the approximation $Y \to \infty$. Then $\varepsilon_1 \approx \varepsilon_2 \approx 1$ and the medium is uniaxial. See § 19.3, and problem 4.7.

5.6. Classification of refractive index and ray surfaces. C.M.A. type diagrams

Refractive index surfaces can take various forms, depending, for example, on whether points of inflection and resonances are present. The forms depend on the particle concentrations N_e, N_i in the plasma, and on the frequency ω, and change when certain transition values of these parameters are crossed. It is useful to classify these forms of the refractive index surface. This was done for an electron plasma by Clemmow and Mullaly (1955) and by Allis (1959), who used diagrams now known as

Fig. 5.2. C.M.A. type diagram for the ordinary wave in a collisionless electron plasma. Table 5.1 gives the meanings of the symbols.

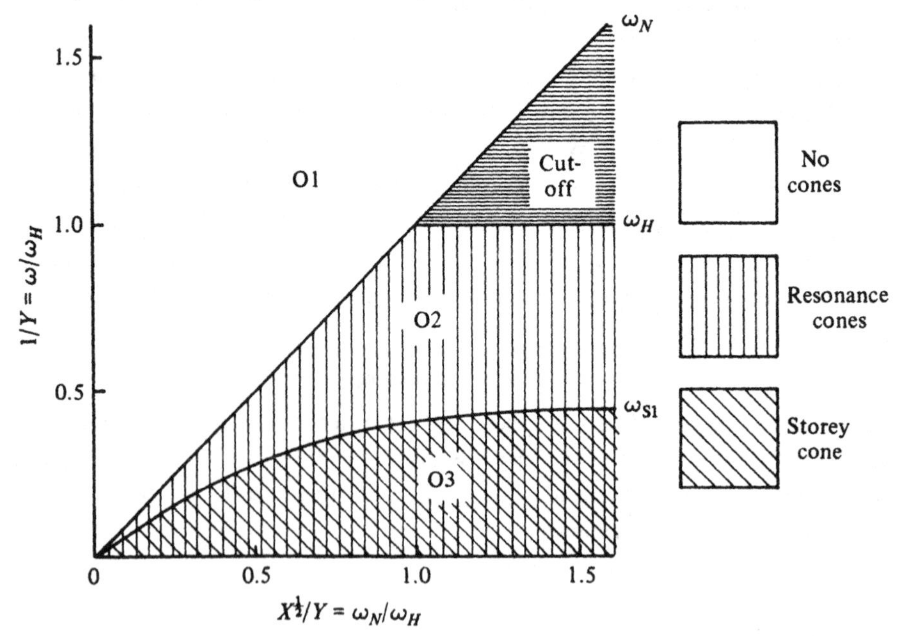

C.M.A. diagrams. When collisions are allowed for, refractive index surfaces are complex and cannot be simply described or classified. A collisional plasma is therefore classified by the real refractive index surfaces that it would have when the collisions are ignored.

In this section we shall consider only a collisionless electron plasma and we shall

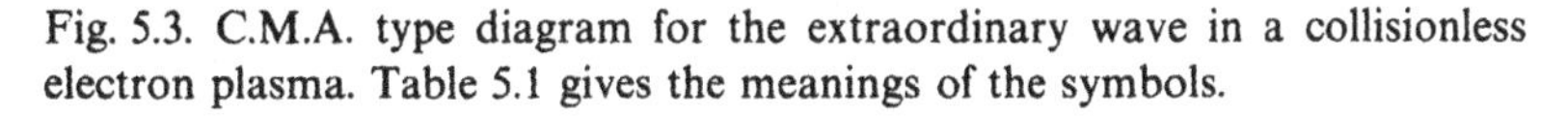

Fig. 5.3. C.M.A. type diagram for the extraordinary wave in a collisionless electron plasma. Table 5.1 gives the meanings of the symbols.

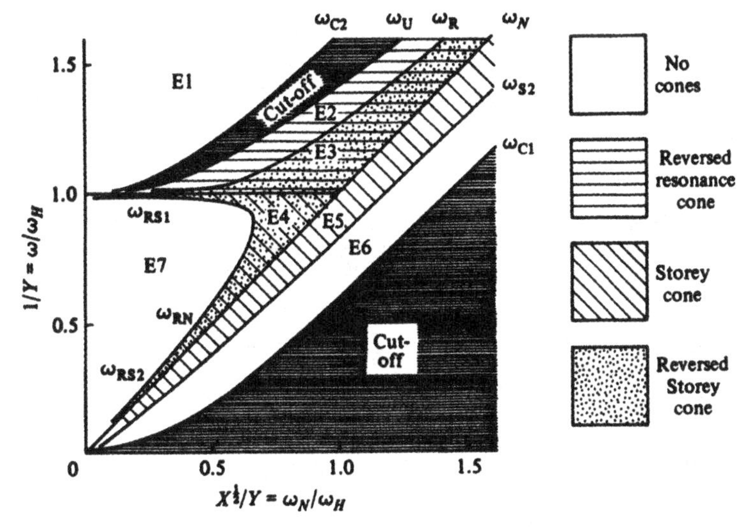

Fig. 5.4. Cross section by a plane ϕ = constant of the refractive index surfaces (left-hand figure) and ray surfaces (right-hand figure). The numbers by the curves are the values of X. In the left figure the curve $X = 0$ is the unit circle and in the right figure it is the circle of radius c. In figures where these circles are drawn as broken lines, they are not part of the system but are shown to give the scale. The same plan is used for figs. 5.5–5.13.

Region O1. Both refractive index surface and ray surface are concave towards the origin. When X is near to 1 the refractive index surface is a narrow cigar shaped surface. In the limit $X \to 1$ it shrinks to a segment of the line $\Theta = 0$ between the window points $n = \pm\{Y/(Y+1)\}^{\frac{1}{2}}$; §§4.11, 4.12, 5.2 item (7). In this example, $Y = 0.5$.

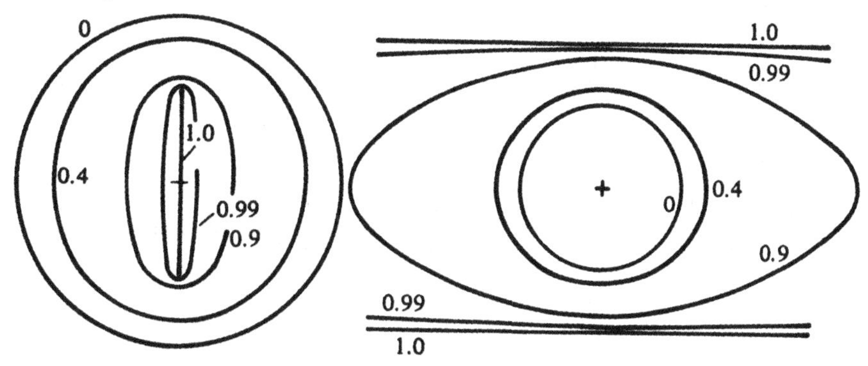

Table 5.1. *'No. of cones' applies to one quadrant* $0 < \Theta < \frac{1}{2}\pi$ or $0 < \beta < \frac{1}{2}\pi$

Region	n^2	Sign of $\partial n/\partial\Theta$	No. of Storey cones	No. of reversed Storey cones	No. of forward resonance cones	No. of reversed resonance cones	Wave mode	Range of ω or ω_N	Fig.
O1	<1	$-$	0	0	0	0	Ord.	$\omega_N \leqslant \omega < \infty$	5.4
O2	>1	$+$	0	0	1	0	Ord. whistler	$\omega_{S1} \leqslant \omega \leqslant \mathrm{Min}(\omega_N, \omega_H)$	5.5
O3	>1	$+$	1	0	1	0	Ord. whistler	$0 < \omega \leqslant \omega_{S1}$	5.6
Cut-off	<0						Ord.	$\omega_H \leqslant \omega \leqslant \omega_N$	—
E1	<1	$+$	0	0	0	0	Ext.	$\omega_{C2} \leqslant \omega < \infty$	5.7
E2	>1	$-$	0	0	0	1	Ext. Z	$\omega_R \leqslant \omega \leqslant \omega_U$	5.8
E3	>1	$-$	0	1	0	1	Ext. Z	$\mathrm{Max}(\omega_N,\omega_H) \leqslant \omega \leqslant \omega_R$	5.9
E4	>1	$-$	1	1	0	0	Ext.	$\omega_{NS} \leqslant \omega_N \leqslant \omega \leqslant \omega_H$	5.10
E5	<1	$+$	1	0	0	0	Ext. Z if $Y=1$	$\omega_{S2} \leqslant \omega \leqslant \omega_N$	5.11
E6	<1	$+$	0	0	0	0	Ext. Z if $Y<1$	$\omega_{C1} \leqslant \omega \leqslant \omega_{S2}$	5.12
E7	>1	$-$	0	0	0	0	Ext.	$\omega_N \leqslant \omega_{NS}$ and $\omega < \omega_H$	5.13
Cut-off	<0						Ext.	$\omega_U \leqslant \omega \leqslant \omega_{C2}$	
Cut-off	<0						Ext.	$0 < \omega \leqslant \omega_{C1}$	

list and classify the various possible forms of the refractive index surface and the associated ray surface. There are now only two parameters, the electron concentration N_e now written simply as N, and ω/ω_H proportional to frequency, so it is possible to use a diagram in two dimensions. Clemmow and Mullaly (1955) used a diagram of Y vs X. In it lines of constant frequency were lines of constant Y, and lines of constant N were parabolae. Later authors have used Y^2 vs X so that lines of

Fig. 5.5. See caption of fig. 5.4. Region O2. Both refractive index surface and ray surface have two sheets and forward resonance cones and both are convex towards the origin. When $\omega_N < \omega_H$ and ω increases and approaches the value ω_N, the refractive index surface is two narrow cone-like surfaces closed at their inner ends and stretching to infinity. In the limit $X \to 1$ they shrink to two segments of the line $\Theta = 0$ beyond the window points $n = \pm\{Y/(Y-1)\}^{\frac{1}{2}}$. See §§5.2 (item (7)) and 4.11, 4.12. When $\omega_N > \omega_H$ and $\omega \to \omega_H$ the shape is similar but the closed inner ends of the tubes are beyond the window points and these points move to infinity in the limit $Y \to 1$. The curves marked T are for $\omega = \omega_{S1}$, and similar curves appear also in fig. 5.6. In this example $Y = 4$.

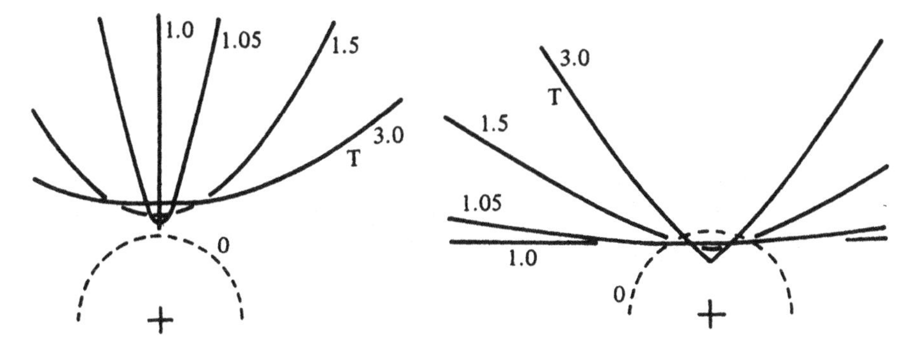

Fig. 5.6. See caption of fig. 5.4. Region O3. Both refractive index surface and ray surface have two sheets, forward resonance cones and forward Storey cones. This is the form for the simplest case of whistlers (Storey, 1953). The ray surface has an optic ray axis at $\beta = 0$; §§ 5.5. The curves marked T are for $\omega = \omega_{S1}$, and similar curves appear also in fig. 5.5. In this example $Y = 22$.

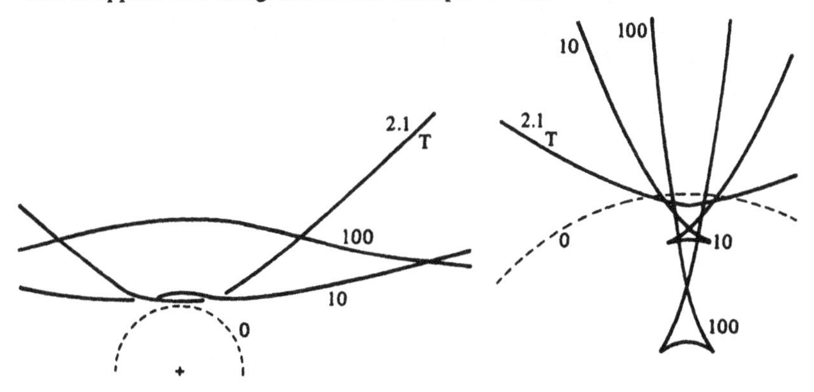

Fig. 5.7. See caption of fig. 5.4. Region E1. Both refractive index surface and ray surface are concave towards the origin. When ω is near to ω_{C2} the refractive index surface is a small ellipsoid (5.7) and the ray surface is a large ellipsoid (5.37). In the limit of cutoff $\omega \to \omega_{C2}$ the refractive index surface shrinks to a point and the ray surface expands to infinity; §§ 5.2, 5.4. In this example $Y = 0.5$.

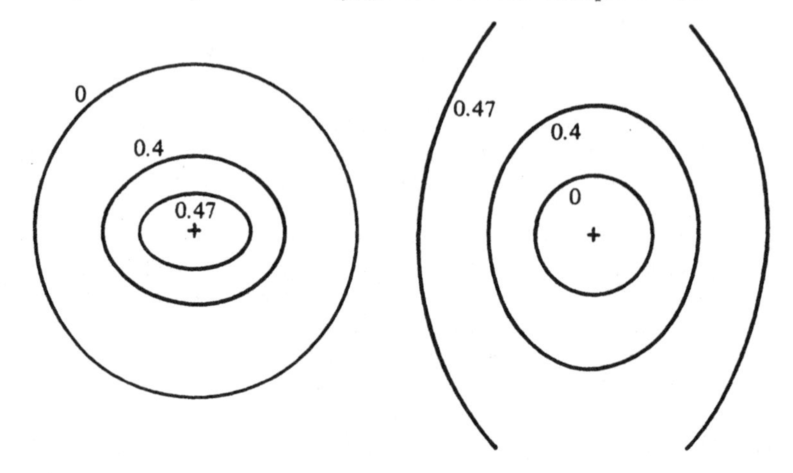

Fig. 5.8. See caption of fig. 5.4. Region E2. Both refractive index surface and ray surface have one sheet and are convex towards the origin. They have reversed resonance cones. This is one of the forms for the Z-mode. The curves marked T are for $\omega = \omega_R$ and the same curves appear in fig. 5.9. In this example $Y = 0.6$.

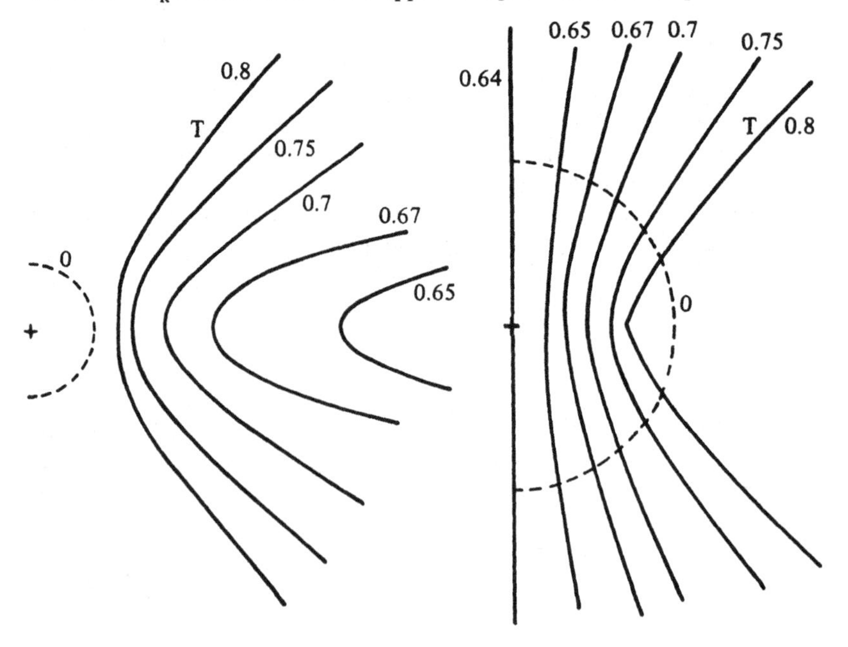

constant N are straight lines through the origin. In this book we use $1/Y$ vs $X^{\frac{1}{2}}/Y = \omega_N/\omega_H$ so that the ordinate is proportional to frequency and independent of N, and the abscissa is proportional to $N^{\frac{1}{2}}$ and independent of ω. Thus for all frequencies a vertical line refers to the same plasma.

In the Appleton–Lassen formula (4.48) or its equivalent forms (4.67), (4.68) (second expression), when collisions are neglected and X is real, the square root can never be zero or infinite for any frequency when $\sin\Theta \neq 0$. Thus the square root never reverses sign when ω or N changes. The positive value when used in (4.48) gives the refractive index n_O for the ordinary wave, as defined in § 4.15, and the negative value gives n_E for the extraordinary wave. The two values n_O, n_E are continuous distinct functions of ω and N for all real ω and N, except where one of them has a resonance. We therefore use separate classification diagrams for the ordinary and extraordinary waves. In the limit $\sin\Theta \to 0$, n_O and n_E can be equal at points in refractive index space called window points, §§ 4.11, 4.12, 5.2 Item (7), but only if

Fig. 5.9. See caption of fig. 5.4. Region E3. Both refractive index surface and ray surface have one sheet, reversed resonance curves and reversed Storey cones. This is another form for the Z-mode. If $\omega_N > \omega_H$ and $\omega \to \omega_N$, $X \to 1$, the limiting form of the refractive index surface is the unit sphere together with the two segments of the line $\Theta = 0$ beyond the points $n = \pm 1$. The limiting form of the ray surface is the sphere $V = c$ together with the two planes $V\cos\beta = \pm c$. If $\omega_N < \omega_H$ and $\omega \to \omega_H$ ($X < 1$, $Y \to 1$), the resonance cones of the ray surface move in towards $\Theta = 0, \pi$. If ω changes to a value less than ω_H the boundary is crossed into region E4 and the refractive index surface is then closed at $\Theta = 0, \pi$ as in fig. 5.10. The curves marked T are for $\omega = \omega_R$ and the same curves appear in fig. 5.8. In this example $Y = 0.6$.

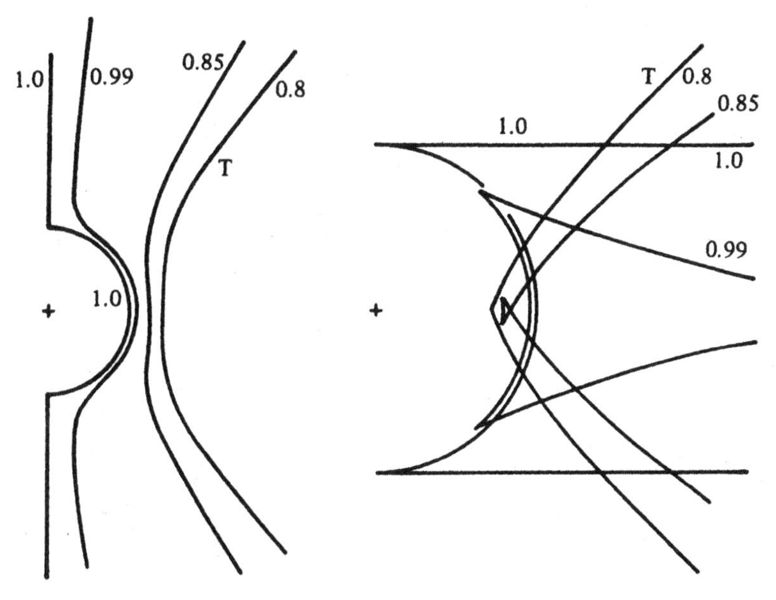

Fig. 5.10. See caption of fig. 5.4. Region E4. Both refractive index surface and ray surface are closed and have a Storey cone and a reversed Storey cone. When the boundary, marked $\omega_{RS1}-\omega_{RN}-\omega_{RS2}$ in fig. 5.3, is reached, these two cones move together as shown by the curves marked T which appear also in fig. 5.13. The cones disappear after the boundary is crossed. In the limit $\omega\rightarrow\omega_N$, $X\rightarrow 1$, the refractive index surface shrinks to the unit sphere together with the two segments of the line $\Theta=0$ between $n=\pm 1$ and the window points $n=\pm\{Y/(Y-1)\}^{\frac{1}{2}}$. The limiting form of the ray surface is the sphere $V=c$ together with the two pairs of planes $V\cos\beta=\pm c\{(Y-1)/Y\}^{\frac{1}{2}}$ and $V\cos\beta=\pm c$. In this example $Y=1.15$.

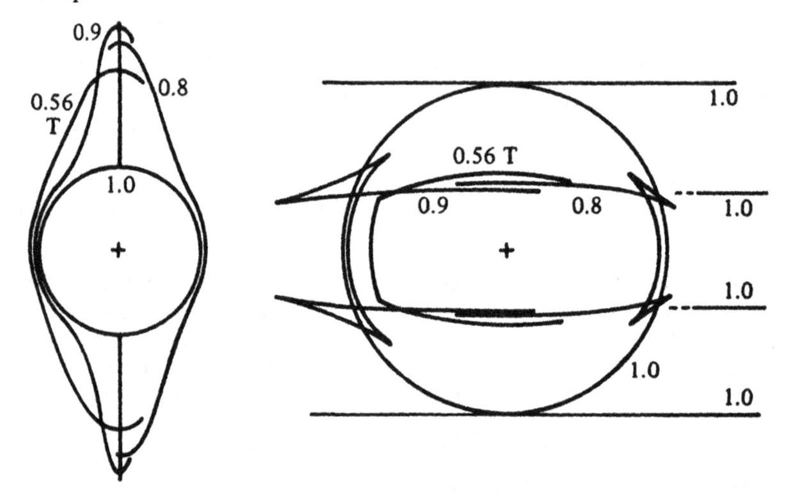

Fig. 5.11. See caption of fig. 5.4. Region E5. Both refractive index surface and ray surface are closed and have a Storey cone. The ray surface has an optic ray axis at $\beta=0$; §§ 5.5. In the limit $\omega\rightarrow\omega_N$, $X\rightarrow 1$ the refractive index surface approaches the unit sphere together with the two segments of the line $\Theta=0$ between the points $n=\pm 1$ and the window points $n=\pm\{Y/(Y+1)\}^{\frac{1}{2}}$. The limiting form of the ray surface is the sphere $V=c$ together with the two pairs of planes $V\cos\beta=\pm c$ and $V\cos\beta=\pm c\{(Y+1)/Y\}^{\frac{1}{2}}$. The curves marked T are for $\omega=\omega_{S2}$ and appear also in fig. 5.12. In this example $Y=1.15$.

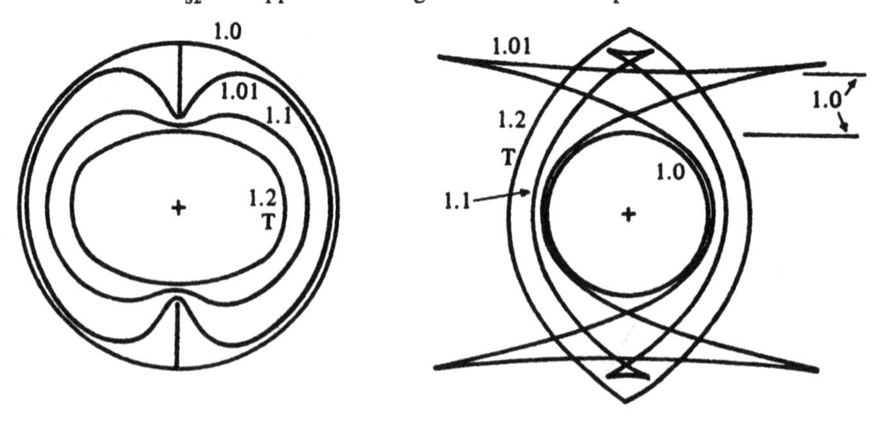

$X = 1$. It is therefore important to note the behaviour of the refractive index surfaces in the limit $X \to 1$ (see captions of figs. 5.4, 5.5, 5.9–5.11).

For the more general collisionless plasma, the square root in (4.51) or (4.68) (first expression) still can never be zero, when $\sin\Theta\cos\Theta \neq 0$, but it can be infinite. The situation is therefore a little more complicated but it turns out that n_O and n_E are still distinct functions for all frequencies for a given plasma.

Fig. 5.2 shows the classification diagram for the ordinary wave in a collisionless electron plasma. Throughout any one of the three regions marked O1, O2, O3, the

Fig. 5.12. See caption of fig. 5.4. Region E6. The topology is the same as for region E1, fig. 5.7. The curves marked T are for $\omega = \omega_{S2}$ and appear also in fig. 5.11. In this example $Y = 0.5$.

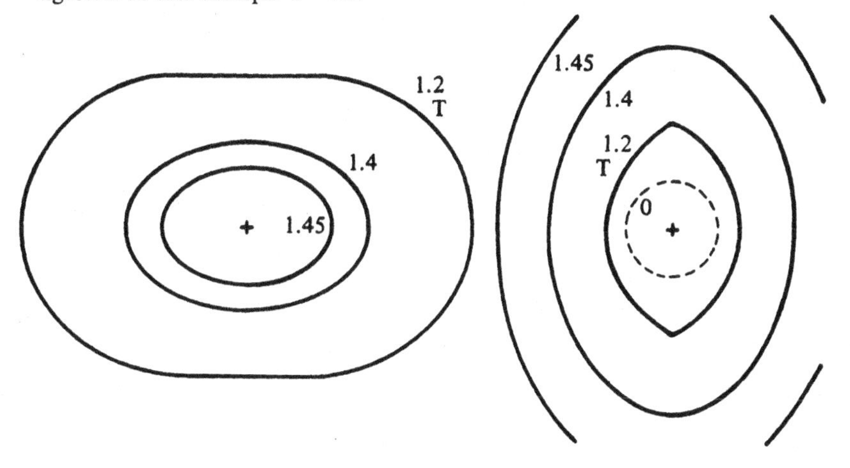

Fig. 5.13. See caption of fig. 5.4. Region E7. The topology is the same as for region O1, fig. 5.4, but the transition curve T is different. It is for the boundary marked $\omega_{RS1} - \omega_{RN} - \omega_{RS2}$ in fig. 5.3, and it appears also in fig. 5.10. In this example $Y = 1.15$.

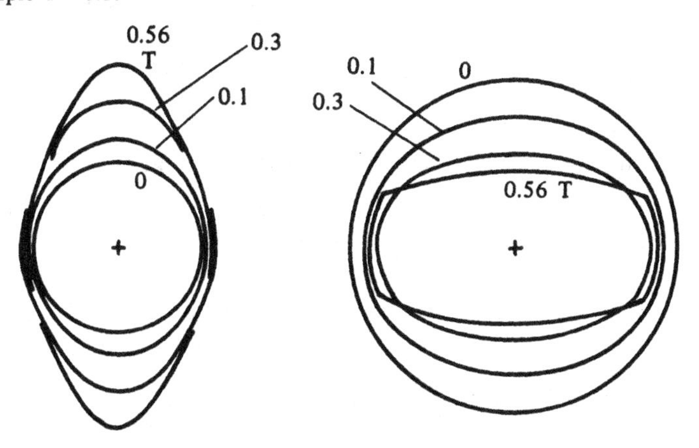

refractive index surface has the same form, figs. 5.4, 5.5, 5.6 respectively. There is one region, shaded and marked 'cut-off' where n_0^2 is negative for all Θ. Similarly fig. 5.3 shows the classification diagram for the extraordinary wave. There are now seven different forms of refractive index surface for regions E1 to E7 and shown in figs. 5.7–5.13 respectively, and there are two separate regions of 'cut-off'. The properties of the ten different forms are summarised in table 5.1.

The regions can be further subdivided depending on the behaviour of the group velocity. Details are not given in this book but see Booker (1975), Walker (1977b).

The boundary curves in figs. 5.2 and 5.3 show where a transition occurs from one type of refractive index and ray surface to another. Examples of the limiting form of these surfaces at a transition are shown in some of the figs. 5.4 to 5.13, and are marked with a T. The boundary curves are labelled with symbols giving the value of ω. Thus, in figs. 5.2, 5.3, ω_N is the plasma frequency and ω_H is the gyro-frequency. In fig. 5.2 the frequency ω_{S1} is where

$$\varepsilon_2 + \varepsilon_3 = 0, \quad X = 2(1 - Y)/(2 - Y) \tag{5.50}$$

(Walker, 1977b; compare (5.15)). When ω decreases through the value ω_{S1}, Storey cones appear at $\Theta = 0$ and π and move to greater values of Θ and $\pi - \Theta$ respectively (figs. 5.5, 5.6).

In fig. 5.3 the frequencies ω_{C1}, ω_{C2} are the cut-off frequencies where $X = 1 + Y$, $1 - Y$ respectively; (3.59), (3.60). The frequency ω_U is the upper hybrid resonance frequency where $X = 1 - Y^2$, (3.62). The frequency ω_{S2} is where (5.14) (5.15) are satisfied. When ω increases through the value ω_{S2}, Storey cones appear at $\Theta = 0$ and π and move to greater values of Θ and $\pi - \Theta$ respectively (figs. 5.12, 5.11). The frequency ω_R is where (5.18), (5.19) are satisfied. When ω decreases through the value ω_R reversed Storey cones appear at $\Theta = \frac{1}{2}\pi$ and move to greater values of $|\frac{1}{2}\pi - \Theta|$ (figs. 5.8, 5.9).

5.7. Dependence of refractive index on frequency

In ch. 4 examples were given, in figs. 4.3–4.6, to show how the two refractive indices of an electron plasma depend on the electron concentration N and the wave normal angle Θ, when the frequency is held constant. It is also of interest to give curves that show how the two values of n^2 depend on frequency for a plasma with a given fixed N. Examples are given for the two cases $\omega_N < \omega_H$, fig. 5.14 and $\omega_N > \omega_H$, fig. 5.15. These figures use the same conventions as in figs. 4.3, 4.5, 4.6. Thus curves for longitudinal propagation $\Theta = 0$ are shown as broken lines and curves for transverse propagation $\Theta = \frac{1}{2}\pi$ are shown as chain lines. The continuous lines are for an intermediate value of Θ. The marked frequencies in these two figures can be identified in the C.M.A. type diagrams, figs. 5.2, 5.3.

Some of the main features to be noted in these figures are as follows:

(1) For any wave normal angle Θ in the range $0 < \Theta < \frac{1}{2}\pi$, the curves of n^2 can never cross those for $\Theta = 0$ or $\Theta = \frac{1}{2}\pi$, except at the cut-off frequencies (proved in § 5.2 item (12)). Hence the curve for the ordinary wave always lies within the region shaded with dots and that for the extraordinary wave always lies in the region with line shading.

(2) The vertical line $\omega/\omega_N = 1$ ($X = 1$) is part of the curves when $\sin\Theta = 0$ (§§ 4.11, 4.12). In fig. 5.14, where $Y > 1$ when $X = 1$, the segment of this vertical line between the two real window points (P, P_2 in fig. 5.14) is for the extraordinary wave, and the remaining segments are for the ordinary wave. As the limit $|\sin\Theta| \to 0$ is approached, the curve for the extraordinary wave makes a steep traverse between regions close to the two window points. In fig. 5.15, where $Y < 1$ when $X = 1$, one of the two window points is where n^2 is negative. The segment of the line $\omega/\omega_N = 1$ between these window points is for the ordinary wave, and the remaining segments are for the extraordinary wave. As the limit $|\sin\Theta| \to 0$ is approached, the curve for the ordinary wave makes a steep traverse between regions close to the two window points; compare § 4.11.

Fig. 5.14. Curves showing how the square n^2 of the refractive index, for a cold electron magnetoplasma, depends on frequency when collisions are neglected, $U = 1$. The plasma frequency $\omega_N/2\pi$ and the electron gyro-frequency $\omega_H/2\pi$ are held constant and in this example $\omega_N/\omega_H = \frac{2}{3}$. The continuous curves are for $\Theta = 30°$. The broken curves are for longitudinal propagation $\sin\Theta = 0$, and the chain curves are for transverse propagation $\cos\Theta = 0$. For any other Θ the curves for the ordinary wave must lie within the regions shaded with dots, and those for the extraordinary wave within the regions shaded with lines. Compare figs. 4.3–4.6.

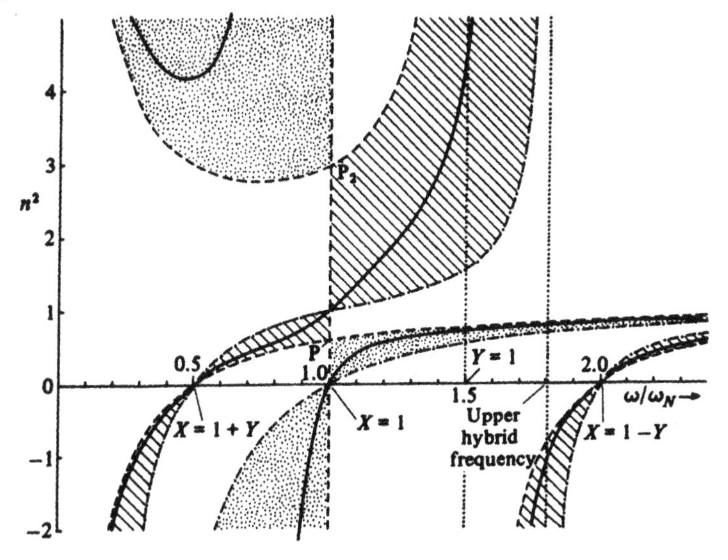

(3) The ordinary wave shows cut-off where $\omega/\omega_N = 1 (X = 1)$ for all Θ.

(4) The extraordinary wave shows cut-off where

$$\frac{\omega}{\omega_N} = \left\{\frac{1}{4}\left(\frac{\omega_H}{\omega_N}\right)^2 + 1\right\}^{\frac{1}{2}} \mp \frac{1}{2}\frac{\omega_H}{\omega_N} \tag{5.51}$$

for $X = 1 \pm Y$ respectively, for all Θ.

(5) Let ω_C denote any one of the cut-off frequencies mentioned in (3) and (4) above. Then the n^2 that is cut off is given by

$$n^2 = K(\omega - \omega_C) + O(\omega - \omega_C)^2 \tag{5.52}$$

where K is a positive constant (compare § 5.2 item (10)). When $X \approx 1 + Y$ it can be shown from (5.7) that

$$K = (2 + Y)/\{(1 + Y)(1 + \cos^2\Theta)\}. \tag{5.53}$$

When $X \approx 1 - Y$, (5.7) with the sign of Y reversed gives

$$K = (2 - Y)/\{(1 - Y)(1 + \cos^2\Theta)\}. \tag{5.54}$$

For positive X this can only occur where $Y < 1$. When $X \approx 1$, ε_3 is small and a formula analogous to (5.7) then gives $n^2 \approx \varepsilon_3 \operatorname{cosec}^2\Theta$. Hence for the cut-off at $X = 1$

$$K = 2\operatorname{cosec}^2\Theta \quad (\Theta \neq 0) \tag{5.55}$$

Thus the curve of n^2 versus ω always has a positive slope at cut-off. But n is proportional to $(\omega - \omega_C)^{\frac{1}{2}}$ so that near cut-off

$$\frac{dn}{d\omega} \approx \frac{K}{2n} \propto (\omega - \omega_C)^{-\frac{1}{2}}. \tag{5.56}$$

This tends to infinity as cut-off is approached.

Fig. 5.15. Similar to fig. 5.14 but $\omega_N/\omega_H = 1.5$ and the continuous curves are for $\Theta = 45°$. Note that the ordinate scale is different for the two parts of the figure.

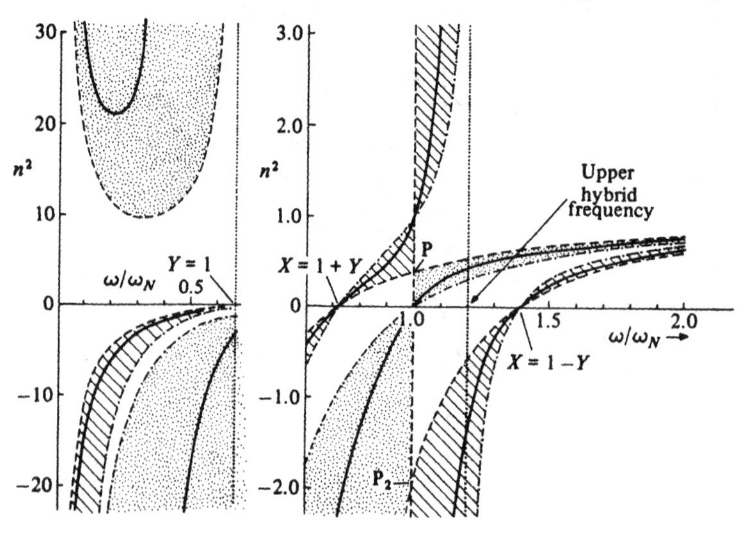

(6) The ordinary wave shows a resonance where

$$\left(\frac{\omega}{\omega_N}\right)^2 = \frac{1}{2} + \frac{1}{2}\frac{\omega_H^2}{\omega_N^2} - \left\{\frac{1}{4}\left(\frac{\omega_H^2}{\omega_N^2} + 1\right)^2 - \frac{\omega_H^2}{\omega_N^2}\cos^2\Theta\right\}^{\frac{1}{2}} \tag{5.57}$$

which is derived from (4.76). When $\sin\Theta = 0$ this is where $\omega = \mathrm{Min}(\omega_H, \omega_N)$, that is ω_N in fig. 5.14, and ω_H in fig. 5.15. When $\cos\Theta \to 0$ this resonance occurs where $\omega/\omega_N \to 0$. For intermediate Θ this resonance is between these limits. It occurs in the branch of the curves for the whistler mode, and the refractive index surfaces are shown in fig. 5.6. In the limit $X \to \infty$, that is $\omega \to 0$, these surfaces extend round towards $\Theta \to \pm\frac{1}{2}\pi$.

(7) The extraordinary wave shows a resonance given by (5.57) when the first $-$ sign is changed to $+$. When $\sin\Theta = 0$ this is where $\omega = \mathrm{Max}(\omega_H, \omega_N)$ that is ω_H in fig. 5.14 and ω_N in fig. 5.15. When $\cos\Theta \to 0$ this resonance occurs at the upper hybrid resonance frequency (3.62), that is

$$\omega \to \omega_U = (\omega_N^2 + \omega_H^2)^{\frac{1}{2}}. \tag{5.58}$$

For intermediate Θ the resonance is between these limits. It occurs in the branch of the curves for the Z-mode, and the refractive index surfaces are shown in figs. 5.8, 5.9.

(8) Let ω_∞ denote either of the resonance frequencies mentioned in (6) and (7) above. Then it can be shown, from (4.76) or (4.77) that near a resonance, $n^2(\omega)$ has a simple pole so that it can be expressed as a Laurent series

$$n^2 = \frac{\mathscr{A}}{\omega - \omega_\infty}\{1 + \mathrm{O}(\omega - \omega_\infty)\}. \tag{5.59}$$

It can further be shown that for the collisionless electron plasma

$$\mathscr{A} = \frac{-X(1 - Y^2\cos^2\Theta)^2(1 + Y^2\cos^2\Theta)\omega_\infty}{2\sin^2\Theta\{(1 - X\cos^2\Theta)^2 + X^2\cos^2\Theta\sin^{-2}\Theta\}}. \tag{5.60}$$

Thus when X and ω_∞ are positive, $\mathscr{A}$ is negative, so that n^2 is positive for frequencies slightly less than any ω_∞, or equivalently for values of X slightly greater than the value X_∞ at exact resonance. The property (5.59) also shows that, near resonance

$$\frac{\mathrm{d}n}{\mathrm{d}\omega} \approx -\frac{\mathscr{A}}{2n(\omega - \omega_\infty)^2} \propto (\omega - \omega_\infty)^{-\frac{3}{2}}. \tag{5.61}$$

This tends to infinity as resonance is approached. (Compare the property (5.56) at cut-off).

(9) The two values of n^2 are continuous and bounded functions of ω near $\omega = \omega_H$, that is $Y = 1$, except when $\sin\Theta = 0$.

(10) Both values of n^2 tend to infinity when $\omega \to 0$. It must, however, be remembered that the effect of ions has here been neglected. At low frequencies the ions have an important effect and introduce additional resonances and cut-offs. For an

actual plasma, at frequencies less than the smallest ion gyro-frequency, it can be shown that, when $\omega \to 0$, the two squared refractive indices tend to the bounded values $n^2 = \varepsilon_1$, $\varepsilon_1 \sec^2\Theta$ where ε_1 is given by the first two terms of (3.72).

When collisions are allowed for $n = \mu - i\chi$ is complex and it would be necessary to draw curves of μ and χ versus ω. These would be similar in many of their properties to the curves of μ and χ versus X discussed in §4.16, figs. 4.7–4.11. In particular segments of the vertical line $\omega = \omega_N$ in figs. 5.14, 5.15 would not now form part of the curves when $\sin\Theta = 0$. As Θ increases and passes through the value Θ_t given by (4.96) the curves of $\mu(\omega)$ and $\chi(\omega)$, near $\omega = \omega_N$, would show intersections or bends similar to those in figs. 4.7–4.11. The curves would join exactly when $\Theta = \Theta_t$, as in fig. 4.9. When $\Theta > \Theta_t$, the curves of $\mu(\omega)$ and $\chi(\omega)$ for the ordinary wave would be continuous for all ω, and similarly for the extraordinary wave. When $\Theta < \Theta_t$ the two curves would still be continuous for all ω including $\omega = \omega_N$. But one curve would be ordinary for $\omega < \omega_N$ and extraordinary for $\omega > \omega_N$, and conversely for the other curve. This is just the same change of description as was needed in §4.16 near $X = 1$.

5.8. Group velocity

In radar and in many radio experiments (see § 1.6), pulses of radio waves are used. A pulse of this kind can be thought of as the signal from a transmitter that generates a constant frequency ω_1 and that is switched on for a short time T. The study of the propagation of pulses is useful for introducing the concept of group velocity, but the importance of this concept is not confined to the study of pulses.

A pulsed radio signal can be represented by a Fourier integral thus

$$F(t) = \int_{-\infty}^{\infty} M(\omega)e^{i\omega t}\,d\omega \tag{5.62}$$

where F is some field component in the radio wave. The form of the function $M(\omega)$ depends on the shape of the pulse which is usually a signal of constant frequency ω_1 and roughly constant amplitude lasting for a time T, which, though short, is long compared with $1/\omega_1$. T may be 200 μs or less and $\omega_1/2\pi$ may be anything from 50 kHz upwards. $M(\omega)$ then has maxima at or near $\omega = \pm\omega_1$, and ω_1 is therefore called the predominant frequency. If $T \gg 1/\omega_1$, the maxima are narrow so that only frequencies very close to $\pm\omega_1$ play a part in the propagation of the pulse. Because $F(t)$ is real it follows that $|M(\omega)|$ is symmetric and $\arg M(\omega)$ is antisymmetric about $\omega = 0$. The centre of the pulse is assumed to be at $t = 0$ which ensures that $\arg M(\omega)$ does not vary appreciably in a small range near ω_1. In this chapter the exact form of $M(\omega)$ is unimportant. It is considered in more detail in §11.16.

Suppose now that the signal (5.62) is the field of an infinite plane wave with its wave normal in the z direction, and let it be launched where $z = 0$ into a homogeneous plasma. Then each component frequency gives a wave whose field contains the factor $\exp(-ikn z)$ as in (2.32), and the field at a distance z from the launching point is

$$F(t, z) = \int_{-\infty}^{\infty} M(\omega)\exp\{i(\omega t - \omega n z/c)\}\,d\omega. \tag{5.63}$$

This integral is the sum of contributions from frequencies in narrow ranges near $\pm\omega_1$. When these contributions have the same phase, the integral has its greatest value. Thus the pulse reaches the distance z at a time t when the phase of the exponential term in (5.63) is stationary with respect to changes of ω. This gives

$$t - \frac{\partial(\omega n)}{\partial\omega}\frac{z}{c} = 0 \text{ for } \omega = \omega_1 \tag{5.64}$$

so that the velocity $\mathscr{U}_z = z/t$ of the pulse in the z direction is given by

$$c/\mathscr{U}_z = n'(\omega_1) = \partial(\omega n)/\partial\omega \quad (\omega = \omega_1). \tag{5.65}$$

If the velocity of the pulse has any component perpendicular to the z direction, this could not be detected in the present example since the pulse was assumed to be an infinite plane wave. To study the x and y components of the velocity it would be necessary to consider a pulse of limited extent in the x and y directions, that is a 'wave packet' and this is done later in this section. It is shown that for an isotropic medium the wave packet travels in the direction of the wave normal, that is the z direction, so that $\mathscr{U}_z$ is the true velocity of the pulse. But for an anisotropic medium the pulse travels in the direction of the ray, that is at an angle α to the z axis as shown in § 5.3, and given by (5.33).

The true velocity $\mathscr{U}$ of a wave packet is called the 'group velocity'. Thus for an isotropic medium (5.65) gives the value $\mathscr{U}_z = \mathscr{U}$ of the group velocity. For a collisionless isotropic electron plasma, (4.10) gives

$$n = (1 - \omega_N^2/\omega^2)^{\frac{1}{2}}, \quad n' = \partial(\omega n)/\partial\omega = (1 - \omega_N^2/\omega^2)^{-\frac{1}{2}} = 1/n. \tag{5.66}$$

In this case $\mathscr{U} = c/n'$ from (5.65), and the phase velocity $v = c/n$ is the same as the ray velocity V, (5.34), (5.35) so that

$$\mathscr{U}V = c^2. \tag{5.67}$$

It is interesting that several other types of wave motion obey a relation similar to (5.67). For example, it is true for one mode in a loss-free wave guide. In wave mechanics, for the waves associated with particles of zero potential energy but various total energies, when treated relativistically, it can be shown that the group velocity $\mathscr{U}$ is the velocity of the particle, and (5.67) is true when V is the phase velocity. But for waves in a plasma it is not true except in the special case (5.66).

For a general anisotropic plasma it was shown in § 5.3 that for a plane wave the direction of energy propagation is the direction of the ray and makes an angle α with the wave normal, (5.33). It is shown below that the velocity $\mathscr{U}$ of a wave packet, that is the group velocity, is in the direction of the ray, so that $\mathscr{U}_z = \mathscr{U}\cos\alpha$ is its component in the direction of the wave normal. Thus from (5.65)

$$\mathscr{U} = c/(n'\cos\alpha). \tag{5.68}$$

It must be emphasised that in general the group velocity is not equal to c/n'.

To find the velocity of a wave packet consider again the expression (5.29) for the field F emitted from a source at the origin. It represents the field as an angular spectrum of plane waves and was derived by expressing it as a Fourier integral in space. We now suppose that the emitted signal is a pulse, like (5.62) so that the amplitude A in (5.29) is also a function of the frequency ω, and the integral is a triple integral. In (5.29) the two variables of integration are n_ξ, n_η and they determine the direction of the wave normal of the wave represented by the integrand, but any two combinations of n_ξ, n_η, n_ζ could have been used. We now have to choose three combinations of the four variables ω, n_ξ, n_η, n_ζ to determine the direction of the wave normal and the frequency. It is convenient to use the three components of the wave vector

$$\boldsymbol{\kappa} = (\kappa_\xi, \kappa_\eta, \kappa_\zeta) = \frac{\omega}{c}(n_\xi, n_\eta, n_\zeta) \tag{5.69}$$

(compare (2.31). In many branches of physics $\boldsymbol{k}$ is used instead of $\boldsymbol{\kappa}$). Then the field of the signal is

$$F(t, \xi, \eta, \zeta) = \int_{-\infty}^{\infty}\int_{-\infty}^{\infty}\int_{-\infty}^{\infty} A(\omega, \boldsymbol{\kappa})\exp\{\mathrm{i}(\omega t - \xi\kappa_\xi - \eta\kappa_\eta - \zeta\kappa_\zeta)\}\,\mathrm{d}\kappa_\xi \mathrm{d}\kappa_\eta \mathrm{d}\kappa_\zeta. \tag{5.70}$$

Now it is assumed that the signal is a pulse as in (5.62) so that A is appreciable only near $\omega = \pm\omega_1$ where ω_1 is the predominant frequency. It is further assumed that the signal is in a narrow beam so that the wave vectors κ are all near a predominant value $\boldsymbol{\kappa}_1$. Thus A is appreciable only near $\boldsymbol{\kappa} = \pm\boldsymbol{\kappa}_1$. The signal is therefore a wave packet, that is a pulse travelling along the beam which is its path. We now invoke again the principle of stationary phase which was used to derive (5.31) and (5.64). The signal F in (5.70) is appreciable only for those values of t, ξ, η, ζ for which the phase, that is the exponent of the integrand, is stationary for small changes of κ_ξ, κ_η, κ_ζ, when $\omega = \omega_1$ and $\boldsymbol{\kappa} = \boldsymbol{\kappa}_1$. This gives

$$\xi/t = \partial\omega/\partial\kappa_\xi, \quad \eta/t = \partial\omega/\partial\kappa_\eta, \quad \zeta/t = \partial\omega/\partial\kappa_\zeta \quad (\omega = \omega_1, \boldsymbol{\kappa} = \boldsymbol{\kappa}_1). \tag{5.71}$$

These are just the three components of the velocity of the wave packet, that is of the group velocity $\mathscr{U}$. The result (5.71) is usually written in the shorthand notation

$$\mathscr{U} = \partial\omega/\partial\boldsymbol{\kappa}. \tag{5.72}$$

It is well known and used in many branches of physics.

To use (5.72) it is necessary to express ω as a function of $\boldsymbol{\kappa} = \omega_1\boldsymbol{n}/c$. The equation

$\omega(\boldsymbol{\kappa}) = \omega_1$ is an equation relating $(\omega/c)(n_\xi, n_\eta, n_\zeta)$. It is just the equation of the refractive index surface. On this surface $\delta\omega$ must be zero so that

$$\frac{\partial\omega}{\partial\boldsymbol{\kappa}}\cdot\delta\boldsymbol{n} = \boldsymbol{\mathscr{U}}\cdot\delta\boldsymbol{n} = 0. \tag{5.73}$$

But any allowed $\delta\boldsymbol{n}$ lies in the refractive index surface. Thus $\boldsymbol{\mathscr{U}}$ is normal to this surface and therefore has the same direction as the ray.

For a plasma it is algebraically complicated to express ω as a function of $\boldsymbol{\kappa}$ or equivalently of $\boldsymbol{n}$. The dispersion relation (4.65) for an electron plasma is of sixth degree in ω and when ions are allowed for it is of still higher degree, and it would be necessary to solve this equation. It is much easier to use the result (5.68), which is studied in the following section. It can be shown (see problem 5.6) that the results (5.68) and (5.72) are the same.

5.9. Properties of the group velocity

In this section we study the dependence of group velocity, in an electron plasma, on electron concentration and on frequency. The dispersion relation (4.65) with (4.66) is used. The A used in these equations has no connection with the A used in § 5.8. Differentiation of (4.65) with respect to ω gives

$$4n(n^2A - B)\partial n/\partial\omega + n^4\partial A/\partial\omega - 2n^2\partial B/\partial\omega + \partial C/\partial\omega = 0. \tag{5.74}$$

Hence from (5.65)

$$n' = n + \omega\partial n/\partial\omega = \frac{n^4\left(A - \frac{1}{4}\omega\frac{\partial A}{\partial\omega}\right) - n^2\left(B - \frac{1}{2}\omega\frac{\partial B}{\partial\omega}\right) - \frac{1}{4}\omega\frac{\partial C}{\partial\omega}}{n(n^2A - B)}. \tag{5.75}$$

This is the group refractive index and is equal to $c/\mathscr{U}_z$ where $\mathscr{U}_z$ is the component of the group velocity in the direction of the wave normal. It is important in the theory of the probing of a horizontally stratified ionosphere by vertically incident pulses of radio waves. When a vertically incident wave packet enters the ionosphere it is deviated laterally (see § 10.12 and fig. 10.11) but the wave normal remains vertical. It travels up to a reflection level and returns by the same path. To find its travel time it is necessary to know only the vertical component $\mathscr{U}_z = c/n'$ of its velocity. This was at one time called the group velocity and some authors still use this term. The true group velocity is $\mathscr{U} = \mathscr{U}_z \sec\alpha$. Curves of n' versus X and tables have been published (Shinn and Whale, 1952; Shinn, 1955). Curves of group velocity versus frequency for both cold and warm plasmas were given by Aubry, Bitoun and Graff (1970).

When collisions are neglected, (4.66) shows that

$$\begin{aligned} A - \tfrac{1}{4}\omega\partial A/\partial\omega &= 1 - \tfrac{3}{2}X - Y^2(\tfrac{3}{2} - 2X) - 2XY^2\sin^2\Theta \\ B - \tfrac{1}{2}\omega\partial B/\partial\omega &= (1 - X)(1 - 3X) - Y^2(2 - 3X) - \tfrac{3}{2}XY^2\sin^2\Theta \\ -\tfrac{1}{4}\omega\partial C/\partial\omega &= -\tfrac{3}{2}X(1 - X)^2 - Y^2(\tfrac{1}{2} - X). \end{aligned} \tag{5.76}$$

These were used in (5.75) to calculate the curves of figs. 5.16, 5.17. They show examples of how n' depends on X and are for the same conditions as figs. 4.3, 4.6 respectively for n^2 vs X.

Equations (5.56) and (5.61) show that $\partial n/\partial\omega$ and therefore n' is infinite when either cut-off or resonance is approached and this can be seen in fig. 5.16, 5.17. It is important to know how n' approaches infinity in these two cases. In (5.75) near any cut-off condition all terms except n in the denominator are analytic functions of X.

Fig. 5.16. Shows how the group refractive index depends on X for a cold collisionless electron plasma. In this example $Y = \frac{1}{2}$. The continuous curves are for $\Theta = 23.27°$. The broken curves are for longitudinal propagation, $\sin\Theta = 0$. The broken curve for the extraordinary wave when $X < 0.5$ is extremely close to the continuous curve E and therefore is not shown. The chain curves are for transverse propagation $\cos\Theta = 0$. Note the change of name from O ordinary, to E extraordinary, where the broken curve crosses the line $X = 1$.

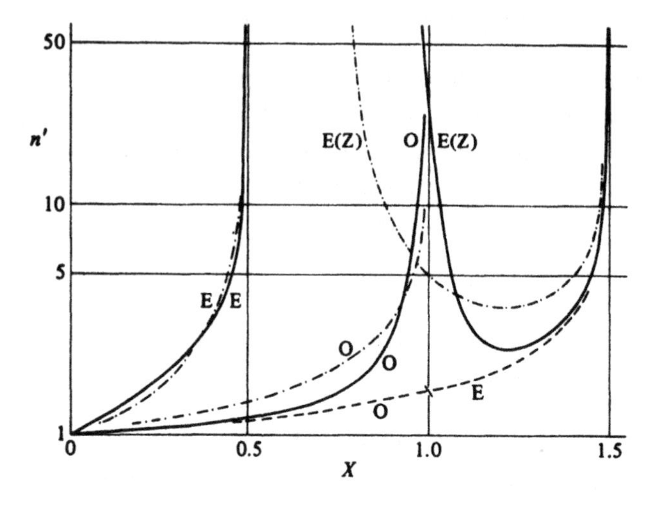

Fig. 5.17. Similar to fig. 5.16, but in this example $Y = 2$.

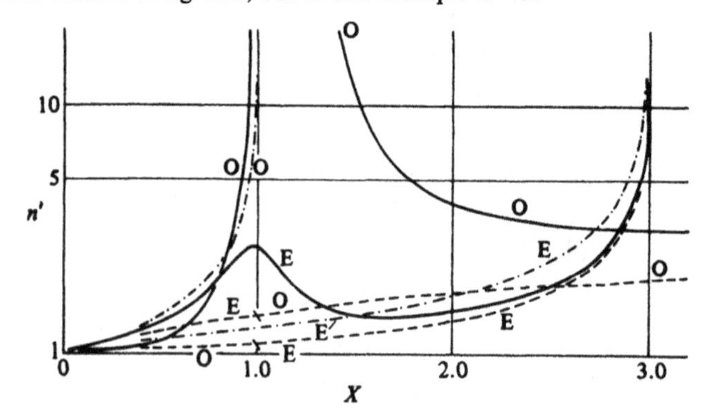

In them we may therefore give X its value at cut-off and set $n^2 = 0$, to give

$$n' \approx \tfrac{1}{4}\omega\frac{\partial C}{\partial\omega}/Bn. \tag{5.77}$$

For the ordinary wave near cut-off at $X = 1$, this gives $n' = \operatorname{cosec}^2\Theta/n$. Now (4.65) (4.66), with $U = 1$, show that $n^2 \approx (1 - X)\operatorname{cosec}^2\Theta$. Hence

$$n' \approx \operatorname{cosec}\Theta(1 - X)^{-\frac{1}{2}}. \tag{5.78}$$

Similarly for the extraordinary wave cut-off at $X = 1 + Y$, (5.77) gives $n' = (1 + \frac{1}{2}Y)/\{(1 + Y)(1 - \frac{1}{2}\sin^2\Theta)n\}$, and n^2 is given by (5.7). Hence

$$n' \approx (1 + \tfrac{1}{2}Y)\{(1 + Y)(1 + Y - X)\}^{-\frac{1}{2}}. \tag{5.79}$$

For the cut-off at $X = 1 - Y$ the result is (5.79) with the sign of Y reversed. In (5.78) and (5.79) it is the power $-\frac{1}{2}$ that is important and it is mentioned again in § 19.6.

The results (5.78), (5.79) show that for every real cut-off condition, n' is real when X is slightly less than its value at exact cut-off. This behaviour can be seen in figs. 5.16, 5.17. The same property holds for the refractive index n; § 5.2 item (10), § 5.7 item (5).

For a resonance a similar argument can be used. It is more complicated because A is zero at a resonance. But it can be shown from (5.75) after a little algebra, that near resonance

$$n' \approx -\tfrac{1}{4}\omega\frac{\partial A}{\partial\omega}n^3/B. \tag{5.80}$$

Now (4.65) (4.66), with $U = 1$, show that

$$n^2 \approx 2B/A = 2B(Y^2\cos^2\Theta - 1)^{-1}(X - X_\infty)^{-1} \tag{5.81}$$

where X_∞ is the value of X at resonance. Hence finally

$$n' \approx Y|\sin\Theta|\{\sin^2\Theta + \cos^2\Theta(1 - Y^2)\} \\ \times (1 + Y^2\cos^2\Theta)^{\frac{1}{2}}|1 - Y^2\cos^2\Theta|^{-3}X_\infty^{\frac{1}{2}}(X - X_\infty)^{-\frac{3}{2}} \tag{5.82}$$

Here it is the factor $(X - X_\infty)^{-\frac{3}{2}}$ that is important. See § 19.6.

For every real resonance, (5.82) shows that n' is real when X slightly exceeds its value X_∞ at exact resonance. This behaviour can be seen in figs. 5.16, 5.17. The same property holds for the refractive index n; § 5.7 item (8).

To find the true group velocity $\mathscr{U} = c/n'\cos\alpha$ it is necessary to know the angle α. It can be found by differentiating (4.65) with respect to Θ, and is given, from (5.33), by

$$\tan\alpha = \frac{1}{n}\frac{\partial n}{\partial\Theta} = \frac{XY^2\sin\Theta\cos\Theta(n^2 - 1)}{2(An^2 - B)}. \tag{5.83}$$

Figs. 5.18–5.20 show an example of how $n'\cos\alpha = c/\mathscr{U}$ and α depend on frequency ω for a fixed value of the ray direction β. The advantage of plotting $c/\mathscr{U}$ versus frequency is that it is proportional to the time of travel of a pulsed signal over a fixed range in the given direction β. If the scale of $c/\mathscr{U}$ were linear, the curves would be the same shape as the record that would be obtained from an 'on-line' spectral analyser of the type commonly used for studying whistlers (§ 13.8), when the source is an impulse. In figs. 5.18, 5.20, however, the scale of $c/\mathscr{U}$ is logarithmic. It should also be remembered that these results are for a homogeneous plasma and differ in some respects from observations made for an inhomogeneous plasma such as the ionosphere or the magnetosphere. On the path of a whistler the angle β is not constant.

The importance of the angle α may be illustrated by considering two space vehicles immersed in the homogeneous plasma. A radio signal is to be sent from one to the other by a directed radio beam. Suppose that the transmitting aerial has a large aperture which is energised with constant phase. If it were in free space this aerial would radiate a single narrow beam in the direction of its axis, normal to the aperture. When it is in the plasma the wave normal of the beam is still parallel to the aerial's axis, but there are two rays, ordinary and extraordinary, which leave the aerial in different directions, because they have different values of the angle α. If one of these rays is to reach the second vehicle, whose true direction makes an angle β with the earth's magnetic field, the aerial would have to be pointed in a direction that makes an angle $\Theta = \beta + \alpha$ with this magnetic field. Now α is itself a function of Θ. Thus when the aerial axis is rotated uniformly in a plane parallel to the magnetic field, the two beams do not rotate uniformly with it. There are cases where the ray direction $\Theta - \alpha$ decreases when Θ is increasing, for a small range of Θ, and it is then possible to find as many as four different directions of the transmitting aerial that will send a signal towards the second vehicle, with different times of travel. Examples

Fig. 5.18. Shows how $n' \cos\alpha = c/\mathscr{U}$ depends on frequency for a cold collisionless electron plasma, when the direction of the ray is fixed. In this example $\omega_N/\omega_H = \frac{2}{3}$ and the ray direction is $\beta = 10°$. The symbols by the curves indicate the regions of the CMA type diagrams, figs. 5.2, 5.3. Different abscissa scales are used in different parts of the range, to show up the details. The transition frequencies indicated by arrows are defined in §§5.6, 5.7. The point marked SC is where a Storey cone crosses the ray direction. Similarly RSC is where a reversed Storey cone crosses the ray direction. Note the change from ordinary to extraordinary at the two window points P, P_2 on the line $X = 1$,

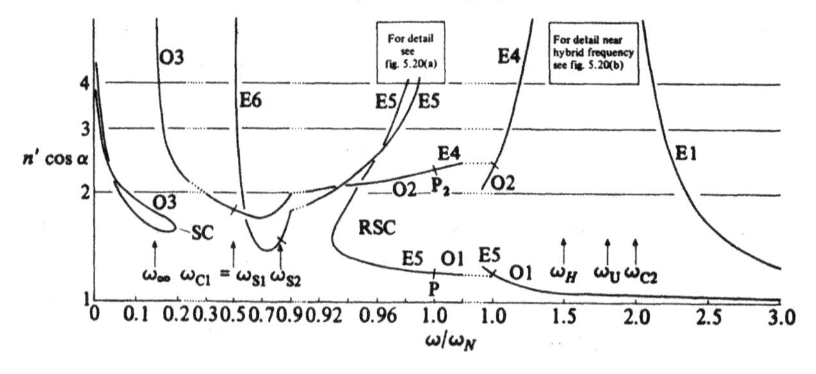

can be seen in fig. 5.19. For part of the branches O3 there are three different real values of α, and for part of the branches E5 there are four different values.

Some of the main features to be noted in figs. 5.18–5.20 are the following. It is assumed that $|\Theta| < \frac{1}{2}\pi$.

(1) The symbols by the various sections of the curves show the regions of the C.M.A. type diagrams, figs. 5.2, 5.3 that are applicable. They can be used, with table 5.1 and figs. 5.4–5.13, to find the shape of the refractive index surface and ray surface for each section.

(2) Two cut-off frequencies ω_{C1} and ω_{C2} appear in fig. 5.19. These would be the same for any other β. The group velocity is real when ω is slightly greater than each ω_C, and tends to ∞ when $\omega \to \omega_C$; see especially (5.80). In this limit $\tan\Theta = 2\tan\beta$ (see (5.37)) and $\alpha = \Theta - \beta$ so that, with $\beta = 10°$ in this example, $\alpha \to 9.42°$.

(3) In figs. 5.18, 5.19 there is one resonance frequency ω_∞ in the branch O3. This is where the frequency dependent resonance cone angle β_R is equal to the ray direction β. Thus the value of ω_∞ depends on β. This resonance is a forward resonance and the limiting value of α is $-90°$ (fig. 5.19), so that $\cos\alpha \to 0$. It was shown in § 5.9, equation (5.82) that at resonance $n' \to \infty$. It can also be shown (Budden and Stott, 1980, § 4.3) that $n'\cos\alpha \to \infty$, as in fig. 5.18. In fig. 5.20(b) there is another ω_∞ for the branch E2. This is a reversed resonance and $\alpha \to +90°$. Note that, for both these resonances $n'\cos\alpha$ is real when ω slightly *exceeds* ω_∞, whereas (5.82) shows that n' is real when $\omega < \omega_\infty$. It must be remembered that, when β is held constant, both Θ and α change when ω changes. The reader should satisfy himself that these changes will explain the behaviour of the curves near resonance.

(4) For the branch E4, $n'\cos\alpha \to \infty$ when $\omega \to \omega_H$ and in this limit $\alpha = -\beta$, so that $\Theta = 0$.

(5) The extreme left-hand curve O3 in figs. 5.18, 5.19 is for the whistler signal. The

Fig. 5.19. Shows how the angle α between the ray and the wave normal depends on frequency for the same example as in fig. 5.18. The wave normal direction is $\Theta = \beta + \alpha$ and here $\beta = 10°$.

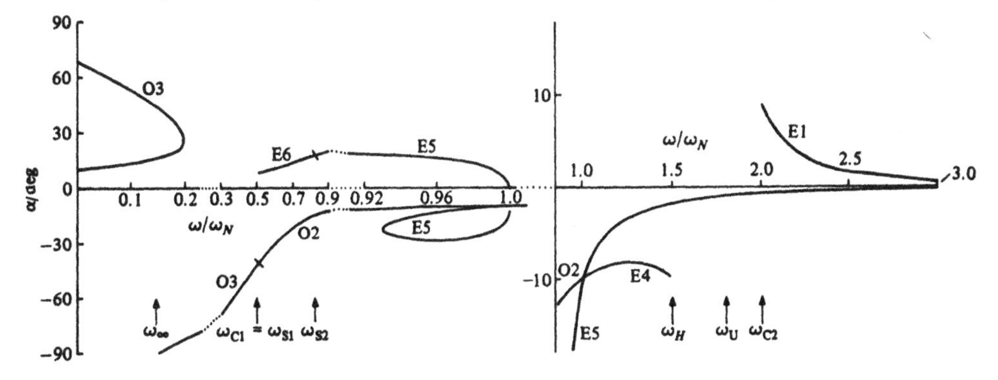

two branches of the curve can be identified in the ray surface, fig. 5.6(b). In the limit $\omega \to 0$ it can be shown (see problem 5.5) that $\tan\Theta = 2\tan\alpha$ whence it follows that in this example for $\beta = 10°$, the two limiting values of α are 10.7° and 69.3° as in fig. 5.19. But it should be remembered that at extremely low frequencies the form of the curves is very different when the effect of ions is allowed for; see Budden and Stott (1980). The frequency near $\omega/\omega_N = 0.17$ where $n'\cos\alpha$ is a minimum, corresponds to the 'nose frequency' of a whistler; see § 13.8.

(6) In fig. 5.18 there is a Storey cone, marked SC, in the whistler branch O3, and a reversed Storey cone, marked RSC, in the Z mode branch E5. In fig. 5.20(a) there is a reversed Storey cone, marked RSC, in the Z mode branch E4. These three cones can be identified in the ray surfaces, figs. 5.6(b), 5.11(b) and 5.10(b) respectively. These ray surfaces have cusps at values β_S of β which cross the actual ray direction β where the curves in figs. 5.18, 5.20 have a vertical tangent. Note that there are no cusps here for the group velocity $\mathscr{U}$ or its reciprocal.

(7) For frequencies very close to the electron plasma frequency (window frequency), two branches of the curves of $n'\cos\alpha$ tend to infinity as shown in fig. 5.20(a). They are associated with the part of the refractive index surface very close to the points $n = 1$, $\sin\Theta = 0$. This behaviour is interesting and needs further study. A start on this has been made by Budden and Stott (1980).

Another example of the dependence of $n'\cos\alpha = c/\mathscr{U}$ on frequency has been given

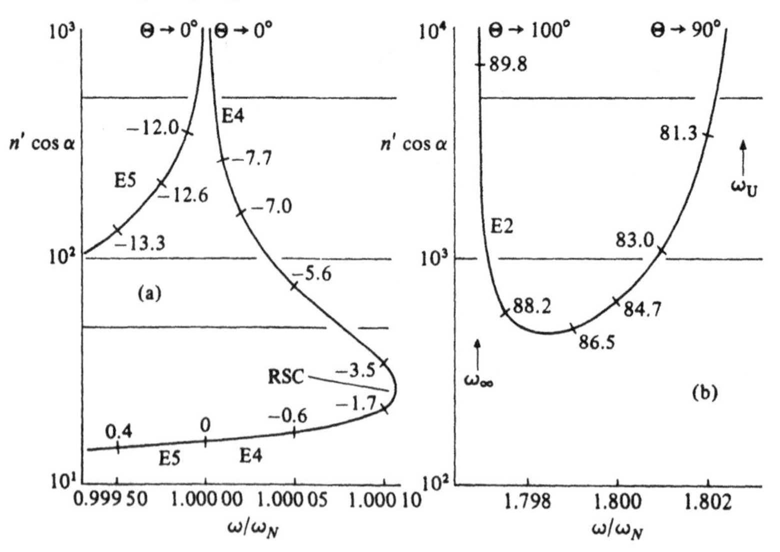

Fig. 5.20. Detail for fig. 5.18. Values of α in degrees are shown as numbers by the curves. (a) is for frequencies extremely close to the electron plasma frequency and is associated with the region of refractive index space very near to $n = 1$, $\Theta = 0$. (b) is for the Z-mode at frequencies extremely close to the upper hybrid resonance frequency ω_U.

by Budden and Stott (1980) who allowed for the presence of ions. Curves of group velocity $\mathscr{U}$ versus ray direction β when the frequency is held constant have been given by Booker (1975, 1984), and Walker (1977a).

5.10. Effect of electron collisions on the group refractive index

The position of a wave packet at a given time was found in § 5.8 by expressing the condition that the phase must be stationary for small variations of frequency, and this led to the formula (5.65). If electron collisions are allowed for, the refractive index n is complex and its real part μ and imaginary part $-\chi$ are both frequency dependent. The dependence of χ on frequency leads to broadening and distortion of the wave packet; see, for example, Stratton (1941) ch. V; Gibbons and Rao (1957). But if the electron collision frequency v is not too great, χ is small for a wave whose squared refractive index would be real and positive if collisions were neglected. Then the phase of the wave is determined mainly by μ, and if χ is neglected the travel time of a wave packet can be found exactly as in § 5.8. If u_z is the component of the wave packet's velocity in the direction of the wave normal, equation (5.65) is replaced by

$$c/u_z = \mu' = \text{Re}(n') = \mu + \omega\frac{\partial\mu}{\partial\omega} = \text{Re}\left(n + \omega\frac{\partial n}{\partial\omega}\right). \tag{5.84}$$

Here n' is now a complex number, and its real part μ' is often called the group refractive index. It is close to the real n' that would be obtained if collisions were

Fig. 5.21. Dependence on X of Re(n')–continuous curves, and Im(n')–broken curves, for an electron plasma at a fixed frequency. In this example $Y=4$ and $\Theta = 23.27°$. The numbers by the curves are the values of Z. These curves are for the ordinary wave.

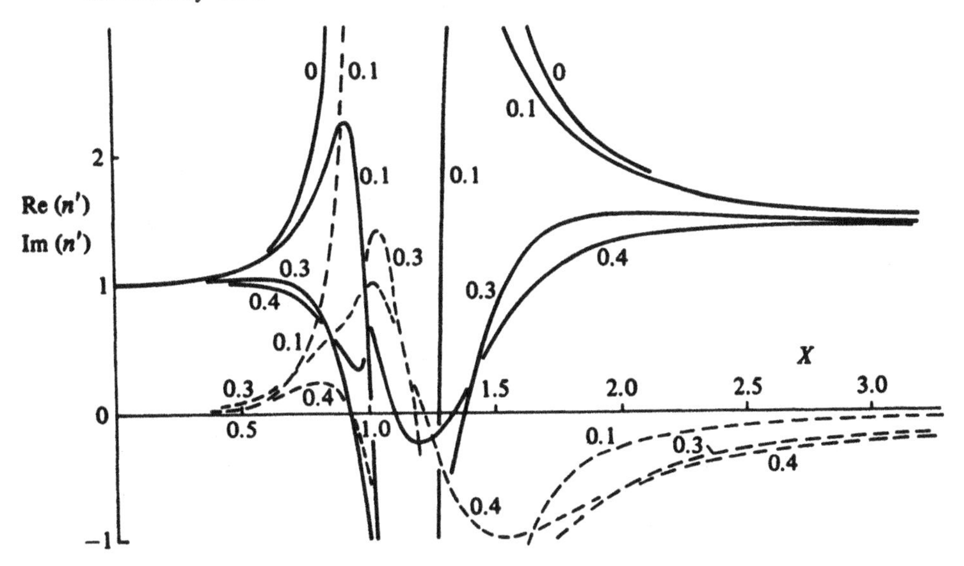

neglected, but the small difference can sometimes be important. For example in the vertical sounding of the ionosphere by radio pulses, the ionosonde technique § 1.7, it is sometimes necessary, when high accuracy is required, to allow for the effect of collisions on the group refractive index. This is especially true at lower frequencies because the effect depends on $Z = \nu/\omega$. It can be especially important at frequencies less than the electron gyro-frequency so that $Y > 1$.

To calculate μ' we first calculate n' and then take the real part. The formula (5.75) may still be used, where A and B are now complex numbers given by (4.66) and the complex n is given by (4.47). The factors (5.76) now have additional terms containing Z, as follows

$$\left.\begin{aligned} A - \tfrac{1}{4}\omega\partial A/\partial\omega &= 1 - \tfrac{3}{2}X - Y^2(\tfrac{3}{2} - 2X) - 2XY^2\sin^2\Theta + \tfrac{1}{2}Z^2(4X - 9) \\ &\quad + \tfrac{1}{4}\mathrm{i}Z(14X + 7Y^2 + 7Z^2 - 15) \\ B - \tfrac{1}{2}\omega\partial B/\partial\omega &= (1 - X)(1 - 3X - 6Z^2) - Y^2(2 - 3X) - \tfrac{3}{2}XY^2\sin^2\Theta \\ &\quad + \tfrac{1}{2}\mathrm{i}Z(20X - 7X^2 + 5Y^2 + 5Z^2 - 9) \\ -\tfrac{1}{4}\omega\partial C/\partial\omega &= -\tfrac{3}{2}X(1 - X)^2 - Y^2(\tfrac{1}{2} - X) \\ &\quad - \tfrac{3}{2}Z^2 + 3XZ^2 + \tfrac{3}{4}\mathrm{i}Z(6X - 5X^2 + Y^2 + Z^2 - 1). \end{aligned}\right\} \quad (5.85)$$

Figs. 5.21, 5.22 show some typical results, for the ordinary and extraordinary waves respectively. For further examples see Millington (1938a), Titheridge (1961b). The curves show how the real and imaginary parts of n' depend on X for a fixed frequency, and various values of Z. In this example Z_t, (4.27), is 0.3398. Thus the values 0.1, 0.3 of Z are less than Z_t and the curves in each figure are continuous, although some of them make large deviations near $X = 1$. By contrast the value $Z = 0.4$ is greater than Z_t. The curves for the ordinary wave when $X < 1$, fig. 5.21, are continuous with those for the extraordinary wave when $X > 1$, fig. 5.22, and vice

Fig. 5.22. The same as fig. 5.21, but for the extraordinary wave. The curve of Re(n') for $Z = 0$ is extremely close to that for $Z = 0.1$, and therefore only a part of it can be shown.

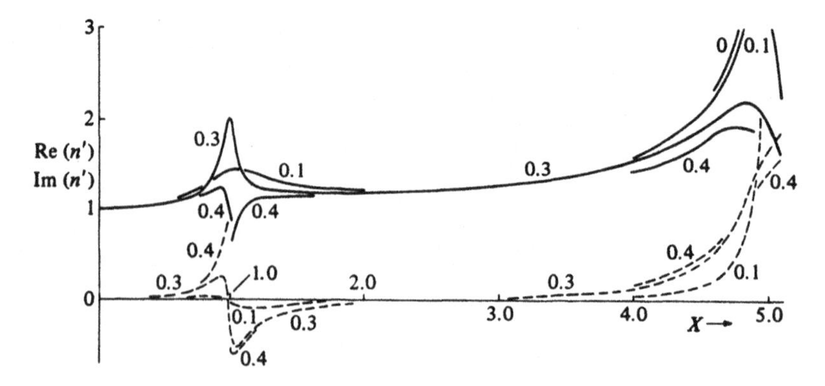

versa. This apparent discontinuity is simply the interchange of the names ordinary and extraordinary as described in §4.16.

When Z is equal to Z_t there is a coupling point on the real X axis at $X = 1$, as explained in §4.16. It can be shown that here both $\mathrm{Re}(n')$ and $\mathrm{Im}(n')$ are infinite. The proof is left to the reader (show that the denominator of (5.75) is zero).

PROBLEMS 5

5.1. The polar curves $n(\Theta)$ and $V(\beta)$ are the cross sections, by a plane through the axis, of a refractive index surface and its associated ray surface for a collisionless plasma. The angle between corresponding radii for the two curves is $\alpha = \Theta - \beta$. A Storey cone in either diagram may be defined as the cone $\Theta = \Theta_s$ or $\beta = \beta_s$ where $\partial\beta/\partial\Theta = 0$. Show that on a Storey cone $n\partial^2 n/\partial\Theta^2 - 2(\partial n/\partial\Theta)^2 - n^2 = 0$, and that this is the condition for a point of inflection of the refractive index surface. Show also that $\partial(1/V)/\partial\beta = n \sin\alpha/c$. Since this is bounded, the ray surface has a cusp at the Storey cone. If $\mathcal{U}$ is the group velocity show that $\partial\mathcal{U}/\partial\beta$ is infinite at a Storey cone. Hence show that a polar plot of $\mathcal{U}$ versus β does not have a cusp but a loop, at a Storey cone. (For examples see Walker, 1977a; Booker, 1975, 1984).

5.2. For the polar curves of problem 5.1 show that, at corresponding points, the product of the curvatures is $\cos^3\alpha/c$. Hence show that for $\Theta = \beta = 0$ in an electron plasma, the ray surface can have zero curvature only in the three conditions (a) $X = 1$ (What is then the value of V? See figs. 5.4, 5.5, 5.9, 5.10), (b) cut-off $X = 1 \pm Y$ (trivial because $V \to \infty$), and (c) $Y = 1$ (trivial because $V \to 0$). Show also that for $\Theta = \beta = \frac{1}{2}\pi$ the ray surface for the ordinary wave never has zero curvature, and the ray surface for the extraordinary wave has zero curvature only at cut-off $X = 1 \pm Y$ (trivial because $V \to \infty$) and at the upper hybrid frequency $X = 1 - Y^2$ (trivial because $V \to 0$).

5.3. For an electron plasma, there is an optic ray axis in the direction $\beta = 0$, so that the ray surfaces there have the form of two opposite cones with apices touching. Show that the wave normals make with the axis an angle Θ given by

$$\tan\Theta = \frac{\{P(1+Y)+1\}\{P(1-Y)+1\}(X-1)}{(P+1)\{P(1-Y^2-X)+1-X\}}$$

where

$$P = \pm\{(1-X)/(1-X-Y^2)\}^{\frac{1}{2}}.$$

(One method is to use equations (28), (11), (12) of Budden and Daniell, 1965). Alternatively it can be done by using (5.12), (5.33) with $\beta = 0$). Show in a similar way that when two ray surfaces cross on the plane $\beta = \frac{1}{2}\pi$, their wave normals make with the axis $\beta = 0$ an angle Θ given by

$$\cot^2\Theta = -\tfrac{1}{2} \pm \frac{Y^2 - 1 + X}{2(1-X)(1-Y^2)^{\frac{1}{2}}}.$$

5.4. Evaluate $(1/n)\,dn/d\Theta$ for the general collisionless plasma with ions included. Hence find the condition that $dn/d\Theta = 0$ when $\sin\Theta \neq 0$, $\cos\Theta \neq 0$.

5.5. Show that for a collisionless electron plasma at extremely low frequency ω, one of the values n of the refractive index is given by $n^2 \approx (\omega_N^2/\omega\omega_H)\sec\Theta$ where Θ is the angle between the wave normal and the superimposed magnetic field. Hence find the values of Θ for a given ray direction β, in the limit $\omega \to 0$. (Note that this is not realistic in practice because the ions have been ignored.)

5.6. In a homogeneous magnetoplasma a wave has ray velocity $V = c/(n\cos\alpha)$ and group velocity $\mathscr{U} = c/(n'\cos\alpha)$. ($\alpha$ and n' defined in (5.33) and (5.65) respectively.) Prove that $1/\mathscr{U} = \{(\partial/\partial\omega)(\omega/V)\}_\beta$ where the ray direction β is held constant for the differentiation. (Note that, when β is constant, both n and α change when ω changes.) Hence show that the two expressions $c/(n'\cos\alpha)$ and $|\partial\omega/\partial\boldsymbol{\kappa}|$ for $\mathscr{U}$ are the same.

5.7. Show that, for frequencies very close to cut-off, the angles Θ between the axis and the wave normal, and β between the axis and the ray, are related by $\tan\Theta = 2\tan\beta$.

5.8. Show that for a given ray direction β, when the frequency approaches a resonance frequency, $n'\cos\alpha \to \infty$ so that $\mathscr{U} \to 0$. (For the solution see Budden and Stott, 1980, § 4.3).

5.9. For a collisionless electron plasma, a family of refractive index surfaces is drawn for a sequence of values of X, with a common origin and axis. Prove that when $Y > 1$ the envelope of these surfaces intersects their common axis at an angle $\arctan\{2(Y-1)\}^{\frac{1}{2}}$.

5.10. For a collisionless electron plasma show that, for propagation with the wave normals parallel to the superimposed magnetic field, the two values of $n' = \partial(\omega n)/\partial\omega$ are $\{1 - \frac{1}{2}XY(1+Y)^{-2}\}/\{1 - X/(1+Y)\}^{-\frac{1}{2}}$ and the same expression with the sign of Y reversed. Show that the two values are equal when $X = \{2(3Y^2-1) \pm 2(1+Y^2)(2Y^4-2Y^2+1)^{\frac{1}{2}}\}/\{Y^2(3+Y^2)\}$. Show that this gives propagated waves (real n and n') only when $Y > 1$.

5.11. For longitudinal propagation in a collisionless electron plasma one of the refractive indices n is given by $n^2 = 1 - X/(1-Y)$. Show that the group refractive index n' is given by $nn' = 1 + \frac{1}{2}XY/(1-Y)^2$. Hence or otherwise show that for the whistler mode ($Y > 1$), when f_N is the plasma frequency and f_H is the electron gyro-frequency, n' has a minimum value when the frequency f satisfies

$$(f/f_H)^4 - (f/f_H)^3 - (f_N/f_H)^2(f/f_H - \tfrac{1}{4}) = 0.$$

Show that this gives $f/f_H \approx \frac{1}{4}$ when f_N/f_H is large, and that in all cases $f/f_H < \frac{1}{4}$.

5.12. Show that for an isotropic electron plasma the complex refractive index n and group refractive index n' satisfy $nn' = 1 + \frac{1}{2}\mathrm{i}XZ(1-\mathrm{i}Z)^{-2}$, Show also that $1/\mathrm{Re}(n) - \mathrm{Re}(n') = \frac{1}{8}XZ^2(3X-8)(2X-1)(1-X)^{-\frac{5}{2}} + O(Z^3)$.

6

Stratified media. The Booker quartic

6.1. Introduction

The earlier chapters have discussed the propagation of a plane progressive radio wave in a homogeneous plasma. It is now necessary to study propagation in a medium that varies in space, and this is the main subject of the rest of this book. The most important case is a medium that is plane stratified, and the later chapters are largely concerned with the earth's ionosphere that is assumed to be horizontally stratified. Because of the earth's curvature the stratification is not exactly plane, but for many purposes this curvature can be neglected. In cases of curved stratification it is often possible to neglect the curvature and treat the medium as locally plane stratified. Some cases where the earth's curvature is allowed for are discussed in §§ 10.4, 18.8.

The theory of this chapter is given in terms of the earth's ionosphere. A Cartesian coordinate system x, y, z is used with x, y horizontal and z vertically upwards. These coordinates are not in general the same as the x, y, z of chs. 2–4. The composition of the plasma, that is the electron and ion concentrations N_e, N_i are assumed to be functions of z only. For the study of radio propagation the frequency is usually great enough for the effect of the ions to be ignored. When radio waves are reflected in the ionosphere the important effects occur in a range of height that is small enough for the spatial variation of the earth's magnetic induction $\boldsymbol{B}$ to be ignored. The vector $\boldsymbol{Y}$ (3.22), antiparallel to $\boldsymbol{B}$, is therefore assumed to be constant in magnitude and direction and to have direction cosines l_x, l_y, l_z.

The origin $z = 0$ is usually taken to be at the ground. Below the ionosphere there is a region of free space which extends up to a height of about $z = 70$ km or more.

In the theory of the reflection of radio waves from the ionosphere it is usually assumed that the incident wave is a plane wave in the free space, and travelling obliquely upwards. In practice the incident wave is nearly always a spherical wave coming from a transmitter of small dimensions, but at a range of 70 km or more the curvature of the wave front is small so that the wave can be treated as plane to a first approximation. The validity of this is examined in § 11.14. Alternatively the incident wave can be expressed as an integral, (6.11) below, whose integrand represents a plane wave. It is thus represented as an 'angular spectrum of plane waves' (Booker and Clemmow, 1950; Clemmow, 1966; see also §§ 5.3, 10.2, 10.3, 11.14). The propagation of one plane wave component of the incident wave is first studied, and the properties for the spherical wave are then found by an integration. For details see §§ 11.14, 11.15. Here we shall give the theory for an incident plane wave.

6.2. The variable q

Let a plane wave be incident on the ionosphere from below, and in the free space let its wave normal have direction cosines S_1, S_2, C so that it makes an angle θ with the vertical, where $\cos\theta = C$, $\sin^2\theta = S_1^2 + S_2^2$, and each field component contains the factor

$$\exp\{-ik(S_1x + S_2y + Cz)\}. \tag{6.1}$$

Now imagine that the ionosphere is replaced by a number of thin discrete strata, in each of which the medium is homogeneous. By making these strata thin enough and numerous enough we may approximate as closely as we please to the actual ionosphere. The incident wave is partially reflected and partially transmitted at the lower boundary of the first stratum. The transmitted wave is partially reflected and transmitted at the second boundary, and so on. In any one stratum there are upgoing and downgoing waves which are the resultants of all the partially reflected and transmitted waves entering the stratum. Because the medium is doubly refracting, there are actually two obliquely upgoing and two obliquely downgoing waves. At each boundary the field components must satisfy some boundary conditions so that they must all depend on x and y in the same way on the two sides of each boundary. Hence the waves depend on x and y only through the factor

$$\exp\{-ik(S_1x + S_2y)\} \tag{6.2}$$

in all the strata.

In the rth stratum let the wave normal of one of the four resultant waves make an angle θ_r with the z axis, and let the refractive index for this wave be n_r. Then each field component contains the factor

$$\exp[-ik\{S_1x + S_2y + n_r(\cos\theta_r)z\}]. \tag{6.3}$$

Further

$$n_r \sin\theta_r = (S_1^2 + S_2^2)^{\frac{1}{2}} = \sin\theta \tag{6.4}$$

which may be regarded as the result of applying Snell's law of refraction in succession at each of the boundaries in the region between the rth stratum and the free space. Now n_r in general depends on the direction of the wave normal, that is on

S_1, S_2 and θ_r. The unknown angle θ_r could be found from (6.4) if n_r were known, but it depends on the angle Θ between the wave normal and $\boldsymbol{Y}$ and this in turn depends on θ_r. Hence (6.4) cannot immediately be used to give θ_r. To overcome this difficulty introduce the quantity

$$q = n_r \cos\theta_r \tag{6.5}$$

so that

$$n_r^2 = S_1^2 + S_2^2 + q^2. \tag{6.6}$$

The expression (6.3) now becomes

$$\exp\{-ik(S_1 x + S_2 y + qz)\} \tag{6.7}$$

and it appears as a factor in all field quantities of the wave being considered. There are in general four such waves, and it is shown in the following section that the four values of q are the roots of a quartic equation, the Booker quartic.

The total field in the stratum is the sum of four terms like (6.7) with the common factor (6.2) but with different values of q and with amplitudes that do not depend on x and y. In any one stratum these amplitudes do not depend on z, but they change in going from one stratum to another because of the partial reflections and transmissions at the boundaries. In the limit of infinitesimally thin strata, that is in a continuous stratified medium, the four amplitudes depend on z but not on x and y.

For the special case where the medium is isotropic, n_r is independent of θ_r and is known at all heights z, from §4.2. Then (6.6) can be used at once to give q. The four values of q then consist of two equal pairs given by

$$q = \pm(n_r^2 - S_1^2 - S_2^2)^{\frac{1}{2}}. \tag{6.8}$$

Objections are sometimes raised to the assertion that the ionosphere is equivalent to the system of discrete homogeneous strata, perhaps because, however thin and numerous we make the strata, there are discontinuities at their boundaries that are not present in the actual ionosphere. This difficulty can be overcome by using the differential equations (2.45), in which the variables x, y, z are chosen to be the same as in §6.1. The electric displacement is $\boldsymbol{D} = \varepsilon_0 \boldsymbol{\varepsilon} \boldsymbol{E}$ and the elements of $\boldsymbol{\varepsilon}$ are functions of z only. To find solutions of this set of six partial differential equations the method of 'separation of the variables' may be used. We seek solutions in which each of the six dependent variables E_x, E_y, E_z, $\mathscr{H}_x$, $\mathscr{H}_y$, $\mathscr{H}_z$ is a product of the form

$$E_x = \mathscr{X}(x)\mathscr{Y}(y)\mathscr{Z}(z) \tag{6.9}$$

with similar expressions for the other five variables. These are substituted into (2.45). It is then clear that the factors $\mathscr{X}(x)$ and $\mathscr{Y}(y)$ can be cancelled from the equations, provided that the functions $\mathscr{X}(x)$ and $\mathscr{Y}(y)$ are the same for all six variables, and that

$$\frac{1}{\mathscr{X}}\frac{d\mathscr{X}}{dx} = \text{constant}; \quad \frac{1}{\mathscr{Y}}\frac{d\mathscr{Y}}{dy} = \text{constant}. \tag{6.10}$$

In the differential equations that remain, the variables and the coefficients are functions only of z. These equations are given and discussed in § 7.13, where it is shown how they lead to the Booker quartic.

The two constants in (6.10) are called 'separation constants', and they are the same as $-ikS_1$, $-ikS_2$ in (6.2), (6.3), (6.7). For each plane $z =$ constant, the fields may be expressed as a double Fourier integral thus

$$E_x(x, y, z) = \iint \check{E}_x(S_1, S_2, z) \exp\{-ik(S_1 x + S_2 y)\} \mathrm{d}S_1 \mathrm{d}S_2 \tag{6.11}$$

with similar expressions for the other five field components. The process of separating the variables, described above only in outline, shows that the differential equations for any set of Fourier amplitudes $\check{E}_x(S_1, S_2, z)$ etc., for given S_1, S_2, are independent of those for any other pair S_1, S_2. The use of the incident plane wave (6.1) is equivalent to studying just one of the Fourier amplitudes. This technique is so widely used that the Fourier interpretation is rarely mentioned. The field components are written simply as $E_x(z)$ etc. instead of $\check{E}_x(S_1, S_2, z)$ etc. The functional dependence on S_1 and S_2 is omitted, and the spatial dependence factor (6.2) is also omitted in the same way as the time factor $\exp(i\omega t)$ is omitted (§ 2.5). These conventions are used in this book, except where otherwise stated.

6.3. The Booker quartic. Derivation

Consider again one of the four plane waves with the properties (6.5), (6.6), in the rth stratum of § 6.2. The directions cosines of its wave normal are

$$S_1/n_r,\ S_2/n_r,\ q/n_r \tag{6.12}$$

so that the angle Θ between the wave normal and the vector $\boldsymbol{Y}$ (direction cosines l_x, l_y, l_z) is given by

$$n_r \cos\Theta = (l_x S_1 + l_y S_2 + l_z q). \tag{6.13}$$

Now (6.6) and the square of (6.13) are substituted into the form (4.62) or (4.65) of the dispersion relation. This gives an equation for finding q, and inspection shows that is a quartic equation.

We use, first, the form (4.65) for an electron plasma. It gives

$$\begin{aligned}
&(q^2 + S_1^2 + S_2^2)^2\{U^2(U - X) - UY^2\} \\
&+ (q^2 - C^2)(l_x S_1 + l_y S_2 + l_z q)^2 XY^2 \\
&- (q^2 + S_1^2 + S_2^2)\{2U(U - X)^2 - 2Y^2(U - X) - XY^2\} \\
&+ (U - X)^3 - Y^2(U - X) = 0
\end{aligned} \tag{6.14}$$

where the property $1 - S_1^2 - S_2^2 = C^2$ has been used. This may be written

$$F(q) \equiv \alpha q^4 + \beta q^3 + \gamma q^2 + \delta q + \varepsilon = 0 \tag{6.15}$$

where

$$\left.\begin{aligned} \alpha &= U(U^2 - Y^2) + X(Y^2 l_z^2 - U^2), \\ \beta &= 2(l_x S_1 + l_y S_2) l_z X Y^2, \\ \gamma &= -2U(U - X)(C^2 U - X) + 2Y^2(C^2 U - X) \\ &\quad + X Y^2(1 - l_z^2 C^2 + l_x^2 S_1^2 + l_y^2 S_2^2), \\ \delta &= -2C^2(l_x S_1 + l_y S_2) l_z X Y^2, \\ \varepsilon &= (U - X)(C^2 U - X)^2 - C^2 Y^2 (C^2 U - X) - (l_x S_1 + l_y S_2)^2 C^2 X Y^2. \end{aligned}\right\} \quad (6.16)$$

The equations (6.15), (6.16) were given by Booker (1936, 1939, 1949) in three papers which are about the most important papers in the whole of the theory of radio wave propagation. The notation used here is almost the same as Booker's original notation. The coefficients α, β must not be confused with the angles α, β used in ch. 5.

A similar process can be used with (4.62) for the more general plasma when ions are allowed for, or when the electron collision frequency is velocity dependent, for example with the Sen–Wyller formulae, § 3.12. The notation is as in § 3.10 including (3.51) for G and J. We use $S^2 = S_1^2 + S_2^2$. Then (6.16) is replaced by

$$\left.\begin{aligned} \alpha &= \tfrac{1}{2}(\varepsilon_1 + \varepsilon_2) - l_z^2 G, \\ \beta &= -2(l_x S_1 + l_y S_2) l_z G, \\ \gamma &= -\tfrac{1}{2}\varepsilon_3(\varepsilon_1 + \varepsilon_2) - \varepsilon_1\varepsilon_2 + S^2(\varepsilon_1 + \varepsilon_2 - l_z^2 G) - l_z^2 J - (l_x S_1 + l_y S_2)^2 G, \\ \delta &= -2(l_x S_1 + l_y S_2) l_z (J + S^2 G), \\ \varepsilon &= \{\varepsilon_1\varepsilon_2 - \tfrac{1}{2}(\varepsilon_1 + \varepsilon_2) S^2\}(\varepsilon_3 - S^2) - (l_x S_1 + l_y S_2)^2 (J + S^2 G). \end{aligned}\right\} \quad (6.17)$$

When the values (3.50) for the electron plasma are used in (6.17), and the coefficients are all multiplied by $U(U^2 - Y^2)$, it can be shown that they are then the same as (6.16).

There are several other methods of deriving the Booker quartic. An important method is given in § 7.13. Another method of some interest is worth mentioning briefly. It has the slight advantage that it makes direct use of Maxwell's equations and does not use the dispersion relation. It is required to find q for one of the four waves, in the general stratum of § 6.2, whose fields have the spatial dependence (6.7), so that for all field variables

$$\frac{\partial}{\partial x} \equiv -ikS_1, \quad \frac{\partial}{\partial y} \equiv -ikS_2, \quad \frac{\partial}{\partial z} \equiv -ikq. \quad (6.18)$$

Then Maxwell's equations (2.44) may be written in matrix notation

$$\mathbf{\Gamma E} = \boldsymbol{\mathscr{H}}, \quad \mathbf{\Gamma}\boldsymbol{\mathscr{H}} = -(\mathbf{1} + \mathbf{M})\mathbf{E}, \quad (6.19)$$

compare (4.70), where the matrix $\mathbf{\Gamma}$ is

$$\begin{pmatrix} 0 & -q & S_2 \\ q & 0 & -S_1 \\ -S_2 & S_1 & 0 \end{pmatrix}. \quad (6.20)$$

In (6.19) **1** is the unit 3 × 3 matrix and $\mathbf{1}+\boldsymbol{M}$ is the permittivity expressed in the coordinate system of § 6.1. For an electron plasma the susceptibility matrix $\boldsymbol{M}$ is given by (3.35). Elimination of $\mathscr{H}$ gives

$$(\boldsymbol{\Gamma}^2+\boldsymbol{M}+\mathbf{1})\boldsymbol{E}=0. \tag{6.21}$$

If this set of three equations is to have a self-consistent solution $\boldsymbol{E}$, it is necessary that the determinant of the matrix shall be zero. This gives

$$\begin{vmatrix} 1-q^2-S_2^2+M_{xx} & S_1S_2+M_{xy} & S_1q+M_{xz} \\ S_1S_2+M_{yx} & 1-q^2-S_1^2+M_{yy} & S_2q+M_{yz} \\ S_1q+M_{zx} & S_2q+M_{zy} & C^2+M_{zz} \end{vmatrix}=0. \tag{6.22}$$

When this is multiplied out and (3.35) is used, it gives the Booker quartic equation (6.15), (6.16).

6.4. Some properties of the Booker quartic

The Booker quartic is a generalised form of the dispersion relation for a cold magnetoplasma. It was obtained in § 6.3 from the dispersion relation by rotating the Cartesian axes in refractive index space, so that the axis of $n_z=q$ is perpendicular to a given fixed plane. This plane is usually chosen to be parallel to the strata in a medium such as the ionosphere. The numbers S_1, S_2, q are the Cartesian components of the vector $\boldsymbol{n}$. The quartic is therefore the equation of the system of two refractive index surfaces in these new axes.

This choice of axes still leaves us free to rotate the axis system about the n_z axis. The components $n_x=S_1$ and $n_y=S_2$ were determined by the incident plane wave (6.1) in free space. It is nearly always convenient to choose the x and y axes so that the wave normal of this wave is in the x–z plane and therefore $S_2=0$. This is done for nearly all the problems discussed in this book. We also write S for S_1, so that $C^2 = 1-S^2$. The coefficients (6.16) for an electron plasma then become

$$\left.\begin{aligned} \alpha &= U(U^2-Y^2)+X(Y^2l_z^2-U^2), \\ \beta &= 2l_xl_zSXY^2, \\ \gamma &= -2U(U-X)(C^2U-X)+2Y^2(C^2U-X)+XY^2(1-l_z^2C^2+l_x^2S^2), \\ \delta &= -2l_xl_zSC^2XY^2, \\ \varepsilon &= (U-X)(C^2U-X)^2-C^2Y^2(C^2U-X)-l_x^2S^2C^2XY^2. \end{aligned}\right\} \tag{6.23}$$

Similar results for the more general plasma can be derived from (6.17). Some authors, for example Pitteway (1965) have preferred to retain S_1, S_2 and to choose the axes so that $l_y=0$. This means that the vector $\boldsymbol{Y}$ is in the x–z plane. The resulting formulae can easily be derived from (6.16) or (6.17) but are not needed in this book.

To illustrate the properties of the quartic it is useful first to ignore collisions as was done when studying the refractive indices in §§ 4.11, 4.12, 5.2. Then $U=1$ and the coefficients (6.23) are

$$\left.\begin{aligned}
\alpha &= 1 - X - Y^2 + XY^2 l_z^2,\\
\beta &= 2l_x l_z SXY^2,\\
\gamma &= -2(1-X)(C^2-X) + 2Y^2(C^2-X) + XY^2(1 - l_z^2C^2 + l_x^2S^2),\\
\delta &= -2l_x l_z SC^2XY^2,\\
\varepsilon &= (1-X)(C^2-X)^2 - C^2Y^2(C^2-X) - l_x^2S^2C^2XY^2.
\end{aligned}\right\} \tag{6.24}$$

Some typical curves of q versus X for this case are given in §§6.7–6.9.

In some problems S, C, X can take complex values but here we consider only the simple situation where they are all real. Then the coefficients (6.24) are all real. The four roots of the quartic (6.15) are then either (a) all real, or (b) two are real and two form a conjugate complex pair, or (c) the roots are in two conjugate complex pairs.

This property can be related to the geometry of the refractive index surfaces. Imagine that in a refractive index space with axes n_x, n_y, n_z parallel to the axes x, y, z of § 6.1, the two refractive index surfaces are drawn. Fig. 6.1 shows a cross section of these surfaces by the plane $n_y = 0$. Now let a line be drawn parallel to the n_z axis with $n_x = S$, $n_y = 0$. The values of n_z where it cuts the refractive index surfaces are the four roots q of the quartic. Three possible positions for this line are shown in fig. 6.1, for the three cases (a), (b), (c). For case (a) the line may cut both refractive index surfaces

Fig. 6.1. Cross section by the plane $n_y = 0$ of the two refractive index surfaces. Each of the three lines (a), (b), (c) is for the fixed values $n_x = S$, $n_y = 0$. The points where one of these lines cuts the surfaces give the values of $q = n_z$. The radius from the origin has the direction of the wave normal and its length is the refractive index n. The arrows show the projections on to the plane of the diagram of the normals to the surfaces, that is the ray directions, at the points of intersection. This example is typical of a loss-free electron plasma with $Y < 1$, $X < 1 - Y$.

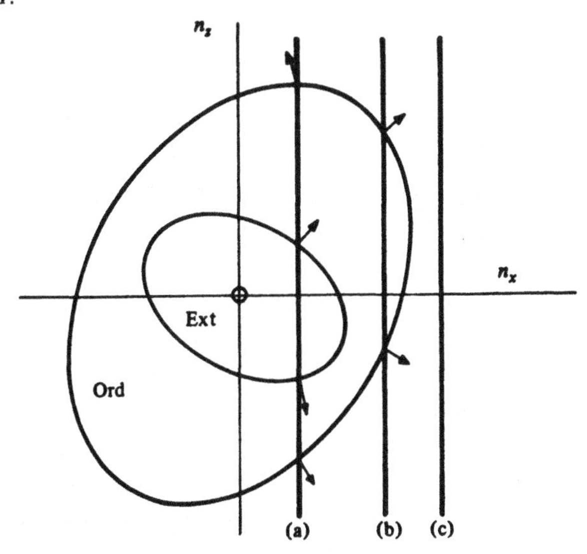

at two real points each, as shown in the figure, or it may cut one refractive index surface at four real points and not cut the other at all. This last possibility can occur, for example, when one refractive index surface has points of inflection as in figs. 5.10 or 5.11. In case (b) the line cuts only one of the refractive index surfaces, at two real points. In case (c) the line does not cut either refractive index surface at real points.

In case (a) imagine that the outward normals to the refractive index surface are drawn at the four real points of intersection. These normals give the directions of the four rays, which are also the directions of energy flux (§ 5.3). They are not in the plane of the diagram except in special cases. Their projections are shown in fig. 6.1. It is clear from the form of the refractive index surfaces, figs. 5.4–5.13, that two of these normals must make acute angles with the n_z axis and the other two must make obtuse angles with it. Thus two of the four waves are obliquely upgoing and the other two are obliquely downgoing. When two of the qs form a complex conjugate pair, it can be shown, see § 7.14(8), that for both the associated waves the component Π_z of the energy flux is zero. The wave for which $\mathrm{Im}(q)$ is negative is attenuated as z increases. If now we imagine that the collision frequency is given a very small value, energy must be absorbed from the wave at a rate H, say, per unit volume per unit time, and therefore Π_z must now have a small positive value so that

$$d\Pi_z/dz = 2k\,\mathrm{Im}(q)\Pi_z = -H. \tag{6.25}$$

This wave is therefore an upgoing wave. The other wave has positive $\mathrm{Im}(q)$ and is attenuated as z decreases. By a similar argument it must be a downgoing wave. Thus in all cases two of the four qs correspond to upgoing waves and two to downgoing waves. This result is still true when the collision frequency is large, provided that S is real. The proof follows.

If the reference line used in fig. 6.1 just touches one of the refractive index surfaces, the two points of intersection have moved together so that the quartic has two equal roots. This is associated with reflection of the wave and is described in detail in § 10.6.

When Z is not zero, so that the form (6.23) of the coefficients is used, no root of the quartic can be real, when S and X are real. For if q is real, (6.5), (6.6) show that n_r and Θ_r are both real. But it was shown in § 4.13 that when the wave normal is in a real direction the refractive index cannot be purely real unless $Z = 0$.

When Z is not zero and S, X are real, two of the roots of the quartic must have negative imaginary parts and the other two must have positive imaginary parts. This can be proved as follows. Consider some set of fixed values of X, U, Y, l_x, l_y, l_z for which the coefficient α is not zero. Starting with $S = 0$, let S be slowly increased (so that C^2 decreases). No root of the quartic can be real for any real value of S, so the imaginary parts of the roots cannot change sign. Now when $S = 0$, so that $q = n_z = n$, the four values of q are the four values of n given by the dispersion relation, and n^2 always has a negative imaginary part (§ 4.13). Hence two values of n have negative imaginary parts, for the upgoing waves, and two have positive imaginary parts, for the downgoing waves. This must also apply for other real values of S and therefore always applies to the four roots of the quartic.

A wave for which $\mathrm{Im}(q)$ is negative is attenuated as z increases, and it is an (obliquely) upgoing wave. Similarly when $\mathrm{Im}(q)$ is positive, the wave is downgoing. It will be shown later that for an upgoing wave, $\mathrm{Re}(q)$ may be either positive or negative. A negative value of $\mathrm{Re}(q)$ means that the wave fronts are travelling obliquely downwards. This apparent contradiction arises because the directions of the wave normal and the ray are in general different (§ 5.3).

When Z is zero some of the roots q may be real. To decide whether a given real q refers to an upgoing or downgoing wave, it is sometimes convenient to give Z a small non-zero value and then to examine the sign of $\mathrm{Im}(q)$. If it is negative, the wave is an upgoing wave.

In some physical problems a complex value of S is used. This occurs, for example, in the study of guided waves by the layers of a stratified system. It is then no longer necessarily true that two waves are upgoing and two are downgoing. It is possible to find a complex S such that three waves are upgoing and one downgoing, or three downgoing and one upgoing. But the subject of guided waves is beyond the scope of this book.

One root of the quartic is infinite when the coefficient $\alpha = 0$. For the collisionless electron plasma, (6.24), this gives

$$X = X_{\infty} \equiv (1 - Y^2)/(1 - Y^2 l_z^2) \tag{6.26}$$

which is independent of S, C, that is of the angle of incidence Θ. The remaining solutions are now given by a cubic equation for q, with real coefficients when the plasma is loss-free. Then at least one of these solutions must be real.

One root of the quartic is zero when the coefficient $\varepsilon = 0$. From (6.23) or (6.24) this gives a cubic equation for X. Its solutions for the collisionless case (6.24) have been studied by Booker (1939), Budden (1961a). When $S^2 \leqslant 1$, $l_x^2 \leqslant 1$ the solutions X are all real. By studying the function $\varepsilon(X)$ it can be shown that one solution is in the range $X \geqslant 1$, one is in the range $C^2 \leqslant X \leqslant 1$, and one in the range $X < C^2$. This last solution is positive if $Y^2 < 1$.

It must be remembered that the levels where $q = 0$ are not necessarily levels of reflection; see § 10.6.

When $X = 1$, the coefficients (6.24) all contain Y^2 as a factor. This may be cancelled from the quartic, which then becomes

$$\begin{aligned}(1 - l_z^2)q^4 - 2Sl_x l_z q^3 + (S^2 - C^2 + C^2 l_z^2 - S^2 l_x^2)q^2 + 2SC^2 l_x l_z q \\ - S^2C^2(1 - l_x^2) = 0.\end{aligned} \tag{6.27}$$

Hence at the level where $X = 1$, the solutions q depend on the direction of the earth's magnetic field, but not on its magnitude. Equation (6.27) factorises and its four solutions are

$$q = \pm C, \quad q = S(l_x l_z \pm \mathrm{i} l_y)/(1 - l_z^2). \tag{6.28}$$

The last two are real only when $l_y = 0$, that is for propagation from magnetic north to south or south to north. Then the quartic has a double root equal to Sl_z/l_x, since $1 - l_z^2 = l_x^2$.

It is useful to display the solutions of the quartic as curves showing how q depends on X; see §§ 6.7–6.9. Since the quartic (6.15) is valid for all values of X, we must have $\mathrm{d}F/\mathrm{d}X = 0$, which gives

$$\frac{\partial F}{\partial q}\frac{\partial q}{\partial X} + \frac{\partial \alpha}{\partial X}q^4 + \frac{\partial \beta}{\partial X}q^3 + \frac{\partial \gamma}{\partial X}q^2 + \frac{\partial \delta}{\partial X}q + \frac{\partial \varepsilon}{\partial X} = 0 \tag{6.29}$$

whence

$$\frac{\partial q}{\partial X} = -\left(\frac{\partial \alpha}{\partial X}q^4 + \frac{\partial \beta}{\partial X}q^3 + \frac{\partial \gamma}{\partial X}q^2 + \frac{\partial \delta}{\partial X}q + \frac{\partial \varepsilon}{\partial X}\right)\Bigg/\frac{\partial F}{\partial q}. \tag{6.30}$$

The denominator $\partial F/\partial q$ is zero where the quartic has a double root. In general the numerator is not also zero so that $\partial q/\partial X$ is infinite. Then the curves of q vs X have a vertical tangent where there is a double root. For examples see figs. 6.2, 6.4–6.7.

The numerator of (6.30) can also be zero at a double root, in four special cases. When $X = 1$, $l_y = 0$ there is a double root given by the second equation (6.28), where $q = Sl_z/l_x$, and the numerator of (6.30) is then

$$\{Y^2(S^2 - l_x^2)^2 - S^4\}/l_x^4. \tag{6.31}$$

This is zero if

$$S = \pm l_x\{Y/(Y \pm 1)\}^{\frac{1}{2}}, \tag{6.32}$$

which gives four values of S. Only two of them are real if $Y < 1$. They are the values of S for which the reference line of fig. 6.1 passes through one of the four window points. For a double root at $X = 1$ where (6.32) is satisfied, two curves of q vs X intersect, and their tangents are not vertical. There is an example in fig. 6.6(c). The condition (6.32) is also related to the phenomena of the 'Spitze', § 10.9, and of radio windows, §§ 17.6–17.9.

It follows from (6.28) that three roots of the quartic are equal at the level $X = 1$ if $l_y = 0$ and $C = Sl_z/l_x$. This means that the wave normal of the incident wave (6.1) in free space is either parallel or antiparallel to the earth's magnetic field. It can be shown that for this condition the numerator of (6.30) is not zero. For three roots to be equal it is necessary that both $\partial F/\partial q$ and $\partial^2 F/\partial q^2$ are zero. Thence it can be shown, by multiplying (6.29) by $\partial X/\partial q$ and then differentiating with respect to q, that $\partial^2 X/\partial q^2$ is zero. Hence the curve of q vs X has a vertical tangent and a point of inflection where $X = 1$. There is an example in fig. 6.6(e).

For the illustrative examples given later in this chapter it is assumed that: (a) l_z is positive. Thus the results apply for the northern hemisphere. (b) S is positive, and l_x, when non-zero, is positive. This means that the wave normal of the incident wave (6.1) makes an acute angle with the vector $\mathbf{Y}$, and the wave normal of any reflected wave makes an obtuse angle with $\mathbf{Y}$.

If the sign of either S or l_x is reversed, the only effect is to reverse the signs of the coefficients β, δ in (6.23) or (6.24). The four solutions of the resulting quartic are the same as before except that all four signs are reversed. Similarly if the sign of l_z is reversed, the only effect is to reverse the signs of β and δ, with consequent sign reversal of all four qs. Thus all cases of interest are covered by studying positive values of S, l_x, l_z.

6.5. Some special cases of the Booker quartic

When the coefficients β and δ in (6.23) or (6.24) are both zero, the quartic reduces to a quadratic equation for q^2. This happens when $X = 0$ so that the quartic is applied to a region of free space. It is then easy to show that the solutions are in two equal pairs

$$q = \pm C. \tag{6.33}$$

It happens in three other important special cases (Booker, 1939). The first is when $S = 0$ so that the waves are vertically incident on the ionosphere. Then $q = n_z = n$ and the quartic is simply the dispersion relation (4.65). Its solutions are the four values of n and their properties have been discussed in chs. 4, 5.

The second case is when $l_x = 0$, which means that the earth's magnetic field is in the y–z plane. The plane containing the wave normal is the x–z plane so that the waves travel from magnetic east to west or west to east. The solutions of the quadratic are then

$$q^2 = \frac{(C^2U - X)\{U(U-X) - Y^2\} - \frac{1}{2}XY^2(1 - C^2l_z^2) \pm XY\{\frac{1}{4}Y^2(1 - C^2l_z^2)^2 + l_z^2(C^2U - X)(U - X)\}^{\frac{1}{2}}}{U(U^2 - Y^2) + X(Y^2l_z^2 - U^2)}$$

$$= C^2 - \frac{X(U - X)}{U(U-X) - \frac{1}{2}Y^2(1 - C^2l_z^2) \pm Y\{\frac{1}{4}Y^2(1 - C^2l_z^2)^2 + l_z^2(C^2U - X)(U - X)\}^{\frac{1}{2}}}. \tag{6.34}$$

Some of their properties, for $U = 1$, are illustrated in §6.7.

The third special case is when $l_z = 0$ so that the earth's magnetic field is horizontal. This therefore applies for propagation at the magnetic equator. The solutions of the quadratic are then

$$q^2 = \frac{(C^2U - X)\{U(U-X) - Y^2\} - \frac{1}{2}XY^2(1 + l_x^2S^2) \pm XY\{\frac{1}{4}Y^2(1 - l_x^2S^2)^2 + l_x^2S^2U(U - X)\}^{\frac{1}{2}}}{U^2(U - X) - UY^2}$$

$$= C^2 - \frac{X(U - X)}{U(U-X) - \frac{1}{2}Y^2(1 - l_x^2S^2) \pm Y\{\frac{1}{4}Y^2(1 - l_x^2S^2)^2 + l_x^2S^2U(U - X)\}^{\frac{1}{2}}}. \tag{6.35}$$

If in addition $l_x = 0$ so that propagation is east–west or west–east at the magnetic equator, the solutions (6.35) become

$$q^2 = C^2 - \frac{X}{U}, \quad q^2 = C^2 - \frac{X(U-X)}{U(U-X)-Y^2}. \tag{6.36}$$

The first is independent of Y, which means that the waves are unaffected by the earth's magnetic field. It can be shown that in this case the electric field is horizontal and parallel to the y axis, so that the electrons are moved only in directions parallel to the earth's magnetic field, and are unaffected by it. The wave is an ordinary wave.

The solutions (6.35) are not discussed further here. For a detailed discussion see Booker (1939), Chatterjee (1952).

6.6. The discriminant of the Booker quartic

The condition that two roots of the quartic shall be equal is very important. It may be associated with reflection, § 10.6 or with coupling between two of the four waves, chs. 16, 17. The condition is that

$$\partial F/\partial q = 4\alpha q^3 + 3\beta q^2 + 2\gamma q + \delta = 0. \tag{6.37}$$

From this equation and the quartic (6.15), q may be eliminated to give a condition that uses only the five coefficients α to ε. The details are given in books on algebra. See, for example, Barnard and Child (1936), or Burnside and Panton (1912). The result is

$$\Delta = 0 \tag{6.38}$$

where

$$\Delta = 4\mathscr{I}^3 - \mathscr{J}^2; \quad \mathscr{I} = 12\alpha\varepsilon - 3\beta\delta + \gamma^2,$$

$$\mathscr{J} = 72\alpha\gamma\varepsilon + 9\beta\gamma\delta - 27(\alpha\delta^2 + \varepsilon\beta^2) - 2\gamma^3. \tag{6.39}$$

Here Δ is called the discriminant of the quartic. It can also be written as a determinant

$$\Delta = \frac{-27}{4}\begin{vmatrix} 4\alpha\varepsilon + 2\beta\delta - \gamma^2 & 6\beta\varepsilon - \gamma\delta & 6\gamma\delta - \beta\gamma \\ 6\beta\varepsilon - \gamma\delta & 8\gamma\varepsilon - 3\delta^2 & 16\alpha\varepsilon - \beta\delta \\ 6\alpha\delta - \beta\gamma & 16\alpha\varepsilon - \beta\delta & 8\alpha\gamma - 3\beta^2 \end{vmatrix}. \tag{6.40}$$

The four roots q of the quartic are functions of the height z because the coefficients α to ε depend on X and Z which are functions of z. For many purposes it is useful (see §§ 7.19, 8.21, 16.5, 16.7) to allow z and thence X, Z, to take complex values. Points in the complex z plane where (6.38) is satisfied are called 'coupling points'. They include 'reflection points' because reflection is a special case of coupling. These points have their greatest influence on radio propagation when they are on or near the real z axis.

The solutions of (6.38) for real X and Z were studied by Pitteway (1959), who used the form (6.40) for Δ. He showed that Δ has a factor X^4. Thus there are four coincident coupling points at the origin of the complex X plane. Here the medium is free space and the roots q are equal in two pairs (6.33). These coupling points are important in the study of limiting polarisation, § 17.10, 17.11. The remaining factor of Δ is of degree eight in X so that there are eight other coupling points of possible interest in the complex X plane. They were studied by Jones and Foley (1972) and by Smith, M.S. (1974a) who used real Z but complex X. They are discussed in § 16.5.

For vertical incidence, $S = 0$, the four roots q have the values $\pm n_{\mathrm{O}}$, $\pm n_{\mathrm{E}}$ where $n_{\mathrm{O}} n_{\mathrm{E}}$ are the refractive indices of the ordinary and extraordinary waves, respectively. One coupling point is where $n_{\mathrm{O}} = 0$ and is associated with the condition $X = U$ of (4.74), for reflection of the ordinary wave. It is labelled O in Jones and Foley's (1972) and in Smith's (1974a) diagrams, and in figs. 6.2, 6.4, 6.5. Two are where $n_{\mathrm{E}} = 0$ and they correspond to the conditions $X = U \pm Y$ in (4.74) for reflection of the extraordinary wave. They are labelled $\mathrm{E}_+, \mathrm{E}_-$ respectively in Smith's diagrams. A fourth coupling point is also a reflection point labelled R. It coincides with the resonance (4.76) when $S = 0$. The remaining four coupling points are labelled $\mathrm{C}_1, \mathrm{C}_2, \mathrm{C}_3, \mathrm{C}_4$ in Smith's diagrams. They are where the square root in (4.47) or (4.67) is zero, that is where

$$X = U \pm \tfrac{1}{2}\mathrm{i} Y \sin^2 \Theta / \cos \Theta. \tag{6.41}$$

This should be compared with (4.27) where a real value of Θ was used. Here at C_i, $i = 1$ to 4, q is in general complex so that (6.41) uses a complex Θ. Each sign in (6.41) has two different C_is and Θs, one for upgoing waves and one for downgoing. This explains why there are four C_is.

When S changes, the coupling points move in the complex X plane, and those that are coincident move apart. But they can retain the labels assigned for vertical incidence, and can thus be identified. Their behaviour, when S takes successively greater real values, is fully described in Smith's paper (1974a). In later chapters the properties of these reflection and coupling points are referred to as they are needed. See § 16.5 and fig. 16.1

6.7. The Booker quartic for east–west and west–east propagation

To illustrate the properties of the solutions q for east–west or west–east propagation it is useful to plot curves of q^2 against X for the special case where collisions are neglected as was done for n^2 in §§4.11–4.12. Then $U = 1$ and the solutions (6.34) are

$$q^2 = \frac{(C^2 - X)(1 - X - Y^2) - \tfrac{1}{2} X Y^2 (1 - C^2 l_z^2) \pm X Y \{\tfrac{1}{4} Y^2 (1 - C^2 l_z^2)^2 + l_z^2 (C^2 - X)(1 - X)\}^{\frac{1}{2}}}{1 - Y^2 - X(1 - Y^2 l_z^2)}. \tag{6.42}$$

This is to be studied for real values of l_z and of C.

If the wave (6.1) is vertically incident, $C = 1$ and $S = 0$. Then $l_z = \cos\Theta$ where Θ is the angle between the wave normal (vertical) and the vector $\mathbf{Y}$. The square root in (6.42) is the same as that in (4.67) with $U = 1$. It can never be zero for any real value of X (§ 5.2 item (5)). But if $C < 1$, the square root in (6.42) is not the same as in (4.67). It is now zero when

$$X^2 - X(1 + C^2) + C^2 + \tfrac{1}{4}Y^2(1 - C^2 l_z^2)/l_z^2 = 0 \tag{6.43}$$

and this has two real solutions X if

$$l_z^2 S^4 > Y^2(1 - C^2 l_z^2)^2 \tag{6.44}$$

(Booker, 1939). A case where this is satisfied is shown in fig. 6.2. When X is between the two solutions of (6.43) the square root is purely imaginary and the two values of q^2 are complex conjugates. Their real and imaginary parts are shown in fig. 6.2. If X is outside this range, both values of q^2 are real.

When the two values of q^2 are equal, that is where (6.43) is satisfied, they may be positive or negative. If they are positive they must occur for two coincident points on the same refractive index surface. For (6.6) shows that the two values of n^2 are then equal for the same real direction of the wave normal. It was shown in § 5.2 item (5) that the two refractive index surfaces cannot intersect. In the example of fig. 6.2, q^2 is negative at both values of X that satisfy (6.43).

When q^2 is positive the wave normal has a real direction (6.12) and the waves can be identified as ordinary or extraordinary, as indicated in fig. 6.1. But when q^2 is negative, q is imaginary and the wave normal direction (6.12) is complex. There can then be an ambiguity in the terms 'ordinary' and 'extraordinary' similar to the one that occurs when collisions are allowed for, as described in § 4.16. Cases could arise

Fig. 6.2. Shows how q^2 depends on X for propagation from magnetic east to west, or west to east. In this example $Y = \tfrac{1}{2}$, $l_z = 0.91868$, $C = 0.9$ and electron collisions are neglected. One solution is infinite where $X = 0.945$. The square root in the solution (6.42) is zero where $X = 0.865$ and 0.945. When X is between these values the two solutions are complex conjugates. The real part of q^2 is shown as a broken line and the imaginary part as a dotted line. The coupling points E_-, E_+, O, $C_1 - C_4$ are explained in § 6.6.

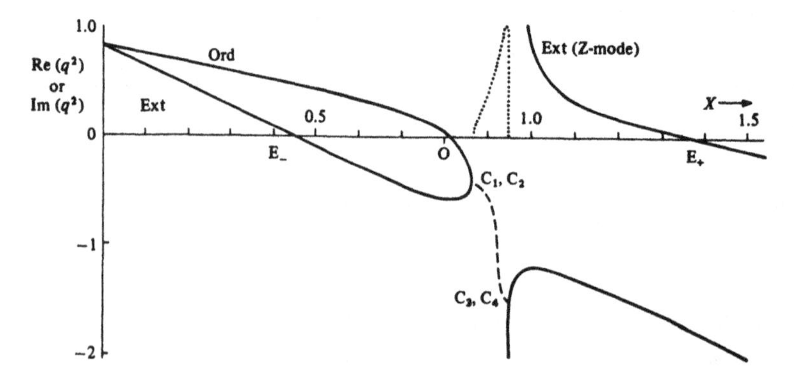

where the adjectives 'ordinary' and extraordinary' have to be interchanged. But for complex Θ there is no universally accepted convention for this.

One value of q^2 is infinite where (6.26) is satisfied. Here two roots of the Booker quartic are infinite. In the example of fig. 6.2 this X is very close to the greater of the two solutions of (6.43).

One value of q^2 is zero where the coefficient ε in (6.24), with $l_x^2 = 0$, is zero. One of the factors of ε gives

$$X = C^2. \tag{6.45}$$

This is associated with reflection of the ordinary wave. It gives the coupling point with the label O as explained in §6.6 and marked in fig. 6.2. The remaining factor of ε gives

$$X = 1 - \tfrac{1}{2}S^2 \pm (C^2 Y^2 + \tfrac{1}{4}S^4)^{\frac{1}{2}}. \tag{6.46}$$

For $C = 1$, $S = 0$ these go over to the cut-off conditions $X = 1 \pm Y$ for the extraordinary wave. They correspond to the coupling points E_+, E_- of §6.6, and are associated with reflection of the extraordinary wave.

Where X is the lesser of the two solutions of (6.43) there are two coincident coupling points C_1, C_2, and for the greater solution there are again two coincident coupling points C_3, C_4. The labelling of §6.6 is not unique, however, and an alternative labelling is possible in which one of C_3, C_4 is replaced by R. (See Smith, 1974a).

For propagation from magnetic east to west or west to east, the ray is in general deviated out of the plane of incidence. This is called 'lateral deviation' and is explained in § 10.13.

For further discussion and results for east–west and west–east propagation see Booker (1939, 1949), Millington (1951, 1954), Chatterjee (1952), Budden (1961a).

6.8. The Booker quartic for north–south and south–north propagation

For propagation from magnetic north to south or south to north, the direction cosine $l_y = 0$ so that $l_x^2 + l_z^2 = 1$. This does not give any simplification of the expressions (6.24) for the coefficients α to ε, but the quartic has some properties of special interest. This is because the vector $\boldsymbol{Y}$ is in the plane of incidence. It is useful to draw a diagram like fig. 6.1 containing cross sections of the refractive index surfaces, and in it the direction of $\boldsymbol{Y}$ can be shown as a line through the origin. Fig. 6.3 gives an example. On this line are the four window points P, Q, P_2, Q_2 where $n = \pm\{Y/(Y \pm 1)\}^{\frac{1}{2}}$, though only two are real if $Y < 1$. On it also are the two points K_1, K_2 where $n = \pm 1$, $\Theta = 0$. These six points are all on the refractive index surface when $X = 1$.

In this diagram a reference line is now drawn parallel to the n_z axis, where $n_x = S$. For a given value of X the cross sections of the two refractive index surfaces are then drawn. Where they are cut by the reference line, the four values of n_z are the required values of q. To plot curves of q versus X we must imagine two sequences of these cross sections to be drawn, one for the ordinary and one for the extraordinary wave, for a sequence of values of X. The form of the q–X curves depends on the value of S chosen for the reference line. Transition cases occur when it goes through any of the six points P, Q, P_2, Q_2, K_1, K_2. Here we shall consider only positive values of S. The transition values, marked in fig. 6.3, are:

$$\begin{aligned} S_A &= \{Y/(Y+1)\}^{\frac{1}{2}} l_x \quad &&\text{point P;} \\ S_B &= l_x &&\text{point } K_1; \\ S_C &= \{Y/(Y-1)\}^{\frac{1}{2}} l_x &&\text{point } P_2. \end{aligned} \tag{6.47}$$

S_C is real only if $Y > 1$. Its value can then be > 1.

For a wave (6.1) incident on the ionosphere from below, the angle of incidence is usually real so that S is real and < 1. A wave with S real and > 1 could not easily be launched from the ground. It could be emitted by a transmitter on a space vehicle within the ionosphere. If the transmitter is in a region where one value of n^2 is > 1, it could emit a wave with $1 < S^2 < n^2$, whose wave normal would have a real direction. Curves of q vs X for $S > 1$ are therefore of some interest but they are not considered further here. An example of a radio signal with $S > 1$ is the subprotonic whistler; see end of § 13.8 and Walker (1968a, b).

Fig. 6.4 shows curves of q vs X for a case where $1 > S > S_B$, and $Y < 1$. For $X = 0$

Fig. 6.3. Similar to fig. 6.1. Shows the plane of incidence $n_y = 0$ in refractive index space for the special case when the earth's magnetic field is in the plane of incidence. The four points P, Q, P_2, Q_2 are window points as described in § 4.11, 4.12, 5.2 item (7). The general form of the curves of q vs X depends on the value of S and changes when S passes through any of the transition values S_A, S_B, S_C.

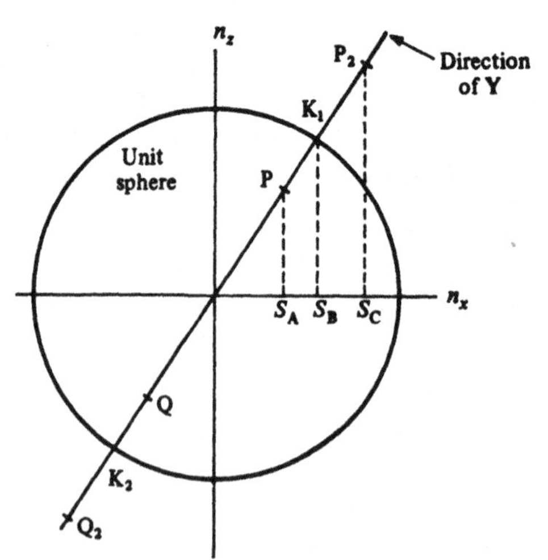

both refractive index surfaces are the unit sphere and the four values of q are in two pairs $q = \pm C$. As X increases, both refractive index surfaces shrink but that for the extraordinary wave shrinks faster (compare fig. 4.3). A value of X is reached where it just touches the reference line. Here the two qs for the extraordinary wave are equal and the tangent to the q–X curve is vertical as shown by (6.30). This is the coupling point E_- of § 6.6. For slightly larger X the two qs for the extraordinary wave are a complex conjugate pair. Similarly, for a still larger X, the reference line touches the ordinary refractive surface and the two qs for the ordinary wave are equal. This is the coupling point O of § 6.6.

The equal qs at E_- and at O have opposite signs, because the ordinary refractive index surfaces are prolate, fig. 5.4, and the extraordinary are oblate, fig. 5.7, as is shown by item (12) of § 5.2.

For $X > 1 - Y^2$ the refractive index surface for the extraordinary wave is again real, and $n > 1$. This is for the Z-mode. The surface is open and has reversed resonance cones (§ 5.2 item (11), figs. 5.8, 5.9). When X reaches the value X_∞ given by (6.26) where one root q is infinite, the reference line is parallel to the generator of the resonance cone, that is to the asymptote of the cross section of the refractive index surface in the magnetic meridian plane. The reference line then cuts the surface at

Fig. 6.4. Shows how the four values of q depend on X for propagation from magnetic north to south or south to north. In this example $Y = 0.5$, $l_x = 0.5$, $S = 0.7071$ and electron collisions are neglected. One q is infinite where $X = 0.9231$. Where two qs are a complex conjugate pair, the real part is shown as a broken line and the imaginary part as a dotted line. The points C, $-C$ are where $X = 1, q = \pm C$. The point D is the double root where $X = 1, q = Sl_z/l_x$ ($= 1.225$ in this example). The curve for X near to 1 is shown in more detail in fig. 6.6(f).

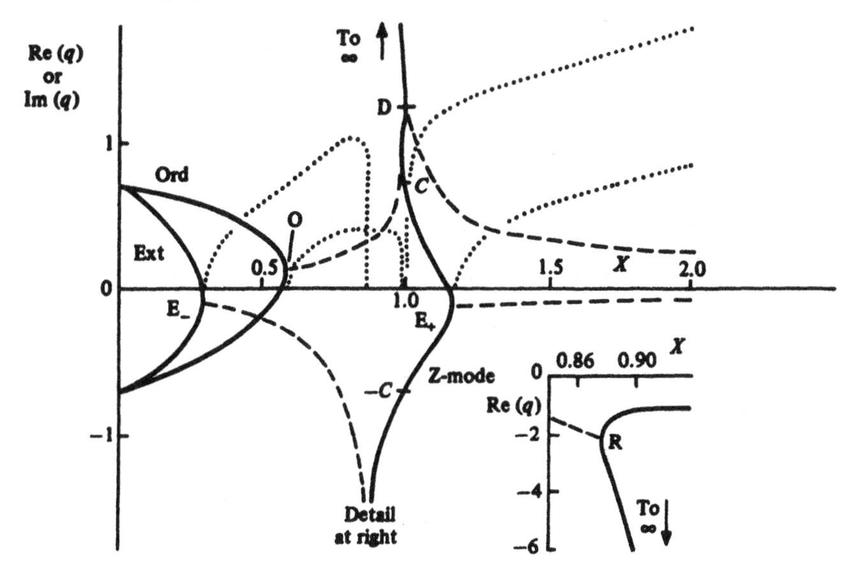

infinity and in at least one other point. This follows very simply from the geometry of the surface and its asymptotes. It was also proved in § 6.4 just after (6.26). Hence, when X is decreased, the refractive index surface for the Z-mode must cut the reference line at two points giving two real values of q for some range of X less than X_∞. This can be seen in figs. 6.4, 6.5.

Fig. 6.5 shows curves of q vs X for a case where $S < S_A$ ((6.47) and fig. 6.3), and $Y < 1$. The reference line now crosses the line segment PQ (fig. 6.3) and this is the limiting form of the refractive index surface for the ordinary wave when $X = 1$ (§ 5.2 item (7) and fig. 5.4). Hence the q–X curve for the ordinary wave extends to the value $X = 1$, and here it has a vertical tangent because there is a double root. This applies for any S in the range $-S_A < S < S_A$, but only when $l_y = 0$. For any other l_y the curve for the ordinary wave has a vertical tangent where $X < 0$.

The behaviour of the curves for X near to 1 depends on S and is shown in the six diagrams of fig. 6.6. This is for $Y < 1$ and for positive l_x. The quartic has three positive roots. There is one negative root, not shown in fig. 6.6, but visible in figs. 6.4, 6.5.

For $Y > 1$, the $q - X$ curve for the ordinary wave where $X < 1$ is similar to that for $Y < 1$. But the curves for the extraordinary wave are different in form and there is another branch of the curves for the ordinary wave corresponding to the whistler mode. The transition point S_C, (6.47) is now real. A full description would need many

Fig. 6.5. The same as fig. 6.4 except that the value of S is 0.258 819. The curve for the ordinary wave now extends to $X = 1$, and the line $X = 1$ is a tangent to it at the point O. Here there is a double root where $q = Sl_z/l_x$ ($= 0.448$ in this example). The curve for X near to 1 is shown in more detail in fig. 6.6(a).

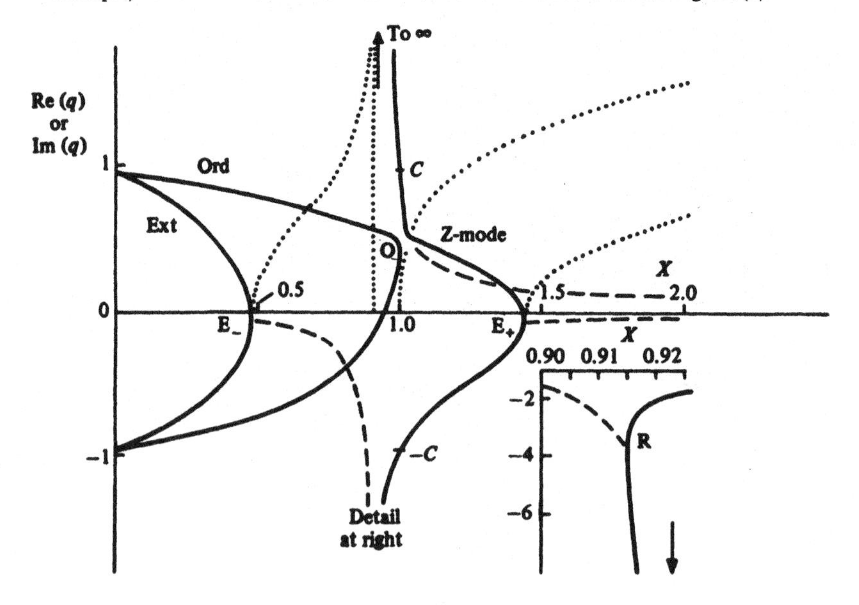

Fig. 6.6. Shows how q depends on X for various values of S and for X in a small range near $X = 1$. The other parameters are the same as in figs. 6.4, 6.5. Only three of the four qs are shown. The remaining q is real and negative in all six cases. The point C is where $q = C$, $X = 1$. The point D is where $q = Sl_z/l_x$, $X = 1$. Here there is a double root, and the tangent to the curve is vertical. In (e) the points C and D coincide and there is a triple root. The transition values (6.47) are $S_A = 0.2887$, $S_B = 0.5$. In (a) $S_A > S = 0.258\,819$ as in fig. 6.5. In (b) $S_A > S = 0.28$. In (c) $S = S_A$. The reference line goes through the window point P. In (d) $S = 0.3$; $S_A > S > S_B$. In (e) $S = S_B$. In (f) $S_B < S = 0.7071$ as in fig. 6.4. The alternative labels O, C_1, C_3 are the coupling points described in § 16.5.

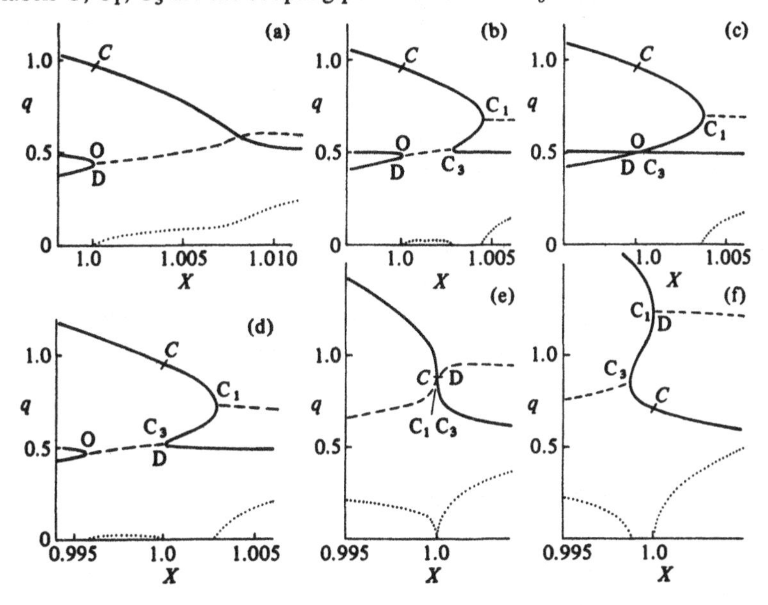

Fig. 6.7. Shows how the four values of q depend on X for propagation from magnetic north to south or south to north. In this example $Y = 2$, $l_x = 0.707$, $S = 0.9$ and electron collisions are neglected. One q is infinite where $X = 2.998$ (not shown here). Other details are as in fig. 6.4. Here $Sl_z/l_x = 0.9$. There is another branch of the curve for the whistler wave beyond the bottom right corner.

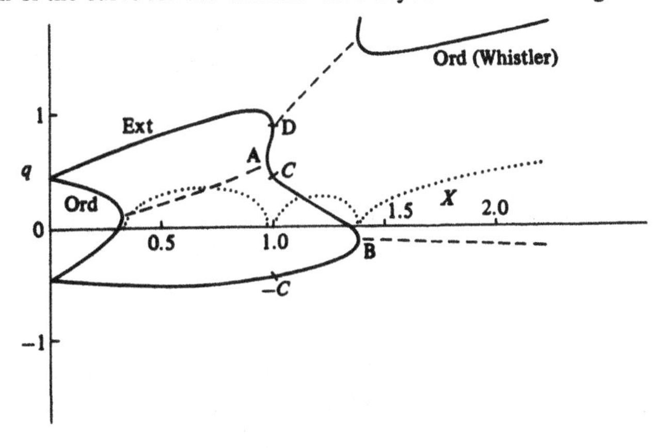

curves and would require too much space. For some examples see Booker (1939), Chatterjee (1952), Millington (1951, 1954), Budden (1961a). Here we must be content with just one example, fig. 6.7.

When $0 < X < 1$ and $Y > 1$ the refractive index for the extraordinary wave is greater than unity (see §4.12 and figs. 4.5, 4.6). Hence the curve of q vs X for the extraordinary wave is outside that for the ordinary, and $|q|$ increases at first when X increases from zero. The curve always extends to $X = 1$ and beyond it, provided that $S < 1$. The points where $X = 1, q = \pm C$ always lie on it. If the reference line of fig. 6.3 cuts the line segment P P_2, that is $S_A < S < S_B$, (6.47), there is a point D on the curve where $X = 1, q = Sl_z/l_x$ and here there is a double root and the tangent to the curve is vertical. This case is illustrated in fig. 6.7. For a small range of X slightly less than 1 all four roots q are real and on the curve for the extraordinary wave. This is possible because here the refractive index surface has two points of inflection, fig. 5.10, and can be cut four times by the reference line. In this case, therefore, the extraordinary wave curve has three points where the tangent is vertical, representing three levels of reflection. It is used later, fig. 10.2, for illustration. If S were equal to S_B, (6.47), the points C and D would coincide and there would be a triple root where $X = 1$.

For larger X the reference line again cuts the refractive index surface for the ordinary wave where $n^2 > 1$, that is the whistler mode. One branch of the resulting $q - X$ curve is in fig. 6.7. If $S < S_A$, the ordinary wave curve for $X < 1$ extends up to $X = 1$ and the double root at D lies on it. If $S > S_C$, the branch for the whistler mode extends down to $X = 1$ and the double root at D lies on it.

6.9. Effect of electron collisions on solutions of the Booker quartic

Figs. 6.8, 6.9 are examples of how the four qs depend on X when the effect of electron collisions is allowed for. Fig. 6.8 is for the same conditions as in fig. 6.4 except that now $Z \neq 0$. The qs are now all complex when $X \neq 0$ so that curves of both Re(q) and Im(q) are shown. Only real values of X are used. When $X \neq 0$ no value of Im(q) is ever zero. Two values are negative and apply to upgoing waves; the curves are labelled 1, 2 in figs. 6.8, 6.9. The other two values are positive and apply to downgoing waves; the curves are labelled 3, 4. This property must always hold when X is real, as was proved in §6.4.

Each of the coupling points E_-, E_+, O is a reflection point where two qs are equal, one for an upgoing and one for a downgoing wave. This cannot now occur for real X because the two qs must have non-zero Im(q) with opposite signs. The point of resonance $X = X_\infty$, now given by $\alpha = 0$ in (6.23), is no longer on the real X axis. Thus there are no values of Re(X) where one q is infinite, and in fig. 6.8 there are no values where two qs are equal, and no vertical tangents.

It is possible, however, for two qs to be equal when X and Z have real and non-zero values X_c, Z_c respectively. This is because the discriminant (6.38) can be zero for real

X, Z. The two equal qs must then be either both for upgoing or both for downgoing waves. This condition occurs when one of the coupling points C_1 to C_4 of § 6.6 lies on the real X axis. The values of X_c, Z_c were fully described by Pitteway (1959) who gave curves for various values of S, Y and l_x, l_y, l_z. Fig. 6.9 gives an example of q–X

Fig. 6.8. Shows how the four values of Re(q) and Im(q) depend on X when electron collisions are allowed for. This example is for propagation from magnetic north to south or south to north, and $Y = 0.5$, $l_x = 0.5$, $S = 0.7071$ as in fig. 6.4. But here $Z = 0.05$. The curves are identifiable by the numbers 1 – 4 which are the same for Re(q) and Im(q). If collisions were neglected, one q would be infinite where $X = 0.9231$.

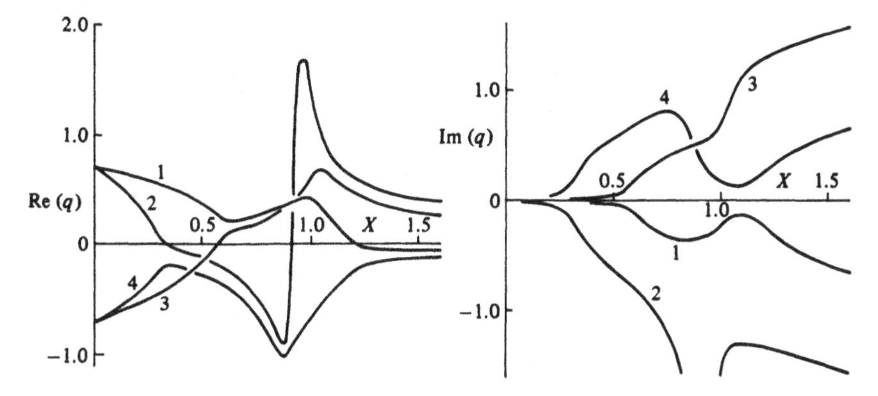

Fig. 6.9. Shows how the four values of Re(q) and Im(q) depend on X when electron collisions are allowed for. In this example $Y = 0.5$, $S = 0.5$, $l_x = l_y = 0.353\,553$, and $Z = 0.274 = Z_c$. If collisions were neglected one q would be infinite where $X = 0.9231$ but the effect of this resonance is masked by the large Z. Two values of q, for the two downgoing waves, are equal where $X = X_c = 0.711$. Here the tangents to curves 3 and 4 are vertical. The values of X_c, Z_c used here were found from fig. 3 of Pitteway (1959).

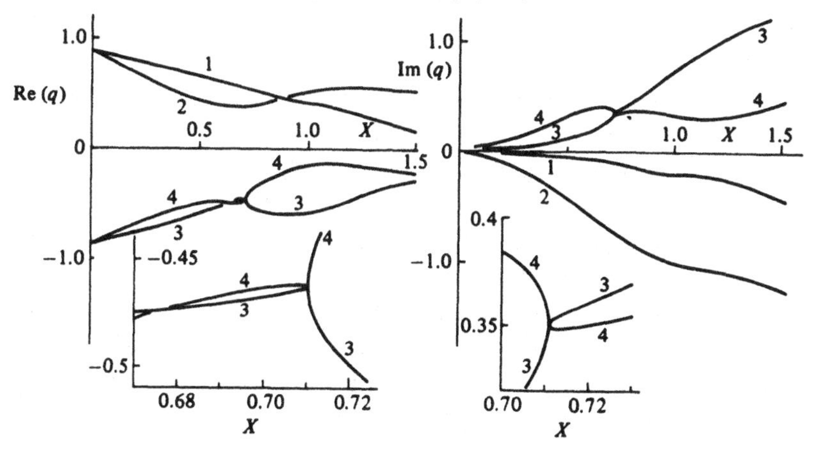

curves for one of these cases, in which $Z_c = 0.274$, $X_c = 0.711$. It can be seen that the tangents to the curves are vertical for both Re(q) and Im(q) at the coupling point, as they must be because $\partial q/\partial X \to \infty$.

6.10. The electromagnetic fields

Consider one of the thin strata described in § 6.2. In it there are in general four progressive waves corresponding to the four values of q. They are called the four 'characteristic waves' for the level z of the stratum. The fields of all these waves depend on x and y only through the factor (6.2) and we now assume that $S_2 = 0$, $S_1 = S$. Hence for all field components at all heights z

$$\frac{\partial}{\partial x} \equiv -ikS, \quad \frac{\partial}{\partial y} \equiv 0. \tag{6.48}$$

For any one of these waves, we have the further property

$$\frac{\partial}{\partial z} \equiv -ikqz \tag{6.49}$$

but the qs are in general different for the four waves. Maxwell's equations (2.45) and the constitutive relation $\boldsymbol{D} = \varepsilon_0 \boldsymbol{\varepsilon} \boldsymbol{E}$ may now be applied to any one of the waves to give

$$qE_y = \mathscr{H}_x, \quad qE_x - SE_z = \mathscr{H}_y, \quad SE_y = \mathscr{H}_z \tag{6.50}$$

$$\left.\begin{aligned} q\mathscr{H}_y &= \varepsilon_{xx}E_x + \varepsilon_{xy}E_y + \varepsilon_{xz}E_z \\ q\mathscr{H}_x - S\mathscr{H}_z &= \varepsilon_{yx}E_x + \varepsilon_{yy}E_y + \varepsilon_{yz}E_z \\ S\mathscr{H}_y &= \varepsilon_{zx}E_x + \varepsilon_{zy}E_y + \varepsilon_{zz}E_z. \end{aligned}\right\} \tag{6.51}$$

These six equations determine the ratios of the six components of $\boldsymbol{E}$ and $\boldsymbol{\mathscr{H}}$. From them E_z and $\mathscr{H}_z$ may be eliminated to give the ratios of the four components

$$E_x, \; -E_y, \; \mathscr{H}_x, \; \mathscr{H}_y. \tag{6.52}$$

The reason for using $-E_y$ is explained in §§ 7.14(3), (8). Explicit expressions for these ratios are given later, § 7.16, equations (7.118), (7.134). They depend on q so that there are four sets of the ratios (6.52) and they are written as the columns of a 4×4 matrix $\mathbf{S}$. The four columns are linearly independent provided that the four qs are distinct, and then $\mathbf{S}$ is non-singular; see § 7.14(1). Equations (6.50), (6.51) determine only the ratios of (6.52) so that each column of $\mathbf{S}$ may be multiplied by any constant. The choice of the constants depends on the particular problem but there is one choice that has special advantages. This is explained in § 7.14(7), (8).

Suppose now that, at any height z, the set of values (6.52) is given for the total field. Let it be written as a column matrix $\mathbf{e}$. Then it can always be expressed as the sum of the fields of the four characteristic waves, with amplitudes f_1, f_2, f_3, f_4; these are written as a column matrix $\mathbf{f}$. Thus

$$\mathbf{e} = \mathbf{S}\mathbf{f}; \quad \mathbf{f} = \mathbf{S}^{-1}\mathbf{e}. \tag{6.53}$$

From a knowledge of **S** and **e**, the amplitudes **f** can be found. This still applies even when we do not use the concept of discrete strata. If the fields (6.52) are given for any z in a continuous stratified medium, the amplitudes **f** can be found from (6.53). This is equivalent to considering a fictitious homogeneous medium with the properties of the actual medium at the height z, and finding what amplitudes **f** of the four characteristic waves would be needed to give the actual fields. The discussion of this topic is resumed in §§ 7.14, 16.2.

The amplitudes **f** play an important part in the theory in this book, and frequent reference is made in later chapters to the transformations (6.53).

PROBLEMS 6

Note. Problems 6.1, 6.2 below may interest those readers who like to pursue algebraic details. The results are useful when tracing curves of q or q^2 vs X. They have no particular practical application in radio propagation problems.

6.1. For a collisionless electron plasma, when $X = 1$, the Booker quartic has two solutions $q = \pm C$. Show that here the curve of q vs X has slope

$$\mathscr{S} = \partial q/\partial X = \mp\{2Y^2C(Cl_x \mp Sl_z)^2\}^{-1}.$$

For propagation from magnetic north to south or south to north ($l_y = 0$), S is given the value $l_x\{Y/(Y+1)\}^{\frac{1}{2}}$ (as in fig. 6.6(c)) so that two branches of the $q - X$ curves intersect where $X = 1$. Show that their slopes $\mathscr{S}$ are solutions of the quadratic equation

$$Yl_x^2\mathscr{S}^2 - 4l_z\{Y(Y+1)\}^{\frac{1}{2}}\mathscr{S} - 2 = 0.$$

6.2. For propagation from magnetic east to west or west to east, the Booker quartic is a quadratic equation for q^2 and, when collisions are neglected, q^2 is zero where $X = C^2$. Show that here the slope $\partial(q^2)/\partial X = -(1 - l_z^2C^2)^{-1}$. The angle of incidence is now chosen so that the quadratic equation has equal roots for two coincident real values of X. Show that $l_z^2S^4 = Y^2(1 - C^2l_z^2)^2$ and that equal roots q^2 occur where $X = \frac{1}{2}(1 + C^2)$, $q^2 = -\frac{1}{2}S^2(1 + l_z^2C^2)/l_y^2$. Here two curves of q^2 vs X intersect. Show that their slopes $\partial(q^2)/\partial X$ are $(1 - l_z^2C^2)/l_y^2$ and $-(1 - l_z^2C^2)\{2(1 + l_z^2C^2) + l_y^2\}/l_y^4$.

6.3. Show that, for propagation near the magnetic pole, the Booker quartic is a quadratic equation for q^2 and find its solutions for a collisionless electron plasma. Show that the two solutions are equal when $X = 1 - \frac{1}{2}S^2\{1 \mp (1 - Y^2)^{\frac{1}{2}}\}$ and that then $q^2 = \frac{1}{2}(1 + C^2)\{1 \pm (1 - Y^2)^{-\frac{1}{2}}\}$. Note that these are real only when $Y < 1$, and then only the upper signs give a positive value of q^2. Sketch the refractive index surface for this case and show that it applies for the Z-mode. Sketch the curves of q^2 vs X.

6.4. Show that if the ionosphere is an electron plasma and X is very small, the two values of q for obliquely upgoing waves are given by $q - C \approx bX$ where b is a solution of the quadratic equation

$$4C^2U(U^2 - Y^2)b^2 + 2\{CY^2(l_zC + l_xS)^2 - CY^2 + 2CU^2\}b + U = 0.$$

Check that this gives the correct results for the special cases (a) $S = 0$, $l_z = 0$, and (b) $S = 0$, $l_z = 1$.

7

Slowly varying medium. The W.K.B. solutions

7.1. Introduction

This chapter continues the discussion of radio waves in a stratified ionosphere and uses a coordinate system x, y, z as defined in §6.1. Thus the electric permittivity $\varepsilon(z)$ of the plasma is a function only of the height z. It is assumed that the incident wave below the ionosphere is a plane wave (6.1) with $S_2 = 0$, $S_1 = S$ so that wave normals, at all heights z, are parallel to the plane $y = \text{constant}$, and for all field components (6.48) is satisfied. It was shown in § 6.10 that if the four components $E_x, E_y, \mathscr{H}_x, \mathscr{H}_y$ of the total field are known at any height z, they can be expressed as the sum of the fields of the four characteristic waves, with factors $\mathfrak{f}_1, \mathfrak{f}_2, \mathfrak{f}_3, \mathfrak{f}_4$. In a homogeneous medium the four waves would be progressive waves, and these factors would be $\exp(-ikq_i z)$, $i = 1, 2, 3, 4$. We now enquire how they depend on z in a variable medium. This question is equivalent to asking whether there is, for a variable medium, any analogue of the progressive characteristic waves in a homogeneous medium. The answer is that there is no exact analogue. There are, however, approximate solutions, the W.K.B. solutions, which have many of the properties of progressive waves.

This problem has proved to be of the greatest importance in all those branches of physics concerned with wave propagation. It has intrigued not only physicists but many mathematicians, and the subject has been taken to levels of complication that are beyond the interests of most physicists. The studies have been devoted almost entirely to wave fields that satisfy differential equations of the second order. In the radio problem this usually means that the plasma medium must be assumed to be isotropic. An anisotropic ionosphere leads, in general, to the equivalent of a fourth order differential equation. Here the studies have been less extensive.

Most of this chapter is devoted to the isotropic ionosphere, that is to the solutions of second order differential equations. The anisotropic ionosphere is discussed in §§ 7.13 onwards.

7.2. The differential equations for an isotropic ionosphere

For an isotropic ionosphere the earth's magnetic field is neglected and the electric permittivity of the plasma is a scalar

$$\varepsilon(z) = n^2 = 1 - X/U \tag{7.1}$$

given by (3.17). Then $\boldsymbol{D} = \varepsilon_0 n^2 \boldsymbol{E}$, (2.29), and, from (6.6), the values of q are in two equal pairs

$$q = \pm(n^2 - S^2)^{\frac{1}{2}} = \pm(C^2 - X/U)^{\frac{1}{2}}. \tag{7.2}$$

When using Maxwell's equations (2.45) we may still use (6.48), but (6.49) must not be used because it is the z dependence of the fields that is to be studied. Thus (2.45) give

$$\frac{\partial E_y}{\partial z} = \mathrm{i}k\mathscr{H}_x, \quad SE_y = \mathscr{H}_z, \quad \frac{\partial \mathscr{H}_x}{\partial z} = \mathrm{i}k(n^2 E_y - S\mathscr{H}_z) \tag{7.3}$$

and

$$\frac{\partial E_x}{\partial z} = -\mathrm{i}k(\mathscr{H}_y + SE_z), \quad \frac{\partial \mathscr{H}_y}{\partial z} = -\mathrm{i}kn^2 E_x, \quad S\mathscr{H}_y = -n^2 E_z. \tag{7.4}$$

The equations have here been deliberately separated into the two sets. The first (7.3) depend only on E_y, $\mathscr{H}_x$, $\mathscr{H}_z$, and the second (7.4) only on E_x, E_z, $\mathscr{H}_y$. Thus if one of these sets is zero at any height z, it remains zero at all heights. The equations separate into two independent sets (7.3) and (7.4).

The set (7.3) is the simpler and will be discussed first. The electric field at all levels has only the component E_y, which is horizontal. The waves are sometimes said to be 'horizontally polarised' at all levels. The magnetic field has no $\mathscr{H}_y$ component so that it is in the plane of incidence. For the other set (7.4), E_y is everywhere zero. The waves are said to be 'vertically polarised' because $\boldsymbol{E}$ lies in the vertical plane of incidence $y =$ constant. This case is a little more complicated and is discussed in §7.12.

From (7.3), $\mathscr{H}_z$ may be eliminated, and (7.2) is used to give

$$\frac{\partial E_y}{\partial z} = \mathrm{i}k\mathscr{H}_x, \quad \frac{\partial \mathscr{H}_x}{\partial z} = \mathrm{i}k\{q(z)\}^2 E_y. \tag{7.5}$$

It is at once clear that there is no solution of (7.5) that represents a 'progressive' wave. For in a progressive wave all field quantities vary with z through the same factor $\exp\{\mathrm{i}\phi(z)\}$. Now the first equation (7.5) would require that $\mathrm{d}\phi/\mathrm{d}z$ is a constant while the second would require that it is proportional to $\{q(z)\}^2$, which is only possible if q is a constant. In spite of this, it is possible to derive *approximate* solutions of (7.5) which have many of the properties of progressive waves. These are the W.K.B. solutions. At every place where q is varying, the reflection process is going on, and it is this which prevents the occurrence of a true progressive wave. Except in certain places, however, the reflection process is very weak, so that the

W.K.B. solutions are often very good approximations. This argument is illustrated in more detail in later sections.

Elimination of $\mathscr{H}_x$ from (7.5) gives

$$\frac{\mathrm{d}^2 E_y}{\mathrm{d}z^2} + k^2 q^2 E_y = 0. \tag{7.6}$$

This is a very important differential equation satisfied by the electric field E_y. It is assumed that the factor $\exp\{\mathrm{i}(\omega t - kSx)\}$ is omitted so that E_y is a function of z only. We therefore use the total derivative sign $\mathrm{d}/\mathrm{d}z$ instead of the partial derivative sign $\partial/\partial z$.

Equation (7.6) is of the form which occurs in the study of nearly all kinds of wave propagation. For example if ψ is the wave function of a particle of mass m and of constant total energy E, moving in a one-dimensional potential field $V(z)$, the Schrödinger equation is

$$\frac{\mathrm{d}^2\psi}{\mathrm{d}z^2} + \frac{2m}{\hbar^2}\{E - V(z)\}\psi = 0 \tag{7.7}$$

so that q^2 in (7.6) is analogous to the classical kinetic energy $E - V$ in (7.7). Again if p is the excess pressure in a plane sound wave of small amplitude travelling in a gas of molecular weight M and temperature T, the wave equation is

$$\frac{\mathrm{d}^2 p}{\mathrm{d}z^2} + \frac{\omega^2 M}{\gamma R T} p = 0 \tag{7.8}$$

where R is the gas constant, and γ is the ratio of specific heats; M, γ and T may be functions of z. Further examples will be found in Problems 7 at the end of this chapter. Results which will now be derived for (7.6) can be adapted to apply to these other equations.

There are other kinds of wave propagation where the differential equation is less simple than (7.6) and may be of higher order than the second, as occurs, for instance, in the study of waves in anisotropic media but in nearly all cases, the W.K.B. method of 7.6 can still be used.

7.3. The phase memory concept

Suppose that in (7.5) and (7.6) the variation of $q(z)$ with z is so slow that a range of z can be chosen where q can be treated as constant. Consider a progressive wave in this region travelling in the direction of positive z. Then its field components would be given by

$$E_y = A\exp(-\mathrm{i}kqz), \quad \mathscr{H}_x = -Aq\exp(-\mathrm{i}kqz) \tag{7.9}$$

where A is constant to a first approximation. Now the first equation (7.9) ought to satisfy (7.6) but by direct substitution it can be seen that the fit is not very good unless $\mathrm{d}q/dz$ and $\mathrm{d}^2q/\mathrm{d}z^2$ are very small.

If q were real, the part kqz of the exponents in (7.9) would be a real angle called the 'phase' of the wave. Since q is in general complex, the angle kqz is complex but it is still called the (complex) phase. Its real part is the same as the phase angle as ordinarily understood, and its imaginary part gives an additional term in the exponent. This term is real because of the factor $-\mathrm{i}$, and therefore affects only the amplitude of the wave. When the wave represented by (7.9) passes through an infinitesimal thickness δz of the medium, the change in its (complex) phase is $kq\delta z$, which therefore depends on q. When it passes through a finite thickness, from $z=0$ to $z=z_1$, say, the change of phase is $k\int_0^{z_1} q\,\mathrm{d}z$. This suggests that a better solution than (7.9) might be

$$E_y = A\exp\left(-\mathrm{i}k\int_0^z q\,\mathrm{d}z\right), \quad \mathscr{H}_x = -Aq\exp\left(-\mathrm{i}k\int_0^z q\,\mathrm{d}z\right). \tag{7.10}$$

These satisfy the first equation (7.5) exactly, but the second is approximately satisfied only when $\mathrm{d}q/\mathrm{d}z$ is small. The agreement is better, however, than for (7.9). The integral in (7.10) constitutes what is sometimes called the 'phase memory' concept. It expresses the idea that a change of phase is cumulative for a wave passing through a slowly varying medium. The solution (7.10) is adequate for the discussion of many wave propagation problems. An extension of it to three dimensions forms the basis of much of the theory of geometrical optics or ray theory, ch. 14. But for some purposes a still more accurate solution is needed. This is the W.K.B. solution, derived in the following sections.

7.4. Loss-free medium. Constancy of energy flow

To obtain a solution that is more accurate than (7.10) it is convenient to treat A as a function of z. One method of deriving an approximation to this function can be used if the medium is loss-free. We therefore now consider the special case where n^2 is real, and we study a range of z where q^2 is positive. Then, in a progressive wave, the average flow of energy in the z direction must be the same for all z. It is proportional to the z component Π_z of the time averaged Poynting vector, (2.63), that is to

$$[\mathrm{Re}(\boldsymbol{E}\wedge\mathscr{H}^*)]_z = -\tfrac{1}{2}(E_y\mathscr{H}_x^* + E_y^*\mathscr{H}_x). \tag{7.11}$$

If q is positive, the exponents in (7.10) are purely imaginary and (7.11) is proportional to qAA^* which must be constant. Hence A is proportional to $q^{-\frac{1}{2}}$, so that the solution becomes

$$E_y = A_1 q^{-\frac{1}{2}}\exp\left(-\mathrm{i}k\int_0^z q\,\mathrm{d}z\right), \quad \mathscr{H}_x = -A_1 q^{\frac{1}{2}}\exp\left(-\mathrm{i}k\int_0^z q\,\mathrm{d}z\right) \tag{7.12}$$

where A_1 is a constant. The argument applies also to a wave travelling in the direction of negative z, so that another solution is

$$E_y = A_2 q^{-\frac{1}{2}} \exp\left(ik\int_0^z q\,dz\right), \quad \mathscr{H}_x = A_2 q^{\frac{1}{2}} \exp\left(ik\int_0^z q\,dz\right) \tag{7.13}$$

where A_2 is another constant.

These two pairs are the two W.K.B. solutions. They are approximate because the reflection process, here neglected, is actually occurring at all values of z where $dq/dz \neq 0$, so that the assumption that the energy flow is constant is not strictly true.

The above derivation applies only when q is real and positive, but it is useful because it illustrates the physical significance of W.K.B. solutions. In § 7.6 another derivation is given which applies for any value of q including complex and purely imaginary values.

The argument used here for a loss-free system may be applied to other kinds of wave. Consider, for example, the Schrödinger equation (7.7) for a particle in a potential field $V(z)$. Let the wave function $\psi(z)$ represent a progressive wave travelling in the positive z direction. Then the first equation (7.10) is replaced by:

$$\psi = \Psi \exp\left[i\int_0^z \{(E-V)2m/\hbar^2\}^{\frac{1}{2}} dz\right] \tag{7.14}$$

(in accordance with the invariable practice in wave mechanics it is here assumed that the suppressed time factor is $e^{-i\omega t}$). If $V(z)$ varies sufficiently slowly, the flux of the particles must be approximately constant so that

$$\psi\frac{d\psi^*}{dz} - \psi^*\frac{d\psi}{dz} = \text{constant} \tag{7.15}$$

(see, for example, Mott, 1952, § II.5). Substitution from (7.14) gives $\Psi\Psi^* \propto (E-V)^{-\frac{1}{2}}$ which suggests that

$$\psi = \psi_0 (E-V)^{-\frac{1}{4}} \exp\left[i\int_0^z \{(E-V)2m/\hbar^2\}^{\frac{1}{2}} dz\right]. \tag{7.16}$$

This is one W.K.B. solution. The other has a minus sign in the exponent and represents a particle travelling in the negative z direction.

7.5. W.K.B. solutions

The solutions that are here called W.K.B. solutions were used in wave-mechanical problems independently by Wentzel (1926), Kramers (1926) and Brillouin (1926). The name W.K.B. is derived from the initials of these three authors. These solutions had been studied earlier by Jeffreys (1924) and some authors refer to them as J.–W.K.B. or W.K.B.–J. solutions. There are, however, still earlier accounts of this type of solution. They were used by Gans (1915) in the study of radio waves in the ionosphere, and by Lord Rayleigh (1912) in the study of sound waves. Jeffreys and Jeffreys (*Methods of Mathematical Physics*, 1972) call them 'asymptotic approximations of Green's type' because Green used them in 1837 in the study of waves on the surface of water. They also draw attention to the use of this kind of approximation by Carlini in 1817. Some authors, for example Olver (1974), now refer to these solutions as L.G. approximations, after Liouville (1837) and Green (1837) (see § 7.9).

The term 'W.K.B. solutions' will be used here because it is well known and widely used. For the history of the subject and further references, see Olver (1974), Heading (1962a).

There seems to be no generally accepted *precise* definition of the term 'W.K.B. solution' that covers all cases. Many of the mathematical treatments study ordinary differential equations of the second order (Jeffreys and Jeffreys, 1972; Heading, 1962a; Olver, 1974) and it is then often possible to formulate a precise definition. But equations of higher order have not received such detailed study. For equations of the fourth order the subject is discussed in §§ 7.13–7.17.

7.6. The W.K.B. method

The W.K.B. solutions of the differential equation (7.6) are now to be derived by a method which is not restricted to real values of q. When q is complex, the convention is adopted, in this section, that its real part is positive.

In a homogeneous medium there are two solutions of (7.6) which represent progressive waves. In both of them the field E_y depends on z, only through a factor $\exp\{i\phi(z)\}$ where $\phi(z) = \mp kqz$. The function $\phi(z)$ is called the generalised phase of the wave and it is in general complex because q is complex. In a homogeneous medium q is independent of z so that $d\phi/dz = \mp kq$, and $d^2\phi/dz^2 = 0$. This suggests that we try to find the function $\phi(z)$ for a slowly varying medium, and we may expect that $d^2\phi/dz^2$ is very small, so that its square and higher powers may be neglected. This idea is the basis of the following method. The function $\phi(z)$ is an example of the eikonal function (Sommerfeld, 1954, §§ 35, 48; see also § 14.2).

Let

$$E_y = A \exp\{i\phi(z)\} \tag{7.17}$$

where A is a constant. Then

$$\frac{d^2E_y}{dz^2} = A\left\{i\frac{d^2\phi}{dz^2} - \left(\frac{d\phi}{dz}\right)^2\right\}e^{i\phi}. \tag{7.18}$$

When this is substituted in the differential equation (7.6) it shows that $\phi(z)$ must satisfy:

$$\left(\frac{d\phi}{dz}\right)^2 = k^2q^2 + i\frac{d^2\phi}{dz^2}. \tag{7.19}$$

This is a non-linear differential equation which would be very difficult to solve exactly, but an approximate solution may be found as follows. Since $d^2\phi/dz^2$ is small, it may be neglected to a first approximation. Then

$$\frac{d\phi}{dz} \approx \mp kq \tag{7.20}$$

and thence

$$\frac{d^2\phi}{dz^2} \approx \mp k\frac{dq}{dz}. \tag{7.21}$$

The approximate value (7.21) is now substituted in (7.19) which gives as a second approximation

$$\frac{d\phi}{dz} \approx \left\{k^2q^2 \mp ik\frac{dq}{dz}\right\}^{\frac{1}{2}}. \tag{7.22}$$

The square root is expanded by the binomial theorem, and since the second term is small, only two terms of the expansion need be retained. The result is

$$\frac{d\phi}{dz} \approx \mp kq\left\{1 \mp \frac{i}{2kq^2}\frac{dq}{dz}\right\}. \tag{7.23}$$

Now the second minus sign in (7.23) was obtained from the minus sign in the first approximation (7.20) and for this case the first sign in (7.23) must be minus. Similarly, if the second sign is plus, the first sign must be plus also. Hence (7.23) becomes

$$\frac{d\phi}{dz} \approx \mp kq + \frac{i}{2q}\frac{dq}{dz}. \tag{7.24}$$

which may be integrated at once to give

$$\phi \approx \mp k\int^{z} q\,dz + i\ln(q^{\frac{1}{2}}). \tag{7.25}$$

The constant of integration may be made zero by a suitable choice of the lower limit of the integral, which is left unspecified here, since its value does not affect the way in which $\phi(z)$ depends on z. Substitution of (7.25) into (7.17) gives

$$E_y \approx Aq^{-\frac{1}{2}}\exp\left\{\mp ik\int^{z} q\,dz\right\}. \tag{7.26}$$

These two solutions are the W.K.B. solutions of (7.6).

If (7.26) is substituted in the first equation (7.5) it gives

$$\mathscr{H}_x = A\left[\mp q^{\frac{1}{2}} - \tfrac{1}{2}q^{-\frac{3}{2}}\frac{dq}{dz}\right]\exp\left\{\mp ik\int^{z} q\,dz\right\}. \tag{7.27}$$

Here the second term in the square brackets is small compared with the first, for a slowly varying medium, and can usually be neglected.

It is stressed that this derivation of the W.K.B. solutions for an isotropic stratified medium is valid for the most general case where q is a complex variable function.

The process can be continued to higher orders of approximation to obtain a still more accurate solution. This is algebraically complicated, however, and it is usually more profitable to use other methods to get a more accurate solution; see, for example, § 8.11 equation (8.30) onwards.

7.7. Discrete strata

The physical significance of W.K.B. solutions is further illustrated by the following alternative derivation, which was used by Bremmer (1951). See also Lord Rayleigh (1912), Brekhovskikh (1960, § 16), Bremmer (1949). It is supposed that the slowly varying medium is divided, by planes $z=$ constant, into thin strata in each of which n and q are constant (compare § 6.2). By making the strata sufficiently thin and numerous it is possible to approximate as closely as desired to the actual medium. Suppose that for $z<0$ the medium is homogeneous with $q=q_0$. Let the strata be of small thickness h and let them be numbered consecutively for $z>0$, so that the values of q are q_1, q_2, etc. Let a plane wave

$$E_y=\exp(-ikq_0z), \quad \mathscr{H}_x=-q_0\exp(-ikq_0z) \tag{7.28}$$

(compare (7.9)) be incident from the direction of negative z. When it reaches the first boundary plane at $z=0$ it gives rise to a reflected wave

$$E_y=R_0\exp(ikq_0z), \quad \mathscr{H}_x=R_0q_0\exp(ikq_0z) \quad \text{for } z<0 \tag{7.29}$$

and a transmitted wave

$$E_y=T_1\exp(-ikq_1z), \quad \mathscr{H}_x=-T_1q_1\exp(-ikq_1z) \quad \text{for } 0<z<h. \tag{7.30}$$

On crossing a boundary the components of electric and magnetic intensity parallel to it must be continuous. Hence, for the boundary at $z=0$

$$1+R_0=T_1, \quad q_0(1-R_0)=q_1T_1 \tag{7.31}$$

whence

$$R_0=(q_0-q_1)/(q_0+q_1), \quad T_1=2q_0/(q_0+q_1). \tag{7.32}$$

The transmitted wave now impinges on the next boundary at $z=h$, and is there partially reflected and partially transmitted. The reflected wave from this boundary is

$$E_y=R_1T_1\exp(-ikq_1h)\exp\{ikq_1(z-h)\} \quad \text{for } 0<z<h \tag{7.33}$$

and the transmitted wave is

$$E_y=T_1T_2\exp(-ikq_1h)\exp\{-ikq_2(z-h)\} \quad \text{for } h<z<2h \tag{7.34}$$

where, by analogy with (7.32):

$$R_1=(q_1-q_2)/(q_1+q_2), \quad T_2=2q_1/(q_1+q_2). \tag{7.35}$$

The medium is assumed to be slowly varying so that the differences q_0-q_1, q_1-q_2, etc. for adjacent strata are very small. Thus, provided that q_0, q_1, q_2 etc. are not too small, the reflected waves have small amplitude.

The reflected waves now impinge on the boundaries and are again reflected. These second reflections add on to the waves travelling in the direction of positive z. They have extremely small amplitude, however, and as a first approximation they can be neglected. The reason is as follows. By making the thickness h of the strata and the differences $q_{i+1} - q_i$ small enough, the reflection coefficients R_0, R_1 etc. can be made indefinitely small. But this also makes the reflecting boundaries indefinitely numerous, so that the numerous first reflections have to be added. The addition must take account of the amplitude and phase of the contributions and is conveniently done by an amplitude–phase diagram of the kind used in physical optics (see, for example, Lipson and Lipson, 1969, § 6.4.5; or Jenkins and White, 1976, ch. 12). When the wave (7.33) reflected from the boundary at $z = h$ reaches the origin it is given by

$$E_y = R_1 T_1 \exp(-2ikq_1 h) \tag{7.36}$$

where the phase $2kq_1h$ contains the factor 2 because the wave has made two traverses of the stratum $0 \leqslant z \leqslant h$. Similarly when the wave reflected from $z = 2h$ reaches the origin it is given by

$$E_y = R_2 T_1 T_2 \exp\{-2ikh(q_1 + q_2)\} \tag{7.37}$$

and in general, for the boundary at $z = mh$

$$E_y = R_m T_1 T_2 \dots T_m \exp\{-2ikh(q_1 + q_2 + \dots + q_m)\}. \tag{7.38}$$

Thus there is a phase difference $2khq_1$ between successive contributions to this sequence, so that the amplitude–phase diagram is a polygon resembling a tightly wrapped spiral. The resultant amplitude of all the first reflections is therefore very small for two reasons: (i) the contributing vectors such as (7.38) are small because R_m is small, and (ii) the phase difference between successive contributions gives the amplitude–phase diagram a large curvature. Consequently the resultant amplitude of the twice reflected waves is smaller still.

Neglect of the twice reflected waves means that the transmitted wave in the mth stratum is given by

$$E_y = A \exp\left[-ik\{h(q_1 + q_2 + \dots + q_{m-1}) + q_m(z - h)\}\right] \tag{7.39}$$

where

$$A = T_1 T_2 \dots T_m, \quad T_m = 2q_{m-1}/(q_{m-1} + q_m). \tag{7.40}$$

In going from the $m - 1^{\text{th}}$ stratum to the mth stratum the increment of $\ln A$ is

$$\Delta(\ln A) = \ln T_m = \ln\left\{\left(1 + \frac{q_m - q_{m-1}}{2q_{m-1}}\right)^{-1}\right\} \approx -\tfrac{1}{2}\frac{\Delta q}{q}. \tag{7.41}$$

If it is assumed that the strata are so thin that this relation can be integrated, it gives

$A = Kq_m^{-\frac{1}{2}}$ and because $A = 1$ when $z < 0$, the constant K is $q_0^{\frac{1}{2}}$. The sum in the exponent of (7.39) can similarly be expressed as an integral so that (7.39) gives

$$E_y = Kq^{-\frac{1}{2}} \exp\left\{ -ik \int_0^z q\,\mathrm{d}z \right\} \tag{7.42}$$

which is one W.K.B. solution.

It may happen that the wave (7.42) continues on to a region where q_m is small. Then the reflection coefficients $R_m = (q_m - q_{m+1})/(q_m + q_{m+1})$ are no longer small. Similarly the phase change $2khq_m$ between waves reflected from adjacent boundaries is very small, so that successive contributions augment each other. In this region the reflection process is strong and the W.K.B. solutions fail. A reflected wave is generated which travels in the negative z direction. After it has left this region of reflection, its field is given by the other W.K.B. solution.

For further discussion of the use of discrete strata see §18.3.

7.8. Coupling between upgoing and downgoing waves

The discussion in § 7.7 showed that a wave represented by one W.K.B. solution may give rise, as it travels, to another wave represented by the other W.K.B. solution, but that this process is negligible provided that q varies slowly enough with z, and is not small. This idea is illustrated by the following alternative derivation of the W.K.B. solutions.

In a homogeneous medium a progressive wave travelling obliquely in the direction of increasing z has field components E_y, $\mathscr{H}_x$ related by $\mathscr{H}_x = -qE_y$, from (7.9), and a progressive wave travelling obliquely in the direction of decreasing z has $\mathscr{H}_x = qE_y$. For an inhomogeneous medium these relations may be used to define the analogues of the progressive waves. Hence let

$$E_y = E_y^{(1)} + E_y^{(2)}, \quad \mathscr{H}_x = \mathscr{H}_x^{(1)} + \mathscr{H}_x^{(2)} \tag{7.43}$$

where

$$\mathscr{H}_x^{(1)} = -qE_y^{(1)}, \quad \mathscr{H}_x^{(2)} = qE_y^{(2)}. \tag{7.44}$$

These are used in the Maxwell equation (7.5) to give

$$\left.\begin{aligned} \frac{\mathrm{d}E_y^{(1)}}{\mathrm{d}z} + \frac{\mathrm{d}E_y^{(2)}}{\mathrm{d}z} &= ikq\mathscr{H}_x^{(1)} + ikq\mathscr{H}_x^{(2)}, \\ -q\frac{\mathrm{d}E_y^{(1)}}{\mathrm{d}z} + q\frac{\mathrm{d}E_y^{(2)}}{\mathrm{d}z} &= ikq^2E_y^{(1)} + ikq^2E_y^{(2)} + E_y^{(1)}\frac{\mathrm{d}q}{\mathrm{d}z} - E_y^{(2)}\frac{\mathrm{d}q}{\mathrm{d}z}. \end{aligned}\right\} \tag{7.45}$$

Divide the second of these by q and subtract it from, or add it to, the first. This gives the two equations

$$\frac{\mathrm{d}E_y^{(1)}}{\mathrm{d}z} + ikqE_y^{(1)} + \frac{1}{2q}\frac{\mathrm{d}q}{\mathrm{d}z}E_y^{(1)} = \frac{1}{2q}\frac{\mathrm{d}q}{\mathrm{d}z}E_y^{(2)}, \tag{7.46}$$

$$\frac{dE_y^{(2)}}{dz} - ikqE_y^{(2)} + \frac{1}{2q}\frac{dq}{dz}E_y^{(2)} = \frac{1}{2q}\frac{dq}{dz}E_y^{(1)}. \tag{7.47}$$

These are coupled equations in the sense of the definition given later in § 16.1. They may be solved by successive approximations. The right-hand sides both contain the factor $\frac{1}{2}q^{-1}\,dq/dz$ which is small in a slowly varying medium except near where $q = 0$. As a first approximation, therefore, the right-hand sides are neglected. This is in effect the same assumption as was made in (7.20) where $d^2\phi/dz^2$ was neglected. Then (7.46), (7.47) give two independent differential equations for $E_y^{(1)}$, $E_y^{(2)}$ whose solutions are

$$\left.\begin{matrix} E_y^{(1)} \\ E_y^{(2)} \end{matrix}\right\} = q^{-\frac{1}{2}}\exp\left(\mp ik\int^z q\,dz\right). \tag{7.48}$$

These are simply the W.K.B. solutions. This method of deriving them shows how they arise by neglecting the coupling process, which in this case is a reflection process. The method is extended later, § 7.15, to deal with the anisotropic ionosphere.

The solutions (7.48) may be substituted in the right-hand sides of (7.46), (7.47) and the equations solved to give a better approximation. An example where this is useful is as follows. Suppose that a plane radio wave travels obliquely upwards in the earth's troposphere. Here the refractive index is very close to unity at all levels, but there can be small variations and there can be stratified layers caused by water vapour or by temperature changes with height. The upgoing wave, (7.48) upper sign, would travel right through such a layer, and any resulting downgoing wave would have such an extremely small amplitude that the right-hand side of (7.46) could be neglected at all levels. But the very weak reflection from a layer in the troposphere can sometimes be strong enough to be detected when it reaches the ground. It is therefore important to be able to calculate its amplitude. The solution (7.48) for $E_y^{(1)}$ is substituted in the right-hand side of (7.47). The resulting inhomogeneous first order differential equation for $E_y^{(2)}$ can then be solved at once by using an integrating factor, to give the amplitude of this downgoing wave below the reflecting layer. The formula was given by Wait (1962, § IV.6) who applied it to some special cases and gave further references.

7.9. Liouville method and Schwarzian derivative

The form (7.26) of the W.K.B. solutions is rather complicated because of the integrals $\mp\int q\,dz$ in the exponents. These represent the phases of the two waves, and it would be useful if they could be expressed in simpler form. This suggests that we should try to use a new independent variable ζ, instead of z. It is equivalent to postulating a fictitious 'comparison' medium with a new 'height' coordinate ζ, instead of z. The idea is now to be applied to the differential equation (7.6) and it is convenient to omit the subscript y. A prime $'$ is used to denote $d/d\zeta$. Then

$$E' = \frac{dE}{dz}z', \quad E'' = \frac{d^2E}{dz^2}(z')^2 + \frac{dE}{dz}z'' \tag{7.49}$$

and with (7.6) this gives

$$E'' - E'z''/z' + Ek^2(z')^2q^2 = 0. \tag{7.50}$$

To remove the term in E', the dependent variable is changed to F, thus

$$E = (z')^{\frac{1}{2}}F \tag{7.51}$$

so that

$$F'' + F[(kqz')^2 - (z')^{\frac{1}{2}}\{(z')^{-\frac{1}{2}}\}''] = 0. \tag{7.52}$$

So far there are no approximations. From § 7.2 we know that there are no solutions that represent true progressive waves. We therefore now examine the conditions that will make the second term in the square brackets of (7.52) negligible compared to the first. If, then, there are true progressive wave solutions, the first term must be a constant. Hence let

$$qz' = 1, \text{ so that } \zeta = \int^z q\,dz. \tag{7.53}$$

The lower limit of the integral affects only the origin of ζ and may be assigned later. With this choice of ζ, the 'comparison' medium is free space and the (approximate) progressive wave solutions of (7.52) are

$$F = \exp(\mp ik\zeta). \tag{7.54}$$

From this and (7.53), (7.51)

$$E = q^{-\frac{1}{2}}\exp(\mp ik\int^z q\,dz) \tag{7.55}$$

which are the two W.K.B. solutions.

Liouville (1837) obtained solutions of this form and he used a change of independent variable similar to that used here. In the same year Green (1837) obtained similar solutions. He did not change the independent variable and his method was similar to that of § 7.6.

The neglected term in (7.52) is

$$-(z')^{\frac{1}{2}}\{(z')^{-\frac{1}{2}}\}'' = \tfrac{1}{2}\{(z')^{-1}z''' - \tfrac{3}{2}(z')^{-2}z''\} = \tfrac{1}{2}\mathscr{D}(z;\zeta) \tag{7.56}$$

and $\mathscr{D}$ is called the Schwarzian derivative of z with respect to ζ. Some of its properties are described by Heading (1975c). For the approximation to be justified it is necessary that $\frac{1}{2}\mathscr{D}$ is very small compared with the first term in the square brackets in (7.52). In terms of q and z, from (7.53), this gives

$$|\tfrac{1}{2}\mathscr{D}| = \left|\frac{3}{4}\left(\frac{1}{q^2}\frac{dq}{dz}\right)^2 - \frac{1}{2q^3}\frac{d^2q}{dz^2}\right| = \left|\frac{5}{16q^6}\left\{\frac{d(q^2)}{dz}\right\}^2 - \frac{1}{4q^4}\frac{d^2(q^2)}{dz^2}\right| \ll k^2. \tag{7.57}$$

This is sometimes used as a condition of validity of the W.K.B. solutions (7.55). An alternative way of deriving it is to substitute (7.55) into the differential equation (7.6). It does not fit exactly, and the expression that remains must be small compared with either term of (7.6). When the term k^2q^2E is used, the result is (7.57).

It must be stressed that the criterion (7.57) applies only when the differential equation has the simple form (7.6). It does not apply for vertical polarisation (7.4) (see § 7.12) nor for an anisotropic plasma (see §§ 7.13–7.16).

A criterion such as (7.57) is not entirely satisfactory because although it requires that $|\frac{1}{2}\mathscr{D}|$ is small, it does not say how small, and it does not tell us how good the approximation is when $|\frac{1}{2}\mathscr{D}|$ is known. One possible exact solution of (7.6) may be written

$$E = q^{-\frac{1}{2}}\exp\left(-\mathrm{i}k\int^z q\,\mathrm{d}z\right)\{1+\mathscr{E}(z)\}. \tag{7.58}$$

The great importance of the Liouville method is that it can be extended so as to specify an upper bound for the fractional error $|\mathscr{E}|$ of the W.K.B. solution. The subject is algebraically complicated and beyond the scope of this book. A very clear account is given by Olver (1974, ch. 6) who is the main originator of the subject.

7.10. Conditions for the validity of the W.K.B. solutions

For the W.K.B. solutions (7.26) to be good approximations, the criterion is (7.57). To get an estimate of how small the left-hand side must be, it is useful to study the special case where q^2 is a linear function of z, as in (8.2). Then an exact solution of (7.6) is the Airy integral function $\mathrm{Ai}(\zeta)$, ch. 8, where, from (8.6), $\zeta = -q^2(k/a)^{\frac{2}{3}} = (k^2a)^{\frac{1}{3}}(z-z_0)$. (This ζ is not the same as that in § 7.9). If these are substituted in the left-hand side of (7.57) it gives $\frac{5}{16}k^2/|\zeta^3|$. The W.K.B. solutions in this case are the same as the asymptotic forms for $\mathrm{Ai}(\zeta)$ and it is shown in § 8.10 that if $|\zeta| \geqslant 1$, the error in using them is less than about 8 to 9%. To achieve this degree of accuracy the criterion (7.57) gives

$$\frac{1}{k^2}\left|\frac{3}{4}\left(\frac{1}{q^2}\frac{\mathrm{d}q}{\mathrm{d}z}\right)^2 - \frac{1}{2q^3}\frac{\mathrm{d}^2q}{\mathrm{d}z^2}\right| \leqslant \frac{5}{16}. \tag{7.59}$$

It must be stressed that if $q^2(z)$ is not approximately linear the error in using W.K.B. solutions may exceed 8 to 9% even when (7.59) is satisfied.

In deriving the W.K.B. solutions (7.26) it was assumed that, in the binomial expansion of (7.22)

$$\frac{1}{k}\left|\frac{1}{q^2}\frac{\mathrm{d}q}{\mathrm{d}z}\right| = \frac{1}{k}\left|\frac{\mathrm{d}}{\mathrm{d}z}\left(\frac{1}{q}\right)\right| \ll 1. \tag{7.60}$$

This is a necessary condition, but not sufficient because (7.59) shows that there is a restriction on $\mathrm{d}^2q/\mathrm{d}z^2$ that is not included. But (7.60) is simple, and adequate for most practical applications. It can be made more quantitative by using again the special case where q^2 is a linear function of z. Then for $|\zeta| \geqslant 1$ it can be shown that (7.60) gives

$$\frac{1}{k}\left|\frac{\mathrm{d}}{\mathrm{d}z}\frac{1}{q}\right| \leqslant 0.5. \tag{7.61}$$

This result can be extended so as to apply in an anisotropic ionosphere; see end of § 16.3.

Either (7.59) or (7.60) is a quantitative definition of the term 'slowly varying'. They require that the derivatives $\mathrm{d}q/\mathrm{d}z$ and, for (7.59), $\mathrm{d}^2q/\mathrm{d}z^2$ shall be sufficiently small, and that q is not too small. Because of the factor $1/k$ or $1/k^2$ the condition is most easily satisfied at high frequencies, but no matter how great the frequency nor how small the derivatives, the condition is certain to fail near values of z where q^2 passes through a zero. Then one W.K.B. solution can generate some of the other, and this constitutes the process of 'reflection'. The approximations made in §§ 7.6, 7.8 no longer apply, and to study the reflection process it is necessary to use more accurate solutions of the differential equations. It is these that are supplied by solutions of the Stokes equation, ch. 8, or other equations, ch. 15.

The condition (7.59) fails if the derivatives $\mathrm{d}q/dz$ and $\mathrm{d}^2q/\mathrm{d}z^2$ are large, even when q is not small. An extreme example of this is the reflection at the sharp boundary between two homogeneous media, where the derivatives are infinite. It is shown in § 15.16 that reflection at a steep gradient and reflection at a zero of q can be regarded as different aspects of the same phenomenon.

7.11. Properties of the W.K.B. solutions

The two W.K.B. solutions (7.26) were derived by starting with the idea of a progressive wave in a small region of an inhomogeneous stratified medium. Equation (7.6) has no exact solution that represents a purely progressive wave, except when q^2 is independent of z. In an inhomogeneous medium the W.K.B. solutions represent waves that are the analogues of the progressive waves travelling obliquely in the positive and negative z directions. Equation (7.27) shows that, if the small second term is neglected

$$\mathscr{H}_x = \mp qE_y \tag{7.62}$$

and (7.3) gives

$$\mathscr{H}_z = SE_y. \tag{7.63}$$

$S = \sin\Theta$ can in general be complex, but it is here assumed to be real. These relations are the same as for progressive waves in a homogeneous medium, for all values of q, except that E_y and q depend on z. Thus for either wave, and for all q, the horizontal components of the time averaged Poynting vector $\mathbf{\Pi}_{\mathrm{av}}$ are

$$\Pi_x = \tfrac{1}{2}Z_0^{-1}S|E_y^2|, \quad \Pi_y = 0 \tag{7.64}$$

so Π_x has the same sign as S and is zero only for vertical incidence, $S=0$. If q is real and positive, the minus sign in (7.62) gives $\Pi_z = \frac{1}{2}Z_0^{-\frac{1}{2}} q|E_y^2|$ so that the wave is obliquely upgoing. For the plus sign in (7.62), Π_z has a minus sign and the wave is obliquely downgoing.

In the lower ionosphere, if N increases monotonically with z, and if collisions are neglected, $q^2 = n^2 - S^2 = C^2 - X$ is at first real and decreases monotonically as z increases. Then (7.26) shows that the electric field E_y of either wave increases because of the factor $q^{-\frac{1}{2}}$, and (7.62) shows that the magnetic field decreases, because of the factor $q^{\frac{1}{2}}$. As the value z where $q=0$ is approached, the electric field would become indefinitely large according to (7.26), but before that happens, a place is reached where the W.K.B. solutions can no longer be used because the condition (7.59) is violated. When this region is passed, however, for still larger z, (7.59) is again satisfied and the solutions (7.26) are again valid. Hence two W.K.B. solutions are possible above the level of reflection. Here q^2 is negative and q is purely imaginary so that one W.K.B. solution is given by

$$E_y = Aq^{-\frac{1}{2}} \exp(-k \int^z |q| \mathrm{d}z), \quad \mathscr{H}_x = -qE_y. \tag{7.65}$$

These fields are now in quadrature so that Π_z is zero. Thus there is no average flow of energy in the z direction but the horizontal flow Π_x, (7.64) is still present if $S \neq 0$. There is, stored in the electromagnetic field, some energy which pulsates back and forth in the z direction with twice the wave frequency. Both fields (7.65) decrease as z increases because of the exponential, and for both of them the phase is independent of z. The wave is inhomogeneous, § 2.15, if $S \neq 0$, and evanescent, § 2.14, if $S = 0$.

The other W.K.B. solution where q^2 is negative is

$$E_y = Aq^{-\frac{1}{2}} \exp(k \int^z |q| \mathrm{d}z), \quad \mathscr{H}_x = qE_y. \tag{7.66}$$

It has similar properties to the wave in (7.65) but both fields increase indefinitely as z increases. This wave, therefore, could not be excited by waves incident from the negative z direction, but it could occur if there were sources of waves where z is very large and positive.

In lossy systems where n^2 and q^2 are complex it often happens that q never becomes exactly zero for any real value of z. In many cases of interest, however, there is still a region of the real z axis where $|q|$ is so small that the condition (7.59) is violated. Then the W.K.B. solutions fail and some reflection occurs. If (7.59) is valid for all real values of z, the reflection is extremely small and the wave incident from the negative z direction continues to travel in the direction of positive z until it is absorbed. Its fields are given with good accuracy by the W.K.B. solution, for all real z.

It is often important to know the relation between the amplitudes of the two

W.K.B. solutions where z is negative and where z is positive, with a range of z in between where (7.59) is violated and there is failure of the W.K.B. solutions. Such a relation is called a 'circuit relation'. The main purpose of the papers by Wentzel, Kramers and Brillouin cited in § 7.5 was a study of circuit relations rather than the derivation of the W.K.B. solutions. A knowledge of the circuit relations is essential in the calculation of reflection coefficients, and the subject is therefore studied in some detail in later chapters. See §§ 8.19, 15.13.

7.12. W.K.B. solutions for oblique incidence and vertical polarisation

When the electric field is parallel to the $x-z$ plane, the appropriate differential equations are (7.4), and they are not quite so simple as the set (7.3) which has been discussed in the earlier sections. The field component E_z may be eliminated from (7.4) to give

$$\frac{\mathrm{d}E_x}{\mathrm{d}z} = -\mathrm{i}k\frac{q^2}{n^2}\mathscr{H}_y, \quad \frac{\mathrm{d}\mathscr{H}_y}{\mathrm{d}z} = -\mathrm{i}kn^2E_x \tag{7.67}$$

which corresponds to (7.5) for horizontal polarisation. It is not now possible to obtain a second order differential equation of the simple form (7.6). Elimination of E_x from (7.67) gives

$$\frac{\mathrm{d}^2\mathscr{H}_y}{\mathrm{d}z^2} - \frac{1}{n^2}\frac{\mathrm{d}(n^2)}{\mathrm{d}z}\frac{\mathrm{d}\mathscr{H}_y}{\mathrm{d}z} + k^2q^2\mathscr{H}_y = 0. \tag{7.68}$$

Some important properties of this equation were given by Heading (1970a). The W.K.B. solutions may be found by the method of § 7.6 Let

$$\mathscr{H}_y = \exp\{\mathrm{i}\phi(z)\}, \tag{7.69}$$

compare (7.17), where $\phi(z)$ is the complex phase. Then the differential equation that ϕ must satisfy is

$$\mathrm{i}\frac{\mathrm{d}^2\phi}{\mathrm{d}z^2} - \left(\frac{\mathrm{d}\phi}{\mathrm{d}z}\right)^2 - \frac{\mathrm{i}}{n^2}\frac{\mathrm{d}(n^2)}{\mathrm{d}z}\frac{\mathrm{d}\phi}{\mathrm{d}z} + k^2q^2 = 0, \tag{7.70}$$

compare (7.19). Since the medium is slowly varying, the first and third terms are small and hence to a first approximation

$$\frac{\mathrm{d}\phi}{\mathrm{d}z} = \mp kq, \quad \frac{\mathrm{d}^2\phi}{\mathrm{d}z^2} = \mp k\frac{\mathrm{d}q}{\mathrm{d}z}. \tag{7.71}$$

These are inserted in the first and third terms of (7.70) which gives for the second approximation

$$\frac{\mathrm{d}\phi}{\mathrm{d}z} = \mp\left\{k^2q^2 \mp \mathrm{i}k\left(\frac{\mathrm{d}q}{\mathrm{d}z} - \frac{2q}{n}\frac{\mathrm{d}n}{\mathrm{d}z}\right)\right\}^{\frac{1}{2}} \tag{7.72}$$

where either the upper or the lower signs must be used throughout, for the same reasons as in § 7.6. The last two terms on the right of (7.72) are small. Then expanding by the binomial theorem gives

$$\frac{\mathrm{d}\phi}{\mathrm{d}z} \approx \mp kq + \frac{1}{2}\frac{\mathrm{i}}{q}\frac{\mathrm{d}q}{\mathrm{d}z} - \frac{\mathrm{i}}{n}\frac{\mathrm{d}n}{\mathrm{d}z} \tag{7.73}$$

which may be integrated, and the result inserted in (7.69). This gives

$$\mathscr{H}_y = nq^{-\frac{1}{2}}\exp\left(\mp \mathrm{i}k\int^z q\,\mathrm{d}z\right). \tag{7.74}$$

Now E_x may be found from the second equation (7.67). If derivatives of q and n are neglected

$$E_x = \pm n^{-1}q^{\frac{1}{2}}\exp\left(\mp \mathrm{i}k\int^z q\,\mathrm{d}z\right). \tag{7.75}$$

The expressions (7.74), (7.75) are the W.K.B. solutions for vertical polarisation. They contain a factor $\exp\{\mathrm{i}(\omega t - kSx)\}$ which is omitted as usual; §7.2.

A useful criterion for the validity of (7.74) rests on the use in (7.73) of only two terms of the binomial expansion of (7.72). For this it is necessary that

$$\frac{1}{k}\left|\frac{\mathrm{d}}{\mathrm{d}z}\left(\frac{1}{q}\right) + \frac{2}{qn}\frac{\mathrm{d}n}{\mathrm{d}z}\right| \ll 1, \tag{7.76}$$

compare (7.60). This fails near the level where $q = 0$, no matter how small are the derivatives $\mathrm{d}q/\mathrm{d}z$, $\mathrm{d}n/\mathrm{d}z$. Hence $q = 0$ gives the level of reflection. The condition (7.76) also fails near the level where $n = 0$ which is not a level of reflection if $S \neq 0$. If the electron concentration N increases monotonically with height z, the level where $n = 0$ is above the reflection level where $q = 0$. If these two levels are well separated, the reflection process is unaffected by the failure of (7.76) at the level $n = 0$, and the reflection coefficient is given by (7.152) just as for horizontal polarisation. But if the two levels where $q = 0$ and $n = 0$ are close together, the reflection coefficient may be affected, and a more detailed study of the differential equation is needed to find its true value. This is given in §§15.5–15.7.

7.13. Differential equations for anisotropic ionosphere

When the earth's magnetic field is allowed for, the differential equations (2.45) cannot be separated into two independent sets such as (7.3) and (7.4). They must be treated together as one set. As in the earlier sections it is assumed that the x and y dependence of all field components is through the factor $\exp(-\mathrm{i}kSx)$, so that (6.48) is used. Then Maxwell's equations (2.45) give

$$\frac{\partial E_x}{\partial z} = -\mathrm{i}k(\mathscr{H}_y + SE_z), \quad \frac{\partial E_y}{\partial z} = \mathrm{i}k\mathscr{H}_x, \quad SE_y = \mathscr{H}_z \tag{7.77}$$

and

$$\frac{\partial \mathscr{H}_x}{\partial z} = \mathrm{i}k(\varepsilon_0^{-1}D_y - S\mathscr{H}_z), \quad \frac{\partial \mathscr{H}_y}{\partial z} = -\mathrm{i}k\varepsilon_0^{-1}D_x, \quad S\mathscr{H}_y = \varepsilon_0^{-1}D_z. \tag{7.78}$$

Now

$$\boldsymbol{D} = \varepsilon_0 \boldsymbol{\varepsilon} \boldsymbol{E} = \varepsilon_0 \boldsymbol{E} + \boldsymbol{P} = \varepsilon_0(\boldsymbol{E} + \boldsymbol{M}\cdot\boldsymbol{E}). \tag{7.79}$$

For an electron plasma in which the collision frequency is not velocity dependent, the susceptibility matrix $\boldsymbol{M}$ is as given in § 3.9. It may be used in (7.78) to express $\boldsymbol{D}$ in terms of $\boldsymbol{E}$. In (7.77) and (7.78) the only field components that appear in derivatives are the four horizontal components E_x, E_y, $\mathscr{H}_x$, $\mathscr{H}_y$. We may therefore use the last equations in (7.77) and (7.78) to eliminate $\mathscr{H}_z$ and E_z. This leads to four homogeneous first order differential equations for the horizontal field components. These can very conveniently be written in matrix form. Let $\mathbf{e}$ be the column matrix with four elements E_x, $-E_y$, $\mathscr{H}_x$, $\mathscr{H}_y$. The reason for using $-E_y$ is given in § 7.14 (3), (8). Then the equations (7.77), (7.78) become

$$\mathrm{d}\mathbf{e}/\mathrm{d}z = -ik\mathbf{T}\mathbf{e} \tag{7.80}$$

where $\mathbf{T}$ is the 4×4 matrix

$$\mathbf{T} = \begin{pmatrix} -\dfrac{SM_{zx}}{1+M_{zz}} & \dfrac{SM_{zy}}{1+M_{zz}} & 0 & \dfrac{C^2+M_{zz}}{1+M_{zz}} \\ 0 & 0 & 1 & 0 \\ \dfrac{M_{yz}M_{zx}}{1+M_{zz}} - M_{yx} & C^2 + M_{yy} - \dfrac{M_{yz}M_{zy}}{1+M_{zz}} & 0 & \dfrac{SM_{yz}}{1+M_{zz}} \\ 1+M_{xx} - \dfrac{M_{xz}M_{zx}}{1+M_{zz}} & \dfrac{M_{xz}M_{zy}}{1+M_{zz}} - M_{xy} & 0 & \dfrac{-SM_{xz}}{1+M_{zz}} \end{pmatrix}. \tag{7.81}$$

The algebraic steps of the derivation of (7.80), (7.81) were given in more detail by Budden (1961a).

If the electron collision frequency is assumed to be velocity dependent, as for example in the Sen–Wyller formulae § 3.12, or if the effect of heavy ions is allowed for, the form (7.81) can still be used but the elements of $\boldsymbol{M}$ are not as given in § 3.9. In these cases they can be found from (3.55) which gives $\boldsymbol{\varepsilon} = \boldsymbol{M} + 1$. The elements of $\boldsymbol{\varepsilon}$ are written ε_{ij}, $i, j = x, y, z$. Then $\mathbf{T}$ is given (Walker and Lindsay, 1975) by:

$$\varepsilon_{zz}\mathbf{T} = \begin{pmatrix} -S\varepsilon_{zx} & S\varepsilon_{zy} & 0 & \varepsilon_{zz} - S^2 \\ 0 & 0 & \varepsilon_{zz} & 0 \\ \varepsilon_{yz}\varepsilon_{zx} - \varepsilon_{yz}\varepsilon_{zz} & \varepsilon_{yy}\varepsilon_{zz} - \varepsilon_{yz}\varepsilon_{zy} - S^2\varepsilon_{zz} & 0 & S\varepsilon_{yz} \\ \varepsilon_{xx}\varepsilon_{zz} - \varepsilon_{xz}\varepsilon_{zx} & \varepsilon_{xz}\varepsilon_{zy} - \varepsilon_{xy}\varepsilon_{zz} & 0 & -S\varepsilon_{xz} \end{pmatrix}. \tag{7.82}$$

Equations (7.80), (7.81) were first given by Clemmow and Heading (1954). An almost identical set for the application to optical waves was given by Berreman (1972).

The equations (7.80) are a set of four simultaneous linear first order differential equations. It is possible to eliminate three of the four dependent variables $\mathbf{e}$, and the result would be one linear fourth order differential equation for the remaining variable. Thus we expect (7.80) to have four linearly independent solutions. If it is studied in a range of z where the plasma is homogeneous, the elements of $\mathbf{T}$ are then all independent of z. In such a region, as shown in ch. 6, there are four solutions representing obliquely travelling progressive waves, associated with the four roots of the Booker quartic. For this case, therefore, we seek a progressive wave solution of (7.80) in which all field components depend on z only through a factor $\exp(-ikqz)$. Then it gives

$$(\mathbf{T}-q)\mathbf{e}=0. \tag{7.83}$$

This is a set of four simultaneous algebraic equations for finding the elements of $\mathbf{e}$. It can only have a non-trivial solution if the determinant of the coefficients is zero, that is

$$\det(\mathbf{T}-q)=0. \tag{7.84}$$

This is a fourth degree equation for finding q, and is simply the Booker quartic. It can be verified that it is the same as the various forms of the quartic given in ch. 6. It shows that the four qs are the eigen values of the matrix $\mathbf{T}$, and when they are inserted in (7.83), the resulting solutions for $\mathbf{e}$ are their associated eigen columns.

7.14. Matrix theory

We now derive further properties of the matrix form (7.80) of the differential equations. A column matrix with four elements is denoted by a bold lower case sans serif letter, for example $\mathbf{e}$. A 4×4 square matrix is denoted by a bold face capital sans serif letter, for example $\mathbf{T}$. The transpose of a matrix is indicated by a superscript T. Thus $\mathbf{e}^{\mathrm{T}}$ is a row matrix with four elements.

In this section some important results are derived and it is useful, for future reference, to set them out in a numbered sequence.

(1) *Eigen columns and eigen rows of* $\mathbf{T}$

Let the four values of q be q_i, $i=1,2,3,4$, and let the associated eigen columns of $\mathbf{T}$ be $\mathbf{s}_i$. The four $\mathbf{s}_i$ are used as the four columns of a 4×4 matrix $\mathbf{S}$. For the present it is assumed that the four q_i are distinct, so that the four $\mathbf{s}_i$ are linearly independent and the matrix $\mathbf{S}$ is non-singular. The case where two qs approach equality is dealt with in § 16.3. From (7.83)

$$\mathbf{T}\mathbf{s}_i=q_i\mathbf{s}_i, \quad \mathbf{T}\mathbf{S}=\mathbf{S}\mathbf{Q} \tag{7.85}$$

where $\mathbf{Q}$ is the 4×4 diagonal matrix with elements q_i. Multiplication by $\mathbf{S}^{-1}$ on both sides gives

$$\mathbf{S}^{-1}\mathbf{T}=\mathbf{Q}\mathbf{S}^{-1}, \tag{7.86}$$

which shows that the rows of $\mathbf{S}^{-1}$ are eigen rows of $\mathbf{T}$ with the eigen values q_i.

If any $\mathbf{s}_i$ is multiplied by a constant, it remains an eigen column of $\mathbf{T}$. To complete the definitions of $\mathbf{s}_i$ and $\mathbf{S}$ it is necessary to impose some condition on the field amplitudes. In other words the $\mathbf{s}_i$ must be normalised. One possible rule for this, used in most of this book, is given in (7) below, and the reason for choosing it is explained in (8). Other normalisation methods are possible and this is discussed in §§ 16.2, 16.3.

(2) *Integrating factor*

In the study of differential equations an integrating factor is often used. The technique is described in books on differential equations, for example Ince (1927, § 2.2), Birkhoff and Rota (1962, chs. I. 5, II. 5). It is applied here to the matrix equation (7.80).

We seek, as an integrating factor, a row matrix $\mathbf{g}^{\mathrm{T}}$ that multiplies (7.80) on the left so that the result is a perfect differential, thus

$$\mathbf{g}^{\mathrm{T}}\mathrm{d}\mathbf{e}/\mathrm{d}z + ik\mathbf{g}^{\mathrm{T}}\mathbf{T}\mathbf{e} = \mathrm{d}W/\mathrm{d}z = 0 \tag{7.87}$$

where W is a scalar. This is satisfied if

$$\mathrm{d}\mathbf{g}^{\mathrm{T}}/\mathrm{d}z = ik\mathbf{g}^{\mathrm{T}}\mathbf{T}, \tag{7.88}$$

and

$$W = \mathbf{g}^{\mathrm{T}}\mathbf{e}. \tag{7.89}$$

Then W is called the bilinear concomitant.

The transpose of (7.88) is

$$\mathrm{d}\mathbf{g}/\mathrm{d}z = ik\mathbf{T}^{\mathrm{T}}\mathbf{g}. \tag{7.90}$$

(3) *Symmetry properties of* $\mathbf{T}$

The matrix $\mathbf{T}$ uses the elements of $\boldsymbol{M}$ (3.35). If the vector $\boldsymbol{Y}$ is reversed in direction, that is if l_x, l_y, l_z are all reversed in sign, $\boldsymbol{M}$ is changed to a new matrix $\bar{\boldsymbol{M}}$ which is simply $\boldsymbol{M}^{\mathrm{T}}$. When this change is made in (7.81), let the resulting matrix be denoted by $\bar{\mathbf{T}}$. Then inspection of (7.81) shows that $\bar{\mathbf{T}}$ is obtained by transposing $\mathbf{T}$ about its trailing diagonal, thus

$$\bar{\mathrm{T}}_{ij} = \mathrm{T}_{5-j,5-i} \tag{7.91}$$

(Budden and Clemmow, 1957). This property still holds when the more general form (7.82) of $\mathbf{T}$ is used. To write it in matrix notation introduce the 4×4 matrix used by Lacoume (1967) and Bennett (1976):

$$\mathbf{B} = \begin{pmatrix} 0 & 0 & 0 & 1 \\ 0 & 0 & 1 & 0 \\ 0 & 1 & 0 & 0 \\ 1 & 0 & 0 & 0 \end{pmatrix} \tag{7.92}$$

with the property

$$\mathbf{B}^{\mathrm{T}} = \mathbf{B}^{-1} = \mathbf{B}. \tag{7.93}$$

Then

$$\bar{\mathbf{T}} = \mathbf{B}\mathbf{T}^{\mathrm{T}}\mathbf{B}, \quad \mathbf{T}^{\mathrm{T}} = \mathbf{B}\bar{\mathbf{T}}\mathbf{B}. \tag{7.94}$$

This property of $\mathbf{T}$ would not have held if E_y had been given a plus sign in the definition of $\mathbf{e}$, § 7.13. This is one reason why E_y has a minus sign.

(4) *Adjoint matrix. Adjoint differential equation*

Now $\mathbf{T}^{\mathrm{T}}$ from (7.94) is substituted in (7.90) and the result is multiplied on the left by $\mathbf{B}$ to give

$$\mathrm{d}\bar{\mathbf{e}}/\mathrm{d}z = \mathrm{i}k\bar{\mathbf{T}}\bar{\mathbf{e}} \tag{7.95}$$

where

$$\bar{\mathbf{e}} = \mathbf{B}\mathbf{g}. \tag{7.96}$$

Here $\bar{\mathbf{e}}$, $\bar{\mathbf{T}}$ are called the adjoints of $\mathbf{e}$, $\mathbf{T}$, and (7.95) or its equivalent form (7.88) is the differential equation adjoint to (7.80). These adjoints, and the bilinear concomitant W, were introduced into the theory of radio propagation by Suchy and Altman (1975a) in a most important paper. For further discussion see § 14.12.

Let $\mathbf{m}$ be a square matrix with elements m_{ij} and let M_{ij} be the cofactor of m_{ij}. Then the matrix $\mathbf{M}$ whose elements are M_{ji} is commonly called the 'adjoint' of $\mathbf{m}$, and less commonly the 'adjugate'. Note that it is the transpose of the matrix of the cofactors. Here, as in Jeffreys and Jeffreys (1972, p. 117), the term 'adjugate' will be used. The adjoint form $\bar{\mathbf{T}}$ of $\mathbf{T}$ as defined by (7.94) is not the same as the adjugate of $\mathbf{T}$.

The elements of $\bar{\mathbf{e}}$ will be written $\bar{E}_x$, $-\bar{E}_y$, $\bar{\mathscr{H}}_x$, $\bar{\mathscr{H}}_y$. Then (7.95) shows that $\bar{\mathbf{e}}$ gives these field components in a fictitious 'adjoint' medium in which the direction of $\boldsymbol{Y}$, and therefore of the earth's magnetic field $\boldsymbol{B}$, is reversed, and the sign of $k = \omega/c$ is reversed. The sign of ω is reversed in the factor $\exp(\mathrm{i}\omega t)$, see (7.100) below, but not in $Z = \nu/\omega$ and not where it appears in ε_1, ε_2, ε_3. The adjoint fields $\bar{\mathbf{e}}$ are introduced only to assist the theory and need not satisfy any physical conditions. For example in a range of z where the medium is homogeneous with losses, (7.95) has a solution proportional to $\exp(+\mathrm{i}qkz)$ where q is any root of the Booker quartic. If it applies to an upgoing wave, $\mathrm{Im}(q)$ is negative, so the adjoint fields *increase* as z increases.

(5) *The bilinear concomitant*

From (7.89), (7.96):

$$W = \bar{\mathbf{e}}^{\mathrm{T}}\mathbf{B}\mathbf{e} = \bar{\mathscr{H}}_y E_x - \bar{\mathscr{H}}_x E_y - \bar{E}_y \mathscr{H}_x + \bar{E}_x \mathscr{H}_y. \tag{7.97}$$

This expression is very similar to (2.65) for $4Z_0\Pi_z$ where Π_z is the z component of the time averaged Poynting vector. The only difference is that, instead of using complex conjugates E_x^* etc., we use the adjoints $\bar{E}_x$ etc. The adjoints resemble the complex conjugates in that neither of them satisfies Maxwell's equations nor any physical conditions. But they are needed in the definitions of W and Π_z respectively.

It is shown later § 14.12, that W is the z component of a vector $\boldsymbol{W}$, the bilinear concomitant vector, which is a generalisation of the time averaged Poynting vector Π_{av}. From its definition (2.64), Π_{av} is necessarily real. But $\boldsymbol{W}$ is in general complex. This applies particularly in a medium with losses.

(6) *Eigen columns and eigen values of* $\bar{\mathbf{T}}$

Let $\mathbf{r}$ be an eigen column of $\bar{\mathbf{T}}$ with eigen value q, so that

$$q\mathbf{r} = \bar{\mathbf{T}}\mathbf{r}. \tag{7.98}$$

Transpose this, multiply on the right by $\mathbf{B}$, and use (7.94). Then

$$q\mathbf{r}^{\mathrm{T}}\mathbf{B} = \mathbf{r}^{\mathrm{T}}\bar{\mathbf{T}}^{\mathrm{T}}\mathbf{B} = \mathbf{r}^{\mathrm{T}}\mathbf{B}\mathbf{B}\bar{\mathbf{T}}^{\mathrm{T}}\mathbf{B} = \mathbf{r}^{\mathrm{T}}\mathbf{B}\mathbf{T} \tag{7.99}$$

which shows, first, that the eigen values q of $\bar{\mathbf{T}}$ are the same as those of $\mathbf{T}$. This also follows by transposition of the determinant (7.84) about its trailing diagonal. Second, (7.99) shows that $\mathbf{r}^{\mathrm{T}}\mathbf{B}$ is an eigen row of $\mathbf{T}$. Then (7.86) shows that we are free to choose $\mathbf{r}$ so that $\mathbf{r}^{\mathrm{T}}\mathbf{B}$ is a row of $\mathbf{S}^{-1}$. The eigen columns of $\bar{\mathbf{T}}$ are the rows of $\mathbf{S}^{-1}$ with the order of the elements reversed.

The field components given by the elements of $\mathbf{S}$ all depend on x, y and t through a factor $\exp\{\mathrm{i}(\omega t - kSx)\}$ which is usually omitted; § 6.2. This suggests that the adjoint fields given by the elements of $\mathbf{S}^{-1}$ should contain a factor $\exp\{-\mathrm{i}(\omega t - kSx)\}$. Then in the adjoint of a homogeneous medium, (7.95) is satisfied and the four progressive wave solutions are

$$(\mathbf{S}^{-1})_j \exp\{\mathrm{i}(-\omega t + kSx + kq_j z)\} \tag{7.100}$$

where the first factor is the jth row of $\mathbf{S}^{-1}$. In all terms of the exponent the signs of k and ω have therefore been reversed.

For a progressive wave in a homogeneous medium choose the jth column of $\mathbf{S}$, and for the adjoint solution $\mathbf{g}^{\mathrm{T}}$ of (7.88) choose (7.100). Then the bilinear concomitant W is $+1$.

(7) *Normalisation of* $\mathbf{S}$

In this section it is assumed that the exponential factors $\exp\{\pm\mathrm{i}(\omega t - kSx)\}$ are omitted. Specific expressions for $\mathbf{S}$ and $\mathbf{S}^{-1}$ are given by (7.118), (7.121) in §7.16, and it is there shown that for any column of $\mathbf{S}$ and the same numbered row of $\mathbf{S}^{-1}$, here taken as the jth column and row,

$$(\mathbf{S}^{-1})_{j3}/\mathrm{S}_{2j} = (\mathbf{S}^{-1})_{j2}/\mathrm{S}_{3j} \quad \text{(not summed).} \tag{7.101}$$

Hence we now choose for the normalisation of $\mathbf{S}$:

$$\mathrm{S}_{2j} = (\mathbf{S}^{-1})_{j3}, \quad \mathrm{S}_{3j} = (\mathbf{S}^{-1})_{j2}. \tag{7.102}$$

The reason for this choice is given in (8) below.

(8) *Bilinear concomitant, and Poynting vector*

Consider now the special case where the ionosphere is loss-free, so that the qs are either real or in complex conjugate pairs; see § 6.4. Then inspection of (7.81) with (3.35) or (3.55) shows that reversal of the direction of Y is just the same as changing the sign of i. The adjoint $\tilde{\mathbf{T}}$ is the same as the complex conjugate $\mathbf{T}^*$.

First let q be real and let the elements of its associated column in $\mathbf{S}$ be E_x, $-E_y$, $\mathscr{H}_x$, $\mathscr{H}_y$. This is an eigen column of $\mathbf{T}$. Suppose that it applies for an upgoing wave. The elements of the corresponding row of $\mathbf{S}^{-1}$ are those of the eigen column of $\tilde{\mathbf{T}}$, that is of $\mathbf{T}^*$, in reversed order, and any multiple of this may be chosen as the adjoint solution

$$\mathbf{g}^{\mathrm{T}} = \mathscr{H}_y^*, \mathscr{H}_x^*, -E_y^*, E_x^* \tag{7.103}$$

of (7.88). Then the bilinear concomitant is

$$W = E_x\mathscr{H}_y^* - E_y\mathscr{H}_x^* - E_y^*\mathscr{H}_x + E_x^*\mathscr{H}_y = 4Z_0\Pi_z, \tag{7.104}$$

where (2.64) has been used. For the upgoing wave on the real z axis this is real and positive. For a downgoing wave we may choose, for $\mathbf{g}^{\mathrm{T}}$, the set (7.103) with the signs on the right reversed, and this would make W real and negative. The normalisation in (7) above was deliberately chosen so that, for a progressive wave with a real q in a loss-free ionosphere, the adjoint set (7.103) is the complex conjugate of the horizontal field components.

Next let q and q^* be two conjugate complex solutions of the quartic and let them be associated with the first and second columns of $\mathbf{S}$. Let the elements of these two columns be

$$\mathbf{s}_1 = E_x(q), -E_y(q), \mathscr{H}_x(q), \mathscr{H}_y(q): \quad \mathbf{s}_2 = E_x(q^*), -E_y(q^*), \mathscr{H}_x(q^*), \mathscr{H}_y(q^*) \tag{7.105}$$

respectively. These are identical except for the change from q to q^*. Let the two corresponding rows of $\mathbf{S}^{-1}$ be $\mathbf{r}_1^{\mathrm{T}}$, $\mathbf{r}_2^{\mathrm{T}}$. They are not the complex conjugates of $\mathbf{s}_1$, $\mathbf{s}_2$ respectively, with the elements in reversed order, because in forming them we do not replace q by q^* or vice versa. But clearly

$$\mathbf{r}_2^{\mathrm{T}} = (\mathbf{s}_1^{\mathrm{T}}\mathbf{B})^*, \quad \mathbf{r}_1^{\mathrm{T}} = (\mathbf{s}_2^{\mathrm{T}}\mathbf{B})^*. \tag{7.106}$$

The product $\mathbf{r}_2^{\mathrm{T}}\mathbf{s}_1$ is $4Z_0\Pi_z$ for the wave associated with q, and $\mathbf{r}_1^{\mathrm{T}}\mathbf{s}_2$ is $4Z_0\Pi_z$ for that associated with q^*. But both these products are zero. This confirms the statement made in § 6.4, that for waves associated with complex conjugate qs in a loss-free medium, the z components of the time averaged Poynting vectors are zero.

The association of the bilinear concomitant product (7.104) with the Poynting vector is another consequence of the use of $-E_y$ in the definition of $\mathbf{e}$ in § 7.13, and in (7.103).

7.15. W.K.B. solutions for anisotropic ionosphere

Two different methods have been given for deriving the W.K.B. solutions when the earth's magnetic field is neglected. In the first, § 7.6, the differential equation (7.6) for

E_y was formulated and the change of variable $E_y = A\exp\{i\phi(z)\}$ (7.17) was then made, leading to a non-linear differential equation (7.19) for ϕ. To use the same method in the more general case it would be necessary to formulate the differential equation for one field variable. We have already seen, near the end of § 7.13, that this is of fourth order, and it is too complicated to be worth deriving specifically, although in principle the method could be used. It is better, however, to extend the second method, § 7.8, in which the fields were expressed as the sum of partial fields for the upgoing and downgoing waves. In this way the differential equations were expressed in coupled form (7.46), (7.47). Neglect of the coupling terms on the right-hand sides gave two separate differential equations whose solutions were the W.K.B. solutions for the upgoing and downgoing waves. This method will now be extended to the general case. The fields are to be expressed as the sum of the partial fields for the four characteristic waves, and the differential equations are expressed in the general coupled form (7.109) below. Neglect of the coupling terms gives four separate equations whose solutions are the four W.K.B. solutions.

Suppose that the ionospheric plasma varies slowly with the height z so that $\mathbf{T}$ and $\mathbf{S}$ are functions of z. At any height z a solution $\mathbf{e}(z)$ of (7.80) can be expressed as a linear combination of the four characteristic waves, that is the four columns of $\mathbf{S}$, with factors f_j as in (6.53), § 6.10. In a homogeneous ionosphere the four f_j would be proportional to $\exp(-ikq_j z)$. We now enquire how they vary with z in a slowly varying ionosphere. The following method is the exact analogue of the 'coupling' method used in § 7.8 for an isotropic ionosphere. It was first given by Clemmow and Heading (1954). In the following theory the summation convention, § 2.11, for repeated suffixes is not used.

A prime ′ is now used to denote $k^{-1}\,\mathrm{d}/\mathrm{d}z$. Then insertion of (6.53) into (7.80) gives

$$(\mathbf{Sf})' = -\mathrm{i}\mathbf{TSf} \tag{7.107}$$

so that

$$\mathbf{f}' + \mathrm{i}\mathbf{S}^{-1}\mathbf{TSf} = \mathbf{\Gamma f}, \quad \text{where} \quad \mathbf{\Gamma} = -\mathbf{S}^{-1}\mathbf{S}' \tag{7.108}$$

whence from (7.85)

$$\mathbf{f}' + \mathrm{i}\mathbf{Qf} = \mathbf{\Gamma f} \tag{7.109}$$

where $\mathbf{Q}$ is the 4×4 diagonal matrix whose elements are the four qs. The terms on the right correspond to the terms on the right of (7.46), (7.47). They are called 'coupling terms', and $\mathbf{\Gamma}$ is called the 'coupling matrix'; see ch. 16. In a homogeneous medium they are zero because $\mathbf{S}'$ and thence $\mathbf{\Gamma}$ are zero. In a sufficiently slowly varying medium they are small and as a first approximation they are neglected. Then (7.109) is four separate first order differential equations

$$\mathrm{f}_j' + \mathrm{i}q_j\mathrm{f}_j = 0 \tag{7.110}$$

with the solutions

$$\mathsf{f}_j = \mathsf{K}_j \exp\left(-ik \int_0^z q_j \mathrm{d}z \right) \quad j = 1, 2, 3, 4 \tag{7.111}$$

where K_j is a constant. The contribution of one of these terms to the solution $\mathbf{e}$ is, from (6.53),

$$\mathbf{e} = \mathsf{K}_j \mathbf{s}_j \exp\left(-ik \int_0^z q_j \mathrm{d}z \right) \tag{7.112}$$

where $\mathbf{s}_j$ is the jth column of $\mathbf{S}$. These are here taken to be the four W.K.B. solutions for the anisotropic ionosphere. The full expression is given later at (7.136).

In an exactly similar way, let the adjoint solution $\bar{\mathbf{e}}^{\mathrm{T}}$ be given by

$$\bar{\mathbf{e}}^{\mathrm{T}} = \bar{\mathbf{f}}^{\mathrm{T}} \mathbf{S}^{-1}. \tag{7.113}$$

It must satisfy the adjoint differential equation (7.88), whence it can be shown that, if coupling terms are neglected,

$$\bar{\mathsf{f}}_j' - iq_j \bar{\mathsf{f}}_j = 0 \tag{7.114}$$

with the solution

$$\bar{\mathsf{f}}_j = \bar{\mathsf{K}}_j \exp\left(ik \int_0^z q_j \mathrm{d}z \right) \tag{7.115}$$

where $\bar{\mathsf{K}}_j$ is another constant.

Thus the adjoint of (7.112) is

$$\bar{\mathbf{e}} = \bar{\mathsf{K}}_j \mathbf{r}_j \exp\left(ik \int_0^z q_j \mathrm{d}z \right) \tag{7.116}$$

where $\mathbf{r}_j$ is the jth row of $\mathbf{S}^{-1}$. The W.K.B. solution (7.112) and its adjoint (7.116) give a bilinear concomitant $W = \mathsf{K}_j \bar{\mathsf{K}}_j$ that is independent of z. This means that for the W.K.B. solution in the special case of a loss-free ionosphere and a progressive wave, the z component Π_z of the time averaged Poynting vector is independent of z (compare § 7.4). It is easily verified that for an isotropic ionosphere with losses, the bilinear concomitant W is independent of z for the W.K.B. solutions (7.12) and (7.13) (horizontal polarisation) and for (7.74), (7.75) (vertical polarisation).

Equation (7.110) for f_j should, strictly speaking, have a term on the right $\Gamma_{jj}\mathsf{f}_j$ (not summed) from the diagonal element of $\mathbf{\Gamma}$ in (7.109). Then (7.111) would become

$$\mathsf{f}_j = \mathsf{K}_j \exp(-ik \int^z q_j \mathrm{d}z + k \int^z \Gamma_{jj} \mathrm{d}z) \quad \text{(not summed).} \tag{7.117}$$

The exponential in (7.111) has, in § 7.3, been called the phase memory term, and the new factor $\exp(k\int^z \Gamma_{jj} \mathrm{d}z)$ is called 'additional memory' (Budden and Smith, 1976). In some important special cases the elements Γ_{jj} are zero; see § 16.2. When they are not zero they ought to be included in the W.K.B. solutions (7.111). Some consequences of including them are described in § 16.9.

Suppose that the matrix $\mathbf{S}$ used to give $\mathbf{s}_j$ in (7.112) has been normalised in a different way so that the column $\mathbf{s}_j$ is replaced by $b(z)\mathbf{s}_j$. Then it can be shown that the element Γ_{jj} now has an extra added term $-b'(z)/b(z)$ and the additional memory term $\exp(k\int^z \Gamma_{jj} dz)$ therefore has a factor $1/b(z)$ that cancels the $b(z)$ multiplying $\mathbf{s}_j$. Thus, if the additional memory term is included, the W.K.B. solution (7.112) is unaffected by the choice of normalisation of $\mathbf{S}$.

7.16. The matrices S and $\mathbf{S}^{-1}$

Any column of $\mathbf{S}$ is a solution of the set of four homogeneous equations (7.85); see §6.10 and end of §7.13. It is proportional to the cofactors of any row of the matrix $\mathbf{T}-q_j\mathbf{1}$, where q_j applies to the jth column of $\mathbf{S}$. Because of the zeros in the second row and third column of $\mathbf{T}$, (7.81), it is simplest to choose the third row. Then

$$S_{1j}:S_{2j}:S_{3j}:S_{4j} \propto a_3 q_j + a_4 : A_j : q_j A_j : a_5 q_j + a_6 \tag{7.118}$$

where

$$\left.\begin{aligned} A_j = q_j^2 + a_1 q_j + a_2, \\ a_1 = -(T_{11}+T_{44}), \quad a_2 = T_{11}T_{44} - T_{14}T_{41}, \quad a_3 = T_{12}, \\ a_4 = T_{14}T_{42} - T_{12}T_{44}, \quad a_5 = T_{42}, \quad a_6 = T_{12}T_{41} - T_{11}T_{42}. \end{aligned}\right\} \tag{7.119}$$

Similarly, from §7.14(6), the jth row of $\mathbf{S}^{-1}$ is proportional to the jth row of the adjoint matrix $\bar{\mathbf{T}}-q_j\mathbf{1}$, with the order reversed. The adjoints $\bar{a}_k$ of the a_k are found from (7.119) by using $\bar{\mathbf{T}}$, that is $\mathbf{T}$ transposed about its trailing diagonal. Thus

$$\begin{aligned} &\bar{a}_1 = a_1, \quad \bar{a}_2 = a_2, \quad \bar{A}_j = A_j \\ &\bar{a}_3 = T_{34}, \quad \bar{a}_4 = T_{14}T_{31} - T_{34}T_{11}, \bar{a}_5 = T_{31}, \bar{a}_6 = T_{34}T_{41} - T_{44}T_{31}. \end{aligned} \tag{7.120}$$

Hence

$$(\mathbf{S}^{-1})_{j1}:(\mathbf{S}^{-1})_{j2}:(\mathbf{S}^{-1})_{j3}:(\mathbf{S}^{-1})_{j4} \propto \bar{a}_5 q_j + a_6 : q_j A_j : A_j : \bar{a}_3 q_j + \bar{a}_4. \tag{7.121}$$

For the electron plasma, from (7.81), the expressions for $a_1 - a_6$ are as follows. Let

$$D = X/\{U(U^2-Y^2) - X(U^2 - l_z^2 Y^2)\}. \tag{7.122}$$

Then

$$\left.\begin{aligned} a_1 &= 2SY^2 l_x l_z D, \\ a_2 &= -C^2 + \{U(U-X) - Y^2(l_x^2 C^2 + l_z^2 S^2)\}D, \\ a_3 &= S(l_y l_z Y^2 - \mathrm{i} l_x U Y)D, \\ a_4 &= -\{C^2 l_x l_y Y^2 + \mathrm{i} l_z Y(UC^2 - X)\}D, \\ a_5 &= -\{l_x l_y Y^2 + \mathrm{i} l_z Y(U-X)\}D, \\ a_6 &= S\{l_y l_z Y^2 - \mathrm{i} l_x Y(U-X)\}D. \end{aligned}\right\} \tag{7.123}$$

For the more general plasma, from (7.82) the expressions are (G and J are given by (3.51)):

$$\left.\begin{aligned}
a_1 &= -2Sl_xl_zG/\varepsilon_{zz},\\
a_2 &= [S^2\{\varepsilon_3 + (l_y^2 + l_z^2)G\} - \{\tfrac{1}{2}\varepsilon_3(\varepsilon_1 + \varepsilon_2) - l_yJ\}]/\varepsilon_{zz},\\
a_3 &= -S\{l_yl_zG + \tfrac{1}{2}il_x(\varepsilon_1 - \varepsilon_2)\}/\varepsilon_{zz},\\
a_4 &= [S^2\{-l_xl_yG + \tfrac{1}{2}il_z(\varepsilon_1 - \varepsilon_2)\} - \{l_xl_yJ + \tfrac{1}{2}il_z\varepsilon_3(\varepsilon_1 - \varepsilon_2)\}]/\varepsilon_{zz},\\
a_5 &= -\{l_yl_zJ + \tfrac{1}{2}il_z\varepsilon_3(\varepsilon_1 - \varepsilon_2)\}/\varepsilon_{zz}\\
a_6 &= S\{l_yl_zJ - \tfrac{1}{2}il_z\varepsilon_3(\varepsilon_1 - \varepsilon_2)\}/\varepsilon_{zz}
\end{aligned}\right\} \tag{7.124}$$

where

$$\varepsilon_{zz} = \tfrac{1}{2}(\varepsilon_1 + \varepsilon_2) - l_z^2G. \tag{7.125}$$

The sets (7.118), (7.121) must be normalised by dividing them by suitable factors. Here we give the result of applying the rule in §7.14(7). In (7.118), (7.121) the condition (7.102) is already satisfied so that, for both, the required normalising factor N_j is the same. It is found from the condition that the diagonal elements of $\mathbf{S}^{-1}\mathbf{S}$ are unity. Hence

$$N_j^2 = (a_3q_j + a_4)(\bar{a}_5q_j + \bar{a}_6) + 2q_jA_j^2 + (a_5q_j + a_6)(\bar{a}_3q_j + \bar{a}_4). \tag{7.126}$$

Now let

$$F_1 = (q_1 - q_2)(q_1 - q_3)(q_1 - q_4) \tag{7.127}$$

and let F_2, F_3, F_4 be obtained from this by cyclic permutation of the suffixes. The quartic found by multiplying out the determinant (7.84) is

$$(q^2 - \mathrm{T}_{32})(q^2 + a_1q + a_2) - (a_3\bar{a}_5 + \bar{a}_3a_5)q - (a_4\bar{a}_5 + a_6\bar{a}_3) = 0 \tag{7.128}$$

from which F_j is found by differentiating with respect to q and then setting $q = q_j$. This gives

$$F_j = 2q_jA_j + (q_j^2 - \mathrm{T}_{32})(2q_j + a_1) - (a_3\bar{a}_5 + \bar{a}_3a_5) \tag{7.129}$$

whence

$$\begin{aligned}
A_jF_j &= 2q_jA_j^2 + (2q_j + a_1)\{(a_3\bar{a}_5 + \bar{a}_3a_5)q_j + a_4\bar{a}_5 + a_6\bar{a}_3\} - (a_3\bar{a}_5 + \bar{a}_3a_5)A_j\\
&= 2q_jA_j^2 + q_j^2(a_3\bar{a}_5 + \bar{a}_3a_5) + 2q_j(a_4\bar{a}_5 + a_6\bar{a}_3)\\
&\quad + a_1(a_4\bar{a}_5 + a_6\bar{a}_3) - a_2(a_3\bar{a}_5 + \bar{a}_3a_5).
\end{aligned} \tag{7.130}$$

Now from (7.119), (7.120) it can be shown that

$$a_4\bar{a}_5 + a_6\bar{a}_3 = a_3\bar{a}_6 + a_5\bar{a}_4 \tag{7.131}$$

$$a_1(a_4\bar{a}_5 + a_6\bar{a}_3) - a_2(a_3\bar{a}_5 + \bar{a}_3a_5) = a_4\bar{a}_6 + \bar{a}_4a_6 \tag{7.132}$$

whence from (7.130), (7.126)

$$A_j F_j = N_j^2. \tag{7.133}$$

Then we finally get for a column of $\mathbf{S}$ from (7.118)

$$(\mathsf{S}_{1j}, \mathsf{S}_{2j}, \mathsf{S}_{3j}, \mathsf{S}_{4j}) = (A_j F_j)^{-\frac{1}{2}}(a_3 q_j + a_4, A_j, q_j A_j, a_5 q_j + a_6) \tag{7.134}$$

and for a row of $\mathbf{S}^{-1}$ from (7.121)

$$\{(\mathbf{S}^{-1})_{j1}, (\mathbf{S}^{-1})_{j2}, (\mathbf{S}^{-1})_{j3}, (\mathbf{S}^{-1})_{j4}\} = (A_j F_j)^{-\frac{1}{2}}(\bar{a}_5 q_j + \bar{a}_6, q_j A_j, A_j, \bar{a}_3 q_j + \bar{a}_4). \tag{7.135}$$

With (7.134) it is now possible to write specific expressions for the horizontal field components of the W.K.B. solution (7.112), thus

$$(E_x, -E_y, \mathscr{H}_x, \mathscr{H}_y)_j = (A_j F_j)^{-\frac{1}{2}}(a_3 q_j + a_4, A_j, q_j A_j, a_5 q_j + a_6)\exp\left(-\mathrm{i}k\int^z q_j \mathrm{d}z\right) \tag{7.136}$$

where the omission of the lower limit of the integral is equivalent to the inclusion of an arbitrary constant multiplier, such as K_j in (7.112).

The factor $F_j^{-\frac{1}{2}}$ in (7.136) is infinite, from (7.127), when q_j is equal to any other q. It is in this condition that $\mathbf{S}$ is singular and the arguments of § 7.14 (1) onwards must fail. Thus we expect the W.K.B. solution (7.136) to be invalid in regions of the complex z plane near where q_j is equal to any other q. It does not necessarily fail when q_j is zero, nor when two qs are equal if neither of them is q_j.

One important case of this failure is when $X \to 0$ so that the medium is free space. Then the four qs are in two equal pairs $\pm C$. It can be shown that for small X the denominator $(A_j F_j)^{\frac{1}{2}}$ in (7.136) is proportional to X but the four elements in brackets are also proportional to X. Thus each of the four field expressions (7.136) tends to a bounded non-zero limit when $X \to 0$. The limits are functions of $U = 1 - \mathrm{i}Z$ and may vary with height z when no electrons are present. This behaviour is of interest in the study of the phenomenon of 'limiting polarisation', discussed in §§ 17.10, 17.11.

7.17. W.K.B. solutions for vertical incidence

For vertical incidence, $S = 0$, the axes may be chosen so that $l_y = 0$. They are then the same as the x, y, z axes of ch. 4. For the electron plasma (7.123) shows that

$$\begin{aligned} a_1 = a_3 = \bar{a}_3 = a_6 = \bar{a}_6 = 0 \\ a_2 = -1 + \{U(U-X) - l_x^2 Y^2\}D \\ a_4 = a_5 = -\bar{a}_4 = -\bar{a}_5 = -\mathrm{i}l_z Y(U-X)D. \end{aligned} \tag{7.137}$$

The roots q are the refractive indices n and the quartic (7.128) is

$$q^4 + q^2(a_2 - \mathsf{T}_{32}) + a_4^2 - a_2\mathsf{T}_{32} = 0 \tag{7.138}$$

which is the same as the dispersion relation (4.65). Its solutions may be taken as

$$q_1 = -q_2 = n_\mathrm{O}, \; q_3 = -q_4 = n_\mathrm{E} \tag{7.139}$$

where n_O, n_E are the refractive indices for the ordinary and extraordinary waves.

Then

$$A_1 = A_2 = n_O^2 + a_2, \quad A_3 = A_4 = n_E^2 + a_2, \quad A_1 A_3 = a_4^2 \tag{7.140}$$

$$F_1 = -F_2 = 2n_O(n_O^2 - n_E^2), \quad F_3 = -F_4 = -2n_E(n_O^2 - n_E^2). \tag{7.141}$$

The expression (7.134) for a column of $\mathbf{S}$ may now be written down. In particular

$$(S_{11}, S_{21}, S_{31}, S_{41}) = (A_1 F_1)^{-\frac{1}{2}}\{a_4, n_O^2 + a_2, n_O(n_O^2 + a_2), a_4 n_O\} \tag{7.142}$$

which refers to the upgoing ordinary wave and shows that for this wave

$$\frac{E_y}{E_x} = -\frac{\mathscr{H}_x}{\mathscr{H}_y} = -\frac{n_O^2 + a_2}{a_4} = -\frac{A_1}{a_4} = \rho_O \tag{7.143}$$

from the definition (4.13), (4.14) of ρ. The same result is obtained for the downgoing ordinary wave (see end of § 4.4 for remarks on the sign of ρ). Similarly for the upgoing or downgoing extraordinary waves

$$\frac{E_y}{E_x} = -\frac{n_E^2 + a_2}{a_4} = -\frac{A_3}{a_4} = \rho_E. \tag{7.144}$$

Equations (7.140) now show that $\rho_O \rho_E = 1$, in agreement with (4.29). The field components E_x, $-E_y$, $\mathscr{H}_x$, $\mathscr{H}_y$ given by the four W.K.B. solutions (7.136) are then

Upgoing ordinary wave

$$-\{2(\rho_O^2 - 1)n_O\}^{-\frac{1}{2}}(1, -\rho_O, -\rho_O n_O, n_O)\exp\left(-ik\int^z n_O \mathrm{d}z\right). \tag{7.145}$$

Downgoing ordinary wave

$$-\{-2(\rho_O^2 - 1)n_O\}^{-\frac{1}{2}}(1, -\rho_O, \rho_O n_O, -n_O)\exp\left(ik\int^z n_O \mathrm{d}z\right). \tag{7.146}$$

Upgoing extraordinary wave

$$-\{2(\rho_O^2 - 1)n_E\}^{-\frac{1}{2}}(\rho_O, -1, -n_E, \rho_O n_E)\exp\left(-ik\int^z n_E \mathrm{d}z\right) \tag{7.147}$$

Downgoing extraordinary wave

$$-\{-2(\rho_O^2 - 1)n_E\}^{-\frac{1}{2}}(\rho_O, -1, n_E, -\rho_O n_E)\exp\left(ik\int^z n_E \mathrm{d}z\right). \tag{7.148}$$

7.18. Ray theory and 'full wave' theory

In this chapter it has been shown that the W.K.B. solutions may be used at nearly all levels in a slowly varying stratified medium. They fail near certain levels where reflection or coupling between the characteristic waves is occurring.

Every W.K.B. solution is of the form (7.136). It is the product of the exponential which expresses the 'phase memory' concept, and the other terms, which vary much more slowly than the exponential. The exponent is equal to $-\mathrm{i}$ multiplied by the generalised or complex phase of the wave, which is calculated as though the wave travelled at each level according to the laws of geometrical optics. A W.K.B. solution is therefore often said to be a mathematical expression of 'ray' theory. A more precise meaning for the term 'ray' is given in ch. 10.

When the W.K.B. solutions fail, it is necessary to make a further study of the differential equations, and to find a solution which cannot in general be interpreted in terms of geometrical optics. Such a solution is sometimes called a 'full wave' solution. For high frequencies, greater than about 1 MHz, most of the important aspects of radio wave propagation in the ionosphere can be handled by 'ray theory'. At levels of reflection or coupling, the W.K.B. solutions fail, but an investigation of the reflection or coupling process by 'full wave' theory shows that in general the 'ray theory' can still be considered to hold, with only trivial modification. An exception to this is the phenomenon of partial penetration and reflection for frequencies near the penetration frequency of an ionised layer (§ 15.10). Here the ray theory is inadequate, and a 'full wave' treatment is necessary.

For lower frequencies, less than about 1 MHz, and especially for very low frequencies, less than about 100 kHz, the ionosphere can change appreciably within a distance of one wavelength, and cannot always be regarded as a slowly varying medium. Then the W.K.B. solutions fail to be good approximations, and a full wave solution is required for nearly every problem. Chs. 10 to 14 deal with problems that can be treated by ray theory methods. The later chapters are almost entirely concerned with 'full wave' solutions.

7.19. The reflection coefficient

Consider again the W.K.B. solutions (7.26) for horizontally polarised waves in an isotropic stratified ionosphere, as discussed in §§ 7.2–7.11. Suppose that an upgoing radio wave of unit amplitude is generated at the ground, $z = 0$. Then its W.K.B. solution

$$E_y = C^{\frac{1}{2}} q^{-\frac{1}{2}} \exp\left(-ik \int_0^z q\mathrm{d}z \right) \tag{7.149}$$

gives the complex amplitude at any other level where q is not near to zero. The constant $C^{\frac{1}{2}}$ is included because, in the free space at the ground, $q = C$. It is required to find the amplitude R of the resulting reflected wave when it reaches the ground. Its field at other levels is given by its W.K.B. solution

$$E_y = RC^{\frac{1}{2}} q^{-\frac{1}{2}} \exp\left(ik \int_0^z q\mathrm{d}z \right). \tag{7.150}$$

These W.K.B. solutions fail where $q = 0$, and the reflection occurs there. If, nevertheless, it is supposed that (7.149) and (7.150) tend to equality when $q \to 0$, then clearly

$$R = \exp\left(-2ik \int_0^{z_0} q\mathrm{d}z \right) \tag{7.151}$$

where z_0 is the value of z that makes $q = 0$. A more detailed study of the differential equations in the region near $z = z_0$, given in § 8.20, shows that this formula needs a small modification. The right-hand side must be multiplied by a factor i. This represents a phase advance of $\frac{1}{2}\pi$ that is not predicted by the cumulative phase change given by the integral. For many purposes it is unimportant, but will be included for completeness. Thus the formula (7.151) should read

$$R = \mathrm{i}\exp\left(-2\mathrm{i}k \int_0^{z_0} q\mathrm{d}z \right). \tag{7.152}$$

In a similar way (7.150) shows that the complex amplitude of the reflected wave at a height $z < z_0$ within the ionosphere is

$$E_y = \mathrm{i}C^{\frac{1}{2}} q^{-\frac{1}{2}} \exp\left(-\mathrm{i}k \int_0^{z_0} q\mathrm{d}z + \mathrm{i}k \int_{z_0}^{z} q\mathrm{d}z \right). \tag{7.153}$$

The two integrals together express the total change of the complex phase of the wave in its passage from the ground to the reflection level z_0 and then back to the level z. They may be combined into a single integral called the 'phase integral'. If the effect of collisions is included, the condition $q = 0$ does not hold for any real value of z. Often, however, q is a known analytic function of z and is zero for a complex value $z = z_0$. Then the phase integral is a contour integral in the complex z plane. This is the basis of the 'phase integral method'; see §§ 8.21, 16.7.

The same property applies for the more general anisotropic ionosphere. The definition of the reflection coefficient R is then more complicated and depends on the wave polarisations; see ch. 11. But it can be shown in a similar way that if q_1 refers to the upgoing incident wave and if q_2 refers to the reflected wave, then the reflection coefficient at the ground contains a factor

$$\mathrm{i}\exp\left\{ -\mathrm{i}k \int_0^{z_0} (q_1 - q_2)\mathrm{d}z \right\} \tag{7.154}$$

where the reflection level z_0 is now the value of z, possibly complex, for which $q_1 = q_2$. Similarly for some other $z < \mathrm{Re}(z_0)$ the field of the reflected wave contains a factor

$$\mathrm{i}\exp\left\{ -\mathrm{i}k\left(\int_0^{z_0} q_1\mathrm{d}z + \int_{z_0}^{z} q_2\mathrm{d}z \right) \right\}. \tag{7.155}$$

These results, without the first factor i, are used in the theory of ray tracing, ch. 10.

PROBLEMS 7

7.1. Formulate the differential equation governing the transverse displacement y for waves of constant frequency on a string, stretched with constant tension T, whose mass per unit length $m(x)$ is a function of position, x. Find the W.K.B. solutions and deduce how the amplitude of a 'progressive' wave depends on m.

7.2. A perfect gas is stratified so that the ambient pressure P and temperature T are slowly varying functions of z. For a very weak plane sound wave of angular frequency ω, with its wave normal parallel to the z axis, find the differential equations satisfied by the excess pressure p and the displacement ζ. Find the W.K.B. solutions for the cases when (a) P is independent of z but T varies, (b) T is independent of z but P varies. In each case find how the amplitudes of p and ζ depend on P or T, and verify that, for a progressive wave, the average energy flow is constant.

7.3. A simple pendulum has a bob of mass m. The string passes through a fixed hole and its length L is made to vary slowly in time t. The pendulum swings with small amplitude. How do the amplitude a, the period T and the energy E of the oscillations depend on L?

7.4. A radio wave travels upwards into the horizontally stratified ionosphere and its wave normal is vertical. The coordinate z is vertically upwards. The earth's magnetic field is allowed for and the vector $\boldsymbol{Y}$ is in the x–z plane. Collisions are neglected. The wave is an ordinary wave with real and positive refractive index n_{O} and polarisation $\rho_{\mathrm{O}} = E_y/E_x$. Suppose that the ionosphere is so slowly varying that no appreciable reflection is occurring. Find an expression for the z component Π_z of the time averaged Poynting vector. Show that if this is independent of z, the field components must be given by the W.K.B. solution (7.145).

In the same problem, but with collisions allowed for, show that for the characteristic polarisations ρ_{O}, ρ_{E} of the ordinary and extraordinary waves, the adjoints $\bar{\rho}_{\mathrm{O}}$, $\bar{\rho}_{\mathrm{E}}$ are $-\rho_{\mathrm{O}}$, $-\rho_{\mathrm{E}}$ respectively. Hence show that, for a W.K.B. solution the bilinear concomitant W, (7.97), is independent of height z.

7.5. A radio wave travels obliquely upwards into an ionosphere simulated by a Chapman layer with scale height $H = 10\,\mathrm{km}$. The earth's magnetic field and collisions are neglected. The wave is reflected at height $z = z_0$. In a range $|z - z_0|$ near this the W.K.B. solutions fail to be good approximations. Use the criterion (7.61) to find an expression for this range. Find its minimum value, that is the value when z_0 is where $\mathrm{d}N/\mathrm{d}z$ is a maximum. What is this minimum value if the frequency is 1 MHz?

8

The Airy integral function and the Stokes phenomenon

8.1. Introduction

In ch. 7 it was shown that at most levels in a slowly varying stratified ionosphere, and for radio waves of frequency greater than about 100 kHz, the propagation can be described by approximate solutions of the differential equations, known as W.K.B. solutions. The approximations fail, however, near levels where two roots q of the Booker quartic equation approach equality. In this chapter we begin the study of how to solve the differential equations when this failure occurs. For an isotropic ionosphere there are only two values of q and they are equal and opposite, and given by (7.1), (7.2). The present chapter examines this case. It leads on to a detailed study of the Airy integral function, which is needed also for the solution of other problems. Its use for studying propagation in an anisotropic ionosphere where two qs approach equality is described in § 16.3.

For an isotropic ionosphere, a level where $q = 0$ is a level of reflection. In § 7.19 it was implied that the W.K.B. solution for an upgoing wave is somehow converted, at the reflection level, into the W.K.B. solution for a downgoing wave with the same amplitude factor, and this led to the expression (7.151) for the reflection coefficient R. The justification for this assertion is examined in this chapter and it is shown in § 8.20 to require only a small modification, as in (7.152).

8.2. Linear height distribution of electron concentration and isolated zero of *q*

The coordinate system used here is as defined in § 6.1.

If electron collisions are neglected, q^2 is related to the electron concentration N thus

$$q^2 = C^2 - X, \tag{8.1}$$

where X is given by (3.5) and is directly proportional to N, and $C = \cos\theta$ where θ is the angle between the incident wave normal and the vertical. Thus q depends on z through X, and is zero when $X = C^2$.

The simplest example of a zero of q occurs when X is a slowly increasing monotonic function of z, as shown in fig. 8.1(a). Then q^2 is a decreasing monotonic function of z as shown in fig. 8.1(b). It is zero where $z = z_0$, and to a first approximation the variation of q^2 with z may be taken as linear near z_0, so that

$$q^2 \approx -a(z - z_0), \tag{8.2}$$

where a is a constant. The right side of (8.2) may be regarded as the first term in the Taylor expansion for q^2 about the point z_0. It was shown in § 7.10 that the W.K.B. solutions in this case are good approximations provided that $|z - z_0|$ exceeds a certain minimum value, M, say. In the present chapter up to § 8.19 it is assumed that (8.2) may be used for q^2 for all values of $|z - z_0|$ from zero up to M. Within this range the solution of the differential equations can be expressed in terms of Airy integral functions, and outside the range the W.K.B. solutions are so chosen that they fit continuously to the solution within the range.

If q^2 is given exactly by (8.2) there is only one value of z, namely z_0 which makes

Fig. 8.1. Dependence on the height z of (a) X (proportional to electron concentration) and (b) q^2, for a slowly varying ionosphere when q^2 has an 'isolated' zero.

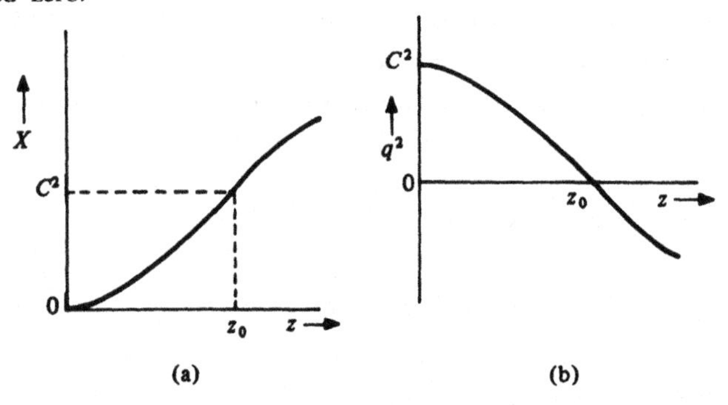

Fig. 8.2. Dependence on the height z of (a) X (proportional to electron concentration) and (b) q^2, when q^2 has two zeros close together.

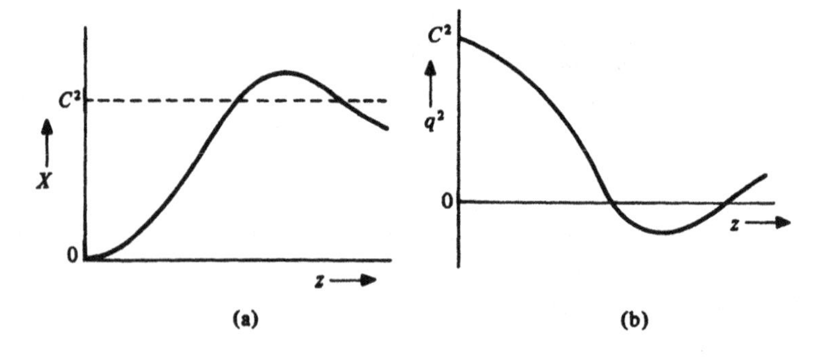

$q = 0$. If, however, the line in fig. 8.1(b) is slightly curved where $z = z_0$ then, in a range of z near z_0, (8.2) may be replaced by

$$q^2 \approx -a(z - z_0) + b(z - z_0)^2, \tag{8.3}$$

where b determines the curvature at $z = z_0$. Now q is zero both at $z = z_0$ and at $z = z_0 + a/b$. This case is illustrated in fig. 8.2(a, b), and could occur, for a frequency less than the penetration frequency, when the electron concentration has a maximum value. If b is very small in (8.3), the second zero is at a great distance a/b from the first, provided that a is not small.

Hence the condition that the variation of q^2 with z shall be nearly linear, near a zero of z, is equivalent to saying that this zero is at a great distance from the next nearest zero. In other words, the zero of q must be isolated. A more exact statement of this condition is given by (8.75), § 8.20.

If a is very small, the two zeros of (8.3) can be close together even when b is small. This case could occur when the frequency is only slightly less than the penetration frequency. Here the W.K.B. solutions fail for a range of z which includes both zeros. It is then necessary to study the differential equations when q^2 varies according to the parabolic law. This leads to the phenomena of partial penetration and reflection, and is discussed in § 15.10. For frequencies slightly greater than the penetration frequency, q has no zeros at any real value of z, but there are zeros in the complex z plane. This case also is discussed in §§ 15.9, 15.10.

Another form of variation of q with z that will be encountered (§ 19.6) is

$$q^2 = \alpha \frac{z - z_0}{z - z_1}, \tag{8.4}$$

where α, z_0, z_1 are constants. Here q^2 has only one zero where $z = z_0$, but it has an infinity where $z = z_1$. The curvature of the curve of q^2 versus z at $z = z_0$ is $-2\alpha/(z_0 - z_1)^2$. If this is to be small, $|z_0 - z_1|$ must be large, so that the infinity and the zero must be well separated. When the expression (8.2) is used for q^2, therefore, it implies that the zero of q at $z = z_0$ is at a great distance both from other zeros and from infinities. A method of solving the differential equations that allows for a small curvature of the function $q^2(z)$ near its zero is described in § 8.20.

8.3. The differential equation for horizontal polarisation and oblique incidence

The differential equations to be satisfied by the wave fields for oblique incidence were derived in § 7.2 for the case when the earth's magnetic field is neglected. It was shown that they separate into two sets, one for fields in which the electric vector is everywhere horizontal, and the other for fields in which the electric vector is everywhere in the plane of incidence. These two cases are usually said to apply to horizontal and vertical polarisation respectively. For horizontal polarisation the equations are (7.3) or (7.5) or (7.6), where q is given by (7.2). If the effect of electron collisions is neglected, so that $Z=0$, then (7.2) reduces to (8.1). If $S^2=0$ so that $C^2=1$, (7.2) gives $q^2=n^2$, and the differential equations apply for vertical incidence. The solution for vertical incidence is therefore a special case of the solution for oblique incidence with horizontal polarisation.

When (8.2) is inserted in (7.6) it gives

$$\frac{\mathrm{d}^2E_y}{\mathrm{d}z^2}-k^2a(z-z_0)E_y=0. \tag{8.5}$$

This is a very important differential equation which governs the behaviour of the wave fields near a zero of q. Its properties are given in the following sections.

For 'vertical' polarisation the differential equations are (7.4) or (7.68). When (8.2) is inserted in (7.4) it gives an equation which is more complicated than (8.5); the discussion of this case is postponed until §§ 15.5–15.7.

8.4. The Stokes differential equation

The theory in the rest of this chapter refers only to horizontal polarisation, so that E_y is the only non-zero component of the electric field. The subscript y will therefore be omitted. In (8.5) it is convenient to use the new independent variable

$$\zeta=(k^2a)^{\frac{1}{3}}(z-z_0), \tag{8.6}$$

where the value of $(k^2a)^{\frac{1}{3}}$ is taken to be real and positive. Thus ζ is a measure of the height. Then (8.5) becomes

$$\frac{\mathrm{d}^2E}{\mathrm{d}\zeta^2}=\zeta E. \tag{8.7}$$

This is known as the Stokes differential equation. The same name is sometimes given to the equation $\mathrm{d}^2E/\mathrm{d}x^2+xE=0$ which is easily converted to (8.7) by the substitution $x=(-1)^{\frac{1}{3}}\zeta$.

In this book the name Stokes is used for two distinct purposes. First, it is used for the differential equation (8.7) or its equivalent. Second, it is used for the Stokes phenomenon, § 8.12. This was discovered by Stokes originally for solutions of (8.7) but it can occur for any function with an irregular singularity. The terms 'Stokes lines' and 'anti-Stokes lines' and 'Stokes diagram' are used as part of the description of the Stokes phenomenon and are explained in §§ 8.13, 8.14. Many of the functions of theoretical physics have irregular singularities at infinity and there are examples in ch. 15. They display the Stokes phenomenon but they are not necessarily connected with solutions of (8.7).

Equation (8.7) has no singularities when ζ is bounded and its solution must therefore be bounded and single valued, except possibly at $\zeta=\infty$. It is necessary to study the properties of these solutions for both real and complex values of ζ.

Solutions of (8.7) can be found as series in ascending powers of ζ, by the standard method. Assume that a solution is $E=a_0+a_1\zeta+a_2\zeta+\dots$. Substitute this in (8.7) and equate powers of ζ. This gives relations between the constants a_0, a_1, a_2, etc., and leads finally to

$$E = a_0\left\{1 + \frac{\zeta^3}{3\cdot 2} + \frac{\zeta^6}{6\cdot 5\cdot 3\cdot 2} + \frac{\zeta^9}{9\cdot 8\cdot 6\cdot 5\cdot 3\cdot 2} + \dots\right\}$$
$$+ a_1\left\{\zeta + \frac{\zeta^4}{4\cdot 3} + \frac{\zeta^7}{7\cdot 6\cdot 4\cdot 3} + \frac{\zeta^{10}}{10\cdot 9\cdot 7\cdot 6\cdot 4\cdot 3} + \dots\right\}, \tag{8.8}$$

which contains the two arbitrary constants a_0 and a_1, and is therefore the most general solution. The series are convergent for all ζ, which confirms that every solution of (8.7) is bounded, continuous and single valued. Series for the derivative $\mathrm{d}E/\mathrm{d}\zeta$ are easily found from (8.8). The two series (8.8) separately have no particular physical significance. The constants a_0 and a_1 for the functions $\mathrm{Ai}(\zeta)$ and $\mathrm{Bi}(\zeta)$ can be found from (8.16), (8.17) below or from their derivatives with respect to ζ, by putting $\zeta = 0$. Their values are as follows:

For $\mathrm{Ai}(\zeta)$:

$$a_0 = 3^{-\frac{2}{3}}/(-\tfrac{1}{3})! = 0.355\,03,$$
$$a_1 = -3^{-\frac{1}{3}}/(-\tfrac{2}{3})! = -0.258\,82. \tag{8.9}$$

For $\mathrm{Bi}(\zeta)$:

$$a_0 = 3^{-\frac{1}{6}}/(-\tfrac{1}{3})! = 0.614\,93,$$
$$a_1 = 3^{\frac{1}{6}}/(-\tfrac{2}{3})! = 0.448\,29.$$

These values, with the series (8.8) and their derivatives, may be used for computing for real and complex ζ up to about $|\zeta| = 5.0$, though in radio propagation problems they are rarely needed for values of $|\zeta|$ greater than about 2.0.

8.5. Qualitative discussion of the solutions of the Stokes equation

Equation (8.7) shows that, if E and ζ are real, $\mathrm{d}^2E/\mathrm{d}\zeta^2$ is real. If $\mathrm{d}E/\mathrm{d}\zeta$ is also real, than E must be real for all real values of ζ. This is also apparent from (8.8). It is of interest to trace the curve of E versus ζ when ζ and E are both real. If ζ is positive, (8.7) shows that the curvature of the curve has the same sign as E. Hence the curve is convex towards the line $E = 0$. If the curve is traced step by step from $\zeta = 0$ upwards, there are three possibilities which are illustrated in fig. 8.3. First, if the initial slope is sufficiently negative, the curve can cross the line $E = 0$ (curve A). When it does so, the sign of the curvature changes, and for higher values of ζ the magnitude of the slope must increase indefinitely. Hence the curve moves indefinitely further from the line $E = 0$ and can never cross it again. Secondly, if the initial slope is positive or only slightly negative, the slope can become zero before the curve reaches the line $E = 0$ (curve B). Thereafter the curve moves indefinitely further from this line and can never cross it. In both these cases E ultimately becomes indefinitely large as ζ increases. The third possibility occurs for one particular negative value of the initial slope. The curve then approaches the line $E = 0$, and never actually reaches it, but gets closer and closer to it (curve C). The slope must always have the opposite sign to E, and E must become smaller and smaller as ζ increases. This last case is of particular importance, and the solution $\mathrm{Ai}(\zeta)$, described later, has this property.

When ζ is negative, the curvature has the opposite sign to E. Hence the curve is concave towards the line $E = 0$. If the curve is traced step by step from $\zeta = 0$ towards increasingly negative values of ζ, it must always curve towards the line $E = 0$ and eventually cross it. The sign of the curvature then changes so that the curve again bends towards the line $E = 0$, and crosses it again. In all cases therefore the function E is oscillatory and as ζ becomes more negative the curvature for a given E increases, so that the oscillation gets more rapid and its amplitude gets smaller. This is illustrated for all three curves in fig. 8.3.

8.6. Solutions of the Stokes equation expressed as contour integrals

A useful form of the solutions of (8.7) can be found in the form of a contour integral. Let

$$E = \int_a^b e^{\zeta t} h(t)\,dt, \tag{8.10}$$

where t is a complex variable and the integral is evaluated along some path in the complex t plane, whose end points a and b are to be specified later. An integral of the form (8.10) is said to be of Laplace type. This form can always be used for a linear differential equation whose coefficients are linear functions of the independent variable. Since (8.10) must satisfy (8.7), it is necessary that

$$\int_a^b (t^2 - \zeta)h(t)e^{\zeta t}\,dt = 0. \tag{8.11}$$

Fig. 8.3. Behaviour of solutions $E(\zeta)$ of the Stokes equation.

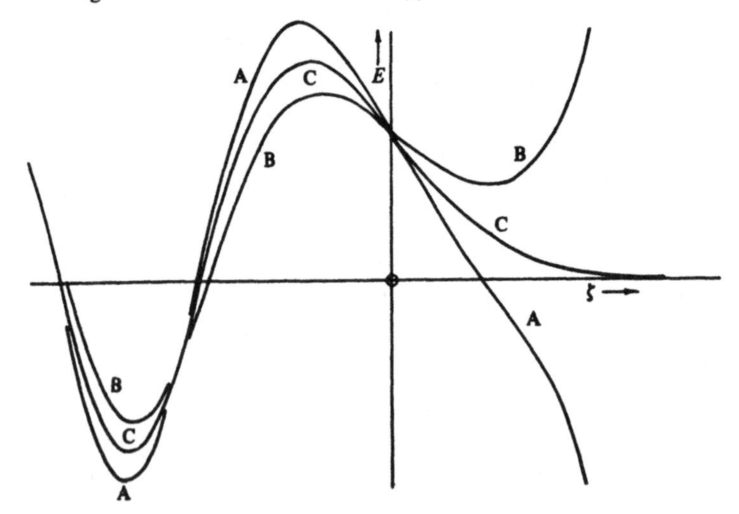

The second term can be integrated by parts, which gives

$$-\left.e^{\zeta t}h(t)\right|_a^b + \int_a^b \left\{t^2 h(t) + \frac{dh(t)}{dt}\right\} e^{\zeta t}\,dt = 0. \tag{8.12}$$

The limits a, b are to be chosen so that the first term vanishes at both limits. Then (8.11) is satisfied if

$$\frac{dh(t)}{dt} + t^2 h(t) = 0, \tag{8.13}$$

that is if

$$h(t) = A\exp(-\tfrac{1}{3}t^3) \tag{8.14}$$

where A is a constant. The limits a and b must therefore be chosen so that $\exp(-\frac{1}{3}t^3 + \zeta t)$ is zero for both. This is only possible if $|t| \to \infty$ and $2\pi r - \frac{1}{2}\pi < 3\arg t < 2\pi r + \frac{1}{2}\pi$, where r is an integer. Fig. 8.4 is a diagram of the complex t plane, and a and b must each be at infinity in one of the shaded sectors. They cannot both be in the same sector, for then the integral (8.10) would be zero. Hence the path may be chosen in three ways, as shown by the three curves C_1, C_2, C_3. This might appear at first to give three independent solutions of (8.7). But the path C_1 can be distorted so as to coincide with the two paths $C_2 + C_3$, so that

$$\int_{C_1} = \int_{C_2} + \int_{C_3}, \tag{8.15}$$

and therefore there are only two independent solutions.

Fig. 8.4. The complex t plane, with possible paths for the integral representation (8.10) of solutions of the Stokes equation.

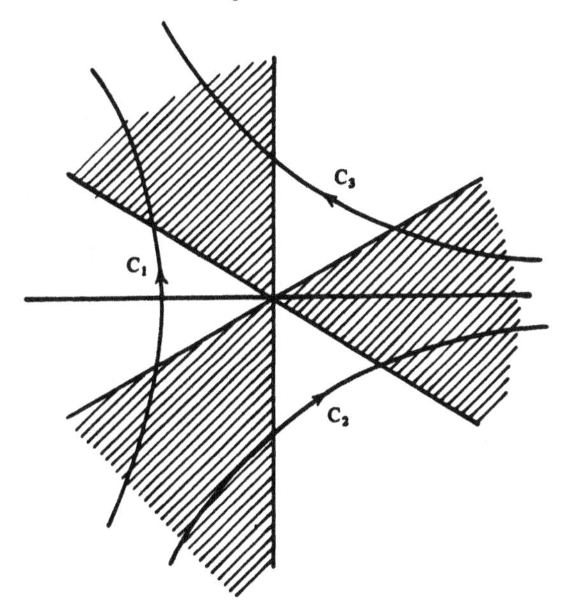

Jeffreys and Jeffreys (1972) define the two functions $\mathrm{Ai}(\zeta)$ and $\mathrm{Bi}(\zeta)$ as follows:

$$\mathrm{Ai}(\zeta)=\frac{1}{2\pi\mathrm{i}}\int_{C_1}\exp(-\tfrac{1}{3}t^3+\zeta t)\,\mathrm{d}t, \tag{8.16}$$

$$\mathrm{Bi}(\zeta)=\frac{1}{2\pi}\int_{C_2}\exp(-\tfrac{1}{3}t^3+\zeta t)\,\mathrm{d}t-\frac{1}{2\pi}\int_{C_3}\exp(-\tfrac{1}{3}t^3+\zeta t)\,\mathrm{d}t. \tag{8.17}$$

In (8.16) the path C_1 can be distorted so as to coincide with the imaginary t axis for almost its whole length. It must be displaced very slightly to the left of this axis at its ends. Let $t=\mathrm{i}s$. Then (8.16) becomes

$$\mathrm{Ai}(\zeta)=\frac{1}{2\pi}\int_{-\infty}^{\infty}\exp\{\mathrm{i}(\zeta s+\tfrac{1}{3}s^3)\}\,\mathrm{d}s. \tag{8.18}$$

Here the imaginary part of the integrand is an odd function of s, and contributes nothing to the integral, which may therefore be written

$$\mathrm{Ai}(\zeta)=\frac{1}{\pi}\int_{0}^{\infty}\cos(\zeta s+\tfrac{1}{3}s^3)\,\mathrm{d}s. \tag{8.19}$$

Apart from a constant, this is the same as the expression used by Airy (1838, 1849). It is known as the Airy integral, and $\mathrm{Ai}(\zeta)$ is called the Airy integral function.

The functions $\mathrm{Ai}(\zeta)$ and $\mathrm{Bi}(\zeta)$ are shown in fig. 8.5.

8.7. Solutions of the Stokes equation expressed as Bessel functions

The Stokes equation (8.7) can be converted into Bessel's equation by changing both the dependent and independent variables. First let

$$\tfrac{2}{3}\zeta^{\frac{3}{2}}=\mathrm{i}\xi. \tag{8.20}$$

Fig. 8.5. The functions $\mathrm{Ai}(\zeta)$ continuous curve, and $\mathrm{Bi}(\zeta)$ broken curve.

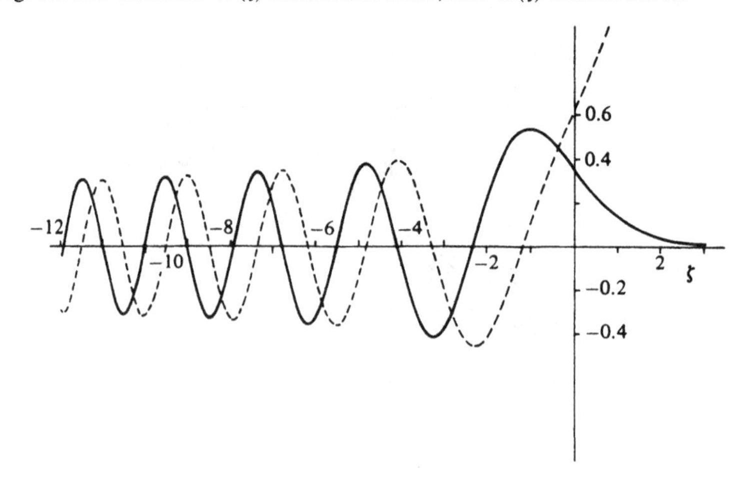

Then (8.7) becomes

$$\frac{\mathrm{d}^2E}{\mathrm{d}\xi^2}+\frac{1}{3\xi}\frac{\delta E}{\mathrm{d}\xi}+E=0. \tag{8.21}$$

Next let

$$E=\xi^{\frac{1}{3}}\mathscr{E}=(-\tfrac{2}{3}\mathrm{i})^{\frac{1}{3}}\zeta^{\frac{1}{2}}\mathscr{E}. \tag{8.22}$$

Then (8.21) becomes

$$\frac{\mathrm{d}^2\mathscr{E}}{\mathrm{d}\xi^2}+\frac{1}{\xi}\frac{\mathrm{d}\mathscr{E}}{\mathrm{d}\xi}+\left(1-\frac{1}{9\xi^2}\right)\mathscr{E}=0. \tag{8.23}$$

This is Bessel's equation of order one-third. The transformations (8.20) and (8.22) are complicated and introduce ambiguities. For example, (8.20) does not define ξ completely, since there is an ambiguity of the sign of the term on the left. Consequently the use of the Bessel function solutions can lead to errors. Moreover, one solution of Bessel's equation (8.23) has a singularity where $\xi=0$, whereas the solutions of the Stokes equation have no singularities. The transformations (8.20) and (8.22) in fact introduce compensating singularities. The relations between the Bessel functions and Ai(ζ), Bi(ζ) are given by Watson (1944), Miller (1946), Olver (1974), Abramowitz and Stegun (1965).

8.8. Tables of the Airy integral functions. Computing

Many tables of the functions Ai(ζ), Bi(ζ) and related functions are available, and a list is given by Abramowitz and Stegun (1965). These are nearly all for real ζ only. One of the most useful is that of Miller (1946) who also gives a summary of the properties of these functions.

Tables for complex ζ are given by Woodward and Woodward (1946). Tables of the functions

$$h_1(z)=(12)^{\frac{1}{6}}\exp(-\tfrac{1}{6}\pi\mathrm{i})\{\mathrm{Ai}(-z)-\mathrm{iBi}(-z)\}$$
$$h_2(z)=(12)^{\frac{1}{6}}\exp(\tfrac{1}{6}\pi\mathrm{i})\{\mathrm{Ai}(-z)+\mathrm{iBi}(-z)\} \tag{8.24}$$

and their derivatives, for complex z, are given by the staff of the Computation Laboratory at Cambridge, Mass. (1945). From them Ai(z) and Bi(z) and their derivatives can be found.

Subroutines for computing Ai(ζ), Bi(ζ) and their derivatives are available in the libraries of many of the larger computers, although in some cases only for real values of ζ. For real or complex ζ the series (8.8) with (8.9) can be used for $|\zeta|$ less than about 4 or 5. For larger $|\zeta|$ the asymptotic forms (8.52)–(8.61) below can be used, and, if greater accuracy is needed, these forms can be combined with asymptotic expansions (8.30)–(8.32).

8.9. Zeros and turning points of Ai(ζ) and Bi(ζ)

From the discussion of § (8.5) it is clear that Ai(ζ), Ai′(ζ), Bi(ζ) and Bi′(ζ) have no zeros when ζ is real and positive. Each of them has an infinite sequence of zeros when ζ is real and negative; compare fig. 8.5. It can be shown that $\mathrm{Ai}(\zeta)$ and $\mathrm{Ai}'(\zeta)$ have no other zeros, but $\mathrm{Bi}(\zeta)$ and $\mathrm{Bi}'(\zeta)$ have an infinite sequence of zeros on each of the lines $\arg\zeta = \pm\pi/3$. The proof of these results is given by Olver (1974, §§ 11.5, 11.8).

8.10. The W.K.B. solutions of the Stokes equation

Approximate solutions of the Stokes equation (8.7) can be found by the W.K.B. method of § 7.6. Thus in (8.7) let

$$E = A\exp\{\mathrm{i}\phi(\zeta)\}. \tag{8.25}$$

Then the method gives (compare (7.18), (7.19)):

$$\left(\frac{\mathrm{d}\phi}{\mathrm{d}\zeta}\right)^2 = -\zeta + \mathrm{i}\frac{\mathrm{d}^2\phi}{\mathrm{d}\zeta^2}. \tag{8.26}$$

The steps of (7.19)–(7.26) are now followed through and lead to the two W.K.B. solutions

$$E = \zeta^{-\frac{1}{4}}\exp(-\tfrac{2}{3}\zeta^{\frac{3}{2}}) \tag{8.27}$$

and

$$E = \zeta^{-\frac{1}{4}}\exp(\tfrac{2}{3}\zeta^{\frac{3}{2}}). \tag{8.28}$$

For large enough $|\zeta|$ any solution of (8.7) is a linear combination of (8.27), (8.28) but not the same combination for all $\arg\zeta$; see § 8.12. The combinations for $\mathrm{Ai}(\zeta)$, $\mathrm{Bi}(\zeta)$ are given later at (8.52), (8.53) and (8.56)–(8.58) respectively. A condition that these are good approximations is that $|\zeta|$ shall be large enough. It is useful to study them when $|\zeta| = 1$. The function $\mathrm{Ai}(\zeta)$ has only the one term (8.27) in its asymptotic approximation when $0 < \arg\zeta < \frac{2}{3}\pi$, and this range runs from one Stokes line to the next; see § 8.13. The function $\mathrm{Ai}(\zeta)$ and its asymptotic approximation from (8.52) were computed and compared. The minimum error was 7.0% for $\arg\zeta = 0$ where the term is subdominant. The maximum error was 14.3% for $\arg\zeta = \frac{2}{3}\pi$ where the term is dominant. On the anti-Stokes line, $\arg\zeta = \frac{1}{3}\pi$, the error was 8.2%. The average error over this whole range was 8.9%. For $|\zeta| = 2$ the corresponding figures were 3.1%, 5.3%, 3.4%, 3.7%. These results suggest that if

$$|\zeta| \geqslant 1, \tag{8.29}$$

the error in using the asymptotic forms alone is less than about 8 to 9%. This criterion is used as a rough guide in this book.

8.11. Asymptotic expansions

In many books on linear differential equations there is a discussion of the properties of the solutions at various points in the complex plane of the independent variable. The points are classified into (a) ordinary points, (b) regular singular points or regular singularities, and (c) irregular singular points or irregular singularities.

Regular singularities are not used in this section but are mentioned later, §§ 15.5, 15.12 ff. Every differential equation must have regular or irregular singular points and for nearly all the differential equations of physics it is found that the point at infinity is an irregular singularity. An exception is the hypergeometric equation, discussed in § 15.12 ff. At an ordinary point, say $\zeta = \zeta_0$, it is always possible to find two convergent series solutions of the differential equation in ascending integer powers of $\zeta - \zeta_0$, and the radius of convergence is the distance from ζ_0 to the nearest singular point. For the Stokes equation (8.7) every point of the complex ζ plane except infinity is an ordinary point. The series (8.8) are the convergent expansions about the origin $\zeta = 0$, and since the only singular point is at infinity, they are convergent for all ζ.

For the region near an irregular singularity the series method can sometimes be used, but the series is divergent and might not use integer powers. This applies to the Stokes equation, as shown below, and for most other irregular singularities that arise in physical problems. A divergent series cannot define any function but it can still be used to give important properties of the solution. Thus we have to deal with two important things: (i) solutions near an irregular singularity can be studied by using divergent series; (ii) the most commonly occurring irregular singularity is at infinity. These two ideas are really quite separate. The fact that they usually occur together can sometimes lead to confusion.

To study the behaviour at infinity it is sometimes convenient to change the independent variable to $x = 1/\zeta$ so that the point $\zeta = \infty$ is transformed to the origin of the complex x plane, and we study the resulting differential equation near its irregular singularity at $x = 0$. Then by using a trial series in ascending powers of x and substituting in the differential equation, we can formally find the coefficients but the series is found to be divergent. There is, however, no need to make this transformation. The series is the same as a series in descending powers of ζ. We can use such a trial series in the original differential equation.

For the Stokes equation (8.7), the W.K.B. solutions (8.27), (8.28) are good approximations, according to (8.29), when $|\zeta|$ is large. But these functions cannot be expanded in descending powers of ζ, and indeed it is not possible to find a solution of (8.7) in descending powers of ζ. But we may take as a trial solution

$$E \sim \zeta^{-\frac{1}{4}} \exp(-\tfrac{2}{3}\zeta^{\frac{3}{2}})\{1 + a_1\zeta^{-r} + a_2\zeta^{-2r} + \cdots\} \tag{8.30}$$

where r is positive. If this is substituted in (8.7), a factor $\exp(-\frac{2}{3}\zeta^{\frac{3}{2}})$ may be cancelled from every term. If then the coefficients of the various powers of ζ in succession are equated to zero, it is found that $r = \frac{3}{2}$ and the coefficients a_1, $a_2 \cdots$ can be found. Similarly the trial solution

$$E \sim \zeta^{-\frac{1}{4}} \exp(\tfrac{2}{3}\zeta^{\frac{3}{2}})\{1 + b_1\zeta^{-r} + b_2\zeta^{-2r} + \cdots\} \tag{8.31}$$

may be used. The same process again yields $r = \frac{3}{2}$, and b_1, b_2 can be found. The values are

$$b_m = (-1)^m a_m = \frac{(3m-\frac{1}{2})!}{(2m)!(-\frac{1}{2})!3^{2m}} = \frac{(6m-1)(6m-3)\cdots(2m+1)}{m!(144)^m}. \quad (8.32)$$

The standard convergence tests show that the series in (8.30), (8.31) are divergent for all values of ζ.

These are examples of the type of series known as 'asymptotic expansions'. Many physicists find that this is a difficult subject, and the reader who wants to master it must expect to devote much effort and time to it. For accounts of it see Olver (1974), Jeffreys and Jeffreys (1972), Jeffreys (1962), Dingle (1973). Here we give only a brief summary of how these expansions are used.

To avoid repeatedly writing the fractional power $r = \frac{3}{2}$, let

$$s = \tfrac{2}{3}\zeta^{\frac{3}{2}}. \quad (8.33)$$

Then the series representation (8.30) may be written

$$E\zeta^{\frac{1}{4}}e^s \equiv Y(s) \sim 1 + c_1/s + c_2/s^2 + \cdots + c_m/s^m + \cdots. \quad (8.34)$$

To study the function $Y(s)$ we must have some definition of it. It is often expressible as a contour integral. In the present example, E in (8.34) is proportional to (8.16). If the series in (8.34) does not converge it does not define any number and so the = sign cannot be used. We use the sign $\sim$ whose meaning is explained below. Even if (8.34) is divergent, it can always be made to 'converge at first' by making $|s|$ large enough. For example we can ensure that the terms decrease steadily as m increases up to the Nth term, by taking $|s| > Q$ where Q is the greatest value of $|c_{m+1}/c_m|$ for $m < N$. If $|s|$ is made extremely large, some of the terms after the first will be extremely small. In series such as (8.31), however, the much later terms get large no matter how large $|s|$ is. But it is clear that there must be a smallest term. Which numbered term it is depends on $|s|$.

Let $S_m(s)$ denote the sum of the first $m+1$ terms of the series

$$S_m(s) = 1 + c_1/s + \cdots + c_m s^m. \quad (8.35)$$

Then the difference

$$R_m(s) = Y(s) - S_m(s) \quad (8.36)$$

is called the 'remainder'. For fixed s, $|R_m(s)|$ will not get small when m gets large because the series is divergent. But it is possible that, for fixed m, $|R_m(s)|$ gets small when $|s|$ gets large. This leads to the Poincaré definition. If for every fixed value of m

$$s^m R_m(s) \to 0 \text{ as } |s| \to \infty \quad (8.37)$$

then we say that the series (8.34) is an asymptotic expansion for $Y(s)$. This gives the meaning of the $\sim$ sign.

Note that the Poincaré definition is concerned only with what happens when $|s| \to \infty$. It is not concerned with non-infinite values of s.

To use a divergent asymptotic expansion the series must be truncated in some way. It can be shown that the error is least when the series is stopped at or just before the smallest term. Then the modulus of the remainder is less than the last retained term of the series.

If $|\zeta|$ is large enough in (8.30), many terms of the series after the first are small. In many practical problems, including most of those discussed in this book, these terms are neglected and we use only the first term. The resulting expression is called an 'asymptotic form'. All W.K.B. solutions are asymptotic forms in this sense. The series and the Poincaré definition are needed for reference in the discussions of later sections. Occasionally it is useful to use a few terms of the asymptotic series when the W.K.B. solutions alone are not accurate enough.

8.12. The Stokes phenomenon of the 'discontinuity of the constants'

The expressions (8.27) and (8.28) are multiple valued functions, whereas any solution of (8.7) is single valued. Hence a solution of (8.7) cannot be represented by the same combination of (8.27) and (8.28) for all values of ζ. To illustrate this, consider the function $\mathrm{Ai}(\zeta)$. It was mentioned in § 8.5 that when ζ is real and positive, this function decreases steadily as ζ increases (fig. 8.3, curve C), but never becomes zero. Hence the W.K.B. approximation for $\mathrm{Ai}(\zeta)$ when ζ is real and positive must be given by

$$\mathrm{Ai}(\zeta) \sim A\zeta^{-\frac{1}{4}}\exp(-\tfrac{2}{3}\zeta^{\frac{3}{2}}) \quad (\text{for } \arg\zeta = 0), \tag{8.38}$$

where A is a constant and the positive values of $\zeta^{-\frac{1}{4}}$, $\zeta^{\frac{3}{2}}$ are used. It cannot include a multiple of (8.28) for this would make the function increase indefinitely for large ζ. Now let $|\zeta|$ be kept constant, and let $\arg\zeta$ increase continuously from 0 to π, so that ζ becomes real and negative. Then the expression (8.38) becomes

$$Ae^{-i\frac{1}{4}\pi}|\zeta^{-\frac{1}{4}}|\exp\{\tfrac{2}{3}i|\zeta^{\frac{3}{2}}|\} \quad (\text{for } \arg\zeta = \pi), \tag{8.39}$$

which is a complex function. But $\mathrm{Ai}(\zeta)$ is real for all real values of ζ, so that (8.39) cannot represent $\mathrm{Ai}(\zeta)$ when ζ is real and negative, and the correct representation must include multiples of both (8.27) and (8.28). It is shown later that an additional term should be added to (8.38) when $\arg\zeta > \frac{2}{3}\pi$.

The W.K.B. approximation to the most general solution of the Stokes equation is

$$A\zeta^{-\frac{1}{4}}\exp(-\tfrac{2}{3}\zeta^{\frac{3}{2}}) + B\zeta^{-\frac{1}{4}}\exp(\tfrac{2}{3}\zeta^{\frac{3}{2}}), \tag{8.40}$$

but this can apply only to a part of the complex ζ plane. If ζ moves out of this part, one of the arbitrary constants A or B must be changed. This phenomenon was discovered by Stokes (1858), and is called the 'Stokes phenomenon of the discontinuity of the arbitrary constants.' See the note in § 8.4 about the two distinct uses of the name Stokes in this book.

8.13. Stokes lines and anti-Stokes lines

The exponents of both the exponentials in (8.40) are real if

$$\arg\zeta = 0, \tfrac{2}{3}\pi, \text{ or } \tfrac{4}{3}\pi, \tag{8.41}$$

and then if $|\zeta|$ becomes indefinitely large, one exponential becomes indefinitely small, and the other indefinitely large. Moreover, if $|\zeta|$ is kept constant, and $\arg\zeta$ is varied, both exponentials have maximum or minimum values when $\arg\zeta$ is given by (8.41), which defines three lines radiating from the origin of the complex ζ plane. These are known as the 'Stokes lines'.

The exponents have equal moduli if

$$\arg\zeta = \tfrac{1}{3}\pi,\ \pi,\ \text{or}\ \tfrac{5}{3}\pi, \tag{8.42}$$

and then if $|\zeta|$ becomes indefinitely large, the moduli of both exponentials remain equal to unity, but the terms oscillate more and more rapidly. The radial lines defined by (8.42) are called the 'anti-Stokes lines'. This nomenclature is used here and is found in many papers on radio propagation. It is not accepted by all authors, however, and for example Olver (1974, p. 518) recommends a different nomenclature.

For all values of $\arg\zeta$ except those in (8.42) one exponential must have modulus greater than unity, and the other must have modulus less than unity. The larger exponential remains so, as long as ζ lies in the 120° sector between two anti-Stokes lines, and the corresponding term in (8.40) is called the 'dominant' term. The other term, containing the smaller exponential, is called the 'subdominant' term. Each term changes from dominant to subdominant, or the reverse, when ζ crosses an anti-Stokes line.

It was shown in § 8.11 that one of the constants A and B must change when ζ crosses some line in the complex ζ plane. It is fairly clear that the constant in the dominant term cannot change, for this would give a detectable discontinuity in the function (8.40), which would mean that it would not even approximately satisfy the differential equation. Hence the constant which changes must be that in the subdominant term, and the most likely place for the change to occur is on a Stokes line, for there the ratio of the subdominant to the dominant term is smallest.

The solution (8.40) is approximate. The two terms are W.K.B. solutions and both are subject to error. To get a more accurate solution, these terms may each be multiplied by a divergent asymptotic expansion, § 8.11. To get the greatest accuracy this must be truncated just before or just after the smallest term. This smallest term then gives an upper bound for the error. It was shown by Stokes (1858) that when the multiplier of the subdominant term changes on a Stokes line, the resulting discontinuity is less than the error bound for the dominant term.

In the following two sections it will be assumed that the arbitrary constant in the subdominant term may change on a Stokes line. These sections are descriptive and are intended to help towards an understanding of the Stokes phenomenon. A more formal mathematical proof of the results is given in § 9.7.

8.14. The Stokes diagram

In any 120° sector between two Stokes lines, a solution of the Stokes equation may contain either one or both of the terms in (8.40). To indicate the nature of the W.K.B. or asymptotic approximation to a particular solution, Stokes (1858) used a diagram constructed as follows. Radial lines are drawn from a fixed point O (fig. 8.6) in the directions of the Stokes and anti-Stokes lines, and labelled S or A respectively. A circle is drawn with centre O. In this diagram radial directions indicate values of $\arg\zeta$ but radial distances do not indicate values of $|\zeta|$. The plane of the diagram, therefore, is not the complex ζ plane. One or two thick lines are drawn in the diagram, inside and outside the circle, and they cross the circle on the anti-Stokes lines. For some value of $\arg\zeta$ a radial line is imagined to be drawn in the corresponding direction. If it crosses a thick line inside the circle, this shows that there is a subdominant term in the asymptotic approximation. If it crosses a thick line outside the circle, this shows that there is a dominant term.

Where the thick line crosses a Stokes line inside the circle, the constant multiplying the associated subdominant term may change, and this is indicated by a break in the thick line. The thick line crosses the circle on the anti-Stokes lines because there the associated terms change from dominant to subdominant or the reverse. The thick line must remain unbroken except where it meets a Stokes line inside the circle, for only there can the associated constant change. Moreover, this change can only occur if the dominant term is present. If it were absent, a change in the constant of the subdominant term would give a detectable discontinuity, since there is now no dominant term to mask it.

These properties are illustrated in fig. 8.6(a), which is the Stokes diagram for the function $\mathrm{Ai}(\zeta)$. In the sector $-\frac{2}{3}\pi < \arg\zeta < \frac{2}{3}\pi$ there is only one term in the asymptotic approximation, and this is subdominant in the sector $-\frac{1}{3}\pi < \arg\zeta < \frac{1}{3}\pi$. The constant cannot change on the Stokes line at $\arg\zeta = 0$ because there is no dominant term. The same term is dominant on the Stokes line at $\arg\zeta = \frac{2}{3}\pi$, and for greater values of $\arg\zeta$ there is also a subdominant term, which becomes dominant

Fig. 8.6. Stokes diagrams for the functions (a) $\mathrm{Ai}(\zeta)$, (b) $\mathrm{Bi}(\zeta)$, (c) $\mathrm{Ai}\{\zeta\exp(\frac{4}{3}i\pi)\} = \mathrm{Ai}\{\zeta\exp(-\frac{2}{3}i\pi)\}$.

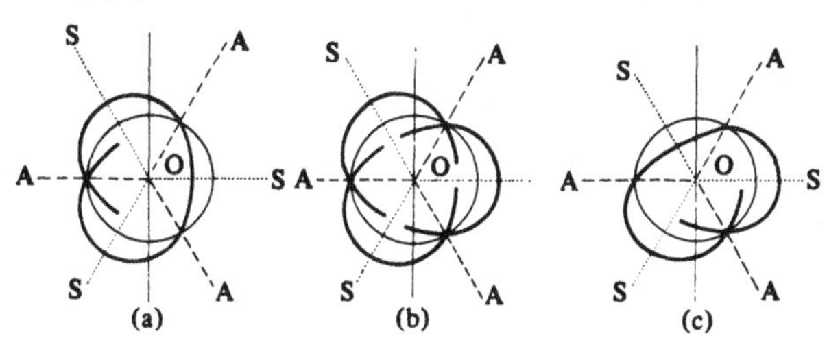

when $\arg\zeta > \pi$, and here the original term again becomes subdominant. This term disappears beyond the Stokes line at $\arg\zeta = \frac{4}{3}\pi$ or $-\frac{2}{3}\pi$.

Another example is given in fig. 8.6(b), which is the Stokes diagram for the function $\mathrm{Bi}(\zeta)$. Here there are both dominant and subdominant terms for all values of $\arg\zeta$, and the constant in the subdominant term changes on all three Stokes lines.

8.15. Definition of the Stokes multiplier

It is now necessary to determine by how much the constant in the subdominant term changes when a Stokes line is crossed. Suppose that in the sector $0 < \arg\zeta < \frac{2}{3}\pi$ (sector I of fig. 8.7) a given solution of the Stokes equation has the asymptotic approximation

$$\zeta^{-\frac{1}{4}}[A_1 \exp(-\tfrac{2}{3}\zeta^{\frac{3}{2}}) + B_1 \exp(\tfrac{2}{3}\zeta^{\frac{3}{2}})]. \tag{8.43}$$

On the Stokes line at $\arg\zeta = \frac{2}{3}\pi$ (S_2 in fig. 8.7) the first term is dominant, and hence for the sector $\frac{2}{3}\pi < \arg\zeta < \frac{4}{3}\pi$ (sector II) the asymptotic approximation is

$$\zeta^{-\frac{1}{4}}[A_1 \exp(-\tfrac{2}{3}\zeta^{\frac{3}{2}}) + B_2 \exp(\tfrac{2}{3}\zeta^{\frac{3}{2}})]. \tag{8.44}$$

The constant in the subdominant term has changed by $B_2 - B_1$. Now this change is zero if A_1 is zero. It cannot depend on B_1, for it would be unaltered if we added to (8.43) any multiple of the solution in which $A_1 = 0$. Since the differential equation is linear, $B_2 - B_1$ must be proportional to A_1, so that

$$B_2 - B_1 = \lambda_2 A_1, \tag{8.45}$$

where λ_2 is a constant called the 'Stokes multiplier' for the Stokes line at $\arg\zeta = \frac{2}{3}\pi$. It gives the change in the constant for the subdominant term when the Stokes line is crossed in an anticlockwise direction. If the crossing is clockwise, the Stokes multiplier has the opposite sign. Stokes multipliers can be defined in a similar way for the other two Stokes lines. It will be shown in §§ 8.16, 9.7, that for the Stokes equation (8.7) all three Stokes multipliers are equal to i.

8.16. Furry's derivation of the Stokes multipliers for the Stokes equation

The following derivation of the Stokes multipliers seems to have been used first by Furry (1947). The two terms in (8.43) are multiple valued functions of ζ with a branch point at the origin. Hence we introduce a cut in the complex ζ plane from 0 to $-\infty$ along the real axis, and take $-\pi \leqslant \arg\zeta \leqslant \pi$. Consider a solution whose asymptotic approximation is (8.43) in sector I of fig. 8.7. Then in the top part of sector II it is

$$\zeta^{-\frac{1}{4}}[A_1 \exp(-\tfrac{2}{3}\zeta^{\frac{3}{2}}) + (B_1 + \lambda_2 A_1) \exp(\tfrac{2}{3}\zeta^{\frac{3}{2}})]. \tag{8.46}$$

On the Stokes line S_1 the second term of (8.43) is dominant, and the constant in the first term changes so that in sector III the asymptotic approximation is

$$\zeta^{-\frac{1}{4}}[(A_1 - \lambda_1 B_1) \exp(-\tfrac{2}{3}\zeta^{\frac{3}{2}}) + B_1 \exp(\tfrac{2}{3}\zeta^{\frac{3}{2}})], \tag{8.47}$$

where λ_1 is the Stokes multiplier for the Stokes line S_1. On the Stokes line S_3 the first term of (8.47) is dominant, and the constant in the second term changes so that in the lower part of sector II the asymptotic approximation is

$$\zeta^{-\frac{1}{4}}[(A_1-\lambda_1 B_1)\exp(-\tfrac{2}{3}\zeta^{\frac{3}{2}})+\{B_1-\lambda_3(A_1-\lambda_1 B_1)\}\exp(\tfrac{2}{3}\zeta^{\frac{3}{2}})], \tag{8.48}$$

where λ_3 is the Stokes multiplier for the Stokes line S_3. Now the cut passes through the middle of sector II. The solution must be continuous across the cut, and hence (8.48) and (8.46) must agree. On crossing the cut from top to bottom, $\zeta^{-\frac{1}{4}}$ changes by a factor $e^{\frac{1}{2}i\pi}$, and $\zeta^{\frac{3}{2}}$ changes sign. Hence by equating coefficients of the exponentials in (8.46) and (8.48) we obtain

$$i(A_1-\lambda_1 B_1)=B_1+\lambda_2 A_1, \tag{8.49}$$

$$i\{B_1-\lambda_3(A_1-\lambda_1 B_1)\}=A_1. \tag{8.50}$$

Now this argument must apply whatever the values of A_1 and B_1. If $A_1=0$, (8.49) shows that $\lambda_1=i$. If $B_1=0$, (8.49) gives $\lambda_2=i$, and (8.50) gives $\lambda_3=i$. Hence

$$\lambda_1=\lambda_2=\lambda_3=i. \tag{8.51}$$

From the differential equation (8.7) it could have been predicted that all three Stokes multipliers are the same, for this equation is unaltered if ζ is replaced by $\zeta e^{\frac{2}{3}\pi i}$, and it therefore has the same properties on each Stokes line.

8.17. The range of validity of asymptotic approximations

The discussion in §§ 8.12–8.16, and the Stokes diagram fig. 8.6 (a) for the function $\mathrm{Ai}(\zeta)$, show that for the range $\frac{2}{3}\pi \leqslant \arg\zeta \leqslant \frac{4}{3}\pi$, the asymptotic approximation for $\mathrm{Ai}(\zeta)$ has multiples of the two terms (8.27), (8.28) that contain the exponential factors $\exp(\pm\frac{2}{3}\zeta^{\frac{3}{2}})$. For all $\arg\zeta$ in this range, except the anti-Stokes line $\arg\zeta=\pi$, one of the two terms is dominant and the other is subdominant. For a fixed $\arg\zeta\neq\pi$ in this range, let $|\zeta|$ increase indefinitely. Then the subdominant term tends to zero and the ratio of $\mathrm{Ai}(\zeta)$ to the dominant term tends to unity as $|\zeta|$ tends to ∞. Hence the asymptotic approximation to $\mathrm{Ai}(\zeta)$ as given by the Poincaré definition of § 8.11 need only contain the dominant term. It is only on the anti-Stokes line, where $\arg\zeta=\pi$ exactly, that it is necessary to include both terms, according to the Poincaré criterion.

Fig. 8.7. The complex ζ plane.

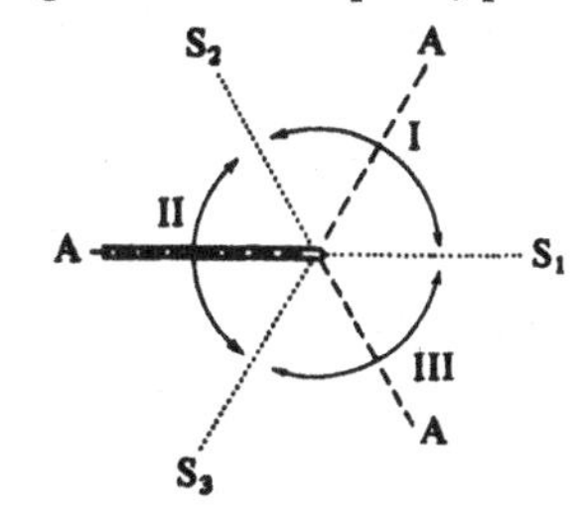

The Poincaré definition thus applies only to the limiting behaviour when $|\zeta| \to \infty$. When $|\zeta|$ is finite, it may clearly be insufficient to include only the dominant term. For example if $\arg \zeta = \pi$ the two terms are equal. If $|\zeta|$ is held constant and $\arg \zeta$ is made to differ slightly from π, the dominant term begins to increase and the subdominant term to decrease but at first the two terms are nearly equal, and in computing $\mathrm{Ai}(\zeta)$ it would be necessary to include both. But the Poincaré definition allows the subdominant term to be omitted. Care is therefore needed when specifying the range of $\arg \zeta$ over which a given asymptotic approximation is valid. For $\mathrm{Ai}(\zeta)$ the best way of doing this is as follows:

$$\mathrm{Ai}(\zeta) \sim \tfrac{1}{2}\pi^{-\frac{1}{2}}\zeta^{-\frac{1}{4}} \exp(-\tfrac{2}{3}\zeta^{\frac{3}{2}}) \quad \text{for} \quad -\tfrac{2}{3}\pi \leqslant \arg \zeta \leqslant \tfrac{2}{3}\pi, \tag{8.52}$$

$$\mathrm{Ai}(\zeta) \sim \tfrac{1}{2}\pi^{-\frac{1}{2}}\zeta^{-\frac{1}{4}}\{\exp(-\tfrac{2}{3}\zeta^{\frac{3}{2}}) + \mathrm{i}\exp(\tfrac{2}{3}\zeta^{\frac{3}{2}})\} \quad \text{for} \quad \tfrac{2}{3}\pi \leqslant \arg\zeta \leqslant \tfrac{4}{3}\pi, \tag{8.53}$$

where a fractional power ζ^f always means $|\zeta^f| \exp\{\mathrm{i}f \arg \zeta\}$. The constant $\frac{1}{2}\pi^{-\frac{1}{2}}$ is the same as A in (8.38). It is found from the contour integral (8.16) which is evaluated by the method of steepest descents as explained in § 9.7. In (8.52), (8.53) the ranges of $\arg \zeta$ end on Stokes lines, and this convention is used in the present book. According to the Poincaré definition the range in (8.52) could be written $-\pi < \arg \zeta < \pi$, and in (8.53) it could be written $\frac{1}{3}\pi < \arg \zeta < \frac{5}{3}\pi$. These ranges end just short of anti-Stokes lines, and the inequality signs are $<$ and not $\leqslant$, so that the two anti-Stokes lines are just outside the range at the ends. The two ranges of $\arg \zeta$ used in (8.52), (8.53) do not overlap. The larger ranges allowed by the Poincaré definition do overlap. They have common parts where both dominant and subdominant terms are present. In the Poincaré sense it does not then matter whether the subdominant term is included or not.

In most textbooks of mathematics that give asymptotic approximations, the range of $\arg \zeta$ is assessed according to the Poincaré definition, and this can sometimes lead to errors. An example of this is given in § 15.10, for the parabolic cylinder function. There are some further comments on this topic in § 15.8.

The asymptotic approximations for $\mathrm{Ai}'(\zeta)$ will also be needed later. They can be found from those for $\mathrm{Ai}(\zeta)$ by differentiating the asymptotic expansion, which gives new expansions for $\mathrm{Ai}'(\zeta)$. The first terms of these are as follows:

$$\mathrm{Ai}'(\zeta) \sim -\tfrac{1}{2}\pi^{-\frac{1}{2}}\zeta^{\frac{1}{4}} \exp(-\tfrac{2}{3}\zeta^{\frac{3}{2}}) \quad \text{for} \quad -\tfrac{2}{3}\pi \leqslant \arg\zeta \leqslant \tfrac{2}{3}, \tag{8.54}$$

$$\mathrm{Ai}'(\zeta) \sim \tfrac{1}{2}\pi^{-\frac{1}{2}}\zeta^{\frac{1}{4}}\{-\exp(-\tfrac{2}{3}\zeta^{\frac{3}{2}}) + \mathrm{i}\exp(\tfrac{2}{3}\zeta^{\frac{3}{2}})\} \quad \text{for} \quad \tfrac{2}{3}\pi \leqslant \arg\zeta \leqslant \tfrac{4}{3}\pi. \tag{8.55}$$

8.18. The choice of a fundamental system of solutions of the Stokes equation

Since the Stokes differential equation (8.7) is of the second order, it has two independent solutions which may be chosen in various ways, and these may be called the fundamental solutions. Any other solution is then a linear combination of them. There is no absolute criterion for these solutions and their choice is a matter of convenience. An obvious choice for one fundamental solution is the Airy integral function $\mathrm{Ai}(\zeta)$. As a second solution Jeffreys and Jeffreys (1972) use the function $\mathrm{Bi}(\zeta)$ defined in § 8.6. The choice of $\mathrm{Ai}(\zeta)$ and $\mathrm{Bi}(\zeta)$ has the advantage that both functions are real when ζ is real. The asymptotic approximation for $\mathrm{Bi}(\zeta)$ may be found from the contour integrals (8.17) which are evaluated by the method of steepest descents as explained in § 9.7. The result is:

$$\mathrm{Bi}(\zeta) \sim \tfrac{1}{2}\pi^{-\frac{1}{2}}\zeta^{-\frac{1}{4}}\{\mathrm{i}\exp(-\tfrac{2}{3}\zeta^{\frac{3}{2}}) + 2\exp(\tfrac{2}{3}\zeta^{\frac{3}{2}})\} \text{ for } 0 \leqslant \arg\zeta \leqslant \tfrac{2}{3}\pi, \tag{8.56}$$

$$\mathrm{Bi}(\zeta) \sim \tfrac{1}{2}\pi^{-\frac{1}{2}}\zeta^{-\frac{1}{4}}\{\mathrm{i}\exp(-\tfrac{2}{3}\zeta^{\frac{3}{2}}) + \exp(\tfrac{2}{3}\zeta^{\frac{3}{2}})\} \text{ for } \tfrac{2}{3}\pi \leqslant \arg\zeta \leqslant \tfrac{4}{3}\pi, \tag{8.57}$$

$$\mathrm{Bi}(\zeta) \sim \tfrac{1}{2}\pi^{-\frac{1}{2}}\zeta^{-\frac{1}{4}}\{2\mathrm{i}\exp(-\tfrac{2}{3}\zeta^{\frac{3}{2}}) + \exp(\tfrac{2}{3}\zeta^{\frac{3}{2}})\} \text{ for } \tfrac{4}{3}\pi \leqslant \arg\zeta \leqslant 2\pi, \tag{8.58}$$

where a fractional power of ζ has the meaning given in the preceding section. An alternative form of (8.58) is

$$\mathrm{Bi}(\zeta) \sim \tfrac{1}{2}\pi^{-\frac{1}{2}}\zeta^{-\frac{1}{4}}\{-\mathrm{i}\exp(-\tfrac{2}{3}\zeta^{\frac{3}{2}}) + 2\exp(\tfrac{2}{3}\zeta^{\frac{3}{2}})\} \text{ for } -\tfrac{2}{3}\pi \leqslant \arg\zeta \leqslant 0. \tag{8.59}$$

In these formulae the ranges of $\arg\zeta$ end on Stokes lines, as explained in the preceding section. There are both dominant and subdominant terms in all three ranges of $\arg\zeta$ and the asymptotic behaviour of $\mathrm{Bi}(\zeta)$ is therefore considerably more complicated than for $\mathrm{Ai}(\zeta)$. This is illustrated by the Stokes diagram for $\mathrm{Bi}(\zeta)$ given in fig. 8.6(b).

It is easily shown that the Stokes equation (8.7) is unaltered when ζ is replaced by $\zeta\exp(\frac{2}{3}\mathrm{i}\pi)$ or $\zeta\exp(\frac{4}{3}\mathrm{i}\pi)$. Hence $\mathrm{Ai}(\zeta e^{\frac{2}{3}\mathrm{i}\pi})$ and $\mathrm{Ai}(\zeta e^{\frac{4}{3}\mathrm{i}\pi})$ are solutions of the Stokes equation. For some purposes it is convenient to use $\mathrm{Ai}(\zeta e^{\frac{2}{3}\mathrm{i}\pi})$ as the second fundamental solution, instead of $\mathrm{Bi}(\zeta)$. It readily follows from (8.52), (8.53) that the asymptotic approximation for $\mathrm{Ai}(\zeta e^{\frac{2}{3}\pi\mathrm{i}})$ is

$$\mathrm{Ai}(\zeta e^{\frac{2}{3}\pi\mathrm{i}}) \sim \tfrac{1}{2}\pi^{-\frac{1}{2}}\zeta^{-\frac{1}{4}}e^{-\frac{1}{6}\pi\mathrm{i}}\exp(\tfrac{2}{3}\zeta^{\frac{3}{2}}) \text{ for } -\tfrac{4}{3}\pi \leqslant \arg\zeta \leqslant 0, \tag{8.60}$$

$$\mathrm{Ai}(\zeta e^{\frac{2}{3}\pi\mathrm{i}}) \sim \tfrac{1}{2}\pi^{-\frac{1}{2}}\zeta^{-\frac{1}{4}}e^{-\frac{1}{6}\pi\mathrm{i}}\{\exp(\tfrac{2}{3}\zeta^{\frac{3}{2}}) + \mathrm{i}\exp(-\tfrac{2}{3}\zeta^{\frac{3}{2}})\} \text{ for } 0 \leqslant \arg\zeta \leqslant \tfrac{2}{3}\pi. \tag{8.61}$$

The Stokes diagram for $\mathrm{Ai}(\zeta e^{\frac{4}{3}\pi\mathrm{i}})$ is shown in fig. 8.6(c) and is obtained from that for $\mathrm{Ai}(\zeta)$ (fig. 8.6(a)) by rotation through 120° clockwise.

It must be possible to express $\mathrm{Ai}(\zeta e^{\frac{2}{3}\pi\mathrm{i}})$ and $\mathrm{Ai}(\zeta e^{\frac{4}{3}\pi\mathrm{i}})$ as linear combinations of $\mathrm{Ai}(\zeta)$ and $\mathrm{Bi}(\zeta)$. It can be shown (see, for example, Miller, 1946), that

$$2e^{-\frac{1}{6}\pi\mathrm{i}}\mathrm{Ai}(\zeta e^{\frac{2}{3}\pi\mathrm{i}}) = \mathrm{Ai}(\zeta) - \mathrm{i}\,\mathrm{Bi}(\zeta), \quad 2e^{\frac{1}{6}\pi\mathrm{i}}\mathrm{Ai}(\zeta e^{\frac{4}{3}\pi\mathrm{i}}) = \mathrm{Ai}(\zeta) + \mathrm{i}\,\mathrm{Bi}(\zeta) \tag{8.62}$$

and

$$\mathrm{Ai}(\zeta) + e^{\frac{2}{3}\pi\mathrm{i}}\mathrm{Ai}(\zeta e^{\frac{2}{3}\pi\mathrm{i}}) + e^{\frac{4}{3}\pi\mathrm{i}}\mathrm{Ai}(\zeta e^{\frac{4}{3}\pi\mathrm{i}}) = 0 \tag{8.63}$$

8.19. Connection formulae, or circuit relations

Equations (8.52) and (8.53) show that in the special case when ζ is real, the asymptotic approximations for $\mathrm{Ai}(\zeta)$ are different according as ζ is positive or negative. A formula which gives one asymptotic approximation when the other is known is sometimes called a 'connection formula' or 'circuit relation'. For example, the connection formula for $\mathrm{Ai}(\zeta)$ is

$$\zeta^{-\frac{1}{4}}\{\exp(-\tfrac{2}{3}\zeta^{\frac{3}{2}}) + \mathrm{i}\exp(\tfrac{3}{2}\zeta^{\frac{3}{2}})\} \leftrightarrow \zeta^{-\frac{1}{4}}\exp(-\tfrac{2}{3}\zeta^{\frac{3}{2}}) \tag{8.64}$$

($\arg\zeta = \pi$ on left, and 0 on right). The use of the double arrow was introduced by Jeffreys (1924). The terms on the left are for negative ζ and those on the right for positive ζ. Clearly, in order to give a connection formula it is necessary to know the value of the Stokes multiplier.

The connection formula for $\mathrm{Bi}(\zeta)$ is written slightly differently thus

$$\zeta^{-\frac{1}{4}}\{\mathrm{i}\exp(-\tfrac{2}{3}\zeta^{\frac{3}{2}}) + \exp(\tfrac{2}{3}\zeta^{\frac{3}{2}})\} \rightarrow 2\zeta^{-\frac{1}{4}}\exp(\tfrac{2}{3}\zeta^{\frac{3}{2}}). \tag{8.65}$$

Since the line $\arg\zeta = 0$ is a Stokes line it does not matter whether the subdominant term is present or not. But the dominant term alone, for positive ζ, is insufficient to determine the asymptotic behaviour for negative ζ. On the other hand, the asymptotic behaviour for negative ζ determines completely the behaviour for positive ζ. Hence, following Langer (1934), we use only a single arrow in (8.65). For a discussion of connection formulae, see for example Heading (1970a, 1977a, 1979).

8.20. Stratified ionosphere. Uniform approximation

The earlier sections of this chapter started with the assumption that in the ionosphere the function $q^2(z)$, (8.2) is an exactly linear function of z and this led to the Stokes equation (8.7). This assumption may often be approximately true for a small range of z, but it is never exactly true in practice. It is important, therefore, to study solutions of the differential equation (7.6) when q^2 has an isolated zero as in fig. 8.1(b), but is not exactly linear. This can be done by using the method of § 7.9. We seek a fictitious 'comparison' medium with a new 'height' variable ζ, instead of z. The transformation is given by (7.49)–(7.52). The second term in the square brackets of (7.52) is $\frac{1}{2}\mathscr{D}(z;\zeta)$ where $\mathscr{D}$ is the Schwarzian derivative, given by (7.56). The new variable ζ is to be chosen so that $\mathscr{D}$ is zero when $q^2(z)$ is exactly linear, and small when q departs slightly from linearity. The derivative $z' = \mathrm{d}z/\mathrm{d}\zeta$, used in (7.56), must therefore exist and be bounded. This means that ζ is an analytic function of z. Thus there are two conditions. First, for the equation (7.52) to reduce to the Stokes equation, the first term in the square brackets must be given by

$$(kqz')^2 = -\zeta \tag{8.66}$$

which leads to

$$\pm\tfrac{2}{3}\mathrm{i}\zeta^{\frac{3}{2}} = k\int^{z} q\,\mathrm{d}z. \tag{8.67}$$

Second, since ζ is an analytic function of z, the lower limit of the integral must be at $z = z_0$ where $q = 0$. Hence

$$\zeta = \left(\frac{3}{2}\mathrm{i}k\int_{z_0}^{z} q\,\mathrm{d}z\right)^{\frac{2}{3}}. \tag{8.68}$$

If q^2 is given exactly by the linear function (8.2), then (8.68) is the same as (8.6). The sign and the fractional power in (8.68) are chosen so that ζ is positive when $\mathrm{Re}(z-z_0)$ is positive. Let $w=\zeta(k^2a)^{-\frac{1}{3}}$. If q^2 is given by (8.3), in which b gives the curvature of the $q^2(z)$ curve, then it can be shown that

$$\left.\begin{aligned} z-z_0 &= w+\frac{1}{5}\frac{b}{a}w^2+\frac{88}{700}\left(\frac{b}{a}\right)^2 w^3+\cdots, \\ w &= z-z_0-\frac{1}{5}\frac{b}{a}(z-z_0)^2-\frac{32}{700}\left(\frac{b}{a}\right)^2(z-z_0)^3-\cdots. \end{aligned}\right\} \tag{8.69}$$

Now (8.69) is substituted in (7.56) to give

$$\left.\begin{aligned} |\tfrac{1}{2}\mathscr{D}| &= \frac{1}{k^2}\left|\frac{3\zeta}{4q^4}\left(\frac{\mathrm{d}q}{\mathrm{d}z}\right)^2-\frac{\zeta}{2q^3}\frac{\mathrm{d}^2q}{\mathrm{d}z^2}+\frac{5k^2}{16\zeta^2}\right| \\ &= \frac{1}{k^2}\left|\frac{5\zeta}{16q^6}\left\{\frac{\mathrm{d}(q^2)}{\mathrm{d}z}\right\}^2-\frac{\zeta}{4q^4}\frac{\mathrm{d}^2(q^2)}{\mathrm{d}z^2}+\frac{5k^2}{16\zeta^2}\right| \end{aligned}\right\} \tag{8.70}$$

(compare (7.57)). Although the separate terms in (8.70) are infinite where $\zeta=0$, $q^2=0$, the combination of them is bounded. It can be shown that (8.70) is zero if $q^2(z)$ is exactly linear as in (8.2). If q^2 is given by (8.3) then it can be shown that

$$|\tfrac{1}{2}\mathscr{D}| = -\left(\frac{b}{a}\right)^2(k^2a)^{-\frac{2}{3}}\left\{\frac{9}{35}+\frac{809}{1050}\frac{b}{a}(z-z_0)+\cdots\right\}. \tag{8.71}$$

This confirms that when the curvature factor b of (8.3) is small, $|\tfrac{1}{2}\mathscr{D}|$ is very small, of order b^2.

If $\mathscr{D}$ is neglected, (7.52) is the same as the Stokes equation (8.7) with F for E. Its solution F must be some combination of $\mathrm{Ai}(\zeta)$ and $\mathrm{Bi}(\zeta)$ chosen to satisfy the physical conditions. When $z-z_0$ and thence ζ are large and positive $\mathrm{Ai}(\zeta)$ and $\mathrm{Bi}(\zeta)$ may be replaced by their asymptotic approximations. That for $\mathrm{Ai}(\zeta)$ is given by (8.52) and includes only the subdominant term $\zeta^{-\frac{1}{4}}\exp(-\frac{2}{3}\zeta^{\frac{3}{2}})$, which gets indefinitely smaller as ζ increases. That for $\mathrm{Bi}(\zeta)$ is given by (8.56) or (8.59) and includes the dominant term $\zeta^{-\frac{1}{4}}\exp(\frac{2}{3}\zeta^{\frac{3}{2}})$ which gets indefinitely larger as ζ increases. If the solution included any multiple of $\mathrm{Bi}(\zeta)$ the electric field would get larger and larger as the height z increased and the energy in the field would be extremely large at very great heights. This obviously could not happen in a field produced by radio waves incident on the ionosphere from below. Hence the solution cannot contain any multiple of $\mathrm{Bi}(\zeta)$ and so the physical conditions of the problem show that F is some multiple of $\mathrm{Ai}(\zeta)$. Further discussions of how to choose the correct solution are given in chs. 15, 18, 19.

The solution (7.51) is then

$$E\approx(z')^{\frac{1}{2}}\mathrm{Ai}(\zeta)\propto\zeta^{\frac{1}{4}}q^{-\frac{1}{2}}\mathrm{Ai}(\zeta) \tag{8.72}$$

where (8.66) has been used. For a level far enough below $z=z_0$, where ζ is negative, the asymptotic form (8.53) for $\mathrm{Ai}(\zeta)$ can be used, provided that $|\zeta|\gtrsim 1$ (compare

(8.29)). Now ζ is given by (8.68). Hence

$$E \approx q^{-\frac{1}{2}}\left\{\exp\left(-\mathrm{i}k\int_{z_0}^{z} q\,\mathrm{d}z\right) + \mathrm{i}\exp\left(\mathrm{i}k\int_{z_0}^{z} q\,\mathrm{d}z\right)\right\} \tag{8.73}$$

where a constant multiplying factor has been omitted. Thus the two terms in (8.73) are just the two W.K.B. solutions (7.26). The approximate solution (8.72) can be used near the level $z = z_0$ where the W.K.B. solutions fail, but it goes over continuously into the two W.K.B. solutions in a region where its asymptotic approximations can be used. It is an example of a 'uniform approximation'. It was first given, for the case of an isolated zero of q^2, by Langer (1937). Other examples of uniform approximations are mentioned in §§ 16.3, 17.3, 17.4.

The non-linearity of q^2 near its zero is important principally in the exponential terms of (8.73). These are the rapidly varying terms of the W.K.B. solutions, especially when k is large. If $q^2(z)$ is exactly linear, z' is independent of height z and therefore of ζ. When $q^2(z)$ is slightly non-linear as in the example of (8.3), where b is small, it can be shown from (8.69) that

$$|z'| = (k^2/a)^{\frac{1}{3}}\{1 + \tfrac{2}{5}b(a^2k)^{-\frac{2}{3}}\zeta + \ldots.\}. \tag{8.74}$$

The use of (8.66) in (7.52), and thence of (8.72), rests on the assumption that z' does not change appreciably for values of $|\zeta|$ up to about unity. This requires that

$$|b(a^2k)^{-\frac{2}{3}}| \ll 1. \tag{8.75}$$

For larger values of $|\zeta|$ the W.K.B. solutions can be used. Another way of deriving (8.75) was given by Budden (1961a, § 16.8).

The solution (8.72) is often needed in a form where the upgoing component wave, that is the first term in (8.73), has unit amplitude at the ground $z = 0$. This is achieved by using a constant multiplying factor with the second expression (8.72), thus

$$E = 2\zeta^{\frac{1}{4}}(\pi C/q)^{\frac{1}{2}}\mathrm{Ai}(\zeta)\exp\left(-\mathrm{i}k\int_{0}^{z_0} q\,\mathrm{d}z\right). \tag{8.76}$$

The results of this section and particularly (8.73) have now supplied the justification for the argument in § 7.19 that led to the expression (7.152) for the reflection coefficient, including the factor i which comes from the second term of (8.73). This means that the phase of the wave is advanced by $\frac{1}{2}\pi$ during the reflection process. An alternative explanation of this phase advance is given in the following section.

8.21. The phase integral method for reflection

In the formula (7.152) for the reflection coefficient, the integral is a contour integral along a path in the complex z plane, beginning at the ground $z = 0$ and ending at the reflection point $z = z_0$. It may be written in other forms. The function $q(z)$ is a two-valued function (7.2) with a branch point at z_0, and the two values will now be written $q_1(z)$, $q_2(z)$. The value used in (7.152) is $q_1(z)$, chosen so that its real part is positive on the real z axis. It is therefore associated with the upgoing wave (7.149). The other value is $q_2(z) = -q_1(z)$, and if inserted in the exponential of (7.149) it would give the downgoing wave (7.150). Now (7.152) may be written

$$R = \mathrm{i}\exp\left\{-\mathrm{i}k\int_0^{z_0} q_1(z)\,\mathrm{d}z - \mathrm{i}k\int_{z_0}^{0} q_2(z)\,\mathrm{d}z\right\}. \tag{8.77}$$

The contours for these two integrals need not be the same. They can be moved in different ways provided that no singularities are crossed.

Now let a branch cut be inserted in the complex z plane, running from z_0 to infinity as shown in fig. 8.8. Its exact position is not important. Then on crossing the cut q_1 and q_2 both change sign. The value of q_1 at a point on one side of the cut is the same as the value of q_2 at the adjacent point on the other side and vice versa. At the branch point z_0 both q_1 and q_2 are zero. The two contours in (8.77) can now be replaced by one contour C_1, fig. 8.8(a). The first part of C_1 runs from $z = 0$ to a point P very close to z_0 and the integrand is q_1. The next part is a very small circle that encircles the point z_0 clockwise from P to Q. Before the cut is crossed the integrand is q_1 and afterwards it is q_2. Since both q_1 and q_2 are small, the value of the integral on this circle can be made as small as desired by choosing the radius to be small enough. The third part of C_1 runs from Q back to $z = 0$ and the integrand is q_2. Then (8.77) becomes

$$R = \mathrm{i}\exp\left(-\mathrm{i}k\int_{\mathrm{C}} q\,\mathrm{d}z\right) \tag{8.78}$$

where C is C_1, and q, now written without a subscript, means q_1 before the cut is crossed and q_2 afterwards. The integrand in (8.78) is a continuous analytic function of z at all points on the contour. The contour C can therefore be distorted so that it is not very near the point z_0, as shown at C_2, fig. 8.8(b), provided that no singularities of q are crossed. The values of the integral and of R are unaffected. In particular it can be moved out into a region where the W.K.B. solutions are good approximations at all points on it, provided that this is not prevented by singularities of q that are too

Fig. 8.8. The complex z plane showing possible contours $\mathrm{C}_1, \mathrm{C}_2$ for use with the phase integral formula (8.78). The shaded region in (b) is where $|\zeta| \lesssim 1$.

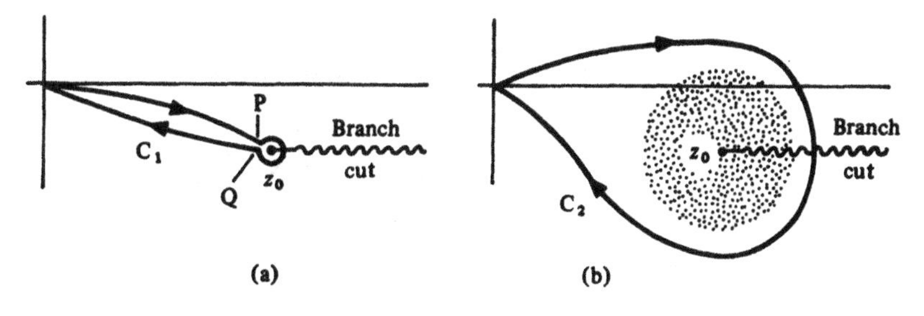

near to z_0. In the example of fig. 8.8(b) the region shaded with dots is where $|\zeta| \lesssim 1$. Outside this region the W.K.B. approximations can be used and this applies for all points on C_2. The integral now expresses the cumulative change of phase in the phase memory term of one W.K.B. solution as we proceed from the transmitter at the ground, $z = 0$, along the contour C_2 in the complex z plane and back to a receiver at $z = 0$. The integral is the phase integral and (8.78) is called the phase integral formula for reflection.

This process can be expressed in another way. The Stokes diagram for the function $\mathrm{Ai}(\zeta)$ in the uniform approximation solution (8.72) is shown in fig. 8.6(a). The Stokes lines $\arg\zeta = 0, \pm\frac{2}{3}\pi$, and the anti-Stokes lines $\arg\zeta = \pm\frac{1}{3}\pi, \pi$, radiate from the point $z = z_0$. They can be drawn in a diagram of the complex z plane as in fig. 8.9(a). If $q^2(z)$ were exactly linear they would be straight but in general they are curved. The Stokes diagram fig. 8.6(a) is shown again in fig. 8.9(b). Now consider the upgoing wave (7.149) near $z = 0$. It is one of the asymptotic forms of the uniform approximation solution (8.72). Its modulus gets larger as z gets more negative and therefore it is the dominant term. It is indicated by the label 'incident wave' in fig. 8.9(b). Now let the upper limit z of the integral in (7.149) move clockwise along the contour C_2 of fig. 8.8(b). The wave is at first the dominant term, but becomes subdominant after the anti-Stokes line A1 is crossed. The Stokes diagram fig. 8.9(b) shows that here there is no dominant term and for this wave the Stokes phenomenon does not occur near the Stokes line S2. After crossing the anti-Stokes line A2, the wave is again dominant. It becomes subdominant again after A3 is crossed, just before the real z axis is reached, and before the Stokes line S1 is crossed. Now it has been converted to the downgoing wave, indicated as 'reflected wave' in fig. 8.9(b). On the whole path the expression is a continuous analytic function of z that does not undergo the discontinuity associated with the Stokes phenomenon. This explains why the contour, in this example, must encircle the branch point in the clockwise sense. The upgoing and downgoing waves near the ground are thus linked by this continuous process. In some other applications of this method the branch point

Fig. 8.9. (a) is the complex z plane and shows the Stokes lines S and anti-Stokes lines A for the function $\mathrm{Ai}(\zeta)$. (b) is the Stokes diagram for this function.

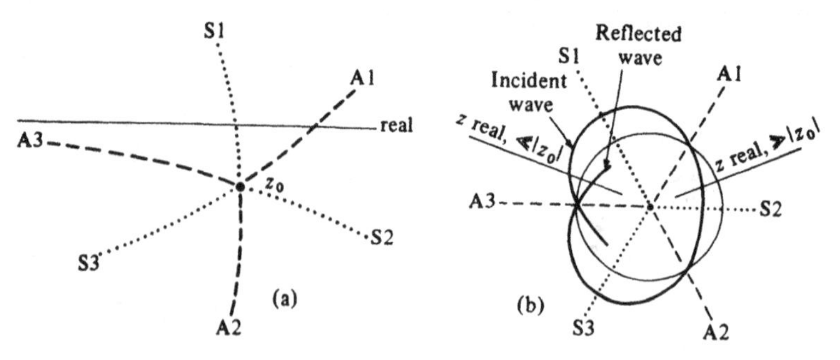

must be encircled anticlockwise. In each problem of this kind the correct sense must be found by examining the relevant Stokes diagram.

Finally consider the factor $q^{-\frac{1}{2}}$ in (7.149). Now q^2 is analytic and may be represented by a series whose first two terms are (8.3). For the clockwise contours $\arg(q^2)$ decreases by 2π and so $\arg(q^{-\frac{1}{2}})$ increases by $\frac{1}{2}\pi$. Thus on returning to the initial point, the final value $q_2^{-\frac{1}{2}}$ is i times the initial value $q_1^{-\frac{1}{2}}$. Now in the expression (7.150) for the downgoing wave q means q_1 and $q^{-\frac{1}{2}}$ is therefore $-\mathrm{i}q_2^{-\frac{1}{2}}$. Thus the reflection coefficient R as given by (7.151) must be multiplied by i. This factor i is included in (7.152) and thence in (8.77), (8.78). It was used there because it appears in (8.73) as the Stokes multiplier of the Stokes equation. It has now been shown that it occurs as a direct consequence of the continuity of the W.K.B. solutions for the incident and reflected waves when the reflection branch point is encircled.

When the earth's magnetic field is allowed for, q_1 and q_2 are solutions of the Booker quartic, ch. 6, and in general $q_1 \neq -q_2$. The reflection point z_0 is where $q_1 = q_2$ but in general they are not zero here. For this case there is a uniform approximation solution (16.29)–(16.32), and the phase integral formula (8.78) can still be used. It can be applied to either the ordinary or the extraordinary wave. For a discussion of this see §§ 16.3, 16.4.

The validity of the phase integral formula for the reflection coefficient rests on the use of the function $\mathrm{Ai}(\zeta)$ in the uniform approximation solution (8.72) because the Stokes diagram, fig. 8.9(b), for this function contains a single continuous curve. This in turn implies that at real heights above the reflection level z_0 there is no downgoing wave. This might not be true if $q(z)$ has another singularity that is too close to z_0. For example, the reflection level z_0 where q is zero may be just below the maximum of an ionospheric layer. Then there is another point z_{00} just above the maximum where q is again zero. These two points are both branch points of $q(z)$. If they are too close together there may be no region between them where the W.K.B. solutions are good approximations, so that it cannot be asserted that there is no downgoing wave in the region above z_0 but below z_{00}. It is not then possible to find a contour C_2 as in fig. 8.8(b). This situation occurs when the electron concentration $N(z)$ has a parabolic distribution and the problem is discussed in §§ 15.9–15.11. Another example occurs for vertical polarisation and near vertical incidence, when q must be replaced by an 'effective' value Q, (15.27) that has a pole just above z_0. This case is discussed in §§ 15.5–15.7.

The first to use phase integral methods for radio propagation problems was T.L. Eckersley. His first main paper on the subject (Eckersley, 1931) is entitled 'On the connection between ray theory of electric waves and dynamics'. In the preceding years the analogy between the motion of a particle in a potential field, and the motion of a wave packet in a medium whose refractive index varies in space, had been used to develop the science of wave mechanics. It was postulated that a moving particle is represented by a progressive wave whose phase must be a single valued function of the space coordinates. Thus when it goes round a closed orbit, the total change of phase, that is the 'phase integral', must be an integer times 2π. This is the origin of the term 'phase integral method'. Ray theory is closely analogous to classical mechanics with the addition of the concept of phase, leading to some of the quantum conditions.

Eckersley's first group of papers (1931, 1932a, b) on the subject was concerned with guided waves. The simplest kind of wave guide has two plane reflecting boundaries with free space between them. A single wave guide mode can be considered as two plane progressive waves in this space, with their wave normals making angles θ, $\pi - \theta$ with the normal to the boundary planes (Brillouin, 1936; Chu and Barrow, 1938). For a self-consistent mode, the twice reflected wave must be identical with the original wave. This requires that the total change of complex phase in the double traverse of the guide width and at the two reflections must be an integer times 2π. It leads to the well known 'mode condition', and is a very simple example of the use of the phase integral. It is analogous to the phase integral condition for the wave associated with a particle in a closed orbit.

This simple example is exceptional because the reflections at the sharp boundary planes cannot be dealt with by ray theory. But in some other problems considered by Eckersley the reflections occurred in continuous slowly varying media, such as the ionosphere, where it was possible to solve the whole problem by ray theory. In this way the phase integral method developed into a method of calculating the reflection coefficient of a continuously varying stratified medium, as described earlier in this section, even when no mode condition is imposed.

For a review of the subject see Budden (1975). For discussion of the mathematical details see Heading (1962a, c, 1976, 1977b), Wait (1962), Evgrafov and Fedoryuk (1966).

The W.K.B. solutions, and the phase integral formula for reflection (7.152) or equivalently (8.78) are the basis of ray tracing methods, chs. 10, 12–14, and are normally used for studying propagation and reflection of high frequency waves in the ionosphere and magnetosphere. In practice this means frequencies greater than about 100 to 200 kHz, or possibly less in the upper ionosphere and magnetosphere when the medium changes little within one free-space wavelength. For high frequencies the methods of ray theory are the most convenient, and the only sensible way of studying propagation. It is shown in §§ 10.17–10.19 that a ray can be traced right through a region near where it touches a caustic surface, and this is effectively a region of reflection. The use of ray theory in this way is possible when the reflection points are well isolated, that is well separated in complex space from others. The only additional feature of the phase integral formula is the factor i in (7.152), (8.78) but this is not important in ray theory. Coupling points, like reflection points, are points in the complex z plane where two roots of the Booker quartic are equal. It is shown in § 16.7 that coupling between two characteristic waves, ordinary and extraordinary, can be treated by the phase integral method, and if the coupling point is sufficiently isolated, the process is very similar to the phase integral method for reflection. Mathematically, the two processes are the same, and the term 'coupling point' is used to mean either a coupling or a reflection point.

In this sense, therefore, the phase integral method is a widely used physical principle that is essentially the same as ray theory. The main purpose of discussing it in this book is to see how far it can be extended and, most important of all, to see when it fails and why.

When two coupling points are close together but can be studied separately from others, this introduces ideas that are the basis of new physical principles going beyond simple ray theory; for example, partial penetration and reflection §§ 15.10, 17.3, radio windows §§ 17.6–17.9, limiting polarisation §§ 17.10, 17.11. But for computing it is now simpler and safer to use numerical integration of the basic differential equations; chs. 18, 19. The phase integral method uses transformations of the variables that introduce, at the coupling points, singularities that are not present in the original basic differential equations. These transformations are useful only as a means of helping to explain the results in terms of physical processes. They are best avoided in numerical integration of the differential equations.

When three or more coupling points need to be considered together, the phenomena are now too complicated to be interpretable in terms of simple physical processes. It is still possible to use the phase integral method in some special cases; see, for example, Heading (1977b). But a general rule that will cover all cases has not been formulated. For very low frequencies, therefore, and for other cases where ray theory is suspect because of the proximity of coupling points, it is safest and also easiest to use full wave numerical solutions.

8.22. The intensity of light near a caustic

The integral (8.19) was originally derived by Airy (1838, 1849) in a study of the variation of the intensity of light near a caustic. Caustics can also be formed by radio waves, see §§ 10.17–10.23, and it is therefore useful to give the theory. Huyghens's principle is used to construct 'amplitude–phase' diagrams for a series of neighbouring points. Each diagram is a spiral, somewhat analogous to Cornu's spiral, and its vector resultant gives the amplitude and phase of the light at the point considered.

Consider a parallel beam of light incident from the left on a convex lens, fig. 8.10. For simplicity the lens is assumed to be cylindrical, and the problem is treated as though it is in two dimensions only. It is well known that the rays of the beam are tangents to a caustic curve XY. It is required to find the intensity of the light at the points of a line AB perpendicular to the caustic. Let Q be a typical point of this line.

Consider a plane wave front of the beam (RS in fig. 8.10), just before it reaches the lens. Imagine this wave front to be divided into infinitesimal strips of equal width δs, which all radiate cylindrical waves of the same intensity. On arrival at Q these waves all have the same amplitude, but their phases depend on the position, s, of the strip from which each originates. If Q is inside the caustic, as shown in fig. 8.10, two geometrical rays pass through it. Assume that one of these, RQ, is included in the narrow pencil between rays 1 and 2, and the other, SQ, in the pencil between rays 4 and 5.

The pencil enclosed by the rays 1 and 2 comes to a focus at X. The point Q is within this pencil, and the light reaches it before it reaches the focus X. It is then easily shown that for the ray RQ within this pencil Fermat's principle is a principle of least time. The variation with s of the phase $\phi(s)$ therefore has a minimum for the Huyghens wavelets that originate near R. The pencil enclosed by the rays 4 and 5 comes to a focus at Y. The point Q is within the pencil and beyond the focus, and it can then similarly be shown that for the ray SQ Fermat's principle is a principle of greatest time, so that the phase $\phi(s)$ has a maximum for Huyghens wavelets that originate near S. The function $\phi(s)$ has no other turning points (fig. 8.11, curve Q). If Q is outside the caustic, at T say, no geometrical rays pass through it. Then $\phi(s)$ has no turning points, and varies as shown in fig. 8.11, curve T. If Q is on the caustic, at W say, the two turning points coincide (fig. 8.11, curve W).

Let the zero of s be chosen midway between the two turning points, or at the point of least slope when there are no turning points. Then the form of the curves in fig. 8.11 is given by the cubic law

$$\phi = K(\zeta s + \tfrac{1}{3}s^3), \tag{8.79}$$

where ζ depends on the distance of Q from the caustic, and K is a constant. Clearly the scale of s can be chosen so that $K = 1$. When Q is on the illuminated side of the caustic, as in fig. 8.10, ζ is negative, and when Q is outside the caustic ζ is positive.

The value of ζ can be found as follows. Let the distance QW $= u$, and let the radius

Fig. 8.10. Caustic formed by a beam of light incident on a lens.

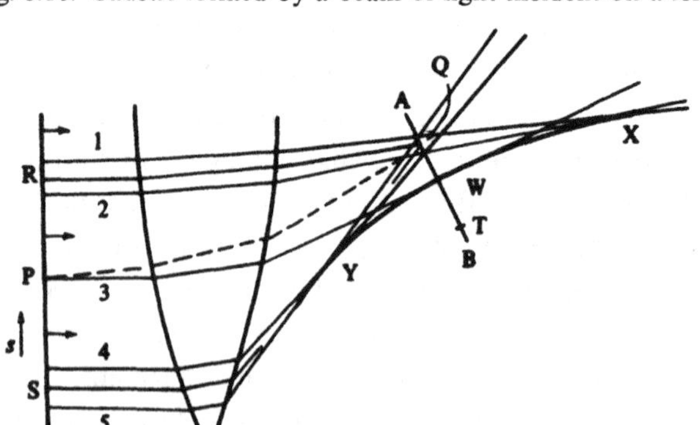

of curvature of the caustic be R. For curve 1, in fig. 8.11, ζ is negative, and the difference between the maximum and minimum values of ϕ is $\frac{4}{3}\zeta^{\frac{3}{2}}$ (with $K = 1$). This is the difference between the phases of the light arriving at Q via the ray pencils 4–5 and 1–2. Now any two rays that are close together, such as 4 and 5, have the same phase where they meet on the caustic. Hence the phase difference between the waves arriving at X and Y is just the phase difference that would arise from the curved path YWX along the caustic, namely $4\pi R\theta/\lambda$ where 2θ is the angle subtended by the arc XY at its centre of curvature. Thus the difference between the phases of the light arriving at Q via the two ray pencils is

$$\frac{2\pi}{\lambda}\{\mathrm{YQ} + \mathrm{QX} - \mathrm{arc(YWX)}\} = 4\pi R(\tan\theta - \theta)/\lambda \approx \frac{4\pi R\theta^3}{3\lambda}, \tag{8.80}$$

so that $\zeta^3 \approx \pi^2 R^2 \theta^6/\lambda^2$. Now $\mathrm{QW} = u \approx \frac{1}{2}R\theta^2$, so that when u is small

$$\zeta = 2u\left(\frac{\pi^2}{R\lambda^2}\right)^{\frac{1}{3}}. \tag{8.81}$$

Clearly ζ could be expressed by an ascending power series in $u = \mathrm{QW}$, and if ζ is small enough, powers higher than the first can be neglected.

For each position of the point Q an amplitude–phase diagram may now be constructed. The curvature of this at a given point is proportional to the slope at the corresponding point in the ϕ–s curve. Since this slope eventually increases indefinitely both for s increasing and for s decreasing, the amplitude–phase diagrams end in spirals. Typical diagrams are shown in fig. 8.12, corresponding to the three curves of fig. 8.11. In each case the closing vector of the diagram is marked with an arrow. It gives the resultant amplitude of the light that reaches Q. There is an obvious analogy between these diagrams and Cornu's spiral, for which the ϕ–s curve is a parabola. Cornu's spiral always has the same form. In diffraction problems whose solution involves a Cornu spiral, variations of intensity occur because

Fig. 8.11. Dependence of phase $\phi(s)$ of Huyghens wavelet on position s of source on initial wavefront.

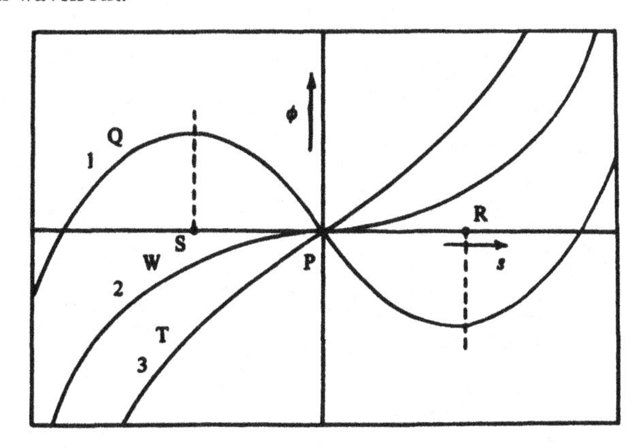

different fractions of the spiral are used to determine the intensity reaching different points. In the present problem however, the shape of the spiral is different for different values of ζ. For all points of the line AB, the resultant of the whole of the spiral is found, but the intensity varies because the shape of the spiral varies from point to point.

In fig. 8.12, let the x axis be chosen parallel to the element corresponding to $s = 0$. Then the diagrams are symmetrical about the y axis, and the resultant vector is always parallel to the x axis. When ζ is positive (first diagram of fig. 8.12), the curvature of the spiral always has the same sign and the length of the resultant decreases monotonically as ζ increases. When ζ is negative (third diagram of fig. 8.12), the middle section of the spiral has opposite curvature from the rest, and as ζ decreases the limiting points of the spiral repeatedly move completely round the origin. The resultant thus alternates in sign, and in the third diagram of fig. 8.12 it is negative.

The contribution of an element δs of the wave front to the resultant vector is proportional to $\delta s \cos\phi$, and hence the resultant vector is proportional to

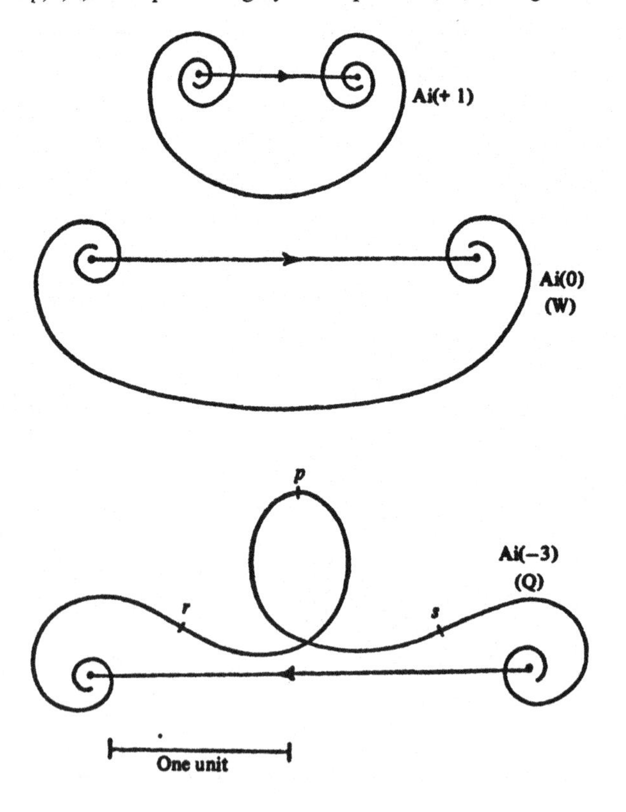

Fig. 8.12. Spirals associated with the Airy integral function. In the third spiral the points p, r, s, correspond roughly to the points P, R, S in figs. 8.10 and 8.11.

$$\int_{-\infty}^{\infty} \cos\phi \, ds = \int_{-\infty}^{\infty} \cos(\zeta s + \tfrac{1}{3}s^3) \, ds. \tag{8.82}$$

Apart from a constant, this is the standard expression (8.19) for the Airy integral function, $\mathrm{Ai}(\zeta)$.

When Q is well beyond the caustic on the illuminated side (ζ large and negative), the turning points of the ϕ–s curve are widely separated, and near each turning point the curve is approximately a parabola. Hence the portion of the amplitude–phase diagram corresponding to the neighbourhood of each turning point is approximately a Cornu spiral. The two Cornu spirals curve opposite ways, and together constitute the more complex spiral whose resultant is the Airy integral function; see fig. 8.13.

PROBLEMS 8

8.1. A linear second order differential equation with independent variable z has two solutions $S_1(z)$, $S_2(z)$. A prime $'$ is used to denote $\mathrm{d}/\mathrm{d}z$. The function $S_1S_2' - S_2S_1' = W(S_1, S_2)$ is called the Wronskian for these solutions. Prove that if the differential equation has no first derivative term, any Wronskian is a constant.

For the Stokes differential equation $\mathrm{d}^2E/\mathrm{d}z^2 = zE$ prove that $W(\mathrm{Ai}, \mathrm{Bi}) = 1/\pi$. If S_1, S_2 are the two series solutions in (8.8) find (a) $W(S_1, S_2)$, (b) $W(S_1, \mathrm{Ai})$, (c) $W(S_2, \mathrm{Ai})$. (Express (b) and (c) in terms of $\mathrm{Ai}(0)$, $\mathrm{Ai}'(0)$).

8.2. The functions $\mathrm{Ai}(z)$ and $\mathrm{Bi}(z)$ (fig. 8.5) are both real and oscillatory when z is real and negative. They can be thought of as 'standing waves', analogous in some ways to $\cos kz$, $\sin kz$. It is sometimes more convenient to use solutions which are analogues of progressive waves. This suggests the use of $\mathrm{C}_+ = \mathrm{Ai} + \mathrm{i}\,\mathrm{Bi}$, $\mathrm{C}_- = \mathrm{Ai} - \mathrm{i}\,\mathrm{Bi}$. Prove that $\mathrm{C}_+(z) = -2\Omega^2\,\mathrm{Ai}(\Omega^2 z)$, and $\mathrm{C}_-(z) = -2\Omega\,\mathrm{Ai}(\Omega z)$ where $\Omega = \exp(\tfrac{2}{3}\pi\mathrm{i})$. (Use the contour integrals (8.16), (8.17)). Show that for real negative z their asymptotic forms are proportional to $z^{-\frac{1}{4}}\exp(\mp\tfrac{2}{3}\mathrm{i}|z|^{\frac{3}{2}})$ respectively (upper sign for C_+). Hence confirm that they represent progressive waves.

Fig. 8.13. Spiral associated with Airy integral function of large negative argument, showing resemblance to two Cornu spirals. The part of the left half where overlapping occurs is not shown.

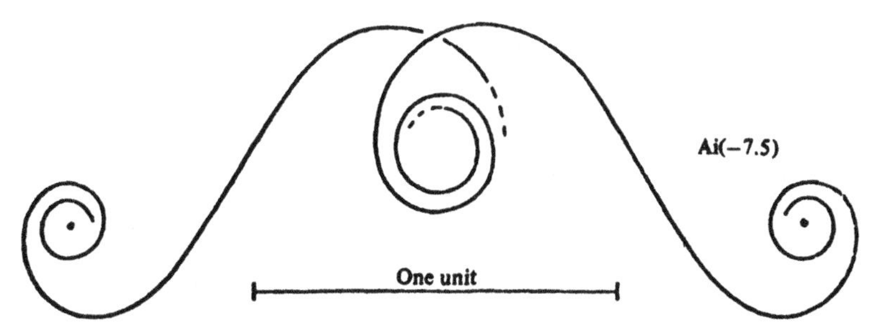

8.3. Find solutions $S(z)$ of the Stokes equation which, if possible, satisfy the following conditions. Express $S(z)$ in terms of $\mathrm{Ai}(z)$ and $\mathrm{Bi}(z)$, whose Wronskian is $1/\pi$. When there is no possible solution explain why. (a is a number that makes $\mathrm{Ai}(-a)=0$.)

(a) $S(0)=0, \quad S'(0)=-1.$

(b) $S(0)=1, \quad S'(0)=0.$

(c) $S(-1)=0, \quad S(0)=1.$

(d) $S(0)=0, \quad S(1)=0.$

(e) $S(0)=1, \quad S(\infty)=0.$

(f) $S(0)=0, \quad S(\infty)=0.$

(g) $S(-a)=0, \quad S(\infty)=0.$

(h) $S(-a)=0, S'(-a)=1.$

(i) $S(-a)=1, \quad S(\infty)=0.$

8.4. The function $y(z)$ satisfies $\mathrm{d}^2y/\mathrm{d}z^2=(z-a)y$ where a is an adjustable parameter, constant for any one solution. Find those values of a for which solutions are possible in the following cases, and express the solutions in terms of Ai and Bi. In each case state whether the given boundary conditions are homogeneous or inhomogeneous.

(a) $y(0)=0$, $y(\infty)$ bounded. y has no zeros for $z>0$.

(b) $y(0)=0, y(1)=0.$

(c) $y(0)=1$, $y(\infty)$ bounded.

(d) $y'(0)=0$, $y(\infty)$ bounded.

(e) $y'(0)=0, y'(1)=0.$

(f) $y'(0)=1, y(1)=0.$

8.5. Find a second order differential equation satisfied by $u=\mathrm{d}y/\mathrm{d}z$ where y can be any solution of the Stokes equation.

8.6. The differential equation $\mathrm{d}^2y/\mathrm{d}z^2=\frac{1}{4}z^2y$ is a special case of Weber's equation. Find (a) the W.K.B. solutions, (b) the Stokes lines, (c) the anti-Stokes lines and (d) the Stokes multipliers.

9

Integration by steepest descents

9.1. Introduction

In some problems in radio propagation and in other branches of physics it is necessary to evaluate integrals of the form

$$I = \int_C g(t) \exp\{f(t)\} \mathrm{d}t \tag{9.1}$$

where C is some path in the complex plane of the variable t. This can sometimes be done by the method of steepest descents. In its simplest form it is an approximate method but it is possible to increase the accuracy by using extensions and refinements of it. One of its most important uses is in the evaluation of the contour integral representations of special functions, for example the solutions (8.16), (8.17) of the Stokes equation. These are discussed in detail in § 9.7, as an illustration of the method. The account of the method given in this chapter is only an outline. For fuller mathematical details the reader should consult a standard treatise such as Jeffreys and Jeffreys (1972) or Morse and Feshbach (1953). See also Baños (1966).

9.2. Some properties of complex variables and complex functions

A complex variable t may be expressed as the sum of its real and imaginary parts thus

$$t = x + \mathrm{i}y \tag{9.2}$$

where x and y are real. A function $f(t)$ may similarly be written

$$f(t) = \phi(x, y) + \mathrm{i}\psi(x, y) \tag{9.3}$$

where ϕ and ψ are real. Then they satisfy the Cauchy relations

$$\partial\phi/\partial x = \partial\psi/\partial y, \quad \partial\phi/\partial y = -\partial\psi/\partial x. \tag{9.4}$$

Let grad ϕ be the vector whose components are $\partial\phi/\partial x$, $\partial\phi/\partial y$ in the x–y plane, that is the complex t plane. Then (9.4) shows that

$$\operatorname{grad}\phi\cdot\operatorname{grad}\psi = 0. \tag{9.5}$$

Thus the vectors grad ϕ and grad ψ are at right angles, which means that, in the complex t plane, lines on which ϕ is constant are also lines on which ψ changes most rapidly, and vice versa. The function $\exp\{f(t)\}$ has constant modulus on lines of constant ϕ, which are called 'level lines' for this function. The same function has constant argument or phase on lines for which ψ is constant. On these lines ϕ is changing most rapidly, so they are called lines of 'steepest descent' (or 'steepest ascent'). Equation (9.5) shows that the level lines and lines of steepest descent are at right angles. A diagram of the complex t plane, in which level lines for various values of $|\exp f(t)|$ are drawn, is like a contour map. The contours are the level lines, and their orthogonal trajectories are the lines of steepest descent or ascent. In this chapter frequent use is made of this kind of diagram. The value of $|\exp f(t)|$ can never be negative so that there are no contours at levels below zero. On the other hand $|\exp f(t)|$ can become indefinitely large, so that the contour maps may show some 'infinitely high' mountains. This kind of 'contour' must not be confused with a 'contour of integration', which is not, in general, a level line. In this chapter the word 'contour' is used in the sense of a level line in a contour map, and the word 'path' is used for a contour or path of integration.

If the two equations in (9.4) are again differentiated with respect to x, y (or y, x) respectively and combined, the result is

$$\frac{\partial^2\phi}{\partial x^2}+\frac{\partial^2\phi}{\partial y^2}=0, \quad \frac{\partial^2\psi}{\partial x^2}+\frac{\partial^2\psi}{\partial y^2}=0. \tag{9.6}$$

Hence both ϕ and ψ behave like potential functions in two-dimensional space, and neither can have a maximum or minimum value at any point where (9.6) holds. For example, if ϕ had a maximum at some point, this would require that $\partial^2\phi/\partial x^2$ and $\partial^2\phi/\partial y^2$ are both negative, which would violate (9.6). A point where the conditions (9.6) and the Cauchy relations (9.4) do not hold is called a 'singular point' of the function $f(t)$. A line of steepest descent is one on which ϕ decreases or increases at the greatest possible rate. It is now clear that ϕ must continue to decrease or increase until the line meets a singular point, for there is no other way in which ϕ can have a maximum or minimum value. Thus the contour maps will never show anything resembling a mountain with a top at a finite height. If a level line forms a closed curve, there must be inside this curve some singularity of $f(t)$, though not necessarily of $\exp\{f(t)\}$. This could be an infinitely high mountain, where $\phi\to+\infty$, or a zero of $\exp f(t)$, where $\phi\to-\infty$.

Clearly, a line of steepest descent (or ascent) can never form a closed curve. The term 'line of steepest descents' (plural) through a saddle point has a slightly different meaning, explained in the following section. Such a line sometimes can form a closed curve but there are no examples in this book.

9.3. Saddle points

The Cauchy relations (9.4) arise from the requirement that $f(t)$ shall have a unique derivative, which is given by

$$\frac{\mathrm{d}f(t)}{\mathrm{d}t}=\frac{\partial\phi}{\partial x}+\mathrm{i}\frac{\partial\psi}{\partial x}=\frac{\partial\psi}{\partial y}-\mathrm{i}\frac{\partial\phi}{\partial y}. \tag{9.7}$$

At a point where $\mathrm{d}f/\mathrm{d}t$ is zero, both grad ϕ and grad ψ are zero so that both the modulus and the argument of $\exp\{f(t)\}$ are stationary. Such a point is called a 'saddle point'. Let $f(t)$ have a saddle point where $t=t_0$ and let

$$f(t_0)=f_0,\quad \mathrm{d}^2f/\mathrm{d}t^2=A\mathrm{e}^{\mathrm{i}\alpha} \tag{9.8}$$

at the saddle point, where A and α are real, but f_0 is in general complex. Then Taylor's theorem gives

$$f(t)=f_0+\tfrac{1}{2}(t-t_0)^2A\mathrm{e}^{\mathrm{i}\alpha}+\mathrm{O}\{(t-t_0)^3\}. \tag{9.9}$$

Let $t-t_0=s\mathrm{e}^{\mathrm{i}\theta}$ where s and θ are real. Then

$$|\exp f(t)|=|\exp f_0|\exp\{\tfrac{1}{2}As^2\cos(2\theta+\alpha)+\mathrm{O}(s^3)\} \tag{9.10}$$

which shows that at the saddle point there are four directions (two crossing lines) for which $|\exp f(t)|$ is independent of s. These are given by

$$2\theta+\alpha=\pm\tfrac{1}{2}\pi,\ \pm\tfrac{3}{2}\pi. \tag{9.11}$$

Hence, through every saddle point there are two level lines crossing at right angles. Further

$$\arg\{\exp f(t)\}=\arg f_0+\tfrac{1}{2}As^2\sin(2\theta+\alpha)+\mathrm{O}(s^3) \tag{9.12}$$

which shows that there are four directions (two crossing lines) for which the argument or phase of $\exp f(t)$ is independent of s. These are given by

$$2\theta+\alpha=0,\ \pm\pi,\ 2\pi \tag{9.13}$$

and if one of these directions is followed, in moving away from a saddle point, it becomes a line of steepest descent, or steepest ascent. Hence through every saddle point there are two such lines crossing at right angles, and their directions are at 45° to the level lines.

If any straight line through a saddle point is traversed, $|\exp f(t)|$ has a maximum or minimum value at the saddle point, according as $\cos(2\theta+\alpha)$ in (9.10) is negative or positive. If the line is tangential to a line of steepest descent or ascent, the rate of change of $|\exp f(t)|$ is greatest. In a contour map of $|\exp f(t)|$ a saddle point appears as a mountain pass, or col, shaped like a saddle.

If both $\mathrm{d}f/\mathrm{d}t$ and $\mathrm{d}^2f/\mathrm{d}t^2$ are zero at some point, then two saddle points coalesce and the point is called a 'double saddle point'. To study the behaviour of the functions, it is then necessary to include the third derivative of $f(t)$. The results are not needed in this book, but see problem 9.5.

Fig. 9.1 shows part of a contour map for $\exp\{f(t)\}$ near a saddle point. The line a is a line of steepest descent which passes very close to the saddle point but not through it. It must turn quite quickly through a right angle away from the saddle point, as shown. The line b is also a line of steepest descent which passes slightly to the other side of the saddle point; it also turns sharply through a right angle but in the opposite direction. The line c is a line of steepest descent which reaches the saddle point. To continue the descent it is necessary to turn through a right angle to either the right or left. If, instead, the line is continued straight through the saddle point, the value of $|\exp f(t)|$ is a minimum at the saddle point, and increases again when it is passed. Thus a true line of steepest descent turns sharply through a right angle when it meets a saddle point. The term 'line of steepest descents through a saddle point', however, is used in a slightly different sense to mean the line for which $|\exp f(t)|$ decreases on both sides of the saddle point at the greatest possible rate (shown as a broken line in fig. 9.1). The word 'descents' is now in the plural, because there are two descents, one

Fig. 9.1. Contour map of the region near the saddle point S. The thin continuous lines are contours, that is level lines. The thick lines are lines of steepest descent and the arrows point downhill. The dotted lines are the pair of level lines through the saddle point. The broken line is the line of steepest descents through the saddle point. Notice how the contours crowd together at the sides of the valleys.

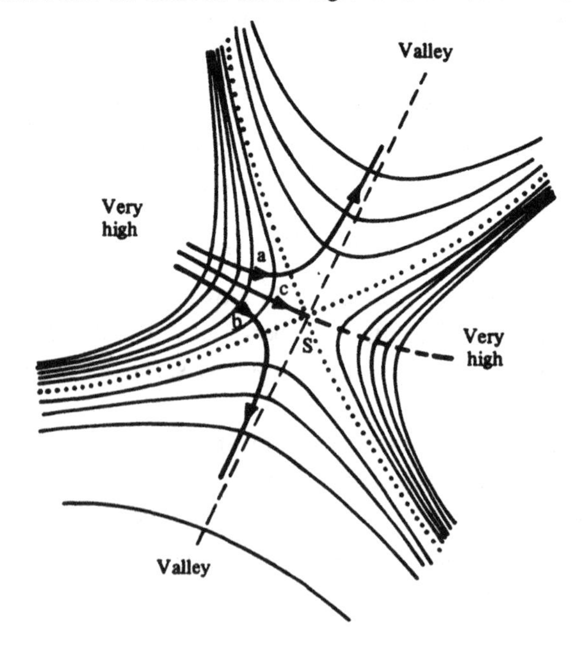

on each side. This line passes smoothly through the saddle point and does not turn through a right angle. Crossing it at right angles at the saddle point is another line on which $|\exp f(t)|$ increases on both sides at the greatest possible rate. This is sometimes called a 'line of steepest ascents through the saddle point'.

The symbol customarily used to mark a saddle point on a contour map can be seen at S_1 and S_2 in figs. 9.5–9.10. The two reversed arcs enclose the line of steepest descents.

9.4. Error integrals and Fresnel integrals

The integral

$$I_0 = \int_{-\infty}^{\infty} \exp(-t^2)\mathrm{d}t = \pi^{\frac{1}{2}} \tag{9.14}$$

may be thought of as a contour integral in the complex t plane, in which the path is the real axis. The integrand is greatest near $t = 0$ and decreases rapidly as $|t|$ increases. When $|t|$ is large the integrand is very small, so that if the infinite limits are replaced by finite values $\pm L$, the error is small when L is large. The value of the integral when modified in this way is

$$\int_{-L}^{L} \exp(-t^2)\mathrm{d}t = 2\int_{0}^{L} \exp(-t^2)\mathrm{d}t = \pi^{\frac{1}{2}}\,\mathrm{erf}(L) \tag{9.15}$$

where erf denotes the 'error function'. The fractional error introduced by the modification is therefore

$$1 - \mathrm{erf}(L) = \mathrm{erfc}(L) = 2\pi^{-\frac{1}{2}} \int_{L}^{\infty} \exp(-t^2)\,\mathrm{d}t. \tag{9.16}$$

The error function and its complement erfc are tabulated (for example by Abramowitz and Stegun, 1965, pp. 310 ff.). The tables show that (9.16) is about 1/100 when $L = 1.82$ and about 1/10 when $L = 1.16$. This idea that the main contribution

Fig. 9.2. The complex t plane. Path of integration for the integral I_0.

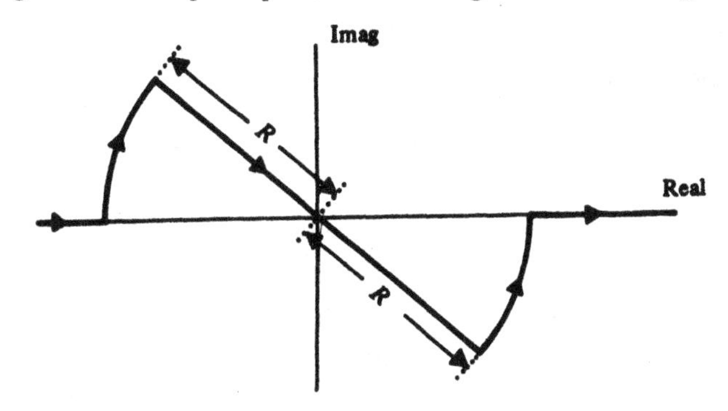

to I_0 comes from small values of $|t|$ less than about 2 is of great importance in the method of steepest descents.

The integrand of I_0 has no singularities (except at ∞) so the path can be distorted, without affecting the value of the integral, provided that it begins and ends as required by the limits of the integral. For example it can be formed by a straight line through the origin with $\arg t = -\frac{1}{4}\pi$, as shown in fig. 9.2 provided that the ends of this line are joined to the real axis. This can be done by arcs of circles of large radius R. We have already seen that the remote parts of the real axis contribute very little to the integral, and it can be proved that their contribution tends to zero as R tends to infinity. Consider now the contribution to I_0 from the 45° circular arc from $\arg t = -\frac{1}{4}\pi$ to 0. Let $t = R\exp(i\theta)$. Then the contribution is

$$iR\int_{-\frac{1}{4}\pi}^{0} \exp\{-R^2(\cos 2\theta + i\sin 2\theta) + i\theta\}\,d\theta. \qquad (9.17)$$

Now for $-\frac{1}{4}\pi \leqslant \theta \leqslant 0$ it is easy to show that $\cos 2\theta \geqslant 1 + 4\theta/\pi$ so that the modulus of the integrand of (9.17) is less than or equal to $\exp\{-R^2(1 + 4\theta/\pi)\}$. This expression can be integrated with respect to θ and the result shows that the modulus of (9.17) is less than $\{1 - \exp(-R^2)\}\pi/4R$. But this tends to zero as R tends to infinity so that the contribution (9.17) also tends to zero. The same applies to the other 45° arc in fig. 9.2. Thus the line at $\theta = -\frac{1}{4}\pi$ in fig. 9.2 could be supposed to extend to infinity. It need not return to the real axis. The limit $+\infty$ in I_0 (9.14) can be supposed to mean that $|t| \to \infty$ with $-\frac{1}{4}\pi \leqslant \arg t \leqslant \frac{1}{4}\pi$, and similarly for the limit $-\infty$.

Fig. 9.3. The complex t plane showing a contour map for the function $\exp(-t^2)$. The numbers by the curves are the values of $\exp(-t^2)$. The arrows point downhill.

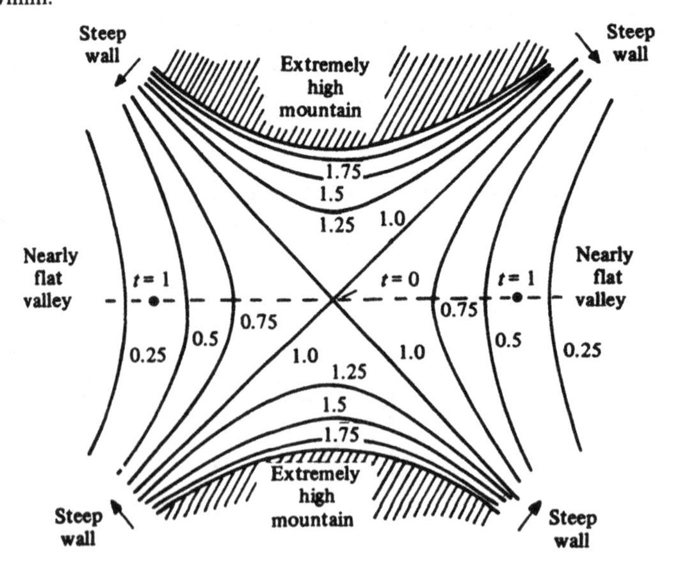

The contours or level lines for the integrand $\exp(-t^2)$ of I_0 are shown in fig. 9.3. It is clear that where $|t|$ is large there are wide valleys in which the level is almost zero, where $-\frac{1}{4}\pi < \arg t < \frac{1}{4}\pi$ and $\frac{3}{4}\pi < \arg t < \frac{5}{4}\pi$. The sides of these valleys are near the lines $\arg t = \frac{1}{4}\pi, \frac{3}{4}\pi, \frac{5}{4}\pi, \frac{7}{4}\pi$, and here the gradient is extremely steep and gets steeper as $|t|$ increases. For large $|t|$ the sides of the valleys are almost like vertical walls of extremely great height. The two ends of the path of integration for I_0 can be anywhere at infinity in these two valleys, but the walls form a boundary, beyond which the ends cannot go. In the sectors $\frac{1}{4}\pi < \arg t < \frac{3}{4}\pi$ and $\frac{5}{4}\pi < \arg t < \frac{7}{4}\pi$ there are extremely high mountains.

A complex integral like I_0 can be thought of as the sum of many complex numbers $\exp(-t^2)\mathrm{d}t$ each of which can be represented as a vector in the Argand diagram. This gives a diagram sometimes called an 'amplitude–phase' diagram or vector diagram, of the kind used in the study of optical diffraction (for examples see figs. 8.12, 8.13, 9.4). The value of the integral is then the 'resultant' or closing vector of this diagram. The form of the diagram depends on the choice of path. The resultant must be unaffected by the choice of path because the integrand is analytic.

Suppose now that the path for I_0 is the real t axis. This is the line of steepest descents through the saddle point, $t = 0$, of the integrand $\exp(-t^2)$. The vectors $\exp(-t^2)\mathrm{d}t$ now all have zero argument or phase, but for equal elements $|\mathrm{d}t|$ of the path their magnitudes decrease rapidly as $|t|$ increases. The vector diagram is therefore a straight line, fig. 9.4(a), whose length is $\pi^{\frac{1}{2}}$. Next suppose that the path is

Fig. 9.4. Amplitude–phase diagrams for various different choices of integration path in the complex t plane for the integral I_0, (9.14). (a) The real axis or line of steepest descents. Parallel vectors not of equal length. (b) A level line through the saddle point. Vectors of equal length but variable direction, giving Cornu spiral. (c) Intermediate case. Vectors vary in both length and direction.

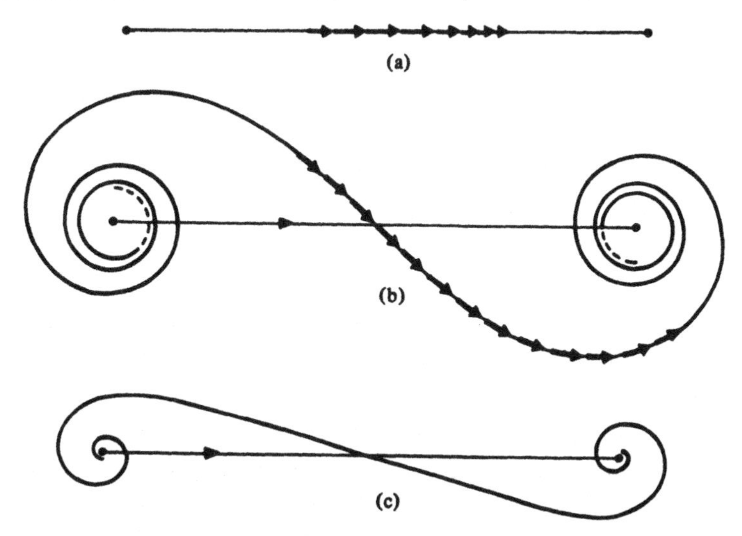

the line $\arg t = -\frac{1}{4}\pi$. Let

$$t = s\exp(-\tfrac{1}{4}\mathrm{i}\pi) \tag{9.18}$$

so that $\mathrm{d}t = \mathrm{d}s\exp(-\frac{1}{4}\mathrm{i}\pi)$, where s is real. The path is now a level line through the saddle point at $t=0$. The integrand is $\exp(\mathrm{i}s^2 - \frac{1}{4}\mathrm{i}\pi)\mathrm{d}s$. For equal elements $\mathrm{d}s$ the vectors in the vector diagram now all have the same length, $\mathrm{d}s$, but their phase is $\varphi = s^2 - \frac{1}{4}\pi$. The amplitude–phase diagram is a Cornu spiral, as shown in fig. 9.4(b). For small s the phase φ is near $-\frac{1}{4}\pi$ and changes only slowly with s, so that the vectors near the middle of the diagram have nearly constant direction. The direction of the change reverses where $s=0$, for then $\mathrm{d}\varphi/\mathrm{d}s = 0$, and this is called a point of stationary phase. For large $|s|$, however, the phase changes very rapidly and then the arms of the spiral are tightly wrapped. They must converge to limit points in such a way that the resultant has length $\pi^{\frac{1}{2}}$ and argument zero, as before.

Finally suppose that the path is the line $\arg t = -\gamma$ where $\gamma \neq 0$ or $\frac{1}{4}\pi$. This still goes through the saddle point but is neither a level line nor a line of steepest descent. Clearly now for constant $|\mathrm{d}t|$ the vectors $\exp(-t^2)\mathrm{d}t$ change both in magnitude and in phase as t varies. The vector diagram is a double spiral which converges to the limit points more rapidly than the Cornu spiral. This is illustrated in fig. 9.4(c) which shows the spiral for $\gamma \approx \frac{1}{8}\pi$. Again the resultant must have length $\pi^{\frac{1}{2}}$ and argument zero. Other paths could be chosen which are not straight and do not pass through the saddle point. They could give very complicated vector diagrams. These paths must begin and end in the correct valleys, and then the resultant vector must in all cases be the same.

Consider again the integral I_0 when the path is the level line as given by (9.18). Then the integral is

$$\begin{aligned} I_0 = \pi^{\frac{1}{2}} &= \exp(-\tfrac{1}{4}\mathrm{i}\pi)\int_{-\infty}^{\infty}\exp(\mathrm{i}s^2)\mathrm{d}s \\ &= (2\pi)^{\frac{1}{2}}\exp(-\tfrac{1}{4}\mathrm{i}\pi)\left[\int_0^{\infty}\cos(\tfrac{1}{2}\pi u^2)\mathrm{d}u + \mathrm{i}\int_0^{\infty}\sin(\tfrac{1}{2}\pi u^2)\mathrm{d}u\right]. \end{aligned} \tag{9.19}$$

The two integrals in square brackets may be written $\mathrm{Lim}_{x\to\infty}C(x)$ and $\mathrm{Lim}_{x\to\infty}S(x)$ respectively where

$$C(x) = \int_0^x \cos(\tfrac{1}{2}\pi u^2)\mathrm{d}u \quad S(x) = \int_0^x \sin(\tfrac{1}{2}\pi u^2)\mathrm{d}u. \tag{9.20}$$

These are the Fresnel integrals. The locus of a point in the C–S plane is a Cornu spiral which converges to the limit points $C = S = \frac{1}{2}$, and $C = S = -\frac{1}{2}$, when $x \to \pm\infty$. The two functions (9.20) are tabulated (Abramowitz and Stegun, 1965). If the limits $\pm\infty$ in (9.19) are replaced by $\pm L$ (compare (9.15)) the fractional error can be calculated from these tables and it is found to be about 1/100 when $L = 45$ and about 1/10 when $L = 4.5$ (compared with 1.82 and 1.16 for (9.15)). Thus the form (9.19) with the level line path converges much more slowly than the form (9.14) with the path of steepest descents. This shows why the path of steepest descents has special importance.

9.5. Contour maps

In the contour map for $\exp(-t^2)$, fig. 9.3, the line of steepest descents through the saddle point and the level lines through the saddle point are straight. On either level line the slope of the hillside increases as distance from the saddle point increases. The valleys, which are nearly flat and at nearly zero altitude when $|t|$ is large, extend on the two opposite sides of the saddle point. They contain the line of steepest descents and they are bounded by the level lines. The contour map for $\exp(-\beta t^2)$ has similar features. If the constant β is real and positive the only change is a change of scale of the map by a factor $\beta^{\frac{1}{2}}$. If β is complex, the line of steepest descents through the saddle point $t = 0$ is where $\arg t = -\frac{1}{2}\arg\beta$, and the level lines through the saddle point are where $\arg t = -\frac{1}{2}\arg\beta \pm \frac{1}{4}\pi$. Thus the contour map is obtained from that for $\exp(-t^2)$ simply by a rotation through an angle $-\frac{1}{2}\arg\beta$ and a scale change by a factor $|\beta|^{\frac{1}{2}}$.

For a more general function $f(t)$, the contour map for $\exp\{f(t)\}$ has more than one saddle point in nearly all cases of practical importance. The contour map near one saddle point S_1 that is remote from singularities of $f(t)$ and from other saddle points is similar to that for $\exp(-\beta t^2)$, as was described in § 9.3. But now, in general, the level lines and lines of steepest descents through S_1 are curved; see fig. 9.1 for example. The most important feature still is that valleys extend in two sectors from S_1. They are bounded by the level lines through S_1 and they contain the line of steepest descents.

Now consider another saddle point S_2 that is the nearest to S_1. There are several possibilities. (a) If S_2 is where $|\exp\{f(t)\}|$ is greater than at S_1, it is in one of the two mountain regions extending from S_1, and the line of steepest descents through S_1 cannot go near it. (b) If S_2 is where $|\exp\{f(t)\}|$ is less than at S_1 it must be in one of the valleys extending from S_1. Adjoining S_2 are two mountain regions and S_1 is in one of these. The other, remote from S_1, must have a contour at the same level as S_1 because all mountains are infinitely high (see § 9.2). Beyond this contour there might be other saddle points or singularities. This mountain divides the valley of S_1 into two parts, V_1 and V_2. The line of steepest descents for S_1 runs down into one part, say V_1. When using an integration path that is required to run down into V_2, we let the path first run from S_1 along the line of steepest descents into valley V_1 and thence through S_2 along its line of steepest descents into valley V_2. Thus there is a contribution to the integral from near S_2 as well as from near S_1. If a path from S_1 running directly into V_2 were used, the integrand could not decrease as fast as it does on the line of steepest descents, so that there are contributions to the integral not only from near S_1 but from more remote parts of the path. These contributions are taken care of by first using the line of steepest descents for S_1 and then adding in the contribution from near S_2. (c) If S_2 and S_1 are at the same level, this is really a special case of (b). The steepest descents contributions from S_1 and S_2 are now of comparable order of magnitude. An example of this case is given in § 9.7, fig. 9.11.

For further understanding of how to choose the integration path it is helpful to study actual examples. These can be found in §§ 9.7, 10.18–10.20.

9.6. Integration by the method of steepest descents

Consider first a simplified form of (9.1) in which $g(t) = 1$ so that

$$I = \int_C \exp\{f(t)\}\mathrm{d}t. \tag{9.21}$$

To evaluate this there are two steps. First, the path C is deformed so that it goes through one or more saddle points of the integrand into the valleys on either side. Then the contributions to the integral from near each used saddle point are evaluated and added.

In problems of physics it nearly always happens that each end of the path C is at a singularity of $f(t)$ where

$$\mathrm{Re}\{f(t)\} \to -\infty, \ |\exp\{f(t)\}| \to 0. \tag{9.22}$$

Then the path begins and ends in a valley. To deform the path we first find the valleys for all the saddle points of $f(t)$. Then from the valley where C begins we let the path run over a saddle point along the line of steepest descents into another valley. If the end point of C is not in this valley we now go along the line of steepest descents of another saddle point into a third valley. The process is repeated until we reach the valley where the end point of C is known to lie. Thus the path is distorted to follow one or a succession of lines of steepest descents through saddle points. This process is best illustrated by the specific examples given later.

Suppose that one of the saddle points on the deformed path is at t_0. On the line of steepest descents through it, $\mathrm{Im}\{f(t)\}$ is constant, so the equation of this line is

$$\mathrm{Im}\{f(t)\} = \mathrm{Im}\{f(t_0)\}. \tag{9.23}$$

This is also the equation of the line of steepest ascents through t_0, and it may have other branches which do not pass through t_0. The branch of (9.23) that is to be used for the path must be found by inspection (see § 9.7, for some examples), but as already explained it is not necessary to know its exact position.

When the path of integration is moved, certain precautions must be taken. The following remarks apply also for the more general form (9.1) with a factor $g(t)$ in the integrand.

If, in moving the path of integration, a pole of the integrand is crossed, it must be allowed for. If it is remote from the contributing saddle points, this can be done by appropriately adding or subtracting $2\pi i$ times its residue. If it is near a saddle point, a process for removing the pole must be used before applying the method of steepest descents (Baños, 1966). If a branch point is crossed, a suitable branch cut must be inserted, and the contribution from the integral around this cut must be included. Both these cases occur in radio propagation problems, the pole in the theory of surface waves (Sommerfeld, 1909; Baños, 1966), and the branch cut in the theory of guided waves and lateral waves (Ott, 1942; Felsen, 1967) but they are beyond the scope of this book.

The contribution to the integral (9.21) from near t_0 is now found as follows. The exponent $f(t)$ is expanded in a Taylor series about t_0, so that the integrand becomes

$$\exp\{f(t)\} = \exp(f_0)\exp\left\{\frac{1}{2!}(t-t_0)^2 f_0'' + \frac{1}{3!}(t-t_0)^3 f_0''' + \cdots\right\} \tag{9.24}$$

where f_0, f_0'', f_0''' are the values of f, d^2f/dt^2, d^3f/dt^3 at the saddle point. On the line of steepest descents through t_0 the last exponent (the series) in (9.24) is real because of (9.23). It is zero at t_0 and negative elsewhere. Hence let

$$\frac{1}{2!}(t-t_0)^2 f_0'' + \frac{1}{3!}(t-t_0)^3 f_0''' + \cdots = -s^2 \tag{9.25}$$

where s is real. The sign of s is to be chosen so that s is negative when the point t moves along the path towards the saddle point, and positive after t has passed the saddle point and is moving away from it. At the two ends of the line of steepest descents $\mathrm{Re}\{f(t)\} = -\infty$ so that s runs from $-\infty$ to $+\infty$. Now if $|t-t_0|$ is small enough the second and later terms of (9.25) are small compared with the first. Thus

$$dt \approx \{f_0''/2!\}^{-\frac{1}{2}} i\, ds. \tag{9.26}$$

Let

$$f_0'' = A e^{i\alpha} \quad (0 \leqslant \alpha < 2\pi). \tag{9.27}$$

Then (9.26) gives

$$dt = \pm(2/A)^{\frac{1}{2}} \exp\{i\tfrac{1}{2}(\pi-\alpha)\}\, ds = P\, ds. \tag{9.28}$$

The sign depends on the angle Θ between the path and the real t axis at the saddle point. It is positive if $-\frac{1}{2}\pi < \Theta \leqslant \frac{1}{2}\pi$ and negative otherwise.

The integrand of (9.21) contains the factor $\exp(-s^2)$ which is appreciable only near $s=0$, that is when $|t-t_0|$ is small. If we use the approximation (9.28), the contribution to the integral from the neighbourhood of t_0 is

$$\exp(f_0)P\int_{-\infty}^{\infty} \exp(-s^2)\, ds = \pm(2\pi/A)^{\frac{1}{2}} \exp(f_0)\exp\{\tfrac{1}{2}i(\pi-\alpha)\}. \tag{9.29}$$

The sign depends on Θ and must be decided by inspection of the path (see the examples in § 9.7).

Formula (9.29) is the main result of the method of steepest descents. It is approximate because of (9.26) but the error is often very small. More accurate formulae are derived later, § 9.9, and these provide an estimate of the error. It can be seen from (9.25), however, that since the term $\frac{1}{6}f_0'''(t-t_0)^3$ and higher terms have been neglected, they ought to be small compared with $\frac{1}{2}(t-t_0)^2 f_0''$ for values of s up to about 2; see section 9.4 just after (9.16). We shall expect the method to fail if f_0'' is too small. But this case can be dealt with by using Airy integral functions, which is perhaps the most important use of these functions; §§9.9 (end) and 10.20.

Consider now the more general integral (9.1) with the factor $g(t)$ restored. It can be written

$$\int_C \exp\{f(t) + \ln g(t)\}\,\mathrm{d}t \tag{9.30}$$

and the foregoing method could in principle be applied to it. Complications arise however, if $g(t)$ has a pole or a zero near a saddle point of $\exp\{f(t)\}$. It commonly happens that $g(t)$ is a 'slowly varying' function of t near a saddle point of $\exp\{f(t)\}$. Then in the region of the t plane where the integrand of (9.1) is appreciable $g(t)$ may be taken as a constant, $g(t_0)$. The contribution to (9.1) from near the saddle point t_0 of $\exp\{f(t)\}$ is then

$$\pm(2\pi/A)^{\frac{1}{2}} g(t_0)\exp(f_0)\exp\{\tfrac{1}{2}\mathrm{i}(\pi-\alpha)\}. \tag{9.31}$$

When $g(t)$ cannot be treated as constant, it is expanded in powers of s and included as a factor in the integrand of (9.29). If terms after the first on the left of (9.25) can be neglected, s is proportional to $t-t_0$, so that $g(t)$ is simply expanded in powers of $t-t_0$:

$$g(t) = g(t_0) + (t-t_0)g_0' + \frac{1}{2!}(t-t_0)^2 g_0'' + \dots \tag{9.32}$$

where $g_0', g_0'' \cdots$ are $\mathrm{d}g/\mathrm{d}t$, $\mathrm{d}^2g/\mathrm{d}t^2 \cdots$ at $t=t_0$. On integration, the odd powers of s or $t-t_0$ give zero. The third term of (9.32) gives a contribution which must be added to (9.31), and when (9.27) is used, this is

$$\begin{aligned} &-(g_0''/f_0'')\exp(f_0)P\int_{-\infty}^{\infty}\exp(-s^2)s^2\,\mathrm{d}s \\ &= \mp\tfrac{1}{2}g_0''(2\pi)^{\frac{1}{2}}A^{-\frac{3}{2}}\exp(f_0)\exp\{\tfrac{1}{2}\mathrm{i}(\pi-3\alpha)\}. \end{aligned} \tag{9.33}$$

This term may be important if $g(t)$ has a zero at or near t_0, so that $g(t_0)$ in (9.31) is zero or small.

When terms after the first, on the left of (9.25), cannot be neglected, the more elaborate treatment of § 9.9 must be used.

For the methods of (9.31)–(9.33) to work there is one fairly obvious restriction on the function $g(t)$. Suppose that the saddle points and lines of steepest descents for $\exp\{f(t)\}$ have been found. On these lines this function contains a factor $\exp(-s^2)$

whose properties are used in the integral (9.29). But this would be wrong if $g(t)$ contained a factor that behaved like $\exp(Ks^2)$ where K is real and greater than unity. The path would not now be on a line of steepest descents and the extra factor would make the integral in (9.29) diverge. Thus $g(t)$ must be 'devoid of exponential behaviour' (quoted from Baños, 1966). In fact the extra factor would not matter if K were less than unity, though its presence might reduce the accuracy of the steepest descents result.

9.7. Application to solutions of the Stokes equation

The simplest possible integral that can be evaluated by steepest descents is the error integral, I_0 in (9.14). When the integrand is the exponential of any quadratic function, it can be changed to the form I_0 simply by a change of variable. There is only one saddle point, where $t = 0$, and the value $\pi^{\frac{1}{2}}$ in (9.14) is exact.

The next simplest possible integral is when the integrand is the exponential of a cubic function of t. This case is therefore very useful for illustration. It is very important in physics because the solutions (8.16), (8.17) of the Stokes differential equation can be expressed in this form. We shall therefore study the integral (8.16) for $\mathrm{Ai}(\zeta)$.

The exponent in (8.16) has two saddle points where

$$t = t_0 = \pm \zeta^{\frac{1}{2}} \tag{9.34}$$

and the lines of steepest descents through them will first be found. The path of integration is then distorted to coincide with one or both of them. The configuration of these lines depends on the value of $\arg \zeta$, and it will be shown that sometimes one and sometimes both of these lines must be used. The transition from one case to the other gives the Stokes phenomenon, § 8.12.

Let $t = u + \mathrm{i}\, v$ where u and v are real and let

$$\zeta = m^2 \mathrm{e}^{\mathrm{i}\theta} \tag{9.35}$$

where m and θ are real and positive. Then the two saddle points are where

$$t = t_0 = \pm m \exp(\tfrac{1}{2}\mathrm{i}\theta). \tag{9.36}$$

The exponent $\zeta t - \frac{1}{3}t^3$ in (8.16) at either saddle point has the value

$$f_0 = \tfrac{2}{3}\zeta t_0 \tag{9.37}$$

and its second derivative is

$$f''_0 = -2t_0. \tag{9.38}$$

On the lines of steepest descent and ascent through a saddle point the imaginary part of the exponent is constant. Hence the equation of these lines is

$$\mathrm{Im}(\zeta t - \tfrac{1}{3}t^3) = \mathrm{Im}(f_0) \tag{9.39}$$

(compare 9.23) which leads to

$$\tfrac{1}{3}v^3 - u^2 v + vm^2 \cos\theta + um^2 \sin\theta = \pm\tfrac{2}{3}m^3 \sin(\tfrac{3}{2}\theta). \tag{9.40}$$

Suppose first that ζ is real and positive so that $\theta = 0$. Then the saddle points lie on the real t axis at $\pm m$ and (9.40) becomes

$$v(\tfrac{1}{3}v^2 - u^2 + m^2) = 0. \tag{9.41}$$

The curves are shown in fig. 9.5 which is a contour map of the integrand. The line $v = 0$ is a line of steepest ascent through the saddle point S_1 and a line of steepest descent through the saddle point S_2. The other factor is the equation of a hyperbola whose asymptotes are given by $v = \pm u\sqrt{3}$. The left branch is a line of steepest descents through the saddle point S_1, and it begins and ends where $\arg t \approx \mp\frac{2}{3}$, that is within the valleys where the path C must begin and end. Hence the path of integration may be distorted to coincide with the left-hand branch of the hyperbola as indicated by arrows in fig. 9.5. The right-hand branch of the hyperbola, shown as a chain line in fig. 9.5, is a line of steepest ascents through the other saddle point S_2,

Fig. 9.5. Complex t plane. Contour map of the function $\exp(t\zeta - \frac{1}{3}t^3)$, when ζ is real and positive. The thin continuous lines are contours, and the numbers on them are the values of the altitude $\exp\{\mathrm{Re}(t\zeta - \frac{1}{3}t^3)\}$. In this example and in figs. 9.6–9.11, $|\zeta| = 1.0$. The thick continuous line is the line of steepest descents through the saddle point S_1 and the arrows show the path for the contour integral (8.16). The broken lines are at $\pm 60°$ to the real axis and are asymptotes of the lines of steepest descent and ascent.

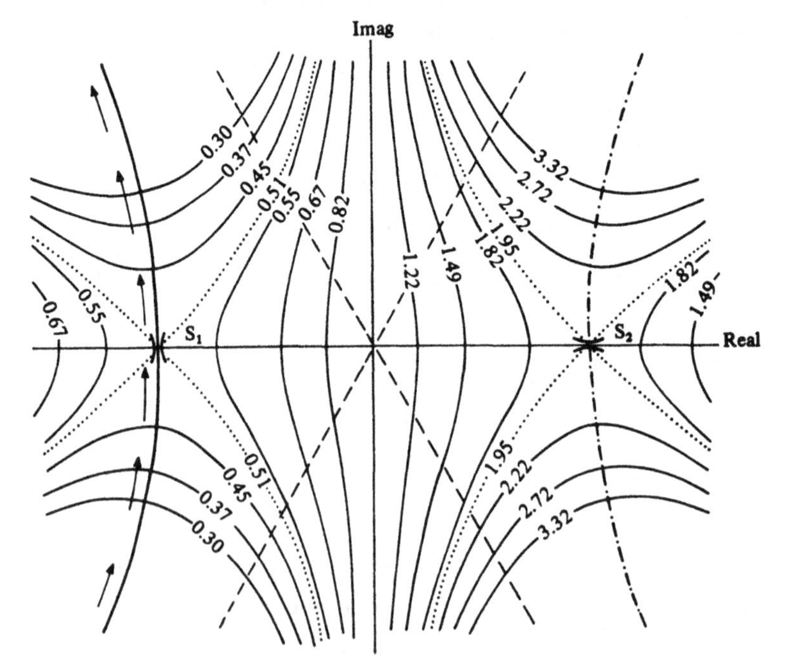

and is not part of the path, which therefore passes through only the one saddle point S_1 at $t_0 = -m$. Equations (9.37), (9.38) then show that $f_0 = -\frac{2}{3}\zeta^{\frac{3}{2}}$ and $f''_0 = 2\zeta^{\frac{1}{2}}$, so in (9.27), $\alpha = 0$, $A = 2\zeta^{\frac{1}{2}}$, where the positive value of $\zeta^{\frac{1}{2}}$ is used.

The formula (9.29), with the factor $1/(2\pi i)$ from (8.16), may now be used to give the approximate value of the integral. The direction Θ of the contour is $\frac{1}{2}\pi$ so the sign in (9.28) is $+$, whence

$$\mathrm{Ai}(\zeta) \sim \tfrac{1}{2}\pi^{-\frac{1}{2}}\zeta^{-\frac{1}{4}}\exp(-\tfrac{2}{3}\zeta^{\frac{3}{2}}) \tag{9.42}$$

where the real positive values of $\zeta^{-\frac{1}{4}}$ and $\zeta^{\frac{3}{2}}$ are used. This is the asymptotic form for $\mathrm{Ai}(\zeta)$ when ζ is real and positive (compare (8.52)). It is the same as the subdominant W.K.B. term (8.27) with a multiplying constant. This illustrates a most important use of the first order steepest descents formula (9.29). When applied to a contour integral representation of a solution of a differential equation, the contribution from one saddle point, if sufficiently isolated, is one of the W.K.B. solutions.

The unused saddle point S_2 in fig. 9.5 is at a much greater altitude than S_1. For other solutions of the Stokes equation, for example $\mathrm{Bi}(\zeta)$, it does give a contribution, proportional to the dominant term, (8.28).

Next suppose that $\theta(=\arg\zeta)$ is in the range $0 < \theta < \frac{1}{3}\pi$. The expression (9.40) now gives two separate equations. The plus sign gives the lines of steepest ascent and descent through the saddle point S_2 at $t_0 = me^{\frac{1}{2}i\theta}$, shown as a chain line in fig. 9.6, and the minus sign gives those through the other saddle point S_1, shown as continuous lines. Each equation is now a cubic. The curve for the minus sign has two intersecting branches which are the lines of steepest descent and ascent through the saddle point S_1 at $t_0 = -me^{\frac{1}{2}i\theta}$, and a third branch which does not pass through either saddle point. The asymptotes are the lines $v = 0$, $v = \pm u\sqrt{3}$. The line of steepest descents begins and ends in the correct sectors as before, and is used as the path C_1, which is indicated by arrows in fig. 9.6, and which therefore passes

Fig. 9.6. Similar to fig. 9.5 but for $0 < \arg\zeta < \frac{1}{3}\pi$. In this and figs. 9.7–9.11 the contours (level lines) are not shown.

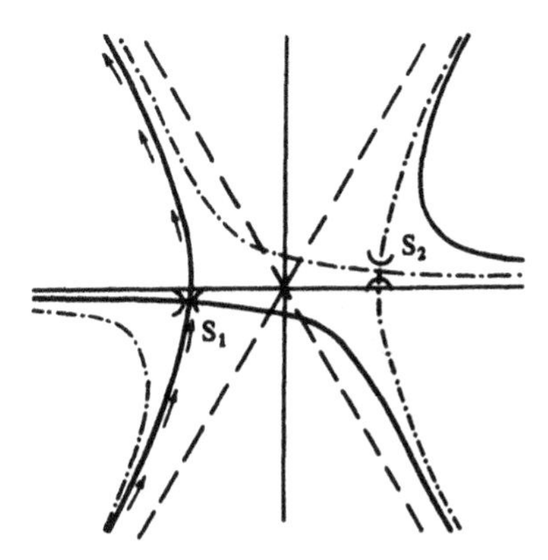

through only the one saddle point at $t_0 = -me^{\frac{1}{2}i\theta}$. Equations (9.37) and (9.38) show that $f_0 = -\frac{2}{3}\zeta^{\frac{3}{2}}$ (where $\arg \zeta^{\frac{3}{2}}$ is $\frac{3}{2}\theta$) and $f_0'' = 2me^{\frac{1}{2}i\theta}$, so that $\alpha = \frac{1}{2}\theta$, $A = 2m$. The direction Θ of the path at the saddle point is $\frac{1}{2}\pi - \frac{1}{4}\theta$, so the + sign is used in (9.29) which gives

$$\mathrm{Ai}(\zeta) = \tfrac{1}{2}\pi^{-\frac{1}{2}}|m^{-\frac{1}{2}}|e^{-\frac{1}{4}i\theta}\exp(-\tfrac{2}{3}\zeta^{\frac{3}{2}}). \tag{9.43}$$

This is the same as (9.42) if $\zeta^{-\frac{1}{4}}$ takes the value which goes continuously into $|\zeta^{-\frac{1}{4}}|$ as θ decreases to zero.

The curves given by (9.40) can be drawn in a similar way for other values of $\theta = \arg \zeta$. Fig. 9.7 shows them for $\theta = \frac{1}{3}\pi$, which is an anti-Stokes line, and fig. 9.8 shows them for $\frac{1}{3}\pi < \theta < \frac{2}{3}\pi$. In both these cases the path is similar to those in figs. 9.5 and 9.6, and the approximate value of the integral is given by (9.42) as before. In

Fig. 9.7. $\operatorname{Arg} \zeta = \frac{1}{3}\pi$. Anti-Stokes line.

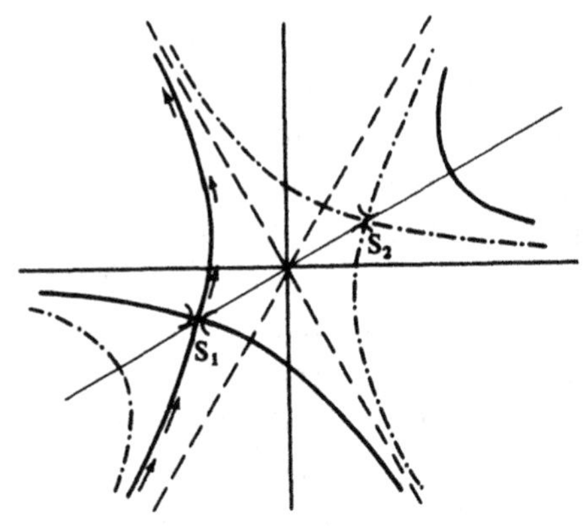

Fig. 9.8. $\frac{1}{3}\pi < \arg \zeta < \frac{2}{3}\pi$. See captions for figs. 9.5, 9.6.

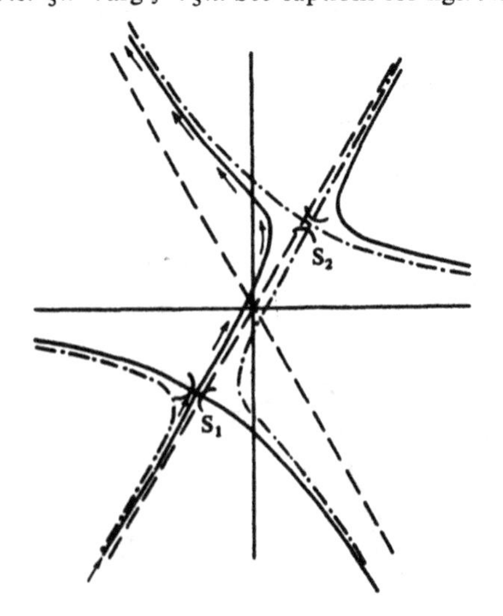

fig. 9.8 the path approaches the saddle point S_2 at $t_0 = me^{\frac{1}{3}i\theta}$, but turns fairly sharply through nearly a right angle to the left, and ends in the correct sector.

Now consider the curves shown in fig. 9.9 for $\theta = \frac{2}{3}\pi$ which is a Stokes line. They are like those for $\theta = 0$, but rotated through 60°. The line of steepest descents through the saddle point S_1 at $t_0 = -me^{\frac{1}{3}i\theta}$ is a straight line which begins in the correct sector (at $\arg t = -\frac{2}{3}\pi$), ascends to the saddle point S_1 and then descends and reaches the other saddle point S_2. To continue the descent, it is necessary to turn through a right angle either to the right or to the left. For the integration path we choose a left turn as indicated by the arrows, so that the path ends in the correct sector ($\arg t = \frac{2}{3}\pi$), and the approximate value of the integral is (9.42) as before.

Fig. 9.9. $\operatorname{Arg} \zeta = \frac{2}{3}\pi$. Stokes line.

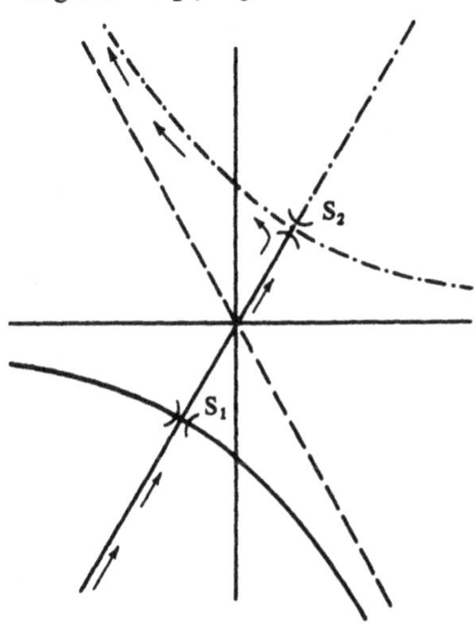

Fig. 9.10. $\frac{2}{3}\pi < \arg \zeta < \pi$. See captions for figs. 9.5, 9.6.

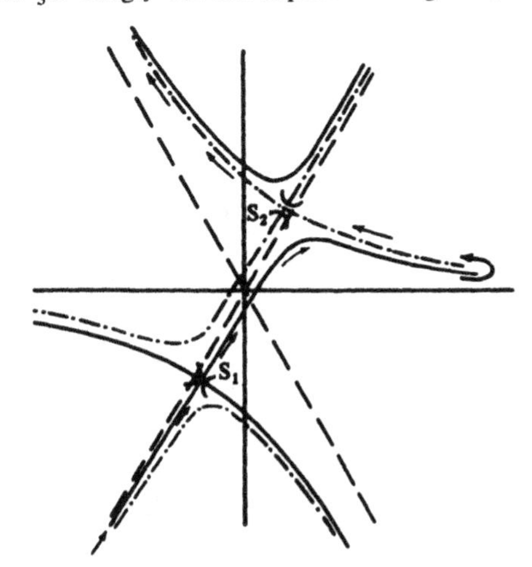

Fig. 9.10 shows the curves for $\frac{2}{3}\pi < \theta < \pi$. The line of steepest descents through the saddle point begins in the correct sector, passes through the saddle point, but when it approaches the other saddle point S_2 it now turns fairly sharply through a right angle to the right, and ends near the asymptote $v = 0$, which is not in the correct sector for the end of the path. The line of steepest descents through the saddle point S_2, however, begins near the asymptote $v = 0$, and ends where $\arg t = \frac{2}{3}\pi$. The path is therefore distorted so as to coincide with both lines of steepest descent, as shown by the arrows in fig. 9.10, and, when the integral is evaluated, the contributions from both saddle points must be included.

The saddle point S_2 is where $t_0 = me^{\frac{1}{2}i\theta}$, and equations (9.37) and (9.38) show that $f_0 = \frac{2}{3}\zeta^{\frac{3}{2}}$ (where $\arg \zeta^{\frac{3}{2}}$ is $\frac{3}{2}\theta$) and $f_0'' = -2me^{\frac{1}{2}i\theta}$, so that $\alpha = \pi + \frac{1}{2}\theta$, $A = 2m$. The direction Θ of the contour at the saddle point is $\pi - \frac{1}{4}\theta$, so that the minus sign is used in (9.29). It gives for the approximate contribution to the integral at the saddle point S_2:

$$\tfrac{1}{2}\pi^{-\frac{1}{2}}|m^{-\frac{1}{2}}|\exp\{i(\tfrac{1}{2}\pi - \tfrac{1}{4}\theta)\}\exp(\tfrac{2}{3}\zeta^{\frac{3}{2}}). \tag{9.44}$$

The contribution from the saddle point S_1 is (9.42) as before, and the two terms together give

$$\mathrm{Ai}(\zeta) \sim \tfrac{1}{2}\pi^{-\frac{1}{2}}\zeta^{-\frac{1}{4}}\{\exp(-\tfrac{2}{3}\zeta^{\frac{3}{2}}) + i\exp(\tfrac{2}{3}\zeta^{\frac{3}{2}})\}, \quad (\tfrac{2}{3}\pi \leqslant \arg\zeta \leqslant \tfrac{4}{3}\pi) \tag{9.45}$$

where $\arg(\zeta^{\frac{3}{2}}) = \frac{3}{2}\theta$; $\arg(\zeta^{-\frac{1}{4}}) = -\frac{1}{4}\theta$. The factor i in (9.45) is the Stokes multiplier.

The above argument shows how the second term comes in discontinuously at the Stokes line $\theta = \frac{2}{3}\pi$. When $\theta - \frac{2}{3}\pi$ is positive but small, the value of the integrand $\exp(\zeta t - \frac{1}{3}t^3)$ is much smaller at S_1 than at S_2, and the second term in (9.45) is small; it is the subdominant term. As $\arg\zeta$ approaches π, however, the two terms become more nearly equal, and on the anti-Stokes line at $\theta = \pi$ they have equal moduli. The curves for $\theta = \pi$ are shown in fig. 9.11. They are like those for $\theta = \frac{1}{3}\pi$, but rotated

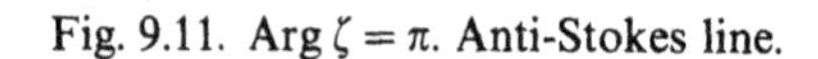
Fig. 9.11. Arg $\zeta = \pi$. Anti-Stokes line.

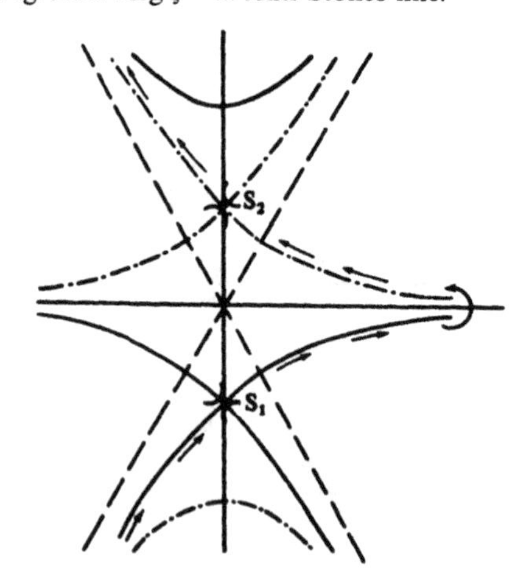

through 60°. Here ζ is real and negative, and the two saddle points are on the imaginary axis. The curves could be traced in a similar way for other values of θ. For the Stokes line at $\theta = \frac{4}{3}\pi$ they would be like those for $\theta = \frac{2}{3}\pi$, but rotated through 60°. For $\frac{4}{3}\pi < \theta < 2\pi$ it is again possible to choose the contour so that it coincides with a line of steepest descents through only one saddle point, and in this range there is only one term in the asymptotic approximation. When $\theta = 2\pi$, the curves again become as in fig. 9.5. When θ increases from 0 to 2π, the two saddle points move only through an angle π in the complex t plane, so that their positions are interchanged. This is illustrated by the fact that the line in the Stokes diagram (fig. 8.6) for $\mathrm{Ai}(\zeta)$ is not a closed curve.

9.8. The method of stationary phase

Instead of using the method of steepest descents, it is possible to evaluate integrals of the type (9.1) by the method of stationary phase. In this method the path is distorted so as to pass through the same saddle points as for the method of steepest descents, but instead of following the lines of steepest descents, it is made to coincide with the level lines through the saddle points. The contributions from the neighbourhood of each saddle point used are then evaluated and added.

Whenever the path passes through a saddle point it must run anywhere within the two valleys on each side. For those parts of a valley which are very distant from the saddle point, the valley has steep walls, as explained in § 9.4, and a level line through the saddle point runs in one of these walls. The path can be moved up to either of these level lines. At a place remote from all saddle points, the steep wall of a valley may contain the level lines from several different saddle points, and their separation in the complex t plane is small because of the steepness of the wall. If, therefore, more than one saddle point is used, the level lines through them must be linked together, and this is done at points in steep hillsides remote from the saddle points, where the lengths of the links are extremely small, so that their contributions to the integral are negligible. The choice of the path for the method of stationary phase usually requires considerably more care than for steepest descents, partly because in any one valley there is only one line of steepest descent through the saddle point, whereas there are two level lines through it.

A level line may be chosen to run continuously through a saddle point or to turn through a right angle there. It is now assumed to run continuously. The case where there is a right angle turn can be dealt with by a simple extension of the following method.

The contribution to (9.21) from near one saddle point, at $t = t_0$, is found as follows. The exponent $f(t)$ is expanded in a Taylor series about $t = t_0$, and, since $\mathrm{d}f/\mathrm{d}t$ is zero there, the result leads to (9.24) as before. On a level line through the saddle point the last exponent in curly brackets in (9.24) is purely imaginary, and is zero at the saddle point. There are two level lines through the saddle point, and for one the phase is a minimum and for the other it is a maximum. Suppose that we choose the one on which it is a minimum. Then we take

$$\frac{1}{2!}(t-t_0)^2 f_0'' + \frac{1}{3!}(t-t_0)^3 f_0''' + \cdots = \mathrm{i}\tau^2 \tag{9.46}$$

where τ is real. Now if $|t-t_0|$ is small enough the second and later terms of (9.46) are small compared with the first. Thus

$$\mathrm{d}t \approx \pm(f_0''/2)^{-\frac{1}{2}}\exp(\tfrac{1}{4}\mathrm{i}\pi)\mathrm{d}\tau. \tag{9.47}$$

The sign depends on the angle Θ between the path and the real t axis at the saddle point. It is positive if $-\frac{3}{4}\pi < \Theta \leqslant \frac{1}{4}\pi$. The integral of (9.24) thus contains the factor

$$\int_{-\infty}^{\infty} \exp(\mathrm{i}\tau^2)\,\mathrm{d}\tau = \pi^{\frac{1}{2}}\exp(\tfrac{1}{4}\mathrm{i}\pi) \tag{9.48}$$

from (9.19) so that the contribution to (9.21) becomes

$$\pm(2\pi/A)^{\frac{1}{2}}\exp(f_0)\exp\{\tfrac{1}{2}\mathrm{i}(\pi-\alpha)\} \tag{9.49}$$

where $f_0'' = A\mathrm{e}^{\mathrm{i}\alpha}$ and $0 \leqslant \alpha < 2\pi$. This is exactly the same as the result (9.29) obtained by the method of steepest descents. If the other level line through the saddle point is used, so that the phase is a maximum there, it can be shown that again the result is the same.

In the method of steepest descents the integral is reduced to the form in (9.29) which is the error integral (9.14), and the integrand is always real. In the method of stationary phase the integral is reduced to the complex Fresnel integral (9.48) (compare (9.19)), in which the integrand has modulus unity, but its argument or phase varies along the contour and is proportional to τ^2. Its value is thus the resultant of a Cornu spiral, fig. 9.4(b). The main contribution is from near the saddle point where small contributing vectors have nearly the same direction because the phase is stationary there. On receding from the saddle point the phase changes more and more rapidly, so that the curve spirals in more and more quickly.

A good illustration of the use of the method of stationary phase is provided by the contour integral (8.16) for $\mathrm{Ai}(\zeta)$ when ζ is real and negative. Then the saddle points $t_0 = \pm\zeta^{\frac{1}{2}}$ lie on the imaginary t axis, which is a level line through both saddle points. The contour may be distorted to coincide with the imaginary axis, and it is convenient to let $t = \mathrm{i}s$. This was done in § 8.6 and was shown to lead to Airy's expression (8.19). For the method of stationary phase, however, we use the form (8.18). The saddle points are where $s = \pm|\zeta^{\frac{1}{2}}|$. When $s = +|\zeta^{\frac{1}{2}}|$ the phase $\zeta s + \frac{1}{3}s^3$ is a minimum, and the Cornu spiral is as in fig. 9.4(b). When $s = -|\zeta^{\frac{1}{2}}|$ the phase is a maximum, and the Cornu spiral has the opposite curvature to that in fig. 9.4(b). The asymptotic approximation to $\mathrm{Ai}(\zeta)$ is therefore the sum of the resultants of two equal Cornu spirals with opposite curvature. This was illustrated in more detail in § 8.22 and fig. 8.13.

The contributions from the two saddle points may each be expressed in the form (9.49). It can be shown that when they are added together with the correct signs, they lead to the same expression (9.45) as was obtained by the method of steepest descents.

9.9. Higher order approximation in steepest descents

In the method of steepest descents, as described in §9.6, it was assumed that the second and later terms of the series (9.25) could be neglected, and this led to the approximation (9.26) and to the approximate value (9.29) for the contribution to the integral. We must now examine what happens when the approximation is not made. This leads to a more accurate value for the contribution (9.29) or (9.31) or (9.33) and to a quantitative criterion to show when the approximation (9.26) can be used.

The change of variable (9.25) can be written

$$s^2 = f_0 - f(t) \tag{9.50}$$

and then the integral (9.1) becomes

$$I = \exp(f_0) \int_{-\infty}^{\infty} \exp(-s^2)\phi(s)\,\mathrm{d}s \tag{9.51}$$

where

$$\phi(s) = g(t)\,\mathrm{d}t/\mathrm{d}s. \tag{9.52}$$

From (9.25), s can be expressed as a series in powers of $t - t_0$, and then, by reversion (Abramowitz and Stegun, 1965, p. 16), $t - t_0$ can be expressed as a series in powers of s. The first term is found from (9.25) simply by neglecting the third and higher powers of $t - t_0$. Hence

$$t - t_0 = \pm \mathrm{i}s\sqrt{2}(f_0'')^{-\frac{1}{2}}\{1 + \tfrac{1}{2}C_1 s + \tfrac{1}{3}C_2 s^2 + \cdots\}. \tag{9.53}$$

The sign depends on the direction Θ of the path in the t plane at the saddle point. It is $+$ if $-\frac{1}{2}\pi < \Theta \leqslant \frac{1}{2}\pi$. The $+$ sign is used here. Then it can be shown that

$$C_1 = \frac{\mathrm{i}\sqrt{2}}{3} f_0'''(f_0'')^{-\frac{3}{2}}$$

$$C_2 = \{\tfrac{1}{4} f_0'''' f_0'' - \tfrac{5}{12}(f_0''')^2\}(f_0'')^{-3}. \tag{9.54}$$

Now suppose, first, that only the first term of (9.25) is appreciable. Then $f(t) - f_0$ is proportional to $(t - t_0)^2$ so that $\exp\{f(t)\}$ has only one saddle point, and no singularities except at infinity. We may say that its saddle point is 'isolated'. We also assume that $g(t)$ is slowly varying and can be taken as a constant $g(t_0)$, as was done in (9.31). Then (9.52), (9.53) give

$$\phi(s) = \pm \mathrm{i}\{\tfrac{1}{2} f_0''\}^{-\frac{1}{2}} g(t_0) \tag{9.55}$$

which is a constant, and was used to get the simple first order result (9.31).

Next suppose that the first two terms of (9.25) are appreciable but the others are negligible. Now $f(t)$ has two saddle points, one where $t = t_0$ and the other where

$$t = t_0 - 2f''_0/f'''_0. \tag{9.56}$$

If f'''_0 is very small, the saddle points are a long way apart, and we may still say that the one at t_0 is 'isolated'. Then (9.55) is still a good approximation for $\phi(s)$ and we can expect to use (9.31) as before. But if f'''_0 is not small, $\phi(s)$ is not a constant and we must use more than just the first term of (9.53).

If more than two terms of (9.25) are appreciable, $f(t)$ has more than two saddle points. If the coefficients of the second and third terms are small enough, the other saddle points are a long way from t_0.

Now from (9.52), (9.50)

$$\phi(s) = -2sg(t)/f'(t). \tag{9.57}$$

But $f'(t)$ is zero at a saddle point. For $t = t_0$, s is also zero and ϕ is bounded. But for any other saddle point ϕ is infinite, so that the series (9.53) cannot converge. Hence we have the very important result: for all functions $f(t)$ except a simple quadratic function, the series (9.53) cannot converge for all values of s. It must have a finite radius of convergence R. This is always less than the distance from t_0 to the next nearest saddle point, or to the next nearest singularity of $f(t)$.

If the saddle point at t_0 is sufficiently 'isolated', the radius of convergence R may be large, and a few of the coefficients $C_1, C_2 \cdots$ after the first in (9.53) may be small. But the series must ultimately diverge where $|s| > R$. In spite of this divergence of (9.53), we insert its derivative with respect to s in (9.52), (9.51), and integrate term by term. The odd powers of s then give zero, but the even powers give

$$I \sim \exp(f_0)\, P\pi^{\frac{1}{2}}(1 + \tfrac{1}{2}C_2 + \tfrac{3}{4}C_4 + \cdots) \tag{9.58}$$

where P is given by (9.28), (9.29). This series is the contribution to I from near the saddle point at t_0. It also is divergent, but nevertheless is useful when the first few terms after the first get successively smaller. It can be shown that it is an asymptotic series in the sense of Poincaré; see § 8.11. The proof of this is known as Watson's lemma. It is a difficult subject and beyond the scope of this book.

The distance from t_0 to the next nearest saddle point depends on some parameter in the integral (9.1) or (9.51). For example in the Airy integral (8.16), the separation of the two saddle points is $2|\zeta^{\frac{1}{2}}|$, and is large when ζ is large. Here ζ is the parameter. When $|\zeta|$ is large the two saddle points are isolated and the simple form of the method of steepest descents should have good accuracy.

For this case

$$f(t) = \zeta t - \tfrac{1}{3}t^3 \tag{9.59}$$

whence it can be shown, from (9.54), that

$$\tfrac{1}{2}C_2 = -\tfrac{5}{48}\zeta^{-\frac{3}{2}} \tag{9.60}$$

and this is the same as the second term of the asymptotic series in (8.30), as given by (8.32), with $m = 1$. When this term is $\ll 1$, it is sufficiently accurate to use only the first order steepest descents formula (9.29) at each contributing saddle point. If it is required that the modulus of (9.60) shall be less than about 1/10, then the condition is $|\zeta| \geqslant 1$. We have already seen, § 8.10, equation (8.29), that this is also the condition that the W.K.B. solutions (8.27), (8.28) shall be good approximations to solutions of the Stokes equation.

The derivation of the series (9.58) via (9.54) is extremely cumbersome algebraically and this method is rarely used in actual computations. When an asymptotic series is needed it is usually easier to derive it in some other way, such as by the method of §8.11. The main use of C_2 in (9.54) is to test the accuracy of the first order steepest descents result (9.29) as illustrated by the example of (9.59), (9.60).

When two saddle points of (9.1) are close together but remote from all other saddle points and singularities, there is another method for finding their combined contribution. In their neighbourhood the exponent $f(t)$ cannot be treated as a quadratic function as was done in (9.26). Instead it is equated to a cubic function $\zeta s - \frac{1}{3}s^3$ which replaces the $-s^2$ in (9.25). It is arranged that the two saddle points where $\mathrm{d}f/\mathrm{d}t$ is zero in the t plane map into the two turning points $s = \pm\zeta^{\frac{1}{2}}$ of the cubic. Then, instead of the error integral in (9.29), an Airy integral appears. The resulting contribution is thus expressed in terms of Airy integral functions. In this way an analogue of the method of steepest descents can be constructed for two saddle points. It can be carried to higher orders of approximation in analogy with (9.58). It is then found that there are contributions from $\mathrm{Ai}(\zeta)$ and $\mathrm{Ai}'(\zeta)$, each multiplied by a series as in (9.58). The full theory was given in a most important paper by Chester, Friedman and Ursell (1957). The method has been used in a radio propagation problem (not described in this book) by Al'pert, Budden, Moiseyev and Stott (1983). An extension of the method is mentioned in § 10.20.

9.10. Double steepest descents

In wave propagation problems it is often necessary to evaluate double integrals of the form

$$\iint g(t, u)\exp\{f(t, u)\}\mathrm{d}t\,\mathrm{d}u \tag{9.61}$$

where t and u are complex variables. See, for example, (10.2). The methods of the earlier sections may be applied to the two variables t and u in turn. Suppose that the t integration is done first. Then u is held constant, the saddle points t_s where

$$\partial f/\partial t = 0 \tag{9.62}$$

are found, and the formula (9.31) is used to give the first order steepest descents contribution at each saddle point. It contains the factor $(\partial^2 f/\partial t^2)^{-\frac{1}{2}}$ as shown by

(9.27). From the physical conditions it is usually easy to decide which saddle points contribute to the integral. Quite often there is only one. In this integration it is assumed that $g(t, u)$ is slowly varying near the saddle point, so that it is treated as a constant $g(t_s, u)$; see § 9.6.

Next the u integration is done. The saddle points in the u plane are where $\mathrm{d}f\{t_s(u), u\}/\mathrm{d}u = 0$. But (9.62) remains true so that this is the same as the two conditions

$$\partial f/\partial t = 0, \quad \partial f/\partial u = 0. \tag{9.63}$$

These give a double saddle point $t = t_s$, $u = u_s$ in the complex t and u planes. The formula (9.31), with u for t, is again applied, and again the factors other than the exponential are treated as constants. Then it can be shown that the first order steepest descents contribution to (9.61) from the double saddle point is

$$\pm 2\pi g(t_s, u_s)\Delta^{-\frac{1}{2}} \exp\{f(t_s, u_s)\} \tag{9.64}$$

where

$$\Delta = \frac{\partial^2 f}{\partial t^2}\frac{\partial^2 f}{\partial u^2} - \left(\frac{\partial^2 f}{\partial t \partial u}\right)^2 \text{ at } t = t_s,\ u = u_s. \tag{9.65}$$

The important feature of the result (9.64) is that the exponential in (9.61) simply takes its value at the double saddle point. The paths of integration in the t and u planes can be thought of as lines of steepest descents through the double saddle point, but in physical problems it is often more convenient to take them as level lines. The condition (9.63) then shows that the phase of $\exp\{f(t, u)\}$ is stationary with respect to t and u at the double saddle point.

Integrals of the type (9.61) occur in §§ 10.2, 10.18, 10.19, 11.14. The use of double steepest descents in radio propagation problems has been discussed by Budden and Terry (1971), Budden (1976).

PROBLEMS 9

9.1. Evaluate $\int_{-\infty}^{\infty} \exp\{a(1 - \cosh x)\}\mathrm{d}x$ approximately by the two following methods, given that $\mathrm{Re}(a) > 0$. (a) Expand the exponent in powers of x and ignore powers greater than x^2. (b) Put $a(1 - \cosh x) = -u^2$. Express $\mathrm{d}x/\mathrm{d}u$ as a series of ascending powers of u and integrate two or three terms. Hence show that in method (a) the fractional error is about $1/8a$, that is, very small when $|a|$ is large.

Try the same methods for $\int_{-\frac{1}{2}\pi}^{\frac{1}{2}\pi} \exp\{a(1 - \sec\theta)\}\,\mathrm{d}\theta$.

9.2. Evaluate $\int_{-\frac{1}{2}\pi}^{\frac{1}{2}\pi} \exp(-a\tan^2 x)\mathrm{d}x$ as an asymptotic expansion given that $\mathrm{Re}(a) > 0$.

9.3. The Hankel function $H_0^{(1)}(z)$ is given by the Sommerfeld contour integral $(1/\pi)\int_C \exp(\mathrm{i}z\cos u)\mathrm{d}u$ where the path C runs from $u = -\frac{1}{2}\pi + \mathrm{i}\infty$ to $u = \frac{1}{2}\pi - \mathrm{i}\infty$.

Evaluate the integral approximately by the method of steepest descents and hence show that the asymptotic form for $H_0^{(1)}(z)$ is $(2/\pi z)^{\frac{1}{2}} \exp\{i(z - \frac{1}{4}\pi)\}$.
(See Sommerfeld, 1949, § 19).

9.4. The factorial function is defined by the integral $z! = \int_0^\infty t^z e^{-t} dt$. Use the method of steepest descents to show that when $|z|$ is large an approximate value of $z!$ is $(2\pi z)^{\frac{1}{2}} z^z e^{-z}$ (first term of Stirling's formula). Sketch the paths used in the method of steepest descents when arg z is (a) zero, (b) $\frac{1}{2}\pi$, (c) $\pi - \varepsilon$ where ε is small.

9.5. The function $\exp\{f(t)\}$ has a double saddle point at $t = t_0$ (where both $df/dt = 0$ and $d^2f/dt^2 = 0$). Show that through this point there are six level lines (three crossing lines) with angles $\frac{1}{3}\pi$ between adjacent lines. Show also that there are three lines of steepest descent and three lines of steepest ascent, forming three crossing lines, and that these are at angles $\frac{1}{6}\pi$ to the level lines. Sketch the contour map for $\exp\{f(t)\}$ near t_0.

9.6. Use the method of steepest descents to find an approximation, valid when $a \gg 1$, for $F(a, m) = \int_0^\infty \exp(ax^m - x) dx$ where m is a real constant, and $0 < m < 1$. In particular show that $F(a, \frac{1}{2}) \sim a\pi^{\frac{1}{2}} \exp(\frac{1}{4}a^2)$.
(Natural Sciences Tripos 1971. Part II. Theoretical Physics).

10

Ray tracing in a loss-free stratified medium

10.1. Introduction

The term 'ray' was used in § 5.3 when discussing the field of a radio wave that travels out from a source of small dimensions at the origin of coordinates ξ, η, ζ in a homogeneous medium. The field was expressed as an integral (5.29) representing an angular spectrum of plane waves. The main contribution to the integral was from 'predominant' values of the components n_ξ, n_η of the refractive index vector $\boldsymbol{n}$ such that the phase of the integrand was stationary for small variations δn_ξ, δn_η. For any point ξ, η, ζ there were one or more predominant values of the refractive index vector $\boldsymbol{n}$, such that the line from the origin to ξ, η, ζ was normal to the refractive index surface. Each predominant $\boldsymbol{n}$ defines a progressive plane wave that travels through the whole of the homogeneous medium.

Instead of selecting a point ξ, η, ζ and finding the predominant $\boldsymbol{n}$s, let us choose a fixed $\boldsymbol{n}$ and find the locus of points ξ, η, ζ for which this $\boldsymbol{n}$ is predominant. For a homogeneous medium this locus is called the 'ray'. The arguments leading to (5.31) still apply and show that it is the straight line through the origin, in a direction normal to the refractive index surface at the point given by the chosen $\boldsymbol{n}$. It was shown in § 5.3 that for a loss-free medium the direction of the ray is the same as the direction of the energy flux as given by the time averaged Poynting vector $\boldsymbol{\Pi}_{\text{av}}$, for the progressive plane wave that has the chosen $\boldsymbol{n}$.

These ideas may now be extended to apply to a plane stratified medium such as the ionosphere. We shall now therefore use the coordinate system x, y, z of § 6.1, with the z axis perpendicular to the strata. As in § 6.2 it is useful to imagine that the ionosphere is replaced by a number of thin discrete strata in each of which the medium is homogeneous. For a given progressive plane wave in any one stratum, the direction of the ray can be found. When this wave crosses a boundary, the x and y direction cosines S_1 and S_2 of its wave normal do not change; this is Snell's law, see § 6.2. But its value of q changes, so that in the next stratum the direction of the ray is different. In each stratum the ray is normal to the refractive index surface, but this gives only its direction. To find the actual curved ray path, the ray must be traced step by step starting from its source. In the limit when the strata are indefinitely thin and numerous, this leads to the integrals (10.3) below. The tracing of rays in this way has many uses of which the principal ones are as follows.

(a) When the direction of a ray reaching a receiver is known, the associated predominant $\boldsymbol{n}$ can be found and this gives the direction of the wave normal and therefore of the arriving wave front.
(b) For a loss-free medium the rays have the direction of the energy flux. By studying the configuration of adjacent rays in a ray pencil, the energy flux at the receiver can be found. This is discussed in §§ 10.17–10.20, 14.8.
(c) For a plasma with a very small collision frequency, the effect of the collisions on the real part of the refractive index $\boldsymbol{n}$ can often be neglected. The rays are traced using only the real part of $\boldsymbol{n}$ and are unaffected by the collisions. The attenuation of the signal is then found by suitably integrating the imaginary part of $\boldsymbol{n}$ along the ray path; see § 10.15.
(d) When pulsed signals are used, the pulse travels along the ray with the group velocity $\mathscr{U}$. The time of travel is found by integrating $1/\mathscr{U}$ along the ray path. See §§ 10.3, 10.16, 14.7.

10.2. The ray path

The radio transmitter may be thought of as a source of small dimensions in free space, at the origin of coordinates x, y, z, and emitting waves of fixed angular frequency $\omega = kc$, whose wave fronts are approximately spherical. Its radiated field is expressed, as in § 6.2, by an integral representing an angular spectrum of plane waves. In the free space below the ionosphere any one of the field components is then given by

$$F(x, y, z, t) = \int_{-\infty}^{\infty}\int_{-\infty}^{\infty} A(S_1, S_2)\exp\{ik(ct - S_1x - S_2y - Cz)\}\,dS_1\,dS_2 \quad (10.1)$$

where the function $A(S_1, S_2)$ depends on the kind of transmitting aerial and on the field component used. In many cases of importance A is a very slowly varying function, and for small ranges of S_1, S_2 it can be treated as a constant when compared with the exponential in (10.1). This is a much more rapidly varying function of S_1, S_2, especially when kx, ky, kz are large. An important example where $A(S_1, S_2)$ is not slowly varying is described in § 10.3.

The integrand of (10.1) represents a plane progressive wave that travels obliquely upwards and enters the ionosphere. Within the ionosphere, for each S_1, S_2, there are two waves that can travel obliquely upwards, and these have different polarisations and different values of q, and they are reflected at different levels. When the incident wave enters the ionosphere, therefore, it divides into two waves that are propagated independently, and their relative amplitudes depend on the state of polarisation of the incident wave. It is possible to choose this so that only one magnetoionic component is present in the ionosphere. For this purpose, at moderate or high magnetic latitudes, and for near vertical incidence, the incident wave must be nearly circularly polarised. In practice it is more usual for the incident wave to be linearly polarised, and it then gives the two magnetoionic components with roughly equal amplitudes. This division of the incident wave into two magnetoionic components is related to the phenomenon called 'limiting polarisation', and the theory is given in §§ 17.10–17.11.

When the transmitted signal is a pulse of radio waves, as in the ionosonde technique, the incident pulse splits into two pulses that travel over different paths with different group velocities. After reflection they reach the receiver at different times so that the received signal or echo is said to be 'split'. The phenomenon is sometimes called magnetoionic splitting.

We now study one of the two magnetoionic component waves in the ionosphere. Let q be its associated root of the Booker quartic equation (ch. 6). In the free space below the ionosphere, $q = C$. It is assumed that, for levels z in the ionosphere far enough below the reflection level, the W.K.B. solution is sufficiently accurate. Below the ionosphere this is simply the factor $\exp(-ikCz)$ in (10.1). Within the ionosphere this is replaced by a multiple of the W.K.B. solution derived from (7.136). It consists, as explained in § 7.18, of a phase memory factor $\exp(-ik\int_0^z q\,dz)$ and a more slowly varying factor that depends on which field component is used for F in (10.1). Thus (10.1) is replaced, within the ionosphere, by

$$F(x, y, z, t) = \int_{-\infty}^{\infty}\int_{-\infty}^{\infty} B(S_1, S_2, z)\exp\left\{ik\left(ct - S_1x - S_2y - \int_0^z q\,dz\right)\right\}dS_1\,dS_2. \tag{10.2}$$

Here the slowly varying part of the W.K.B. solution has been combined with the factor A in (10.1) to give the new slowly varying factor B.

The formula (10.2) fails if z is near a level of reflection, but it can be used for an obliquely downgoing wave that has been reflected at some greater value of z. This is done by arranging that the range of the integral $\int q\,dz$ extends from the ground up to the level of reflection (in general different for each component wave) and then down again to the receiving point, with q replaced by a new q for the downgoing wave. This technique was explained in § 7.19.

The double integral (10.2) is of the type (9.61) and may be evaluated by the method of double steepest descents, § 9.10. There is a double saddle point where the partial derivatives of the exponent with respect to S_1 and S_2 are both zero; compare (9.63). The exponent in (10.2) may be denoted by $i\varphi$ where φ is the phase. Then these conditions are equivalent to saying that the phase must be stationary with respect to variations of S_1 and S_2 (compare § 5.3). The conditions are

$$-\partial\varphi/\partial S_1 \equiv x + \int_0^z (\partial q/\partial S_1)\mathrm{d}z = 0, \quad -\partial\varphi/\partial S_2 \equiv y + \int_0^z (\partial q/\partial S_2)\mathrm{d}z = 0. \tag{10.3}$$

For any given position (x, y, z) of the receiver these might be solved for S_1, S_2 and would give $S_1 = S_{10}$, $S_2 = S_{20}$ say. Since B is slowly varying it may be replaced by $B(S_{10}, S_{20}, z)$. It is important only for small ranges of S_1 and S_2 near S_{10} and S_{20} respectively. If B fell to zero outside this range, the signal at the receiver would not be appreciably affected.

The value of the integral (10.2) is now

$$F(x, y, z, t) \approx G(S_{10}, S_{20}, z)\exp\left\{\mathrm{i}k\left(ct - S_{10}x - S_{20}y - \int_0^z q\mathrm{d}z\right)\right\} \tag{10.4}$$

where G is a slowly varying function of z, and q takes its value for $S_1 = S_{10}$, $S_2 = S_{20}$. This gives the predominant progressive wave near the point (x, y, z).

We now trace the path traversed by the energy arriving at the receiver. Since this comes only from component waves with S_1 and S_2 near S_{10} and S_{20} respectively, we must find a series of points for which the signal is a maximum when $S_1 = S_{10}$, $S_2 = S_{20}$. The locus of these points is called the 'ray path' or simply the 'ray'. For each point on it the signal is a maximum when the phase is stationary for variations of S_1 and S_2. Hence (10.3) are the equations of the ray, provided that in $\partial q/\partial S_1$, $\partial q/\partial S_2$ we put $S_1 = S_{10}$, $S_2 = S_{20}$ after differentiation.

If (10.3) are differentiated with respect to z, they give

$$\mathrm{d}x/\mathrm{d}z = -\partial q/\partial S_1, \quad \mathrm{d}y/\mathrm{d}z = -\partial q/\partial S_2 \quad \text{for } S_1 = S_{10},\ S_2 = S_{20} \tag{10.5}$$

and these give the direction of the ray at each point. Now S_{10}, S_{20}, q are the coordinates of the point on the refractive index surface for the predominant progressive wave. Thus (10.5) confirms that the ray path is normal to the refractive index surface.

In general it is not easy to solve (10.3) for S_{10} and S_{20} when x, y, z are given. The more usual procedure is to assume a pair of values S_{10}, S_{20} and trace the ray corresponding to them. Successive pairs are tried until a pair is found that gives a ray passing through a given receiving point.

10.3. Wave packets

On a given ray the signal is the resultant of waves whose original direction cosines S_1, S_2 are within narrow ranges near S_{10}, S_{20} respectively. Suppose now that the transmitter radiates only these waves. Then the function $B(S_1, S_2, z)$ in (10.2) is no longer nearly independent of S_1, S_2, but has a narrow maximum near $S_1 = S_{10}$,

$S_2 = S_{20}$. It is no longer a slowly varying function, and the stationary phase condition (10.3) must be replaced by

$$x + \int_0^z \frac{\partial q}{\partial S_1} dz + \frac{i}{k}\frac{\partial(\ln B)}{\partial S_1} = 0, \quad y + \int_0^z \frac{\partial q}{\partial S_2} dz + \frac{i}{k}\frac{\partial(\ln B)}{\partial S_2} = 0. \tag{10.6}$$

If in these we put $S_1 = S_{10}$, $S_2 = S_{20}$, they give the equation of the ray. But the ray must pass through the transmitter, which is at the origin. Hence

$$\frac{\partial(\ln B)}{\partial S_1} = \frac{\partial(\ln B)}{\partial S_2} = 0 \tag{10.7}$$

when $S_1 = S_{10}$, $S_2 = S_{20}$. Therefore, although B is not now slowly varying, this adds nothing to the stationary phase condition. The resulting signal leaving the transmitter is in a narrow pencil of radiation with its maximum in the direction (S_{10}, S_{20}). This pencil enters the ionosphere and then has a curved path which is the ray path.

Suppose, further, that the signal is a pulse like that described in § 5.8. Then it must contain a range of frequencies, with maximum amplitude near the centre of the range. Since $k = 2\pi f/c$ is proportional to the frequency f, the pulsed signal may be expressed by assuming that B is a function also of k, and the expression becomes

$$F(x, y, z, t) = \int_{-\infty}^{\infty}\int_{-\infty}^{\infty}\int_{-\infty}^{\infty} B(k, S_1, S_2, z)\exp\left\{ik\left(ct - S_1 x - S_2 y - \int_0^z q\,dz\right)\right\} dk\,dS_1\,dS_2 \tag{10.8}$$

where B has a narrow maximum near

$$k = k_0, \quad S_1 = S_{10}, \quad S_2 = S_{20} \tag{10.9}$$

and these will be called the 'predominant' values of k, S_1, S_2. This represents a signal confined to a narrow pencil and of short duration. It is therefore a small packet of waves, and at a given time t it has maximum amplitude at some point (x, y, z). As t increases, the locus of this point is the path of the wave packet. The wave packet is assumed to leave the transmitter when $t = 0$ and this requires that $\partial(\ln B)/\partial k = 0$ when (10.9) is satisfied.

The signal (10.2) is a maximum when the phase is stationary with respect to variations of k, S_1, S_2 at the predominant values (10.9). This gives the two equations (10.3), and in addition

$$ct - S_1 x - S_2 y - \int_0^z \frac{\partial(kq)}{\partial k} dz = 0. \tag{10.10}$$

Here t is the time taken for the wave packet to travel from the origin to the point (x, y, z) where x and y are given in terms of z by (10.3). In this time the wave packet could travel a distance $P' = ct$ in free space, and this is called the 'equivalent path'.

Thus (10.10) can be rewritten

$$P' = S_1 x + S_2 y + \int_0^z \frac{\partial(fq)}{\partial f} \mathrm{d}z \tag{10.11}$$

where (10.9) is satisfied.

Although most transmitters do not radiate wave packets, we have shown that the problem of finding the path of the energy reaching a given receiver is the same as that of finding the path of a wave packet. In future, therefore, we shall speak of a ray and of the path of a wave packet as the same thing.

10.4. Equations of the ray path

It has been shown that for a ray path in a plane stratified medium, S_1 and S_2 have constant values S_{10} and S_{20}. In the discussion of rays that follows, the second subscript 0 will be omitted, and we use simply S_1, S_2 for the constant values on a ray.

The coordinates x, y, z of a point on a ray in a stratified medium are related by the two differential equations (10.5), or by the two equations in integrated form, (10.3). These were first given by Booker (1939) and are now widely known as Booker's ray tracing equations.

Assume first that the ionospheric plasma is isotropic. Then q is given, from (6.6), by

$$q^2 = n^2 - S_1^2 - S_2^2 \tag{10.12}$$

where n is the refractive index and is independent of S_1 and S_2. It was shown in § 10.3 that a ray is the same as the path of a wave packet and that on it S_1 and S_2 are constant. Hence (10.5) give

$$\mathrm{d}x/\mathrm{d}z = S_1/q, \quad \mathrm{d}y/\mathrm{d}z = S_2/q \tag{10.13}$$

and

$$x = S_1 \int_0^z \frac{\mathrm{d}z}{q}, \quad y = S_2 \int_0^z \frac{\mathrm{d}z}{q}. \tag{10.14}$$

These show that on a ray the ratio x/y is a constant so that the ray remains in the same vertical plane. It is nearly always convenient to choose the axes so that this is the plane $y = 0$, and $S_2 = 0$.

To find $\partial q/\partial S_1$ and $\partial q/\partial S_2$ for the general anisotropic plasma we make this choice of axes from the start, so that $S_2 = 0$, $S_1 = S$. The Booker quartic (6.15) must be true for all values of S_1 and S_2. Hence

$$\mathrm{d}F(q)/\mathrm{d}S_1 = \mathrm{d}F(q)/\mathrm{d}S_2 = 0. \tag{10.15}$$

Now

$$\frac{\mathrm{d}F(q)}{\mathrm{d}S_1} = \frac{\partial F(q)}{\partial q}\frac{\partial q}{\partial S_1} + \frac{\partial \alpha}{\partial S_1} q^4 + \frac{\partial \beta}{\partial S_1} q^3 + \frac{\partial \gamma}{\partial S_1} q^2 + \frac{\partial \delta}{\partial S_1} q + \frac{\partial \varepsilon}{\partial S_1} = 0 \tag{10.16}$$

and there is a similar equation with S_1 replaced by S_2. Equation (6.16) shows that $\partial\alpha/\partial S_1 = \partial\alpha/\partial S_2 = 0$. Hence, for a ray path, from (10.5)

$$\left.\begin{aligned}\frac{\mathrm{d}x}{\mathrm{d}z} &= \left(\frac{\partial\beta}{\partial S_1}q^3 + \frac{\partial\gamma}{\partial S_1}q^2 + \frac{\partial\delta}{\partial S_1}q + \frac{\partial\varepsilon}{\partial S_1}\right)\Bigg/\frac{\partial F(q)}{\partial q}\\ \frac{\mathrm{d}y}{\mathrm{d}z} &= \left(\frac{\partial\beta}{\partial S_2}q^3 + \frac{\partial\gamma}{\partial S_2}q^2 + \frac{\partial\delta}{\partial S_2}q + \frac{\partial\varepsilon}{\partial S_2}\right)\Bigg/\frac{\partial F(q)}{\partial q}\end{aligned}\right\}\text{ when } S_1 = S \text{ and } S_2 = 0. \tag{10.17}$$

Curves have been given by Booker (1949) to show how these quantities depend on X in some special cases.

The second expression in (10.17) is not in general zero even when $S_2 = 0$. Thus a ray that starts out in the vertical plane $y = 0$ does not necessarily remain in that plane. It may be laterally deviated. Some examples of this lateral deviation are given in §§ 10.12–10.14.

In some radio problems it is necessary to allow for the earth's curvature. Consider therefore a spherically stratified isotropic ionosphere, so that the refractive index $n(r)$ is a function of distance r from the centre of the earth. At radius r let the wave normal of a progressive wave make an angle ψ with the radius. Because the medium is now assumed to be isotropic, this is also the direction of the ray. Now ψ satisfies Bouger's law

$$rn\sin\psi = K \tag{10.18}$$

where K is a constant. This replaces Snell's law (6.4) for a plane stratified medium. Equation (10.18) is easy to prove by simple geometry (see problem 10.3). If r, θ are the polar coordinates of any point on a ray path

$$\frac{r\mathrm{d}\theta}{\mathrm{d}r} = \tan\psi = \frac{K}{rn\cos\psi} = \frac{K/r}{q} \tag{10.19}$$

where

$$q^2 = n^2 - K^2/r^2. \tag{10.20}$$

Equation (10.19) is the differential equation of a ray path in various forms. It should be compared with (10.13). An observer who thought that the earth was flat would treat $\mathrm{d}r$ as an element of height $\mathrm{d}z$, and $r\mathrm{d}\theta$ as an element of horizontal distance $\mathrm{d}x$. Then (10.19), (10.20) are the same as the first equation (10.13), and (10.12) with $S_2 = 0$, provided that we take

$$S_1 = K/r. \tag{10.21}$$

This shows that, for the curved earth, the ionosphere can be treated as though it is plane stratified provided that we use (10.21), so that S_1 is no longer constant on a ray. This device is called an 'earth flattening method'. In the present example, with an isotropic ionosphere, it is exact. For an anisotropic ionosphere it can also be used but then involves approximations, and we speak of an 'earth flattening approximation'. This topic is discussed further in §18.8.

Equations (10.19), (10.20) can also be written

$$r\mathrm{d}\theta/\mathrm{d}r = S_0/\bar{q}, \quad \bar{q}^2 = \bar{n}^2 - S_0^2, \quad \bar{n} = nr/r_0 \tag{10.22}$$

where r_0 is some fixed value of r, for example at the earth's surface, and S_0 is a constant equal to the value of $n \sin\psi$ where $r = r_0$. Thus (10.22) provides an alternative 'earth flattening' method, in which S_0 is constant on a ray. Instead of n it uses the modified value $\bar{n}$. This is very similar to the modified refractive index used by Booker and Walkinshaw (1946) in one of the pioneering papers on guided radio waves. For a further study of equations (10.18)–(10.20) see § 10.23.

10.5. The reversibility of the path

A wave packet leaves the transmitter, is deviated by the ionosphere and returns to the ground. Suppose now that every component plane wave in the downcoming wave packet is reversed in direction. This means that both S_1 and S_2 are reversed in sign, and (6.16) shows that the only effect on the quartic is to reverse the signs of the coefficients of q and q^3. Thus q also changes sign, so that $\partial q/\partial S_1$ and $\partial q/\partial S_2$ remain unchanged and (10.5) shows that the direction of the path is the same as before. The wave packet therefore simply retraces its original path back to the transmitter. The state of polarisation of the wave in a wave packet can be found from (6.21), which gives the ratios of the three components E_x, E_y, E_z. These depend on the elements of the 3×3 matrix in (6.21), (6.22), which remain unaltered when q, S_1 and S_2 all change sign. Hence the polarisation referred to fixed axes x, y, z is the same for the original and the reversed wave packet. This has an important bearing on the study of reciprocity; §§ 14.13, 14.14.

The equations (10.5) were based on the assumption that the ray is a feature of a wave system of the form (10.2) in which the z dependence of the exponent is given entirely by the factor $\int_0^z q\,\mathrm{d}z$. Here q refers to one of the four magnetoionic components and it is implied that this is propagating independently of the other three. A wave can, however, give rise as it travels to some of the other components, and these react back on the original wave and modify its properties. The subject is studied in ch. 16 on coupled wave equations, where it is shown, § 16.9, that q must be replaced by $q + \mathrm{i}\Gamma_{jj}$. In many practical cases the extra term is negligible. In cases where Γ_{jj} cannot be neglected, the ray tracing equations (10.5) must be modified. It is then found that the rays are not reversible. This subject has been studied by Smith, M.S. (1975, 1976) who gives some examples.

10.6. The reflection of a wave packet

In an isotropic ionosphere a wave is reflected at the level where the ray and the wave normal are horizontal, that is where $q = 0$. Here the integrands in (10.14) are infinite. These integrals, however, always converge to a finite limit when z is at the reflection level. To illustrate this consider a simple example where the electron height distribution is linear so that $X = pz$. Let $S_2 = 0$, $S_1 = S$. Then

$$q^2 = n^2 - S^2 = C^2 - X = C^2 - pz. \tag{10.23}$$

Let a ray leave the origin in this medium. To evaluate the integral (10.14) it is often simplest to change to q as the variable of integration. This gives

$$x = 2S \int_C^q \frac{dq}{d(q^2)/dz} = \frac{2S}{p}(C - q). \tag{10.24}$$

Clearly this converges to the limit $2SC/p$ at the reflection level where

$$q = 0, \quad z = z_0 = C^2/p. \tag{10.25}$$

The ray path (10.24) is the parabola

$$S^2 z = SCx - \tfrac{1}{4}px^2. \tag{10.26}$$

It was shown in §§7.19, 8.20 that when an upgoing wave represented by the integrand of (10.2) reaches the level of reflection $z = z_0$, it is converted to a downgoing wave, whose fields are given simply by using the factor (7.155) in the integrand. This gives the phase memory for the ray path up to the reflection level and then down again after reflection. Thus the ray can be traced right through the reflection level as a continuous curve. The phase advance factor i mentioned at the end of § 8.20 does not affect the ray path. In the present example the parabolic path (10.26) extends beyond its maximum to the descending part.

For an anisotropic ionosphere at a reflection level, two roots of the Booker quartic, say q_1 and q_2, are equal. The denominators of the ray equations (10.17) are $\partial F(q)/\partial q$ which has a factor $q_1 - q_2$. Thus integration of (10.17) uses an integral with an infinite integrand at the reflection level. But again it can be shown that the integral must converge to a limit. This is because $(q_1 - q_2)^2$ is an analytic function of z at $z = z_0$, like q^2 in (10.23). For the proof see § 16.3. Thus in an anisotropic medium also, a ray can be traced right through a reflection level.

It is possible for the quartic to have two equal roots which differ from zero. Where this happens the path of the ray is horizontal, but the direction of the wave normal is not horizontal. The wave packet does not in general travel in the direction of the wave normal. Similarly, at a level where one value of q is zero, the associated wave packet is not necessarily travelling horizontally, although the wave normal is horizontal. Thus the level where $q = 0$ is not a level of reflection, except in the special case where there are two roots which are both zero at the same level.

The first expression (10.17) is dx/dz for the ray path. It can change sign at a level where the numerator is zero and the path of the wave packet is then parallel to the y–z plane, and the x component of its motion is reversed. Some examples of this can be seen in fig. 10.10. The sign of the vertical component of its motion is unaltered, however, and can only be reversed where $\partial F(q)/\partial q$ is zero.

In § 6.4 it was shown that when S_1, S_2 are real two of the four waves are obliquely upgoing waves and two are obliquely downgoing.

10.7. An example of a ray path at oblique incidence

Suppose that the frequency is greater than the gyro-frequency, and consider a ray that leaves the earth obliquely in the magnetic meridian plane. It is shown later, § 10.14, that there is no lateral deviation in this case, so the ray remains in the magnetic meridian throughout its path. We take $S_2 = 0$ so that the x–z plane is the plane of incidence and also the magnetic meridian plane. Fig. 10.1(a) shows an example of how the four roots of the quartic depend on X in this case. When the ray enters the ionosphere it splits into ordinary and extraordinary rays as described in § 10.2. The direction of the ordinary ray is given at each level by (10.17), first equation, and the ray can thus be plotted as shown in fig. 10.1(b). When the level B is reached, two roots of the quartic are equal and the ray is horizontal, but q is not zero and the wave normal is still directed obliquely upwards because q is positive. When the downgoing ray reaches the level where $q = 0$, at A in figs. 10.1(a, b), the wave normal is horizontal. Thereafter, on the rest of the downward path, the wave normal is obliquely downwards. The behaviour of the extraordinary ray is similar, but the level of reflection is lower than for the ordinary ray, and the wave normal becomes horizontal before reflection occurs.

Another example of a ray path is shown in fig. 10.2. It is for a frequency less than the electron gyro-frequency, $Y = 2$, and again is for a ray in the magnetic meridian plane. The dependence of q on X is shown by the curve marked Ext in fig. 6.7. There are vertical tangents at the three points D, A, B and the ray path is horizontal at the

Fig. 10.1. Finding ray paths at oblique incidence. In this example the incident ray is in the magnetic meridian plane and travelling southwards. (a) shows how the four roots q of the Booker quartic depend on X. (b) shows the path of the ordinary ray and the arrows show the direction of the wave normal at each level.

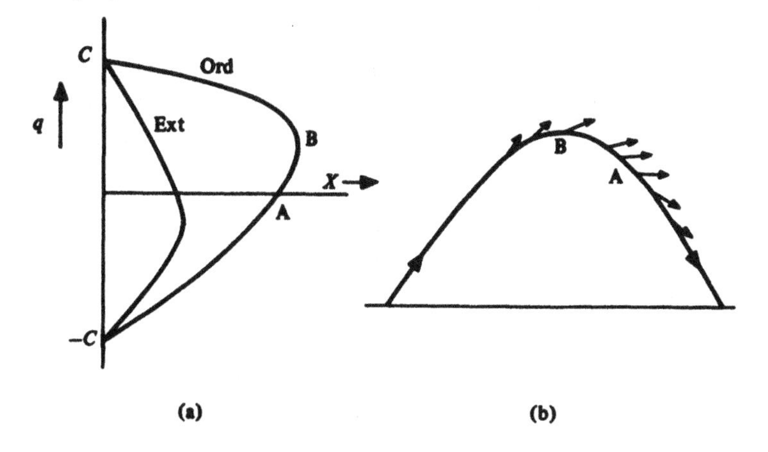

three corresponding levels, as shown in fig. 10.2. The wave normal is horizontal where $q = 0$ at a level just below B, as shown by the arrows.

10.8. Poeverlein's construction

A method for finding the general shape of a ray path in a stratified ionosphere was given by Poeverlein (1948, 1949, 1950). In the free space below the ionosphere the incident ray is in the x–z plane at an angle θ to the z axis. It enters the ionosphere and there splits into two rays, ordinary and extraordinary. We study one of these, say the ordinary ray. A diagram is drawn with axes n_x, n_z parallel to the x, z axes of ordinary space. It is thus a cross section of refractive index space by a plane parallel to the plane of incidence. A sequence of values of X is now selected, and for each X the cross section of the refractive index surface for the ordinary wave is drawn in the diagram. An example is shown in fig. 10.3. In general the earth's magnetic field is not in the plane of this diagram, but its projection is shown in the figure. The outermost curve is for the free space below the ionosphere and is a circle of unit radius. As the electron concentration increases the curves get smaller until they shrink to a point at the origin when $X = 1$.

Now let a vertical line, AB in fig. 10.3, be drawn in the $n_x - n_z$ plane at a distance $S = \sin\theta$ from the origin; compare fig. 6.1. It is here called the reference line. For some level let it cut the refractive index surface at the two points C and D, and let OC make an angle ψ with the n_z axis. Then OC is the refractive index n for a wave whose normal is in the direction OC, and $\text{OC}\sin\psi = n\sin\psi = S$, which is simply Snell's law. Hence OC is one possible direction for the wave normal of the ordinary wave. Another possible direction is OD, and two other directions are found from the intersection of the line AB with the refractive index surface for the extraordinary wave. It is also clear from the figure that $\text{CE} = n\cos\psi = q$.

Now it was shown in § 5.3 that the ray direction is perpendicular to the refractive index surface. Hence the two possible ray directions for the ordinary wave are the perpendiculars to the surface at C and D. Clearly the one at C must be inclined

Fig. 10.2. An example of a ray path in the magnetic meridian plane. The dependence of q on X for this example is shown by the curve marked Ext in fig. 6.7.

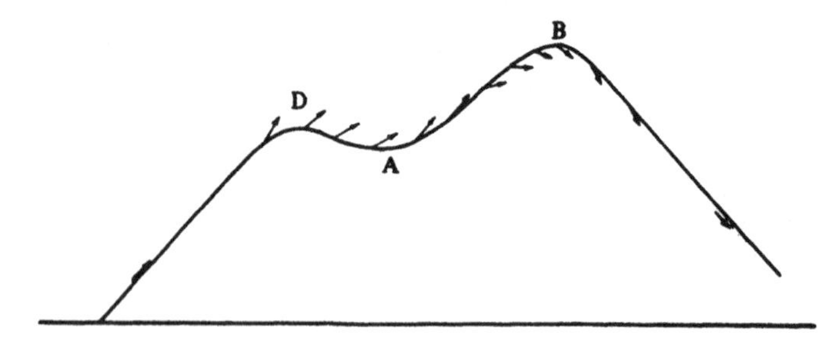

upwards and that at D downwards. These ray directions are not in general in the plane of the figure, that is the n_x–n_z plane. For example, that at C must be coplanar with OC and the earth's magnetic field.

The reference line cuts the circle of unit radius at A so that OA is the direction in which the ray enters the ionosphere. As the wave packet travels upwards, the refractive index surface changes, so that the direction of the wave normal and the ray also change. The wave normal is always in the plane of the diagram, but the ray may not be. The points C and D move closer together until a level is reached where the reference line just touches a refractive index surface at F in fig. 10.3. Then the ray direction is horizontal and the two associated values of q are equal. This is the level of reflection, and thereafter the wave packet travels downwards and we consider the same series of surfaces as before, but in reverse order. Finally the surface of unit radius is again reached, and the wave packet leaves the ionosphere with the ray and wave normal both in the direction OB.

It is clear that in general the ray is directed out of the plane of the diagram. The only exception to this is for propagation from (magnetic) north to south or south to north, and this case is discussed separately in the following sections.

If Poeverlein's construction is applied to a real ray that originates in the free space

Fig. 10.3. To illustrate Poeverlein's construction. Cross section by the plane of incidence of a sequence of refractive index surfaces for the ordinary ray, for various values of X.

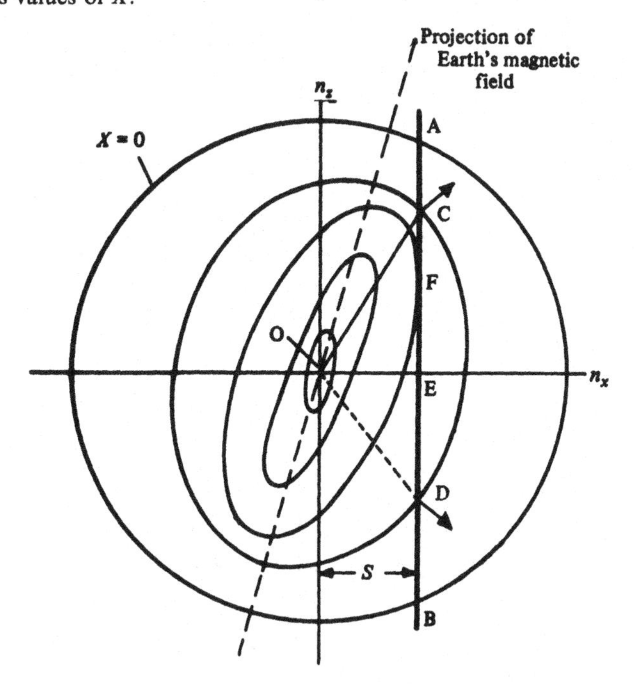

below the ionosphere, the angle of incidence θ is real so that S must be real and $|S| < 1$. But in a plasma the refractive index can exceed unity and then fig. 10.3 might contain refractive index surfaces where $|n| > 1$. Some of these could be real surfaces in the region where $|S| > 1$. There are examples in figs. 10.7, 10.9. A reference line AB in fig. 10.3 with $S > 1$ would cut such a surface at real points and here the wave normal and the ray would have real directions. Such a ray could be propagated in a horizontally stratified ionosphere, although it could not originate from a ray with real θ in free space. It might come from a transmitter in a space vehicle within the ionosphere, or from outside the ionosphere by refraction where there are horizontal variations. Signals of this kind have been observed. An example is the subprotonic whistler described at the end of § 13.8.

10.9. Propagation in magnetic meridian plane. The 'Spitze'

Fig. 10.3 shows a series of cross sections of the refractive index surfaces by the plane of incidence, that is the n_x–n_z plane. When this plane is also the magnetic meridian, these cross sections, for the ordinary ray, have a slightly different form, and an example is shown in fig. 10.4. As the electron concentration increases, the curves get smaller, as in fig. 10.3, but instead of shrinking to a point when $X = 1$, they shrink to the line PQ. This shows that the refractive index n for the ordinary wave is zero when $X = 1$ for all directions except the direction of the earth's magnetic field. For this case

Fig. 10.4. Similar to fig. 10.3 but the plane of incidence is now the magnetic meridian. In this example $Y = \frac{1}{2}$ (compare fig. 5.4, left-hand figure). The numbers by the curves are the values of X.

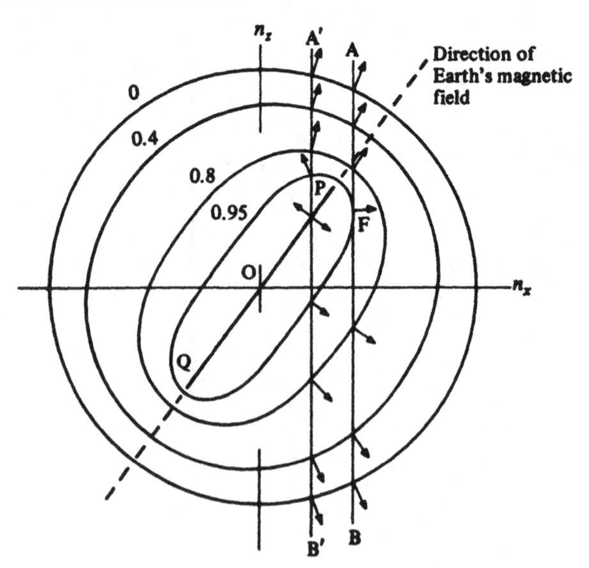

of purely longitudinal propagation there is an ambiguity in the value of n when $X = 1$, mentioned in §4.11.

Poeverlein's construction may now be used as described in the previous section. A vertical reference line is drawn at a distance S from the origin. Suppose it is at AB in fig. 10.4. Then the ray directions are given by the perpendiculars to the refractive index surfaces where the line cuts them. They are now always in the plane of incidence, so there is no lateral deviation. Suppose that a wave packet travels upwards in an ionosphere in which the electron concentration increases monotonically. Eventually a level is reached where the reference line just touches a refractive index surface, as at F in fig. 10.4. Here the ray is horizontal and this is the level of reflection. The phenomenon is similar to that described in the previous section. The surface that the line just touches gives the value of X at the level of reflection. If the angle of incidence, arcsin S, increases, the reference line moves further from the origin and the surface which is touched corresponds to a smaller value of X. The reflection level therefore decreases as the angle of incidence increases.

Suppose, however, that the reference line is at A′B′ in fig. 10.4. The ray directions are given as before by the perpendiculars to the refractive index surfaces where the line cuts them, but there is now no surface which is touched by the line. Instead, the line A′B′ cuts the line PQ, which refers to the level where $X = 1$. Above PQ the outward perpendicular to the refractive index surface is directed obliquely upwards; below PQ the perpendicular is obliquely downwards. On the line PQ the ray therefore reverses its direction abruptly, and this is the level of reflection. Here the ray path is perpendicular to the line PQ, that is to the earth's magnetic field. The ray path never becomes horizontal, but has a cusp at the level $X = 1$, called by Poeverlein the 'Spitze'. Fig. 10.5 shows some typical ray paths for various angles of incidence. It is similar to a diagram given by Poeverlein (1950). As long as the angle

Fig. 10.5. Typical ray paths for the ordinary ray in the magnetic meridian, showing the Spitze.

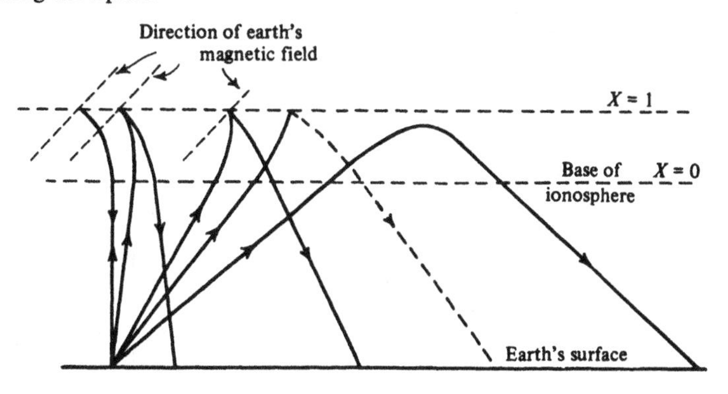

of incidence is such that the line A′B′ cuts the line PQ, there is always a Spitze, and the reflection level is where $X = 1$ for all angles of incidence.

If the reference line is drawn where S is given by (6.32), plus sign, it goes through P. Then the upgoing ray reaches the level where $X = 1$ but is not now reflected. It is converted to an extraordinary ray in the Z-mode, above the level where $X = 1$, and its path is described in § 10.10, fig. 10.8(a). The incident ray is in the magnetic meridian at an angle θ_w to the vertical. For directions near this in a small cone whose width is a few degrees, the upgoing ray is partially reflected and partially converted to a Z-mode ray. This small cone of angles is known as a 'radio window'. The phenomenon cannot be explained by simple ray theory alone and a full wave treatment is needed, given in §§ 17.6–17.9. There is a similar effect if the reference line in fig. 10.4 goes through Q; see fig. 10.8(b). For this reason the points P and Q in refractive index space are known as window points.

If the plane of incidence is turned very slightly from the magnetic meridian, the line PQ moves out of the plane of the diagram in fig. 10.4. The line A′B′ then touches one of the refractive index surfaces, for X slightly less than unity. This is a long cigar-shaped surface close to PQ. The perpendicular to it is a horizontal line almost perpendicular to the plane of the diagram. The ray path is then a twisted curve in three dimensions, and fig. 10.6(a, b) shows how it would appear when seen from the west and the north respectively. The receiver is not in the plane of incidence through the transmitter, for the reasons explained in § 10.14. A ray path with a Spitze is a limiting case, when the width of the curve of fig. 10.6(b) in the east–west direction becomes infinitesimally small.

10.10. Ray paths for the extraordinary ray when $Y < 1$

When $Y < 1$ so that the frequency is greater than the gyro-frequency, one set of refractive index surfaces, for the extraordinary ray, extends from $X = 0$ to $X = 1 - Y$; see fig. 5.7 left-hand diagram. The outermost curve for $X = 0$ is a circle of unit radius. The curves get smaller as X increases and eventually shrink to a point

Fig. 10.6. Projections of the ray path for the ordinary ray when the plane of incidence is slightly inclined to the magnetic meridian. (a) is projected on to the plane of incidence and (b) on to a vertical plane at right angles to it.

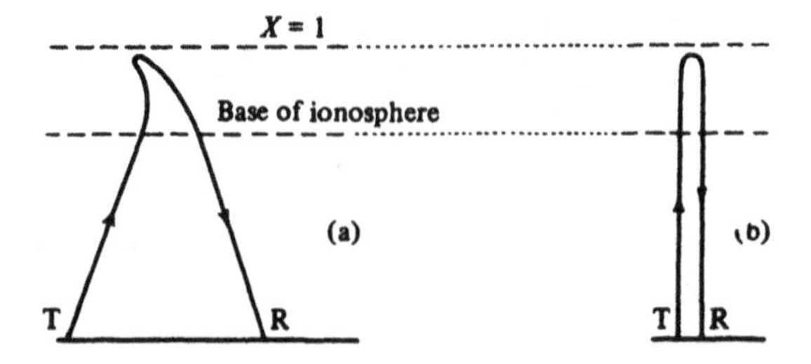

when $X = 1 - Y$. They never shrink to a line as in fig. 10.4 so that there is no phenomenon of the Spitze for the extraordinary ray when $X < 1$. The value of X at reflexion decreases as S increases, for the whole range $0 \leqslant S \leqslant 1$.

The refractive index for the extraordinary ray is imaginary when $1 - Y < X < 1 - Y^2$, but for $X > 1 - Y^2$ it is again real, and another set of refractive index surfaces can be constructed. The cross section of these by the magnetic meridian is shown in fig. 10.7. Now some values of n are greater than unity, so that there are some curves outside the unit circle. There is no curve for $X = 0$. The curves shrink to zero when $X = 1 + Y$. When $X = 1$ (see figs. 5.9, 5.11), the curve includes the unit circle, but it also includes the two straight line segments PP_1 and QQ_1 which are along the direction of the earth's magnetic field. This is because for purely longitudinal propagation there is an ambiguity in the value of n when $X = 1$. The points P and Q in fig. 10.7 are the window points and are the same as those in fig. 10.4. The two sets of surfaces have these points in common.

In general, for a wave packet incident from below, the extraordinary ray could not reach the levels depicted in fig. 10.7, for it would be reflected at or below the level where $X = 1 - Y$. There is a special case, however, when the angle of incidence is such that the reference line in fig. 10.4 goes through the point P. This figure refers to the ordinary wave, but at the level where $X = 1$ the wave is converted into an extraordinary wave, and the remainder of its path is found from fig. 10.7. The line AB

Fig. 10.7. Cross section, by the magnetic meridian plane, of the second set of refractive index surfaces for the extraordinary ray when $Y = \frac{1}{2}$. Compare figs. 5.8, 5.9. The numbers by the curves are the values of X.

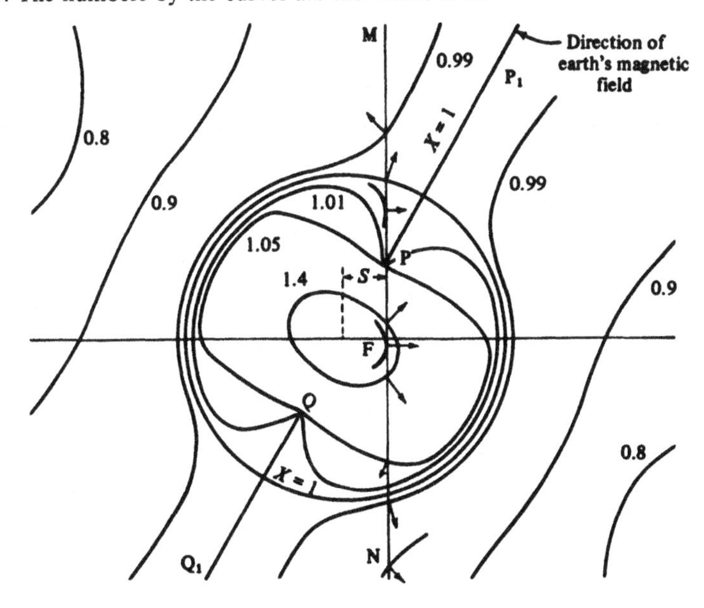

of fig. 10.4 must then be continued in fig. 10.7 and is there shown as the line MN. The two parts PM and PN correspond to different rays. The part PM near to P gives a ray travelling downwards and arriving at the level $X = 1$ from above, and so this ray cannot be present. The part PN gives a ray in the Z-mode that travels up to the level where the line PN just touches a refractive index surface at F. Here the ray is horizontal so this is a level of reflection. The ray then comes down again and crosses the level where $X = 1$, but can never leave the ionosphere because the level $X = 0$ does not appear in fig. 10.7. The line PN successively cuts curves of increasing radius and the ray becomes more and more horizontal. The wave packet therefore travels nearly horizontally just above the level where $X = 1 - Y^2$. In fact its energy would ultimately be absorbed because of electron collisions, which cannot be neglected in this case. The ray path is shown in fig. 10.8(a). The behaviour is similar when the line in figs. 10.4 and 10.7 goes through the point Q. The ray path for this case is shown in fig. 10.8(b). For a more detailed study of these ray paths see Poeverlein (1950, p. 157).

10.11. Extraordinary ray when $Y > 1$

When $Y > 1$, the cross sections by the magnetic meridian plane of the refractive index surfaces for the extraordinary ray are as shown in fig. 10.9 which is a combination of the left-hand diagrams of figs. 5.10–5.13. As X approaches $1 + Y$, the curves get smaller and eventually shrink to a point. The unit circle is the curve for $X = 0$. The curve for $X = 1$ is also the unit circle, but in addition it includes the two straight line segments PP_2, QQ_2 which are along the direction of the earth's magnetic field. The four points P, Q, P_2, Q_2 are window points, and P, Q are the same as in fig. 10.4. For

Fig. 10.8. Ray path for an incident ordinary ray in the magnetic meridian plane, at the transitional angle of incidence. It is converted to an extraordinary ray in the Z-mode at a window point. In (a) the lines AB in fig. 10.4 and MN in fig. 10.7 pass through the window point P. In (b) these lines pass through the window point Q.

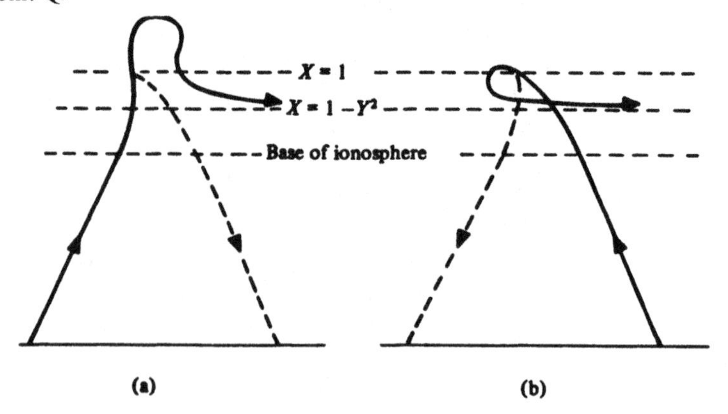

$X < 1$ the refractive index curves have an envelope shown in the inset of fig. 10.9 (see problem 5.9).

When the frequency is small enough to make $Y > 1$, it is not usually permissible to neglect electron collisions, so that the curves of fig. 10.9, in which collisions are neglected, cannot be applied to the actual ionosphere. It is, nevertheless, instructive to use fig. 10.9 to plot ray paths in the magnetic meridian, for a fictitious ionosphere in which collisions are absent. This is done by Poeverlein's construction exactly as in §§ 10.8–10.10. A vertical reference line AB is drawn at a distance S from the origin. Six examples are shown in fig. 10.10. For the line A_1B_1 (fig. 10.9) the ray path is as in fig. 10.10(a), and for the line A_2B_2 it is as in fig. 10.10(b). In both these cases when the upgoing ray passes the level where $X = 1$, its direction is parallel to the incident ray, and when the downgoing ray passes this level its direction is parallel to the emergent ray below the ionosphere. Fig. 10.10(c) shows the ray path for the line A_3B_3 in fig. 10.9. This crosses the straight segment PP_2 to the right of the point R and to the left of the right branch of the envelope, so that it does not intersect this branch. When the ray first reaches the level where $X = 1$, the corresponding point is S in

Fig. 10.9. Cross section, by the magnetic meridian plane, of the refractive index surfaces for the extraordinary ray when $Y = 2$. The numbers by the curves are the values of X. Inset is an enlarged view of part of the diagram.

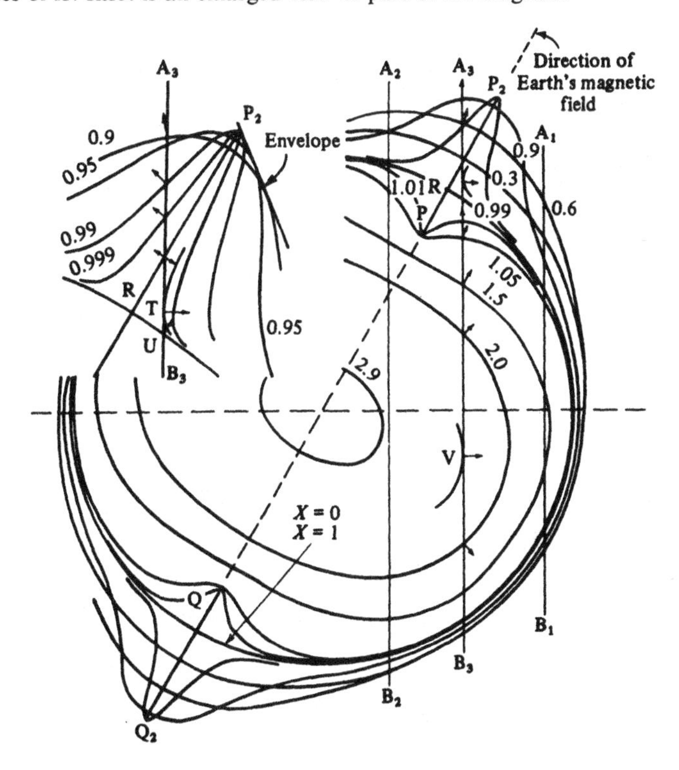

fig. 10.10(c). The ray direction is here suddenly reversed, in exactly the way described in § 10.8 for the ordinary ray, and there is a 'Spitze' in the ray path. The ray next goes down to a level where it is horizontal, corresponding to the point T in the figures. Then it rises again, passes through the level where $X = 1$ (point U, figs. 10.9 and

Fig. 10.10. Ray paths for the extraordinary ray in the magnetic meridian plane when $Y > 1$ and electron collisions are neglected. In each diagram the angle marked θ is equal to the angle of incidence. In (c) and (d) the line B shows the direction of the earth's magnetic field. The position of the line AB in fig. 10.9 is as follows. For (a) at A_1 B_1, for (b) at A_2 B_2, and for (c) at A_3 B_3. For (d) it is between P and R, for (e) it passes through R and for (f) it passes through the window point P.

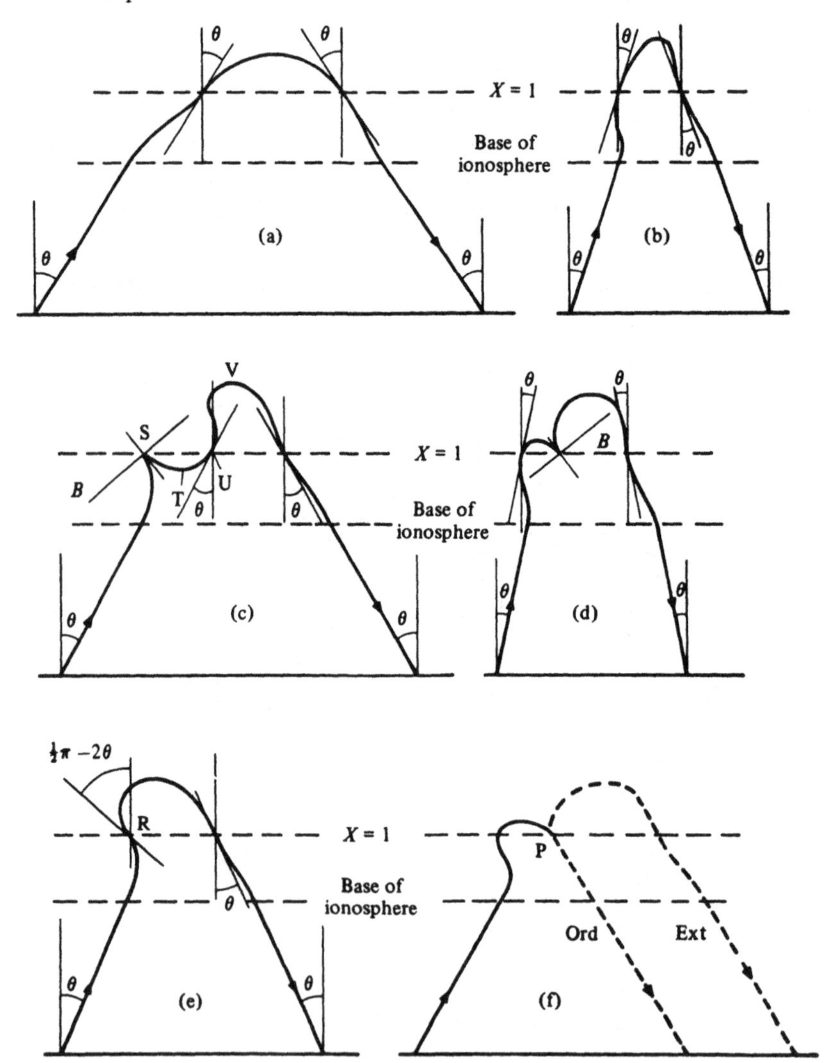

10.10(c)), and again becomes horizontal at the level corresponding to point V. Thereafter it travels downwards, again passing the level where $X = 1$, and eventually reaches the ground.

Fig. 10.10(d) shows the ray path when A_3B_3 passes between P and R in fig. 10.9, and fig. 10.10(e) is the path when A_3B_3 passes through R. The detailed tracing of these rays may be left as an exercise for the reader, who should verify that at the level where $X = 1$, corresponding to the point R, the ray is inclined to the vertical at an angle $\frac{1}{2}\pi - 2\theta$ where θ is the angle of incidence.

When the line AB in fig. 10.9 passes through the point P the ray path can be traced as before until the representative point reaches P. The extraordinary wave can here be converted into an ordinary wave, for the point P is common to the surfaces in fig. 10.9 and 10.4. The rest of the ray path is then found from fig. 10.4, and the ray returns to earth as an ordinary wave although it started out as an extraordinary wave. The complete ray path in this case is as shown in fig. 10.10(f). The wave would split into an ordinary and extraordinary wave for a cone of angles of incidence near the critical value, as explained in § 10.9.

The case where the reference line goes through the window point P_2 (or Q_2) is not given in detail here. This point is common to fig. 10.9 and to the set of refractive index surfaces for the ordinary wave when $Y > 1, X > 1$, that is for the whistler mode, fig. 5.5 left diagram. The points P_2, Q_2 are window points sometimes referred to as the 'second window' (Jones, 1976b; Budden, 1980); § 17.9.

10.12. Lateral deviation at vertical incidence

The Booker quartic may be used to find the path of a wave packet that is vertically incident on the ionosphere from below. It is convenient to choose the x axis to be in the magnetic meridian, pointing north, so that the direction cosine $l_y = 0$. The partial derivatives with respect to S_2 of all five coefficients α to ε in (6.16) are zero when $S_2 = 0$, and

$$\left.\begin{aligned} \partial\alpha/\partial S_1 = \partial\gamma/\partial S_1 = \partial\varepsilon/\partial S_1 = 0, \\ \partial\beta/\partial S_1 = -\,\partial\delta/\partial S_1 = 2l_x l_z X Y^2, \end{aligned}\right\} \quad \text{when } S_1 = 0. \tag{10.27}$$

Hence $\partial q/\partial S_2$ and thence $\mathrm{d}y/\mathrm{d}z$, (10.17), are zero, which shows that the wave packet remains in the x–z plane, that is in the magnetic meridian.

At vertical incidence q is the same as the refractive index n. We neglect collisions so that $U = 1$ in (6.16). Then (10.17) gives

$$\frac{\mathrm{d}x}{\mathrm{d}z} = \frac{l_x l_y X Y^2(n^2 - 1)}{2\alpha n^2 + \gamma}. \tag{10.28}$$

This is the tangent of the angle of inclination of the ray path to the vertical. By using the appropriate value of n it may be applied to either the ordinary or the extraordinary ray. Now n must satisfy $\alpha n^4 + \gamma n^2 + \varepsilon = 0$ which shows that the denominator in (10.28) is $\pm(\gamma^2 - 4\alpha\varepsilon)^{\frac{1}{2}}$ and

$$\frac{\mathrm{d}x}{\mathrm{d}z} = \pm \frac{l_x l_z Y(1-n^2)}{\{4l_z^2(1-X)^2 + l_x^4 Y^2\}^{\frac{1}{2}}} \tag{10.29}$$

where the minus sign applies to the ordinary ray and the plus sign to the extraordinary ray. This has the same values for upgoing and downgoing wave packets so the upward and downward paths are the same. In the northern hemisphere l_z is positive and l_x is negative so that $\mathrm{d}x/\mathrm{d}z$ is positive for the ordinary ray. In the southern hemisphere l_x and l_z are both negative and $\mathrm{d}x/\mathrm{d}z$ is negative for the ordinary ray. Hence the ordinary ray is always deviated towards the nearest magnetic pole and the extraordinary ray towards the equator. For frequencies greater than the electron gyro-frequency, $n_\mathrm{O}^2 > n_\mathrm{E}^2$ when $X < 1 - Y$. Hence $(1 - n_\mathrm{O}^2) < (1 - n_\mathrm{E}^2)$ and the inclination of the ray path to the vertical is smaller for the ordinary ray.

The ordinary ray is reflected when $X = 1$ and there $\mathrm{d}x/\mathrm{d}z = l_z/l_x = -\cot\Theta$ where Θ is the angle between the earth's magnetic field and the vertical. Hence at reflection the ordinary ray path is perpendicular to the earth's magnetic field, as was shown by Poeverlein's construction, § 10.9, figs. 10.4, 10.5. It should be remembered, however, that the wave normal is always vertical. It was explained in ch. 5 that the ray path and the wave normal are not in general parallel. See problem 10.7 on the polarisation of a wave at its reflection level.

Fig. 10.11. The paths of vertically incident wave packets in the northern hemisphere for a frequency greater than the electron gyro-frequency. The observer is looking towards the (magnetic) west.

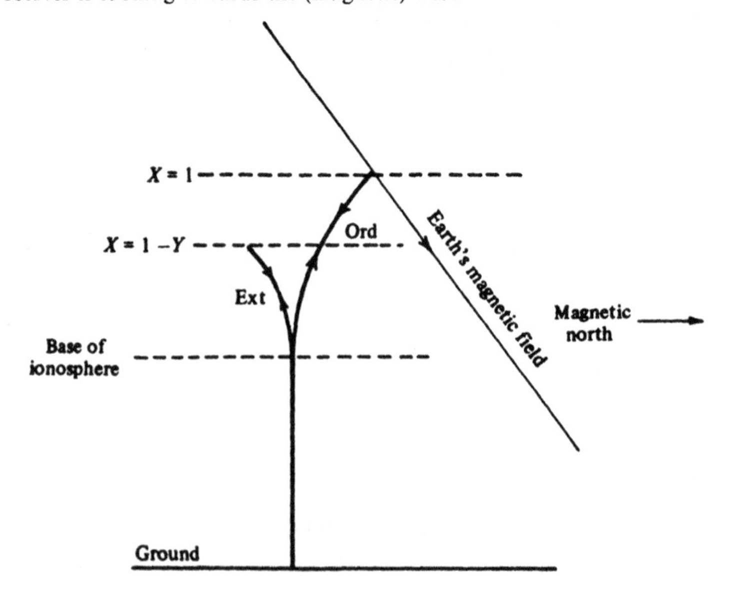

Fig. 10.11 shows the form of the two ray paths for vertical incidence. An important result of this lateral deviation is that the reflection points for the ordinary and extraordinary rays are at different latitudes. The gyro-frequency is sometimes assessed by measuring the difference of the penetration frequencies of an ionospheric layer for the two rays. If the electron density at the maximum of the layer varies appreciably with latitude the measurement will give a wrong value for the gyro-frequency (discussed by Millington, 1949, 1951).

10.13. Lateral deviation for propagation from (magnetic) east to west or west to east

In § 6.7 it was shown that when propagation is from (magnetic) east to west or west to east, the coefficients β and δ in the Booker quartic are zero and the quartic is a quadratic equation for q^2. In finding the path of a wave packet, however, it is important to notice that β and δ must not be set equal to zero until after the differential coefficients in (10.17) have been found. This is because a wave packet may include some component plane waves whose normals are not in the x–z plane.

In (6.16) put $l_x = 0$, and $U = 1$. Then

$$\frac{\partial\alpha}{\partial S_2} = 0, \frac{\partial\beta}{\partial S_2} = 2l_y l_z XY^2, \frac{\partial\gamma}{\partial S_2} = 0, \frac{\partial\delta}{\partial S_2} = -2Cl_y l_z XY^2, \frac{\partial\varepsilon}{\partial S_2} = 0, \tag{10.30}$$

when $S_2 = 0$, and (10.17) gives

$$\frac{\mathrm{d}y}{\mathrm{d}z} = -\left(\frac{\partial q}{\partial S_2}\right)_{S_2=0} = \frac{l_y l_z SY^2(q^2 - C^2)}{2\alpha q^2 + \gamma}. \tag{10.31}$$

This shows that the wave packet is deviated out of the x–z plane. If its path is projected on to the y–z plane then $\mathrm{d}y/\mathrm{d}z$ is the tangent of the angle of inclination of

Fig. 10.12. The paths of wave packets in the northern hemisphere when the plane of incidence is (magnetic) west to east.

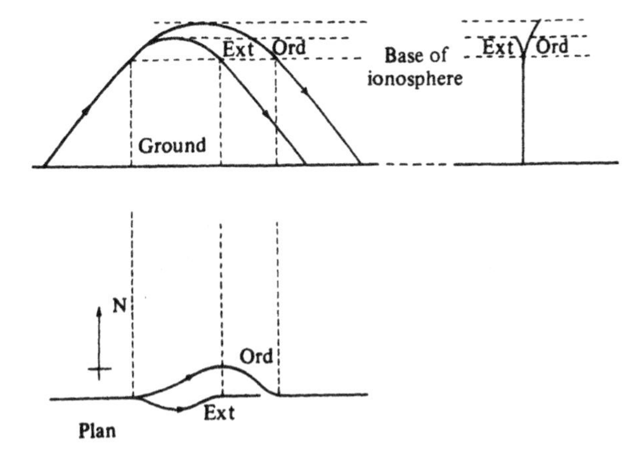

the projected path to the vertical. By using the appropriate value of q, (10.31) may be applied to either the ordinary or the extraordinary ray. It is the same for an upgoing or downgoing wave packet so that the upward and downward projected paths are the same. Thus a wave packet may leave the x–z plane on its upward journey but returns to it again on its downward journey. This is illustrated in fig. 10.12 which shows the projection of the ray paths on to the horizontal plane and on to two vertical planes.

Now q must satisfy the equation $\alpha q^4 + \gamma q^2 + \varepsilon = 0$ which shows that the denominator in (10.31) is $\pm(\gamma^2 - 4\alpha\varepsilon)^{\frac{1}{2}}$, and

$$\frac{\mathrm{d}y}{\mathrm{d}x} = \pm\frac{l_y l_z Y^2(C^2 - q^2)}{\{Y^4(1 - C^2 l_z^2)^2 + 4Y^2 l_z^2(1 - X)(C - X)\}^{\frac{1}{2}}}. \tag{10.32}$$

This is similar to (10.29) to which it reduces when $C = 1$ (with y for x and l_y for l_x). (See Millington, 1951.)

10.14. Lateral deviation in the general case

In the general case the derivatives with respect to S_2 of the coefficients (6.16) of the Booker quartic, with $U = 1$, are

$$\left.\begin{aligned}\frac{\partial\alpha}{\partial S_2} = 0, \quad \frac{\partial\beta}{\partial S_2} = 2l_y l_z XY^2, \quad \frac{\partial\gamma}{\partial S_2} = 2l_x l_y SXY^2,\\ \frac{\partial\delta}{\partial S_2} = -2C^2 l_y l_z XY^2, \quad \frac{\partial\varepsilon}{\partial S_2} = 2C^2 l_x l_y SXY^2,\end{aligned}\right\} \tag{10.33}$$

when $S_2 = 0$, where S is written for S_1. The formulae (10.17) now give

$$\frac{\mathrm{d}y}{\mathrm{d}z} = -\left(\frac{\partial q}{\partial S_2}\right)_{S_2=0} = \frac{2l_y XY^2(q^2 - C^2)(l_z q + l_x S)}{4\alpha q^3 + 3\beta q^2 + 2\gamma q + \delta}. \tag{10.34}$$

Fig. 10.13. The paths of wave packets in the northern hemisphere when the plane of incidence is not in the direction of one of the four (magnetic) cardinal points. The double and triple arrows show the apparent direction of the transmitter when using the ordinary and extraordinary rays respectively.

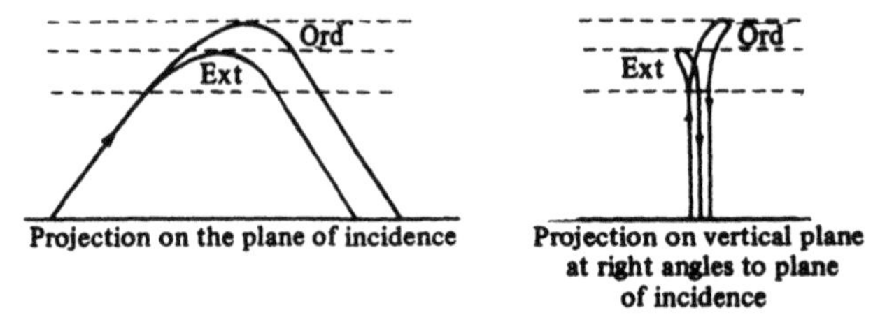

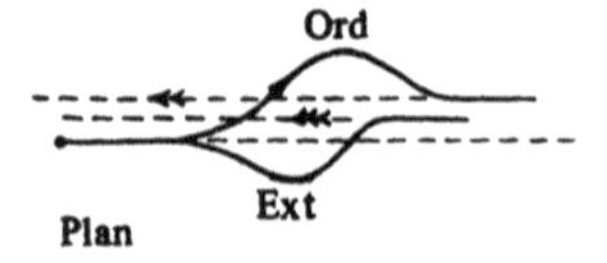

If the ray path is projected on the y–z plane then $\mathrm{d}y/\mathrm{d}z$ is the tangent of the angle of inclination of the projected path to the vertical. Now (10.34) contains odd powers of q and, at a given level, is not the same for the upward and downward paths. Hence the projections of the upward and downward paths are different. A wave packet leaves the x–z plane on its upward path and does not return to it again, or crosses it and returns to earth in a different plane. Below the ionosphere $\mathrm{d}y/\mathrm{d}z$ is zero, so that a wave packet returns to earth in a plane parallel to the x–z plane but not necessarily coincident with it. The only exceptions to this are for propagation from (magnetic) north to south or south to north, when the wave packet always remains in the x–z plane, and for propagation from (magnetic) east to west or west to east, when the wave packet moves out of the x–z plane but returns to it when it leaves the ionosphere.

The form of the ray paths in the general case is illustrated in fig. 10.13. The lateral deviation in the general case may mean that a signal reaching a receiver from a sender which is not in the direction of one of the four (magnetic) cardinal points, arrives in a vertical plane different from that containing the receiver and sender. The wave normals of this signal are in the plane of arrival, so that if a direction-finding aerial is used, the apparent bearing of the sender will differ from the true bearing. This theory has been discussed by Booker (1949) and Millington (1954), who showed that the bearing error would not be serious except in unusual conditions.

10.15. Calculation of attenuation, using the Booker quartic

Although this chapter is mainly concerned with loss-free media, so that q is purely real at all points of a ray, it is here of interest to consider the effect of a small electron collision frequency ν. The quantity q depends on ν only through $U = 1 - \mathrm{i}Z$. Let q_0 be its value when $Z = 0$. Then

$$q = q_0 - \mathrm{i}Z\frac{\partial q}{\partial U} - \frac{1}{2}Z^2\frac{\partial^2 q}{\partial U^2} + \cdots, \tag{10.35}$$

where the derivatives have their values for $Z = 0$. It can easily be shown that $\partial q/\partial U$ is then real when q is real. If Z is small, only two terms in (10.35) need be used. Then the real part of q is the same as if there were no collisions, and determines the path of the wave packet, which is the same as before. Alternatively the third term of (10.35) may be retained. This gives a slightly changed value of Re(q) so that the path of the wave packet is modified. There is now an imaginary part $-\mathrm{i}Z(\partial q/\partial U)$ of q, which gives attenuation of the wave. Equation (10.4) shows that the amplitude is reduced by a factor

$$\exp\left\{-k\int_0^z Z\frac{\partial q}{\partial U}\mathrm{d}z\right\}, \tag{10.36}$$

where the range of integration extends over the path of the wave packet. For a ray reflected from the ionosphere the range would extend from the ground to the level of reflection and back to the ground, and the integrand would in general be different for the upward and downward paths.

The quantity $\partial q/\partial U$ can be found from the quartic (6.15) using the form (6.16) for the coefficients. Since the quartic is satisfied for all values of U, we have $\mathrm{d}F(q)/\mathrm{d}U = 0$, whence

$$\frac{\partial q}{\partial U} = -\frac{q^4\dfrac{\partial \alpha}{\partial U} + q^2\dfrac{\partial \gamma}{\partial U} + \dfrac{\partial \varepsilon}{\partial U}}{4q^3\alpha + 3q^2\beta + 2q\gamma + \delta} \tag{10.37}$$

(β and δ do not contain U and therefore do not appear in the numerator). Booker (1949) has given curves showing how $\partial q/\partial U$ depends on X in some special cases. For another method of finding the attenuation see §§14.7, 14.10.

In § 10.5 it was shown that when every component wave in a wave packet is reversed in direction, the wave packet simply retraces its original path back to the transmitter. At each point of the path q, S_1 and S_2 are reversed in sign. Now (10.37) shows that $\partial q/\partial U$ is also simply reversed in sign, but the path of integration used in determining the attenuation is also reversed so that the total attenuation is the same for the original and the reversed paths. This result has an important bearing on the study of reciprocity (§§ 14.13, 14.14).

10.16. Phase path. Group or equivalent path

Let the vector $\mathrm{d}\boldsymbol{s}$, with components $\mathrm{d}x, \mathrm{d}y, \mathrm{d}z$, be an element of the path of a ray, and let the predominant wave in the ray have a refractive index vector $\boldsymbol{n}$ with components S_1, S_2, q (compare (10.4); the second subscripts 0 are now omitted). The change of phase in the path element $\mathrm{d}\boldsymbol{s}$ is

$$k\boldsymbol{n}\cdot\mathrm{d}\boldsymbol{s} = k(S_1\mathrm{d}x + S_2\mathrm{d}y + q\mathrm{d}z). \tag{10.38}$$

Suppose the ray runs from the origin to the point x, y, z. In this section of its path the change of phase is

$$k\int \boldsymbol{n}\cdot\mathrm{d}\boldsymbol{s} = kP = k\left(S_1x + S_2y + \int_0^z q\mathrm{d}z\right) \tag{10.39}$$

and P is called the phase path. Thus P/c is the time it would take a feature of the wave, such as a wave crest, to travel from the origin to the point x, y, z. The time of travel P'/c of a wave packet over the same path was found in § 10.3 and given by (10.11), whence

$$P' = \partial(fP)/\partial f = \partial(kP)/\partial k \tag{10.40}$$

($k = 2\pi f/c$). A wave packet travels with the group velocity and so P' is called the group path, or equivalent path.

The formulae (10.39) and (10.11), (10.40) are used in this book mainly for vertically incident waves, §§ 13.3–13.6, and for oblique and vertical incidence in an isotropic ionosphere, §§ 12.2–12.8. Some results from (10.39) for oblique incidence and an anisotropic ionosphere have been given by Booker (1949).

For a medium with losses, q is complex as shown in § 10.15, and so P, (10.39), is also complex. Then (10.36) shows that the amplitude of the signal is reduced by a factor

$$\exp\{k\,\mathrm{Im}(P)\} \tag{10.41}$$

as it travels along the ray. The group path P' is also complex. The physical significance of this is discussed in § 11.16.

10.17. Ray pencils

The equations (10.3) give the path of one ray in a stratified medium. But one ray cannot exist in isolation. At any point on it there must be a wave field of the type (10.4) which extends laterally, and at nearby points not on the original ray there must be other rays. Thus any ray is just one of a family or pencil of rays. The intensity of the signal on a ray depends on the convergence or divergence of neighbouring rays, but the ray tracing equations such as (10.3) give no information about this. For example in fig. 10.14 the ray shown as a thick line is exactly the same in the two figures, but the neighbouring rays have very different configurations.

Within a family of rays we can select a set that lie in the surface of a narrow tube. Since, in a loss-free medium, the energy flux at each point is in the direction of the ray, it follows that no energy can flow across the surface of such a tube. For a continuous wave system with a fixed single frequency the energy flux does not vary with time. Then the power crossing any cross section of a ray tube must be the same for all cross sections along its whole length. This property can be used to find how the signal intensity varies along a ray path. The intensity is affected by convergence or divergence of neighbouring rays in a ray pencil. The effect of convergence is

Fig. 10.14. Typical families of rays entering the ionosphere from below. In (a) the transmitter is a point source at O. In (b) the incident wave is a plane wave in the free space below the ionosphere. In practice the true situation is usually more complicated than in (a); see § 10.21 and fig. 10.16.

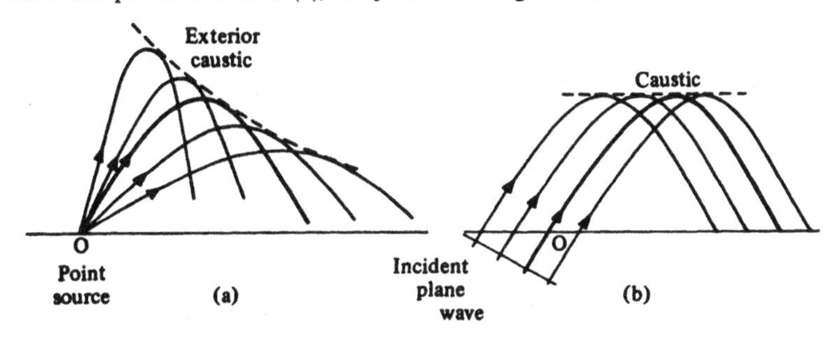

important near a focus, or near the edge of a 'skip' zone, § 10.22, or near a caustic surface, § 10.18. Various aspects of the problem have been studied, mainly for loss-free media, by Försterling and Lassen (1933), Inston and Jeffs (1968), Kelso (1964) and Rawer (1948, 1952), who gives some further references.

Equations (10.5) show that the direction cosines of a ray are proportional to

$$-\partial q/\partial S_1, \quad -\partial q/\partial S_2, \quad 1. \tag{10.42}$$

Now let

$$K_{ij} = \int_0^z \frac{\partial^2 q}{\partial S_i \partial S_j} \mathrm{d}z \quad (i,j = 1,2), \tag{10.43}$$
$$\Delta = \det(K_{ij}).$$

It will be necessary to apply these formulae both to upgoing rays and to downgoing rays that have been reflected at a higher level. It has been shown in §§ 8.20, 7.19 that for these downgoing rays the integral in (10.4) must be regarded as extending from $z = 0$ up to the reflection level $z = z_0$ and thence down to the level z; see (7.153), (7.155). This same extension is implied in the integral of (10.43) when it is used for downgoing rays.

Consider a narrow ray pencil of rectangular cross section emerging from the origin into free space, and for the four rays at its corners let S_1 and S_2 be given by (S_1, S_2), $(S_1 + \delta S_1, S_2)$, $(S_1, S_2 + \delta S_2)$, $(S_1 + \delta S_1, S_2 + \delta S_2)$. Let $\mathscr{N}_0$ be the power per unit solid angle in the pencil. Then the power flowing in the pencil is

$$\mathscr{N}_0 \delta S_1 \delta S_2 / C. \tag{10.44}$$

Since the medium is loss-free, this is the same for all cross sections of the pencil.

The pencil enters the ionosphere and cuts any plane $z = \text{constant}$, in a parallelogram whose sides are, from (10.3), (10.43), the two vectors in this plane with components

$$\left.\begin{aligned} \delta x = K_{11}\delta S_1, \quad \delta y = K_{21}\delta S_1, \\ \text{and} \qquad \delta x = K_{12}\delta S_2, \quad \delta y = K_{22}\delta S_2. \end{aligned}\right\} \tag{10.45}$$

Thus the area of the parallelogram is

$$\delta S_1 \delta S_2 \Delta \tag{10.46}$$

and the power (10.44) crosses this area obliquely in the direction given by (10.42). The time average of the energy flux within the pencil (power per unit area) in the direction of the ray is

$$\begin{aligned} \Pi &= \frac{\mathscr{N}_0 \delta S_1 \delta S_2}{C} \Bigg/ \frac{\delta S_1 \delta S_2 \Delta}{\{1 + (\partial q/\partial S_1)^2 + (\partial q/\partial S_2)^2\}^{\frac{1}{2}}} \\ &= \mathscr{N}_0 \{1 + (\partial q/\partial S_1)^2 + (\partial q/\partial S_2)^2\}^{\frac{1}{2}} / C\Delta. \end{aligned} \tag{10.47}$$

This is the required result, that allows for the divergence or convergence of neighbouring rays in a pencil. If the medium has small losses, (10.47) must be multiplied by the square of (10.41).

For an isotropic plasma q is given by (10.12). We can take $S_2 = 0$ without loss of generality. Then (10.43), with (10.14) gives

$$K_{11} = -\int_0^z \frac{n^2}{q^3}\mathrm{d}z, \quad K_{12} = K_{21} = 0, \quad K_{22} = -\int_0^z \frac{\mathrm{d}z}{q} = -x/S_1 \tag{10.48}$$

and (10.47) is

$$\Pi = \mathcal{N}_0 n/(qCK_{11}K_{22}) = \mathcal{N}_0 nS_1/(-xqCK_{11}). \tag{10.49}$$

The expressions (10.47), (10.49) are functions of the coordinates x, y, z. They can be infinite in certain conditions, and then

$$\Delta = 0, \quad \text{or } K_{11} = 0 \tag{10.50}$$

respectively. These are the equations of a surface which is the envelope of the system of rays leaving the origin. It is not difficult to prove this directly by differential geometry. At any point on this envelope, the rays in a pencil have converged to a line or point focus. Near here the methods of ray theory fail and a more elaborate full wave treatment is needed to study the signal intensity. Some examples are given in the following section.

The source may not always be at the origin. An important case is when a plane wave is obliquely incident from free space, as would occur from an infinitely distant point source. The ray pencil at the origin is then formed from parallel rays. Let the power per unit area in the direction of the ray here be Π_1. This pencil now cuts any plane $z = \text{constant}$ in an area that is independent of z, so the energy flux in the direction of the ray is, from (10.42)

$$\Pi = \Pi_1 C\{1 + (\partial q/\partial S_1)^2 + (\partial q/\partial S_2)^2\}^{\frac{1}{2}}. \tag{10.51}$$

For an isotropic medium this reduces to

$$\Pi = \Pi_1 nC/q. \tag{10.52}$$

10.18. Caustics

A surface forming the envelope of a system of rays is called a caustic surface. An example was given in § 8.22. The objective of the present section is to study how the signal from a radio transmitter varies in space when a caustic surface is present. The theory is given only for an isotropic ionosphere. The theory for an anisotropic ionosphere has not yet been fully worked out. The following version gives only the main points of the theory. For more details see Budden (1976), Maslin (1976a, b). The subject has also been studied by Kravtsov (1964a, b) Ludwig (1966), and Ziolkowski and Deschamps (1984).

Fig. 10.14(b) shows a situation where the incident wave is plane, and all rays have the same shape and the same value of S_1. They all touch a horizontal plane at the 'reflection level' where $q=0$ and the rays are horizontal. This plane is the caustic surface. According to (10.52) the signal intensity would here be infinite, but it has been shown by a 'full wave' study in § 8.20 that the height dependence of the signal is actually given by an Airy integral function, as in (8.72). Above it, although there are no real rays, the signal does not fall abruptly to zero but decays with increasing height in accordance with (7.65). The field here is an inhomogeneous wave, § 2.15, and can be thought of as containing complex rays; see § 14.11. The region where the rays are real is called the 'illuminated' side of the caustic, and the region where there are no real rays is called the 'dark' side.

Fig. 10.14(a) is an example where the rays come from a point transmitter at the origin. They have various different values of S_1 and therefore become horizontal at different levels, in each case where $q=0$. Their envelope is shown as a broken line in the figure but it is actually the cross section by the plane $y=0$ of a surface, the caustic surface. In this example it is called an exterior caustic because it is an outer boundary of the region of space where the rays are real. Another type, called an interior caustic, is possible and discussed in § 10.21, fig. 10.16.

For a point source at the origin the emitted field is given by the angular spectrum of plane waves (10.1). Here the exponential is an obliquely upgoing wave of unit amplitude. A downgoing wave reflected from higher levels may also be present but is not included in (10.1). Suppose, now, that the transmitter is a vertical Hertzian magnetic dipole of unit amplitude. Then the electric field is everywhere horizontal. In the plane $y=0$ studied here, the only non-zero electric field component is E_y. If this is chosen for F in (10.1), it can be shown (Clemmow, 1966, § 2.25) that

$$A(S_1, S_2) = S_1/C \tag{10.53}$$

where a constant multiplier $\mathrm{i}ck^3/8\pi^2$ has been omitted. Similarly, if the transmitter is a vertical Hertzian electric dipole, the only non-zero magnetic field component in the plane $y=0$ is $\mathscr{H}_y$. If $-\mathscr{H}_y$ is chosen for F in (10.1), then (10.53) is again true. Other types of transmitter and other field components can be dealt with by using the appropriate A in (10.53). In the following theory F can be interpreted either as E_y or as $-\mathscr{H}_y/n$. The need for the factor $1/n$ here is shown by (7.74).

It is now assumed that in the ionosphere the electron concentration $N(z)$ increases monotonically with height, so that each plane wave in (10.1) is reflected. The case where some waves penetrate an ionospheric layer is considered later, § 10.22. The exponential in (10.1) must be replaced by the uniform approximation (8.72) for one component plane wave. This gives, with $y=0$,

$$F = \iint 2S_1\zeta^{\frac{1}{4}}(\pi/Cq)^{\frac{1}{2}}\mathrm{Ai}(\zeta)\exp\left\{-\mathrm{i}k\left(S_1 x + \int_0^{z_0} q\,\mathrm{d}z\right)\right\}\mathrm{d}S_1\,\mathrm{d}S_2 \tag{10.54}$$

where ζ is given by (8.68). If the asymptotic form for $\mathrm{Ai}(\zeta)$ is used, this gives

$$F \approx \iint S_1(Cq)^{-\frac{1}{2}} \exp(-ikS_1x)\left\{\exp\left(-ik\int_0^z q\,dz\right) + i\exp\left(-ik\oint_0^z q\,dz\right)\right\} dS_1\,dS_2. \tag{10.55}$$

The first term is the upgoing wave, and the integral in the exponent goes from 0 up to z. The second term is the reflected wave. The symbol $\oint$ implies that the integral extends from $z = 0$ to $z = z_0$ where $q = 0$ and the ray is horizontal, and thence down to the level z with the sign of q reversed, as in (7.153). A caustic surface is nearly always an envelope of the downgoing rays, as in the example of fig. 10.14(a). Near it, therefore, the second term of (10.55) usually has to be studied. But here we shall study both terms. Note that the factor i from (8.73) is retained in (10.55).

The S_1 and S_2 integrals in (10.54), (10.55) are now to be evaluated by the method of steepest descents, ch. 9. We study first (10.55). Either of its two terms may be written

$$F = \iint S_1(Cq)^{-\frac{1}{2}} \exp(-ikp)\,dS_1\,dS_2, \quad p = S_1x + \int_0^z q\,dz \tag{10.56}$$

and the integral sign in p may be either $\int$ or $\oint$. The double integral for F requires the method of double steepest descents, §9.10, but the process is simple in this case because the saddle point for the S_2 integration is at $S_2 = 0$ and independent of S_1. The second derivative $\partial^2 p/\partial S_1 \partial S_2$ is zero. Thus the S_2 integration is done first. It uses

$$\frac{\partial^2 p}{\partial S_2^2} = K_{22} = -x/S_1 \tag{10.57}$$

from (10.43), (10.48). This integration is perhaps more conveniently thought of as using the method of stationary phase; compare §10.2, especially (10.3) second equation. The result is, from (9.31)

$$F = (2\pi)^{\frac{1}{2}} \exp\left(\tfrac{1}{4}i\pi\right) \int S_1^{3/2}(kxCq)^{-\frac{1}{2}} \exp(-ikp)\,dS_1 \tag{10.58}$$

where C, q, p now take their values for $S_2 = 0$.

Finally the S_1 integration must be done. When the saddle points in the S_1 plane are well separated it is sufficiently accurate to use the first order steepest descents result (9.31) at each. This is equivalent to using ray theory. Where two saddle points coincide there is a caustic and for points near it an extension of the method of steepest descents is needed, as in §10.19. Where three saddle points coincide there is a cusp; see §10.21. These properties were used by Maslin (1976a, b) to trace the caustic curves in the plane $y = 0$ and to find the cusps. He showed that, for a monotonic electron height distribution $N(z)$ and for the first term in (10.55), that is for upgoing waves, there are no caustics or cusps. These occur only for the second term, that is for downgoing waves. They can occur, however, for upgoing waves that have penetrated a lower ionospheric layer. The number of contributing saddle points in (10.58) depends on the coordinates x, z. Maslin (1976a, b) showed that it can be as many as five in practical cases.

We now find the contribution to (10.58) from one isolated saddle point. It must occur where $\partial p/\partial S_1 = 0$ which gives $x = S_1 \int_0^z q^{-1} \mathrm{d}z$. This is the same as (10.14). It shows that the saddle point value of S_1 is the value for the ray through the point (x, z). Let this value be S_a and let $q = q_\mathrm{a}$, $C = C_\mathrm{a}$ for $S_1 = S_\mathrm{a}$. The second derivative of p when $S_1 = S_\mathrm{a}$ is, from (10.43), (10.49)

$$\partial^2 p/\partial S_1^2 = K_{11} = -\int_0^z (n^2/q_\mathrm{a}^3)\,\mathrm{d}z. \tag{10.59}$$

Then the contribution to (10.58) from the saddle point is, from (9.31):

$$F = 2\pi \mathrm{i} k^{-1} S_\mathrm{a}^{\frac{3}{2}} (-C_\mathrm{a} q_\mathrm{a} K_{11} x)^{-\frac{1}{2}} \exp\left\{-\mathrm{i}k\left(S_\mathrm{a} x + \int_0^z q_\mathrm{a}\,\mathrm{d}z\right)\right\}. \tag{10.60}$$

This is the field F on a ray that is not near a caustic.

A typical situation is sketched in fig. 10.15. The curve PQRST is a ray that touches the caustic at S. At P it is obliquely upgoing so that (10.60) uses the first term of (10.55). Here q_a is real and positive and (10.59) shows that K_{11} is real and negative. Thus the factor $(-q_\mathrm{a} K_{11})^{-\frac{1}{2}}$ in (10.60) is real and positive. For the second derivative (9.27) of the exponent, $\alpha = \frac{1}{2}\pi$ so that the last factor of (9.29) is $\exp(\frac{1}{4}\mathrm{i}\pi)$. This has been combined with the same factor in (10.58) to give the factor i in (10.60).

The approximation for $\mathrm{Ai}(\zeta)$ used in (10.55) fails near $q = 0$ so that it might be expected that (10.60) fails at Q where $q_\mathrm{a} = 0$. Here (10.59) shows that K_{11} is infinite. Although q_a and K_{11} both have branch points where $q_\mathrm{a} = 0$, it is easy to show that the product $q_\mathrm{a} K_{11}$ is analytic and non-zero there. When Q is passed and we move to R, both q_a and K_{11} have changed sign so q_a is negative and K_{11} is positive. Now in

Fig. 10.15. The continuous curves are two of a family of rays that touch a caustic surface, shown as a broken line.

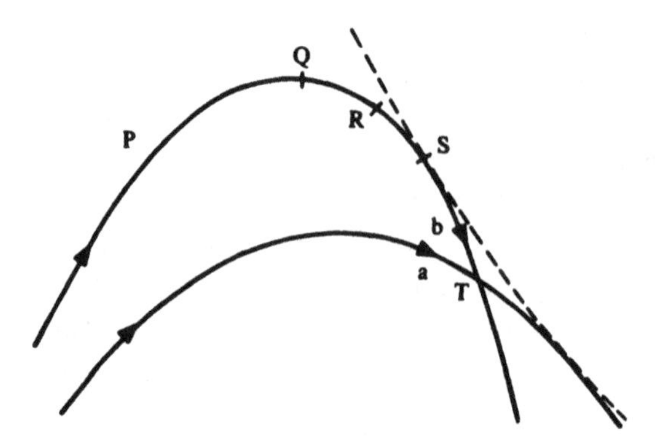

(9.27) α is $3\pi/2$ and the last factor of (9.31) is $\exp(-\frac{1}{4}\mathrm{i}\pi)$. But now the ray is obliquely downgoing and the factor i from the second term of (10.55) must be included. This finally gives a factor $\exp(+\frac{1}{4}\mathrm{i}\pi)$ as before, and (10.60) is still correct for points such as R. It is continuous for points near Q and it turns out that it is still correct there, in spite of the failure of (10.55). This can be proved by the method outlined in § 10.19.

At S in fig. 10.15, two saddle points S_a coincide. This requires that $\partial^2 p/\partial S_1^2 = 0$ so that $K_{11} = 0$ from (10.59); compare (10.50). But q_a is bounded so (10.60) is infinite. Here the simple form of steepest descents fails, so that (10.60) is not valid even though (10.55) is a good approximation.

At T on the same ray K_{11} is again negative. The result (10.60) is valid provided that the extra factor i from the second term of (10.55) is included. It is not now cancelled because of the last term of (9.29). It represents the $\frac{1}{2}\pi$ phase advance that occurs in a ray that has previously touched a caustic surface.

10.19. The field where the rays are horizontal

When the approximation used for (10.55) does not apply, that is near $q = 0$, the more accurate form (10.54) must be used. The algebra for this case, given in full by Budden (1976), is lengthy and only the main results are given here.

For the function Ai in (10.54) we use its integral representation (8.16). This gives

$$F = -\mathrm{i}\iiint S_1 \zeta^{\frac{1}{4}}(\pi C q)^{-\frac{1}{2}} \exp\left\{-\mathrm{i}k\left(S_1 x + \int_0^{z_0} q\,\mathrm{d}z\right) + \zeta t - \frac{1}{3}t^3\right\}\mathrm{d}S_1\,\mathrm{d}S_2\,\mathrm{d}t. \tag{10.61}$$

This is a triple integral but, as for (10.56), the S_2 integration is simple and can be done first by the method of stationary phase; compare (10.58). The result is

$$F = \exp(-\tfrac{1}{4}\mathrm{i}\pi)(2/kx)^{\frac{1}{2}}\iint S_1^{3/2}\zeta^{\frac{1}{4}}(Cq)^{-\frac{1}{2}}\exp(-\mathrm{i}kP + \zeta t - \tfrac{1}{3}t^3)\mathrm{d}S_1\,\mathrm{d}t \tag{10.62}$$

where

$$P = S_1 x + \int_0^{z_0} q\,\mathrm{d}z. \tag{10.63}$$

The integral (10.62) is done by the method of double steepest descents, §9.10. The double saddle points are where $\partial/\partial S_1$ and $\partial/\partial t$ of the exponent are zero. It can be shown that these lead to the first equation (10.14) for finding S_1. This therefore has the value S_a for the ray that goes through the point (x, z).

If the double saddle points are well separated in S_1–t space, the first order double steepest descents formula (9.64) can be used for each. It can thence be shown that the contribution from one double saddle point is (10.60). This result now applies even for a point such as Q in fig. 10.15 where the ray is horizontal. It supplies the proof mentioned near the end of § 10.18.

The W.K.B. solutions of ch. 7 were for the fields in a stratified ionosphere when a plane wave was incident from below it. These solutions failed at a level where the rays were horizontal and touched a horizontal caustic surface. In (10.60) we now have a very important new result. It is a generalisation of the W.K.B. solution and applies when the waves originate from a point source not at infinity. Like the conventional W.K.B. solution it is approximate because its derivation used the first order steepest descents formulae. It fails where the rays touch a caustic surface. But it remains valid where the rays are horizontal provided this occurs at points not near a caustic.

10.20. The field near a caustic surface

Near a caustic surface the integrand of (10.62) has two double saddle points close together so the first order double steepest descents formula cannot be used. For a single integral such as (9.1) with two saddle points close together, the method of Chester, Friedman and Ursell (1957), §9.9, can be used. For the double integral (10.62) an extension of this has been suggested (Budden, 1976). The full theory has been discussed by Ursell (1980). The results are as follows.

Consider a point such as T in fig. 10.15 where two rays intersect. Let subscripts a, b be used to indicate values at T from these two rays, a for the ray that has not yet touched the caustic and b for the ray that has touched it. Thus P_a, P_b are the two phase paths, from (10.39) with $y=0$, and S_a, S_b are the two values of S_1, etc. Now define

$$\xi = \{\tfrac{3}{4}ik(P_a - P_b)\}^{\frac{2}{3}} \tag{10.64}$$

where the fractional power is chosen so that ξ is real and negative when $P_b - P_a$ is real and positive. This should be compared with (8.68) for ζ. If the source is at infinity so that the incident rays a and b are parallel and $S_a = S_b$, then $\xi = \zeta$. Let

$$g_a = ik^{-1}S_a^{3/2}(-C_a q_a K_{11a} x)^{-\frac{1}{2}}, \quad g_b = ik^{-1}S_b^{3/2}(C_b q_b K_{11b} x)^{-\frac{1}{2}}. \tag{10.65}$$

$$V = \xi^{\frac{1}{4}}(g_b + g_a), \quad W = \xi^{-\frac{1}{4}}(g_b - g_a). \tag{10.66}$$

Note that the factor g_a appears in (10.60). Then it can be shown that

$$F = 2\pi^{\frac{1}{2}}\{V\,\mathrm{Ai}(\xi) + W\,\mathrm{Ai}'(\xi)\}\exp\{-\tfrac{1}{2}ik(P_a + P_b)\}. \tag{10.67}$$

If the point T is beyond the caustic in fig. 10.15, there are no real rays through it. But the quantities S_a, S_b, P_a, P_b, etc. can still be defined. They are now complex and we say that complex rays can reach points beyond the caustic; see §14.11. It can be shown that ξ is a continuous real function for points x, z on both sides of the caustic. It is zero on the caustic and positive for points beyond it. Thus (10.67) is uniformly valid in this region. It is the analogue of the 'uniform approximation' solution (8.72) that applied for an incident plane wave.

Finally, for a point on the illuminated side of the caustic where rays are real and ξ is negative, suppose that $|\xi| \gtrsim 1$ and use the asymptotic forms (8.53), (8.55) for Ai, Ai′

in (10.67), and use (10.64) for ξ. Then

$$F \approx 2\pi\{g_a \exp(-ikP_a) + ig_b \exp(-ikP_b)\}. \tag{10.68}$$

The first term is the same as (10.60) for ray a and the second is (10.60) for ray b. The factor i shows the phase advance of $\frac{1}{2}\pi$ already explained at the end of § 10.18.

For a point on the dark side of the caustic the asymptotic forms (8.52), (8.54) would be used, and there would be only one term in (10.68). It would represent the inhomogeneous wave mentioned in § 10.18, and its amplitude would decrease rapidly with increasing distance form the caustic.

For a discussion of the above theory and for results of some calculations closely related to it, see Warren, de Witt and Warber (1982).

10.21. Cusps. Catastrophes

When a family of rays enters the ionosphere from below, the configuration is often more complicated than implied in fig. 10.14(a). A typical situation is sketched in fig. 10.16. The caustic surface is in two parts. One part is the same as the exterior caustic of fig. 10.14(a). But there is another branch which is touched by the rays on the side remote from the transmitter, and it is therefore called an interior caustic. The two branches meet in a point in fig. 10.16, but this is a cross section. The caustics are actually surfaces and they meet in a line, called a cusp line. It can be shown that near this line three saddle points of the integral (10.58) are close together. Both the simple

Fig. 10.16. Example of rays from point transmitter entering the ionosphere and shown as continuous curves. Their envelope, shown as a broken curve, is a caustic with interior and exterior branches meeting in a cusp.

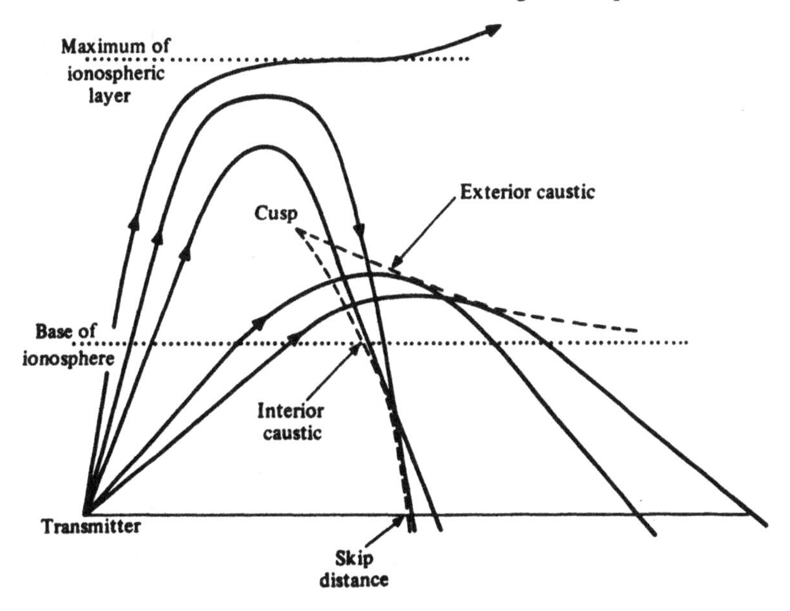

ray theory and the theory of §§ 10.18–10.20 are inadequate and a still more elaborate theory is needed. This has been given for the radio wave problem by Budden (1976). It is found that the Airy integral used near a caustic must now be replaced by a more complicated function given by Pearcey (1946), and sometimes called the 'cusp function'; (see also Maslin, 1976b). Caustics and cusps have played some part in the study of guided waves; see Hartree, Michel and Nicolson (1946), Brekhovskikh (1960).

Caustics and cusps are examples of a class of focusing phenomena that occur in physics, known as catastrophes; see, for example, Poston and Stewart (1978). They are important when ray theory is used to study the propagation of radio waves through a medium containing irregularities of refractive index. The caustic is the only catastrophe that is likely to be often encountered by practical radio engineers, and it is the only one described in any detail in this book.

10.22. The skip distance

Fig. 10.16 is typical for rays from a point transmitter entering an ionospheric layer in which the plasma frequency at the maximum is less than the frequency of the wave. Thus rays near vertical incidence penetrate the layer. If a ray is to return to the ground it must be obliquely incident. It follows that there is a minimum distance D_s from the transmitter within which no ray travelling via the ionosphere can reach the ground. Fig. 10.16 shows that this distance D_s is where the interior caustic meets the ground. Near it we expect the signal amplitude to be given by an Airy integral function, as in (10.67). This distance is called the 'skip' distance. Just beyond it two rays reach the ground and interact to give an interference pattern, expressed by the oscillations of the Airy integral function. These rays are in free space and are straight and the caustic is curved. It was shown in §8.22 that the signal amplitude along a line

Fig. 10.17. Configuration of downcoming rays near the skip distance. The thick curve is the interior caustic. The rays would actually be reflected by the earth, but in this diagram the effect of the earth is ignored.

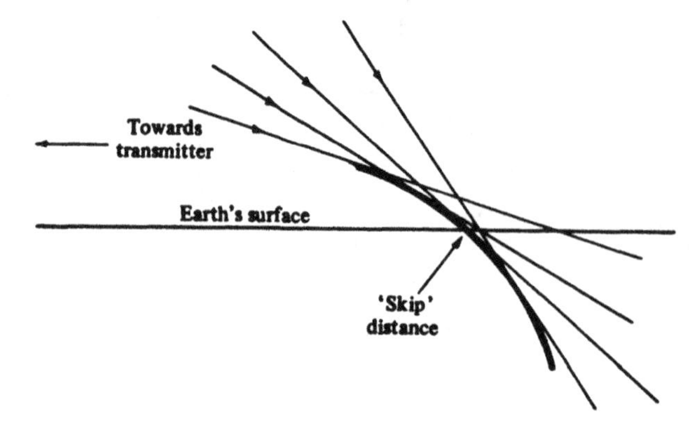

at right angles to the caustic is proportional to

$$\mathrm{Ai}(\zeta) \text{ with } \zeta = 2u(\pi^2/R\lambda^2)^{\frac{1}{3}} \tag{10.69}$$

where u is distance measured perpendicular to the caustic, λ is the wavelength and R is the radius of curvature of the caustic.

The configuration of the rays reaching the ground near the skip distance D_s is sketched in fig. 10.17. Let the particular ray that reaches the gound at range D_s make an angle θ_0 with the vertical. Rays at neighbouring angles θ all reach the ground at greater distances D. It can easily be shown that $R = (\mathrm{d}^2D/\mathrm{d}\theta^2)\cos\theta_0$. If x is distance measured along the ground from the skip distance, towards the transmitter, then $u = x\cos\theta_0$ and the expression in (10.69) for ζ is

$$\zeta = 2x\left(\frac{\pi^2\cos^2\theta_0}{\lambda^2\mathrm{d}^2D/\mathrm{d}\theta^2}\right)^{\frac{1}{3}}. \tag{10.70}$$

Typical values of θ_0 and $\mathrm{d}^2D/\mathrm{d}\theta^2$ can be found from formulae given later, § 12.7, and used to assess the scale of the function $\mathrm{Ai}(\zeta)$. It is found that it is of order $\mathrm{Ai}(0.5x)$ to $\mathrm{Ai}(1.5x)$ where x is in km. Thus the scale of the interference pattern just beyond the skip distance is of order one or two kilometres.

If the receiver is at a fixed distance D from the transmitter, there is some frequency f_M that makes $D_s = D$, and it is called the 'maximum usable frequency'; see §§ 12.8–12.10. For other frequencies, x in (10.70) is $D_s - D$ and approximately

$$x = (f - f_M)\mathrm{d}D_s/\mathrm{d}f. \tag{10.71}$$

If the transmitted frequency f is changed slowly, the received signal amplitude varies as $\mathrm{Ai}(\zeta)$ where

$$\zeta = 2(f - f_M)\frac{\mathrm{d}D_s}{\mathrm{d}f}\left(\frac{\pi^2\cos^2\theta_0}{\lambda^2\mathrm{d}^2D/\mathrm{d}\theta^2}\right)^{\frac{1}{3}}. \tag{10.72}$$

Some properties of this formula have been discussed by Budden (1961a, § 11.12). The value of $\mathrm{d}D_s/\mathrm{d}f$ can be estimated from formulae discussed later, § 12.8.

In practical cases $\mathrm{d}D_s/\mathrm{d}f$ is usually positive. Then if $f < f_M$ two rays reach the receiver, and ζ is negative. If f increases, the signal amplitude oscillates because of the maxima and minima of $\mathrm{Ai}(\zeta)$, which is an interference pattern formed by the two rays. Eventually f may exceed f_M and ζ becomes positive. Then the signal decays and disappears. Similar effects are observed with a fixed frequency because the ionosphere changes with time and f_M changes. The oscillatory effect in the amplitude of a signal that is about to fade right out is well known to radio operators who work near the maximum usable frequency.

10.23. Edge focusing

A radio signal coming in from outside the earth is refracted by the ionosphere. For very high frequencies the refractive index n of the ionosphere is very close to unity and the refraction is negligible, but for smaller frequencies $|n-1|$ can be large enough to be important. The subject is of some interest in radio astronomy. We here consider only the case where the source is effectively at an infinite distance so that the rays reaching the earth from it are all parallel before they enter the ionosphere. For a study of the rays from a source at a finite distance from the earth see, for example, Rawer (1962).

In the top side of the ionosphere, that is above the maximum of the F2-layer, it can be assumed that the electron concentration $N(r)$ is a monotonically decreasing function of the radius r from the centre of the earth. The refraction of the rays is studied by ray tracing in the spherically stratified ionosphere. The frequencies of interest are so great that the earth's magnetic field and electron collisions may be neglected. The equations for the ray path were given in § 10.4. Spherical polar coordinates r, θ, ϕ are used. The receiver is assumed to be in the top side of the ionosphere on the line $\theta = 0$ at radius $r = a$. The plane $\phi = 0$ is chosen so that the direction of the source is in it, at $\theta = \theta_s$.

A ray coming in to the ionosphere is deviated away from the radial direction, and it may become horizontal and then travel outwards again. A possible situation is sketched in fig. 10.18. Here the rays all touch an external caustic surface shown as a broken line. An internal caustic could also occur but is not considered here; see fig. 10.16. For a receiver at a position such as A in fig. 10.18 two different real rays can reach it. If it moves to B, these two rays have moved to coincidence and B is on the caustic. For a receiver at C no real rays can reach it because it is in the shadow zone. Near the caustic on the illuminated side there is a concentration of rays giving a focusing effect, and this has been called 'edge focusing'.

Fig. 10.18. Rays from an infinitely distant source enter the upper ionosphere and touch a caustic surface, shown as a broken line.

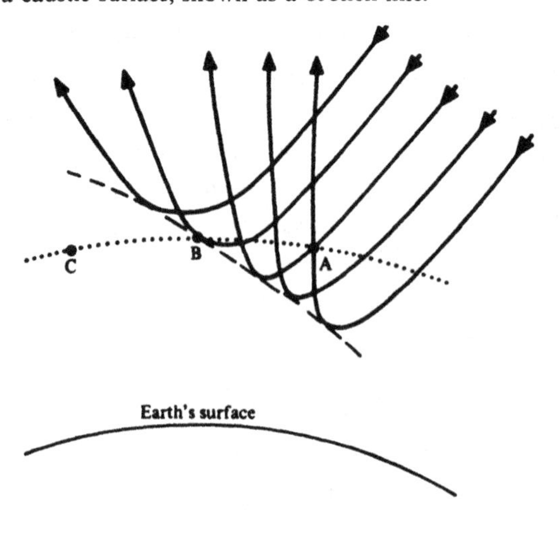

A receiver in the ionosphere must be on a space vehicle that moves, so that the source direction θ_s observed at the receiver changes with time. It is required to find how the signal amplitude depends on θ_s for a fixed value of the receiver height a. The full theory for a collisionless isotropic ionosphere was given by Budden (1961c). Only an outline of the method and the result are given here.

For an inward coming ray in the ionosphere let ψ be the angle between its direction and the inward radius, and let ψ_A be the value of ψ at the receiver. Let n_0 be the refractive index at the receiver. Then Bouger's law (10.18) gives

$$rn \sin\psi = an_0 \sin\psi_a = K. \tag{10.73}$$

The differential equation of a ray path is (10.19). For a ray going through the receiver this may be integrated with respect to r to give

$$\theta_s = an_0 \sin\psi_a \, ⨍_a^\infty \frac{dr}{r^2 q} \tag{10.74}$$

where

$$q = n\cos\psi = (n^2 - n_0^2 \sin^2\psi_a a^2/r^2)^{\frac{1}{2}}. \tag{10.75}$$

If $\cos\psi_a$ is positive, the ray at the receiver is incoming and the integral in (10.74) extends along the real r axis from a to ∞. But if $\cos\psi_a$ is negative, the ray at the receiver is outgoing, because it was horizontal where $r = r_1 < a$. The first part of the path of integration in (10.74) must therefore run from a down to r_1 and on this part $\cos\psi$ and q are negative. The second part runs from r_1 to ∞ and on it $\cos\psi$ and q are positive. The symbol $⨍$ is used to indicate that the correct choice of path must be made. Some properties of the integral (10.74) were discussed by Daniell (1966) who showed that it is convenient to treat it as a contour integral in the complex r plane.

Since n^2 is a known function of r, (10.74) can be evaluated to give an equation relating θ_s and ψ_a. A specific example was given by Budden (1961c). The form of the

Fig. 10.19. Relation between direction θ_s of source, and inclination ψ_a of the ray at the receiver to the inward radius.

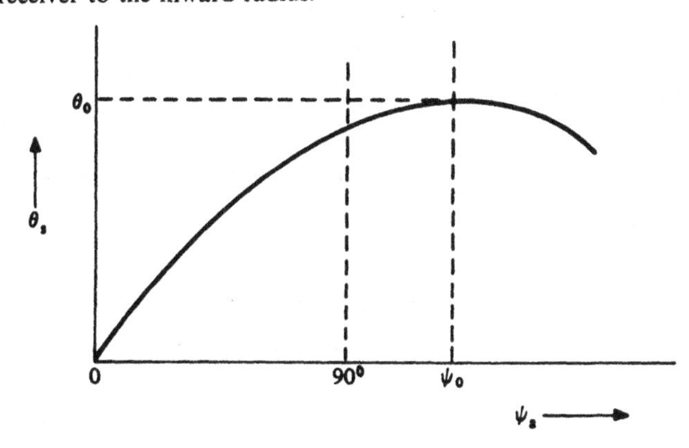

result is sketched in fig. 10.19. It shows that θ_s has a maximum value θ_0, which is the source direction that makes the caustic go through the receiver. It occurs where $\psi_a = \psi_0$. By differentiating (10.74) with respect to ψ_a it can be shown that $\psi_0 > \frac{1}{2}\pi$. This means that only the outgoing parts of the rays can touch the caustic. The corresponding result for a flat earth was mentioned in § 10.18 (Maslin, 1976a, b). Where θ_s is a maximum let

$$d^2\theta_s/d\psi_a^2 = -2\gamma. \tag{10.76}$$

Now let the source emit linearly polarised waves with the electric field E perpendicular to the plane $\phi = 0$. To find the field E in the immediate neighbourhood of the receiver Budden (1961c) assumed that, in a sufficiently small region, the earth's curvature could be neglected and that n^2 varied linearly with radius thus

$$n^2 \approx n_0^2 + \alpha(r - a). \tag{10.77}$$

The differential equation satisfied by E in this region was then formulated and solved. This showed finally that when θ_s is near to θ_0 the signal amplitude is proportional to

$$\mathrm{Ai}\{-(k/\alpha)^{\frac{2}{3}} n_0^2 (\sin\psi_0)^{\frac{1}{3}}(\theta_0 - \theta_s)/\gamma\}. \tag{10.78}$$

It was at one time suggested that edge focusing might be used for detecting a radio star source from the large signal that occurs when θ_s is near θ_0. But the maximum value of $|\mathrm{Ai}|$ is not very large; see fig. 8.5. The term 'focusing' is perhaps misleading. The method has not proved to be practicable.

PROBLEMS 10

10.1. Show that for a radio wave travelling vertically upwards in a horizontally stratified loss-free ionosphere, the time averaged Poynting vector makes with the vertical an angle whose tangent is

$$\frac{\mathrm{i}Y\rho l_x(1 - n^2)}{(1 - X)(\rho^2 - 1)}$$

(use (4.58) for a loss-free plasma). By using ρ from (4.20) show that this is the same as (10.29).

10.2. For an extraordinary ray vertically incident on a horizontally stratified loss-free ionosphere, show that at the reflection level the angle between the ray and the vertical is $\arctan\{\sin 2\Theta/(3 + \cos 2\Theta)\}$ where Θ is the angle between the earth's magnetic field and the vertical.
(See Millington, 1949, 1951).

10.3. In a spherically stratified isotropic ionosphere the refractive index $n(r)$ is a function only of radius r from the earth's centre. A ray at radius r makes an angle ψ with the radius. Prove Bouger's law that $rn \sin\psi$ is constant on the ray.
(Hint: prove it first for the straight rays on the two sides of a spherical boundary

between two homogeneous media. Then extend it to a succession of boundaries and finally to the limit of a continuously varying medium.

Another method is to use the general ray tracing equations (14.33), (14.34).

A third method is to use Fermat's principle.)

10.4. A continuous radio wave from a point source on the ground is vertically incident on a horizontally stratified isotropic loss-free ionosphere. Show that at any height, where the total equivalent path travelled from the ground is P', the signal intensity is proportional to $(P')^{-2}$ for both upgoing and downgoing (reflected) waves.
(Use (10.48), (10.49)).

10.5. An ordinary ray is incident obliquely on a collisionless ionosphere, in the magnetic meridian plane at an angle of incidence arcsin $\{l_x(Y+1)^{\frac{1}{2}}\}$ (see (6.32) and § 10.9). When it reaches the level $X = 1$, it is at a window point. It continues obliquely upwards as an extraordinary ray, but there is also a reflected ordinary ray of infinitesimally small amplitude travelling downwards. (See fig. 10.8(a).) Show that these two rays are inclined to the earth's magnetic field at angles ψ given by

$$\cot\psi = 2\cot\Theta[1 + Y \pm \{(1+Y)^2 + \tfrac{1}{2}(1+Y)\tan^2\Theta\}^{\frac{1}{2}}]$$

where Θ is the angle between the earth's magnetic field and the vertical ($l_x = \sin\Theta$). What happens to these rays (a) at a magnetic pole, (b) at the magnetic equator?

10.6. A ray goes obliquely from a transmitter on the ground at (0, 0, 0) into a horizontally stratified ionosphere. It is reflected and returns to a receiver on the ground at $(x, y, 0)$. When the ionosphere is loss-free the ray, where it leaves the transmitter, has x and y direction cosines S_1, S_2 respectively. The function $q(z)$ on the ray has a branch point at the reflection level $z = z_0$. Show that the phase path for the whole ray is

$$P = S_1 x + S_2 y + \int_C q\,dz.$$

where the contour C is as used in (8.78). When there are no collisions, S_1, S_2 and P are real and there is no attenuation. A collision frequency is now introduced so that Z is small at all heights. Then the ray path is perturbed. Prove that the attenuation, in nepers, from (10.41) is now

$$-k\,\mathrm{Im}(P) \approx k\int_C Z\frac{\partial q}{\partial U}dz.$$

This is the same as if S_1, S_2 are unchanged, and only $q(z)$ is changed. In fact the perturbed ray is complex and S_1, S_2 have complex increments $\delta S_1, \delta S_2$. The proof must take account of this.

10.7. For a radio wave travelling vertically upwards into a horizontally stratified loss-free ionosphere, show that the ordinary wave, at its reflection level, is linearly polarised with its electric vector parallel to the earth's magnetic field $\boldsymbol{B}$, and its magnetic vector horizontal and perpendicular to $\boldsymbol{B}$. Show that for the extraordinary wave, at its reflection level, the electric vector describes a circle in a plane at right angles to $\boldsymbol{B}$.

10.8. A radio ray travels in an isotropic medium whose refractive index n is a function only of the radius r measured from a fixed centre. By using Fermat's principle, or otherwise, show that the path of the ray satisfies

$$\frac{\mathrm{d}r}{\mathrm{d}\theta} = \frac{r}{K}(r^2n^2 - K^2)^{\frac{1}{2}}$$

where K is a constant and r, θ are spherical polar coordinates with respect to an axis in the plane of the ray.

For radio rays of a certain frequency, the refractive index of the earth's outer atmosphere is given by $n^2 = 1 - a^2/r^2$, where a is a constant. A ray approaches the earth along a line which, if produced, would pass at a distance b from the earth's centre. Show that the ray is deviated through an angle

$$\pi\{1 - b(a^2 + b^2)^{-\frac{1}{2}}\}.$$

11

Reflection and transmission coefficients

11.1. Introduction

The later chapters are largely concerned with the derivation and the properties of the reflection coefficient of a stratified ionosphere, and to a lesser extent with the transmission coefficient. The purpose of this chapter therefore, is to set out their definitions and to discuss some of their properties in special cases of importance in radio propagation.

The coordinate system is as defined in §6.1. The z axis is perpendicular to the strata of the ionosphere, so that all properties of the plasma, especially the electron concentration $N(z)$ and collision frequency $\nu(z)$, are functions only of z. The origin is often taken to be at the ground so that z is height above the earth's surface. Sometimes, however, it is convenient to use a different origin.

Up to §11.11 it is assumed that the incident wave is plane and travelling upwards, in general obliquely with its wave normal in the x–z plane at an angle θ to the z axis, measured clockwise. Then the x–z plane is called the 'plane of incidence'. If F is any field component of the incident wave then in the free space below the ionosphere

$$F = F_1 \exp\{-ik(Sx + Cz)\} \tag{11.1}$$

where $S = \sin\theta$, $C = \cos\theta$ and F_1 is a constant, in general complex.

11.2. The reference level for reflection coefficients

Consider, first, a wave vertically incident on the ionosphere, so that $\theta = 0$, and let some field component be given, in the free space below the ionosphere, by

$$F = F_1 \exp(-ikz) \tag{11.2}$$

where z is height above the ground, and F_1 is a complex constant. This gives rise to a reflected wave travelling downwards for which

$$F = F_2 \exp(ikz) \tag{11.3}$$

where F_2 is another complex constant. If the ratio of these two field components is measured at the ground the result is

$$R_0 = F_2/F_1, \tag{11.4}$$

and the complex number R_0 is called the reflection coefficient 'at the ground' or 'with the ground as reference level', or 'referred to the level $z = 0$'. The particular field components F are specified later, § 11.4.

If the ratio were measured at some level $z = z_1$, the result would be

$$R_1 = \{F_2 \exp(ikz_1)\}/\{F_1 \exp(-ikz_1)\} = R_0 \exp(2ikz_1) \tag{11.5}$$

where R_1 is the reflection coefficient referred to the level $z = z_1$. Thus the effect of changing the reference level is to alter the argument but not the modulus of the reflection coefficient. Similarly, the reflection coefficient referred to some other level z_2 is

$$R_2 = R_0 \exp(2ikz_2) = R_1 \exp\{2ik(z_2 - z_1)\}. \tag{11.6}$$

This gives the rule for changing the reference level.

The downgoing wave (11.3) may arise by reflection in an ionosphere in which N varies gradually with height and at a level, $z = z_3$ say, where N is appreciably different from zero, the upgoing and downgoing waves are no longer given by (11.2) and (11.3). Yet it is still possible to use the level $z = z_3$ as the reference level, and to write

$$R_3 = \{F_2 \exp(ikz_3)\}/\{F_1 \exp(-ikz_3)\} = R_0 \exp(2ikz_3). \tag{11.7}$$

Here R_3 is the reflection coefficient referred to the level $z = z_3$ and is calculated *as though this level were in free space*, even though the ratio in (11.7) could never be observed.

For many purposes the ground is the best reference level since this is where the reflection coefficients are measured. But in some cases other levels are more convenient. For example, in §§ 11.7–11.11 where the ionosphere is assumed to be a sharply bounded homogeneous medium, the reference level is taken at the boundary.

Suppose next that the incident wave is oblique so that F is given by (11.1). Then the reflected wave is

$$F = F_2 \exp\{-ik(Sx - Cz)\}. \tag{11.8}$$

Both incident and reflected waves depend on x through the same factor $\exp(-ikSx)$, see § 6.2. The reflection coefficient is defined to be the ratio of these fields *measured at the same point*, whose coordinates are (x, z_1), say, so that

$$R_1 = \frac{F_2}{F_1} \exp(2ikCz_1). \tag{11.9}$$

Thus R_1 is independent of the horizontal coordinate x, but it is still necessary to specify the reference level for the vertical coordinate. If this is changed to a new level z_2 then the new reflection coefficient R_2 is

$$R_2 = R_1 \exp\{2ikC(z_2 - z_1)\}. \quad (11.10)$$

If C is real, the effect of changing the reference level is to alter the argument but not the modulus of the reflection coefficient. Problems sometimes arise in which C is complex, for example when the space between the earth and the ionosphere is treated as a wave guide (Budden 1961b). The incident and reflected waves are then inhomogeneous plane waves (§ 2.15) and changing the reference level can change both the modulus and argument of the reflection coefficient.

In observations at oblique incidence the incident and reflected waves are usually measured at different points on the earth's surface. Suppose that these are separated by a horizontal distance D. Then the ratio of the fields in the reflected and incident waves at the ground is

$$R_D = \{F_2 \exp(-ikSD)\}/F_1 = R_0 \exp(-ikSD). \quad (11.11)$$

11.3. The reference level for transmission coefficients

Suppose that there is free space above the ionosphere and that the wave (11.1) incident from below gives rise to a wave above the ionosphere whose field component F is

$$F = F_3 \exp\{-ik(Sx + Cz)\}. \quad (11.12)$$

These two waves could never be observed at the same point because the ionosphere intervenes between them. The field (11.12), however, can be extrapolated backwards to some level $z = z_1$ below the top of the ionosphere, as though the wave were travelling in free space. Similarly, the field (11.1) can be extrapolated upwards to the same level $z = z_1$. The transmission coefficient T is defined to be the ratio of the fields (11.12) to (11.1) extrapolated in this way to the same point. Thus

$$T = F_3/F_1. \quad (11.13)$$

This ratio is independent of x and z, so that in this example it is unnecessary to specify a reference level for the transmission coefficient.

In some problems however there is, above the ionosphere, a homogeneous medium which is not free space. Then the transmitted field is no longer given by (11.12) but by

$$F = F_3 \exp\{-ik(Sx + qz)\} \quad (11.14)$$

where q is a constant different from C. This field is now extrapolated down to the level $z = z_1$, as though the wave were travelling in the same homogeneous medium throughout, and the transmission coefficient is

$$T = \{F_3 \exp(-ikqz_1)\}/\{F_1 \exp(-ikCz_1)\} = \frac{F_3}{F_1} \exp\{ik(C - q)z_1\}. \quad (11.15)$$

Since this depends on z_1, the reference level must now be specified. Often q is complex even when C is real, so that a change of reference level changes both the modulus and the argument of T.

11.4. The four reflection coefficients and the four transmission coefficients

The most important reflection and transmission coefficients are those deduced when the incident wave is linearly polarised with its electric vector either parallel to the plane of incidence or perpendicular to it. The reflected wave, in general, is elliptically polarised, but may be resolved into linearly polarised components whose electric vectors also are parallel and perpendicular to the plane of incidence. It is convenient to introduce four coefficients $R_{11}, R_{12}, R_{21}, R_{22}$, each equal to the complex ratio of a specified electric field component in the wave after reflection to a specified electric field component in the wave before reflection. The second subscript denotes whether the electric field component specified in the incident wave is parallel, 1, or perpendicular, 2, to the plane of incidence, and the first subscript refers in the same way to the electric field component in the reflected wave.

It is necessary to adopt some convention with regard to sign of the electric fields of the linearly polarised components. For the component with $\boldsymbol{E}$ perpendicular to the plane of incidence the electric field is E_y and its sign is the sign of E_y. For the component with $\boldsymbol{E}$ in the plane of incidence the sign of E is taken to be the sign of $\mathscr{H}_y$.

This sign convention means, for example, that if $R_{11} = 1$ at normal incidence, the electric fields E_x in the incident and reflected waves have opposite directions because the magnetic fields $\mathscr{H}_y$ have the same direction. On the other hand if $R_{22} = 1$ at normal incidence, the electric fields E_y in the incident and reflected waves have the same direction but the magnetic fields $\mathscr{H}_x$ have opposite directions.

These definitions are based on the way in which most radio propagation engineers visualise the incident and reflected waves. They were first formulated in a similar form for the study of the ionospheric reflection of radio waves of very low frequency (Bracewell *et al.*, 1951). When the waves are in the free space below the ionosphere the component of the electric field of one wave in the plane of incidence, that is the x–z plane, is equal to $\mathscr{H}_y$. This is shown from equations (4.2), (4.4), (4.6) by transforming them to the axis system of the present chapter, with $n = 1$. It is therefore useful to use $\mathscr{H}_y$ instead of the electric field component in the plane of incidence. This leads to the following expressions for the reflection coefficients. The superscripts (I), (R), (T) are used to denote the incident, reflected and transmitted waves respectively. Let $E_y^{(I)} = \mathscr{H}_x^{(I)} = 0$. Then

$$R_{11} = \mathscr{H}_y^{(R)}/\mathscr{H}_y^{(I)}, \quad R_{21} = E_y^{(R)}/\mathscr{H}_y^{(I)}. \tag{11.16}$$

Let $E_x^{(I)} = \mathscr{H}_y^{(I)} = 0$. Then

$$R_{12} = \mathscr{H}_y^{(R)}/E_y^{(I)}, \quad R_{22} = E_y^{(R)}/E_y^{(I)}. \tag{11.17}$$

If the incident wave is in a plasma that is not free space, the relation between $\boldsymbol{E}$ and $\mathscr{H}$ is different. For an isotropic plasma, if n is the refractive index, $\mathscr{H}_y$ is n times the component of $\boldsymbol{E}$ in the x–z plane. For an anisotropic plasma the waves are in general combinations of ordinary and extraordinary components and the relation between $\mathscr{H}_y$ and $\boldsymbol{E}$ is complicated and depends on θ. It is convenient and simplest now to retain the definitions (11.16), (11.17), but it must be remembered that they do not now in general use the component of $\boldsymbol{E}$ in the plane of incidence.

In an exactly similar way the four transmission coefficients $T_{11}, T_{12}, T_{21}, T_{22}$ are introduced to indicate the ratios of the field components $\mathscr{H}_y^{(T)}, E_y^{(T)}$ in the transmitted wave to $\mathscr{H}_y^{(I)}, E_y^{(I)}$ in the incident wave. Thus:
Let $E_y^{(I)} = 0$. Then

$$T_{11} = \mathscr{H}_y^{(T)}/\mathscr{H}_y^{(I)}, \quad T_{21} = E_y^{(T)}/\mathscr{H}_y^{(I)}. \tag{11.18}$$

Let $\mathscr{H}_y^{(I)} = 0$. Then

$$T_{12} = \mathscr{H}_y^{(T)}/E_y^{(I)}, \quad T_{22} = E_y^{(T)}/E_y^{(I)}. \tag{11.19}$$

In these definitions, (11.16)–(11.19), the first subscript always refers to the emerging reflected or transmitted wave, and the second subscript refers to the incident wave.

In some older papers on the reflection of radio waves of very low frequency, the subscripts 1, 2, were replaced by $\parallel, \perp$, and the second subscript was placed in front of the symbol. Thus R_{11} was ${}_{\parallel}R_{\parallel}$, R_{12} was ${}_{\perp}R_{\parallel}$ and so on. This notation does not accord with the standard conventions of matrix notation, and has now been almost entirely abandoned.

11.5. Reflection and transmission coefficient matrices

The expressions (11.16)–(11.19) can now be written in matrix notation. Let the field components of the incident wave be written as a column matrix of two elements, and similarly for the reflected and transmitted waves, thus

$$\boldsymbol{e}^{(I)} = \begin{pmatrix} \mathscr{H}_y^{(I)} \\ E_y^{(I)} \end{pmatrix}, \quad \boldsymbol{e}^{(R)} = \begin{pmatrix} \mathscr{H}_y^{(R)} \\ E_y^{(R)} \end{pmatrix}, \quad \boldsymbol{e}^{(T)} = \begin{pmatrix} \mathscr{H}_y^{(T)} \\ E_y^{(T)} \end{pmatrix}. \tag{11.20}$$

Let $\boldsymbol{R}$ and $\boldsymbol{T}$ be the 2×2 square matrices

$$\boldsymbol{R} = \begin{pmatrix} R_{11} & R_{12} \\ R_{21} & R_{22} \end{pmatrix} \tag{11.21}$$

$$\boldsymbol{T} = \begin{pmatrix} T_{11} & T_{12} \\ T_{21} & T_{22} \end{pmatrix}. \tag{11.22}$$

Then (11.16)–(11.19) can be written

$$\boldsymbol{e}^{(R)} = \boldsymbol{R}\boldsymbol{e}^{(I)} \tag{11.23}$$

$$\boldsymbol{e}^{(T)} = \boldsymbol{T}\boldsymbol{e}^{(I)} \tag{11.24}$$

and $\boldsymbol{R}$, $\boldsymbol{T}$ are called the reflection and transmission coefficient matrices respectively.

If a wave is produced by two successive reflection processes whose reflection coefficient matrices are $\boldsymbol{R}_a$, $\boldsymbol{R}_b$, then it follows from (11.23) that the fields in the second reflected wave are $\boldsymbol{R}_b\boldsymbol{R}_a\boldsymbol{e}^{(I)}$. The resultant reflection coefficient matrix is the matrix product of the matrices for the separate reflections. Its value depends on the order in which the reflections occur. This result can clearly be extended to any number of reflections.

11.6. Alternative forms of the reflection coefficient matrix

Instead of resolving the incident and reflected waves into linearly polarised components, they might be resolved into components in a different way to give a different representation of the reflection coefficient matrix. To illustrate this we shall suppose that the waves are resolved into right-handed and left-handed circularly polarised components.

For the incident wave let $E_\parallel$ be the component of $\boldsymbol{E}$ in the plane of incidence. Then (4.4) with $n=1$ shows that $E_\parallel = \mathscr{H}_y$. If the wave has right-handed circular polarisation, an observer looking in the direction of the wave normal sees the electric vector rotating clockwise. This requires that

$$E_{yr} = -\mathrm{i}E_{\parallel r} = -\mathrm{i}\mathscr{H}_{yr} \tag{11.25}$$

where the second subscript r indicates right-handed circular polarisation. Similarly for left-handed circular polarisation

$$E_{yl} = \mathrm{i}\mathscr{H}_{yl}. \tag{11.26}$$

The total fields E_y, $\mathscr{H}_y$ are the sum of (11.25), (11.26). We now use a superscript (I) to indicate the incident wave. Then

$$\mathscr{H}_y^{(I)} = \mathrm{i}E_{yr}^{(I)} - \mathrm{i}E_{yl}^{(I)}, \quad E_y^{(I)} = E_{yr}^{(I)} + E_{yl}^{(I)}. \tag{11.27}$$

Define the two column matrices

$$\boldsymbol{e}_0^{(I)} = \begin{pmatrix} E_{yr}^{(I)} \\ E_{yl}^{(I)} \end{pmatrix}, \quad \boldsymbol{e}_0^{(R)} = \begin{pmatrix} E_{yr}^{(R)} \\ E_{yl}^{(R)} \end{pmatrix}. \tag{11.28}$$

Then (11.27) can be written in matrix form, with (11.20),

$$\boldsymbol{e}^{(I)} = 2^{\frac{1}{2}}\boldsymbol{U}\boldsymbol{e}_0^{(I)} \tag{11.29}$$

where $\boldsymbol{U}$ is the unitary matrix

$$\boldsymbol{U} = 2^{-\frac{1}{2}}\begin{pmatrix} \mathrm{i} & -\mathrm{i} \\ 1 & 1 \end{pmatrix}. \tag{11.30}$$

Similarly for the reflected wave

$$\boldsymbol{e}^{(R)} = 2^{\frac{1}{2}}\boldsymbol{U}\boldsymbol{e}_0^{(R)}. \tag{11.31}$$

Substitution of (11.29) and (11.31) into (11.23) now gives

$$\boldsymbol{U}\boldsymbol{e}_0^{(R)} = \boldsymbol{R}\boldsymbol{U}\boldsymbol{e}_0^{(I)} \tag{11.32}$$

whence

$$e_0^{(R)} = R_0 e_0^{(I)}, \quad R_0 = U^{-1} R U = \begin{pmatrix} R_{rr} & R_{rl} \\ R_{lr} & R_{ll} \end{pmatrix}. \tag{11.33}$$

The reflection coefficient matrix R_0 for circularly polarised components has thus been obtained from R by the unitary transformation (11.33). It is used in § 11.10.

Some of the properties of R and R_0 are illustrated in problems 11.1–11.3.

11.7. Wave impedance and admittance

Let E and H be some components of the electric and magnetic intensities respectively of a wave, at the same point. Then the ratio E/H is a wave impedance. For SI units E is measured in volts per metre and H in amperes per metre so that E/H is measured in ohms and has the dimensions of an electrical impedance. Similarly H/E is a wave admittance and would be measured in siemens. Instead of H we use the measure $\mathscr{H}$ (§ 2.10) for the magnetic intensity. Then the ratios $E/\mathscr{H}$ and $\mathscr{H}/E$ are dimensionless. It will sometimes be necessary to use formulae containing both types of ratio and a single phrase is needed to apply to both. The term 'wave admittance' will be used for this, as is customary. For a discussion of the concept of wave admittance and impedance see Schelkunoff (1938), Booker (1947).

In this book wave admittances are used for a stratified ionosphere with the z axis perpendicular to the strata, and they are therefore defined only in terms of the field components E_x, E_y, $\mathscr{H}_x$, $\mathscr{H}_y$ parallel to the strata. This has the advantage that if the medium has a discontinuity at some plane $z =$ constant, these components are the same on the two sides. Thus the wave admittances are continuous at such a discontinuity.

The concept is now to be applied to a system of waves in the ionosphere whose fields are independent of y and depend on x only through a factor $\exp(-ikSx)$, where $S = \sin\theta$. These might arise from an incident wave in free space with its wave normal in the x–z plane at an angle θ to the z axis. Then in any region where the medium is homogeneous or sufficiently slowly varying there are in general four waves, and the field components of any one of them contain an exponential factor

$$\exp\left\{-ik\left(Sx + \int^z q\,\mathrm{d}z\right)\right\} \tag{11.34}$$

where q is one of the four roots of the Booker quartic equation, ch. 6. As a simple illustration suppose that the medium is isotropic and that only one wave (11.34) is present in a range of height where the medium is homogeneous. Let this wave be a vertically polarised upgoing wave, as in § 7.2, so that $\mathscr{H}_x = E_y = 0$ and from (7.4)

$$q\mathscr{H}_y = n^2 E_x, \quad A_{11} = E_x/\mathscr{H}_y = q/n^2 \tag{11.35}$$

where q is given by (7.2) with the + sign. Here A_{11} is a 'wave admittance'. It is independent of the amplitude of the wave and depends only on θ and on the properties of the medium. For a homogeneous medium it would be independent of z. It is called a 'characteristic admittance' of the medium. If a downgoing vertically polarised wave is also present, the ratio A_{11} is no longer independent of z, and it depends on the relative amplitudes of the two waves.

Next suppose that, in the same height range, the only wave present is a horizontally polarised upgoing wave so that $E_x = \mathscr{H}_y = 0$. Now A_{11} cannot be used, but instead from (7.3) we use

$$-qE_y = \mathscr{H}_x, \quad A_{22} = -\mathscr{H}_x/E_y = q, \tag{11.36}$$

where q is the same as in (11.35). Here A_{22} is another wave admittance. Again, in a homogeneous medium, it is independent of z and is a characteristic admittance of the medium. In these two examples A_{11} is actually an impedance, but A_{22} is an admittance.

The concept of wave admittance must now be generalised so as to apply for an anisotropic medium. For any level consider two different wave systems, with the same θ. Their horizontal field components are

$$E_{x1}, -E_{y1}, \mathscr{H}_{x1}, \mathscr{H}_{y1} \quad \text{and} \quad E_{x2}, -E_{y2}, \mathscr{H}_{x2}, \mathscr{H}_{y2}. \tag{11.37}$$

These are the elements of two column matrices $\mathbf{e}_1$ and $\mathbf{e}_2$ respectively, as in § 7.13. The reason for using $-E_y$ was explained in § 7.14 (3), (8). Now define a 2×2 wave admittance matrix $\boldsymbol{A}$ as follows. Let

$$\boldsymbol{P} = \begin{pmatrix} E_{x1} & E_{x2} \\ -\mathscr{H}_{x1} & -\mathscr{H}_{x2} \end{pmatrix}, \quad \boldsymbol{Q} = \begin{pmatrix} \mathscr{H}_{y1} & \mathscr{H}_{y2} \\ E_{y1} & E_{y2} \end{pmatrix}, \tag{11.38}$$

and let

$$\boldsymbol{A} = \boldsymbol{P}\boldsymbol{Q}^{-1}. \tag{11.39}$$

since $\boldsymbol{A}\boldsymbol{Q} = \boldsymbol{P}$ it follows that $\boldsymbol{A}$ has the property

$$\boldsymbol{A}\begin{pmatrix} \mathscr{H}_y \\ E_y \end{pmatrix} = \begin{pmatrix} E_x \\ -\mathscr{H}_x \end{pmatrix} \tag{11.40}$$

where $E_x, -E_y, \mathscr{H}_x, \mathscr{H}_y$ may be any linear combination of the two sets (11.37). There are other ways of defining an admittance matrix, by placing the elements of (11.37) in a different order and with different signs. See, for example, Barron and Budden (1959), Budden (1961a). The choice (11.40) is made to achieve some algebraic simplicity in later applications §§ 18.9, 18.12.

Some of the main properties of the wave admittance matrix may now be listed.

(1) Consider a range of z where the medium is isotropic and slowly varying. Let the two wave systems (11.37) be upgoing progressive waves, one vertically and the other horizontally polarised. Then the elements of $\mathbf{e}_1$, $\mathbf{e}_2$ are

$$\mathbf{e}_1 = E_x, 0, 0, \frac{n^2}{q}E_x; \quad \mathbf{e}_2 = 0, -E_y, -qE_y, 0 \tag{11.41}$$

and it follows from (11.38), (11.39) that

$$A_{12}=A_{21}=0, \quad A_{11}=q/n^2, \quad A_{22}=q. \tag{11.42}$$

Thus (11.35), (11.36) are the diagonal elements of $\boldsymbol{A}$. The z dependence of both fields (11.41) includes the factor $\exp(-\mathrm{i}k\int^z q\,\mathrm{d}z)$ but this factor cancels and does not appear in $\boldsymbol{A}$.

(2) Consider a range of z where the medium is slowly varying but anisotropic. Let each of the two wave systems (11.37) be a progressive wave given by one of the W.K.B. solutions, §7.15, so that all elements of $\mathbf{e}_1$ depend on z through the same factor $\exp(-\mathrm{i}k\int^z q_1 z)$ and similarly for $\mathbf{e}_2$, with the factor $\exp(-\mathrm{i}k\int^z q_2\,\mathrm{d}z)$. These factors cancel in (11.39). In a homogeneous medium $\boldsymbol{A}$ would be independent of z. It is called a 'characteristic' admittance matrix of the medium. It depends on θ and on which two of the four possible waves are chosen for (11.37).

(3) If more than two of the four possible progressive waves are used in choosing $\mathbf{e}_1$, $\mathbf{e}_2$ in (11.37), then $\boldsymbol{A}$ depends on height z, whether or not the medium is homogeneous. The differential equation satisfied by $\boldsymbol{A}$ is studied later, § 18.9.

(4) Consider the special case where $\theta = 0$, so that the wave normals are all parallel to the z axis. Let the two waves chosen for (11.37) be the upgoing ordinary and extraordinary waves so that $\mathbf{e}_1$ and $\mathbf{e}_2$ are given by (7.145), (7.147). Then it can be shown that

$$\boldsymbol{A}=\frac{1}{n_{\mathrm{O}}-\rho_{\mathrm{O}}^2 n_{\mathrm{E}}}\begin{pmatrix}1-\rho_{\mathrm{O}}^2 & \rho_{\mathrm{O}}(n_{\mathrm{O}}-n_{\mathrm{E}}) \\ \rho_{\mathrm{O}}(n_{\mathrm{O}}-n_{\mathrm{E}}) & n_{\mathrm{O}}n_{\mathrm{E}}(1-\rho_{\mathrm{O}}^2)\end{pmatrix} \tag{11.43}$$

and, as expected from (2), the z dependent factors have cancelled.

(5) $\boldsymbol{A}$ is continuous across any boundary plane $z=$ constant.

(6) $\boldsymbol{A}$ depends on the ratios of field components at one point. This applies also to the reflection coefficient matrix $\boldsymbol{R}$, § 11.5. Thus there is a relation between $\boldsymbol{R}$ and $\boldsymbol{A}$. An example is given at the end of this section. One important use of the continuity property (5) of $\boldsymbol{A}$ is in the derivation of reflection coefficients.

(7) Transmission coefficients depend on the ratios of field components at different points. A knowledge of $\boldsymbol{A}$ at different points is not sufficient to determine the transmission coefficients.

Suppose that $\boldsymbol{A}$ is known in the free space below the ionosphere. Here the fields may be separated into upgoing and downgoing component waves. For fields of the upgoing component we use a superscript I as in §§ 11.4–11.6. Let $\mathscr{H}_y^{(\mathrm{I})}=a$, $E_y^{(\mathrm{I})}=b$ so that $E_x^{(\mathrm{I})}=Ca$, $\mathscr{H}_x^{(\mathrm{I})}=-Cb$ where C is $\cos\theta$. Similarly for the fields of the downgoing component, superscript R, let $\mathscr{H}_y^{(\mathrm{R})}=c$, $E_y^{(\mathrm{R})}=d$, so that $E_x^{(\mathrm{R})}=-Cc$, $\mathscr{H}_x^{(\mathrm{R})}=Cd$. Then the total fields are given by

$$\begin{pmatrix}E_x \\ -E_y \\ \mathscr{H}_x \\ \mathscr{H}_y\end{pmatrix}=\begin{pmatrix}C & 0 & -C & 0 \\ 0 & -1 & 0 & -1 \\ 0 & -C & 0 & C \\ 1 & 0 & 1 & 0\end{pmatrix}\begin{pmatrix}a \\ b \\ c \\ d\end{pmatrix} \tag{11.44}$$

whence by inversion of the 4×4 matrix

$$2\begin{pmatrix} a \\ b \\ c \\ d \end{pmatrix} = \begin{pmatrix} 1/C & 0 & 0 & 1 \\ 0 & -1 & -1/C & 0 \\ -1/C & 0 & 0 & 1 \\ 0 & -1 & 1/C & 0 \end{pmatrix} \begin{pmatrix} E_x \\ -E_y \\ \mathscr{H}_x \\ \mathscr{H}_y \end{pmatrix}. \tag{11.45}$$

From the definition (11.20), (11.23) of the reflection coefficient matrix $\boldsymbol{R}$ it follows that

$$\boldsymbol{R}\begin{pmatrix} a \\ b \end{pmatrix} = \begin{pmatrix} c \\ d \end{pmatrix}. \tag{11.46}$$

Now (11.45) may be partitioned to give

$$\left.\begin{aligned} 2\begin{pmatrix} a \\ b \end{pmatrix} &= \frac{1}{C}\begin{pmatrix} E_x \\ -\mathscr{H}_x \end{pmatrix} + \begin{pmatrix} \mathscr{H}_y \\ E_y \end{pmatrix}, \\ 2\begin{pmatrix} c \\ d \end{pmatrix} &= -\frac{1}{C}\begin{pmatrix} E_x \\ -\mathscr{H}_x \end{pmatrix} + \begin{pmatrix} \mathscr{H}_y \\ E_y \end{pmatrix}. \end{aligned}\right\} \tag{11.47}$$

If (11.46) and (11.40) are now used we obtain

$$\boldsymbol{R}(C + A) = C - A \tag{11.48}$$

where the arbitrary column $(\mathscr{H}_y, E_y)$ on the right has been omitted and C is assumed to be multiplied by the 2×2 unit matrix $\mathbf{1}$. Hence

$$\begin{aligned} \boldsymbol{R} &= (C - A)(C + A)^{-1} = 2C(C + A)^{-1} - \mathbf{1} \\ A &= C(\mathbf{1} - \boldsymbol{R})(\mathbf{1} + \boldsymbol{R})^{-1} = 2C(\mathbf{1} + \boldsymbol{R})^{-1} - C. \end{aligned} \tag{11.49}$$

This shows that if A is transposed then $\boldsymbol{R}$ also is transposed. These results are used in § 18.12. In particular if A and $\boldsymbol{R}$ are diagonal:

$$R_{11} = \frac{C - A_{11}}{C + A_{11}}, \quad A_{11} = C\frac{1 - R_{11}}{1 + R_{11}}; \quad R_{22} = \frac{C - A_{22}}{C + A_{22}}, \quad A_{22} = C\frac{1 - R_{22}}{1 + R_{22}}. \tag{11.50}$$

11.8. Reflection at a sharp boundary 1. Isotropic plasma

The formulae for the reflection and transmission coefficients at a plane boundary between two homogeneous media are useful, even though there are no sharp boundaries in the ionosphere and magnetosphere. For example some authors have adopted a simple model ionosphere consisting of a homogeneous plasma with a lower plane boundary and free space below. This has led with good success to explanations of many of the features of the propagation of radio waves of very low frequency. See for example, Yokoyama and Namba (1932); Rydbeck (1944); Bremmer (1949); Barber and Crombie (1959). The formulae have already been used in § 7.7 to study propagation in a medium comprised of discrete strata.

Suppose that in the region $z > 0$ there is a homogeneous medium with refractive index n_2, and in the region $z < 0$ another homogeneous medium with refractive index

n_1. In this section it is assumed that these media are isotropic. The incident plane wave is in the medium where $z<0$ with its wave normal in the $x-z$ plane at an angle θ to the positive z axis, and $0 \leqslant \theta < \frac{1}{2}\pi$. It gives rise to a reflected plane wave in $z<0$ and to a transmitted plane wave in $z>0$. The field components satisfy some boundary conditions so that all field components must depend on x and y in the same way. This leads at once to the requirement that the wave normals of the reflected and transmitted waves are also in the $x-z$ plane. Let them make angles θ_R, θ_T respectively with the z axis. All three angles θ, θ_R, θ_T are measured clockwise from the positive z axis, for an observer looking in the direction of the positive y axis. Since the x dependence of all fields is the same it then follows that

$$\theta_R = \pi - \theta, \quad n_1 \sin\theta = n_2 \sin\theta_T. \tag{11.51}$$

The first is the reflection law, and the second is Snell's law; compare § 6.2. Note that if the medium in $z<1$ is anisotropic, the refractive index n_R for the reflected wave is in general different from that for the incident wave, n_1. Then the first equation (11.51) must be replaced by

$$n_1 \sin\theta = n_R \sin\theta_R. \tag{11.52}$$

It is now required to find the reflection and transmission coefficients, and the boundary plane $z=0$ will be used as the reference level. They are derived from the boundary conditions that the four components E_x, E_y, $\mathscr{H}_x$, $\mathscr{H}_y$ parallel to the boundary are the same on the two sides of the boundary plane $z=0$. The details are set out in textbooks on optics and on electromagnetic wave theory. They were given, in a notation very similar to that used here, by Budden (1961a), who used a definition of the transmission coefficients different from (11.18), (11.19). In writing the results it is convenient to use.

$$q_I = -q_R = n_1 \cos\theta, \quad q_T = n_2 \cos\theta_T = (n_2^2 - n_1^2 \sin^2\theta)^{\frac{1}{2}}. \tag{11.53}$$

These qs are the same as those defined and used in ch. 6. Then

$$R_{12} = R_{21} = T_{12} = T_{21} = 0, \tag{11.54}$$

$$R_{11} = \frac{n_2\cos\theta - n_1\cos\theta_T}{n_2\cos\theta + n_1\cos\theta_T} = \frac{n_2^2 q_I - n_1^2 q_T}{n_2^2 q_I + n_1^2 q_T}, \tag{11.55}$$

$$T_{11} = \frac{2n_2\cos\theta}{n_2\cos\theta + n_1\cos\theta_T} = \frac{2n_2^2 q_I}{n_2^2 q_I + n_1^2 q_I}, \tag{11.56}$$

$$R_{22} = \frac{n_1\cos\theta - n_2\cos\theta_T}{n_1\cos\theta + n_2\cos\theta_T} = \frac{q_I - q_T}{q_I + q_T}, \tag{11.57}$$

$$T_{22} = \frac{2n_1\cos\theta}{n_1\cos\theta + n_2\cos\theta_T} = \frac{2q_1}{q_1 + q_T}. \tag{11.58}$$

The above formulae for the elements of $\boldsymbol{R}$ are special cases of (11.66) which gives $\boldsymbol{R}$ for the more general case when the medium in $z>0$ is anisotropic. The full derivation is given in § 11.10.

Formulae (11.55), (11.57) and formulae equivalent to (11.56), (11.58) were given by Fresnel. They are often used in textbooks of optics for real n_1 and n_2, but they are not restricted to real variables, and will be applied here to cases where the refractive indices and the angles are in general complex.

They are often needed for normal incidence, $\theta = \theta_T = 0$ and they then become

$$R_{11} = -R_{22} = \frac{n_2 - n_1}{n_2 + n_1}, \tag{11.59}$$

$$T_{11} = \frac{2n_2}{n_1 + n_2}, \qquad T_{22} = \frac{2n_1}{n_1 + n_2}. \tag{11.60}$$

The above formula for R_{11} is based on the use of $\mathscr{H}_y$ as in (11.16). It is sometimes more convenient at normal incidence to use, instead of R_{11}, the ratio of the values of E_x. The resulting reflection coefficient is $-R_{11}$ in (11.59).

11.9. Properties of the Fresnel formulae

In this section the medium where $z < 1$ is assumed to be free space so that $n_1 = 1$, and $q_I = \cos\theta$. The refractive index of the medium where $z > 0$ will be denoted simply by n, with the subscript 2 omitted.

The reflection coefficient R_{11}, (11.55) is zero when $n^2 \cos\theta = q_T$, that is when

$$\theta = \theta_B, \quad \tan\theta_B = n. \tag{11.61}$$

When n is real θ_B is a real angle called the 'Brewster angle'. For real n, when θ is near to θ_B, R_{11} is real. Arg R_{11} changes discontinuously by π when θ increases and passes through θ_B.

When n is complex θ_B is also complex and is called the 'complex Brewster angle'. When $\theta = \theta_B$ the incident and transmitted waves are both inhomogeneous plane waves; § 2.15. When θ is real R_{11} is complex. Then when θ increases, arg R_{11} changes continuously, and the rate of change is greatest, and $|R_{11}|$ is a minimum, when θ is near to $\mathrm{Re}(\theta_B)$.

The formula (11.57) shows that for R_{22} to be zero it is necessary that $\cos\theta = (n^2 - \sin^2\theta)^{\frac{1}{2}}$. This cannot be satisfied for any real or complex θ if $n \neq 1$. Hence in this case there is no Brewster angle.

When $\cos\theta_T$ is zero, $\sin\theta_T$ is unity and Snell's law (11.51) gives

$$\theta = \theta_C, \quad \sin\theta_C = n. \tag{11.62}$$

Then $R_{11} = R_{22} = 1$, and θ_C is called the 'critical angle', by analogy with the corresponding case in optics. If n is complex, the critical angle is also complex.

Suppose that θ and n are real and $\theta_C \leqslant \theta < \frac{1}{2}\pi$, or that θ is real and n is purely imaginary. Then q_T (11.53) is purely imaginary and $|R_{11}| = 1$, $|R_{22}| = 1$, and from (11.55), (11.57)

$$\arg R_{11} = 2\arctan\{(\sin^2\theta - n^2)^{\frac{1}{2}}/n^2\cos\theta\} \tag{11.63}$$

$$\arg R_{22} = 2\arctan\{(\sin^2\theta - n^2)^{\frac{1}{2}}/\cos\theta\} \tag{11.64}$$

where the square root takes its positive value, and the arctangents are between $-\frac{1}{2}\pi$ and $+\frac{1}{2}\pi$. The reader should verify that the signs of (11.63), (11.64) are correct. These are examples where the reflection is total. They are analogous to the phenomenon of 'total internal reflection' in optics.

Curves showing how the coefficients (11.55)–(11.58) depend on the angle θ and on the refractive indices are given in many books. Some examples, where the medium in $z > 0$ is an isotropic electron plasma without or with collisions, were given by Budden (1961a). Since then, however, programmable pocket calculators have become readily available and the interested reader can now quickly plot such curves for himself. See problems 11.5, 11.6, 11.8, 11.9 for some examples of the use of these and related formulae.

11.10. Reflection at a sharp boundary 2. Anisotropic plasma

When one or both of the media on the two sides of the boundary plane $z = 0$ is anisotropic, the derivation of the reflection coefficients is more complicated. In this section the theory is given only for the case where the medium in $z < 0$ is free space, so that $n_1 = 1$. The wave normal of the incident wave is again assumed to be in the $x - z$ plane at an angle θ to the z axis, and we use $S = \sin\theta$, $C = \cos\theta$. In general it gives a reflected wave where $z < 0$ and two transmitted waves, ordinary and extraordinary, in the anisotropic medium where $z > 0$. The quantities associated with these waves will be indicated by superscripts (O), (E), which replace the (T) of § 11.4. They are obliquely upgoing plane waves and their fields are determined by their two values $q^{(O)}$ and $q^{(E)}$ of the roots of the Booker quartic (6.15) with coefficients (6.23). These values of q for upgoing waves are chosen as described in § 6.4. To describe the fields of any of the plane waves, it is convenient to use the column matrix $\mathbf{e}$ with four elements $E_x, -E_y, \mathscr{H}_x, \mathscr{H}_y$ as defined in § 7.13 and used in (7.80). They are given by (7.118) or (7.134). They will now be written $\mathrm{e}_1^{(O)}, \mathrm{e}_2^{(O)}, \mathrm{e}_3^{(O)}, \mathrm{e}_4^{(O)}$ for the ordinary wave, and the same with (O) replaced by (E) for the extraordinary wave. Then the admittance matrix in the medium where $z > 0$ is from (11.38) (11.39)

$$\boldsymbol{A} = \begin{pmatrix} \mathrm{e}_1^{(O)} & \mathrm{e}_1^{(E)} \\ -\mathrm{e}_3^{(O)} & -\mathrm{e}_3^{(E)} \end{pmatrix} \begin{pmatrix} \mathrm{e}_4^{(O)} & \mathrm{e}_4^{(E)} \\ -\mathrm{e}_2^{(O)} & -\mathrm{e}_2^{(E)} \end{pmatrix}^{-1} \tag{11.65}$$

This is independent of z (§11.7 item (2)) and is a characteristic admittance matrix. Now $\boldsymbol{A}$ must be the same on the two sides of the bounday plane $z = 0$, (§ 11.7 item (5)), and hence in the medium where $z < 0$, at the boundary, $\boldsymbol{A}$ is equal to (11.65). But here it is z dependent and not a characteristic admittance matrix. At the boundary, $\boldsymbol{R}$ is found from (11.49), (11.65). Thus the reflection coefficient with reference level at $z = 0$ is

$$\boldsymbol{R} = \begin{pmatrix} C\mathrm{e}_4^{(O)} - \mathrm{e}_1^{(O)} & C\mathrm{e}_4^{(E)} - \mathrm{e}_1^{(E)} \\ -C\mathrm{e}_2^{(O)} + \mathrm{e}_3^{(O)} & -C\mathrm{e}_2^{(E)} + \mathrm{e}_3^{(E)} \end{pmatrix} \begin{pmatrix} C\mathrm{e}_4^{(O)} + \mathrm{e}_1^{(O)} & C\mathrm{e}_4^{(E)} + \mathrm{e}_1^{(E)} \\ -C\mathrm{e}_2^{(O)} - \mathrm{e}_3^{(O)} & -C\mathrm{e}_2^{(E)} - \mathrm{e}_3^{(E)} \end{pmatrix}^{-1} \tag{11.66}$$

A method of finding $\boldsymbol{R}$ that used solutions of the Booker quartic equation, and that is in essentials the same as that used here, was given by Yabroff (1957), and by Johler and Walters (1960). They gave curves to show how the elements of $\boldsymbol{R}$ depend on angle of incidence and on the parameters of the medium.

A completely different method for finding $\boldsymbol{R}$ is described in § 18.10, where it is shown that $\boldsymbol{R}$ is the solution of a matrix equation that does not contain $q^{(\mathrm{O})}$ and $q^{(\mathrm{E})}$. See also Barron and Budden (1959, § 10).

11.11. Normal incidence. Anisotropic plasma with free space below it

For normal incidence $\theta = 0$, two cases will be studied. For the first, in this section, it is assumed, as in § 11.10, that the medium where $z < 1$ is free space, and the homogeneous anisotropic plasma where $z > 0$ is treated as a simple model of the ionosphere with its lower boundary at $z = 0$, and with z measured vertically upwards. The second case, in § 11.12, treats the problem when both media are anisotropic plasmas. For both cases it is convenient to choose the x and y axes so that the vector $\boldsymbol{Y}$, anti-parallel to the earth's magnetic field, is in the x–z plane.

For the field components $\mathbf{e}^{(\mathrm{O})}$, $\mathbf{e}^{(\mathrm{E})}$ in the upper medium we may use, instead of (7.118), the simpler forms given in (7.145), (7.147). Thus the elements are
For $\mathbf{e}^{(\mathrm{O})}$:

$$1, \quad -\rho_{\mathrm{O}}, \quad -\rho_{\mathrm{O}} n_{\mathrm{O}}, \quad n_{\mathrm{O}} \tag{11.67}$$

For $\mathbf{e}^{(\mathrm{E})}$:

$$\rho_{\mathrm{O}}, \quad -1, \quad -n_{\mathrm{E}}, \quad \rho_{\mathrm{O}} n_{\mathrm{E}} \tag{11.68}$$

The subscripts O, E, for ordinary and extraordinary, now replace the superscripts (O), (E). If these are now used in (11.66) with $C = 1$, it gives

$$\boldsymbol{R} = \begin{pmatrix} n_{\mathrm{O}} - 1 & \rho_{\mathrm{O}}(n_{\mathrm{E}} - 1) \\ -\rho_{\mathrm{O}}(n_{\mathrm{O}} - 1) & -(n_{\mathrm{E}} - 1) \end{pmatrix} \begin{pmatrix} n_{\mathrm{O}} + 1 & \rho_{\mathrm{O}}(n_{\mathrm{E}} + 1) \\ \rho_{\mathrm{O}}(n_{\mathrm{O}} + 1) & n_{\mathrm{E}} + 1 \end{pmatrix}^{-1} \tag{11.69}$$

and on multiplying out

$$R_{11} = \frac{1}{\rho_{\mathrm{E}} - \rho_{\mathrm{O}}} \left\{ \rho_{\mathrm{E}} \frac{n_{\mathrm{O}} - 1}{n_{\mathrm{O}} + 1} - \rho_{\mathrm{O}} \frac{n_{\mathrm{E}} - 1}{n_{\mathrm{E}} + 1} \right\}, \tag{11.70}$$

$$R_{22} = \frac{1}{\rho_{\mathrm{E}} - \rho_{\mathrm{O}}} \left\{ \rho_{\mathrm{O}} \frac{n_{\mathrm{O}} - 1}{n_{\mathrm{O}} + 1} - \rho_{\mathrm{E}} \frac{n_{\mathrm{E}} - 1}{n_{\mathrm{E}} + 1} \right\}, \tag{11.71}$$

$$R_{12} = R_{21} = \frac{2}{\rho_{\mathrm{E}} - \rho_{\mathrm{O}}} \left\{ \frac{1}{n_{\mathrm{O}} + 1} - \frac{1}{n_{\mathrm{E}} + 1} \right\}. \tag{11.72}$$

The relation $\rho_{\mathrm{O}} \rho_{\mathrm{E}} = 1$, (4.29), has been used here.

If the earth's magnetic field is horizontal, the wave polarisations are, from (4.82): $\rho_{\mathrm{O}} = 0$ and $\rho_{\mathrm{E}} = \infty$. Then the elements of $\boldsymbol{R}$ are

$$R_{11} = \frac{n_{\mathrm{O}} - 1}{n_{\mathrm{O}} + 1}, \quad R_{22} = -\frac{n_{\mathrm{E}} - 1}{n_{\mathrm{E}} + 1}, \quad R_{12} = R_{21} = 0. \tag{11.73}$$

The first two are just the Fresnel formulae (11.59) for normal incidence with $n_1 = 1$.

Suppose now that the earth's magnetic field is vertical and directed downwards, as at the north magnetic pole. Then the vector $\boldsymbol{Y}$ is directed upwards and its direction cosine is $l_z = +1$ for upgoing waves. For the ordinary wave when $Y < 1$, $\rho_O = -\mathrm{i}$ which corresponds to right-handed circular polarisation. Similarly $\rho_E = +\mathrm{i}$. Hence

$$n_O^2 = 1 - X/(U+Y), \quad n_E^2 = 1 - X/(U-Y). \tag{11.74}$$

If $Y > 1$, the order of the terms ordinary and extraordinary and of the subscripts O, E must be reversed. The elements of R are now

$$\begin{aligned} R_{11} = -R_{22} &= \frac{1}{2}\left\{\frac{n_O - 1}{n_O + 1} + \frac{n_E - 1}{n_E + 1}\right\}, \\ R_{12} = R_{21} &= \mathrm{i}\left\{\frac{1}{n_E + 1} - \frac{1}{n_O + 1}\right\}. \end{aligned} \tag{11.75}$$

These may be simplified by using the alternative representation $\boldsymbol{R}_0$ of § 11.6, in which the incident and reflected waves are resolved into circularly polarised components. As in § 11.6 we use subscripts r, l to denote right-handed and left-handed components, so that

$$\boldsymbol{R}_0 = \begin{pmatrix} R_{rr} & R_{rl} \\ R_{lr} & R_{ll} \end{pmatrix}. \tag{11.76}$$

Then the transformation (11.33), with (11.30), when used with (11.75) gives

$$R_{rr} = R_{ll} = 0, \quad R_{rl} = \frac{1 - n_E}{1 + n_E}, \quad R_{lr} = \frac{1 - n_O}{1 + n_O}. \tag{11.77}$$

These results show that the two waves normally incident on a sharply bounded ionosphere are independently reflected. For example an incident wave that is circularly polarised with a right-handed sense would give a wave in the ionosphere with the same polarisation, and the reflected wave would be circularly polarised with a left-handed sense. This reversal of the sense occurs because the directions of the wave normals are opposite for the incident and reflected waves; compare end of § 4.4. The absolute direction of rotation is the same for all three waves, namely clockwise when seen by an observer looking upwards. Because the wave polarisations are the same for all three waves, the boundary conditions are satisfied with this one polarisation, and no wave with the other polarisation is produced at the boundary.

11.12. Normal incidence. Two anisotropic plasmas

To study the reflection at normal incidence when the media on both sides of the boundary are anisotropic, we shall from the start use a representation in which the waves are resolved, not into linearly polarised components, but into ordinary and extraordinary components. The values of n_O, ρ_O, n_E, ρ_E are different for the two media, and this will be indicated by using superscripts (1) for $z < 0$, and (2) for $z > 0$. Thus the incident wave is resolved into an ordinary component whose fields are the column matrix $\mathbf{e}_{\rm O}^{(I)}$ with elements

$$a_{\rm O}(1, -\rho_{\rm O}^{(1)}, -\rho_{\rm O}^{(1)} n_{\rm O}^{(1)}, n_{\rm O}^{1}) \tag{11.78}$$

and an extraordinary wave whose $\mathbf{e}_{\rm E}^{(I)}$ has elements

$$a_{\rm E}(1, -\rho_{\rm E}^{(1)}, -\rho_{\rm E}^{(1)} n_{\rm E}^{(1)}, n_{\rm E}^{(1)}). \tag{11.79}$$

The reflected wave is similarly resolved into an ordinary component whose $\mathbf{e}_{\rm O}^{(R)}$ has elements

$$b_{\rm O}(1, -\rho_{\rm O}^{(1)}, \rho_{\rm O}^{(1)} n_{\rm O}^{(1)}, -n_{\rm O}^{(1)}) \tag{11.80}$$

and an extraordinary component whose $\mathbf{e}_{\rm E}^{(R)}$ has elements

$$b_{\rm E}(1, -\rho_{\rm E}^{(1)}, \rho_{\rm E}^{(1)} n_{\rm E}^{(1)}, -n_{\rm E}^{(1)}). \tag{11.81}$$

These formulae are given in (7.146), (7.148). The polarisations $\rho_{\rm O}$, $\rho_{\rm E}$ are defined with the same axes for upgoing and downgoing waves and therefore have the same two values. For an explanation of this see end of § 4.4. Finally the fields of the transmitted waves in the upper medium are

$$c_{\rm O}(1, -\rho_{\rm O}^{(2)}, -\rho_{\rm O}^{(2)} n_{\rm O}^{(2)}, n_{\rm O}^{(2)}) \tag{11.82}$$

and

$$c_{\rm E}(1, -\rho_{\rm E}^{(2)}, -\rho_{\rm E}^{(2)} n_{\rm E}^{(2)}, n_{\rm E}^{(2)}). \tag{11.83}$$

The boundary conditions require that the sum of the fields (11.78)–(11.81) is equal to the sum of (11.82) and (11.83). This can be written in matrix form

$$\begin{pmatrix} 1 & 1 & 1 & 1 \\ -\rho_{\rm O}^{(1)} & -\rho_{\rm E}^{(1)} & -\rho_{\rm O}^{(1)} & -\rho_{\rm E}^{(1)} \\ -\rho_{\rm O}^{(1)} n_{\rm O}^{(1)} & -\rho_{\rm E}^{(1)} n_{\rm E}^{(1)} & \rho_{\rm O}^{(1)} n_{\rm O}^{(1)} & \rho_{\rm E}^{(1)} n_{\rm E}^{(1)} \\ n_{\rm O}^{(1)} & n_{\rm E}^{(1)} & -n_{\rm O}^{(1)} & -n_{\rm E}^{(1)} \end{pmatrix} \begin{pmatrix} a_{\rm O} \\ a_{\rm E} \\ b_{\rm O} \\ b_{\rm E} \end{pmatrix} = \begin{pmatrix} 1 & 1 \\ -\rho_{\rm O}^{(2)} & -\rho_{\rm E}^{(2)} \\ -\rho_{\rm O}^{(2)} n_{\rm O}^{(2)} & -\rho_{\rm E}^{(2)} n_{\rm E}^{(2)} \\ n_{\rm O}^{(2)} & n_{\rm E}^{(2)} \end{pmatrix} \begin{pmatrix} c_{\rm O} \\ c_{\rm E} \end{pmatrix}. \tag{11.84}$$

Now the matrices are partitioned thus

$$\boldsymbol{a} = \begin{pmatrix} a_{\rm O} \\ a_{\rm E} \end{pmatrix}, \quad \boldsymbol{b} = \begin{pmatrix} b_{\rm O} \\ b_{\rm E} \end{pmatrix}, \quad \boldsymbol{c} = \begin{pmatrix} c_{\rm O} \\ c_{\rm E} \end{pmatrix}, \tag{11.85}$$

and

$$\boldsymbol{M}_1 = \begin{pmatrix} 1 & 1 \\ -\rho_{\rm O} & -\rho_{\rm E} \end{pmatrix}, \quad \boldsymbol{M}_2 = \begin{pmatrix} -\rho_{\rm O} & -\rho_{\rm E} \\ 1 & 1 \end{pmatrix} \begin{pmatrix} n_{\rm O} & 0 \\ 0 & n_{\rm E} \end{pmatrix} \tag{11.86}$$

where appropriate superscripts (1), (2) are to be attached. Then (11.84) gives

$$\boldsymbol{M}_1^{(1)}(\boldsymbol{a} + \boldsymbol{b}) = \boldsymbol{M}_1^{(2)}\boldsymbol{c}, \quad \boldsymbol{M}_2^{(1)}(\boldsymbol{a} - \boldsymbol{b}) = \boldsymbol{M}_2^{(2)}\boldsymbol{c}. \tag{11.87}$$

Next, by analogy with § 11.4, define reflection and transmission coefficient matrices

$$\boldsymbol{R}_{\rm c} = \begin{pmatrix} R_{\rm OO} & R_{\rm OE} \\ R_{\rm EO} & R_{\rm EE} \end{pmatrix}, \quad \boldsymbol{T}_{\rm c} = \begin{pmatrix} T_{\rm OO} & T_{\rm OE} \\ T_{\rm EO} & T_{\rm EE} \end{pmatrix} \tag{11.88}$$

so that

$$b = R_c a, \quad c = T_c a. \tag{11.89}$$

These are substituted in (11.87). The equations must be true for any a, so the factor a on the right may be omitted, whence

$$M_1^{(1)}(1 + R_c) = M_1^{(2)} T_c, \quad M_2^{(1)}(1 - R_c) = M_2^{(2)} T_c. \tag{11.90}$$

Then elimination of T_c gives

$$(M_1^{(2)})^{-1} M_1^{(1)}(1 + R_c) = (M_2^{(2)})^{-1} M_2^{(1)}(1 - R_c) \tag{11.91}$$

which is a matrix equation for finding R_c. After some reduction it becomes

$$\begin{pmatrix} n_O^{(2)} & 0 \\ 0 & n_E^{(2)} \end{pmatrix} G(1 + R_c) = G \begin{pmatrix} n_O^{(1)} & 0 \\ 0 & n_E^{(1)} \end{pmatrix} (1 - R_c) \tag{11.92}$$

where

$$G = \begin{pmatrix} \rho_O^{(1)} - \rho_E^{(2)} & \rho_E^{(1)} - \rho_E^{(2)} \\ \rho_O^{(2)} - \rho_O^{(1)} & \rho_O^{(2)} - \rho_E^{(1)} \end{pmatrix}. \tag{11.93}$$

Now consider the special case where $\rho_O^{(2)} = \rho_O^{(1)}$, $\rho_E^{(2)} = \rho_E^{(1)}$. Then G is diagonal and is just a constant multiplier $\rho_O - \rho_E$ that can be cancelled from (11.92). It follows that R_c is then diagonal and

$$R_{OO} = \frac{n_O^{(1)} - n_O^{(2)}}{n_O^{(1)} + n_O^{(2)}}, \quad R_{EE} = \frac{n_E^{(1)} - n_E^{(2)}}{n_E^{(1)} + n_E^{(2)}}. \tag{11.94}$$

When both characteristic wave polarisations are the same on the two sides of the boundary there is no intermode conversion at the boundary.

If the lower medium is free space, any wave may be resolved into two waves whose polarisations can be chosen in any way we please; see problem 4.2. They can therefore be chosen to be $\rho_O^{(2)}$, $\rho_E^{(2)}$. Then the formulae (11.94) apply with $n_O^{(1)} = n_E^{(1)} = 1$. Two examples of this have already been given at (11.73) and (11.77).

11.13. Probing the ionosphere by the method of partial reflection

The ionosonde technique, § 1.7, uses vertically incident pulses of radio waves for studying the ionosphere. The main observed reflected pulses come from levels where n_O or n_E is zero, in the E- and F-regions. But the lower parts of the ionosphere, in the height range 60–100 km can contain weak irregularities of electron concentration N that can give discontinuities of N. Thus there can be quite sharp boundaries between two plasmas with slightly different N, and those boundaries that are horizontal or nearly so can give observable reflections. In temperate latitudes, for waves with vertical wave normals, the two wave polarisations ρ_O, ρ_E are very close to $\mp i$, that is circular polarisation, and do not change appreciably when N changes. Thus there is no appreciable intermode conversion at these reflections. The ordinary and extraordinary waves are reflected independently and the reflection coefficients R_{OO}, R_{EE} are given by (11.94) with good accuracy. They are each proportional to the small change δN of N at the reflecting boundary, but the ratio R_{OO}/R_{EE} is independent of δN and depends only on the average composition of the plasma.

The idea that these reflections arise from horizontal sharp boundaries is oversimplified. A better idea is to regard them as produced by scattering from weak random irregularities of N. The theory for this version was given by Booker (1959) who showed that the effective ratio R_{OO}/R_{EE} is the same as for the simpler sharp boundary model.

The values of R_{OO}, R_{EE} are of order 10^{-7} to 10^{-3} so the reflections are extremely weak and it is necessary to make observations in a location where the extraneous electrical noise level is very small. The method was first used by Gardner and Pawsey (1953) and later by numerous other workers, particularly Belrose and Burke (1964), Fejer and Vice (1959a), Holt (1963), Belrose (1970), Belrose, Burke, Coyne and Reed (1972). The frequencies used were typically 2.28 MHz (Gardner and Pawsey), 2.66 and 6.27 MHz (Belrose and Burke).

In some of these experiments the transmitted pulse was linearly polarised. In the ionosphere it splits into two circularly polarised components with opposite senses and equal amplitudes. The reflected pulses were received on two aerials designed to respond to the two circularly polarised components, and the ratio of the amplitudes was measured. In other experiments the transmitting aerial was designed to emit circularly polarised waves and it was arranged that alternate pulses had a right-handed and a left-handed sense.

The reflections occur where N and therefore X is very small so that both refractive indices are close to unity. The ratio of the reflection coefficients (11.94) is then

$$R_{OO}/R_{EE} = \delta(n_O^2)/\delta(n_E^2) \tag{11.95}$$

where δ indicates the small change when the reflecting boundary is crossed. For small X, (4.112) shows that

$$n^2 \approx 1 + X\frac{\frac{1}{2}Y^2\sin^2\Theta - U^2 \pm Y(U^2\cos^2\Theta + \frac{1}{4}Y^2\sin^4\Theta)^{\frac{1}{2}}}{U(U^2 - Y^2)} \tag{11.96}$$

(upper sign for ordinary wave when $Y < 1$), whence

$$\frac{R_{OO}}{R_{EE}} = \frac{\frac{1}{2}Y^2\sin^2\Theta - U^2 + Y(U^2\cos^2\Theta + \frac{1}{4}Y^2\sin^4\Theta)^{\frac{1}{2}}}{\frac{1}{2}Y^2\sin^2\Theta - U^2 - Y(U^2\cos^2\Theta + \frac{1}{4}Y^2\sin^4\Theta)^{\frac{1}{2}}}. \tag{11.97}$$

This depends on the electron collision frequency ν. The main use of the partial reflection method is as a way of measuring ν at heights that are accurately known. For (11.97) it has been assumed that ν is independent of electron velocity. Some workers have used the Sen–Wyller form (3.84) of the constitutive relations so that their version of (11.97) contains the $\mathscr{C}$ integrals (3.83) (Belrose and Burke, 1964).

The waves that travel up to the reflection level and down again are attenuated, and the attenuations are different for the ordinary and extraordinary waves. In measuring R_{OO}/R_{EE} this must be allowed for, and it is particularly important for the greater heights of reflection because the paths through the ionosphere are large. The attenuation depends on X and thence on N so that this method also gives some information about $N(z)$.

11.14. Spherical waves. Choice of reference level

For the definitions of the reflection coefficients given in the preceding sections it was assumed that the incident wave was a plane wave. In practice the incident wave originates at a source of small dimensions, the transmitting aerial, so that the wave front is approximately spherical, but by the time it reaches the ionosphere the radius of curvature is so large that the wave is very nearly a plane wave. In treating it as exactly plane an approximation is used, whose validity must now be examined.

The field emanating from a transmitter of small dimensions at the origin of coordinates can be written as an angular spectrum of plane waves (10.1). It is now assumed that the receiver is in the plane $y = 0$. The field component F in (10.1) may then be E_y when the transmitter is a vertical magnetic Hertzian dipole, or $-\mathscr{H}_y/n$ when it is a vertical electric Hertzian dipole. Then the factor A in (10.1) is given by (10.53) and

$$F = \iint S_1 C^{-1} \exp\{-ik(S_1 x + Cz)\}\, dS_1\, dS_2. \tag{11.98}$$

The factor S_1 gives the angle dependence of the radiated field of the vertical dipole. The integral is proportional to $r^{-1} e^{-ikr} \sin\theta$ where r, θ are the polar coordinates of the receiving point. The limits of both integrals are $\pm\infty$ but they are to be treated as contour integrals in the complex planes of S_1 and S_2. The integrand represents a plane wave whose wave normal has direction cosines S_1, S_2, C. For other types of transmitting aerial, the first factor S_1 is replaced by some other combination of S_1, S_2, C.

Each component plane wave in the integrand of (11.98) is reflected from the ionosphere and gives another plane wave travelling obliquely downwards, whose wave normal has direction cosines S_1, S_2, $-C$. In this wave we consider a field component F_R which may or may not be the same as F. The ratio F_R/F is some element or combination of the elements of $\boldsymbol{R}$ which we denote simply by R. This reflection coefficient R is in general a function of S_1 and S_2, though in simple cases it depends only on $\theta = \arccos C$. Suppose that its reference level (see § 11.2) is where $z = z_1$. Then the reflected field is

$$F_R = \iint R(S_1, S_2) S_1 C^{-1} \exp[-ik\{S_1 x + C(2z_1 - z)\}]\, dS_1\, dS_2. \tag{11.99}$$

This may be evaluated by the method of double steepest descents, § 9.10. Suppose first that $R(S_1, S_2)$ is a very slowly varying function of S_1, S_2, so that the double saddle points S_{10}, S_{20} may be found from the exponential alone. Then R is set equal to $R(S_{10}, S_{20})$ and placed outside the integral. But the integral that remains is the same as (11.98) except that the sign of z is reversed and the source is at the point $x = y = 0$, $z = 2z_1$. This is just the geometrical image point of the origin in the reference plane $z = z_1$.

When (9.63) is applied to (11.99) it gives

$$S_{20} = 0, \quad S_{10}/C = x/(2z_1 - z) = \tan\theta_0. \tag{11.100}$$

These give the direction cosines of the wave normal of the predominant plane wave reaching the receiving point. They show that the reflection is just as for simple geometrical optics from the plane $z = z_1$, with reflection coefficient as for a plane wave at angle of incidence θ_0. In this case therefore, the plane wave reflection coefficient may be used, even though the incident wave front is spherical. The justification for it is that $R(S_1, S_2)$ must be sufficiently slowly varying.

Suppose next that we use a different reference level $z = z_2$, so that, from (11.10), the reflection coefficient is now

$$R_2 = R(S_1, S_2)\exp\{2ikC(z_2 - z_1)\}. \tag{11.101}$$

Then (11.99) is replaced by

$$F_R = \iint R_2 S_1 C^{-1} \exp[-ik\{S_1 x + C(2z_2 - z)\}]\, dS_1\, dS_2. \tag{11.102}$$

If the presence of the exponential factor in (11.101) is not known, and if it is assumed that R_2 is a slowly varying function of S_1, S_2, the method used for (11.99) would now give a wrong result. It would show a wave reaching the receiver with the wrong angle of incidence $\theta_2 = \arctan\{x/(2z_2 - z)\}$ instead of (11.100), it would use the reflection coefficient for $\theta = \theta_2$ instead of θ_0, and it would give the wrong phase for the received signal.

If the exponentials in (11.101) and (11.102) had been combined before finding the saddle point, the correct result (11.100) would have been obtained. It is now clear that the saddle point for (11.99) must be found by writing the integrand

$$S_1 C^{-1} \exp[-ik\{S_1 x + C(2z_1 - z)\} + \ln\{R(S_1, S_2)\}] \tag{11.103}$$

and using this exponent in (9.63). If R is a function only of θ so that it is independent of azimuth, it can be shown that $\partial(\ln R)/\partial S_2$ has a factor S_2 so that still $S_{20} = 0$ as in (11.100). But the equation for finding S_{10} is now

$$-ik\{Cx - (2z_1 - z)S_{10}\} + \frac{1}{R}\frac{\partial R}{\partial\theta} = 0. \tag{11.104}$$

This gives (11.100) only if

$$\left|\frac{1}{R}\frac{\partial R}{\partial\theta}\right| \ll |kCx| \approx |kS_1(2z_1 - z)| \tag{11.105}$$

and this is the condition that must be satisfied when the plane wave reflection coefficient is used for spherical waves. It is amply satisfied in many practical cases of radio reflection from the ionosphere or from the earth's surface. This applies even for radio waves of very low frequency (VLF) for which k is small. It is important that the reference level is correctly chosen so that R does not contain a hidden exponential factor like the one in (11.101)

In general R is complex and

$$\frac{1}{R}\frac{\partial R}{\partial \theta} = \frac{\partial}{\partial \theta}(\ln|R|) + \mathrm{i}\frac{\partial}{\partial \theta}(\arg R). \tag{11.106}$$

The second term is imaginary, and is large if the reference level has been wrongly chosen. Now (11.101) shows that

$$\frac{\partial}{\partial \theta}(\arg R_2) = \frac{\partial}{\partial \theta}(\arg R) - 2kS(z_2 - z_1) \tag{11.107}$$

so that a new reference level z_2 can always be found to make (11.107) zero. This rule would in general make the reference level z_2 depend on θ. In practical studies of reflection coefficients it is usual and simpler to use the same reference level for all θ. It is usually chosen so that (11.107) is zero somewhere within the range of θ that is of interest.

The first term of (11.106) is real so that the solution S_{10} of (11.104) is complex. This topic is discussed further in § 11.15. Very often, however, this term is ignored in (11.102) and S_{10} is taken as real.

The function $R(S_1, S_2)$ in (11.99) in general has poles and zeros and branch points. When a contributing saddle point is near to one of these, the simple form of the method of steepest descents cannot be used, and a more elaborate treatment is needed. Examples of these problems in radio propagation have been mentioned in § 9.6. They are outside the scope of this book.

When the reflection coefficient $R(S_1, S_2)$ in (11.99) is not independent of the azimuth angle, $\arctan(S_2/S_1)$, the integrals can still be evaluated by steepest descents, and it is found that the image transmitter is not at the geometrical image point but displaced from it laterally. This is the phenomenon of lateral deviation. Practical cases of it usually occur at high frequencies and are more easily handled by ray tracing methods. Examples have been given in §§ 10.13, 10.14 and figs. 10.12, 10.13.

11.15. Goos–Hänchen shifts for radio waves

In this section the formula (11.99) is used to give the signal reaching a receiver on the ground $z = 0$, in the plane $y = 0$ at distance x from the transmitter at the origin. It is assumed that the reflection coefficient R is a function only of the angle of incidence $\theta = \arccos C$, as would occur, for example, if the ionosphere were isotropic. To indicate this we write it $R(\theta)$. If it is so slowly varying that it can be set equal to its value at the saddle point and placed outside the integral, the factor that remains is an integral representing a spherical wave coming from a point source at the geometrical image point $x = y = 0$, $z = 2z_1$, of the transmitter in the reference plane $z = z_1$. The line from this image point to the receiver is then the wave normal of the predominant wave front at the receiver. It makes with the z axis an angle θ_0, where

$$\tan\theta_0 = x/2z_1. \tag{11.108}$$

The wave has, in effect, travelled a distance

$$r_0 = (x^2 + 4z_1^2)^{\frac{1}{2}} \tag{11.109}$$

from the image point. If the reflector at $z = z_1$ had been perfect, $R = 1$, the predominant wave at the receiver would contain a factor $e^{-i\varphi}$ where the phase φ is kr_0. If the factor R is allowed for, the complex phase is

$$\varphi_0 = kr_0 + \mathrm{i}\ln R(\theta_0) = k(x\sin\theta_0 + 2z_1\cos\theta_0) - \arg R(\theta_0) + \mathrm{i}\ln|R(\theta_0)|. \tag{11.110}$$

The real part of this now includes the real phase change $-\arg R(\theta_0)$ at reflection. The imaginary part gives an amplitude reduction factor $|R(\theta_0)|$.

These results are modified if the dependence of R on θ is allowed for. The integral (11.99) is evaluated by double steepest descents, §9.10, but the second derivatives $\partial^2/\partial S_1 \partial S_2$ of the exponent are zero so the S_1 and S_2 integrations can be done independently. The double saddle point is where $S_2 = 0$, $\theta = \theta_g$ with

$$\tan\theta_g = \tan\theta_0 - \frac{\sec\theta_g}{2kz_1}\left(\frac{\partial(\arg R)}{\partial\theta} - \frac{\mathrm{i}}{|R|}\frac{\partial|R|}{\partial\theta}\right)_{\theta=\theta_g}. \tag{11.111}$$

This shows that the predominant wave normal at the receiver no longer has the direction θ_0 of the geometrical image point. There is an angular shift $\theta_g - \theta_0$. This is also the shift of the direction of the virtual source point beyond the reference plane at $z = z_1$. Equation (11.111) has more than one solution, so that there are more than one saddle point contributing to the integral (11.99), each corresponding to a different virtual image source. Usually it is only the contribution of greatest amplitude that is important, but cases can occur where there are two reflected waves of approximately equal amplitudes and different directions arriving at the receiver. For an example see the results at the end of this section. The following theory applies for only one of the solution θ_g of (11.111).

The phase φ_g of the wave reaching the receiver is now

$$\varphi_g = k(x\sin\theta_g + 2z_1\cos\theta_g) - \arg R(\theta_g) + \mathrm{i}\ln|R(\theta_g)| \tag{11.112}$$

A term $-\mathrm{i}\ln\{r_0\sin\theta_g/(r_g\sin\theta_0)\}$ should be added to (11.112) but it is very small and is neglected here. Thus there is a shift $\varphi_g - \varphi_0$ of the complex phase. The second derivatives of the exponent of (11.99) are

$$\partial^2/(\partial S_1)^2{:}\ \frac{2\mathrm{i}kz_1}{C^3} + \frac{1}{C}\frac{\partial}{\partial\theta}\left(\frac{1}{CR}\frac{\partial R}{\partial\theta}\right), \tag{11.113}$$

$$\partial^2/(\partial S_2)^2 \colon \frac{2ikz}{C} + \frac{1}{S_1 CR}\frac{\partial R}{\partial \theta}, \tag{11.114}$$

where θ takes the value θ_g. Thus the formulae (9.64), (9.65) show that the double integral (11.99) has a factor $1/r_g$ where

$$r_g^2 = C^{-2}\left\{z_1 - \frac{iC^2}{2k}\frac{\partial}{\partial \theta}\left(\frac{1}{CR}\frac{\partial R}{\partial \theta}\right)\right\}\left\{z_1 - \frac{i}{2kS_1 R}\frac{\partial R}{\partial \theta}\right\} \tag{11.115}$$

and the factor C^{-1} in (11.99) has been included. If R is independent of θ, this gives $r_g = z_1/C$ and since C is now $\cos\theta_0$ from (11.108), this is just r_0 (11.109). The wave front is spherical with its centre at the geometrical image point. But now the two factors of (11.115) show that the S_1 and S_2 integrations respectively give different ranges. The wave front does not come from a single virtual point source and is no longer exactly spherical. The wave normals of the reflected wave front form an astigmatic pencil of rays (Deschamps, 1972). But we shall regard the geometric mean r_g as a new effective range. Thus the angle dependence of R gives a range shift $r_g - r_0$.

Suppose now that there is some range of θ for which $|R| = 1$ when θ is real. This is a case of total reflection, and examples have been given in § 11.9, equation (11.63), (11.64). In optics it occurs when a wave in a transparent medium such as glass is reflected at a boundary where the medium beyond it has a smaller refractive index, and the angle of incidence exceeds the critical angle. It is then known as total internal reflection. In this case $\partial|R|/\partial\theta$ is zero in (11.111) and (11.115). Then θ_g and r_g are both real, and the angular shift $\theta_g - \theta_0$ and the range shift $r_g - r_0$ are real. These two shifts are known as the Goos–Hänchen shifts (Goos and Hänchen, 1947) of the optical image of a point source in a reflecting plane. They have been fully studied in optics. A very complete account has been given in four papers by Lotsch (1970).

For radio waves, the θ dependence of R is important in the study of reflections from the ionosphere, especially for very low frequencies (VLF). Now $\partial|R|/\partial\theta$ is not zero so the angular and range shifts given from (11.111), (11.115) are complex, and there is a shift $\varphi_g - \varphi_0$ of the complex phase, given by (11.110), (11.112). A complex θ_g means that the predominant wave at the receiver is an inhomogeneous plane wave; §2.15. A complex effective range r_g means that the wave front at the receiver has a complex curvature. The physical significance of a wave from a point source whose space coordinates are complex has been discussed by Deschamps (1972) and by Felsen and his co-workers (see, for example, Ra, Bertoni and Felsen, 1973).

In VLF studies measurements are often made of the signal amplitude at the ground for various distances from the transmitter. There are two main contributions to this signal, the direct or ground wave and the wave reflected from the ionosphere. Their relative phase depends on distance so that the curve of signal amplitude versus

Table 11.1

Horizontal Range x (km)	θ_0 (deg)	Complex angle shift $\theta_g - \theta_0$ (deg)	r_0 (km)	Complex range shift $r_g - r_0$ (km)	Complex phase shift $\varphi_g - \varphi_0$ (deg)	$\lvert r_0/r_g \rvert \exp\{\mathrm{Im}(\varphi_g)\}$
100	35.5	0.27 − 0.71i	172.0	− 2.2 + 4.8i	− 0.23 − 0.23i	0.392
200	55.0	1.01 − 1.85i	244.1	− 17.6 + 21.3i	− 1.20 − 2.46i	0.261
		13.25 − 3.88i		− 4.1 − 405.5i	− 63.03 + 72.69i	0.117
250	60.8	2.13 − 3.62i	286.5	− 77.4 + 86.2i	− 0.68 + 0.23i	0.212
		6.40 − 2.81i		− 109.6 − 199.0i	− 0.06 + 0.41i	0.169
260	61.7	2.08 − 4.54i	295.3	− 75.1 + 142.2i	0.23 + 0.28i	0.171
		5.51 − 1.27i		− 77.8 − 146.1i	5.27 + 0.21i	0.168
270	62.6	5.02 − 0.53i	304.1	− 46.0 − 123.1i	9.11 + 0.09i	0.155
		1.68 − 5.29i		− 44.8 + 186.1i	− 0.40 + 0.28i	0.127
300	65.0	4.11 + 0.68i	333.1	2.2 − 92.2i	15.12 − 3.84i	0.143
		0.20 − 6.55i		64.1 + 249.1i	− 5.9 − 141.2i	0.060
400	70.7	2.34 + 1.72i	423.8	39.5 − 40.0i	6.47 + 11.3i	0.175

distance shows an interference effect. It is called a 'Hollingworth interference pattern' since Hollingworth (1926) was one of the first to use this method. Much work of this kind has been done for frequencies of 16 kHz and near (see references in § 19.7). Conversely, for a given model of the ionosphere when $R(\theta)$ is known, it is required to calculate the Hollingworth interference pattern, so that it can be compared with the observed pattern.

It is important to check whether the complex Goos–Hänchen shifts are likely to be important in these calculations. Equations (11.111), (11.115) show that the terms that give these shifts are inversely proportional to k, that is to frequency, so that as the frequency gets smaller, the shifts get more important.

To test this, some calculations were made with $R(\theta)$ for a sharply bounded isotropic medium, given by R_{11} in (11.55) with $n_1 = 1$. The function $R(\theta)$ used is shown in fig. 11.1 and is of similar form to the calculated reflection coefficient for model ionospheres (Budden, 1955b; Deeks, 1966a, b; Pitteway, 1965). This example has, however, been deliberately chosen to have a pronounced minimum of $|R|$ and an associated steep gradient of arg R, so as to illustrate the double image point effect. The reflector was assumed to be at height $z_1 = 70$ km and this was used as the reference level. The receiver on the ground was assumed to be at various distances from the transmitter, and the earth's curvature was neglected. Table 11.1 shows some examples of the calculated shifts for a frequency of 16 kHz. The angle θ_0 and the range r_0 are given by (11.108), (11.109), and the phase angles φ_0, φ_g are given by (11.110), (11.112). The signal amplitude is determined mainly by $|r_0/r_g| \exp\{\text{Im}(\varphi_g)\}$ which appears in the last column.

The table shows that, at distances near 260 km from the transmitter, there are two

Fig. 11.1. Dependence on angle of incidence θ of the modulus and argument of the reflection coefficient R, as given by (11.55), for a sharply bounded medium with refractive index $n = 2.39 - 0.84i$. The electric fields of the waves are in the plane of incidence. This example is used in the adjacent table to study the Goos–Hänchen shifts.

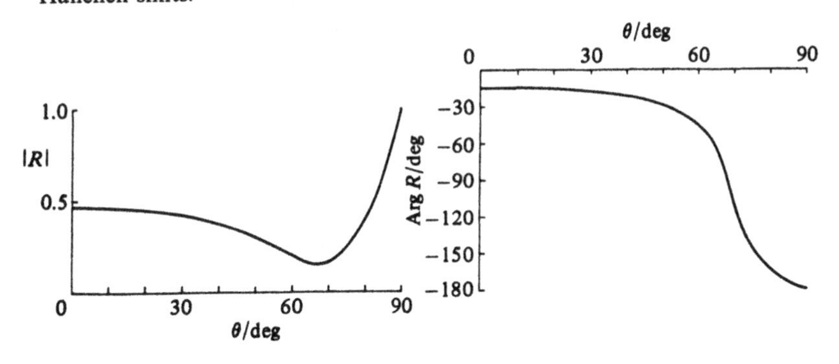

contributing reflected waves with comparable amplitudes. For a function $R(\theta)$ with a less pronounced minimum, however, this effect is absent. These and other similar calculations show that, for 16 kHz, the shifts are small compared with errors of measurement and with uncertainties that arise from variability of the ionosphere. The phase shift of 15° at 300 km, shown in the table, might just be large enough to be important.

11.16. The shape of a pulse of radio waves

In the earlier sections of this chapter it has been assumed that the transmitter emits a continuous wave of constant amplitude and fixed frequency. If the transmitted wave is modulated so that, for example, it consists of pulses, then the emitted signal is a combination of many frequencies, and to find the signal reflected from the ionosphere it is necessary to allow for the frequency dependence of the reflection coefficient. Pulsed radio signals, called wave packets, were studied in §§ 5.8, 10.3 and represented by Fourier integrals (5.62), (10.8). But the exact form of the pulse played no part. It was simply assumed that the pulse had a constant frequency f_1 and roughly constant amplitude, and that it lasted for a time T long compared with $1/f_1$. The spectrum function $M(\omega)$ of (5.62) with $\omega = 2\pi f$, was then large near $f = f_1$ and $f = -f_1$, and became very small at other frequencies. It is now of interest to investigate how the form of a pulse of radio waves is modified by a reflection from the ionosphere. It is convenient to use the actual frequency f instead of the angular frequency ω, or the wave number k.

The signal from the transmitter is to be Fourier analysed into its component frequencies. Each frequency contributes an angular spectrum of plane waves, such as (10.1) or (11.98). After reflection from the ionosphere, the signal reaching the receiver is a predominant wave such as (10.4). If the corresponding wave in the emitted angular spectrum has an electric field proportional to $\exp(2\pi i f t)$, the field of the received signal from this wave is proportional to $\exp\{2\pi i f(t - P/c)\}$ where P is the phase path, defined in § 10.16. Here $2\pi f P/c$ is the total change of phase of the signal as it travels from transmitter to receiver. It may be given by (10.39) for a ray path, or it may be the sum of contributions from the paths up to and down from the ionosphere and the phase change on reflection. See for example the exponent in (11.103). In the examples given later in this section, figs. 11.2, 11.3, $P(f)$ is a real function and is best regarded as arising from a ray path as in (10.39). But in general $P(f)$ has an imaginary part arising from attenuation or because $|R| < 1$.

Let the electric field of the predominant wave in the pulse emitted by the transmitter be

$$E(t) = m(t)\cos(2\pi f_1 t) \tag{11.116}$$

where t is time and $m(t)$ is appreciable only in a range $|t| < T$, and varies very slowly compared with $\cos(2\pi f_1 t)$. If this were delineated on the screen of a cathode ray oscillograph with a linear time base, it would appear as a cosine wave $\cos(2\pi f_1 t)$, the carrier (radio) wave, modulated by the function $m(t)$. This function is therefore called the 'shape' of the pulse, and describes how the envelope of the pulse varies with time.

The function $m(t)$ and its Fourier transform $M(f)$ are related thus

$$m(t) = \int_{-\infty}^{\infty} M(f)\exp(2\pi i f t)\,df, \quad M(f) = \int_{-\infty}^{\infty} m(t)\exp(-2\pi i f t)\,dt \quad (11.117)$$

and $M(f)$ is appreciable only where $|f|$ is very small compared with the carrier frequency f_1. Since $m(t)$ is necessarily real, $|M(f)|$ is a symmetric function of f, and $\arg M(f)$ is antisymmetric. Then a well-known theorem in Fourier analysis shows that

$$E(t) = \mathrm{Re}\int_{-\infty}^{\infty} M(f - f_1)\exp(2\pi i f t)\,df. \quad (11.118)$$

This signal is emitted from the transmitter. The symbol Re is usually omitted but will be retained in this section. Hence the signal reaching the receiver is

$$E_r(t) = \mathrm{Re}\int_{-\infty}^{\infty} M(f - f_1)\exp\{2\pi i f(t - P/c)\}\,df. \quad (11.119)$$

Now $M(f - f_1)$ is only appreciable when $|f - f_1|$ is small. This suggests that the term $P(f)$ in (11.119) should be expanded in a Taylor series about $f = f_1$. Let

$$f - f_1 = \sigma. \quad (11.120)$$

Then, with (10.40),

$$fP(f) = f_1 P(f_1) + \sigma P'(f_1) + \tfrac{1}{2}\sigma^2 P'_1(f_1) + \tfrac{1}{6}\sigma^3 P'_2(f_1) + \cdots \quad (11.121)$$

where P'_1, P'_2, are written for $\partial P'/\partial f, \partial^2 P'/\partial f^2$ respectively when $f = f_1$.

It was shown in § 10.16 that when P' is real, the time for the pulse to travel from transmitter to receiver is P'/c where P' is the equivalent path, given by (10.40). In general P' is complex, so let

$$t = \tau + P'(f_1)/c, \quad \tau = u + iv \quad (11.122)$$

where $u = \mathrm{Re}(\tau)$ is a new measure of time.

The integral (11.119) is now

$$E_r(t) = \mathrm{Re}\exp[2\pi i f_1\{t - P(f_1)/c\}]$$
$$\times \int_{-\infty}^{\infty} M(\sigma)\exp\{2\pi i(\sigma\tau - \tfrac{1}{2}\sigma^2 P'_1/c - \tfrac{1}{6}\sigma^3 P'_2/c \ldots)\}\,d\sigma. \quad (11.123)$$

The first exponential is the high frequency oscillation. If $P(f_1)$ has an imaginary part this gives an amplitude reduction factor $\exp[2\pi f_1\{\mathrm{Im}\,P(f_1)\}/c]$. The integral is the function that modulates the high frequency and therefore gives the pulse shape. Clearly if P'_1, P'_2 and higher derivatives are negligible, and if P' is real, the integral is just $m(\tau)$, from (11.117), that is the original pulse, undistorted but delayed by a time P'/c.

Two special cases are now of interest, both for real P'. The first is when the carrier frequency f_1 is so chosen that the curve of $P'(f)$ has a positive slope and only small curvature. There is an example in fig. 12.5(b) at a frequency f_1; this figure gives $h'(f)$ for vertical incidence, and $P' = 2h'$. Then P'_1 is positive and P'_2 is small. If the original pulse is wide enough, $M(\tau)$ falls quickly to zero as σ increases from zero, and we therefore neglect the σ^3 term in (11.123). Then the pulse shape is given by

$$\int_{-\infty}^{\infty} M(\sigma)\exp(-\pi \mathrm{i} P'_1\sigma^2/c)\exp(2\pi\mathrm{i}\sigma\tau)\,\mathrm{d}\sigma. \tag{11.124}$$

This can be transformed by the convolution theorem of Fourier analysis. The Fourier transform of the first exponential in (11.124) is

$$(P'_1/c)^{-\frac{1}{2}}\exp(-\tfrac{1}{4}\pi\mathrm{i})\exp(\pi\mathrm{i}ct^2/P'_1) \tag{11.125}$$

and hence (11.124) becomes, apart from a constant factor

$$\int_{-\infty}^{\infty} m(\tau - t)\exp(\pi\mathrm{i}ct^2/P'_1)\,\mathrm{d}t. \tag{11.126}$$

Integrals of this type are used in optics in finding the Fresnel diffraction pattern of an aperture. Suppose that a parallel beam of monochromatic light of wavelength λ is incident on a slit which allows light to pass with amplitude $m(x)$ where x is measured at right angles to the length of the slit. A screen is placed at a distance y from the slit, and z is distance measured parallel to x in the plane of the screen. Then the amplitude of light arriving at the screen is proportional to

$$\int_{-\infty}^{\infty} m(z-\xi)\exp(2\pi\mathrm{i}\xi^2/\lambda y)\,\mathrm{d}\xi \tag{11.127}$$

and this is called the Fresnel diffraction pattern of the slit. Hence we may say that when a pulse of radio waves is reflected from the ionosphere, in conditions where the $P'(f)$ curve has a finite slope and small curvature, the pulse shape is modified to that of its own Fresnel diffraction pattern. Fig. 11.2 shows this modification for a pulse which is initially rectangular. The distorted pulses show oscillations whose scale in time depends on the value of P'_1. Further examples are given by Rydbeck (1942b).

The second special case is when the $P'(f)$ curve has zero slope, but appreciable curvature, so that $P'(f)$ is a maximum or a minimum, for example, at a frequency F_N or f_2 in fig. 12.5(b). Then the pulse shape is given by

$$\int_{-\infty}^{\infty} M(\sigma)\exp(-\tfrac{1}{3}\pi\mathrm{i}P'_2\sigma^3/c)\exp(2\pi\mathrm{i}\sigma\tau)\,\mathrm{d}\sigma. \tag{11.128}$$

The Fourier transform of the exponential is

$$\int_{-\infty}^{\infty} \exp\{\pi\mathrm{i}(2\sigma t - \tfrac{1}{3}P'_2\sigma^3/c)\}\,\mathrm{d}\sigma = (2\pi^2 c/P'_2)^{\frac{1}{3}}\mathrm{Ai}\{-2t(c\pi^2/P'_2)^{\frac{1}{3}}\} \tag{11.129}$$

from (8.18) where Ai is the Airy integral function. The convolution theorem may now be used as before and gives for (11.128) (apart from a constant factor)

$$\int_{-\infty}^{\infty} m(\tau - t)\,\mathrm{Ai}\{-2t(c\pi^2/P_2')^{\frac{1}{3}}\}\,\mathrm{d}t. \tag{11.130}$$

This expression has no simple interpretation like the Fresnel diffraction pattern used in the preceding example. Suppose that the original pulse is rectangular, that is $m(t) = 1$ for $0 < t < T$, and $m(t) = 0$ for all other values of t. Then (11.130) becomes

$$\int_{\tau - T}^{\tau} \mathrm{Ai}\{-2t(c\pi^2/P_2')^{\frac{1}{3}}\}\,\mathrm{d}t. \tag{11.131}$$

Fig. 11.3 shows an example of this function. (Another example is given by Rydbeck, 1942b). If P_2' is positive, as at $f = f_2$ in fig. 12.5(b), then the curve is as shown. If P_2' is negative as at $f = F_N$ in fig. 12.5(b), then the curve of fig. 11.3 is reversed in time and the 'tail' becomes a precursor. This occurs because the 'side-band' frequencies arrive before the predominant frequency.

Suppose now that P and P' are complex. Equations (11.119)–(11.126) are still true but τ is complex. The imaginary part of P simply gives attenuation of the pulse as

Fig. 11.2. Distortion of a rectangular pulse after reflection when the $P'(f)$ curve has a constant slope. The original pulse is shown by broken lines. The values of $\mathrm{d}P'/\mathrm{d}f$ in km MHz^{-1} for the three diagrams are: top 4, middle 40, bottom 400.

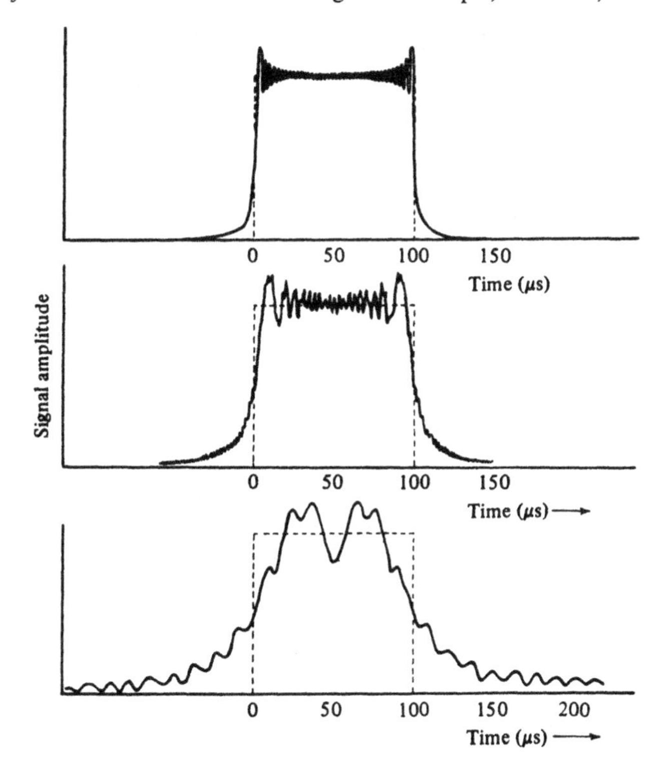

shown by the first factor of (11.123). If P'_1 and P'_2 are small enough to be neglected, (11.124) shows that the received pulse shape is $m(\tau) = m(u + \mathrm{i}v)$ from (11.117). This requires that we interpret the original pulse shape $m(t)$ for complex time t.

Suppose that $m(t) = \exp(-\gamma t^2)$. Then

$$m(\tau) = \exp(-\gamma u^2)\exp(\gamma v^2)\exp(-2\mathrm{i}\gamma vu) \tag{11.132}$$

Here u, given by (11.122), is the new measure of time. The received pulse has maximum amplitude when $u = 0$, that is $t = \mathrm{Re}\{P'(f_1)\}/c$. Thus there is a time delay $\mathrm{Re}(P')/c$. The last factor of (11.132) gives a phase change that is equivalent to a shift of the high frequency carrier to $f_1 - \gamma\,\mathrm{Im}\{P'(f_1)\}/\pi c$. The factor $\exp(\gamma v^2)$ gives an increase of amplitude which must partially compensate the decrease caused by $\mathrm{Im}(P)$.

If $m(t)$ is a rectangular pulse or some similar function with discontinuities, it is not simple to interpret $m(\tau)$ for complex τ. One method now is to use some analytic approximation for $m(t)$. It is then found that $\mathrm{Im}(\tau)$ gives rise to a distortion of the pulse and a frequency shift, but the centre of a symmetric pulse still arrives when $u = 0$, so that the time delay is still $\mathrm{Re}(P')/c$.

If P'_2 is negligible and P'_1 is constant as in (11.124), the result (11.126) for the received pulse shape $m_\tau(u)$ can still be used with complex τ. By changing the variable

Fig. 11.3. Distortion of a rectangular pulse after reflection when the $P'(f)$ curve has a minimum and $\mathrm{d}^2P'/\mathrm{d}f^2 = 2.34 \times 10^{-11}\,\mathrm{km\,s^2} = 23.4\,\mathrm{km\,MHz^{-2}}$. The original pulse is shown by broken lines.

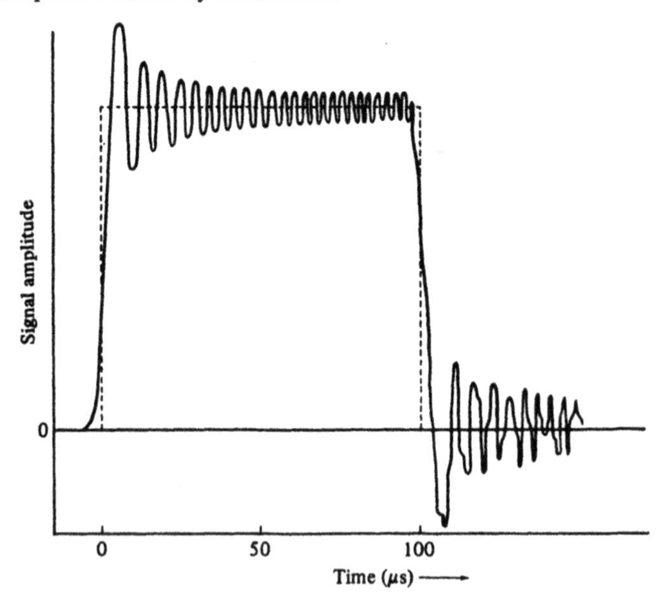

of integration and using (11.122) it can be written

$$m_r(u) = \int_{-\infty}^{\infty} m(s)\exp\{\pi i c(s-u-iv)^2/P'_1\}\,ds. \tag{11.133}$$

Suppose that $m(t)$ is a rectangular pulse

$$m(t) = 1 \text{ for } -\tfrac{1}{2}T \leqslant t \leqslant \tfrac{1}{2}T \tag{11.134}$$

and zero outside this range. Then (11.133) becomes, apart from a constant multiplier,

$$m_r(u) = \int_A^B \exp(-x^2)\,dx, \quad A = \exp(-\tfrac{1}{4}i\pi)(\pi/P'_1)^{\frac{1}{2}}(-\tfrac{1}{2}T - iv - u),$$

$$B = \exp(-\tfrac{1}{4}i\pi)(\pi/P'_1)^{\frac{1}{2}}(\tfrac{1}{2}T - iv - u). \tag{11.135}$$

This integral can be expressed in terms of the error function for a complex argument, and there are published tables of this function.

This discussion has shown that there is no simple physical interpretation for a complex P', but that the time of travel of a wave packet is found from $\mathrm{Re}(P')$. This can be applied to the time of travel T of a wave packet along a ray path of length L in a homogeneous medium. Then $P' = cL/\mathscr{U}$ where $\mathscr{U}$ is the complex group velocity given by (5.68), and

$$T = \mathrm{Re}(P')/c = \mathrm{Re}(n'\cos\alpha)L/c. \tag{11.136}$$

The effective velocity of a wave packet in an attenuating medium is $c/\mathrm{Re}(n'\cos\alpha)$. For further discussion of this topic see Suchy (1972a, b, 1974a, § vii).

PROBLEMS 11

11.1. It is observed that a wave reflected from the ionosphere has the same polarisation no matter what the polarisation of the incident wave. Show that the reflection coefficient matrix $\boldsymbol{R}$ (§ 11.5) is singular, i.e. $\det \boldsymbol{R} = 0$, and find the polarisation of the reflected wave in terms of the elements of $\boldsymbol{R}$. Show that for one polarisation of the incident wave there is no reflection.

11.2. The surface of the sea may be regarded as a perfect conductor, so that the horizontal components of the total electric field close to the surface are zero. Show that for all angles of incidence the reflection coefficient matrix at the surface as reference level, is

$$\boldsymbol{R} = \begin{pmatrix} 1 & 0 \\ 0 & -1 \end{pmatrix}.$$

What is the value of $\boldsymbol{R}_0$ (§11.6)?

11.3. The polarisation of the incident wave is to be adjusted so that the reflected wave has the same polarisation. Show that in general this can be done in two different ways. What condition does $\boldsymbol{R}$ satisfy when there is only one way?

11.4. State Maxwell's electromagnetic equations in vector form for free space. Express them in terms of the components E_x, E_y, E_z, H_x, H_y, H_z of the electric and magnetic fields respectively in a right-handed system of Cartesian coordinates x, y, z, for a plane electromagnetic wave travelling with its normal parallel to the z axis.

A plane wave in free space is moving from left to right along the z axis and is incident on the plane surface $z = a$, where a is positive, and is partially reflected there. In the plane $z = 0$ the components E_x, E_y, H_x, H_y of the total field are measured and found to have the amplitudes and phases given by the following table:

	Amplitude	Phase
E_x	13	arctan (5/12)
E_y	5	0
$Z_0 H_x$	5	0
$Z_0 H_y$	13	− arctan (5/12)

Here Z_0 is the characteristic impedance of free space. Show that the incident wave is linearly polarised, and find its amplitude and the plane of its electric field. Find also the amplitude and polarisation of the reflected wave, and hence deduce the reflection coefficient.

(From Natural Sciences Tripos, 1955. Part II. Physics: Theoretical Option.)

11.5. A plane electromagnetic wave in free space is normally incident on the plane boundary of a homogeneous loss-free medium with magnetic permeability $\mu = 2$ and electric permittivity $\varepsilon = \frac{1}{2}$. Find the reflection coefficient.

(From Natural Sciences Tripos, 1955 Part II. Physics).

11.6. A piece of glass is partially silvered. Find the reflection coefficient for electromagnetic waves incident normally on the surface if the thickness of the silver is negligible compared with one wavelength. Assume that the resistivity of silver is the same at high frequencies as for steady currents. Find how thick the silver must be to make the (intensity) reflection coefficient equal to 0.8. (Refractive index of glass = 1.5. Resistivity of silver = 1.5×10^{-8} ohm m.)

(From Natural Sciences Tripos, 1953 Part II. Physics).

11.7. Find an expression for the reflection coefficient of a sharply bounded homogeneous ionosphere for propagation from (magnetic) east to west or west to east at the magnetic equator, when the electric vector is in the plane of incidence. Show that the expression is not reciprocal, i.e. it is changed when the sign of the angle of incidence is reversed.

(For the solution see Barber and Crombie, 1959).

11.8. As a rough simple model, an ionospheric layer can be pictured as a homogeneous isotropic plasma of refractive index n with two horizontal boundaries a distance d apart. Show that for angle of incidence θ the reflection and transmission coefficients R_{11}, T_{11} are given by

$$R_{11} = \sin(knd\cos\phi)/\sin(\beta - knd\cos\phi)$$

$$T_{11} = \sin\beta\exp(ikd\cos\phi)/\sin(\beta - knd\cos\phi)$$

where $\sin\theta = n\sin\phi$ and $\tan\frac{1}{2}\beta = in\cos\theta/\cos\phi$. The lower boundary is used as the reference level for R_{11}. Show that, if the plasma is loss-free so that n^2 is real, $|R_{11}|^2 + |T_{11}|^2 = 1$.
(For the solution see Budden, 1961a, § 8.10).

11.9. In problem 11.8 for normal incidence $\theta = 0$, show that

$$R_{11} = (n^2 - 1)\tan(knd)/\{(n^2 + 1)\tan(knd) - 2in\}$$

$$T_{11} = -2in\sec(knd)\exp(ikd)/\{(n^2 + 1)\tan(knd) - 2in\}.$$

Suppose that the medium is a collisionless electron plasma so that $n^2 = 1 - \omega_N^2/\omega^2$. Sketch curves to show how $|R_{11}|$ depends on frequency ω, especially when ω is near to ω_N. Show that when $\omega = \omega_N$, $|R_{11}|^2 = (kd)^2/\{4 + (kd)^2\}$, $|T_{11}|^2 = 4/\{4 + (kd)^2\}$.
(For curves giving results see Budden, 1961a, § 8.11.)

11.10. A plane surface has reflection coefficient $R = 1$ for all angles of incidence θ. It is placed at a height $z = z_0$ and there is a transmitter at the origin of coordinates. In calculating the properties of the reflected wave a reference level $z = z_1 < z_0$ is used where $z_0 - z_1$ is small. Thus the effective reflection coefficient is $R_1 = \exp\{2ik(z_1 - z_0)\cos\theta\}$ (see(11.10)). The geometrical image point A of the transmitter in the reference plane is at $x = y = 0$, $z = 2z_1$. Find the Goos–Hänchen shifts of the true image source point from A, in angle and range. Hence show that the true image source is at $x = y = 0$, $z = 2z_0$, that is at the geometrical image point in the actual reflector at $z = z_0$.

12

Ray theory results for isotropic ionosphere

12.1. Introduction

Although the earth's magnetic field has a very important influence on the propagation of radio waves in the ionospheric plasma, it is nevertheless of interest to study propagation when its effect is neglected. This was done in the early days of research in the probing of the ionosphere by radio waves, and it led to an understanding of some of the underlying physical principles; see, for example, Appleton (1928, 1930). In this chapter, therefore, the earth's magnetic field is ignored. For most of the chapter the effect of electron collisions is also ignored, but they are discussed in §§ 12.3, 12.11. The effect both of collisions and of the earth's magnetic field is small at sufficiently high frequencies, so that some of the results are then useful, for example with frequencies of order 40 MHz or more, as used in radio astronomy. For long distance radio communication the frequencies used are often comparable with the maximum usable frequency, §§ 12.8–12.10, which may be three to six times the penetration frequency of the F-layer. They are therefore in the range 10 to 40 MHz. This is large compared with the electron gyro-frequency which is of order 1 MHz, so that here again results for an isotropic ionosphere are useful, although effects of the earth's magnetic field have to be considered for some purposes.

This chapter is largely concerned with the use of pulses of radio waves and the propagation of wave packets, and uses results from ch. 10, especially §§ 10.2–10.6. But the ray direction and the wave normal direction are the same. There is no lateral deviation, §§ 10.12–10.14, and no phenomenon of the Spitze, §§ 10.9–10.11. The first half is largely concerned with probing the ionosphere at vertical incidence, and the second half with oblique incidence and ray paths.

The present chapter is not concerned with the wave polarisation of the waves, and the field of a wave will be denoted simply by E which may be thought of as one component of the electric field, although the arguments could equally well be applied to a component of the magnetic field. The analysis in this chapter will usually be expressed in terms of the wave frequency f rather than the angular frequency $\omega = 2\pi f$.

12.2. Vertically incident pulses

Consider first a horizontally stratified ionosphere in which the electron concentration increases monotonically with height z. A vertically incident pulse of radio waves travels upwards with the group velocity $\mathscr{U} = c/n'$, from (5.66), until it is reflected at some height z_0 where $n = 0$, and returns to the ground $z = 0$. The height $z_0(f)$ is called the 'true height' of reflection. It depends on the carrier frequency f of the pulse. The time of travel of the wave packet is P'/c where P' is the equivalent path or group path § 10.16. For a collisionless plasma, to which (5.66) applies,

$$P'(f) = 2h'(f) = 2c\int_0^{z_0} \frac{dz}{\mathscr{U}}, \tag{12.1}$$

$$h'(f) = \int_0^{z_0} n'\,dz = \int_0^{z_0} \frac{1}{n}\,dz, \tag{12.2}$$

and $h'(f)$ is called the 'equivalent height' of reflection. The f is again used to indicate the carrier frequency of the pulse. The integrand of (12.2) is infinite at the upper limit $z = z_0$, but the integral is bounded when z_0 is not at the maximum of an ionospheric layer, because $1/n$ behaves like $(z_0 - z)^{-\frac{1}{2}}$ and its integral converges. The total change of phase of a component wave of frequency f in the pulse is $2\pi f P/c$ where $P(f) = 2h(f)$ is the phase path. Now

$$h(f) = \int_0^{z_0} n\,dz \tag{12.3}$$

is called the 'phase height' of reflection. On the real part of its path where the wave is propagated, $n \leqslant 1$ and (12.2), (12.3) then show that

$$h(f) < z_0(f) < h'(f). \tag{12.4}$$

Although z_0 is a function of f, this was not allowed for when deriving (12.2). To show that the result is correct it is instructive to derive (12.2) by another method. For any component frequency f in the pulse, the total phase change for the upward and downward path is $4\pi f h(f)/c$, where $h(f)$ is given by (12.3). Thus the field of the reflected wave reaching the ground is, compare (11.119):

$$E(t) = \int_{-\infty}^{\infty} M(f - f_1)\exp\left\{2\pi i f\left(t - \frac{2}{c}\int_0^{z_0(f)} n\,dz\right)\right\} df. \tag{12.5}$$

Here $M(f - f_1)$ is the Fourier transform of the pulse envelope, shifted in frequency, as in (11.117), (11.118), so that it is centred on the carrier frequency of the pulse, here temporarily denoted by f_1. It is a slowly varying function of f. The time τ of arrival of the pulse at the ground is the value of t that makes the phase of the integrand stationary for frequencies near f_1. Hence

$$\left[\frac{\partial}{\partial f}\left\{f\left(\tau-\frac{2}{c}\int_0^{z_0} n\mathrm{d}z\right)\right\}\right]_{f=f_1}=0 \tag{12.6}$$

so that

$$c\tau=2\int_0^{z_0}\left\{\frac{\partial}{\partial f}(fn)\right\}_{f=f_1}\mathrm{d}z+2f_1 n(z_0)\left(\frac{\partial z_0}{\partial f}\right)_{f=f_1} \tag{12.7}$$

Now $n(z_0)$ is zero and $\partial z_0/\partial f$ is bounded except when f_1 is near the penetration frequency of an ionospheric layer, and this case needs special treatment by full wave methods; see § 15.11. Apart from this the last term of (12.7) is zero, and (5.66) shows that the remaining term is the same as (12.2). Thus the dependence of z_0 on frequency does not affect the result. This is because, near the level of reflection, n is close to zero and the wavelength in the medium is very large. Hence small variations of reflection level do not affect the phases of the component waves.

Curves showing the form of the functions $h'(f)$, $z_0(f)$, $h(f)$ in specific cases have been given by Budden (1961a) and others. See also problems 12.1–12.3.

12.3. Effect of collisions on phase height $h(f)$ and equivalent height $h'(f)$

The uniform approximation solution (8.72) for vertically incident waves, and the W.K.B. solutions derived from it, are functions of the height z. The fields they represent can, of course, only be observed at real values of z. But in these solutions we can make z complex and require that the same differential equations are satisfied. This is the process of analytic continuation. It shows that (8.72) is still valid when continued analytically to use complex values of z, provided that the boundary condition at great heights is satisfied. When z is real and large the solution must represent a purely upgoing wave. It can be shown that this is so for the monotonic functions $N(z)$ considered in this section.

When collisions are allowed for the refractive index n is given by (4.9) and is complex. There is no real value z_0 of z that makes $n=0$, but there is a complex z_0. The reflected wave is still given by (7.153) with $C=1$ and $q=n$ for vertical incidence. The integrals are now contour integrals in the complex z plane. The factor i from (8.73) is now included although it is unimportant here. Thus the reflection coefficient measured at the ground is, from (7.151):

$$R=\mathrm{i}\exp\{-2\mathrm{i}kh(f)\} \tag{12.8}$$

where the phase height $h(f)$ is still given by (12.3) but is now complex. In (12.2) n' is no longer equal to $1/n$ so this equation must be rewritten

$$h'(f)=\int_0^{z_0} n'\,\mathrm{d}z=\int_0^{z_0}\frac{\partial}{\partial f}(fn)\mathrm{d}z \tag{12.9}$$

and this defines a complex equivalent height of reflection.

The quantities $2h(f)$ and $2h'(f)$ are special cases of the phase path $P(f)$ and the equivalent path $P'(f)$ respectively. The physical interpretation of complex values of these quantities was discussed in §11.16. It was shown that the time of travel of a wave packet is $\mathrm{Re}\{P'(f)\}/c$ so that the effective equivalent height of reflection is $\mathrm{Re}\{h'(f)\}$.

Usually in radio observations it is the magnitude $|R|$ of the reflection coefficient that is measured. Now (12.8) gives

$$\ln|R| = 2k\,\mathrm{Im}\,(h). \tag{12.10}$$

This is zero if there are no collisions. Introduce a small non-zero collision frequency ν and suppose that it is independent of height z. This is a useful assumption because $\nu(z)$ is much less strongly height dependent than $N(z)$; see fig. 1.2. There is then a small change of the phase height h:

$$\delta h \approx Z(\partial h/\partial Z)_{Z=0} = Z\left\{\int_0^{z_0} (\partial n/\partial Z)_{Z=0}\,\mathrm{d}z + n(z_0)\partial z_0/\partial Z\right\}. \tag{12.11}$$

But $n(z_0)$ is zero, and from (4.9):

$$(\partial n/\partial Z)_{Z=0} = -\tfrac{1}{2}\mathrm{i}X/n = \tfrac{1}{2}\mathrm{i}(n-n') \tag{12.12}$$

so that (12.11), with (12.3), (12.9), gives

$$\delta h \approx \tfrac{1}{2}\mathrm{i}Z(h-h'). \tag{12.13}$$

Then finally, from (12.10), the attenuation is

$$-\ln|R| \approx \frac{\nu}{c}(h'-h). \tag{12.14}$$

This formula was used for estimating the electron collision frequency ν (Appleton, 1935; Farmer and Ratcliffe, 1935). When the earth's magnetic field is allowed for the theory needs some modification. This is discussed in §13.10.

For one useful simple model of the ionosphere, there is free space in the height range $0 \leqslant z \leqslant h_0$ and above this the electron concentration $N(z)$ is proportional to $z-h_0$. The collision frequency is independent of z so that Z is a constant. For this model it can be shown (see problem 12.1) that the real parts of (12.3) and (12.9) are independent of Z. It can further be shown that (12.14) is true for any value of ν and is not restricted to small values.

In another model the electron concentration increases exponentially with height z. The values of $h(f)$ and $h'(f)$ are then as given in problem 12.2. If there is a constant collision frequency, the squared refractive index is

$$n^2 = 1 - F^2\mathrm{e}^{\alpha z}/(1-\mathrm{i}Z). \tag{12.15}$$

It can then be shown that

$$-\ln|R| = k(\arctan Z)\,\mathrm{Re}\{h'(f)-h(f)\}. \tag{12.16}$$

The proof is left to the reader. If Z is small, this is the same as (12.14).

12.4. Equivalent height for a parabolic height distribution of electron concentration

Suppose that the electron concentration is given by the parabolic law

$$N(z) = N_m\{1-(z-z_m)^2/a^2\} \quad \text{for}\,|z-z_m| \leqslant a, \qquad N(z) = 0 \quad \text{for}\,|z-z_m| \geqslant a. \tag{12.17}$$

One curve in fig. 12.1 shows how N depends on z in this case. It has a maximum value N_m where $z = z_m$. The constant a is called the 'half thickness' of the parabolic layer. This example is of particular importance in the study of layers that have a maximum of electron concentration. Some important properties of any layer depend on the curvature of the $N(z)$ curve at the maximum. By studying the properties of a parabolic layer with the same curvature at the maximum it is possible to derive results for other layers.

The plasma frequency that corresponds to the maximum electron concentration N_m will be denoted by f_p and is called the 'penetration frequency' because waves of frequency less than f_p are reflected below z_m and vertically incident waves of greater frequency travel right through the layer without reflection. It is convenient to use a new height variable

$$\zeta = (z-z_m)/a. \tag{12.18}$$

Fig. 12.1. Dependence of f_N^2/f_p^2, proportional to electron concentration $N(z)$, on height z for the parabolic distribution (12.17), the sech^2 distribution (12.26) and the Chapman layer (1.9), (1.10) with $H = \frac{1}{2}a$. All three distributions have a maximum at the same height $z = z_m$ and their curvatures there are the same.

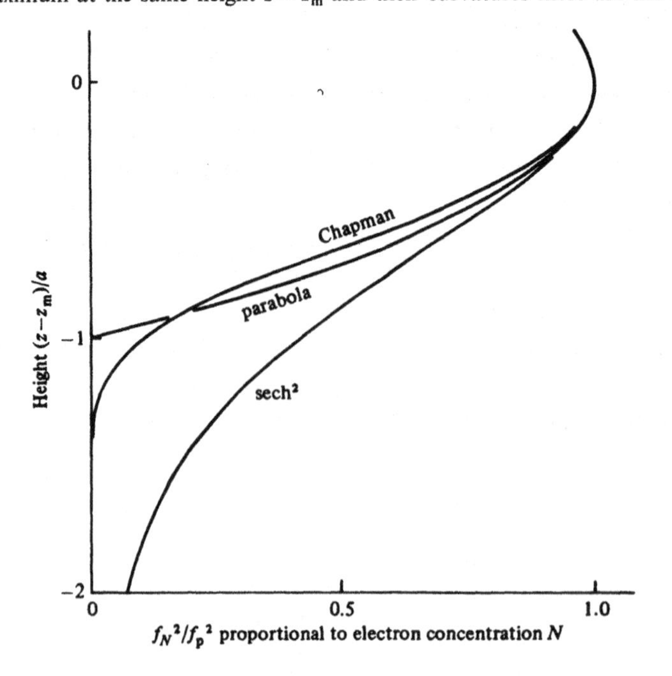

Then (12.17) gives

$$f_N^2 = f_p^2(1-\zeta^2) \text{ for } |\zeta| \leqslant 1. \tag{12.19}$$

Collisions are neglected. The real refractive index n and group refractive index n' are given, from (4.10), (5.66), by

$$n = 1/n' = \{1-(1-\zeta^2)f_p^2/f^2\}^{\frac{1}{2}} \text{ for } |\zeta| \leqslant 1. \tag{12.20}$$

When $|\zeta| \geqslant 1, f_N$ is zero and n, n' are both unity.

Two cases must now be studied. The first is for $f < f_p$ so that reflection occurs at a height z_0 below z_m. The value ζ_0 of ζ at the reflection level is

$$\zeta_0(f) = -\{1-f^2/f_p^2\}^{\frac{1}{2}}. \tag{12.21}$$

The equivalent height of reflection is

$$h'(f) = \int_0^{z_0} n' \,\mathrm{d}z = z_m - a + \tfrac{1}{2}a\frac{f}{f_p}\ln\left(\frac{f_p+f}{f_p-f}\right). \tag{12.22}$$

Similarly the phase height is

$$h(f) = \int_0^{z_0} n \,\mathrm{d}z = z_m - \tfrac{1}{2}a - \tfrac{1}{4}a\left(\frac{f_p}{f} - \frac{f}{f_p}\right)\ln\left(\frac{f_p+f}{f_p-f}\right). \tag{12.23}$$

The dependence on frequency f of the true height from (12.21), equivalent height (12.22) and phase height (12.23) are shown in the left half of fig. 12.2.

The second case is for $f > f_p$ so that the waves travel right through the layer without reflection. They may, however, be reflected from a higher layer, and the parabolic layer can contribute to the equivalent height of reflection. The contribution made by the region between the ground and the top of the parabolic layer will

Fig. 12.2. Left half shows how the equivalent height h', the true height of reflection z_0 and the phase height h depend on frequency f for the parabolic distribution (12.17) of $N(z)$. Right half shows the contributions to h' and h, and the true path for one passage through the layer.

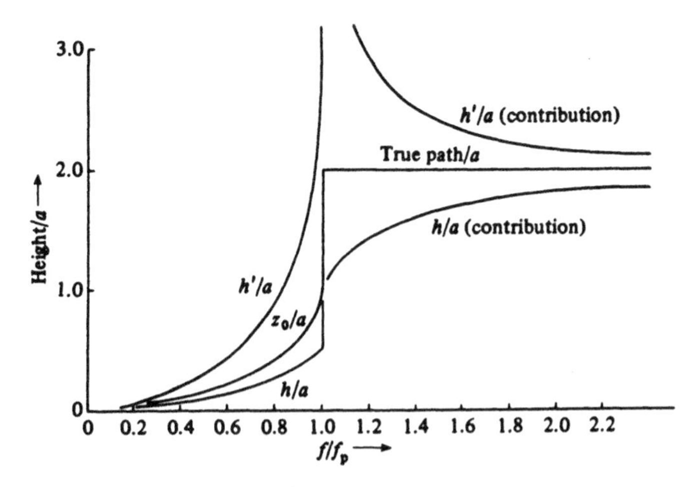

now be calculated. The contribution to the true height is clearly $z_m + a$. The contribution to the equivalent height is

$$h'(f) = \int_0^{z_m+a} n' \mathrm{d}z = z_m - a + \frac{f}{f_p} \ln\left(\frac{f+f_p}{f-f_p}\right). \tag{12.24}$$

The contribution to the phase height is

$$h(f) = \int_0^{z_m+a} n \mathrm{d}z = z_m + \tfrac{1}{2}a\left(\frac{f}{f_p} - \frac{f_p}{f}\right) \ln\left(\frac{f+f_p}{f-f_p}\right). \tag{12.25}$$

The dependence on frequency f of the contribution to the true height, the equivalent height and the phase height are shown in the right half of fig. 12.2.

The above results show that the equivalent height approaches infinity as the frequency f approaches the penetration frequency f_p. The group velocity of a pulse is smallest near the top of its trajectory. For a frequency just below penetration, the region of low group velocity is thicker than it is for lower frequencies, because the gradient of electron concentration is small near the level of reflection. This explains why the equivalent height is large for frequencies close to penetration. The pulse spends a relatively long time near the level of maximum electron concentration, and for a small range of frequencies below f_p the shape of the $h'(f)$ curve is determined predominantly by the curvature of the $N(z)$ curve at the maximum. The shape of the $N(z)$ curve at lower levels is less important for these frequencies, because the pulse spends relatively little time there. This property is illustrated in fig. 12.3 which shows

Fig. 12.3. Curves of $h'(f)$ for the three electron distributions of fig. 12.1. For all three curves, $z_m/a = 6$.

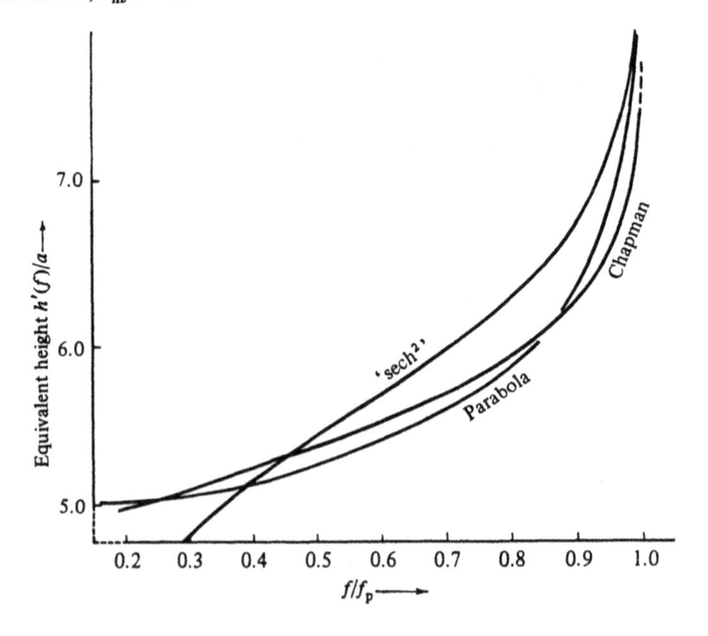

the $h'(f)$ curves for three different electron height distributions, all with the same curvature at the maximum. These are the parabolic layer, the 'sech^2' layer discussed below, and the Chapman layer given by (1.9), (1.10) with $H = \frac{1}{2}a$; see fig. 12.1.

The same property was further illustrated by Appleton and Beynon (1940, appendix). They plotted observed values of $h'(f)$ against $\frac{1}{2}(f/f_p)\ln[(f_p+f)/(f_p-f)]$. For a parabolic layer, (12.22) shows that the graph would be a straight line. For their actual ionospheric layers they fit to a straight line was good for the E-layer, and fairly good but with more scatter of the points for two examples of the F-layer. From graphs of this kind, the slope and the intercept can be used to get estimates of a and $z_m - a$ respectively.

The expression (12.2) for the equivalent height cannot be used for frequencies very close to the penetration frequency, and in this case the problem must be treated by full wave theory. This is done in § 15.11, where it is shown that the effective value of $h'(f)$ near the penetration frequency remains finite but attains a high maximum value. The amplitude of the reflected wave is then small, however, and the maximum is not easy to observe. In most practical cases there is no need to take account of the full wave correction.

The parabolic layer has discontinuities in the gradient of electron concentration at the top and bottom of the layer, $z = z_m \pm a$. In this respect it differs from actual ionospheric layers, in which the transitions are believed to be gradual. A distribution which does not have this drawback is given by

$$f_N^2 = f_p^2 \operatorname{sech}^2\left(\frac{z - z_m}{a}\right). \tag{12.26}$$

One curve in fig. 12.1 shows how the electron concentration N varies with height z in this case. As for the parabola there is a maximum of electron concentration where $z = z_m$, and the penetration frequency is f_p. The electron concentration is never zero; it must have a small non-zero value even at the ground. It is convenient to use the variable ζ defined by (12.18). Then the equivalent height, when $f < f_p$, is given by

$$h'(f) = a \ln\left[\left(\frac{f_p^2}{f^2} - 1\right)^{-\frac{1}{2}} \sinh\frac{z_m}{a} + \left\{\left(\frac{f_p^2}{f^2} - 1\right)^{-1} \sinh^2\left(\frac{z_m}{a}\right) - 1\right\}^{\frac{1}{2}}\right]. \tag{12.27}$$

The curvature of the distribution (12.26) at the maximum is the same as that of the parabolic layer (12.17) for the same values of f_p and a. Fig. 12.3 shows how $h'(f)$ in (12.27) varies with frequency, and the corresponding curve for the parabola is also shown. At high frequencies $h'(f)$ is greater for the 'sech^2' profile. Although the true height of reflection must be lower, the larger number of electrons at low levels causes greater group retardation, which more than compensates for the shorter true path. At low frequencies, however, $h'(f)$ is smaller for the 'sech^2' layer, because the true height falls rapidly as the frequency decreases.

During the daytime there are two main layers in the ionosphere known as E and

F. It is now believed that they are not distinct but that the region between them is ionised with an electron concentration nearly as great as in the E-layer itself; see §1.8 and fig. 1.3. It is nevertheless instructive to assume that these two layers have separate parabolic distributions. Let f_{pE}, z_{mE}, a_E be the penetration frequency, height of maximum and half thickness respectively for the E-layer, and let f_{pF}, z_{mF}, a_F be the corresponding quantities for the F-layer above it. It is assumed that $f_{pF} > f_{pE}$ and that the layers do not overlap. Then if $f < f_{pE}$, $h'(f)$ is given by (12.22) with f_p, z_m and a replaced by f_{pE}, z_{mE} and a_E. If $f_{pE} < f < f_{pF}$,

$$h'(f) = z_{mF} - 2a_E - a_F + a_E \frac{f}{f_{pE}} \ln\left(\frac{f+f_{pE}}{f-f_{pE}}\right) + \tfrac{1}{2} a_F \frac{f}{f_{pE}} \ln\left(\frac{f_{pF}+f}{f_{pE}-f}\right). \quad (12.28)$$

This function is shown plotted for typical values in fig. 12.4. The full wave correction (§ 15.11) near $f_p^{(E)}$ and $f_p^{(F)}$ is not used.

Although the earth's magnetic field has been neglected in deriving (12.28), the curve of fig. 12.4 is very similar in general form to some $h'(f)$ records obtained in practice. A characteristic feature of these records is the infinity at the penetration frequency f_{pE} of the E-layer. When such an infinity appears, it can usually be attributed to penetration of one layer. For frequencies just above it, the waves are being reflected from a higher layer, but are retarded in their passage through the layer that has just been penetrated. For greater frequencies the retardation is less because the group velocity in the lower layer is not so small as for frequencies near the penetration frequency.

12.5. Effect of a 'ledge' in the electron height distribution

A possible form for the electron height distribution $N(z)$ is shown in fig. 12.5(a). Here, for heights below the maximum, $N(z)$ increases monotonically as z increases,

Fig. 12.4. $h'(f)$ curve when the ionosphere consists of two separate parabolic layers. For the lower layer (E), $f_{pE} = 3$ MHz, $z_{mE} = 100$ km, $a_E = 10$ km, and for the upper layer (F), $f_{pF} = 6$ MHz, $z_{mF} = 240$ km, $a_F = 40$ km.

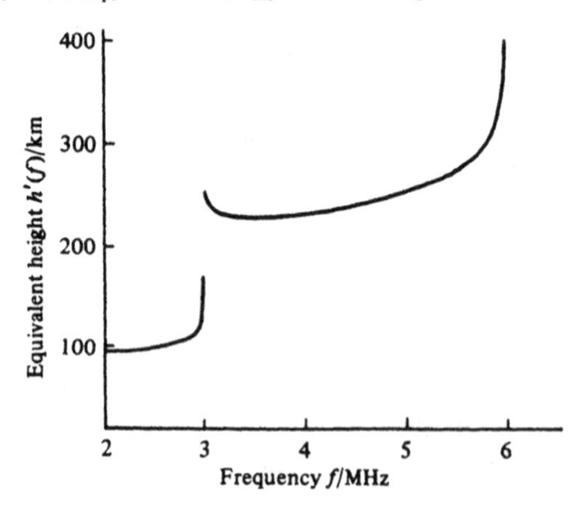

but there is a part of the curve where dN/dz is small. This feature is often called a 'ledge' in the distribution. Suppose that in the middle of the ledge the plasma frequency f_N is denoted by F_N. The form of the $h'(f)$ curve in this case is shown in fig. 12.5(b). For frequencies near F_N the gradient dN/dz is small at the level of reflection, and the group retardation is large, so that $h'(f)$ is large. For greater frequencies, the reflection level is higher and the group retardation in the ledge is less. Consequently $h'(f)$ is less. For still greater frequencies, $h'(f)$ increases again and becomes infinite at the penetration frequency.

The effect of a ledge is therefore to give a maximum in the $h'(f)$ curve. The size of the maximum increases if the gradient dN/dz in the ledge decreases, but $h'(f)$ is always bounded as long as $N(z)$ increases monotonically. Hence, when $N(z)$ is a monotonically increasing function, the corresponding part of the $h'(f)$ curve is bounded and continuous. If $N(z)$ has a maximum, the $h'(f)$ curve shows an infinity at the penetration frequency, and is discontinuous there. When corrected by the full wave theory (§ 15.11), the $h'(f)$ curve does not go to infinity but has a maximum near the penetration frequency, although it is still discontinuous. The maximum is rarely observed on actual $h'(f)$ records, and for practical purposes the curve may be said to go to infinity. When the effect of the earth's magnetic field is allowed for, a discontinuity of a different kind can appear in the $h'(f)$ curve for the extraordinary wave at the gyro-frequency (see § 13.2).

The F1-layer often appears as a ledge below the F2-layer, and the $h'(f)$ curve for the whole F-layer is then similar to fig. 12.5(b). Sometimes the F1-layer may attain an actual maximum, and then the maximum in the $h'(f)$ record will go over into an infinity associated with penetration.

12.6. The calculation of electron concentration $N(z)$, from $h'(f)$

In the preceding sections some examples have been given of the calculation of $h'(f)$ when $N(z)$ is known. Of much greater practical importance is the calculation of $N(z)$

Fig. 12.5. (a) Shows an electron height distribution $N(z)$ with a ledge. (b) The resulting $h'(f)$ curve, not to scale, showing a maximum at a frequency near the plasma frequency F_N in the ledge.

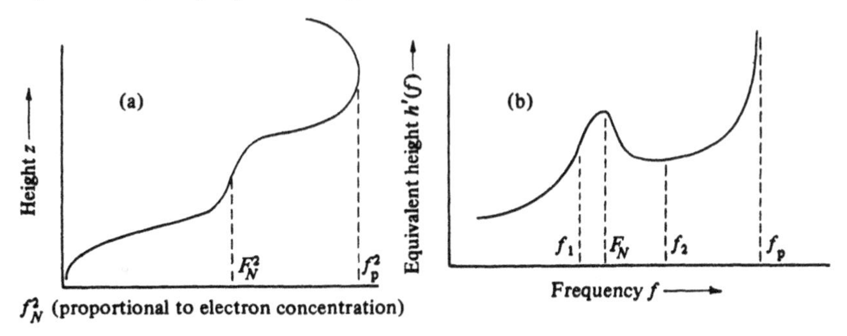

when $h'(f)$ is known, since in the commonest form of radio sounding of the ionosphere it is $h'(f)$ that is observed. The relation between $h'(f)$ and $N(z)$ may be written

$$h'(f) = \int_0^{z_0} n'(N, f)\,\mathrm{d}z. \tag{12.29}$$

The problem is to solve this integral equation for $N(z)$. This can be accomplished analytically in the special case when the earth's magnetic field is neglected. The electron concentration N is proportional to the square of the plasma frequency $f_N(z)$. Hence $f_N(z)$ is to be found. The problem will actually be solved by finding the inverse function $z(f_N)$.

The group refractive index n' is given by (5.66) with ω replaced by f. Hence (12.29) may be written

$$\frac{1}{f}h'(f) = \int_0^{z_0} \frac{\mathrm{d}z}{(f^2 - f_N^2)^{\frac{1}{2}}} \tag{12.30}$$

where z_0 is that value of z for which $f_N = f$. Now it is convenient to use new variables

$$u = f_N^2, \quad v = f^2, \tag{12.31}$$

so that

$$v^{-\frac{1}{2}}h'(v^{\frac{1}{2}}) = \int_0^{z_0} \frac{\mathrm{d}z}{\{v - u(z)\}^{\frac{1}{2}}}. \tag{12.32}$$

Here $u(z)$ is the unknown function that is to be found.

When $u(z)$ is a monotonically increasing function of z, (12.32) can be solved by a standard method (see, for example, Whittaker and Watson, 1927, §11.8). An alternative method using Laplace transforms has been given by Manning (1947, 1949). Multiply both sides by $(1/\pi)(w - v)^{-\frac{1}{2}}$, where

$$w = g^2 \tag{12.33}$$

and g is the value of f_N for which the true height $z(g)$ is to be found. Integrate with

Fig. 12.6. (a) The unknown function $u(z)$ proportional to electron concentration. (b) The same curve plotted in the v–z plane.

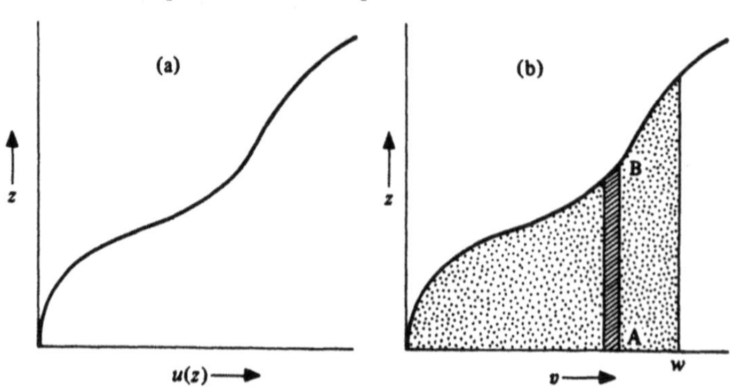

respect to v from 0 to w. Then (12.32) gives

$$\frac{1}{\pi}\int_0^w v^{-\frac{1}{2}}(w-v)^{-\frac{1}{2}}h'(v^{\frac{1}{2}})\,\mathrm{d}v = \frac{1}{\pi}\int_0^w\left[\int_0^{z_0}\frac{\mathrm{d}z}{(w-v)^{\frac{1}{2}}(v-u)^{\frac{1}{2}}}\right]\mathrm{d}v. \tag{12.34}$$

Now suppose that the unknown function $u(z)$ is as shown by the curve of fig. 12.6(a). The integral on the right is a double integral with respect to z and v. It is to be integrated first with respect to z while v is kept constant, and the top limit for z is the value that makes $u=v$. Fig. 12.6(b) is a diagram of the $v-z$ plane, and the curve of fig. 12.6(a) is copied in it. The first integration is over the shaded strip AB. The result is then to be integrated with respect to v from $v=0$ to $v=w$. It is thus clear that the whole integration extends over the dotted region of fig. 12.6(b). The order of integration may now be changed provided that the limits are suitably modified. Then the right side of (12.34) becomes

$$\frac{1}{\pi}\int_0^{z_0(g)}\left[\int_u^w\frac{\mathrm{d}v}{(w-v)^{\frac{1}{2}}(v-u)^{\frac{1}{2}}}\right]\mathrm{d}z, \tag{12.35}$$

where $z_0(g)$ is the value of z that makes $u=w$. The integration with respect to v may be effected by elementary methods (by the substitution $v=w\cos^2\theta+u\sin^2\theta$) and gives simply π. Hence the expressions (12.34) and (12.35) are equal to $z_0(g)$. On the left of (12.34) put $v=w\sin^2\alpha$, which gives

$$z_0(g)=\frac{2}{\pi}\int_0^{\frac{1}{2}\pi}h'(w^{\frac{1}{2}}\sin\alpha)\,\mathrm{d}\alpha. \tag{12.36}$$

Now (12.33) shows that $w^{\frac{1}{2}}=g$. We now put $g=f_N$ so that (12.36) may be written

$$z(f_N)=\frac{2}{\pi}\int_0^{\frac{1}{2}\pi}h'(f_N\sin\alpha)\,\mathrm{d}\alpha, \tag{12.37}$$

which is the required solution.

The equation (12.32) was solved by Abel for a different problem. A particle is projected horizontally with velocity V on to a frictionless slope, for which the height $H(x)$ is an unknown function of the horizontal distance x. The slope is assumed to be so small that the horizontal component of the velocity may be taken to be the same as the actual oblique velocity. The time T taken for the particle to come to rest and then return to its starting point is given by

$$T(V)=2\int_0^{x_0}\frac{\mathrm{d}x}{\{V^2-2gH(x)\}^{\frac{1}{2}}} \tag{12.38}$$

where g is the gravitational acceleration, and x_0 is the value of x for which $H(x)=\frac{1}{2}V^2/g$. The function $T(V)$ can be determined experimentally. Abel's problem was to find the function $H(x)$. Clearly (12.38) is of the same form as (12.32) and can be solved in the same way (see de Groot, 1930). Ratcliffe (1954) has used this analogy by constructing a mechanical model in which a steel ball-bearing is projected up an incline shaped like the electron distribution $N(z)$ in the ionosphere. This exhibits many of the features of the reflection of a pulse of radio waves projected up into the ionosphere.

Suppose now that $u(z)$ is not a monotonically increasing function of z, but has one maximum and one minimum, as shown in fig. 12.7(a). At the maximum let $z = z_1$, $u = u_p$ where

$$u_p = f_p^2, \tag{12.39}$$

and f_p is the penetration frequency. Let $z = z_3$ be the height where u again has the value u_p (see fig. 12.7(a)). It was shown in § 12.4 that the $h'(f)$ curve must have an infinity where $f = f_p$, so that the curve is as in fig. 12.8(a). It is now assumed that this curve is given, and we require to find the curve of fig. 12.7(a). This problem cannot be solved completely. Only the part of the curve below z_1 can be found unambiguously.

The equation (12.32) still holds. To solve it, we apply the method of the last section. Multiply both sides by $(1/\pi)(w - v)^{-\frac{1}{2}}$ where w is given by (12.33). Integrate

Fig. 12.7. (a) Similar to fig. 12.6(a) but the distribution now has a valley. (b) Similar to fig. 12.6(b).

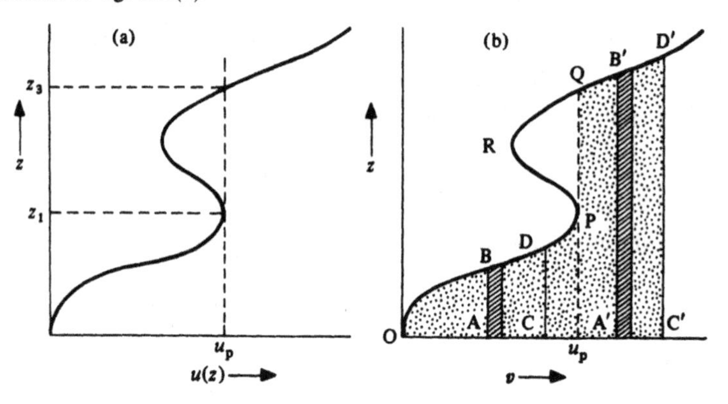

Fig. 12.8. (a) The $h'(f)$ curve from which $u(z)$ is to be found. (b) Result of applying Abel's method to (a). The unknown curve $u(z)$ is shown for comparison, as a broken line.

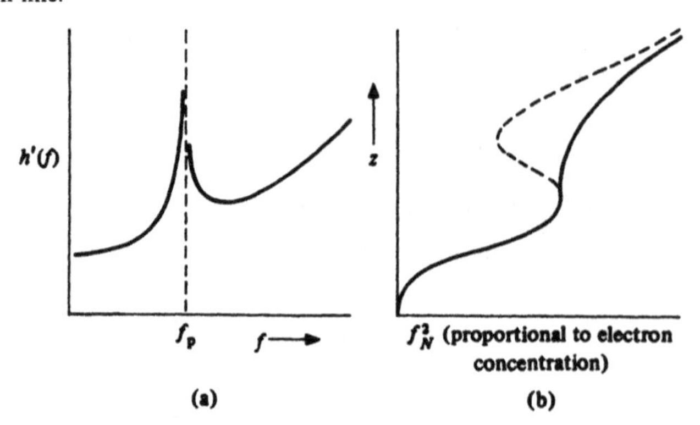

with respect to v from 0 to w, which gives (12.34), whose right side is a double integral with respect to z and v. Fig. 12.7(b) is a diagram of the v–z plane and the curve of fig. 12.7(a) is copied in it. Suppose first that w is less than u_p and is given by the ordinate CD in the figure. Then the double integral extends over the area OACDB. The order of integration may be changed as before and the final result is

$$z(f_N) = \int_0^{\frac{1}{2}\pi} h'(f_N \sin\alpha)\, d\alpha \text{ for } f_N < f_p, \tag{12.40}$$

which is the same as (12.37), and enables the part of the curve below $z = z_1$ in fig. 12.7(a) to be found. Next suppose that w is greater than u_p, and is given by the ordinate C′D′ in fig. 12.7(b). In the double integral on the right of (12.34) the first integration with respect to z extends from $z = 0$ up to the curve. When v is just less than u_p, the upper limit is on the part DP of the curve. When v becomes just greater than u_p, the upper limit suddenly increases to the part QB′ of the curve. It is thus clear that the double integral extends over the dotted area of fig. 12.7(b), and the area PQR is not included. Now the order of integration may be changed, as in the last section, but if the right side of (12.34) were written in the form of (12.35), the integration would extend over the dotted area plus the area PQR. Equation (12.35) is therefore no longer correct. The following integral must be subtracted from it:

$$\frac{1}{\pi}\int_{z_1}^{z_3}\left[\int_u^{u_p} \frac{dv}{(w-v)^{\frac{1}{2}}(v-u)^{\frac{1}{2}}}\right] dz. \tag{12.41}$$

This is the integral over the area PQR. Now (12.35) can be integrated as before, and gives simply $z_0(f_N)$. The integration with respect to v in (12.41) can be effected in the same way as before, and gives for this term

$$\zeta(g) = \frac{1}{\pi}\int_{z_1}^{z_3} \arccos\left(\frac{w - 2u_p + u}{w - u}\right) dz. \tag{12.42}$$

Hence (12.34) becomes

$$z(f_N) = \frac{2}{\pi}\int_0^{\frac{1}{2}\pi} h'(f_N \sin\alpha)\, d\alpha + \zeta(f_N). \tag{12.43}$$

Hence Abel's integral (12.37) gives the correct value for $z(f_N)$ when $f_N < f_p$, but gives a result which is too small by $\zeta(f_N)$ when $f_N > f_p$. This is illustrated in fig. 12.8(b). Thus the expression (12.42) is the error which is contributed by the part of the ionosphere between $z = z_1$ and $z = z_3$.

In the range $z_1 < z < z_3$ it is clear that $u < u_p$. If $w \gg u_p$, the angle arccos (...) in (12.42) must be very small, and hence the error $\zeta(f_N)$ is small. This means that the error is not serious for frequencies which greatly exceed the penetration frequency f_p. In particular the E-layer (for which f_p is of the order 2 to 4 MHz in the daytime) would not seriously affect results for the densest part of the F-layer which may reflect frequencies of order 5 to 10 MHz. There is now good evidence to show that the minimum of the electron density between the E- and F-layers is very shallow, and this would mean that the error ζ is always small.

When $f_N > f_p$, the integral in (12.43) covers a range which includes the infinity at $f = f_p$ in the $h'(f)$ curve (fig. 12.8(a)). This infinity arises from a logarithm term of the form given by (12.22) and (12.24). If the integral is evaluated analytically in this case, the range of integration must be divided into the two parts $0 \leqslant \sin\alpha \leqslant (f_p/f_N) - \varepsilon$ and $(f_p/f_N) + \varepsilon \leqslant \sin\alpha \leqslant 1$ where ε is an arbitrarily small quantity. The two parts of the integral are evaluated separately and added, and ε is then allowed to become indefinitely small. In this way it can be shown that the sum approaches a definite limit known as the 'principal value' of the integral, which is the value to be used in (12.43) (for further details see, for example, Whittaker and Watson, 1935, §4.5).

12.7. Ray paths at oblique incidence

We now study some examples of the paths of obliquely travelling wave packets in a horizontally stratified isotropic ionosphere with negligible collision frequency. Without loss of generality the path can be assumed to be in the plane $y = 0$, and to start from a transmitter at the origin. Then its equation, from (10.14) is

$$x = S\int_0^z \frac{\mathrm{d}z}{q}, \tag{12.44}$$

$$q^2 = n^2 - S^2 \tag{12.45}$$

where $S = \sin\theta$, and θ is the angle between the predominant wave normal and the vertical in free space.

Consider, first, the simple model ionosphere of § 12.3 in which there is free space where $0 \leqslant z \leqslant h_0$, and above this $N(z)$ is proportional $z - h_0$. Let $C = \cos\theta$. Then

$$\begin{aligned} q^2 &= C^2 - \alpha(z - h_0)/f^2 \quad \text{when } z \geqslant h_0, \\ q &= C \quad\quad\quad\quad\quad\quad\quad \text{when } z \leqslant h_0. \end{aligned} \tag{12.46}$$

where α is a constant. Evaluation of the integral in (12.44) is often most easily achieved by changing to q as the variable of integration. If (12.46) is used, it gives, for $z \geqslant h_0$

$$x = h_0\tan\theta + 2\frac{f^2 S}{\alpha}(C - q) = h_0\tan\theta + \frac{f^2\sin 2\theta}{\alpha} - \frac{2f^2\sin\theta}{\alpha}\left\{\cos^2\theta - \frac{\alpha(z - h_0)}{f^2}\right\}^{\frac{1}{2}} \tag{12.47}$$

which shows that within the ionosphere the ray path is a parabola; compare (10.26) which gives the same result for $h_0 = 0$. It is now of interest to find the total horizontal range D traversed by the wave packet when it returns to the ground. The top of its trajectory is where $\mathrm{d}x/\mathrm{d}z = \infty$, that is where $q = 0$ and at this point x is $\frac{1}{2}D$. Hence from (12.47)

$$D = 2h_0 \tan\theta + 2f^2(\sin 2\theta)/\alpha. \tag{12.48}$$

The first term is contributed by the two straight sections of the path below the ionosphere and the second term is the part within the ionosphere.

Fig. 12.9 shows how D/h_0 depends on θ for some typical cases. It can easily be shown that if $f > 4\alpha h_0$ the curve has a maximum and a minimum so that for some values of D (12.48) is satisfied by three different values of θ. This means that wave packets can travel from the transmitter to the receiver by three different paths. This is only possible if $D > 6h_0\sqrt{3}$. The proof is left as an exercise for the reader.

Suppose that the horizontal line D/h_0 = constant in fig. 12.9 just touches the curve at either a maximum or a minimum. Then two of the ray paths have moved to coincidence which means that a caustic surface meets the ground at the range D; see §§ 8.22, 10.18. If the touching point is at a minimum, the range D is a skip distance; see § 10.22 and fig. 10.17. If it is at a maximum, D is a reversed skip distance. Diagrams showing the shape of the rays and of the caustics for this and other cases have been given by Maslin (1976a, b).

Suppose, next, that the electron concentration is given by the parabolic law (12.17), and use ζ from (12.18). Then

$$\begin{aligned} q^2 &= C^2 - (1-\zeta^2)f_p^2/f^2 \quad \text{for } |\zeta| \leqslant 1, \\ q &= C \quad \text{for } |\zeta| \geqslant 1, \end{aligned} \tag{12.49}$$

where f_p is the penetration frequency of the layer. The equation of a ray path within

Fig. 12.9. Shows how the horizontal range D depends on angle of incidence θ for the model ionosphere with free space below the height h_0, and $f_N^2 = \alpha(z - h_0)$ above it. The earth's curvature is neglected. The numbers by the curves are the values of $f^2/\alpha h_0$.

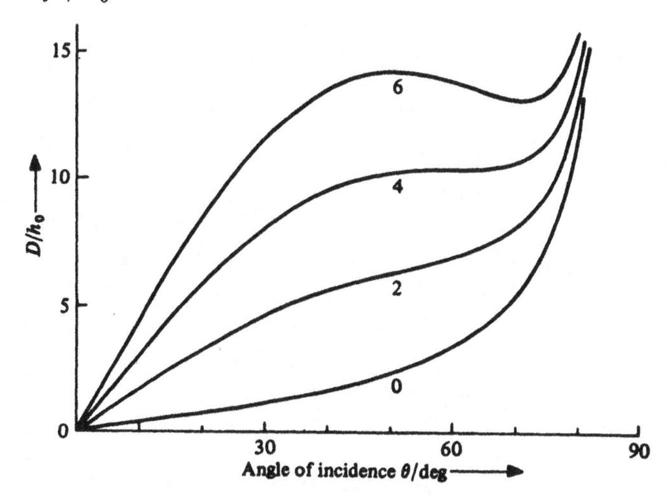

the ionosphere is, from (12.44)

$$x = h_0 \tan\theta + aS\int_{-1}^{\zeta} \{C^2 - (1-\zeta^2)f_p^2/f^2\}^{-\frac{1}{2}}\,d\zeta \tag{12.50}$$

where $h_0 = z_m - a$. The horizontal range D is found, as in (12.48), by finding the part between the transmitter and the top of the trajectory and doubling it. This gives

$$D = 2h_0 \tan\theta + a\sin\theta\frac{f}{f_p}\ln\left(\frac{f_p + f\cos\theta}{f_p - f\cos\theta}\right). \tag{12.51}$$

The first term is contributed by the two straight sections of the path below the ionosphere, and the second term is the part within the ionosphere.

Fig. 12.10 shows how D/h_0 depends on θ for some typical cases. When $f \leqslant f_p$ any wave packet is reflected whatever the value of θ, and D varies between 0 and ∞. When $f > f_p$, however, wave packets that are near vertical incidence can penetrate the ionosphere, and do not return to the ground. For reflection it is necessary that $\cos\theta < f_p/f$ which gives a minimum value of θ, and (12.51) shows that D tends to ∞ as $\cos\theta$ tends to f_p/f and also as $\cos\theta$ tends to zero. This behaviour is clearly shown by the three curves for which $f > f_p$ in fig. 12.10. Between these values there is a minimum value of D, and for any greater value of D wave packets can travel from the transmitter to the receiver by two different paths. The path for which θ is the smaller is known as the Pedersen ray (Pedersen, 1927, § XI 2c). It is shown in § 12.8 that this ray has the greater time of travel.

Fig. 12.10. Shows how the horizontal range D depends on angle of incidence θ for a typical parabolic layer. In this example $a = \frac{1}{5}h_0$. The numbers by the curves are the values of f/f_p.

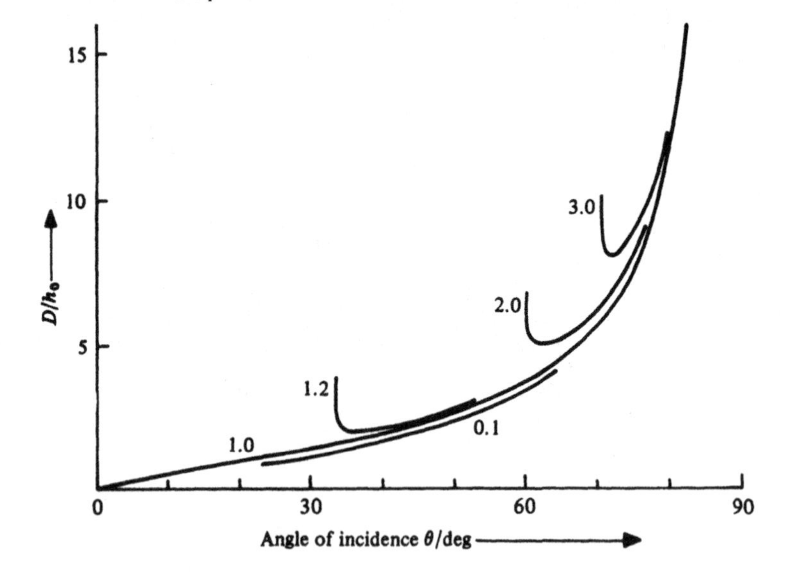

When D has a minimum value, this is an example of a skip distance already discussed in § 10.22. No ray that goes via the ionosphere can reach the ground at a shorter range. But near the skip distance the simple ray theory is inadequate. A caustic surface meets the ground there and the dependence of signal on range is given by an Airy integral function (10.69). The scale of this function is given by (10.70), and uses the curvature $\mathrm{d}^2D/\mathrm{d}\theta^2$ of the $D(\theta)$ curve at its minimum.

12.8. Equivalent path P' at oblique incidence

For a wave packet travelling from the origin to the point $(x, 0, z)$, the equivalent path is given by (10.11) which, with $S_2 = 0$, $y = 0$, $S_1 = S$, becomes

$$P' = Sx + \int_0^z \frac{\partial(fq)}{\partial f}\mathrm{d}z. \tag{12.52}$$

This gives the time of travel $t = P'/c$. Now for the isotropic collisionless plasma, (4.10) and (10.12) show that

$$q^2 = C^2 - f_N^2/f^2 \tag{12.53}$$

from which (12.52) gives

$$P' = Sx + C^2\int_0^z \frac{\mathrm{d}z}{q}. \tag{12.54}$$

Now (12.44) shows that the integral is simply x/S. Hence

$$P' = x/S \tag{12.55}$$

which means that when a wave packet starts out at an angle θ to the vertical and has travelled along some path TA, fig. 12.11, its equivalent path is TB obtained by extending its initial path along a straight line to a point B vertically above A. In particular if the wave packet returns to the ground at a distance $x = D$, the equivalent path is the sum of the two oblique sides TC, CR of the triangle formed by extending the initial and final directions of the wave packet until they intersect at C. Hence

$$P' = D/S. \tag{12.56}$$

Fig. 12.11. Ray path in the ionosphere and the associated triangulated path.

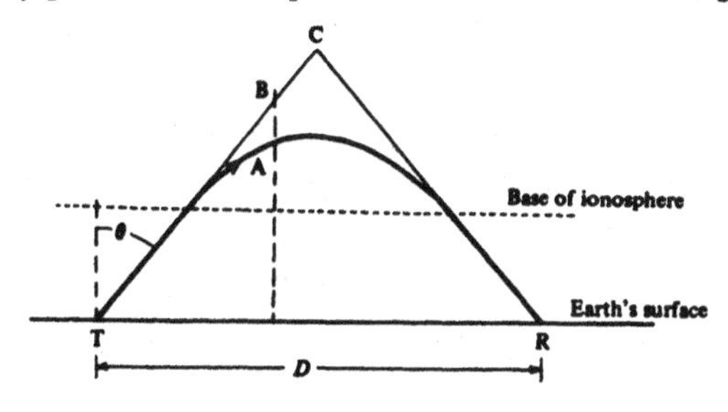

This is Breit and Tuve's theorem (1926). The fictitious path formed by the two straight lines TC, CR is called the triangulated path. The theorem is true only when the earth's magnetic field and curvature are neglected and when the electron collision frequency is very small.

An alternative proof of Breit and Tuve's theorem is based on the result that for an isotropic plasma the ray and the wave normal have the same direction. Hence, if ψ is the inclination of the ray and wave normal to the vertical, Snell's law gives $n \sin\psi = S$. Let δs be an element of the ray path and δx its horizontal projection. Then $\delta x = \delta s \sin\psi = S\delta s/n = S\delta s n'$ from (5.66). The wave packet travels with the group velocity $\mathscr{U} = c/n'$, and hence $n'\delta s = c\delta t$ where δt is the time taken to travel the distance δs. Hence $\delta t = \delta x/Sc$ which on integration gives $P' = x/S$, the same as (12.55).

Equations (12.53), (12.54) may be combined to give

$$P'(f) = \frac{1}{C}\int_0^z \frac{dz}{\{1 - f_N^2/(fC)^2\}^{\frac{1}{2}}} \tag{12.57}$$

where P' is measured at the particular frequency f. Now the integrand is the group refractive index $n'(fC)$ for the frequency fC. A wave of frequency fC vertically incident on the ionosphere would be reflected at the level z_0 where $f_N = fC$ and this is also the level of reflection of the frequency f at oblique incidence. Suppose that the vertically incident wave packet and the wave packet described by (12.57) both return to the ground. Then the upper limit of the integral must be set equal to z_0 and a factor 2 introduced to allow for the upward and downward paths. This gives

$$P'(f) = \frac{2}{C}\int_0^{z_0} n'(fC)\,dz. \tag{12.58}$$

Here the integral is the equivalent height of reflection $h'(fC)$, (12.2), for a frequency fC at vertical incidence. Hence

$$CP'(f) = 2h'(fC). \tag{12.59}$$

In words: 'For wave packets that return to the ground, the equivalent path for a frequency fC at vertical incidence is the vertical projection of the triangulated path for a frequency f at oblique incidence.' This is Martyn's theorem for equivalent path (Martyn, 1935). It must not be confused with Martyn's theorem for attenuation, § 12.11. Some modifications of the theorem, to allow for the earth's curvature, were discussed by Millington (1938b).

The ionosonde technique, § 1.6, uses vertically incident pulses of radio waves for probing the ionosphere and supplies curves of equivalent height $h'(f)$ versus frequency f. It is sometimes used also at oblique incidence, and supplies curves of equivalent path $P'(f)$ versus frequency. It is therefore of interest to calculate $P'(f)$ curves for various models of the ionosphere. For typical observed $P'(f)$ curves at oblique incidence see for example Wieder (1955), Agy and Davies (1959), Davies (1969).

The range D where a wave packet returns to the ground is found from (12.44) by putting $z = z_0$, the height of reflection, and doubling the right-hand side, as explained in § 12.7. This gives D as a function of the angle of incidence θ. Now $\sin\theta = D/P'$ from (12.56), and on substitution for θ this gives an equation for finding P′ when D and f are known.

Consider again the simple model ionosphere of (12.46) with free space below $z = h_0$ and a linear $N(z)$ above this. Then D is given by (12.48) and from (12.56) the required equation is

$$\frac{1}{h_0}(P'^2 - D^2)^{\frac{1}{2}} = 2 + \frac{4f^2}{\alpha h_0}\{1 - (D/P')^2\} \tag{12.60}$$

which may be used to calculate f for various values of P'. Some typical curves are shown in fig. 12.12. It was mentioned in § 12.7 that if $D > 6h_0\sqrt{3}$, three different rays can reach the receiver. This property can be seen in fig. 12.12.

For the parabolic model ionosphere of (12.49), D is given by (12.51). From (12.56) $\sin\theta$ is found and substituted in (12.51) to give an equation relating f/f_p and P'. It is a bit complicated because of the logarithm but can be solved quite quickly on a pocket calculator by using an iterative process. Some typical results are shown in fig. 12.13. The curve for $D = 0$ is the same as for vertical incidence, fig. 12.3, and is included for comparison.

For all the curves, $P' \to \infty$ at the penetration frequency $f = f_p$. For a range of

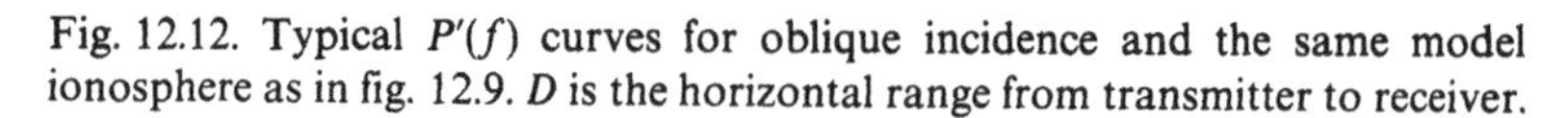

Fig. 12.12. Typical $P'(f)$ curves for oblique incidence and the same model ionosphere as in fig. 12.9. D is the horizontal range from transmitter to receiver.

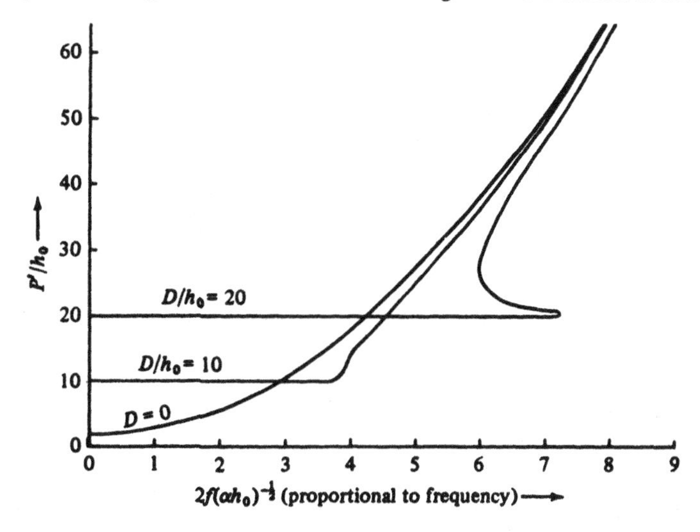

frequency greater than f_p there are two different values of P', which shows that signals arrive at the receiver by two different ray paths. The ray with the longer path and the smaller θ is the 'Pedersen ray' already mentioned in § 12.7. At the maximum frequency of this range, the two curves of P' move together. This frequency is called the 'maximum usable frequency', or by its abbreviation 'MUF', and it is here denoted by f_M. It is where the range D is at the skip distance. It is studied in the following section.

It was shown in § 10.22 that, for frequencies near the MUF, the dependence of signal amplitude on frequency is given by an Airy integral function (10.69), (10.72), and shows an interference effect. But this occurs for a very small range of $f-f_M$ and is not usually seen on oblique incidence ionograms. It is sometimes found that appreciable signals are received at frequencies well beyond the point where the two branches of the $P'(f)$ curve come together, that is at the extreme right of the curves in fig. 12.13. These cannot be explained in terms of wave propagation near the caustic at the skip distance. They might possibly be explained by scattering of the waves from irregularities of N, either at the reflecting level or at some lower level through which the waves pass.

12.9. Maximum usable frequency, MUF

As a simple illustration of the calculation of an MUF, consider the result (12.51) for the parabolic model ionosphere when the earth's curvature is neglected. It gives the range D for a ray that leaves the transmitter at an angle θ to the vertical and returns to the ground. If f is the MUF for a receiver at range D, then D is the skip distance, so that when θ is varied in (12.51), D is a minimum. The condition $\partial D/\partial\theta = 0$ gives

$$2h_0 + a\cos^3\theta \frac{f}{f_p} \ln\left(\frac{f_p + f\cos\theta}{f_p - f\cos\theta}\right) = \frac{2af^2\sin^2\theta\cos^2\theta}{f_p^2 - f^2\cos^2\theta}. \tag{12.61}$$

Fig. 12.13. As for fig. 12.12 but for the parabolic model ionosphere used in fig. 12.10, with $a = \frac{1}{5}h_0$. The numbers by the curves are the values of D/h_0.

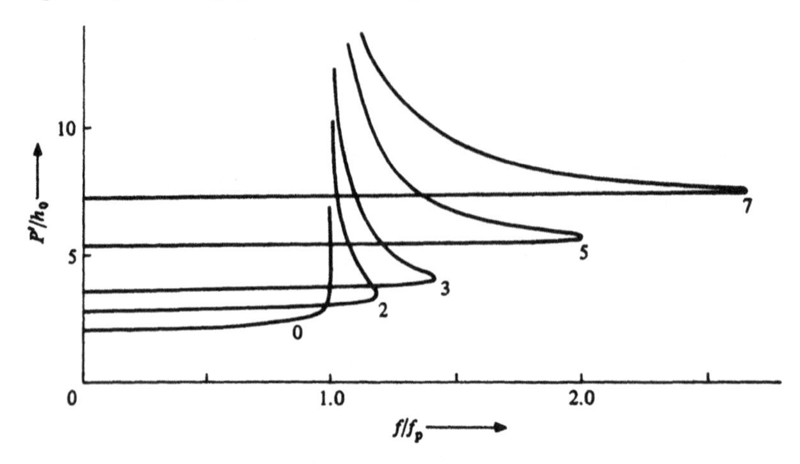

If θ were eliminated from (12.51), (12.61) it would give an equation relating the maximum usable frequency $f = f_M$ and the range D. This, however, is algebraically complicated. Appleton and Beynon (1940) first assumed a value for $\cos\theta f/f_p$ and used it in (12.61) to calculate θ. Thus values of f/f_p and θ are found that satisfy (12.61). These are then used in (12.51) to find D. In this way a curve of D versus $f = f_M$ is plotted.

12.10 The forecasting of MUF

In planning the best use of a given radio communication link, it is important to know in advance what frequencies can be used. We therefore need to know the maximum usable frequency. In forecasting this there are two stages. First, some information is needed about the electron height distribution $N(z)$. This depends on time of day, time of year, position on the earth and sunspot number. A possible method for this is to use the International Reference Ionosphere; see § 1.8. Second, this information is used to estimate the MUF for given positions of the transmitter and receiver. Two methods of doing this have been in use for many years, the method of Appleton and Beynon (1940, 1947), and the method of N. Smith (1937, 1938, 1939, 1941). The problem is complicated so that approximations are necessarily used. In the original forms of both methods the earth's magnetic field was neglected, but corrections that allow for it have been used with later versions. The original forms, however, illustrate well the basic principles used in forecasting MUF and a brief summary is given here. For fuller accounts see the original papers cited above and Millington (1938b), Davies (1969, 1981). More recent methods are described in reports of the CCIR (see end of this section). The author is greatly indebted to Dr P.A. Bradley of the Rutherford Appleton Laboratory, Chilton, England, for information on this subject.

The distance D_G between transmitter and receiver is usually of the order of several thousand kilometres, so that the curvature of the earth cannot be neglected. D_G is measured over the earth's curved surface. It is related to the straight line distance D thus

$$D = 2\rho \sin(D_G/2\rho) \tag{12.62}$$

where ρ is the radius of the earth's surface.

In Appleton and Beynon's (1940, 1947) method the parabolic distribution (12.17) for $N(z)$ is used as an approximation for the forecast $N(z)$. One method of estimating the half thickness a, and the height $h_0 = z_m - a$ of the base, uses the $h'(f)$ curve at vertical incidence and was described in § 12.4. Other methods have been given by Booker and Seaton (1940) and by Ratcliffe (1951). Forecasts of a, z_m and f_p are sometimes available so that a forecast of the complete $N(z)$ is not needed. Because of the earth's curvature, the ray path in the ionosphere must be found by using the Bouger law (10.18) instead of Snell's law (6.4). It leads to an expression for D_G as a function of f/f_p and of the angle i_0 between the ray where it enters the base of the ionosphere and the vertical. This is analogous to (12.51). It is algebraically complicated, but is simplified by assuming that h_0/ρ is small and neglecting squares and higher powers. Now f is the MUF, so that D_G is the skip distance and therefore $\partial D_G/\partial i_0 = 0$. The resulting equation and the equation for D_G are now combined by the method described in §12.9. In this way curves of various kinds can be plotted. In one form of these, for a given value of a/h_0, a set of curves is plotted of f/f_p versus D_G, each curve for a fixed value of $z_m = h_0 + a$. Separate diagrams are used for each value of a/h_0. Thus when a, h_0 and f_p are known, the value of f/f_p for a given D_G can be read off, and f is the required MUF.

Newbern Smith's (1937, 1938, 1939, 1941) method does not use the assumption that $N(z)$ is the parabolic distribution (12.17). Instead it uses directly the $h'(f)$ curve for vertical incidence. This must first be found for the forecast $N(z)$. The method is described here for the case where the earth's curvature is neglected.

A combination of Breit and Tuve's theorem (12.56) and Martyn's theorem (12.59) gives

$$h'(fC) = \tfrac{1}{2}D(f)\cot\theta \tag{12.63}$$

where $D(f)$ is the range of a ray of frequency f that leaves the transmitter at an angle θ to the vertical, and $h'(fC)$ is the equivalent height of reflection at vertical incidence

Fig. 12.14. Simplified version of Newbern Smith's method for finding the MUF when the earth's curvature is neglected. The thick line is an example of an observed $h'(f)$ curve at vertical incidence, showing penetration at 3 MHz. The thin lines are the right-hand side of (12.64) with $D = 1000$ km. In this example the MUF is 6 MHz.

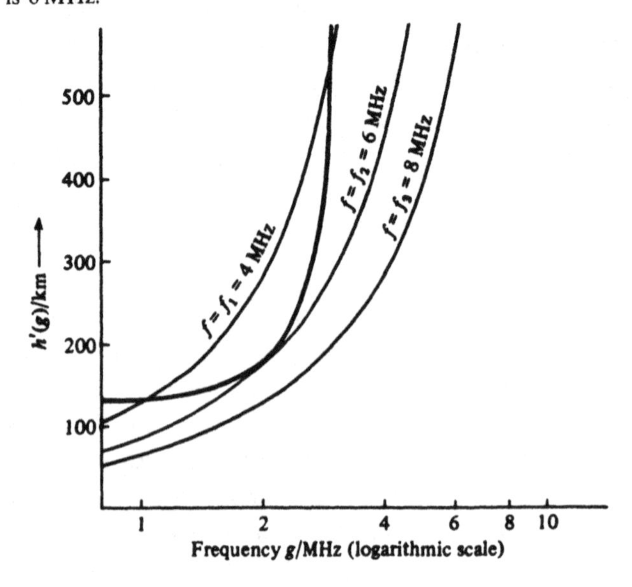

for a frequency fC. Let $fC = g$. Then (12.63) may be written

$$h'(g) = \tfrac{1}{2}D(f)\frac{g/f}{\{1-(g/f)^2\}^{\frac{1}{2}}}. \qquad (12.64)$$

Here g may be called the 'equivalent frequency at vertical incidence'. Now the $h'(g)$ curve is given. Suppose it is plotted as in fig. 12.14. In the same figure let a series of curves be plotted each for a fixed f, and for a chosen value of D, to show how the right-hand side of (12.64), as ordinate, depends on g. Consider the curve for $f = f_1$. It cuts the $h'(g)$ curve at two points, which shows that (12.64) is satisfied for two different frequencies. On the other hand, the curve for the greater frequency $f = f_3$ does not cut the $h'(g)$ curve, so that (12.64) cannot be satisfied. Between them is the curve for an intermediate frequency $f = f_2$, which just touches the $h'(g)$ curve, and therefore f_2 must be the MUF for the chosen D.

This method can be applied more simply as follows. The $h'(g)$ curve is plotted with $\ln g$ as abscissa, and the right-hand side of (12.64) is plotted on a sheet of transparent material with $\ln(g/f)$ as abscissa, so that it has a vertical asymptote where $\ln(g/f) = 0$. The curves are superimposed, so that the horizontal axes coincide, and the transparent sheet is slid horizontally until the two curves just touch. The point where the line $\ln(g/f) = 0$ cuts the g axis of the $h'(g)$ diagram is then noted. At this point $g = f$, so that it gives the MUF, which is simply read off from the horizontal axis. In this way one curve on the transparent sheet can be used for all values of f, but it is necessary to have a different curve for each value of the horizontal range D.

Refinements of this method to allow for the earth's curvature and the earth's magnetic field have been described by Newbern Smith (1938). For a full description of the method see National Bureau of Standards Circular 462 (1948, ch. 6).

When the earth's magnetic field is allowed for, Martyn's theorem and Breit and Tuve's theorem are not valid, and there is no simple method of computing the MUF. It is needed for large values of the range D, so that the earth's curvature cannot be ignored. The problem can be tackled by using general ray tracing methods, ch. 14, and some work on these lines has been done (for example Capon 1962, Herring, 1984). It is necessary first to adopt some model for the ionosphere. In recent work a model that has proved useful is known as the Bradley–Dudeney model (Bradley and Dudeney, 1973; Dudeney, 1978, 1983), and another useful model was proposed by Booker (1977). Then, for a given frequency and bearing, several extraordinary rays are traced, each with a different initial elevation angle, and the horizontal ranges are found. There is, in general, a minimum range or skip distance. The extraordinary ray is used because it gives the smaller skip distance. The calculations are repeated for other frequencies. Then, for a receiving station with the chosen bearing, the MUF is the greatest frequency for which the skip distance is less than the range of the station. For a receiving station with a different bearing, a new set of calculations is needed.

A program of this kind is extremely demanding even for the fastest modern computers, and the uncertainties in predicting the parameters of the chosen model of the ionosphere are so great that it would not be worth while for regular use in forecasting MUFs.

The standardisation of methods for forecasting MUF is one of the functions of the Comité Consultatif International des Radiocommunications (CCIR) which is a committee of the International Telecommunications Union (ITU). The journal of the ITU is the Telecommunications Journal and is published monthly in several languages. It gives addresses, and has bulletins containing data for forecasting purposes. Details of prediction methods are contained in CCIR reports. Usually these are not published in scientific journals, but can be obtained from the ITU.

12.11. Martyn's theorem for attenuation of radio waves

In this chapter, except in § 12.3, electron collisions have been neglected. If the electron collision frequency ν is small enough, most of the foregoing theory can be used without modification and some new results can be established. When electron collisions are allowed for, q from (4.9) and (12.45) is given by

$$q^2 = C^2 - \frac{X}{1+Z^2} - \frac{\mathrm{i}XZ}{1+Z^2} \tag{12.65}$$

so that q is complex and may be written $q_r + \mathrm{i}q_i$ where q_r and q_i are real and q_r is positive. If $Z = \nu/\omega$ is so small that its square and higher powers may be neglected, (12.65) gives

$$q_r^2 - q_i^2 \approx C^2 - X, \quad 2q_r q_i \approx -\mathrm{i}XZ, \tag{12.66}$$

whence

$$q_r \approx (C^2 - X)^{\frac{1}{2}}, \quad q_i \approx -\mathrm{i}\tfrac{1}{2}XZ/q_r. \tag{12.67}$$

Thus q_r is the same as the value of q used when collisions are neglected. The field in a wave packet is given by (10.4). This now includes an extra factor

$$\exp\left(k\int_0^z q_i \,\mathrm{d}z\right). \tag{12.68}$$

Since q_i in (12.67) contains a factor Z and is therefore very small, the equations giving the path and time of travel P'/c of a wave packet are not appreciably affected. The factor (12.68) shows that the wave packet is attenuated. When it returns to the ground its amplitude is reduced by a factor $|R|$ where

$$\ln|R| = 2k\int_0^{z_0} q_i \,\mathrm{d}z. \tag{12.69}$$

When (12.67) for q_i and (3.6), (3.15) for X and Z are used, this gives

$$-\ln|R(f)| = \frac{2\pi}{c}\int_0^{z_0}\frac{f_N^2\nu}{f^2 q_r}\,\mathrm{d}z$$

$$= \frac{2\pi C}{c}\int_0^{z_0}\frac{f_N^2\nu}{C^2f^2(1-f_N^2/C^2f^2)^{\frac{1}{2}}}\,\mathrm{d}z \qquad (12.70)$$

which gives the attenuation for a wave packet of frequency f obliquely incident on the ionosphere at an angle arccos C. If a wave packet of frequency $g = fC$ were vertically incident on the ionosphere, its attenuation would be given by

$$-\ln|R_0(g)| = \frac{2\pi}{c}\int_0^{z_0}\frac{f_N^2\nu}{g^2(1-f_N^2/g^2)^{\frac{1}{2}}}\,\mathrm{d}z \qquad (12.71)$$

where R_0 denotes the reflection coefficient at vertical incidence. The levels of reflection z_0 are the same in the two cases. Hence

$$\ln|R(f)| = C\ln|R_0(fC)|. \qquad (12.72)$$

This is Martyn's theorem for attenuation (Martyn 1935; see also Millington, 1938b; Appleton and Beynon, 1955). It shows that the logarithm of the reflection coefficient for frequency f at oblique incidence is C times the logarithm of the reflection coefficient for frequency fC at vertical incidence. It applies only when the effect of the earth's magnetic field is neglected and when the collision frequency ν is so small that $Z \lesssim 0.1$.

PROBLEMS 12

12.1. In a simple model of the ionosphere there is free space from $z = 0$ to $z = h_0$. Above this the plasma frequency f_N is given by $f_N^2 = \alpha(z - h_0)$. The collision frequency $\nu = 2\pi fZ$ is independent of height. The earth's magnetic field is neglected. Show that for vertically incident waves of frequency f, the phase height, true height of reflection and equivalent height are respectively

$$h = h_0 + \frac{2}{3}\frac{f^2}{\alpha}(1 - \mathrm{i}Z), \quad z_0 = h_0 + \frac{f^2}{\alpha}(1 - \mathrm{i}Z), \quad h' = h_0 + \frac{2f^2}{\alpha}(1 - \tfrac{2}{3}\mathrm{i}Z).$$

Show that the reflection coefficient R satisfies

$$\ln|R| = 2k\,\mathrm{Im}\,(h) = -\frac{4}{3}\frac{\nu}{c}\frac{f^2}{\alpha} = -\frac{\nu}{c}\{\mathrm{Re}\,(h') - \mathrm{Re}(h)\}$$

(compare (12.14)).

12.2. In a model ionosphere the electron concentration increases exponentially with height z above the ground so that $f_N^2 = F^2\mathrm{e}^{\alpha z}$. Collisions are negligible. Show that for vertically incident waves of frequency f, the equivalent height, true height and phase height of reflection are given respectively by

$$h'(f) = \frac{2}{\alpha}\ln\left[\frac{f}{F} + \left\{\left(\frac{f}{F}\right)^2 - 1\right\}^{\frac{1}{2}}\right], \quad z_0 = \frac{2}{\alpha}\ln\frac{f}{F}, \quad h(f) = h'(f) - \frac{2}{\alpha}\left\{1 - \left(\frac{F}{f}\right)^2\right\}^{\frac{1}{2}}.$$

Note that there must be a non-zero electron concentration even at the ground $z=0$, but for a practical model it is very small so that F/f is an extremely small number. Then

$$h'(f)\approx\frac{2}{\alpha}\ln(2f/F),\quad z_0=\frac{2}{\alpha}\ln\frac{f}{F},\quad h(f)\approx\frac{2}{\alpha}\{\ln(2f/F)-1\}.$$

12.3. In problem 12.2 suppose that the electron collision frequency is independent of height. Show that its effect is to increase the equivalent height of reflection $\mathrm{Re}\{h'(f)\}$ by $Z^2/\alpha+O(Z^4)$ where $Z=\nu/2\pi f$.

12.4. What electron height distribution functions $N(z)$ would give the following $h'(f)$ curves if electron collisions and the earth's magnetic field are neglected?
(a)

$$h'(f)=h_0 \text{ when } f\leqslant f_1;\ h'(f)=h_0+\frac{2f}{\alpha}(f^2-f_1^2) \text{ when } f\geqslant f_1.$$

(b)

$$\begin{aligned} h'(f)&=h_0+2f^2/\alpha && \text{when } f\leqslant f_1,\\ &=h_0+2f^2/\alpha+2f(f^2-f_1^2)^{\frac{1}{2}}\left(\frac{1}{\beta}-\frac{1}{\alpha}\right) && \text{when } f\geqslant f_1. \end{aligned}$$

12.5. Suppose that the refractive index n in the ionosphere is given by $n^2=1-\alpha(z-h_0)$ for $z\geqslant h_0$; $n^2=1$ for $z\leqslant h_0$, where α is a real constant, and the earth's magnetic field is neglected. A point transmitter sends up a wave packet at an angle θ to the vertical. Find the distance at which it returns to the earth's surface. (Neglect the earth's curvature.)

The transmitter emits a pulse of radio waves uniformly in all directions. Show that the pulse first returns to the earth at a distance $4h_0(2/h_0\alpha-1)^{\frac{1}{2}}$ from the transmitter, provided that $\alpha<2/h_0$. (From Math. Tripos, 1957, Part III.)

12.6. A radio wave packet with predominant frequency f leaves the ground at an angle θ to the vertical. It is reflected from the ionosphere and returns to the ground at distance D, taking a time $P'(f)/c$. There is free space up to height $z=h_0$ and above this the electron concentration is proportional to $(z-h_0)^2$. Electron collisions and the earth's curvature and magnetic field are to be neglected. Prove that

$$P'(f)\sin\theta=D=2h_0\tan\theta+Kf\sin\theta \text{ where } K \text{ is a constant.}$$

(Adapted from Math. Tripos 1958, Part III.)

12.7. The ionosphere is a very thin layer, for example a sporadic E-layer (E_S), at height h_0, with penetration frequency f_p. Collisions and the earth's magnetic field are to be neglected.
(a) If the earth's curvature is neglected and if the receiver is at range D from the transmitter, show that the maximum usable frequency is $f_M=f_p(1+D^2/4h_0^2)^{\frac{1}{2}}$.

(b) If the earth's surface has radius ρ, and if the range is D_G measured over the curved surface, show that

$$f_M = f_p\{(h_0+\rho)^2+\rho^2-2\rho(\rho+h_0)\cos\Phi\}^{\frac{1}{2}}/(h_0+\rho-\rho\cos\Phi)$$

where $\Phi = \frac{1}{2}D_G/\rho$. Since Φ is small, show that this leads to

$$f_M = f_p\left(1+\frac{D_G^2}{4h_0^2}-\frac{D_G^2}{4\rho h_0}\right)^{\frac{1}{2}}.$$

(For the solution see Appleton and Beynon, 1940, § 4).

13

Ray theory results for anisotropic plasmas

13.1 Introduction

We now have to consider how results such as those of ch. 12 for an isotropic plasma must be modified when the earth's magnetic field is allowed for. The dispersion relation (4.47) or (4.51) is now much more complicated so that, even for simple electron height distribution functions $N(z)$, it is not possible to derive algebraic expressions for such quantities as the equivalent height of reflection $h'(f)$ or the horizontal range $D(\theta)$. These and other quantities must now be evaluated numerically. The ray direction is not now in general the same as the wave normal. Breit and Tuve's theorem and Martyn's theorems do not hold. These properties therefore cannot be used in ray tracing. Some general results for ray tracing in a stratified anisotropic plasma have already been given in ch. 10.

The first part of this chapter §§ 13.2–13.6 is concerned with vertically incident pulses of radio waves on a stratified ionosphere, since this is the basis of the ionosonde technique, §1.7, which is widely used for ionospheric sounding. The transmitted pulses have a radio frequency f which is here called the 'probing frequency'. It is also sometimes called the 'carrier frequency'. The study of this subject involves numerical methods that are important because of their use for analysing ionospheric data and for the prediction of maximum usable frequencies. It has already been explained, § 10.2, that the incident pulse splits into two separate pulses, ordinary and extraordinary, that travel independently. This is the phenomenon of magnetoionic splitting. Ionosonde equipments can be carried in satellites above the maximum of $N(z)$ in the F2-region of the ionosphere, so that they send pulses downwards. This is known as topside sounding and has been extensively used since 1962. The main results are briefly described in § 13.5.

Four other topics come naturally into this chapter, namely Faraday rotation, § 13.7, whistlers and ion cyclotron whistlers §§ 13.8, 13.9, absorption § 13.10, and wave interaction §§ 13.11–13.13.

13.2. Reflection levels and penetration frequencies

A radio wave vertically incident on the ionosphere from below travels upwards until it is reflected at a level where the refractive index $n = 0$. For the ordinary wave, when collisions are neglected, this occurs where $X = 1$, that is $f = f_N$, which is the same condition as for an isotropic plasma. For the extraordinary wave the refractive index is zero where $X = 1 - Y$, which requires that

$$f_N^2 = f^2 - f f_H. \tag{13.1}$$

This is a quadratic equation with the solution

$$f = f_{\mathrm{E}}^{(-)} = \tfrac{1}{2}\{(f_H^2 + 4f_N^2)^{\frac{1}{2}} + f_H\}. \tag{13.2}$$

The other solution is negative and therefore of no practical interest. The refractive index is also zero for the extraordinary wave where $X = 1 + Y$ and this similarly gives

$$f = f_{\mathrm{E}}^{(+)} = \tfrac{1}{2}\{(f_H^2 + 4f_N^2)^{\frac{1}{2}} - f_H\}. \tag{13.3}$$

Thus an extraordinary wave travels upwards until it is reflected at the level where either (13.2) or (13.3) is first satisfied. In the free space below the ionosphere (13.3) is zero and (13.2) is f_H. They both increase when the ionosphere is entered and f_N increases, and (13.2) is always the greater.

If $f > f_H$ then as f_N increases it is (13.2) that first reaches the value f of the probing frequency. Thus it might be expected that reflection at a level given by (13.3) can only be observed if $f < f_H$. This is true according to simple ray theory when the transmitter is at the ground or anywhere in free space. But in topside sounding the transmitter is in the ionosphere and may be at a level where X exceeds the value marked RES in fig. 4.3, so that n^2 is positive for the Z-mode. Then reflection at the level given by (13.3) can be observed even when $f > f_H$; see § 13.5.

When observing from the ground, especially at high latitudes, a reflection from the level given by (13.3) is sometimes seen when $f > f_H$, and the resulting branch of the $h'(f)$ curve is known as the 'Z-trace'. The phenomenon is described later, § 16.8.

The penetration frequency of an ionospheric layer is the frequency that makes $n = 0$ at the maximum of the layer. This can only happen if collisions are neglected, but penetration frequencies are customarily defined by neglecting collisions. For the ordinary wave the penetration frequency is the plasma frequency f_N at the maximum of the layer. It is denoted by $f_{\mathrm{p}}^{\mathrm{O}}$ which is the same as the f_{p} used in ch. 12 for an isotropic ionosphere. When the term 'penetration frequency' is used alone, it normally refers to the ordinary wave.

For the extraordinary wave there are two penetration frequencies. One is where $X = 1 - Y$ at the maximum of the layer. It is given, from (13.2), by

$$f_{\mathrm{p}}^{(\mathrm{X}-)} = \tfrac{1}{2}[\{f_H^2 + 4(f_{\mathrm{p}}^{(\mathrm{O})})^2\}^{\frac{1}{2}} + f_H]. \tag{13.4}$$

This is the one that is normally observed when $f > f_H$. If $f_{\mathrm{p}}^{(\mathrm{O})}$ is large compared with f_H, (13.4) gives

$$f_p^{(X-)} - f_p^{(O)} \approx \tfrac{1}{2} f_H. \tag{13.5}$$

This result was used by Appleton and Builder (1933) to measure f_H in the ionosphere, and to show that it is electrons that are responsible for the reflection of radio waves.

The other extraordinary wave penetration frequency is, from (13.3),

$$f_p^{(X+)} = \tfrac{1}{2}\{(f_H^2 + 4 f_p^{(O)2})^{\frac{1}{2}} - f_H\}. \tag{13.6}$$

The relation between the two penetration frequencies (13.4), (13.6) and $f_p^{(O)}$ is shown in fig. 13.1.

In (13.4) and (13.6) the superscript X is used for 'extraordinary' because the label E is needed to designate the E-layer. In the literature that deals with the reduction of ionograms, the penetration frequencies $f_p^{(O)}$, $f_p^{(X-)}$, $f_p^{(X+)}$ for the E-layer are usually written $F_O E$, $F_X E$, $F_Z E$. For the F-layer the E is replaced by F. The most important penetration frequencies are $f_p^{(O)}$, $f_p^{(X-)}$ for the F2-layer and they are commonly written $F_O F2$, $F_X F2$.

When the probing frequency f increases and approaches a penetration frequency, the equivalent height $h'(f)$ becomes large and tends to infinity. It is in this way that penetration frequencies are identified on ionogram records. It was pointed out

Fig. 13.1. Shows how the two penetration frequencies for the extraordinary ray, shown as continuous and dotted curves, are related to that for the ordinary ray, shown as a chain line.

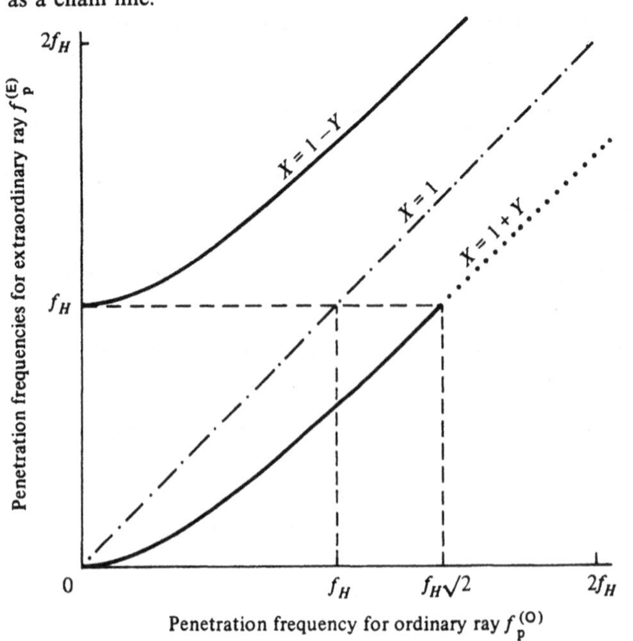

above that, except for the Z-trace, the reflection condition (13.3) is only observed when probing from the ground, if $f < f_H$. The same applies to the penetration frequency (13.6). If it is to be observable from the ground it must be less than f_H, and this requires that

$$f_p^{(O)} < f_H \surd 2. \tag{13.7}$$

For a layer that satisfies this condition, the $h'(f)$ curve for the extraordinary wave shows two infinities, one given by (13.4) where $f > f_H$, and the other by (13.6) where $f < f_H$. Fig. 13.3 shows an example of this. If (13.7) is not satisfied, the $h'(f)$ curve for the extraordinary wave shows only one infinity that is given by (13.4) where $f > f_H$. In either case $h'(f)$ attains large values when $f \to f_H$ but here it tends to a bounded limit; see fig. 13.2. This is not an infinity and is not associated with penetration; see § 13.4.

13.3. The calculation of equivalent height, $h'(f)$

In the first part of this section it is assumed that electron collisions may be neglected. The effect of collisions is discussed at the end of the section.

The equivalent height of reflection for a vertically incident pulse of radio waves is $h'(f) = \int_0^{z_0} n' \, dz$ from (12.2) and this still holds when the earth's magnetic field is

Fig. 13.2. The $h'(f)$ curves for frequencies near the electron gyro-frequency, 1.2 MHz in these examples. The angle between the earth's magnetic field and the vertical is 23.27°. In (a) $f_N^2 = \alpha(z - h_0)$ for $z > h_0$, and $h'(f)$ is measured from the level $z = h_0$. In (b) f_N^2 is a proportional to $e^{\alpha z}$ and $h'(f)$ is measured from the level where $f_N = 0.5$ MHz. For the continuous curves the collision frequency ν is zero. For the broken curves in (a) the numbers are the values of ν in s^{-1}. These collision frequencies make no appreciable difference except for the extraordinary wave at frequencies just less than the gyro-frequency.

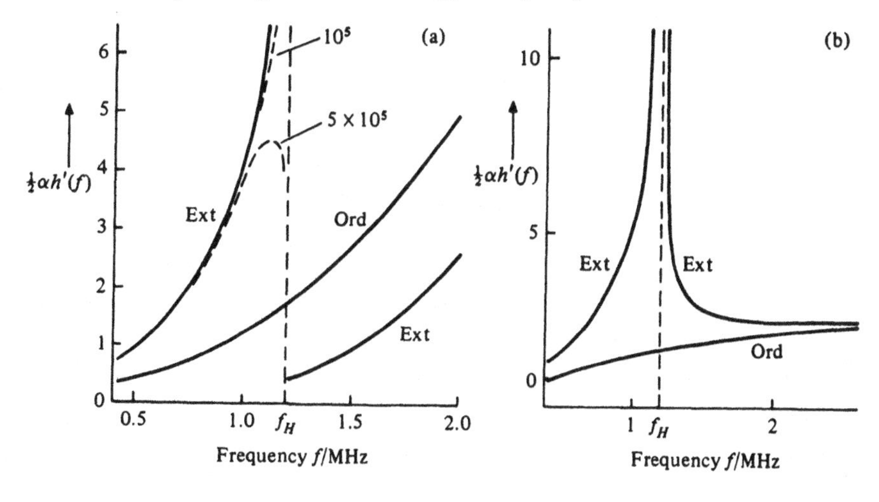

allowed for. It will now be written

$$h'(f)=\int_0^{z_0} n'(f,f_N)\,\mathrm{d}z. \tag{13.8}$$

Here n' is the group refractive index. It is a function of the probing frequency f and the plasma frequency f_N, and it is through f_N that n' depends on z. It also depends on f_H and on the angle Θ between $\boldsymbol{Y}$ and the vertical, but these can be assumed to be constant for any one observing location, and therefore they are not written in (13.8). The reflection level $z=z_0$ is where the refractive index is zero, and here n' tends to infinity; see figs. 5.16, 5.17. For the ordinary wave near the reflection level, n' is given by (5.78) and for the extraordinary wave by (5.79). These can be used to show that the integral (13.8) is bounded even though n' is infinite at the upper limit. The method is the same as in § 12.2.

To evaluate (13.8) it is necessary to use numerical methods. The contribution from the part of the path in free space is simply the length of that part. To find the contribution from the part in the ionosphere it is sometimes convenient to use f_N as the variable of integration, instead of z, thus

$$h'(f)=\int_0^{f_\mathrm{R}} n'(f,f_N)\frac{\mathrm{d}z}{\mathrm{d}f_N}\,\mathrm{d}f_N. \tag{13.9}$$

Here f_R is the value of f_N where reflection occurs. For the ordinary wave $f_\mathrm{R}=f$, and for the extraordinary wave, the two reflection conditions give

$$\text{for } X=1\mp Y:\quad f_\mathrm{R}=(f^2\mp ff_H)^{\frac{1}{2}}. \tag{13.10}$$

The transformation in (13.9) can only be used when $f_N(z)$ is a monotonic function in the range of integration. For numerical integration it is best to use an integrand that does not go to infinity at one limit. This can be achieved by a further transformation. For example, let

$$f_N=f_\mathrm{R}\sin\phi. \tag{13.11}$$

Then (13.9) becomes

$$h'(f)=f_\mathrm{R}\int_0^{\frac{1}{2}\pi} n'(f,f_\mathrm{R}\sin\phi)\frac{\mathrm{d}z}{\mathrm{d}f_N}\cos\phi\,\mathrm{d}\phi. \tag{13.12}$$

It can be shown from (5.78) or (5.79) that $n'\cos\phi$ is now bounded at the upper limit. This can be achieved by other transformations, and the best choice depends on the function $f_N(z)$. Various other forms were used by Millington (1938a) and by Shinn and Whale (1952). An integral such as (13.12) is now suitable for evaluation by one of the standard methods, for example Gaussian quadrature (Abramowitz and Stegun, 1965, table 25.4).

The effect of electron collisions on the equivalent height for an isotropic ionosphere was discussed in § 12.3, and the same general conclusions hold when the earth's magnetic field is allowed for. The process of analytic continuation in the complex z plane is used. The group refractive index n', the height of reflection z_0 and the equivalent height h' are all complex. It was shown in § 11.15 that it is $\mathrm{Re}(h')$ that gives the time of travel of the radio pulse. The complex $h'(f)$ is still given by (13.8) where the integral is now a contour integral in the complex z plane.

The complex reflection height z_0 is where $n=0$, that is where

$$X(z) = 1 + \eta Y - \mathrm{i}Z(z). \tag{13.13}$$

Here, for the ordinary wave $\eta = 0$, and for the extraordinary wave $\eta = -1$ if $f > f_H$ and $+1$ if $f < f_H$. This equation usually has more than one solution z. The required solution z_0 is nearly always the nearest to the real z_0 that would be used when $Z = 0$. The contribution to the integral (13.8) from within the ionosphere can be evaluated by using a transformation similar to (13.11). Let $Z(z_0) = Z_0$ and let

$$f_N = f(1 + \eta Y - \mathrm{i}Z_0)^{\frac{1}{2}} \sin\phi. \tag{13.14}$$

Then the contribution to (13.8) from within the ionosphere is

$$h'(f) = f(1 + \eta Y - \mathrm{i}Z_0)^{\frac{1}{2}} \int_0^{\frac{1}{2}\pi} n'(f, f_N, \nu) \frac{\mathrm{d}z}{\mathrm{d}f_N} \cos\phi \, \mathrm{d}\phi. \tag{13.15}$$

Here the complex group refractive index depends on the collision frequency ν, which may be a function of z. To use this formula it must be possible to express z and $\nu(z)$ as analytic functions of f_N and thence of ϕ. The variable ϕ may be chosen to be real over the range of integration, so that f_N and $\mathrm{d}z/\mathrm{d}f_N$ and ν are complex. At the top limit, $\phi = \frac{1}{2}\pi$, n' is infinite but $\cos\phi$ is zero. It can again be shown, as for (13.12), that the product $n' \cos\phi$ is bounded. To allow for collisions, the proof needs an extension of the argument of § 5.9 that led to (5.78). It is left as an exercise for the reader. The formula (13.14) and extensions of it have been used by Cooper (1961) and by Piggott and Thrane (1966) to study the effect of collisions on the $h'(f)$ curves for a parabolic model and other models of the ionosphere.

It is sometimes necessary to compute $h'(f)$ when $N(z)$ is given not as an algebraic function but as a table of numerical values, usually at equal intervals of z. Then $N(z)$ for any z is found by interpolation in the table. If collisions are neglected, the path for the integral (13.8) is the real z axis, and $h'(f)$ can be found by numerical integration. It is best not to change to f_N as the variable, as was done in (13.9). To transform the integral so that the integrand is bounded at the upper limit, the change of variable

$$z = z_0 \sin^2\phi \tag{13.16}$$

can be used.

When collisions are allowed for it is first necessary to find the complex z_0 where $n = 0$, and this requires analytic continuation of the tabulated $N(z)$ to complex values of z. To do this we first find the real z where $X = 1$ (ordinary wave) or $X = 1 \mp Y$ (extraordinary wave). Then it is assumed that $N(z)$ and therefore $X(z)$ is a linear function in this neighbourhood, whence the complex z_0 that makes $X = 1 - \mathrm{i}Z$, or

$1 \mp Y - iZ$, respectively, can be found. In a few cases studied by the author it was found better to use a quadratic function instead of a linear function. Then the required value of the equivalent height is given by

$$\mathrm{Re}\{h'(f)\} = \int_0^{\mathrm{Re}(z_0)} \mathrm{Re}(n')\,\mathrm{d}z - \int_0^{\mathrm{Im}(z_0)} \mathrm{Im}(n')\,\mathrm{d}s \tag{13.17}$$

where in the second integral $z = \mathrm{Re}(z_0) + is$, and $N(z)$ is calculated by the same linear or quadratic function that was used to find z_0. In the second integral it is also useful to use

$$s = \mathrm{Im}(z_0)\sin^2\phi, \tag{13.18}$$

so that the integrand is bounded at the upper limit; compare (13.16).

Another way of calculating $h'(f)$ is to compute two values h_1, h_2 of $h(f)$ at neighbouring frequencies f_1, f_2 and then to use $h' = \partial(fh)/\partial f \approx (f_1h_1 - f_2h_2)/(f_1 - f_2)$. This method was used by Altman (1965) who included a study of the Z-trace (§16.8).

13.4. Ionograms

'Ionogram' is the name given to the curves of $h'(f)$ obtained with an ionosonde equipment, §1.7. In the following description subscripts will be used to indicate the values $h'_O(f)$, $h'_X(f)$ for the ordinary and extraordinary waves respectively.

Curves of $h'_O(f)$ are of the same general form as those for an isotropic ionosphere. They show similar behaviour near the penetration frequencies of ionospheric layers, and near the plasma frequency in a ledge in the electron height distribution $N(z)$; see §§12.2–12.5 and figs. 12.2–12.4. For frequencies f near to the electron gyro-frequency f_H, the refractive index n and group refractive index n' are bounded and continuous functions of f. The $h'_O(f)$ curves for the ordinary wave show no special features when f is near to f_H.

When $f > f_H$ the extraordinary wave is reflected where $X = 1 - Y$, that is at a lower level than the ordinary wave. Thus $h'_X(f)$ is in general smaller than $h'_O(f)$ but there are exceptions. For example, if f very slightly exceeds the extraordinary penetration frequency of the E-layer, the extraordinary wave is delayed near the maximum of the E-layer and has the larger equivalent height. If the extraordinary wave is reflected in a ledge of $N(z)$, the pulse spends a long time near the reflection level; see §12.5. Then $h'_X(f)$ can be greater than $h'_O(f)$. Examples of this can be seen in the typical $h'(f)$ diagram sketched in fig. 13.4.

Suppose now that the extraordinary wave is reflected from an ionospheric layer with a non-zero value of $G = \mathrm{d}N(z)/\mathrm{d}z$ at its lower edge. Let the frequency f be decreased so that it approaches f_H. The reflection level where $X = 1 - Y$ gets lower as f decreases and in the limit $f \to f_H$ it is where $f_N \to 0$, that is at the base of the layer. But the contribution to $h'_X(f)$ from within the layer does not necessarily decrease. It tends to a non-zero limit and it can be shown that this is proportional to $1/G$. (See problem 13.1.) For actual ionospheric layers $N(z)$ tails off gradually towards zero as z decreases, so that G is effectively zero, or very small. Then $h'_X(f)$ tends to large values as $f \to f_H(+)$. These effects are illustrated by the two examples in fig. 13.2.

Fig. 13.3. Calculated $h'(f)$ curves for a model ionosphere with two parabolic layers. For the lower layer, $f_p^{(O)} = 0.7$ MHz, half thickness $a = 30$ km, height of maximum = 110 km. For the upper layer $f_p^{(O)} = 2.4$ MHz, half thickness $a = 50$ km, height of maximum = 260 km.

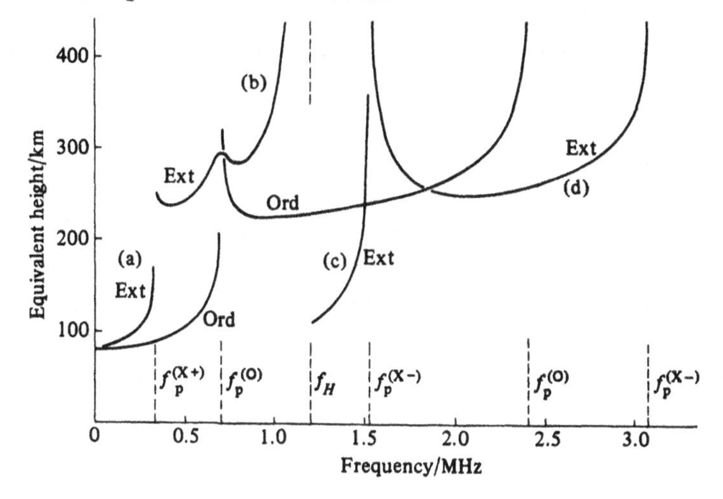

Fig. 13.4. Sketch of a typical ionogram for an undisturbed summer day in temperate latitudes. The trace for the extraordinary wave is shown as a broken line. Note that a logarithmic scale of frequency is used. This is the usual practice in published ionograms.

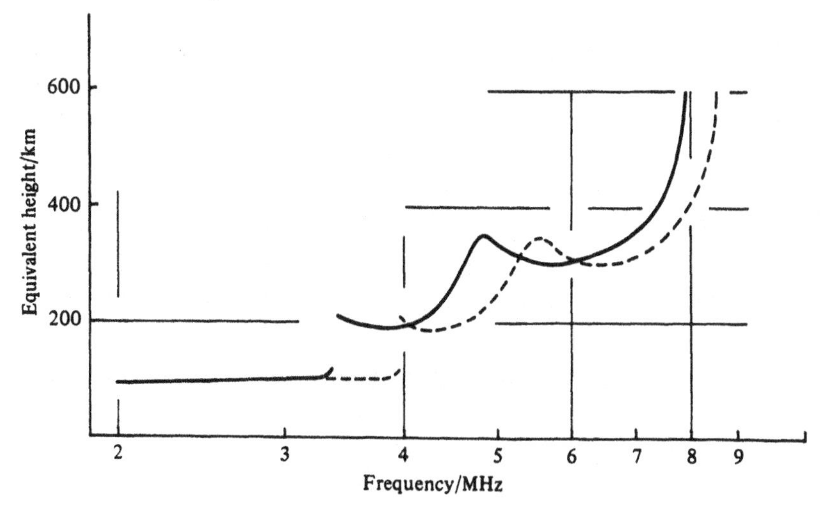

When $f<f_H$ the extraordinary wave is reflected where $X=1+Y$, that is at a higher level than the ordinary wave. Consequently $h'_X(f)$ is in general greater than $h'_O(f)$ but again there can be exceptions. Now let f be increased so that it approaches f_H. Then $h'_X(f)$ gets very large, but this is quite different from the behaviour when f approaches a penetration frequency. It occurs because n' is very large where X and $Y-1$ are small. There are two examples in fig. 13.2.

Fig. 13.3 shows the calculated ionogram for a model of the ionosphere consisting of two parabolic layers simulating the E- and F-layers. The curve $h'_O(f)$ is similar to that obtained when the earth's magnetic field was neglected, § 12.4 and fig. 12.4. The curve $h'_X(f)$ is in four parts. The left branch (a) is for reflection from the lower layer and goes to infinity at the penetration frequency $f_p^{(X+)}$ (13.6). The next branch (b) is for reflection from the upper layer. It shows an upward turn at the left end because the waves have just penetrated the lower layer but are still retarded as they pass through it. At the penetration frequency of the lower layer for the ordinary wave, the extraordinary branch (b) shows a maximum of $h'_X(f)$. This is because, near the maximum $N(z)$ in the lower layer, X is close to unity. Now fig. 5.17 shows that near $X=1$, the group refractive index n' for the extraordinary wave has a large value and this occurs over a large range of height z near the maximum of $N(z)$. Hence the pulse is markedly retarded in the lower layer. For other frequencies, the level where $X \approx 1$ is not near the maximum. It extends over a smaller range of height and the retardation is less. For the branch (b), $h'_X(f)$ tends towards infinity at its right-hand end as the frequency f approaches f_H, but this is not associated with penetration of any layer.

The branch (c) is for f just greater than f_H, and is for reflection from the lower layer. It shows penetration at the frequency $f_p^{(X-)}$, (13.4). The remaining branch (d) is for reflection from the upper layer, and shows a large $h'_X(f)$ at its left-hand end because the waves have then just penetrated the lower layer but are still retarded by it. At the right-hand end it goes to infinity as f approaches the penetration frequency $f_p^{(X-)}$ for the upper layer, (13.4).

Although the two-parabola model used for fig. 13.3 is very simple compared with the actual $N(z)$, the calculated ionogram shows many of the features of observed ionograms. In published ionograms the frequency range covered rarely extends down to the electron gyro-frequency or below it. There are several reasons for this. In the day time radio waves with frequencies of order 1 MHz or less are heavily absorbed in the D-region, so that the reflected amplitudes are small. Ionosonde receivers are subject to interference at these frequencies from transmitters in the broadcast frequency bands. As the frequency is decreased it is necessary to use larger and more elaborate aerials to achieve efficient radiation and to avoid interfering with broadcast reception. The absorption is particularly great for the extraordinary wave near the gyro-frequency so that here its behaviour in ionograms is not easy to observe. At night, however, the absorption is less. Ionograms extending down to 50 kHz have been published (Watts and Brown, 1954; Watts, 1957) that display the maximum in branch (b) of the $h'_X(f)$ curve; fig. 13.3. For frequencies greater than f_H the general form of the ionogram in fig. 13.3 is similar to actual ionograms observed at night. In the day time, however, the penetration frequency of the F-layer is much greater and there is, nearly always, an F1-layer giving a ledge in the function $N(z)$, (see § 12.5 and fig. 12.5) with a resulting maximum in both $h'_O(f)$ and $h'_X(f)$; see fig. 13.4.

The ionosonde technique has been widely used in ionospheric research and the number of published ionograms is very great. For some collections of typical ionograms see, for example, Pickle (1951), Wright and Knecht (1957), Stevens (1961), Rawer and Suchy (1967). For a detailed study of the interpretation of ionograms see Piggott and Rawer (1972, 1978).

The above description applies mainly to the special simple case when the ionosphere is horizontally stratified. In practice ionograms can be very complicated because of disturbing factors. For example when the layers are tilted the transmitted and received rays are oblique. Irregularities in the ionospheric plasma can make the ionogram curves broad and irregular. There is sometimes a thin layer at a height of about 100 km, known as sporadic E, abbreviation E_S, that has a penetration frequency comparable with or greater than that of the F-layer, so that it gives an almost horizontal line at about 100 km on the ionograms, and the reflections from higher levels are masked by it. (See, for example, Smith, E.K. and Matsushita, 1962.) In polar regions the ionosphere is nearly always disturbed and its behaviour differs from that at lower latitudes. These and other disturbances are not considered further in this book. For a general description and references see, for example, Rawer and Suchy (1967), Davies (1969), Ratcliffe (1972).

13.5. Topside sounding

Ionograms are frequently recorded with an ionosonde equipment in a satellite above the maximum of the F2-layer. Then the radio pulses travel downwards and return to the satellite after reflection. The time of travel is $2h'(f)/c$ and $h'(f)$ is often called the 'equivalent depth.' The curves are often presented with the ordinate $h'(f)$ plotted downwards. The following description applies only to the simplest ideal conditions. Fig. 13.6 is a sketch of a typical topside ionogram for this case.

The transmitter and receiver are immersed in the ionospheric plasma. It is here assumed that below them the electron concentration $N(z)$ and thence the plasma frequency $f_N(z)$ increase monotonically as the height z decreases, until the maximum of the F2-layer is reached. There is no intervening maximum and no ledge. It is further assumed that the ionosphere is horizontally stratified so that the wave normals are everywhere vertical. The transmitted pulse splits into ordinary and extraordinary components, and these suffer lateral deviation, as described in § 10.12, but it is the vertical component $\mathscr{U}_z$ of the group velocity that determines the time of travel.

To describe topside ionograms it is useful to plot a diagram, fig. 13.5, with f as abscissa and f_N as ordinate, increasing downwards. Separate diagrams are shown for the ordinary and extraordinary waves.

The horizontal lines AB are drawn where f_N is the plasma frequency $f_s^{(O)}$ at the satellite, and the lines CD where f_N is the penetration frequency $f_p^{(O)}$ of the F2-layer. The continuous lines show where the refractive index $n = 0$ and the waves are reflected, namely $X = 1$ for the ordinary wave and $X = 1 \mp Y$ for the extraordinary wave. The chain line shows where n is infinite. It is given by (4.76) with $U = 1$, and depends on the angle Θ between the earth's magnetic field and the vertical. It corresponds to the value of X marked RES in figs. 4.3, 4.6. The shaded regions between these lines show where the waves are evanescent and cannot be propagated. The various transition values of the frequency f at the satellite are indicated by a subscript s. Thus $f_s^{(O)}$ is where $X = 1$, and $f_s^{(X-)}$, $f_s^{(X+)}$ are where $X = 1 \mp Y$ respectively, and $f_s^{(O\infty)}$ $f_s^{(X\infty)}$ are where n and n' are infinite. The equivalent depth of reflection is given by

$$h'(f) = -\int_{f_s^O}^{f_R} n'(f, f_N) \frac{dz}{df_N} df_N \tag{13.19}$$

compare (13.9), where f_R is the value of f_N at the reflection level.

Consider now the ordinary wave. If $f > f_p^{(O)}$ the wave penetrates the ionosphere. It is reflected at the ground and returns to the satellite. If f only slightly exceeds $f_p^{(O)}$, the group retardation near the maximum of the F-layer is large. If f is slightly less

Fig. 13.5. The f/f_H vs f_N/f_H diagrams used in the study of topside ionograms. The chain curves marked RES are where one refractive index is infinite. They depend on the angle Θ between the earth's magnetic field and the vertical, and in this example $\Theta = 30°$.

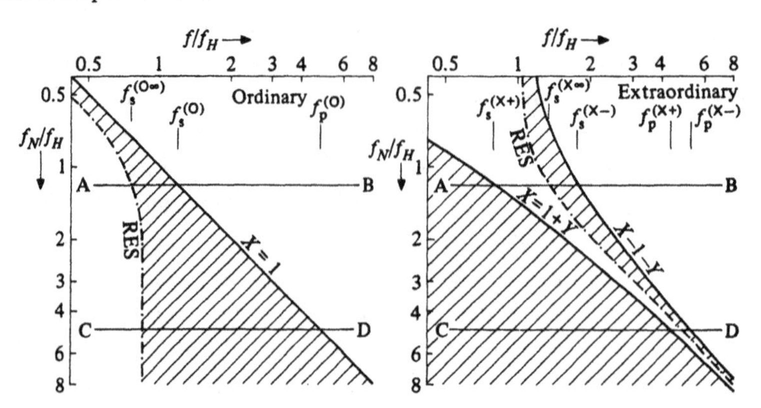

than $f_p^{(O)}$ the wave is reflected just above the maximum of the F-layer and again the group retardation is large, just as in ionograms observed at the ground, fig. 13.3. If f is decreased, the equivalent depth $h'(f)$ gets less. When f slightly exceeds $f_s^{(O)}$ the reflection level is at only a small distance below the satellite and $h'(f)$ is small. It tends to zero when $f \to f_s^{(O)}$. The extraordinary wave is similar. It penetrates the ionosphere and is reflected from the ground when $f > f_p^{(X-)}$. Its $h'(f)$ tends to zero when $f \to f_s^{(X-)}$. These effects are seen in the sketch, fig. 13.6. The ground reflections are at the right-hand end.

An ordinary wave can also be launched from the satellite if $f < f_s^{(O\infty)}$; see fig. 13.5(a). This is where $f < f_H$ and the wave is in the whistler mode; § 13.8. The figure shows that there is no value of f_N where a wave of this frequency can be reflected in the ionosphere. It therefore penetrates through to the lower ionosphere. Here the effect of collisions cannot be ignored for these low frequencies. The wave might be reflected at the ground and return to the satellite but such reflections are rarely seen in topside ionograms.

An extraordinary wave can be launched from the satellite if $f_s^{(X+)} < f < f_s^{(X\infty)}$. It is reflected where $X = 1 + Y$ and its $h'(f)$ curve is called the 'topside Z-trace'. If f slightly exceeds $f_s^{(X+)}$, the value of $h'(f)$ is small and tends to zero when $f \to f_s^{(X+)}$. In the conditions of fig. 13.5(b) this wave could never penetrate the ionosphere. When f increases and approaches the value $f_s^{(X\infty)}$, the value of X near the satellite approaches the value X_∞ that makes the group refractive index n' infinite. It was shown in § 5.9 that n' is then of order $(X - X_\infty)^{-\frac{1}{2}}$; see (5.82). It can thence be shown that the contribution to $h'(f)$, (13.19), from near the lower limit of the integral, tends to

Fig. 13.6. Sketch to show the general form of a topside ionogram.

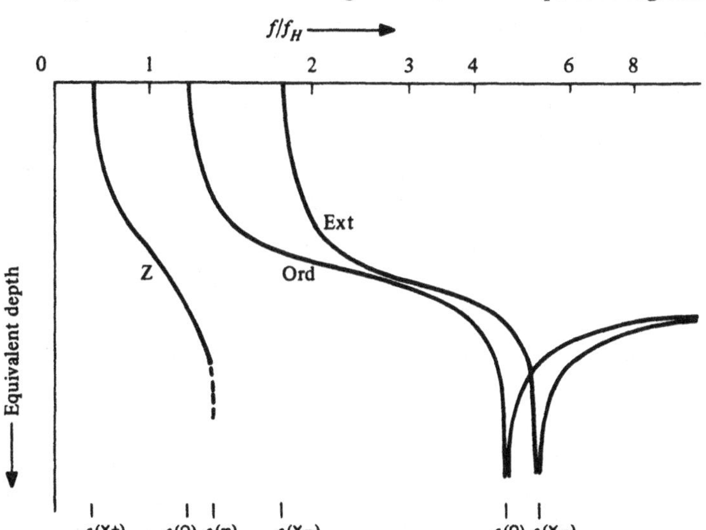

infinity when $f \to f_s^{(X\infty)}$. This is quite different from the behaviour near a reflection level, where $X = X_R$. Then n' is of order $(X - X_R)^{-\frac{1}{2}}$, see (5.78), (5.79), and $h'(f)$ tends to a bounded limit, as was shown in § 13.3. When f increases and approaches $f_s^{(X\infty)}$, therefore, $h'(f)$ tends to infinity because of the large group retardation near the satellite. This infinity is not connected with penetration. In this respect the topside Z-trace is different from the Z-trace in ionograms recorded at the ground, § 16.8, which show $h'(f) \to \infty$ where penetration of the F-layer is approached.

When f is near to $f_s^{(X\infty)}$, conditions are near resonance and it was shown in § 5.4 item (9) that then the ray and the wave normal are nearly at right angles. Since the wave normal is vertical, the path of the wave packet must be nearly horizontal, and it can be shown that it is towards the equator. It can be seen from fig. 13.5(b) that, for the topside Z-trace, the level where $X = 1$ is between the satellite and the reflection level. Here, from § 5.2 item (8), $n = 1$ whence (5.83) shows that $\alpha = 0$, and the path of the wave packet is vertical. At still lower levels the angle α between the ray path and the wave normal has changed sign, so that the downward path here is directed away from the equator. For some examples of the ray paths for all three topside sounder traces see Davies (1969).

Actual topside ionograms are never as simple as is implied by fig. 13.6. The ionosphere may not be horizontally stratified so that there are oblique reflections. It may be disturbed so that the traces are broadened and irregular. The function $N(z)$ may not be monotonic, or may contain a ledge. The most striking thing of all is that the ionograms show resonances or spikes at certain frequencies. These occur because the transmitted energy excites the plasma surrounding the satellite into some kind of resonance that decays in a time comparable with the interval between transmitted pulses. They show in the ionogram as vertical lines extending down from $h'(f) = 0$. They are sometimes so intense that the actual ionogram is partly masked and difficult to read. They occur at the frequencies $f_s^{(O)}$, $f_s^{(X-)}$, $f_s^{(X+)}$, $f_s^{(O\infty)}$, $f_s^{(X\infty)}$, $2f_H$, $3f_H$,.... and less strongly at f_H, and at the upper hybrid frequency $(f_s^{(O)2} + f_H^2)^{\frac{1}{2}}$. These resonances are not within the realm of radio propagation, but they are extremely important in plasma physics and are studied in books on that subject; see, for example, Lockwood (1963). For the theory see Fejer and Calvert (1964), Dougherty and Monaghan (1965), Sturrock (1965), Nuttall (1965a, b).

The details of all aspects of topside sounding, including the resonances, were discussed in a series of papers in a special issue of *Proc. I.E.E.E* in June 1969. For good general accounts see Rawer and Suchy (1967), Davies (1969), Jackson, Schmerling and Whitteker (1980).

13.6. The calculation of electron concentration $N(z)$ from $h'(f)$

Some examples have been given of $h'(f)$ curves calculated when $N(z)$ is known. Of much greater practical importance is the calculation of $N(z)$ when $h'(f)$ is known.

This problem has already been discussed, in § 12.6, for the case when the earth's magnetic field is neglected, and it was shown that the integral equation (12.29) that relates $h'(f)$ to $N(z)$ can be solved analytically in the range of z where $N(z)$ is a monotonically increasing function. When the earth's magnetic field is allowed for, however, the integral equation is more complicated and can only be solved by numerical methods. The integral (13.8) or (13.9) is expressed as a sum of discrete terms that can be written as a matrix product. The problem is then solved by matrix inversion or some equivalent process.

The first method to be described, or variants of it, formed the basis of calculations that were widely used for many years. It is now superseded but it is useful for illustrating the general principles. It is supposed that the ionosphere is divided into discrete laminae, and this type of method is therefore called a 'lamination' method. It is explained here in terms of the ordinary wave and for ionograms observed at the ground, but it can easily be extended to use the extraordinary wave and for topside ionograms. Electron collisions are neglected.

The integral equation to be studied is (13.9) with $f_R = f$. It is to be solved to give the function $z(f_N)$. Suppose that $h'(f)$ is given as a tabulated function at equal intervals Δf of f, and let

$$h'(r\Delta f) = h'_r \tag{13.20}$$

where r is an integer. The range of integration in (13.9) is divided up into discrete intervals Δf so that for the sth interval

$$(s-1)\Delta f < f_N < s\Delta f \tag{13.21}$$

where $s \leqslant r$. It is assumed that, in each interval, $\mathrm{d}z/\mathrm{d}f_N$ is constant and given by

$$\mathrm{d}z/\mathrm{d}f_N \approx (z_s - z_{s-1})/\Delta f \tag{13.22}$$

where z_s means $z(s\Delta f)$. Now let

$$\begin{aligned} M_{rs} &= \Delta f^{-1}\int_{(s-1)\Delta f_N}^{s\Delta f_N} n'(r\Delta f, f_N)\,\mathrm{d}f_N \quad (s \leqslant r), \\ M_{rs} &= 0 \quad (s > r). \end{aligned} \tag{13.23}$$

Then equation (13.9) becomes, for all r

$$h'_r = \sum_{s=1}^{r} M_{rs}(z_{s-1}) \tag{13.24}$$

which may be written

$$\begin{bmatrix} h'_1 \\ h'_2 \\ h'_3 \\ \vdots \\ h'_r \end{bmatrix} = \begin{bmatrix} M_{11} & 0 & 0 & \cdots \\ M_{21}-M_{22} & M_{22} & 0 & \cdots \\ M_{31}-M_{32} & M_{32}-M_{33} & M_{33} & \cdots \\ \vdots & \vdots & \vdots & \\ M_{r1} & M_{r2} & M_{r3} & \cdots \end{bmatrix} \begin{bmatrix} z_1 \\ z_2 \\ z_3 \\ \vdots \\ z_r \end{bmatrix}. \tag{13.25}$$

Let A denote the square matrix on the right. Then the equation may be written in matrix notation and solved by inversion of A, thus

$$\boldsymbol{h}' = \boldsymbol{A}\boldsymbol{z}, \quad \boldsymbol{z} = \boldsymbol{A}^{-1}\boldsymbol{h}'. \tag{13.26}$$

The matrix A is lower triangular, that is, all elements above the leading diagonal are zero. It is therefore possible to find the elements of z by solving the equation (13.25) for $z_1, z_2 \ldots$ in succession. Let

$$B_{rs} = -A_{rs}/A_{rr} \quad \text{for } s \neq r, \quad B_{rr} = 1/A_{rr}. \tag{13.27}$$

Then from (13.25)

$$z_r = \sum_{s=1}^{r-1} B_{rs} z_s + B_{rr} h'_r. \tag{13.28}$$

This process is equivalent to the second equation (13.26) but there is no need to invert A.

The numbers B_{rs} depend only on the properties of the plasma and not on $h'(f)$ or $z(f_N)$. To use the method a table of these numbers must first be computed. In practical cases a value $\Delta f = 0.1$ MHz was used. To cover the frequency range 2 to 6 MHz the table contains 861 numbers. For an example see Budden (1961a). A separate table is needed for each ground observing station. Once computed, it can be used with any set $\boldsymbol{h}'$ to find z. The $h'(f)$ data from ionograms do not extend down to $f = 0$ and in many practical cases the smallest f used is about 1 to 2 MHz. The method can be extended to deal with this. For these and other details, and for examples of the tables see Thomas, Haselgrove and Robbins (1958), Titheridge (1959), Thomas and Vickers (1959), Budden (1961a), Jackson (1969).

In the above version of the method it was supposed that $h'(f)$ is given at equal intervals Δf of the frequency f. Instead of using f, any monotonic function of it may be used. For example King (1957) has shown that there is some advantage in using $\ln f$. For other ways of increasing the accuracy of the method see, for example, Titheridge (1979).

In more recent methods of calculating $N(z)$ from $h'(f)$ the need for a large table of numbers has been avoided. Instead of using a sequence of values of $z(f_N)$ as in (13.25), this function is expressed in some other way. The idea was suggested first by Titheridge (1961a) who used a polynomial expansion for $z(f_N)$ thus

$$z(f_N) = \sum_{j=1}^{r} \alpha_j f_N^{j+1} \tag{13.29}$$

and his method is known as the 'polynomial method'. Here z is measured from the base of the ionospheric layer, where $f_N = 0$. The term in f_N was not included because it would impose the restriction that $\mathrm{d}z/\mathrm{d}N$ is infinite at the base of the layer where $N = 0$, whereas Titheridge assumed that $\mathrm{d}z/\mathrm{d}N$ is bounded here. Substitution of (13.29) into (13.9) with $f_R = f$ gives

$$h'(f) = \sum_{j=1}^{r} \alpha_j(j+1) \int_0^f n'(f, f_N) f_N^j \, \mathrm{d}f_N. \quad (13.30)$$

Suppose now that observed values of $h'(f)$ are given at r different values $f_1, f_2, \ldots, f_r$ of f. Let $h'(f_i)$ be written h_i and let $\boldsymbol{h}'$ be the column matrix of the r values of h_i. Similarly let $\boldsymbol{\alpha}$ be the column matrix of the r values of α_j in (13.29). Then (13.30) gives

$$\boldsymbol{h}' = \boldsymbol{B}\boldsymbol{\alpha} \quad (13.31)$$

where $\boldsymbol{B}$ is a square matrix with elements

$$B_{ij} = (j+1) \int_0^{f_i} n'(f_i, f_N) f_N^j \, \mathrm{d}f_N. \quad (13.32)$$

Equation (13.29) may be written for r different values $f_{N1}, f_{N2}, \ldots, f_{Nr}$ of f_N. Let z_i denote $z(f_{Ni})$ and let $\boldsymbol{z}$ be the column matrix of the z_i's. Then (13.29) gives

$$\boldsymbol{z} = \boldsymbol{A}\boldsymbol{\alpha} \quad (13.33)$$

where $\boldsymbol{A}$ is the square matrix with elements

$$A_{ij} = f_{Ni}^{j+1} \quad (13.34)$$

and on substitution for $\boldsymbol{\alpha}$ from (13.31)

$$\boldsymbol{z} = \boldsymbol{A}\boldsymbol{B}^{-1}\boldsymbol{h}'. \quad (13.35)$$

This is the required solution. The matrix $\boldsymbol{A}\boldsymbol{B}^{-1}$ must be computed first. Titheridge gave examples where $r = 6$ so that the matrix contains only 36 elements. It can be computed very quickly and there is no need for the large table of numbers used with the older method.

The simplest form of the polynomial method described above has some disadvantages which were discussed by Titheridge (1961a) who showed how to modify and extend the method so as to overcome them. In later papers, Titheridge (1967a, b, 1969, 1975), Titheridge and Lobb (1977), the method has been adapted to deal with the extraordinary wave and with topside ionograms, and to give information about $N(z)$ in the lowest parts of the ionosphere where f_N is less than the smallest f used in the $h'(f)$ data.

The expansion (13.29) for z is a single valued function f_N. For values of z near where $f_N(z)$ has a maximum, $z(f_N)$ is necessarily a two-valued function, so that (13.29) cannot show a maximum of $f_N(z)$. In spite of this the method can still be used for z values that are very close to the maximum of a layer.

No method can give the function $N(z)$ completely in a valley region such as that shown between P and Q in fig. 12.7(b), because no ionogram can contain the information. But some properties of the distribution can be found. For example the minimum value of $N(z)$ can be estimated but its height cannot. For a discussion of this type of problem see Becker (1967), Titheridge (1975), Lobb and Titheridge (1977), and the references given below.

Clearly the polynomial expansion (13.29) can be replaced by other expansions. One recent development should be briefly mentioned. From the ionogram the penetration frequency $f_{\mathrm{p}}^{(\mathrm{O})}$ of the F-layer can first be found. Then an expansion of the form

$$z(f_N) = z_{\mathrm{m}} + \{\ln(f_{\mathrm{p}}^{(\mathrm{O})}/f_N)\}^{\frac{1}{2}} \sum_j A_j \mathscr{P}_j(f_N) \tag{13.36}$$

can be used. Here z_{m} is the height, at first unknown, of the maximum of the layer. By using the factor $\{\ln(f_{\mathrm{p}}^{(\mathrm{O})}/f_N)\}^{\frac{1}{2}}$, the difficulty of representing the parabolic shape of $N(z)$ near its maximum by a single valued polynomial is avoided. The $\mathscr{P}_j(f_N)$ are polynomials of some function of f_N. The unknown coefficients are the A_js and z_{m}. These correspond to the α_js in (13.29), and are found by a matrix inversion analogous to $\boldsymbol{B}^{-1}$ in (13.35).

Reinisch and Huang (1983) used an expansion of the type (13.36) for analysing ionograms observed at the ground. For the $\mathscr{P}_j(f_N)$ they used suitably chosen Chebyshev polynomials (Snyder, 1966). Huang and Reinisch (1982) used a similar method for topside ionograms.

13.7. Faraday rotation

When a linearly polarised wave enters an anisotropic plasma it splits into the two characteristic waves, ordinary and extraordinary, and each has its own polarisation $\rho_{\mathrm{O}}, \rho_{\mathrm{E}}$ respectively. The two waves have different phase velocities so that their phase difference changes as they travel. At some point along their path the resultant signal is found by adding the fields of the two waves, but because of the phase difference the resulting polarisation is not the original linear polarisation. For a very simple example of this consider a homogeneous collisionless electron plasma with the superimposed magnetic field parallel to the z axis. Let the wave normal of the incident wave, and therefore also the wave normals of the two characteristic waves, be parallel to the z axis. In these conditions the ray and wave normal directions are the same. Let the incident wave start from $z = 0$ and travel in the direction of positive z and let it be linearly polarised with $\boldsymbol{E}$ parallel to the x axis. Suppose that $\boldsymbol{Y}$ is in the positive z direction, $Y < 1$, and $X < 1 - Y$. Then the ordinary wave is circularly polarised with a right-handed sense; see fig. 4.2. Its electric field has a constant amplitude A_{O}, and where $z = 0$ the components are

$$E_x^{(\mathrm{O})} = A_{\mathrm{O}} \mathrm{e}^{\mathrm{i}\omega t}, \quad E_y^{(\mathrm{O})} = -\mathrm{i} A_{\mathrm{O}} \mathrm{e}^{\mathrm{i}\omega t} \tag{13.37}$$

Similarly the extraordinary wave is circularly polarised with a left-handed sense and its electric field, of amplitude A_{E}, has components

$$E_x^{(\mathrm{E})} = A_{\mathrm{E}} \mathrm{e}^{\mathrm{i}\omega t}, \quad E_y^{(\mathrm{E})} = \mathrm{i} A_{\mathrm{E}} \mathrm{e}^{\mathrm{i}\omega t}. \tag{13.38}$$

The linearly polarised incident wave is the sum of (13.37), (13.38) so that $A_{\mathrm{O}} = A_{\mathrm{E}}$, giving

$$E_x = 2A_{\rm O}e^{i\omega t}, \quad E_y = 0. \tag{13.39}$$

When the wave has travelled a distance z, the fields are, for the ordinary wave

$$E_x^{(\rm O)} = A_{\rm O}\exp\{i(\omega t - kzn_{\rm O})\}, \quad E_y^{(\rm O)} = -\,iE_x^{(\rm O)}. \tag{13.40}$$

and for the extraordinary wave

$$E_x^{(\rm E)} = A_{\rm O}\exp\{i(\omega t - kzn_{\rm E})\}, \quad E_y^{(\rm E)} = iE_x^{(\rm E)}. \tag{13.41}$$

The total fields are the sum of these, namely

$$\left.\begin{matrix} E_x \\ E_y \end{matrix}\right\} = 2A_{\rm O}\exp[i\{\omega t - \tfrac{1}{2}kz(n_{\rm O}+n_{\rm E})\}]\left\{\begin{matrix} \cos\{\tfrac{1}{2}kz(n_{\rm O}-n_{\rm E})\} \\ \sin\{\tfrac{1}{2}kz(n_{\rm O}-n_{\rm E})\} \end{matrix}\right\}. \tag{13.42}$$

Thus the resultant wave is linearly polarised and its plane of polarisation makes an angle

$$\Psi = \tfrac{1}{2}kz(n_{\rm O} - n_{\rm E}) \tag{13.43}$$

with the x axis. This is half the phase difference between the two waves. The plane of polarisation of the resultant wave rotates as z increases. This effect was studied by Faraday (see, for example, Jenkins and White, 1976). It was observed in the laboratory for light waves travelling in a transparent medium with a magnetic field applied in the direction of travel. It occurs not only for the free electrons in a plasma as illustrated here, but for the bound electrons in transparent solids and liquids. It is known as the 'Faraday effect', or 'Faraday rotation'. The spatial rate of rotation is

$$\partial\Psi/\partial z = \tfrac{1}{2}k(n_{\rm O} - n_{\rm E}). \tag{13.44}$$

If the waves travel in the opposite direction, the values $\rho_{\rm O} = -\,i$, $\rho_{\rm E} = i$ for the same axes are unchanged; see end of § 4.4. Thus the Faraday rotation is in the same absolute direction. If the wave system is reflected and retraces its path, the plane of polarisation does not go back to its initial plane, but continues rotating in the same sense as before the reflection.

In this example the refractive indices $n_{\rm O}$, $n_{\rm E}$ are given by (4.84). One important application is for high frequencies so that X and Y are small. If X^2, Y^2 and higher powers can be neglected, this gives

$$\tfrac{1}{2}(n_{\rm O} - n_{\rm E}) \approx \tfrac{1}{2}XY, \quad \tfrac{1}{2}(n_{\rm O} + n_{\rm E}) \approx 1 - \tfrac{1}{2}X. \tag{13.45}$$

The second expression is the average of the two refractive indices and is in the exponential of (13.42). It may be called the 'effective' value of the refractive index, and for this approximation it is the same as for an isotropic plasma. If the first expression is inserted in (13.44) and if (3.5), (3.25) are used for X, Y respectively, then

$$\partial\Psi/\partial z = NBe^3/(8\pi^2\varepsilon_0 cm^2 f^2) \tag{13.46}$$

where B is the strength of the magnetic field. This has been derived for a homogeneous plasma. If N and B are slowly varying functions of z, the equation may be integrated with respect to z to give the total rotation for a given path; see (13.51) below.

Consider again a homogeneous plasma and suppose, as before, that the source is at the origin and the receiver is on the z axis. Let the vector $\boldsymbol{Y}$ be in the x–z plane at an angle β to the z axis. The emitted linearly polarised signal again splits into the two characteristic component waves. Both ray directions are along the z axis but the two wave normal directions are different. They are in the x–z plane and make angles Θ_{O}, Θ_{E} with Y, and angles

$$\alpha_{\mathrm{O}} = \Theta_{\mathrm{O}} - \beta, \quad \alpha_{\mathrm{E}} = \Theta_{\mathrm{E}} - \beta \tag{13.47}$$

with the z axis; compare (5.32) and fig. 5.1. The two electric fields now have components E_z, but we here study only the E_x and E_y components. The two polarisations are no longer exactly circular but if β is small enough they are approximately circular and the two waves have approximately equal amplitudes A_{O}; compare (13.40), (13.41). If they are recombined after a distance z they give a resultant field similar to (13.42) but in general the polarisation is no longer exactly linear. It is an elongated ellipse whose major axis makes an angle Ψ with the x axis. In particular if the phase difference between the waves is $2\pi m$ where m is an integer, their combination has linear polarisation similar to the emitted wave at $z = 0$, but rotated through an angle $\Psi = \pi m$. The rotation angle is half the phase difference, as before.

The two refractive indices n_{O}, n_{E} depend on the wave normal angles Θ_{O}, Θ_{E} respectively. On the z axis the components E_x for the two electric fields are

$$E_x^{(\mathrm{O})} = A_{\mathrm{O}} \exp[\mathrm{i}\{\omega t - kzn_{\mathrm{O}}(\Theta_{\mathrm{O}})\cos\alpha_{\mathrm{O}}\}], \quad E_x^{(\mathrm{E})} = A_{\mathrm{O}} \exp[\mathrm{i}\{\omega t - kzn_{\mathrm{E}}(\Theta_{\mathrm{E}})\cos\alpha_{\mathrm{E}}\}]. \tag{13.48}$$

The two products $n\cos\alpha$ are the ray refractive indices, $\mathscr{M}$; see § 5.3. The spatial rate of Faraday rotation is $\partial/\partial z$ of half the phase difference, that is, compare (13.44)

$$\partial\Psi/\partial z = \tfrac{1}{2}k\{n_{\mathrm{O}}(\Theta_{\mathrm{O}})\cos\alpha_{\mathrm{O}} - n_{\mathrm{E}}(\Theta_{\mathrm{E}})\cos\alpha_{\mathrm{E}}\}. \tag{13.49}$$

Consider again the application to high frequencies so that X and Y are small. Now (5.83) shows that α is of order XY^2. Note that in (5.83) both $n^2 - 1$ and $An^2 - B$ have a factor X. If X^2, Y^2 and higher powers are negligible, both α_{O} and α_{E} are negligibly small so that $\Theta_{\mathrm{O}} = \Theta_{\mathrm{E}} = \beta$. The two refractive indices are now given by (4.113). Hence (13.49) gives

$$\partial\Psi/\partial z \approx \tfrac{1}{2}k\{n_{\mathrm{O}}(\beta) - n_{\mathrm{E}}(\beta)\} \approx \tfrac{1}{2}kXY\cos\beta. \tag{13.50}$$

The only change needed in (13.46) is the insertion of a factor $\cos\beta$. In a slowly varying medium N, B and β may all depend on z. Then integration with respect to z gives for the Faraday rotation

$$\Psi = \frac{e^3}{8\pi^2\varepsilon_0 cm^2 f^2}\int_0^z NB\cos\beta\,\mathrm{d}z. \tag{13.51}$$

This formula is often used in radio astronomy; see, for example Kraus (1966, eqn. 5–76). It is applied to the interstellar and intergalactic plasmas where X and Y are extremely small. The refraction of the rays is then negligible and it can safely be assumed that the two component waves have the same ray path.

Measurements have been made of the Faraday rotation of radio signals received at the ground from a satellite in the upper part of the ionosphere; see for example Titheridge (1966, 1968, 1971a); Garriott, Da Rosa and Ross (1970). The frequencies used were typically 20 MHz to 137 MHz, so that X was small. Hence the refraction of the rays could be neglected, and (13.51) could be used. Now z is used for height and s for distance along the path, so that dz is replaced by ds = dz sec θ where θ is the angle between the path and the vertical. Throughout the ionosphere B and β can be taken as constant to a first approximation, so that (13.51) gives:

$$\Psi = \frac{e^3 B \cos\beta \sec\theta}{8\pi^2 \varepsilon_0 c m^2 f^2} \int_0^z N \, \mathrm{d}z. \tag{13.52}$$

The integral in (13.52) is called the 'total electron content' of the ionosphere up to height z, and Faraday rotation is one of the most important ways of measuring it.

When the approximation, that X and Y are very small, cannot be made, the problem is more complicated. When a linearly polarised wave enters the ionosphere, the two component rays undergo lateral deviation in opposite directions; see §§ 10.12–10.14 and figs. 10.11–10.13. The two ray directions at the transmitter are different and must be chosen so that the two rays both reach the receiver. Thus the two ray paths are different. To find the Faraday rotation for this problem it may be necessary to trace the two rays separately and compute their phase paths P_O, P_E. The Faraday rotation angle is then $\frac{1}{2}k(P_\mathrm{O} - P_\mathrm{E})$. The only case where the two ray paths are the same is for a homogeneous medium.

Table 13.1

X	Y	β/deg	$n_\mathrm{O}(\Theta_\mathrm{O})\cos\alpha_\mathrm{O} - n_\mathrm{E}(\Theta_\mathrm{E})\cos\alpha_\mathrm{E}$	$n_\mathrm{O}(\beta) - n_\mathrm{E}(\beta)$	$XY\cos\beta$
0.1	0.1	10	0.0105	0.0105	0.0098
0.7	0.1	10	0.1296	0.1298	0.0689
0.1	0.1	50	0.0068	0.0068	0.0064
0.7	0.1	50	0.0863	0.0868	0.0450
0.1	0.5	10	0.0706	0.0706	0.0492
0.4	0.5	10	0.4044	0.4054	0.1970
0.1	0.5	50	0.0479	0.0484	0.0321
0.4	0.5	50	0.3102	0.3193	0.1286
0.1	1.1	10	0.431	0.426	0.108
0.7	1.1	10	1.977	1.600	0.758
0.1	1.1	50	0.286	0.257	0.071
0.7	1.1	50	1.210	0.560	0.495

The use of the approximation $n_O(\beta) - n_E(\beta)$ in (13.50) implies that the two wave normal directions are the same as the ray direction β. It is important to know by how much this approximation differs from the correct expression $n_O(\Theta_O)\cos\alpha_O - n_E(\Theta_E)\cos\alpha_E$ from (13.49). Table 13.1 gives some typical computed values for these two quantities and for the approximation $XY\cos\beta$ from (13.50).

13.8. Whistlers

In §4.12, fig. 4.6 the values of n^2 for a collisionless electron plasma were studied and it was shown that for $f < f_H$ the value of n^2 for the ordinary wave is positive for large X and for sufficiently small values of the angle Θ between the wave normal and the earth's magnetic field. The refractive index and ray surfaces are as shown in figs. 5.5, 5.6. This behaviour of the ordinary wave extends to frequencies down in the audible range. This type of wave is known as the 'whistler mode'. To study it we first neglect the effects of heavy positive ions and of collisions. These two effects are considered in the following section. An effect associated with protons is mentioned briefly at the end of this section.

If an audio-frequency amplifier, with headphones or a loud-speaker, is connected to a suitable aerial, such as a long wire as high as possible, various atmospheric noises can be heard and sometimes these include a whistle of falling pitch lasting for several seconds, known as a 'whistler'. The existence of whistlers has been known for many years (Burton and Boardman, 1933; Eckersley, 1935) but the main stimulus to the study of the subject came from the work of Storey (1953). For a full account of all aspects including the history, see Helliwell (1965), and for reviews see Al'pert (1980a, 1983), Carpenter (1983), Walker (1976).

It is now believed that a whistler originates in a lightning flash near the earth's surface. This is an impulse signal that contains a wide range of frequencies including those in the audible range. Some energy in this range can penetrate to the upper ionosphere and magnetosphere. The ray surfaces, fig. 5.6, show that for the low frequencies, that is large X, the ray directions are confined within a cone surrounding the direction of the earth's magnetic field, and Storey (1953) suggested that this causes the rays to be guided roughly in the direction of the magnetic field lines. Thus the signal travels over the equator and returns to earth at a point somewhere near the other end of the magnetic line of force along which it started out. The ray directions in fig. 5.6 may make angles up to 20° or more with the earth's magnetic field, and it is now known that this is not adequate to give the observed guiding. For a study of whistler ray paths see Yabroff (1961), Hashimoto, Kimura and Kumagai (1977). There are ducts in the magnetosphere like tubes containing an enhanced electron concentration, aligned with the magnetic lines of force. These act as wave guides and are now known to cause the guiding.

The two refractive indices are given by (4.46). If collisions are neglected ρ is purely imaginary and $U \approx 1$. For the extremely small frequencies in whistlers X and Y are large, so that the U and the 1 can be neglected. For upgoing waves in the northern hemisphere $\cos\Theta$ is positive. Then (4.46) shows that if n is positive, ρ must be positive imaginary. Thus upgoing whistler waves are polarised with a left-handed sense of rotation in the northern hemisphere. The field vectors $\boldsymbol{E}$, $\boldsymbol{\mathscr{H}}$ rotate in the same direction as free electrons would rotate about the earth's magnetic field. The convention used here to specify the sense of rotation is as explained in §4.3.

Because of the dependence of the group velocity $\mathscr{U}$ on frequency, §5.9, the high frequencies in the audible range travel faster, and the impulse is therefore drawn out into a signal of descending pitch, and is heard as a descending whistle. A whistler that makes one journey over the equator is called a 'short whistler', fig. 13.7(a). When it comes down into the ionosphere it can be reflected and return along roughly the same path to a region near its origin. It is then dispersed twice as much as a short whistler so that its duration is roughly doubled, fig. 13.7(b). It is called a 'long whistler'. The lightning flash that generates the whistler gives an impulsive signal that is often heard as a 'click' preceding a long whistler, and this signal goes directly to the receiver and so is not dispersed. Whistlers can be reflected repeatedly and make many transits over the equator. They are attenuated by collisions, and this would cause the amplitude to fall below the detection level after only a few transits. But

Fig. 13.7. Sketches showing spectrograms of typical whistlers. (a) is a short whistler with $T\sqrt{f} = 50\,\mathrm{s}^{\frac{1}{2}}$. (b) is the same whistler that has made two passages over the equator and returned to near its source. The initiating lightning flash shows as a vertical line at time $t = 0$. $T\sqrt{f} = 100\,\mathrm{s}^{\frac{1}{2}}$. (c) is a nose whistler. (d) is a fractional hop whistler as observed in a satellite, and it has an ion cyclotron (proton whistler) component.

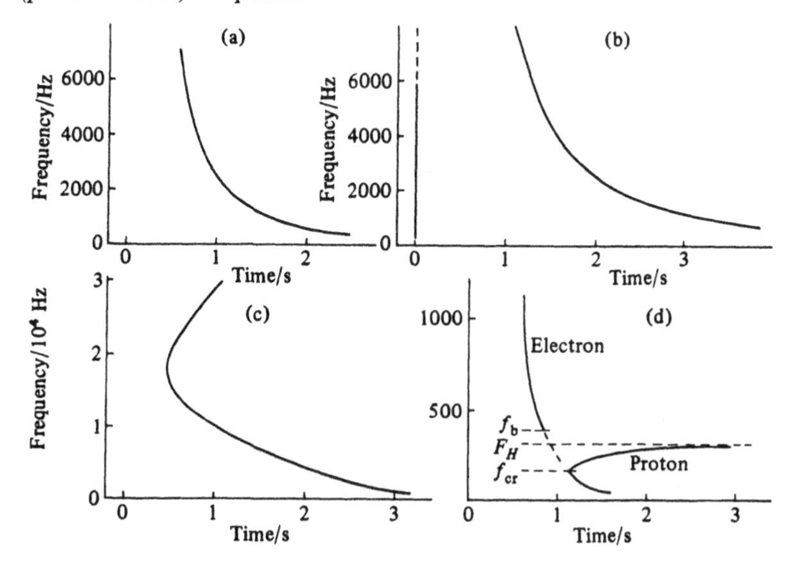

whistler waves can be amplified by interaction with streaming charged particles in the magnetosphere and this compensates for the attenuation. Hence long trains of whistlers are sometimes recorded, all originating from the same lightning flash.

Whistler signals are analysed by feeding them to some form of spectral analyser. This produces a record with frequency as ordinate and time as abscissa and the record is darkened where any frequency is present at a given time. Some typical records are sketched in fig. 13.7. On these records the direct signal from a lightning flash contains all frequencies in the band accepted by the receiver, and appears as a vertical line, fig. 13.7(b).

For the extremely low frequencies of the audible range X may be of order 200 or more, so that the approximation (4.110) can be used. Since Y also is large, the 1 in the denominator can be neglected. Hence the refractive index n for the ordinary wave is given by

$$n \approx \left(\frac{X}{Y}\sec\Theta\right)^{\frac{1}{2}} = f_N \sec^{\frac{1}{2}}\Theta (ff_H)^{-\frac{1}{2}} \qquad (13.53)$$

where Θ is the angle between the wave normal and the earth's magnetic field. Thus for these extremely low frequencies one of the two refractive indices is real and positive and the waves are propagated. It can be shown that if the Lorentz term, § 3.3, is included in the dispersion relation, neither refractive index would be real and positive and there would be no whistlers. The existence of whistlers is perhaps the strongest evidence that the Lorentz term must be omitted (Eckersley, 1935). The angle α between the ray and wave normal is given, from (5.33), by

$$\tan\alpha = \tfrac{1}{2}\tan\Theta. \qquad (13.54)$$

whence the angle $\beta = \Theta - \alpha$ between the ray and the earth's magnetic field is given by

$$\tan\beta = \tan\Theta/(2 + \tan^2\Theta). \qquad (13.55)$$

The maximum value of $\tan\beta = 1/\sqrt{8}$. It occurs where $\Theta = 54°44'$ and gives $\beta(\max) = 19°29'$. This is the semi-vertical angle of the cone within which the ray directions are confined.

The group refractive index n' is given, from (5.65), by

$$n' = \tfrac{1}{2}n \qquad (13.56)$$

so that the group velocity is

$$\mathcal{U} = 2c/(n\cos\alpha). \qquad (13.57)$$

For rays directed along the earth's magnetic field, $\Theta = \alpha = 0$, this gives

$$\mathcal{U} \approx 2c(ff_H)^{\frac{1}{2}}/f_N. \qquad (13.58)$$

For whistlers guided in a duct Θ and α are usually small so that (13.58) is a good approximation. The time of travel T of a whistler from its source to the receiver is the integral of $1/\mathcal{U}$ along the path. Thus for the various frequencies in the signal $T\sqrt{f} = D$ is a constant, called the 'dispersion'. This gives the shape of the spectrogram for low enough frequencies. It is known as the Eckersley law (Eckersley, 1935), and is confirmed by measurements. It would be expected to fail for frequencies that are too great to allow the use of the approximation (13.53), or so small that the effect of positive ions cannot be neglected. This behaviour too is confirmed by the measurements.

When the frequency is increased, the group velocity does not continue to increase in accordance with (13.58) but attains a maximum value where f is slightly less than $\frac{1}{4}f_H$ (see problem 5.11). For this frequency the time of travel is a minimum, and for greater frequencies the time of arrival increases with frequency. Thus the spectrogram has the shape of a nose as in fig. 13.7(c), and the whistler is called a 'nose whistler.' In fact all whistlers would be nose whistlers if the frequency were extended to high enough values. The frequency where T is a minimum is called the nose frequency. It is approximately the average of $\frac{1}{4}f_H$ over the path. A long part of the whistler path is near the top of its trajectory, and it is the electron gyro-frequency here that mainly determines the nose frequency. Typical nose frequencies are of order 20 to 30 kHz. The earth's magnetic field, and therefore f_H, is proportional to the inverse cube of the distance from the earth's centre. Whistlers observed at high latitudes can travel to great heights where f_H is small, and then the nose frequency can sometimes be as small as 3 to 4 kHz.

A lightning flash at the ground emits a spherical pulse of waves into free space. Each component frequency in the pulse can be resolved into an angular spectrum of plane waves as in (10.1), and this includes plane waves with real angles of incidence θ from 0° to 90°. In the ionosphere and above, the refractive index for whistler waves is very large, of order 50 or more. All the plane waves with real θ are therefore refracted so that their wave normals are very nearly vertical. Most theories of whistler paths assume that, where a whistler first enters the topside ionosphere from below, the wave normal is exactly vertical. Above this the medium is no longer horizontally stratified. The whistler enters a duct and is guided as already described.

Whistlers of another kind have been observed with receivers in rockets or satellites in the topside ionosphere. They are known as 'subprotonic whistlers' and were described by Carpenter, Dunckel and Walkup (1964). They were first reported by Barrington and Belrose (1963). Their dispersion is less than for a normal whistler and in a spectrogram they appear as several successive traces at equal intervals, suggesting that the signal travels up and down with repeated reflections at levels of about 100 km and 1000 km. They are confined to lower frequencies less than about 4 kHz. They occur mainly at night and at latitudes greater than 45°.

It is believed that in the topside ionosphere up to about 800 km the positive ions in the plasma are mainly atomic oxygen and helium ions. Above this the relative abundance of protons becomes greater and at 1000 km and above it is the protons that predominate. It has been suggested (Smith, R.L., 1964) that the effect of the protons on the refractive index is great enough to cause the upgoing whistler signals to be refracted into ray paths that become horizontal and are then reflected downwards. The signal thus returns to the ionosphere where the vertical gradients are large and can give an upward reflection. The region above 800 km where the proton concentration becomes appreciable is called the 'protonosphere', and these whistlers are therefore called subprotonic whistlers. Their paths have been studied by Kimura (1966) who used ray tracing methods.

In a horizontally stratified topside ionosphere where the rays become horizontal and reflection is occurring, the wave normals cannot be vertical. These waves cannot, therefore, originate from waves with real θ below the ionosphere if there is horizontal stratification at all levels. The subject has been studied by Walker (1968a, b) who showed that large values of $S = \sin\theta$ of about 10 to 25 are needed to explain the propagation and reflection of subprotonic whistlers. He gave examples of ray paths for various frequencies and values of S, with the wave normal in the magnetic meridian plane, and he was able to explain most of the observed features, assuming that the topside ionosphere is horizontally stratified. But the pulse that initiates these whistlers can only enter this topside ionosphere if there is some region of the lower ionosphere, near the lightning source, where there are strong horizontal variations. One suggestion is that subprotonic whistlers are associated with a strong but localised sporadic E-layer (see §19.8).

13.9. Ion cyclotron whistlers

In the discussion of whistlers in the first part of § 13.8 it was assumed that electrons are the only effective charged particles in the plasma. For frequencies less than about 500 Hz, however, the effect of positive ions cannot be ignored. Hines (1957) was one of the first to consider the effect of heavy positive ions on whistler propagation. This section deals with some effects where the positive ions play an essential part.

When a whistler is received by a satellite above the ionosphere it travels up from the lightning flash at the ground to the receiver. Its dispersion $D = T\sqrt{f}$ is smaller than for a whistler that has travelled over the equator because the distance travelled is smaller. It is called a 'fractional hop' whistler. The spectrogram often shows the effect sketched in fig. 13.7(d). There is a part of the trace for which the frequency increases slowly with time, and approaches a limiting value equal to the proton gyro-frequency F_H at the level of the receiver. Observations have shown that in the northern hemisphere this part of the trace is polarised with a right-handed sense of rotation, and it is called a 'proton whistler'. At lower frequencies still, a similar trace is sometimes observed, associated with helium ions and called a 'helium whistler'. These whistlers are collectively known as ion cyclotron whistlers. Proton whistlers were first reported by Smith, Brice *et al.* (1964), and helium whistlers by Barrington *et al.* (1966).

At higher frequencies the dispersion of the trace is caused mainly by electrons. This part of the upgoing signal is polarised with a left-handed sense in the northern hemisphere, and is an ordinary wave of the type discussed in the previous section. It is called an electron whistler. Part of its trace is absent or weak in the frequency range where the proton whistler appears. This suggests that the incident ordinary wave has changed to an extraordinary wave, with a reversal of the sense of the polarisation.

When the ray and wave normal are parallel or antiparallel to the earth's magnetic field, $\Theta = 0$, the two refractive indices are given by

$$n_{\mathrm{O}}^2 = \varepsilon_2, \quad n_{\mathrm{E}}^2 = \varepsilon_1. \tag{13.59}$$

The frequency dependence of these quantities is shown in figs. 3.1, 3.2. If only electrons are effective then at low enough frequencies ε_2 is positive and ε_1 is negative. The extraordinary wave is evanescent. But if ions are allowed for, there is a range of frequencies less than each ion gyro-frequency, where ε_1 is positive; see fig. 3.2. The frequency dependence of n_{O}^2 and n_{E}^2 is sketched in fig. 13.8. Collisions are neglected so that both are real. In fig. 13.8(a) there is only one ion species present so that n_{O}^2 is never equal to n_{E}^2. In fig. 13.8(b) there are two ion species present and there is now a frequency f_{cr} between the two ion gyro-frequencies where $\varepsilon_1 = \varepsilon_2$, that is $n_{\mathrm{O}}^2 = n_{\mathrm{E}}^2$. This is the 'crossover' frequency discussed in § 3.11 item (15). In both diagrams, for

Fig. 13.8. Sketches, not to scale, to show how n_{O}^2 and n_{E}^2 depend on frequency when collisions are neglected and the wave normal is parallel to the earth's magnetic field. In (a) there is only one species of positive ion, protons, with gyro-frequency F_{H}. In (b) there are two species, protons, and helium ions with gyro-frequency F_{He}, and f_{cr} is the crossover frequency. The dotted curves show how the values change when the wave normal makes a small angle with the earth's magnetic field. RH, right-handed, and LH, left-handed, show the sense of the polarisation for upgoing waves in the northern hemisphere.

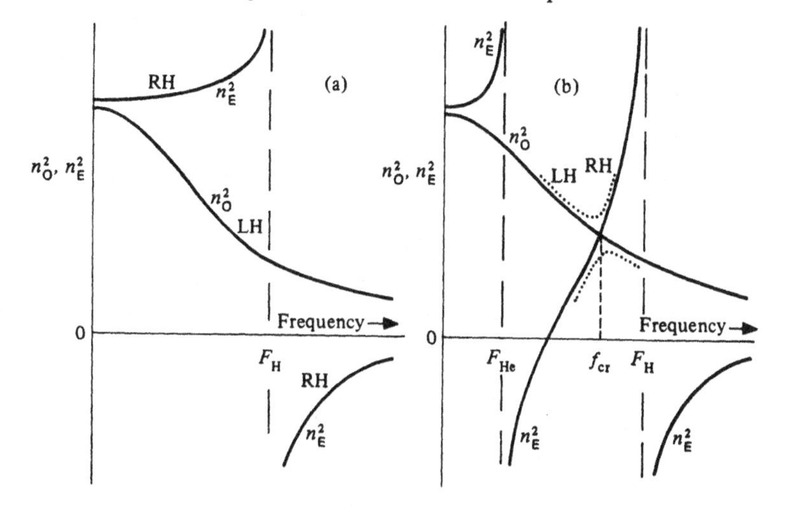

the curves marked n_O^2 the polarisation for upgoing waves in the northern hemisphere is circular with a left-handed sense, and for the curves marked n_E^2 it is circular with a right-handed sense.

If Θ is now made slightly different from zero the square root in (4.51) can never be zero when collisions are neglected. Thus n_O^2 and n_E^2 are never equal at any frequency. The curves cannot intersect, but separate as shown by dotted curves in fig. 13.8(b). The polarisations are no longer exactly circular, but the values of ρ are purely imaginary. When the upper dotted curve of fig. 13.8(b) is followed from left to right, ρ is near $-\mathrm{i}$ for $f < f_{cr}$, giving a left-handed sense as for the electron whistler. When $f = f_{cr}$, ρ has decreased to zero, and when $f > f_{cr}$ it is near to $+\mathrm{i}$ giving a right-handed sense as for the proton whistler.

The crossover frequency f_{cr} depends on the electron and ion concentrations in the plasma, and therefore in the upper ionosphere f_{cr} depends on the height. For a wave of fixed frequency f travelling vertically upwards there may be some height, called the 'crossover level'. where $f = f_{cr}$. Then, according to ray theory, the polarisation changes continuously from left hand to right hand, or vice versa, as the wave travels upwards. Curves showing how ρ/i depends on height for various values of Θ were given by D. Jones (1970). If collisions are negligible this reversal of the sense of polarisation must happen for all waves that encounter a level where $f = f_{cr}$. This change occurs continuously as the height increases. It does not entail any kind of mode conversion. If the wave starts from the ground as an electron whistler, its name changes to 'proton whistler' after it has passed the crossover level. This change of name is similar to the change from 'ordinary' to 'extraordinary' discussed in §4.16. The change occurs at different levels for the various frequencies in the whistler wave.

Crossover cannot occur unless at least two positive ion species are present. When proton whistlers are observed there must be both protons and some heavier positive ions such as singly charged helium ions in the upper ionosphere. Similarly when helium whistlers are observed there must also be some heavier ions such as singly charged atomic oxygen ions.

The following explanation is given in terms of ray theory alone, and this is adequate for the main features of ion cyclotron whistlers. In practice some mode conversion does occur, and it is mentioned again at the end of this section.

The theory of the formation of ion cyclotron whistlers was given by Gurnett, Shawhan *et al.* (1965). They showed that in the full theory the effect of collision damping plays an important part. The following treatment is based on the work of D. Jones (1970, 1969). The second of his papers, though published earlier, is a sequel to the first.

For the extremely low frequencies in whistlers the refractive indices in the upper ionosphere are large, of order 10 to 100; see fig. 13.10 for some examples. Thus the wave normal of a wave entering from free space is refracted so that it is nearly vertical even when the incidence is oblique. We may therefore assume that the wave normal is vertical, and for the range of height up to about 3000 km studied here we assume that the ionosphere is horizontally stratified. Now Θ is the angle between the vector $\boldsymbol{Y}$, antiparallel to the earth's magnetic field, and the vertical.

When collisions are allowed for, ε_1 and ε_2 are complex and are never equal for any real frequency. Thus n_{O} and n_{E} are never equal when $\Theta = 0$. But they are equal for some real transition value Θ_t of Θ; compare §§ 4.4, 4.16. When this occurs the two wave polarisations ρ_{O}, ρ_{E} are also equal. The condition is, from (4.24) with (3.51):

$$\mathfrak{G} = \frac{\varepsilon_3(\varepsilon_1 - \varepsilon_2)}{\varepsilon_1\varepsilon_2 - \frac{1}{2}\varepsilon_3(\varepsilon_1 + \varepsilon_2)} = \pm \mathrm{i}\frac{\sin^2 \Theta_t}{\cos \Theta_t}. \tag{13.60}$$

It requires that

$$\mathrm{Re}(\mathfrak{G}) = 0, \quad \Theta = \Theta_t \tag{13.61}$$

where

$$\sin^2 \Theta_t / \cos \Theta_t = \pm \,\mathrm{Im}(\mathfrak{G}). \tag{13.62}$$

This condition is here called 'exact crossover'. The frequency f_b for which it occurs is here called the 'boundary frequency'. It is a function of the angle Θ. In simple cases there is only one value f_b in the frequency range associated with one ion whistler, but there can be more than one.

In practical cases it is found that, at exact crossover, $\mathrm{Im}(\mathfrak{G})$ is positive. For upgoing waves in the northern hemisphere $\cos \Theta$ is positive, whence (4.24) shows that

$$\rho_{\mathrm{O}} = \rho_{\mathrm{E}} = -1. \tag{13.63}$$

The condition $\mathrm{Re}(\mathfrak{G}) = 0$ alone is called simply 'crossover'. At any given height the frequency for which it occurs is called the 'crossover frequency' for that height, and is denoted by f_{cr}. Then if $\Theta < \Theta_t$, (4.24) shows that the two values of ρ are real, so that the wave polarisations are linear.

To continue this explanation we now need a model of the upper ionosphere for illustration. The model used here is based on suggestions by Gurnett, Shawhan *et al.* (1965) with some modifications used by D. Jones (1970). It applies for heights above 500 km and here the electron concentration decreases monotonically with increasing height. Three positive ion species are present, namely protons, singly charged helium ions and singly charged atomic oxygen ions, and their concentrations depend on height. The collision frequencies also depend on height. The full details are complicated and not needed here, but are given by D. Jones (1970).

The earth's magnetic field and thence the ion gyro-frequencies are proportional to the inverse cube of the distance from the earth's centre. In fig. 13.9 the upper and lower curves show how the proton gyro-frequency F_{H} and the crossover frequency f_{cr}, respectively depend on height. The numbers by the lower curve are the values in degrees of Θ_t as found from (13.62).

Suppose, now, that the satellite is in the northern hemisphere at a height of 2000 km as shown by the vertical line, and let the angle Θ between the vector $\boldsymbol{Y}$ and the vertical be 12°. The waves coming up from the earth are electron whistler waves, polarised with a left-handed sense. The other type of wave polarised with a right-handed sense is heavily attenuated at low heights and does not penetrate the ionosphere. The crossover frequency at the satellite is marked f_{crs} in fig. 13.9. Consider a component frequency f_1 less than f_{crs}. In travelling up to the satellite the wave does not encounter a crossover level and so it remains an electron whistler with left-handed polarisation. This accounts for the part of the trace at the lowest frequencies in fig. 13.7(d). Next consider a frequency $f_2 > f_{\mathrm{crs}}$. The wave now encounters a crossover level below the satellite, at about 1500 km in fig. 13.9. Here Θ_{t} is about 6.9° and Θ is greater than this. Near the crossover level, therefore, the left-handed elliptical polarisation has changed to linear, and at greater heights it becomes elliptical with a right-handed sense; see fig. 13.11 for more details of this. When it arrives at the satellite it is a proton whistler, and appears on the part of the trace marked 'proton' in fig. 13.7(d).

Now consider a frequency f_3 greater than the proton gyro-frequency F_{Hs} at the satellite. This wave encounters a crossover level below the satellite, at about 1200 km in fig. 13.9. Above this it becomes a proton whistler. But before it reaches

Fig. 13.9. Shows how the proton gyro-frequency, upper curve, and the crossover frequency, lower curve depend on height in the upper ionosphere for a typical model. The numbers by the lower curve are the values in degrees of the transition angle Θ_{t} (13.62).

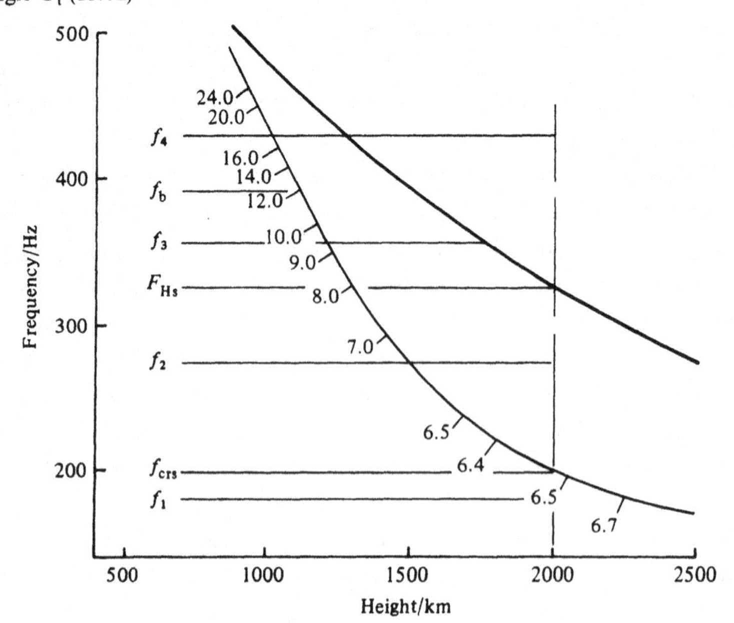

the satellite it attains a level, about 1750 km, where its frequency is equal to the proton gyro-frequency. At greater heights than this Re(n^2) is negative and the wave is almost evanescent. Hence it is not observed at the satellite. Fig. 13.7(d) shows that, for a frequency range slightly greater than F_{Hs}, the trace is absent.

Finally consider a frequency such as f_4, fig. 13.9. The wave now encounters a crossover level at about 1000 km height but here $\Theta < \Theta_t$. Thus the wave remains an electron whistler and the sense of its polarisation does not reverse. It travels on up and is observed at the satellite. This accounts for the part of the trace marked 'electron' in fig. 13.7(d). There is a 'boundary frequency' marked f_b in figs. 13.7(d), 13.9, that separates the two regions where Θ is greater or less than Θ_t. It is the frequency for which the 'exact crossover' condition (13.49), (13.50) is satisfied.

For the other values of Θ and for other compositions of the topside ionosphere the spectrograms of ion cyclotron whistlers can have different forms from that in fig. 13.7(d). For examples see the papers by D. Jones (1970, 1969), where curves similar to fig. 13.9 for helium whistlers are also given.

The dependence of Re(n) and $-$ Im(n) on height near the crossover level for two typical frequencies f is shown in fig. 13.10. For figs. 13.10 (a and b), $f = 220$ Hz and

Fig. 13.10. Height dependence of Re(n) and $-$ Im(n) near crossover. The labels LH, RH show whether the sense of the wave polarisation is left handed (electron whistler) or right handed (proton whistler) respectively. (a) and (b) are for a frequency $f = 220$ Hz so that $f_{crs} < f < f_b$. The upgoing electron whistler changes to a proton whistler. (c) and (d) are for $f = 440$ Hz so that $f > f_b$. At the crossover level $\Theta < \Theta_t$ and the upgoing electron whistler remains an electron whistler above the crossover level.

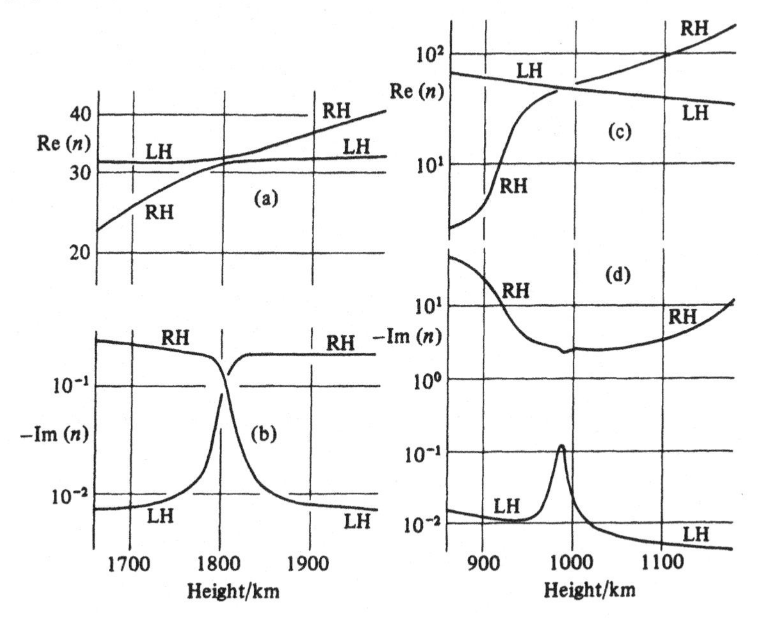

the crossover level is at 1800 km. Here $\Theta_{cr} < \Theta$. The two curves for $-\text{Im}(n)$ cross at this level, and the curves for $\text{Re}(n)$ do not cross. The electron whistler wave, polarised with a left-handed sense, that comes up from lower levels, goes over continuously as the height increases to a proton whistler wave polarised with a right-handed sense. For figs. 13.10(c and d), $f = 440$ Hz and the crossover level is at 987 km. Here $\Theta_{cr} > \Theta$. The two curves for $\text{Re}(n)$ cross, and the curves for $-\text{Im}(n)$ do not cross. The upcoming electron whistler wave remains an electron whistler after it has passed to above the crossover level. It is instructive to compare fig. 13.10 with the sequence figs. 4.7–4.11.

Fig. 13.11 is a diagram of the complex ρ plane and shows how the polarisation ρ of the upcoming electron whistler wave depends on height. For a frequency 440 Hz, curve (b), ρ remains near $+i$. For a frequency 220 Hz, curve (a), ρ changes

Fig. 13.11. The complex ρ plane. The curves show how the wave polarisation ρ depends on height for an upgoing wave that starts as an electron whistler. The numbers by the curves are the heights in km and the underlined number is the crossover height. Curve (a) is for 220 Hz and curve (b) for 440 Hz; see caption of fig. 13.10. Curve (c) is for frequency 390 Hz and curve (d) for 394 Hz. They are near to, and on opposite sides of the boundary frequency 391.8 Hz.

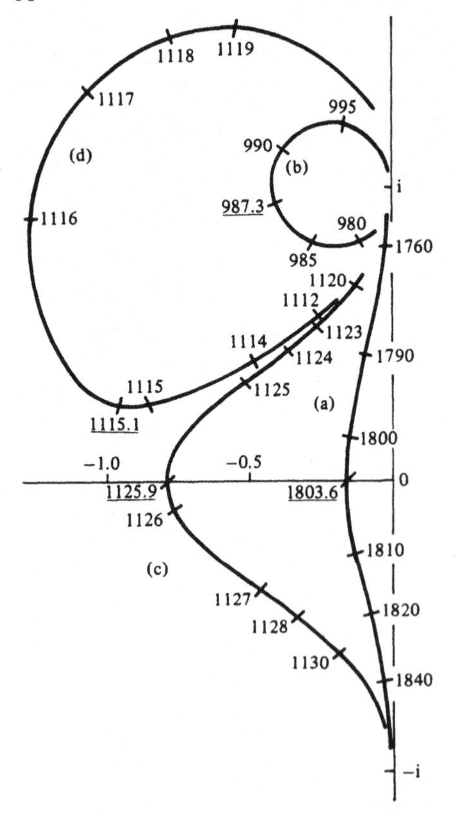

continuously from near $+\mathrm{i}$ through a negative real value at the crossover level to near $-\mathrm{i}$ at greater heights. Curves are also shown for frequencies 390 and 394 Hz, that is just less than and just greater than the boundary frequency $f_{\mathrm{b}} = 391.8$ Hz, for which the exact crossover condition (13.61), (13.62) is satisfied. It is seen that in both cases the value of ρ is near -1 at the crossover level. For 390 Hz $< f_{\mathrm{b}}$, curve (c), it goes over to near $-\mathrm{i}$ and the wave becomes a proton whistler. For 394 Hz $> f_{\mathrm{b}}$, curve (d), it returns to near $+\mathrm{i}$ and the wave remains an electron whistler.

The above description has been given entirely in terms of ray theory, which is the subject of this chapter. Exact crossover occurs for one particular frequency f_{b} when the crossover conditions (13.61), (13.62) are both true at some point on the real z axis. It has been assumed that the approximations of ray theory hold at all points on the real z axis. This, however, is not strictly true. When $f \neq f_{\mathrm{b}}$ the conditions (13.61) (13.62) hold at points $z_{\mathrm{p}}, z_{\mathrm{q}}$ in the complex z plane, known as 'coupling points'. When f is changed, z_{p} and z_{q} move, and when $f = f_{\mathrm{b}}$ one of them is on the real z axis, fig. 13.12(c). The approximations of ray theory break down near a coupling point; this is studied in § 16.3. The two refractive indices are nearly equal there, and we say that the two waves are strongly coupled. One wave as it travels gives rise to some of the other. Thus the transition from an electron whistler to a proton whistler does not occur abruptly for frequencies near to f_{b}.

Consider again the case where collisions are neglected, as in fig. 13.8. Then when z is real, $\varepsilon_1, \varepsilon_2, \varepsilon_3$ and thence $\mathfrak{G}$ are all real. If $\Theta \neq 0$, (13.60), (13.61) cannot be satisfied for any real z, but they are satisfied for two complex values $z_{\mathrm{p}} = z_{\mathrm{r}} + \mathrm{i}z_{\mathrm{i}}$, and $z_{\mathrm{q}} = z_{\mathrm{r}} - \mathrm{i}z_{\mathrm{i}}$, at equal and opposite distances from the real z axis, fig. 13.12(a). These are the two coupling points. At each of them the two upgoing waves are strongly coupled. The real z axis runs between the two coupling points. On it the two refractive indices are real and continuous. For one of them the wave at low heights is an electron whistler and at great heights it has changed its name to proton whistler. The other refractive index similarly is for a proton whistler at low heights and an electron whistler at great heights.

Fig. 13.12. The complex z plane showing the positions of the coupling points z_{p}, z_{q}, for ion cyclotron whistlers (not to scale). Collisions are neglected in (a) but not in (b), (c), (d). (b) is typical of low latitudes where the angle Θ between the earth's magnetic field and the vertical is large. (c) is for exact crossover. (d) is typical of high latitudes where Θ is small and ion whistlers can be produced only by mode conversion. The broken curves in (b) and (d) are possible paths for use with the phase integral method.

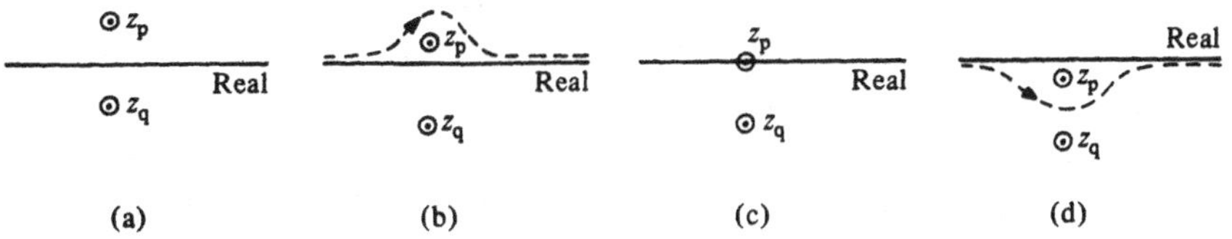

If, now, Θ is made smaller, z_i decreases. The coupling points move closer together, and when $\Theta = 0$ they coalesce on the real z axis. There is then a coalescence C2, § 17.4, and this example is used in § 17.5 for illustration.

In practice collisions cannot be neglected so $\mathfrak{G}$ is complex when z is real. Equations (13.60), (13.61) still give the position of the coupling points z_p, z_q but they are no longer at equal and opposite distances from the real z axis. For the model used earlier in this section it can be shown that they are both displaced in the direction of negative imaginary z. If Θ is small enough, however, they are still on opposite sides of the real z axis, fig. 13.12(b). The two refractive indices now have imaginary parts. Their height dependence is as shown in fig. 13.10(a, b).

For a given Θ there can be some frequency $f = f_b$ for which one of the coupling points is on the real z axis. This is the condition of exact crossover, fig. 13.12(c). It is also called 'critical coupling', § 16.6. For larger Θ both coupling points are on the negative imaginary side of the real z axis, fig. 13.12(d). The coupling points are branch points of the two refractive indices $n(z)$. Now the real z axis runs on the positive imaginary side of the branch point z_p so that on it an electron whistler at low heights remains an electron whistler at great heights, and similarly for a proton whistler. The height dependence of the two refractive indices is now as shown in fig. 13.10(c, d).

In this case, the mode conversion from an electron whistler to a proton whistler can be studied by using, instead of the real z axis, a path in the z plane that runs on the negative imaginary side of z_p, as sketched in fig. 13.12(d). It may be possible to find such a path that does not go too near to z_p so that the W.K.B. solutions are good approximations at all points on it. Then ray theory can still be used by following the ray to complex values of z. The field amplitude where z is large and real is the amplitude of the wave produced by mode coupling. This is an example of the phase integral method for coupling, § 16.7, and may be regarded as a special case of the use of a complex ray, §§ 14.9–14.12.

Similarly, when z_p is on the positive imaginary side of the real z axis, it may be possible to find a good path in the z plane that runs on the positive imaginary side of z_p, as sketched in fig. 13.12(b). This could then be used to find the amplitude at great heights of the electron whistler wave produced by mode coupling from the incident wave that is an electron whistler at low heights.

To study the dependence of the received whistler signal amplitudes on frequency f and angle Θ it is therefore necessary to extend the use of ray theory by using paths in the complex z plane; see §§ 14.9–14.12, 16.7. An alternative method is to use full wave solutions of the governing differential equations, ch. 18. Figs. 13.13, 13.14 show some results calculated in this way for a model of the upper ionosphere similar to that used in figs. 13.9–13.11. The equations (7.80), with (7.82), were integrated through a downward range of height from 1200 km to 950 km for vertically incident waves and using the methods of §§ 18.6, 18.7. For each f and Θ two integrations are needed because there are two different possible upgoing waves at the top. The two solutions are combined so that at the bottom there is an upgoing electron whistler wave of unit amplitude. Fig. 13.13 shows how the amplitudes of the two upgoing waves depend on height z for various values of Θ. The signal amplitudes here are $|\mathbf{f}_i|$ where the two $\mathbf{f}_i$s are the elements of $\mathbf{f}$, (6.53), for the two upgoing waves. It was found that the amplitudes of the downgoing waves were negligible at all heights so there is no appreciable reflection in this part of the ionosphere. The results of figs. 13.13, 13.14 all apply for a frequency of 422.5 Hz, which is less than the proton gyro-frequency F_H for the whole of the height range in the figures. Fig. 13.9 shows that the transition value Θ_t is near 16°, and the crossover height is just above 1000 km. The continuous curves of figs. 13.13, 13.14 apply for the wave that is propagated without mode conversion. The broken curves are the waves produced by mode conversion; these would be absent if the approximations of the simple ray theory of this chapter were used. The label el indicates an electron whistler whose polarisation has a left-handed sense, and pr indicates a proton whistler with a right-handed sense. It can be seen that for $\Theta = 10°$, that is less than Θ_t, the amplitude of the electron whistler is close to unity at all heights, but the growth of the proton whistler, by mode conversion near the coupling level, can be seen. For $\Theta = 20°$, that is greater than Θ_t, the continuous curve represents an electron whistler at low heights and a proton whistler at great heights. Its amplitude decreases with increasing height because of

Fig. 13.13. Dependence of signal amplitude on height for electron and proton whistlers for a frequency 422.5 Hz and various values of Θ as marked by the curves. The model used is slightly different from that of figs. 13.9–13.11. The transition value of Θ is $\Theta_t \approx 15°$ and the crossover height is 1037 km.

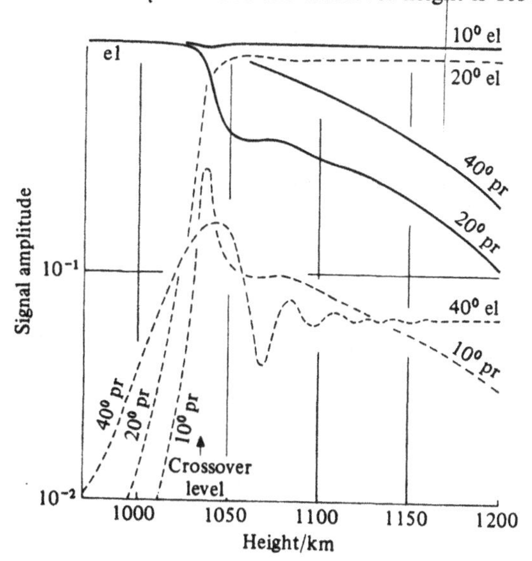

attenuation. The other wave is a proton whistler at low heights and an electron whistler at great heights. The broken curve shows its growth, and it also shows oscillations. These are even more marked for $\Theta = 40^\circ$. They occur because of interference between contributions to the second wave from different levels; their relative phases depend on the difference of the phase velocities of the two upgoing waves.

Fig. 13.14 shows how the amplitudes of the two waves depend on Θ at a height of 1150 km, that is above the crossover level. For small Θ the wave here is mainly an electron whistler and the proton whistler has very small amplitude. This is what is observed at high latitudes. For large Θ, $\approx 40^\circ$, the predominant wave is a proton whistler but the electron whistler is also present. For smaller Θ, although still $> \Theta_t$, the electron whistler formed by mode conversion may have the larger amplitude because the proton whistler is more heavily attenuated. Thus there are ranges of Θ and of frequency f where both electron and proton whistlers can be present with comparable amplitudes. Experimental confirmation of this was given by Rodriguez and Gurnett (1971).

Several workers have used full wave solutions to study ion cyclotron whistlers. Thus Wang (1971) investigated the intermode coupling for vertical incidence when collisions are neglected. This subject is of some theoretical interest; see § 17.5. Fijalkow, Altman and Cory (1973) allowed for collisions and for oblique incidence, and included a study of the propagation of the electron whistler from the ground up

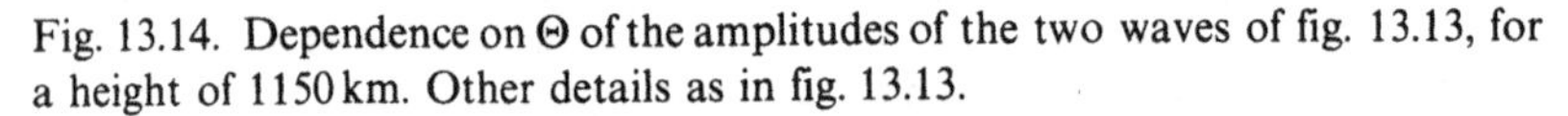

Fig. 13.14. Dependence on Θ of the amplitudes of the two waves of fig. 13.13, for a height of 1150 km. Other details as in fig. 13.13.

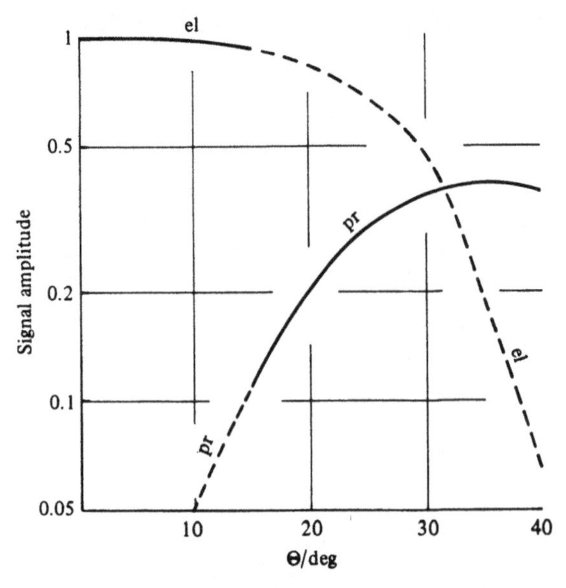

into the topside ionosphere. Their conclusions seem to be generally in agreement with those given here. Arantes and Scarabucci (1975) allowed for collisions and studied vertical incidence. They discussed the physical processes associated with changes of wave polarisation in the region where the coupling is strong. Their computing method was very similar to that used here for figs. 13.13, 13.14.

In this section it has been assumed that the wave normals are vertical. The receiver was assumed to be high in the topside ionosphere somewhere above the source but its exact position was not considered. In a fuller treatment allowance must be made for the lateral deviation of the signals, §§ 10.12–10.14. This means that for a ray to reach a given receiver it must leave the source with the wave normal not vertical. Because collisions are important in this theory, the refractive indices are complex and the rays are refracted through complex angles. Thus the wave normals at the source have complex directions that are different for the electron whistler and the ion whistler. A theory that makes allowance for this was given by Terry (1971, 1978).

13.10. Absorption, non-deviative and deviative

The effect of collisions in a plasma is to cause attenuation of radio waves. This is because some of the electromagnetic energy is absorbed by the medium and causes heating, mainly of the electrons. The present section is concerned with attenuation. Some consequences of heating are studied in §§ 13.11–13.13. The term 'absorption' is commonly used to mean both these aspects.

The effect of electron collisions on the phase height $h(f)$ and the equivalent height $h'(f)$ for wave packets vertically incident on an anisotropic ionosphere has been discussed in § 13.3. The present section also is concerned with vertically incident wave packets. The subject is important in the theory of ionospheric probing using the ionosonde technique. Oblique incidence is discussed in §§ 10.15, 10.16, 14.7, and in problem 10.6.

For an isotropic ionosphere, the effect of electron collisions on the attenuation of a wave packet was discussed in § 12.3 where the important formula (12.14) was derived. There is an analogous relation, (13.67) below, for the anisotropic case. For a fuller treatment of this subject see Dyson and Bennett (1979). The formulae (12.3), (12.9) can still be used. It must be remembered that a vertically incident wave packet is deviated laterally, § 10.12, but the group refractive index n' in (12.9) gives the vertical component c/n' of the group velocity. The predominant wave normal is everywhere vertical. The attenuation in nepers is $-\ln|R| = -2k\,\mathrm{Im}\,\{h(f)\}$, and is given by (12.10), also for the anisotropic case. It is zero if $Z = 0$. It is now instructive to study the effect of a small value of Z, as was done in § 12.3. Then (12.11) can still be used and $(\partial n/\partial Z)_{Z=0}$ must now be found. For $Z = 0$, the refractive index n is a function of X and Y. In the form (4.65) of the dispersion relation let the coefficients A, B, C be divided by U^3. Then each of them is a function of X/U, Y/U. This shows that n is a function of these two variables. It then follows that

$$\frac{\partial n}{\partial Z} = \mathrm{i}X\frac{\partial n}{\partial X} + \mathrm{i}Y\frac{\partial n}{\partial Y} \tag{13.64}$$

where all derivatives have their values for $Z = 0$. Then from (12.10), (12.11)

$$\ln|R| = \frac{1}{c}\int_0^{z_0} \nu\left(X\frac{\partial n}{\partial X} + Y\frac{\partial n}{\partial Y}\right)\mathrm{d}z. \tag{13.65}$$

For $Z = 0$, n depends on frequency f only through X and Y. Now (5.65) shows that $n' = \partial(nf)/\partial f$ whence

$$n' - n = f\frac{\partial n}{\partial f} = -2X\frac{\partial n}{\partial X} - Y\frac{\partial n}{\partial Y} \tag{13.66}$$

and if this is put in (13.65) it gives for the attenuation

$$-\ln|R| = \frac{1}{c}\int_0^{z_0} \tfrac{1}{2}\nu\frac{n' - n}{1 + g}\mathrm{d}z \tag{13.67}$$

where

$$g = \frac{-\tfrac{1}{2}Y\dfrac{\partial n}{\partial Y}}{X\dfrac{\partial n}{\partial X} + Y\dfrac{\partial n}{\partial Y}}. \tag{13.68}$$

Some special cases of this may be noted. For purely longitudinal propagation, $\Theta = 0$, the two values of n^2 are given by (4.84). For the first, (13.68) gives $g = \tfrac{1}{2}Y$. In the ionosphere this may be assumed to be independent of z. If ν also is independent of z, (13.67) gives

$$-\ln|R| = \frac{1}{2}\frac{\nu}{c}\frac{1}{1 + \tfrac{1}{2}Y}(h' - h). \tag{13.69}$$

The wave is ordinary where $X < 1$, extraordinary where $X > 1$ (see §4.16), and is reflected where $X = 1 + Y$. The other value of n^2 in (4.84) gives the same result with the sign of Y reversed. The wave is extraordinary and reflected where $X = 1 - Y$.

For purely transverse propagation, $\Theta = \tfrac{1}{2}\pi$, the value of n^2 for the ordinary wave is the same as for an isotropic medium and is independent of Y. Then $g = 0$ and if ν is independent of z, (13.67) is the same as (12.14). For the extraordinary wave n^2 is given by (4.81), second formula. Then it can be shown that

$$\frac{1}{1 + g} = \frac{Y^2 + (1 - X)^2}{XY^2 + (1 - X)^2}. \tag{13.70}$$

This depends on height z and therefore cannot be taken outside the integral in (13.67).

For the more general case Bennett and Dyson (1978), and Dyson and Bennett (1979) give a very full study of the function g, including curves showing how it depends on X for various Θ.

The electron collision frequency ν decreases with increasing height; see fig. 1.2. It often happens that a vertically incident radio pulse is reflected at a height where ν is small, but travels up and down through a large thickness of ionosphere lower down where ν is larger, but $N(z)$ and X are small. Then most of the attenuation occurs where X is small. For example the pulse may be reflected in the F-region but most of its attenuation occurs in the D- and E-regions. In this type of situation the attenuation occurs where n is close to unity. Here an oblique ray path would be almost straight. The lateral deviation of a vertical ray here is very small. The absorption is then said to be 'non-deviative'. The topic is of some interest for high frequencies that are not near the penetration frequency of any layer. Then Z is negligible near the reflection level and small in the absorbing region. In these conditions (13.65) can be used where the derivatives have their values for $Z = 0$ and small X. For this approximation (4.113) applies and (13.65) gives

$$-\ln|R| = \frac{1}{c}\int_0^{z_0} \nu X(\tfrac{1}{2} \mp Y\cos\Theta)\,dz \tag{13.71}$$

where the upper sign is for the ordinary wave. The condition $|\cos\Theta/\sin^2\Theta| \gg \frac{1}{2}Y$ for (4.112) must be imposed here, so that (13.71) cannot be used for near transverse propagation. In (4.113) it is assumed that Z and Y^2 are negligible. More accurate formulae can be derived if this assumption is not made. For examples see Rawer and Suchy (1967, §9).

The approximation that X is small, used in (13.71), cannot be made for a pulse that is reflected near the maximum of an ionospheric layer. Here dN/dz is small near the reflection level. There is a range of z with nearly constant N where the refractive index is near to zero and the group velocity is very small. The wave packet then spends a long time near the reflection level and most of its attenuation occurs there. This is where the ray is being deviated, and the absorption here is called 'deviative absorption'. For this case also there are approximate formulae for $\mathrm{Im}(n)$ near the reflecting level. For details see Rawer and Suchy (1967, §9) where further references are given.

For discussions of many aspects of absorption, both experimental and theoretical, see *Journal of Atmospheric and Terrestrial Physics* (1962, special volume on absorption).

13.11. Wave interaction 1. General description

The phenomenon of wave interaction, also called cross modulation or the Luxembourg effect, was first observed by Tellegen (1933). He used a receiver in Eindhoven and was receiving a broadcast signal from Beromünster (Switzerland, 652 kHz). It was observed that the radio program from Luxembourg (252 kHz) was heard at the same time. It was established that the modulation of the Luxembourg signal, called the 'disturbing wave', was impressed as amplitude modulation of the

Beromünster signal, called the 'wanted wave', and that the process was occurring within the ionosphere.

This subject has aroused great interest and there are numerous published papers dealing both with experiment and with the theory. It is of practical importance for two main reasons. First, it is a source of interference with broadcast transmissions, which is becoming more serious as more and more transmitters are being built that send disturbing energy into the D- and E-regions. This aspect of the subject has been surveyed by Maslin (1976e). Second, it is a useful means of investigating the D- and E-regions and particularly as a method of measuring the effective collision frequency. It was used in this way most successfully by Fejer (1955) with pulse modulated signals for both the wanted and the disturbing waves. The timing of the pulses could be adjusted so as to choose the height where the interaction occurred. Fejer's method has been used with various elaborations by several groups of later workers (Barrington and Thrane, 1962; Landmark and Lied, 1962; Weisbrod, Ferraro and Lee, 1964).

Most of the earlier theories of wave interaction ignored the earth's magnetic field so that the ionosphere is assumed to be isotropic. An excellent summary, with a detailed discussion of the energy balance of the electrons, is given by Huxley and Ratcliffe (1949). A theory that allows for the earth's magnetic field has been given by Maslin (1974, 1975a, b, 1976c, d, e).

The explanation, in outline, is as follows. The disturbing wave is attenuated by the ionospheric plasma so that some of its energy is converted to heat. It is the electrons that are heated first, so that their average random thermal speed is increased. When the disturbing wave is amplitude modulated the heating rate is greatest when the modulation amplitude is greatest, and varies in step with the modulation envelope of the wave. The heated electrons can lose energy with a time constant of about 1 ms, so that for modulation frequencies less than about 1 kHz the modulation causes the electron temperature to vary. There is a time lag as explained later; see (13.76). Now the collision frequency of the electrons depends on their average speed and is increased when the electron temperature is increased. When a wanted wave goes through the same plasma it is attenuated at a rate that depends on the collision frequency so that its attenuation varies in step with the modulated electron temperature. Thus the wanted wave, through this attenuation, acquires a modulation that originates from the modulation of the disturbing wave.

The changed collision frequency also means that the attenuation of the disturbing wave itself is changed. Usually it increases when the collision frequency increases so that the modulation amplitude is reduced and the disturbing wave is said to be 'self-demodulated'. Observations of this have been made (Aitchison, 1957; Hibberd, 1957; King, 1959). It can happen that if the average value of ν is large enough, an increase of

ν can cause a decrease of attenuation; see, for example, problem 4.6. Then the modulation of the disturbing wave is enhanced.

Wave interaction is, in one sense, a non-linear process, which would place it outside the subject range of this book. But, for radio waves in the ionosphere from communication or broadcasting transmitters, the power is small enough for attenuation to be treated as a linear process, and for the heating to be found by using only the linear form, ch. 3, of the constitutive relations. Attenuation, also called absorption, has been discussed in §§ 10.15, 10.16, 12.3, 12.11, 13.3, 13.10. A brief discussion of the heating process is given in the following section. A full account of wave interaction is included in a text book by Gurevich (1978) on non-linear phenomena.

13.12. Wave interaction 2. Outline of theory

The version of the theory in this section is greatly simplified and gives only the barest essentials. There are many refinements for which the original papers must be consulted.

In any heating process in the ionosphere it is the electrons that acquire energy first, from the heating wave. They already have random thermal speeds of the order of 2×10^5 m s^{-1} (energy of order 0.1 eV). Superimposed on this is an ordered motion caused by the electric field of the heating wave. The ordered speeds are typically about 3×10^3 m s^{-1}, that is very much less than the random speed. This figure applies at 100 km from a 1 MW transmitter of frequency 1 MHz. The ordered motion is an elliptical orbit and is periodic with the period of the heating wave. The resultant orbits, comprising both ordered and disordered motion, are complicated because of the earth's magnetic field, and because the wave is in general elliptically polarised. They have been discussed by Ratcliffe (1959, ch. 14).

The electrons make collisions which, in the ionosphere, are predominantly with neutral particles and may be assumed to be instantaneous. Because the particles that they collide with are massive, the change of the electron's speed in one collision is very small. It is the direction of motion that is changed. Immediately after a collision, the electron's new velocity is the resultant of a new thermal velocity and the same ordered velocity as immediately before the collision. It is easy to show that the average resultant speed is greater than the thermal speed alone. Thus, on the average, a collision occurs when the resultant speed is greater than the thermal speed. The result is that, on the average, the thermal energy after a collision is greater than it was before it. Through collisions, the ordered energy imparted by the heating wave is converted to disordered electron energy, that is to heat.

The collision frequency of an electron in the ionosphere is of order 3×10^3 s^{-1} at 300 km to 10^6 s^{-1} at 80 km. Thus, at the lower levels, an electron makes an average of one or several collisions in one cycle of the heating wave. Although the electron's ordered energy is small, it can be continuously converted to heat by collisions. Let H be the rate at which heat is imparted to the electrons in unit volume by the disturbing wave. Methods of calculating it are discussed below.

In the electron heating process just described it was assumed that an electron does not change its energy when it makes a collision. In fact it does transfer a small fraction of its energy to the heavy particle with which it collides. Thus electrons heated by a disturbing wave can cool down again by transfer of their excess energy to other particles of the plasma. For the electrons in a small volume let Q be the average energy per electron at any fixed time, and let Q_0 be the value that Q would have if the electrons were in thermal equilibrium with the other particles. For thermal equilibrium the electrons would have a Maxwellian velocity distribution with temperature $T_0 = \frac{2}{3}Q_0/K$ where K is Boltzmann's constant. For disturbed electrons the distribution is not in general Maxwellian but we can speak of an effective electron temperature $T = \frac{2}{3}Q/K$. The disturbed electrons tend to return to thermal equilibrium so that on average an electron loses energy $G(Q - Q_0)$ in each collision, where G is a constant. The time dependence of Q is then given by

$$\partial Q/\partial t = -Gv_{av}(Q - Q_0) + H/N \tag{13.72}$$

where v_{av} is the average collision frequency of an electron. The cooling process has a time constant $1/Gv_{av}$. The constant G can be estimated from a study of the mechanics of a collision; see for example, problem 3.7. It is found, however, that the value from these estimates may be too small. Experimental measurements for air show that G is of order 1.3×10^{-3} (Huxley and Ratcliffe, 1949), so that in the lower ionosphere $1/Gv_{av}$ is of order 1 ms. If the disturbing wave is amplitude modulated with a frequency less than about 1 kHz, the electron temperature has time to rise and fall as the heating rate H changes. Thus the modulation is impressed on the electron temperature. For greater modulation frequencies, however, the electrons cannot lose much energy within the short modulation cycles. The electron temperature then simply increases until the average energy loss rate $Gv_{av}(Q - Q_0)$ is equal to the average rate of heating by the disturbing wave.

It was shown in § 2.11 that when a harmonic wave system is present in a plasma with electric intensity $\boldsymbol{E}$ at any point, the heating rate at that point is, from (2.52)

$$H = -\tfrac{1}{2}\omega \,\mathrm{Im}\,(\boldsymbol{E}^* \cdot \boldsymbol{D}) = -\tfrac{1}{2}\omega\varepsilon_0 \,\mathrm{Im}\,(\boldsymbol{E}^* \cdot \boldsymbol{\varepsilon}\boldsymbol{E}). \tag{13.73}$$

Now let $\boldsymbol{E}$ for the disturbing wave have a sinusoidal modulation so that

$$\boldsymbol{E} = \boldsymbol{E}_0(1 + M\cos\Omega t). \tag{13.74}$$

Here Ω is the angular modulation frequency and is very much less than the angular wave frequency ω. Then

$$H = H_0(1 + \tfrac{1}{2}M^2 + 2M\cos\Omega t + \tfrac{1}{2}M^2\cos 2\Omega t), \quad H_0 = -\tfrac{1}{2}\omega_0\varepsilon_0 \,\mathrm{Im}\,(\boldsymbol{E}_0^* \cdot \boldsymbol{\varepsilon}\boldsymbol{E}_0). \tag{13.75}$$

This is now inserted in (13.72). The term $2M\cos\Omega t$ in H then contributes to the solution a term

$$Q = \frac{2MH_0}{N}\{(G\nu_{av})^2 + \Omega^2\}^{-\frac{1}{2}}\cos(\Omega t - \Phi), \quad \tan\Phi = \Omega/G\nu_{av}. \tag{13.76}$$

This shows that the modulation of Q includes a part with angular frequency Ω having a phase lag Φ. There is also a term with angular frequency 2Ω that is not present in the original modulation (13.74). The phase shift Φ is transferred to the modulation imposed on the wanted wave. By measuring Φ for various modulation frequencies Ω, Ratcliffe and Shaw (1948) could locate the region of the ionosphere where the interaction was occurring.

Consider now the method of Fejer (1955) in which the disturbing and wanted waves are radio pulses. In the original version of the method these pulses travel vertically as in the ionosonde technique, § 1.7. It is arranged that the disturbing pulse is upgoing and the wanted pulse is downgoing after reflection at a higher level. The disturbing pulse is transmitted only for alternate wanted pulses so that the received amplitudes of the wanted pulse with and without disturbance can be compared. When the disturbing pulse is present the two pulses first meet at a level $z = z_0$ chosen by the timing of the pulses. The wanted pulse then travels on down through the lower levels recently heated by the disturbing pulse. Let the duration τ of the disturbing pulse be much less than $1/G\nu_{av}$ and for any given height $z_1 < z_0$ let it apply heating at the rate $H(z_1)$, assumed constant during the pulse. Then (13.72) shows that, after the pulse has passed this height, Q has acquired an increment $H(z_1)\tau/N$ that subsequently decays thus

$$Q = \frac{H(z_1)\tau}{N}\exp(-G\nu_{av}t). \tag{13.77}$$

From this the increase of effective temperature and thence the increment $\Delta\nu_{eff}$ of the effective collision frequency for the electrons are found, at the time t when the wanted pulse passes through the level z_1. This depends on how $\nu(v)$ depends on electron velocity v, and on the velocity distribution function of the disturbed electrons. Then by an integration through a height range below z_0 the extra attenuation of the wanted wave, caused by $\Delta\nu_{eff}$ is found and expressed in terms of G, ν_{av} and N. By measuring this extra attenuation estimates can be made of these three quantities, and their height dependence can be studied.

13.13. Wave interaction 3. Kinetic theory

In much published work on the theory of wave interaction it was assumed that an electron's mean free path λ_e is independent of its speed v so that $\nu(v) \approx v/\lambda_e$ and ν is proportional to v. But laboratory experiments have suggested that for electrons of low energy the assumption $\nu \propto v^2$ is better. This topic has been discussed in § 3.12. For slow electrons the correct form for $\nu(v)$ is not known with certainty. But the occurrence of wave interaction shows that $\nu(v)$ must be some increasing function of v to give an increase of the average ν when heating occurs.

When the electrons are in thermal equilibrium it is believed that, at least in the lower ionosphere, their velocity distribution function is Maxwellian. The faster electrons are making more collisions so that, when a heating wave is applied, the fast electrons are heated more than the slow electrons. The velocity distribution is therefore disturbed and no longer Maxwellian. It then changes with time and in the cooling process, when the heating is removed, it must relax back to Maxwellian. In the disturbed plasma the extra attenuation of a wanted wave occurs while the distribution function is not Maxwellian, and a full theory must allow for this. To pursue these problems the Boltzmann distribution function for electrons is used. It satisfies the Boltzmann equation. The study of this is in the realm of plasma kinetic theory. It is a large subject with a very extensive literature. A brief summary of the problem for wave interaction was given by Budden (1983b). For recent work in this field and further references see Garrett (1982a, b, 1983, 1985).

PROBLEMS 13

13.1. If the earth's magnetic field is vertical and electron collisions are neglected, then for waves with vertical wave normals in the ionosphere, the two group refractive indices n' are given (see problem 5.10) by $\{1 + XY/2(1-Y)^2\}\{1 - X/(1-Y)\}^{-\frac{1}{2}}$ and the same expression with the sign of Y reversed. Let $h'(f)$ be the contribution to the equivalent height of reflection from the part of the ray path within the ionosphere. Express it as an integral, for a frequency slightly greater than the electron gyro-frequency f_H, and show that

$$\lim_{\varepsilon\to 0} h'\{f_{\mathrm{H}}(1+\varepsilon)\} = \tfrac{4}{3}/(\mathrm{d}X/\mathrm{d}z)_0$$

where $(\mathrm{d}X/\mathrm{d}z)_0$ is the value of $\mathrm{d}X/\mathrm{d}z$ at the base of the ionosphere. (A useful transformation is $X = \varepsilon(1-t^2)$. See Millington (1938a). The author is indebted to Dr. D.H. Shinn of the G.E.C.-Marconi Research Laboratories (formerly Marconi's Wireless Telegraph Co., Ltd.) for bringing this and some similar problems to his attention.)

13.2. Some substances show natural optical activity. A well known example is a solution of an organic compound whose molecule occurs in two enantiomorphic forms that are mirror images of each other and not identical. For an isotropic medium, containing only one of the two types of molecule, the constitutive relation is

$$\boldsymbol{D} = \varepsilon_0\{\varepsilon\boldsymbol{E} + (\eta/k)\,\mathrm{curl}\,\boldsymbol{E}\}.$$

Note that this displays spatial dispersion (§§ 3.1, 5.5) because the operator curl contains spatial derivatives. Show that when a linearly polarised wave travels in a homogeneous medium of this kind, its plane of polarisation is rotated through an angle $\frac{1}{2}k\eta$ in unit length of its path. Compare and contrast this effect with Faraday rotation in a magnetoplasma. (See Landau and Lifshitz, 1960, p. 337 ff.).

13.3. Let ρ_O and ρ_E be the two values of the wave polarisation ρ for waves of extremely low frequency with vertical wave normals in the ionosphere. The effect of collisions and of heavy positive ions is allowed for. Show that in the complex ρ plane, at the crossover frequency (not exact crossover) the two points that represent ρ_O, ρ_E lie either both on the real axis, or both on a circle of unit radius with centre at the origin.

14

General ray tracing

14.1. Introduction

Ch. 10 studied the problem of ray tracing in a plane stratified medium. The objective of the present chapter is to extend this theory to deal with media that are not plane stratified. For example the composition of the ionosphere nearly always depends on the horizontal coordinates, and there are cases especially near twilight when it cannot be assumed to be horizontally stratified. The structure of the magnetosphere is controlled largely by the earth's magnetic field and the solar wind, see § 1.9, and it is not plane stratified.

The number of published papers on ray tracing methods is very large and the subject has acquired some mathematical interest that often goes beyond the needs of practical radio engineers. In this book no attempt has been made to give all the references, but a few that have come to the author's attention are given at the relevant points in this chapter. The objective here is to present the basic physical ideas of ray tracing and to describe the methods that have mainly been used for radio waves in the ionosphere and magnetosphere.

In § 10.2 the ray path was expressed by integrals (10.3), and for plane stratified media these can often be evaluated in closed form. For the more general plasma, however, this is rarely possible. The ray path was also expressed by its differential equations (10.5) for $\mathrm{d}x/\mathrm{d}z$, $\mathrm{d}y/\mathrm{d}z$. The independent variable was the height z and was measured normal to the strata. In the more general case there is nothing to single out one coordinate in this way. In this chapter the ray will be traced by deriving expressions for $\mathrm{d}x/\mathrm{d}\tau$, $\mathrm{d}y/\mathrm{d}\tau$, $\mathrm{d}z/\mathrm{d}\tau$, or analogous expressions with other coordinates instead of x, y, z. Here τ is some variable that increases as we proceed along a ray. There is some latitude in its choice; see § 14.3 below. It is found that these three equations must be supplemented by three more expressions that give $\mathrm{d}/\mathrm{d}\tau$ of the components of the refractive index vector. The six differential equations for the ray path are then integrated with respect to τ. This is usually done with a computer, using a step-by-step process. The equations will first be derived for Cartesian coordinates x, y, z. They will then be transformed to use other coordinate systems; see § 14.5.

One purpose of ray tracing is to find a ray that goes from a given transmitter to a given receiver. Suppose that the transmitter is at the origin and the receiver is on the ground at the point $(x_0, 0, 0)$. The direction of a ray where it leaves the transmitter may be specified by the polar angles θ, ϕ of its wave normal, and these are not at first known. It is therefore necessary to trace several rays and find where they return to the ground. Simplifications are often possible. For example if it is known that there is no lateral deviation of the rays, then $\phi = 0$. If then a ray that starts with $\theta = \theta_i$ returns where $x = x_i$, for $i = 1, 2$, an estimate of the correct value θ_0 of θ can be made by linear interpolation, sometimes called the 'rule of false position' (Hartree, 1958):

$$\theta_0 \approx \frac{\theta_1 x_2 - \theta_2 x_1 + x_0(\theta_2 - \theta_1)}{x_2 - x_1}. \tag{14.1}$$

To use this formula two rays must first be traced. It may often be necessary to trace two or three more before one is found that goes close enough to the receiver. If there is lateral deviation then both ϕ and θ must be changed for successive rays. A two-dimensional form of the rule of false position can then be devised. To use it, three rays must first be traced; see problem 14.2.

A method of this kind in which successive trials are made is sometimes called a 'shooting method'. If the equations are complicated it can be very demanding on computer time. For this type of problem, therefore, ray tracing is customarily used only when some simplification is possible. Another use of ray tracing is to study the configuration of a whole family of rays that leave a given source. For example Herring (1980) used it in this way for a source in the magnetosphere, in a study of radio emissions in the frequency range 10 to 200 kHz.

The first part of this chapter up to § 14.8, like ch. 10, deals almost entirely with loss-free media. The ray direction is then also the direction of the time averaged Poynting vector; see § 5.3. The rays are the paths of energy flux. The next part, §§ 14.9–14.12, discusses the effect of losses in the medium and introduces the concepts of complex rays and the bilinear concomitant vector. The last part, §§ 14.13–14.14, deals with reciprocity when ray theory is used.

The purpose of ray tracing is nearly always to find out the features of the wave that arrives at a given receiver from a given transmitter. A knowledge of the ray path traversed is of secondary importance. It might be useful if it was desired to interrupt the signal by placing an obstacle where it would screen the receiver. The right position would be somewhere on the ray path.

Four uses of ray tracing, (a)–(d), were listed at the end of § 10.1. For items (c), (d) it is necessary to evaluate integrals along the ray path. This can be done by formulating further differential equations, and integrating them at the same time as the differential equations for the ray path are integrated; see § 14.7.

14.2. The eikonal function

Ray tracing is often used where there are sharp boundaries between different media. Where a ray crosses such a boundary its direction, in general, changes discontinuously, and there may be a partial reflection. There are no sharp boundaries in the ionosphere and magnetosphere, and in most of this chapter it is assumed that the medium is continuous and slowly varying. This means that, at points where ray theory can be used, solutions of Maxwell's equations exist that are the analogues of the W.K.B. solutions, ch. 7, or generalisations of them. It was shown in §§ 7.3–7.7 that a W.K.B. solution represents a progressive wave with an exponential or 'phase memory' factor and a more slowly varying amplitude factor; see for example (7.26). The basic formulae of ray tracing use only the phase memory part. Study of the amplitude needs an extension of the theory as given in §§ 10.17, 14.8.

When ray tracing is used, therefore, it is implied that at each point in space there is a progressive wave in which the electric and magnetic fields are given by

$$\boldsymbol{E} = \boldsymbol{E}_0 \exp(-\mathrm{i}\mathscr{E}), \quad \boldsymbol{\mathscr{H}} = \boldsymbol{\mathscr{H}}_0 \exp(-\mathrm{i}\mathscr{E}) \tag{14.2}$$

where the exponential is the phase memory term. The function $\mathscr{E}$ is called the 'eikonal function' (Sommerfeld, 1954). It depends on the space coordinates. Now let (14.2) be substituted in Maxwell's equations (2.44), and use the constitutive relation (4.8). Assume that the amplitude factors $\boldsymbol{E}_0$, $\boldsymbol{\mathscr{H}}_0$ vary so slowly that their spatial derivatives may be ignored. Then

$$\boldsymbol{\Gamma}\boldsymbol{E}_0 = \boldsymbol{\mathscr{H}}_0, \quad \boldsymbol{\Gamma}\boldsymbol{\mathscr{H}}_0 = -\boldsymbol{\varepsilon}\boldsymbol{E}_0, \tag{14.3}$$

where $\boldsymbol{\Gamma}$ is the matrix

$$\boldsymbol{\Gamma} = \begin{pmatrix} 0 & -\partial\mathscr{E}/\partial z & \partial\mathscr{E}/\partial y \\ \partial\mathscr{E}/\partial z & 0 & -\partial\mathscr{E}/\partial x \\ -\partial\mathscr{E}/\partial y & \partial\mathscr{E}/\partial x & 0 \end{pmatrix}. \tag{14.4}$$

But (14.3) is the same as (4.70) provided that

$$\boldsymbol{n} = \operatorname{grad}\mathscr{E}. \tag{14.5}$$

The components of $\boldsymbol{n}$ satisfy the dispersion relation in the form (4.73). Thus (14.5) is the condition that (14.2) shall satisfy Maxwell's equations. It is called the 'eikonal equation'. It shows that $\mathscr{E}$ is similar to a potential function in space, and its gradient $\boldsymbol{n}$ is an irrotational vector field, so that

$$\operatorname{curl}\boldsymbol{n} = 0. \tag{14.6}$$

If $\mathscr{E}$ is known at any point x_0, y_0, z_0, its value at any other point x, y, z is found by integrating its gradient along any path from A to B. Thus

$$\mathscr{E} = \int_{x_0}^{x} n_x \, \mathrm{d}x + \int_{y_0}^{y} n_y \, \mathrm{d}y + \int_{z_0}^{z} n_z \, \mathrm{d}z + \text{constant} \tag{14.7}$$

where the same path is used for all three integrals.

An example of a progressive wave for a plane stratified medium was given by (10.4). Here S_{10}, S_{20}, q are the same as the components n_x, n_y, n_z of the refractive index vector $\boldsymbol{n}$. The exponent could be written $\exp(-\mathrm{i}\mathscr{E})$ where $\mathscr{E}$ is given by (14.7). In the plane stratified medium Snell's law applies and n_x, n_y in (10.4) are constants. Thus the first two terms of (14.7) give $xn_x + yn_y + \text{constant}$. The remaining term $\int q \, \mathrm{d}z = \int n_z \, \mathrm{d}z$ was called the phase memory term. It was discussed in § 7.3, and used in (10.2). When n_x, n_y are not constant their integrals in (14.7) must be retained. This shows how the eikonal (14.7) is simply an extension to three dimensions of the phase memory concept.

In general the field in the neighbourhood of any point in space is more complicated than the progressive wave (14.2). It may be the sum of two or more progressive waves. It must then be resolved into its component progressive waves, and each ray is associated with one only of these components. The others are studied separately, or ignored. But each component progressive wave is, by itself, a solution of Maxwell's equations.

14.3. The canonical equations for a ray path

The differential equations for a ray path are now to be found. It will be shown that six variables are needed, namely the Cartesian coordinates x, y, z in ordinary space and the components n_x, n_y, n_z of the refractive index vector $\boldsymbol{n}$, in refractive index space. These six variables may be thought of as defining a six-dimensional space.

Let x, y, z be the coordinates of the point where some feature of the wave intersects the ray, at an instant of time $t = \tau$. This feature could be, for example, a wave crest. Then $\mathrm{d}x/\mathrm{d}\tau$, $\mathrm{d}y/\mathrm{d}\tau$, $\mathrm{d}z/\mathrm{d}\tau$ are the components of the ray velocity $\boldsymbol{V}$ that was studied in ch. 5. Here τ is used as the independent variable, and the total derivative sign $\mathrm{d}/\mathrm{d}\tau$ is used to denote changes that occur on the ray path. Other choices of the independent variable are possible, and this is discussed later. Note particularly that τ is not the time of travel of a wave packet. Now $\boldsymbol{V}$ can be found when the dispersion relation is known.

At any point x, y, z, the dispersion relation may be written

$$D(x, y, z; n_x, n_y, n_z) = \text{constant}. \tag{14.8}$$

Various expressions for this were given in § 4.9, for example (4.62)–(4.66) or (4.73), in which the constant is zero, but other forms are possible. Some authors use the form $\omega(x, y, z; \kappa_x, \kappa_y, \kappa_z) = \omega_0$ where ω_0 is the constant frequency of the wave, and they use the vector $\boldsymbol{\kappa} = \omega\boldsymbol{n}/c$ instead of the vector $\boldsymbol{n}$ used here. See for example, Suchy and Paul (1965), Buckley (1982). This form is avoided here because of its algebraic complexity; see end of § 5.8. Possible choices for D in (14.8) are discussed in §§ 14.4, 14.5. At a fixed point x, y, z (14.8) is the equation of the refractive index surface in refractive index space. For a point on this surface, $\boldsymbol{V}$ has the direction of the normal; see § 5.3. Thus

$$\boldsymbol{V} = (\mathrm{d}x/\mathrm{d}\tau, \mathrm{d}y/\mathrm{d}\tau, \mathrm{d}z/\mathrm{d}\tau) = A(\partial D/\partial n_x, \partial D/\partial n_y, \partial D/\partial n_z). \tag{14.9}$$

The multiplier A may be found from (5.36) with x, y, z instead of ξ, η, ζ. Substitution of (14.9) for $\boldsymbol{V}$ shows that

$$A = c/\{n_x \partial D/\partial n_x + n_y \partial D/\partial n_y + n_z \partial D/\partial n_z\}. \tag{14.10}$$

Equations (14.9) with (14.10) are the required differential equations of the ray path. To use them, however, it is necessary to know the correct values of n_x, n_y, n_z. These change as the point x, y, z moves along the ray, and further equations are needed to determine them.

The equations (14.9) depend on D at one point only in space. They take no cognisance of how the medium varies in space. The additional equations must allow for this variation because it is the spatial gradient of the composition of the medium that determines the curvature of the ray. Thus we have to study how D varies in space, not only for points on the ray path but for adjacent points as well.

Equation (14.8) must hold for all points. Consider an infinitesimal displacement δx from a point x, y, z on the ray, with y and z held constant. Then the components n_x, n_y, n_z change but D remains constant. Hence

$$\frac{\delta D}{\delta x} = 0 = \frac{\partial D}{\partial x} + \frac{\partial D}{\partial n_x}\frac{\partial n_x}{\partial x} + \frac{\partial D}{\partial n_y}\frac{\partial n_y}{\partial x} + \frac{\partial D}{\partial n_z}\frac{\partial n_z}{\partial x}. \tag{14.11}$$

Because the displacement δx is infinitesimal, the partial derivatives on the right all take their values on the ray. Now (14.6) shows that

$$\frac{\partial n_y}{\partial x} = \frac{\partial n_x}{\partial y}, \quad \frac{\partial n_z}{\partial x} = \frac{\partial n_x}{\partial z}. \tag{14.12}$$

If this and (14.9) are used in (14.11) it gives:

$$A\frac{\partial D}{\partial x} + \frac{\partial n_x}{\partial x}\frac{\mathrm{d}x}{\mathrm{d}\tau} + \frac{\partial n_x}{\partial y}\frac{\mathrm{d}y}{\mathrm{d}\tau} + \frac{\partial n_x}{\partial z}\frac{\mathrm{d}z}{\mathrm{d}\tau} = 0. \tag{14.13}$$

The last three terms are simply $\mathrm{d}n_x/\mathrm{d}\tau$. A similar argument applies if $\delta D/\delta y$ or $\delta D/\delta z$ is used in (14.11). Hence we obtain the three equations

$$(\mathrm{d}n_x/\mathrm{d}\tau, \mathrm{d}n_y/\mathrm{d}\tau, \mathrm{d}n_z/\mathrm{d}\tau) = -A(\partial D/\partial x, \partial D/\partial y, \partial D/\partial z). \tag{14.14}$$

The six equations (14.9) and (14.14) are the required differential equations for the ray path. They and the arguments that led to them can be found in the work of Sir William R. Hamilton (collected papers 1931; the original papers were written in

1827–32), and he called them the 'canonical' equations of the ray. But the notation he used is quite different from that used here. The ideas permeate a large part of his papers and it is not easy to give a precise page reference. A study of Hamilton's work is essential and rewarding for anyone who aspires to be a specialist in ray theory, but cannot be recommended to the radio engineer who wishes to acquire a practical working knowledge of ray tracing techniques.

Equations (14.14) are a generalisation of Snell's law. For example suppose they are applied to a horizontally stratified ionosphere as studied in ch. 10. Then the dispersion relation is independent of x and y, so that $\partial D/\partial x$, $\partial D/\partial y$ are both zero. Hence n_x and n_y are constant on a ray, and this is Snell's law, as discussed in § 6.2. They are the same as S_1, S_2 respectively.

For a general medium, (14.14) can be used to define an 'effective' local surface of stratification near any point. The vector $\partial D/\partial x$, $\partial D/\partial y$, $\partial D/\partial z$ is the normal to this surface, and will be called the 'effective stratification vector'. For any plane stratified medium it is normal to the strata and its direction is the same for all possible ray directions as given by (14.9). But in general it is different for different ray directions, and there is no unique surface of stratification. For a counter example see problem 14.1.

14.4. Properties of the canonical equations

The coordinates x, y, z of a variable point on a ray may be written as a vector $\boldsymbol{r}$. The operator grad is then often written grad $D = \partial D/\partial \boldsymbol{r}$. With this notation the canonical equations (14.9) (14.14) in vector form are

$$\mathrm{d}\boldsymbol{r}/\mathrm{d}\tau = A\partial D/\partial \boldsymbol{n}, \quad \mathrm{d}\boldsymbol{n}/\mathrm{d}\tau = -A\partial D/\partial \boldsymbol{r}, \tag{14.15}$$

where from (14.10)

$$A = c/(\boldsymbol{n}\cdot\partial D/\partial \boldsymbol{n}). \tag{14.16}$$

For an isotropic medium the ray and the wave normal have the same direction, and the refractive index n is a scalar function of position. The dispersion relation may be written

$$D \equiv (n_x^2 + n_y^2 + n_z^2)^{\frac{1}{2}} - n(x, y, z) = 0. \tag{14.17}$$

Then (14.16) gives $A = c/n$ and the equations (14.15) are

$$\frac{\mathrm{d}\boldsymbol{r}}{\mathrm{d}\tau} = c\frac{\boldsymbol{n}}{n^2}, \quad \frac{\mathrm{d}\boldsymbol{n}}{\mathrm{d}\tau} = \frac{c}{n}\frac{\partial n}{\partial \boldsymbol{r}}. \tag{14.18}$$

For a more general anisotropic medium, when the canonical equations in the form (14.15) are integrated with a step-by-step process they give $\boldsymbol{r}$ and $\boldsymbol{n}$ at successive points along a ray. The value of $\boldsymbol{n}$ ought to satisfy the dispersion relation (14.8), and for an exact solution it would do so because the equations were derived from (14.8). But in numerical integration rounding errors accumulate and lead to a wrong value of $\boldsymbol{n}$ and thence of D and its derivatives. It is advisable, therefore, to use the dispersion relation as the integration proceeds. The rate at which errors accumulate depends on the step size used in the integration. One method that is often used is to carry out the integration for successive groups of steps, and test the dispersion relation at the end of each group. If it is not satisfied with sufficient accuracy, the last group is repeated with a smaller step size.

The form (14.15) is not always the most convenient or simplest and other versions are often used. For example there is one method in which one of the three components of $\mathrm{d}\boldsymbol{n}/\mathrm{d}\tau$ from (14.15) is omitted, and for illustration we shall suppose that it is the equation for $\partial n_z/\partial\tau$. Integration of the remaining two equations gives n_x, n_y, and the dispersion relation is then used to find n_z. There are many other ways of using a modified form of the equations. The choice depends on the particular problem being solved. The methods are best illustrated by describing some examples of particular problems.

Consider the special case of an electron plasma whose composition is independent of y, and let the ray, the wave normal and the vector $\boldsymbol{Y}$ all lie in the plane $y = 0$. Then they remain in this plane and $y = 0$, $n_y = 0$. Only four of the six equations (14.15) are non-trivial. Let χ be the angle between the wave normal and the n_x axis so that

$$\tan\chi = n_z/n_x \tag{14.19}$$

whence

$$\frac{\mathrm{d}\chi}{\mathrm{d}\tau} = \frac{\partial\chi}{\partial n_x}\frac{\mathrm{d}n_x}{\mathrm{d}\tau} + \frac{\partial\chi}{\partial n_z}\frac{\mathrm{d}n_z}{\mathrm{d}\tau} = \frac{1}{n^2}\left(n_x\frac{\mathrm{d}n_z}{\mathrm{d}\tau} - n_z\frac{\mathrm{d}n_x}{\mathrm{d}\tau}\right). \tag{14.20}$$

The refractive index depends on x, z and on the angle χ, and through χ on n_x, n_z. The dispersion relation may be written

$$D \equiv \frac{(n_x^2 + n_z^2)^{\frac{1}{2}}}{n(x, z; \chi)} = 1. \tag{14.21}$$

This form is discussed in more detail in § 14.5. In (14.16) it gives $A = c$. Then the canonical equations (14.15) give, with (14.20)

$$\begin{aligned}
\frac{\mathrm{d}x}{\mathrm{d}\tau} &= \frac{c}{n}\left(\cos\chi + \sin\chi\frac{1}{n}\frac{\partial n}{\partial\chi}\right),\\
\frac{\mathrm{d}z}{\mathrm{d}\tau} &= \frac{c}{n}\left(\sin\chi - \cos\chi\frac{1}{n}\frac{\partial n}{\partial\chi}\right),\\
\frac{\mathrm{d}\chi}{\mathrm{d}\tau} &= \frac{c}{n^2}\left(\cos\chi\frac{\partial n}{\partial z} - \sin\chi\frac{\partial n}{\partial x}\right).
\end{aligned} \tag{14.22}$$

When (14.22) are integrated, the refractive index n can be found at each point, since χ is known. This shows how the two equations for $\mathrm{d}n_x/\mathrm{d}\tau$ and $\mathrm{d}n_z/\mathrm{d}\tau$ may be replaced by a single equation for $\mathrm{d}\chi/\mathrm{d}\tau$ together with the formula for $n(x, z; \chi)$. The equations (14.22) were given by Haselgrove (1954) and were used by her (Haselgrove, 1957) to study rays in the magnetic meridian plane in the ionosphere.

The multiplier A in (14.15) depends on the choice of the function D in the dispersion relation (14.8). Its value does not affect the path of the ray nor the values of $\boldsymbol{n}$ on it. It is often combined with τ to give a new independent variable s, so that equations (14.15) become

$$\mathrm{d}\boldsymbol{r}/\mathrm{d}s = \partial D/\partial \boldsymbol{n}, \quad \mathrm{d}\boldsymbol{n}/\mathrm{d}s = -\partial D/\partial \boldsymbol{r}. \tag{14.23}$$

Suppose now that we wish to trace a ray that goes from free space into the ionosphere, and suppose that (4.62) is chosen for the dispersion relation. It applies for a particular set of axes but in the free space it is the same for all axis systems and reduces to

$$D \equiv n^4 - 2n^2 + 1 = 0. \tag{14.24}$$

Now all elements of $\partial D/\partial \boldsymbol{n}$ are zero, so that A (14.16) is infinite. When ds in (14.23) is finite, dτ must be zero, and d$\boldsymbol{r}$, d$\boldsymbol{n}$ are both zero. We cannot make any progress along the ray. Clearly a different D must be chosen. A possible choice is discussed in the following section.

For a ray that enters the ionosphere from free space there is a further problem. It must be either ordinary or extraordinary, and for both, at the start, $n = 1$. There is nothing in equations (14.15) that will decide which of the two types it is. This may be dealt with by writing, in the dispersion relation,

$$n^2 = 1 + Xp \tag{14.25}$$

and expressing it in terms of p rather than n. Now (4.112) shows that in the limit of free space, $X \to 0$, p tends to a non-zero limit that is different for the ordinary and extraordinary waves. Thus at the start the value of p for the required wave is used. The details are complicated, and are given by Haselgrove and Haselgrove (1960) and by Haselgrove (1963). They use q instead of p but it is avoided here because of possible confusion with Booker's q, ch. 6.

For other versions of the ray tracing equations see, for example, Al'pert (1948), Kimura (1966), Bitoun, Graff and Aubry (1970), Suchy (1972a, b, 1974a, 1981), Rönnmark (1984). See also *Radio Science* (1968, special issue on ray tracing).

If the transmitter or receiver is moving, or if the medium is changing with time, the ray tracing problem is more complicated and the received signal can show a Doppler shift of frequency. The effect is not discussed in this book, but see for example Little and Lawrence (1960), Capon (1961), Bennett (1969).

14.5. The Haselgrove form of the equations

The refractive index n is given by the Appleton–Lassen formula (4.48) or some equivalent form such as (4.51), (4.61). The variables in it are functions of the space coordinates and it also contains the angle Θ, that depends on the direction cosines of the wave normal:

$$\frac{n_x}{(n_x^2+n_y^2+n_z^2)^{\frac{1}{2}}}, \quad \frac{n_y}{(n_x^2+n_y^2+n_z^2)^{\frac{1}{2}}}, \quad \frac{n_z}{(n_x^2+n_y^2+n_z^2)^{\frac{1}{2}}}. \tag{14.26}$$

Thus n may be written $n(x, y, z;\ n_x, n_y, n_z)$ where n_x, n_y, n_z occur only in the combinations (14.26). Now in place of the function D in the dispersion relation (14.8), we choose a new function G so that the dispersion relation is

$$G(x, y, z;\ n_x, n_y, n_z) \equiv \frac{(n_x^2+n_y^2+n_z^2)^{\frac{1}{2}}}{n(x, y, z;\ n_x, n_y, n_z)} = 1. \tag{14.27}$$

If now n_x, n_y, n_z are replaced by $\lambda n_x, \lambda n_y, \lambda n_z$, the combinations (14.26) are unaltered, and G changes to λG. Therefore G is said to be homogeneous of degree one in the set n_x, n_y, n_z, and Euler's theorem for homogeneous functions shows that

$$n_x \partial G/\partial n_x + n_y \partial G/\partial n_y + n_z \partial G/\partial n_z = G. \tag{14.28}$$

The proof of the theorem is not difficult and is given in standard text books (see, for example, Gibson, 1929, p. 412). Now (14.28) and (14.16) with G for D show that, since $G = 1$, $A = c$, and (14.15) becomes:

$$\mathrm{d}\boldsymbol{r}/\mathrm{d}\tau = c\partial G/\partial \boldsymbol{n}, \quad \mathrm{d}\boldsymbol{n}/\mathrm{d}\tau = -c\partial G/\partial \boldsymbol{r}. \tag{14.29}$$

Equations of this form were given by Hamilton (1931) who also used the properties of homogeneous functions. They were derived and used for radio propagation problems by Haselgrove (1954, 1957, 1963), Haselgrove and Haselgrove (1960), and they are commonly called the Haselgrove equations. A simple example (14.22) has already been given. When the equations are expressed in Cartesian coordinates, as used in (14.27), various transformations are possible that greatly simplify the algebra (see the Haselgrove references cited above). This is true even when they are applied in an ionosphere or magnetosphere with spherical or more complicated stratification, because the simplicity of the Cartesian equations outweighs the extra complication of expressing a spherical or more irregular plasma in Cartesian coordinates (Haselgrove, 1963).

The equations are, however, often used in spherical polar coordinates r, θ, ϕ, and their basic form for this case can be derived as follows. At each point in space the components of the refractive index vector are n_r parallel to the radius r in ordinary space, n_θ parallel to the direction $r = \text{const.}$, $\phi = \text{const.}$, and n_ϕ parallel to the direction $r = \text{const.}$, $\theta = \text{const.}$ Thus the refractive index space is a Cartesian space, but the directions of its axes are different for different points in ordinary space. The dispersion relation is

$$G \equiv \mathscr{G}(r, \theta, \phi;\ n_r, n_\theta, n_\phi) \equiv \frac{(n_r^2+n_\theta^2+n_\phi^2)^{\frac{1}{2}}}{\breve{n}(r, \theta, \phi;\ n_r, n_\theta, n_\phi)} = 1. \tag{14.30}$$

Here $\breve{n}$ is numerically the same as n but a different symbol is needed because the functional dependence on the new variables is different. Similarly $\mathscr{G}$ is used instead of G. The equations for transforming the variables are

$$r = (x^2 + y^2 + z^2)^{\frac{1}{2}}, \quad \tan\theta = (x^2 + y^2)^{\frac{1}{2}}/z, \quad \tan\phi = y/x \tag{14.31}$$

$$\left.\begin{aligned} n_r &= (n_x \cos\phi + n_y \sin\phi)\sin\theta + n_z \cos\theta, \\ n_\theta &= (n_x \cos\phi + n_y \sin\phi)\cos\theta - n_z \sin\theta, \\ n_\phi &= -n_x \sin\phi + n_y \cos\phi. \end{aligned}\right\} \tag{14.32}$$

Hence from (14.29), after some algebra

$$\frac{\mathrm{d}r}{\mathrm{d}\tau} = c\frac{\partial\mathscr{G}}{\partial n_r}, \quad \frac{\mathrm{d}\theta}{\mathrm{d}\tau} = \frac{c}{r}\frac{\partial\mathscr{G}}{\partial n_\theta}, \quad \frac{\mathrm{d}\phi}{\mathrm{d}\tau} = \frac{c}{r\sin\theta}\frac{\partial\mathscr{G}}{\partial n_\phi}, \tag{14.33}$$

$$\left.\begin{aligned} \frac{\mathrm{d}n_r}{\mathrm{d}\tau} &= c\left(-\frac{\partial\mathscr{G}}{\partial r} + \frac{n_\theta}{r}\frac{\partial\mathscr{G}}{\partial n_\theta} + \frac{n_\phi}{r}\frac{\partial\mathscr{G}}{\partial n_\phi}\right), \\ \frac{\mathrm{d}n_\theta}{\mathrm{d}\tau} &= c\left(-\frac{1}{r}\frac{\partial\mathscr{G}}{\partial\theta} - \frac{n_r}{r}\frac{\partial\mathscr{G}}{\partial n_\theta} + \frac{n_\phi\cot\theta}{r}\frac{\partial\mathscr{G}}{\partial n_\phi}\right), \\ \frac{\mathrm{d}n_\phi}{\mathrm{d}\tau} &= c\left(-\frac{1}{r\sin\theta}\frac{\partial\mathscr{G}}{\partial\phi} - \frac{n_r + n_\theta\cot\theta}{r}\frac{\partial\mathscr{G}}{\partial n_\phi}\right). \end{aligned}\right\} \tag{14.34}$$

14.6. Fermat's principle

Fermat's principle is usually stated as follows. Consider several different paths running from a fixed point A to another fixed point B. Suppose that a ray can travel along any one of these paths and let T be the time taken for some feature of the wave, such as a wave crest or trough, to go from A to B. It is assumed that this feature moves with the ray velocity in the direction of the path. Then the actual ray that goes from A to B is that path for which T is an extremum.

For an isotropic medium the proof is not difficult. The ray velocity V at each point is c/n in the direction of the wave normal and is the same for all ray directions. This gives at once the first set of the ray tracing equations (14.18), that is

$$\frac{\mathrm{d}x}{\mathrm{d}\tau} = cn_x/n^2 \tag{14.35}$$

and two similar equations with y, n_y, and z, n_z. Any path from A to B is defined by specifying the three functions $x(p)$, $y(p)$, $z(p)$ where p is some parameter that increases monotonically from 0 at A to 1 at B. It is convenient to use the notation

$$x' = \mathrm{d}x/\mathrm{d}p, \quad y' = \mathrm{d}y/\mathrm{d}p, \quad z' = \mathrm{d}z/\mathrm{d}p. \tag{14.36}$$

Then, if $\mathrm{d}s$ is an element of the path,

$$\frac{\mathrm{d}s}{\mathrm{d}p} = \{x'^2 + y'^2 + z'^2\}^{\frac{1}{2}} = \frac{c\,\mathrm{d}\tau}{n\,\mathrm{d}p}. \tag{14.37}$$

The time of travel from A to B is

$$T=\int_{(A)}^{(B)} \mathrm{d}\tau=\frac{1}{c}\int_0^1 n\frac{\mathrm{d}s}{\mathrm{d}p}\mathrm{d}p. \tag{14.38}$$

If, for some path $\mathscr{P}$, this is an extremum, then on using another path infinitesimally different from $\mathscr{P}$ the change δT of T must be zero. This is a standard problem in the calculus of variations. The necessary and sufficient condition that $\delta T=0$ is given by the three Euler equations

$$\frac{\partial}{\partial x}\left\{n\frac{\mathrm{d}s}{\mathrm{d}p}\right\}=\frac{\mathrm{d}}{\mathrm{d}p}\left\{\frac{\partial}{\partial x'}\left(n\frac{\mathrm{d}s}{\mathrm{d}p}\right)\right\} \tag{14.39}$$

and two similar equations for the y and z components. On carrying out the differentiations and using (14.36), (14.37), this gives

$$\frac{\partial n}{\partial x}\frac{\mathrm{d}s}{\mathrm{d}p}=\frac{\mathrm{d}}{\mathrm{d}p}\left\{nx'/\frac{\mathrm{d}s}{\mathrm{d}p}\right\}=\frac{\mathrm{d}}{\mathrm{d}p}\left\{n\frac{\mathrm{d}x}{\mathrm{d}s}\right\}=\frac{\mathrm{d}}{\mathrm{d}p}\left\{\frac{n^2}{c}\frac{\mathrm{d}x}{\mathrm{d}\tau}\right\}. \tag{14.40}$$

Then on rearranging and using (14.36)

$$\frac{\partial n}{\partial x}=\frac{\mathrm{d}n_x}{\mathrm{d}s}=\frac{n}{c}\frac{\mathrm{d}n_x}{\mathrm{d}\tau}. \tag{14.41}$$

This and the two similar equations for the y and z components are the same as the second set of the ray tracing equations (14.18). Thus these equations are the Euler equations that are the equivalent of Fermat's principle.

For an anisotropic medium the problem is more complicated. The magnitude of the ray velocity V now depends on its direction. It is given by $V=c/n\cos\alpha$ where $n\cos\alpha$ is the ray refractive index $\mathscr{M}$, § 5.3. Thus

$$n\cos\alpha=\mathscr{M}(x, y, z;\, V_x, V_y, V_z) \tag{14.42}$$

where the dependence on V_x, V_y, V_z is only through the direction cosines

$$\frac{V_x}{(V_x^2+V_y^2+V_z^2)^{\frac{1}{2}}},\quad \frac{V_y}{(V_x^2+V_y^2+V_z^2)^{\frac{1}{2}}},\quad \frac{V_z}{(V_x^2+V_y^2+V_z^2)^{\frac{1}{2}}} \tag{14.43}$$

of the ray direction. The equation of the ray surface at the point x, y, z may now be written

$$F(x, y, z;\, V_x, V_y, V_z)\equiv\frac{1}{c}(V_x^2+V_y^2+V_z^2)^{\frac{1}{2}}\mathscr{M}(x, y, z;\, V_x, V_y, V_z)=1. \tag{14.44}$$

It is homogeneous of degree 1 in V_x, V_y, V_z and therefore, by Euler's theorem (compare (14.28)),

$$V_x\partial F/\partial V_x+V_y\partial F/\partial V_y+V_z\partial F/\partial V_z=F=1. \tag{14.45}$$

The function F gives the ray surface and is a kind of analogue of the function G, (14.27), that gives the refractive index surface.

The ray surface has the shape of a wave front, see end of § 5.3, and therefore its normal is the wave normal, parallel to $\boldsymbol{n}$. Thus

$$\boldsymbol{n} = c\partial F/\partial \boldsymbol{V} \tag{14.46}$$

where (5.36) and (14.45) have been used to give the factor c.

Now we apply to F an argument similar to that used for D in (14.11)–(14.13). Let δx be an infinitesimal displacement from a point on the ray, with y and z held constant. Then

$$\frac{\delta F}{\delta x} = 0 = \frac{\partial F}{\partial x} + \frac{1}{c}\left(n_x \frac{\partial V_x}{\partial x} + n_y \frac{\partial V_y}{\partial y} + n_z \frac{\partial V_z}{\partial z} \right) \tag{14.47}$$

where (14.46) has been used. The same argument can be applied to G and it gives

$$\frac{\delta G}{\delta x} = 0 = \frac{\partial G}{\partial x} + \frac{1}{c}\left(V_x \frac{\partial n_x}{\partial x} + V_y \frac{\partial n_y}{\partial x} + V_z \frac{\partial n_z}{\partial x} \right) \tag{14.48}$$

where the first equation (14.29) has been used. Addition of (14.47) and (14.48) now gives

$$\frac{\partial}{\partial x}(F + G) = -\frac{1}{c}\frac{\partial}{\partial x}(\boldsymbol{V}\cdot\boldsymbol{n}) = 0 \tag{14.49}$$

from (5.36) and there are similar results for $\partial/\partial y$ and $\partial/\partial z$. Thence from the second equation (14.29)

$$c\partial F/\partial x = \mathrm{d}n_x/\mathrm{d}\tau \tag{14.50}$$

and there are two similar equations with y and z.

Now we again use $x(p)$, $y(p)$, $z(p)$ to specify paths from A to B, and $x' = \mathrm{d}x/\mathrm{d}p$ etc. as in (14.36). Then we have $x' = V_x \mathrm{d}\tau/\mathrm{d}p$ and two similar equations with y and z. There is some latitude in the choice of the parameter p and we choose $p = \tau/T$ so that, for any one path, $\mathrm{d}\tau/\mathrm{d}p$ is a constant. Then

$$\frac{\mathrm{d}s}{\mathrm{d}p} = \{V_x^2 + V_y^2 + V_z^2\}^{\frac{1}{2}}\frac{\mathrm{d}\tau}{\mathrm{d}p}. \tag{14.51}$$

Thus $\mathrm{d}\tau = (\mathscr{M}/c)(\mathrm{d}s/\mathrm{d}p)\mathrm{d}p$ and this is now used in (14.38). It shows that, for the anisotropic medium, the refractive index n is to be replaced by the ray refractive index $\mathscr{M}$. For $\delta T = 0$ the Euler equations, for (14.38) with $\mathscr{M}$ for n, are now

$$\frac{\partial}{\partial x}\left\{\mathscr{M}\frac{\mathrm{d}s}{\mathrm{d}p}\right\} = \frac{\mathrm{d}}{\mathrm{d}p}\left\{\frac{\partial}{\mathrm{d}x'}\left(\mathscr{M}\frac{\mathrm{d}s}{\mathrm{d}p}\right)\right\} \tag{14.52}$$

and two similar equations; compare (14.39). Cancel the constant factor $\mathrm{d}\tau/\mathrm{d}p$ and use (14.44). Then

$$\frac{\partial F}{\partial x} = \frac{\mathrm{d}}{\mathrm{d}p}\left(\frac{\partial F}{\partial x'}\right) = \frac{\mathrm{d}}{\mathrm{d}\tau}\left(\frac{\partial F}{\partial V_x}\right). \tag{14.53}$$

Now (14.50) and (14.46) show that the first and last terms are both equal to $(1/c)(\mathrm{d}n_x/\mathrm{d}\tau)$. Similarly the other two Euler equations are satisfied. Thus Fermat's principle is established. It is again equivalent to the ray tracing equations (14.29) which were used at the stages (14.48) and (14.50).

Fermat's principle is sometimes asserted as an *a priori* postulate, and used to derive the ray tracing equations. This is not satisfactory, because it is necessary first to formulate the definition of a ray and this involves the restriction to a progressive wave as in § 14.2. Moreover although Maxwell's equations must be satisfied for the fields associated with an allowed ray, they cannot be satisfied for all arbitrary fictitious varied ray paths.

The usefulness of Fermat's principle is that it can be proved for Cartesian coordinates, starting from Maxwell's equations, as has been done here in outline. But it must be independent of the coordinate system and it can therefore then be used to derive the equations in other systems. It was used in this way by Haselgrove (1954) to derive the canonical equations in a very general coordinate system.

One further point should be noted. The time of travel T that is an extremum is not the travel time of a wave packet, but of some feature of the wave itself. This prompts the question: when Fermat's principle is asserted *a priori*, why should the time of travel be that of the wave and not of a wave packet? The wave travel time gives the right answer, but to show this it is necessary to have some other basis for Fermat's principle, so that then it is not an *a priori* postulate at all.

For a discussion of Fermat's principle see, for example, Bennett (1969), Suchy (1972b).

14.7. Equivalent path and absorption

It was shown in § 5.8 that a wave packet travels in the direction of the ray with the group velocity $\mathscr{U}$ given by (5.68), (5.75). The time of travel over a given path from A to B is P'/c where P' is called the equivalent path. Hence

$$P' = c\int_{(A)}^{(B)} \mathrm{d}s/\mathscr{U} = \int_{(A)}^{(B)} n' \cos\alpha\,\mathrm{d}s \tag{14.54}$$

where n' is the group refractive index, §§ 5.8, 5.9 and α is the angle between the ray and the wave normal. Now

$$\mathrm{d}s = V\mathrm{d}\tau = \frac{c\mathrm{d}\tau}{n\cos\alpha} \tag{14.55}$$

whence (14.54) becomes

$$P' = c\int_{(A)}^{(B)} \frac{n'}{n}\mathrm{d}\tau = c\int_{(A)}^{(B)}\left(1 + \frac{f}{n}\frac{\partial n}{\partial f}\right)\mathrm{d}\tau \tag{14.56}$$

from (5.75) first equation, where f is the frequency.

If collisions are allowed for, the refractive index vector $\boldsymbol{n}$ is complex and the wave is attenuated as it travels. The general theory for this case is discussed in § 14.9ff. If the collision frequency is small enough, however, $-\operatorname{Im}(\boldsymbol{n})$ is so small that its effect on the ray paths is negligible. The theory of the earlier sections can be applied to Re($\boldsymbol{n}$), and the ray path can be found by ignoring the imaginary part. Once the path is known, the attenuation can be found by an integration along the ray path, as described below.

As we proceed along a ray the signal amplitude may change for two reasons. First, the neighbouring rays in a ray pencil may diverge or converge, so that the energy flux decreases or increases respectively. This was studied for a stratified medium in § 10.17, and it is discussed for the general case in the following section. Second, the signal may be attenuated. The signal amplitude $\mathscr{A}$ is here taken to be the square root of the magnitude of the time averaged Poynting vector (2.63). The following discussion applies only to the attenuation, and not to the divergence or convergence.

Let

$$\boldsymbol{n} = \boldsymbol{\mu} - \mathrm{i}\boldsymbol{\chi}, \quad n = \mu - \mathrm{i}\chi. \tag{14.57}$$

Then χ determines the rate of attenuation in the direction of the wave normal. The attenuation along the ray path is therefore given by

$$\mathscr{A} = \mathscr{A}_0 \exp\left\{ -k \int_{(\mathrm{A})}^{(\mathrm{B})} \chi \cos\alpha \, \mathrm{d}s \right\} \tag{14.58}$$

where $\mathscr{A}_0$ is the signal amplitude at the point A. It is assumed that $\chi \ll \mu$ so that χ can be neglected everywhere except in (14.58). To express $\mathscr{A}$ in terms of the electric field $\boldsymbol{E}$ of the wave, it is assumed that the fields near a point on the ray are given by the progressive wave (14.2) with (14.8). Then it can be shown, from (2.63) and (2.44) first equation that

$$\mathscr{A}^2 = \frac{1}{2Z_0} | \boldsymbol{\mu}(\boldsymbol{E}\cdot\boldsymbol{E}^*) - \mathrm{Re}\,\{\boldsymbol{E}^*(\boldsymbol{\mu}\cdot\boldsymbol{E})\}|. \tag{14.59}$$

Now (14.55) is replaced by

$$\mathrm{d}s \cos\alpha = c\,\mathrm{d}\tau/\mu \tag{14.60}$$

so that (14.58) gives

$$\ln \mathscr{A} = -kc \int_{(\mathrm{A})}^{(\mathrm{B})} \frac{\chi}{\mu} \mathrm{d}\tau + \ln \mathscr{A}_0. \tag{14.61}$$

The six canonical differential equations, such as (14.29), or their equivalent, are used to find the ray path in a step-by-step integration process. At the same time it is convenient to find the equivalent path and the attenuation by expressing (14.56) and (14.61) as additional differential equations, thus

$$\frac{\mathrm{d}P'}{\mathrm{d}\tau} = c\left(1 + \frac{f}{\mu}\frac{\partial\mu}{\partial f}\right), \tag{14.62}$$

$$\frac{\mathrm{d}(\ln \mathscr{A})}{\mathrm{d}\tau} = -kc\frac{\chi}{\mu}. \tag{14.63}$$

The six canonical equations and the two equations (14.62), (14.63) can then be integrated at the same time.

A slightly different method of ray tracing in an anisotropic absorbing medium was used by Suchy (1972a). A wave packet was launched whose shape was initially a Gaussian function in all three directions. As it travelled in real space it was attenuated and it spread, but the point of maximum signal amplitude could be found at successive times. The locus of this point was taken as the effective real ray and its speed was the generalised group velocity, which was shown to be $\mathrm{Re}(\partial\omega/\partial\boldsymbol{\kappa})$ where $\boldsymbol{\kappa}=\omega\boldsymbol{n}/c$. In later papers (Suchy, 1972b, 1974a) these results were related to a form of ray tracing equations of Hamilton's type in real space.

14.8. Signal intensity in ray pencils

One ray does not exist in isolation. It must be one of a family of neighbouring rays, and we can select a set of rays in this family that lie in the surface of a narrow tube. The rays within this tube form a pencil of rays, and the signal intensity depends on the convergence or divergence of the pencil. But the canonical equations apply to one ray only, and give no information about the neighbouring rays. For this some further equations are needed. The problem was discussed in § 10.17 for rays in a stratified medium. It is now extended for the general medium. In this section it is assumed that the medium is loss-free so that the refractive indices and other associated quantities are all real. The method given here is similar to that of Buckley (1982) but there are some differences. Buckley gives references to similar methods used for acoustic waves, where the medium is isotropic and non-dispersive. The author is indebted to Dr. R. Buckley for permission to use his results, and for discussions. The problem for anisotropic dispersive media has also been studied by Harvey (1968) and Bernstein (1975).

In this section we shall need to use some tensor relations, and therefore vectors and tensors are expressed in Cartesian coordinates with the subscript notation. Thus the vector $\boldsymbol{n}$ is now written n_i, where the subscript i takes one of the three values x, y, z, so that the n_i are the components of $\boldsymbol{n}$. The coordinates x, y, z of a point on a ray are the components of the vector $\boldsymbol{r}$ of § 14.4 and are similarly written r_i. In this notation the Haselgrove form (14.29) of the canonical equations is

$$\mathrm{d}r_i/\mathrm{d}\tau = c\partial G/\partial n_i, \quad \mathrm{d}n_i/\mathrm{d}\tau = -c\partial G/\partial r_i. \qquad (14.64)$$

A quantity can have two or three or more subscripts and is then a tensor of rank two, three, etc. When the same subscript appears on two factors in any product of terms, it is implied that that product is the sum of the three products in which the repeated subscript takes each of the three values x, y, z. This is the summation convention and is fully described by Jeffreys (1931) who also gives the definitions and properties of the isotropic tensors δ_{ij} and ε_{ijk} used below.

It is assumed that a ray pencil originates from a point source where $r_i=0$. We consider a ray pencil with the property that its cross section by any plane is a small parallelogram. The pencil is then defined by the four rays at the corners. These rays can be specified by giving the four values of the refractive index n_i at the source. They will be indicated by a capital N, and are taken to be

$$N_i, \quad N_i+\delta N_i^{(1)}, \quad N_i+\delta N_i^{(2)}, \quad N_i+\delta N_i^{(1)}+\delta N_i^{(2)}. \tag{14.65}$$

Here $\delta N_i^{(1)}$, $\delta N_i^{(2)}$ play the same role as δS_1, δS_2 in § 10.17. The ray specified by N_i is the 'main' ray, to be traced by using the canonical equations. The other rays are infinitesimally displaced from it. Let the points where these four rays cut any given plane be

$$r_i, \quad \delta r_i+r_i^{(1)}, \quad r_i+\delta r_i^{(2)}, \quad r_i+\delta r_i^{(1)}+\delta r_i^{(2)}. \tag{14.66}$$

Then the area of the parallelogram where the pencil crosses the plane is a vector A_i in the direction of the normal and given by the vector product of $\delta r_i^{(1)}$ and $\delta r_i^{(2)}$, thus

$$A_i=\varepsilon_{ijk}\delta r_j^{(1)}\delta r_k^{(2)} \tag{14.67}$$

where ε_{ijk} is the isotropic tensor of rank 3. The energy flux at a point in the pencil is the time averaged Poynting vector $\mathbf{\Pi}_{\text{av}}$ which is now written Π_i. The power crossing the parallelogram is then the scalar product $\Pi_i A_i$. This is the power flux in the pencil and is the same for all cross sections, so that

$$\varepsilon_{ijk}\Pi_i\delta r_j^{(1)}\delta r_k^{(2)}=\text{constant}. \tag{14.68}$$

The signal intensity is the magnitude of Π_i and is here written Π, without subscript as in § 10.17. Thus $\Pi^2=\Pi_i\Pi_i$. Equation (14.68) is now to be used to find how Π varies along the pencil.

When N_i is changed by δN_i, the values of r_i, n_i at some point on the main ray change to new values for a neighbouring ray. For this type of change the operator $\mathrm{D}/\mathrm{D}N_i$ will be used, to distinguish it from changes along the same ray, for which $\mathrm{d}/\mathrm{d}\tau$ is used. Then

$$\delta r_i=\frac{\mathrm{D}r_i}{\mathrm{D}N_j}\delta N_j, \quad \delta n_i=\frac{\mathrm{D}n_i}{\mathrm{D}N_j}\delta N_j. \tag{14.69}$$

The two 3×3 tensors

$$J_{ij}=\frac{\mathrm{D}r_i}{\mathrm{D}N_j}, \quad K_{ij}=\frac{\mathrm{D}n_i}{\mathrm{D}N_j} \tag{14.70}$$

must be known at any point where (14.69) is to be used. They are found by formulating differential equations for them. Apply the operator $\mathrm{D}/\mathrm{D}N_j$ to (14.64). This gives

$$\frac{\mathrm{d}J_{ij}}{\mathrm{d}\tau}=c\left\{\frac{\partial^2 G}{\partial n_i\partial r_k}J_{kj}+\frac{\partial^2 G}{\partial n_i\partial n_k}K_{kj}\right\}, \tag{14.71}$$

$$\frac{\mathrm{d}K_{ij}}{\mathrm{d}\tau}=-c\left\{\frac{\partial^2 G}{\partial r_i\partial r_k}J_{kj}+\frac{\partial^2 G}{\partial r_i\partial n_k}K_{kj}\right\}. \tag{14.72}$$

The tensors (14.70) together have 18 elements and so in (14.71), (14.72) there are 18 simultaneous equations. These are integrated at the same time as the canonical equations (14.64), proceeding along the main ray. The function $G(r_i, n_j)$ is known at all points in ordinary space and refractive index space, so its second partial derivatives used in (14.71) (14.72) are known. At the start of the integration, $r_i = 0$ so all elements of J_{ij} are zero, and $n_i = N_i$ so $K_{ij} = \delta_{ij}$. Thus the initial values are known.

Now the first equation (14.69) is used in (14.68) to give

$$\varepsilon_{ijk}\Pi_i J_{jl} J_{km} \delta N_l^{(1)} \delta N_m^{(2)} = \text{const.} \tag{14.73}$$

The K_{ij} are not used here. They are needed only as auxiliary variables in the equations (14.71), (4.72).

We are free to choose the two constant vectors $\delta N_l^{(1)}$, $\mathrm{d}N_m^{(2)}$ in any way we please, except that they must lie in the refractive index surface at the source point. Thus they must be perpendicular to the normal to this surface, that is to $\partial G/\partial N_i$. A slightly modified form of the choice made by Buckley (1982) is used here:

$$\delta N_l^{(1)} = \left(\frac{\partial G}{\partial N_z}, 0, -\frac{\partial G}{\partial N_x}\right)\delta a^{(1)}, \quad \delta N_m^{(2)} = \left(0, \frac{\partial G}{\partial N_z}, \frac{-\partial G}{\partial N_y}\right)\delta a^{(2)}. \tag{14.74}$$

The product $\delta a^{(1)}\delta a^{(2)}$ of the two scalar amplitude factors appears simply as a constant multiplier on the left side of (14.73) and may be omitted. Then the scalar constant (14.73) is ε_{ijk} multiplied by the three vectors

$$\Pi_i, \quad J_{jx}\frac{\partial G}{\partial N_z} - J_{jz}\frac{\partial G}{\partial N_x}, \quad \text{and } J_{ky}\frac{\partial G}{\partial N_z} - J_{kz}\frac{\partial G}{\partial N_y}. \tag{14.75}$$

It is therefore the 3×3 determinant formed by the components of these vectors (Jeffreys, 1931, p. 13). After rearrangement, and cancellation of the constant factor $\partial G/\partial N_z$, it is more concisely written as the 4×4 determinant

$$\Pi \begin{vmatrix} 0 & \partial G/\partial N_x & \partial G/\partial N_y & \partial G/\partial N_z \\ \hat{\Pi}_x & J_{xx} & J_{xy} & J_{xz} \\ \hat{\Pi}_y & J_{yx} & J_{yy} & J_{yz} \\ \hat{\Pi}_z & J_{zx} & J_{zy} & J_{zz} \end{vmatrix} = \text{constant}. \tag{14.76}$$

The vector Π_i has here been written $\Pi\hat{\Pi}_i$ where $\hat{\Pi}_i$ is a unit vector in the direction of the ray, which is known, and Π is the signal intensity.

Equations (14.76), (14.71), (14.72) are the required solution. To use them, the equations (14.71), (14.72) are integrated at the same time as the canonical equations (14.64), along the main ray. Then at any point where Π is required, the determinant in (14.76) is evaluated. The signal intensity is proportional to its reciprocal.

The signal intensity would be infinite where the rays touch a caustic surface, for there the cross-sectional area of the pencil is zero. Then the determinant in (14.76) is zero. The analogous result to this for a stratified medium was discussed in § 10.17; see equation (10.50). But for a ray that touches a caustic there is no failure of the equations. The canonical equations (14.64) and the set (14.71), (14.72) can be integrated right through a caustic region.

The source of a ray pencil is often in free space or in some homogeneous isotropic medium. It is useful to apply the equations to this case as a simple partial check. Then G is given by (14.27) in which n is a constant, and $\partial G/\partial N_i$ is n_i/n^2. The only non-zero coefficient on the right of (14.71), (14.72) is

$$\frac{\partial^2 G}{\partial n_i \partial n_k} = \frac{1}{n^4}(n^2 \delta_{ik} - n_i n_k). \qquad (14.77)$$

Equation (14.72) shows that K_{ij} does not change, but retains its initial value δ_{ij}. Then (14.71) can be integrated. Since J_{ij} is initially zero, it gives

$$J_{ij} = \frac{c\tau}{n^4}(n^2 \delta_{ij} - n_i n_j). \qquad (14.78)$$

The vector $\hat{\Pi}_i$ is n_i/n. These results can be put into (14.76) which gives finally

$$-\Pi c^2 \tau^2 / n^5 = \text{constant}. \qquad (14.79)$$

This shows, correctly, that Π is inversely proportional to R^2 where $R = c\tau/n$ is the distance from the source. The constant is $-1/n^3$ times the power in unit solid angle of the pencil.

For an inhomogeneous isotropic medium the equations (14.76), (14.71), (14.72) take a simpler form, and Buckley (1982) has discussed this case as an illustration. As far as the author knows, at the time of writing no results for an anisotropic medium have been published in which equations of the type (14.71), (14.72) or similar have been used to give computed values of the signal intensity.

14.9. Complex rays. A simple example

The rest of this chapter deals with rays in media for which the losses may be appreciable so that the refractive index $\boldsymbol{n}$ is complex. A method of dealing with this problem when the losses are very small was described in § 10.15 and in § 14.7, equations (14.57)–(14.63). It can be used if $-\text{Im}(\boldsymbol{n})$ is small enough for its effect on the ray path to be neglected, but not for the general case of a lossy medium. The effect of a complex $\boldsymbol{n}$ is that the wave normals, and thence the rays, are refracted so as to have complex directions. Thus it is necessary to consider complex values of the space coordinates x, y, z. The rays must be treated as paths in this complex space. It often happens that a ray path between two real points P and Q is in complex space for almost the whole of its length, and the only points on it that are in real space are the end points P and Q. This idea for radio waves was put forward by Budden and Jull (1964) and a fuller account was given by Jones, R.M. (1970) and by Budden and Terry (1971), and Bennett (1974).

Several other treatments of complex rays have been given. For example Deschamps (1972) introduced the idea of a point source with complex coordinates, and he also studied the energy flux in a ray pencil. Connor and Felsen (1974) were concerned with pulsed waves, or wave packets, and gave examples where it is useful to allow not only the space coordinates but also the time t to take complex values.

When a ray is to be traced through an attenuating medium so as to reach a given receiving point, it is usually necessary to use a 'shooting method', as in § 14.1. Then rays must be traced for successive trial values of the ray direction θ, ϕ at the transmitter. The need to compute with complex variables means that the computing time for any one trial ray is nearly doubled. Since θ and ϕ are now in general complex, their real and imaginary parts comprise four adjustable parameters of the ray, instead of two for real rays. Thus the number of trial complex rays that must be traced is also nearly doubled. The tracing of complex rays has been used for some practical problems, for example radio windows (Budden and Terry, 1971; Budden, 1980; see § 17.8), and ion cyclotron whistlers (Terry 1978; Budden 1983b; see § 13.9). There are ways of partially overcoming some of the complications, but for most practical problems the method is too laborious, and approximate real rays must be used. The importance of complex ray tracing is that the solution it gives is the same as that obtained when the technique of the 'angular spectrum of plane waves' is used, §§ 6.2, 10.1–10.4, and the integrals are evaluated by the method of steepest descents, ch. 9 and §§ 10.2–10.4. For examples see item (7) below and §14.11. The theory in the earlier sections was given in the language of real variables but applies equally well when the coordinates and the refractive index components S_1, S_2, q are complex. Thus the complex ray provides a 'correct' solution that can be used for testing the shorter approximate methods.

Some workers seem to have difficulty in accepting the idea that a ray can go from P to Q without passing through intervening real points. It may therefore be useful first to illustrate the subject of complex rays by a very simple example.

Consider two semi-infinite homogeneous media separated by a plane boundary $z = 0$. The medium where $z < 0$ is free space and it will be called 'medium 1'. The other medium is isotropic with refractive index n and it will be called 'medium 2'. It is required to trace a ray from the source point P, coordinates $x = y = 0, z = -h_1$, to the receiving point Q, coordinates $x = x_2$, $y = 0$, $z = h_2$. See fig. 14.1. In each medium the ray is straight and in the direction of the wave normal, and must lie in the plane $y = 0$. It is in two segments PC and CQ, that make angles θ_1, θ_2 respectively with the z axis. Snell's law gives

$$\sin\theta_1 = n\sin\theta_2 \tag{14.80}$$

and the geometry of the ray path requires that

$$h_1\tan\theta_1 + h_2\tan\theta_2 = x_2. \tag{14.81}$$

The wave is partially reflected at the boundary but the reflected wave is here ignored. Equations (14.80), (14.81) can be converted to an equation of degree 4 in $\tan\theta_1$, but it can be shown that only two of the four solutions are physically acceptable. This subject was studied by Stott (1981) but is not considered further here. If n is real, only one of the physically acceptable solutions has real θ_1 and only this solution will be studied. It corresponds to the ray that would be expected from elementary geometrical optics. Now if n is complex it is easy to see that (14.80), (14.81) can only be satisfied if both θ_1 and θ_2 take complex values. The complex lines PC and CQ in the two media are parts of the two segments of the complex ray from P to Q.

For a ray with given θ_1 in medium 1 the equation of the ray path is

$$x = (z + h_1)\tan\theta_1. \tag{14.82}$$

The boundary is where $z = 0$ and here the ray is refracted. The equation of the ray path in medium 2 is

$$x = h_1 \tan\theta_1 + z\tan\theta_2. \tag{14.83}$$

The equation $y = 0$ is also part of the equation for the ray in both media. In the following discussion it is assumed that always $y = 0$ and the problem is treated as a two-dimensional problem in x and z.

For complex n the two equations (14.82), (14.83) are about the simplest possible example of a complex ray. They exhibit most of the important properties, which will now be summarised. The following list gives examples of the locus of a point that moves along a ray starting from the source point P. In all these examples n is complex.

(1) We start from P, that is $x = 0$, $z = -h_1$ along a ray with real θ_1. The equation is (14.82). If z is kept real, then x is real and the ray is a real ray in the usually understood sense. Let z increase until the value $z = 0$ is reached, that is the point C, fig. 14.1. Here the ray is refracted, and in medium 2 its equation is (14.83). But

Fig. 14.1. Schematic representation of a complex ray PCQ that goes between the two real points P, Q. The point C on the boundary is actually at a complex value of x.

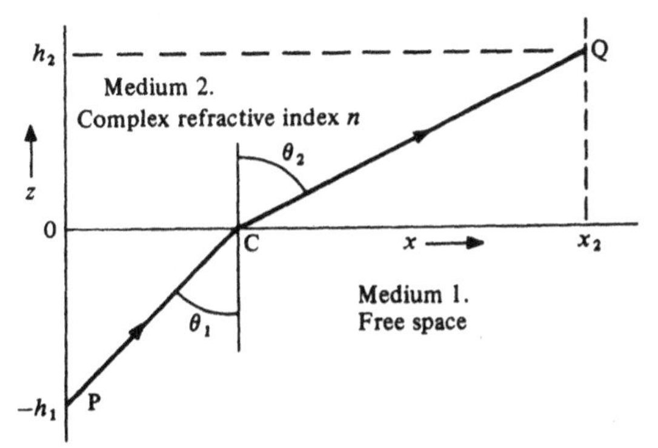

(14.80) shows that θ_2 is complex. If we keep z real and non-zero, x is always complex. This ray can never go through the real point Q.

(2) Start again from P with the same real θ_1, that is on the same ray as in (1), and let z change from its initial value $-h_1$ but let it take a complex value. The equation of the ray is still (14.82) which shows that x also is complex. The point (x, z) is in complex space but it is on the same ray. Equation (14.82) expresses x as a function of z in the usual sense where x and z are complex variables. The function $x(z)$ is represented by a Riemann surface in the four-dimensional space of Re(x), Im(x), Re(z), Im(z). In this example it is a simple plane surface. The real ray of (1), with x and z both real, is a line that is a special cross section of this Riemann surface. In the full sense of the mathematics of complex variables the ray (14.82) is a surface and not a line.

(3) In this example the boundary between the two media is where $z = 0$, and on it z cannot take any other value, real or complex. But x and y can take any values, real or complex. Here we study only the plane $y = 0$. The existence of the boundary does not impose any restriction on x or y.

(4) If the ray in medium 2 is to reach the real point Q, a different ray from that in (1) must be used. Now θ_1 must be complex. If z then increases starting from $-h_1$, and if z is real, x is necessarily complex. When z reaches $z = 0$ the point (x, z) is on the boundary and x has the complex value $h_1 \tan \theta_1$. The complex point $(h_1 \tan \theta_1, 0)$ is the only point that is both on the boundary and on the ray.

(5) Let θ_1 have the same complex value as in (4). Let x increase from zero and remain real. Then z has the complex value $-h_1 + x \cot \theta_1$. There is no real x, except zero, that makes z have a real value. The boundary plane $z = 0$ can never be reached with real x. It is essential to allow x to take complex values.

(6) If we trace the ray by moving along it in small steps of length δs, we have in medium 1

$$\delta x = \delta s \sin \theta_1, \quad \delta z = \delta s \cos \theta_1. \tag{14.84}$$

This ensures that the point remains on the ray (14.82). Now δs is not restricted to real values. We can choose a complex δs to meet any required condition. Since the boundary is at a real value $z = 0$, a natural choice here would be to arrange that z is real so that

$$\arg(\delta s) = -\arg(\cos \theta_1). \tag{14.85}$$

The choice of arg δs is a matter of convenience.

(7) The field at Q from a source at P can be found without recourse to ray theory. The field radiated from P is expressed as an angular spectrum of plane waves. This was done in (10.1) which was for a source at the origin. For a source at P it

becomes

$$F(x, y, z) = \int\int A(S_1, S_2) \exp[-ik\{S_1 x + S_2 y + C(z + h_1)\}] dS_1 dS_2. \quad (14.86)$$

This gives some field component F in medium 1. The time factor $\exp(ikct)$ is omitted. Each plane wave in the integrand is refracted at the boundary, and Snell's law (14.80) applies to it. Thus the field in medium 2 is

$$F(x, y, z) = \int\int B(S_1 S_2) \exp\{-ik(S_1 x + S_2 y + Ch_1 + qz)\} dS_1 dS_2 \quad (14.87)$$

where q is given by (10.12), and $B = TA$. Here $T(S_1, S_2)$ is the transmission coefficient, for example (11.56) or (11.58), at the boundary. Now (14.87) is to be evaluated at the receiving point Q that is $(x_2, 0, h_2)$. The method of double steepest descents §§9.10, 10.2 is used. It is assumed that $B = T(S_1, S_2)\, A(S_1, S_2)$ is slowly varying so that the double saddle point is determined from the exponential only. Thus $\partial/\partial S_2$ and $\partial/\partial S_1$ operating on the exponent must give zero. This leads to

$$S_2 = 0, \quad x_2 - h_1 S_1/C - h_2 S_1 (n^2 - S_1^2)^{-\frac{1}{2}} = 0. \quad (14.88)$$

If now we let

$$S_1 = \sin\theta_1, \quad S_1/n = \sin\theta_2 \quad (14.89)$$

then (14.88) gives (14.81). The predominant wave at Q contains the factor

$$\exp\{-ik(x_2 \sin\theta_1 + h_1 \cos\theta_1 + nh_2 \cos\theta_2)\}. \quad (14.90)$$

But this is exactly the expression that is obtained when the values of θ_1 and θ_2 for the complex ray are used. The complex phase path from P to Q (§ 10.16 and (10.39)) is

$$P = x_2 \sin\theta_1 + h_1 \cos\theta_1 + nh_2 \cos\theta_2. \quad (14.91)$$

This also, when multiplied by k, is the change of complex phase in going from P to Q along any path in the complex ray.

The component plane waves that make up the integrals (14.86) in medium 1 and (14.87) in medium 2 can be thought of as going from P to Q in real space. The ray is the locus of points for which these integrands have double saddle points when (14.89) is satisfied; see § 10.2 after (10.4). This locus is the complex ray, (14.82), (14.83).

(8) The complex ray that goes from P to Q does not go through any intermediate real point. Suppose we select any real point R in medium 2 somewhere between P and Q. There is a complex ray that goes from P to R, but it is a different ray from the one that goes to Q, and it has different complex values of θ and ϕ

(9) Suppose that, instead of being at P, the source is at infinity in the x–z plane, in a direction that makes a real angle Θ with the z axis. Then the wave in medium

1 is a plane wave. The rays are all parallel to the line from the origin to the source, so any one ray has the equation

$$x = z \tan \Theta + x_0. \tag{14.92}$$

Thus for all these rays $\theta_1 = \Theta$ is real. Here x_0 is the value of x at the boundary $z = 0$. It can take any real or complex value, but rays with differing x_0 are different rays; compare (4) above. If after refraction into medium 2 the ray is to go through Q it is necessary that

$$x_0 + h_2 \tan \theta_2 = x_2. \tag{14.93}$$

But $\sin \theta_2 = \sin \theta_1 / n$. Since θ_1 is real, θ_2 is complex. Thus x_0 must be complex. The ray in medium 1 is not in real space but it is still a normal to the incident plane wave.

(10) For more general problems, where the ray is not confined to a plane, it is necessary to let all three coordinates and all three components of n take complex values.

14.10. Real pseudo rays

When losses are very small it is possible at first to ignore $\mathrm{Im}(n)$ and use $\mathrm{Re}(n)$ to trace real rays. The attenuation is then found from the imaginary part of the phase path P that is $\mathrm{Im}(P)$, by integrating $\mathrm{Im}(n)$ along the real ray. This method was described in §§ 10.15, 10.16, 14.7. The question may now be asked: can we use real rays in cases where $\mathrm{Im}(n)$ is not very small?

This question may be examined by using real rays in the simple problem of § 14.9. We therefore suppose that there are rays PC, CQ in fig. 14.1 for which θ_1 and θ_2 have the real values θ_{p1}, θ_{p2}. These rays cannot obey Snell's law (14.80) because n is complex. They will therefore be called 'pseudo rays'. But they can be made to obey a modified form

$$\sin \theta_{p1} = \sin \theta_{p2} \,\mathrm{Re}(n). \tag{14.94}$$

This and (14.81) give real values of θ_{p1}, θ_{p2}. It can be argued that plane waves with normals in the real directions θ_{p1}, θ_{p2} are present in the integrands of (14.86), (14.87) respectively. For these pseudo rays the phase path is

$$P_p = h_1 \sec \theta_{p1} + n h_2 \sec \theta_{p2}. \tag{14.95}$$

Table 14.1 shows some results for the case where medium 2 is a cold isotropic electron plasma with $h_1 = h_2 = \frac{1}{2} x_2$. The values of θ_1, θ_2 were calculated from (14.80), (14.81), P from (14.91), and P_p from (14.95). The real and imaginary parts of all angles are in degrees.

These examples suggest that when $-\arg(n)$ is less than about 10°, the results with real pseudo rays are fairly reliable. For rays in a continuously varying ionosphere this result was confirmed by Budden and Terry (1971) who showed, however,

Table 14.1

X	Z	n	θ_1	θ_2	θ_{p1}	θ_{p2}	P/x_2	P_p/x_2
0.5	0.1	0.711 − 0.035i	34.4 − 3.5i	52.9 + 1.8i	34.5	52.7	1.196 − 0.029i	1.194 − 0.027i
0.5	0.5	0.785 − 0.127i	37.9 − 5.2i	51.2 + 3.3i	35.5	52.2	1.257 − 0.101i	1.213 − 0.014i
10.0	30.0	1.008 − 0.165i	45.5 − 3.1i	44.8 + 3.2i	45.2	44.7	1.425 − 0.116i	1.420 − 0.116i
30.0	30.0	1.086 − 0.460i	53.0 − 8.7i	39.9 + 14.5i	47.2	42.6	1.505 − 0.302i	1.473 − 0.312i
30.0	10.0	1.370 − 1.084i	60.9 − 3.1i	13.5 + 12.4i	52.3	35.3	1.717 − 0.614i	1.656 − 0.664i

that real pseudo rays can give more serious errors in conditions near to penetration of an ionospheric layer. For the last three entries in the table the large values of X and Z are typical for waves of very low frequency, say 10 to 50 kHz in the lower ionosphere. For the last two entries $-\arg(n)$ exceeds 20° and the errors in P_p are more serious. At these frequencies, however, the wavelength is very long and ray theory is rarely used for the ionosphere. It is sometimes used for the magnetosphere, but then Z is much smaller than the values used in the last three entries of the table.

We conclude that the results with real pseudo rays are remarkably good even when $-\arg(n)$ is large, but still it is unwise to use them in cases where there is any doubt, without first checking their accuracy.

14.11. Complex rays in stratified isotropic media

The equations for ray tracing in a stratified ionosphere were derived in §§ 10.2–10.4 and discussed, for an isotropic ionosphere, in § 12.7. Although the discussion was almost entirely in terms of loss-free media and real rays, the equations are still valid when the coordinates and the variables such as n, q, $S = \sin\theta$, take complex values. As an illustration, consider the example used in (12.46). The coordinates are x, y, z with z vertical. The earth's curvature is neglected. There is free space where $0 \leqslant z \leqslant h_0$ and above this $N(z)$ is proportional to $z - h_0$. The plasma is a cold isotropic electron plasma and we now suppose that the collision frequency ν, and thence Z, is independent of height. Then the constant α in (12.46) has a factor $1/(1 - \mathrm{i}Z)$, and is therefore complex. Consider a ray that leaves the transmitter at the origin and returns to a receiver on the ground where $x = D$, $y = 0$, $z = 0$. Thus D is the horizontal range, given by (12.48), and must be real. The equation shows that, for this ray, θ must be complex because α is complex. The first and last parts of the ray are in free space and are straight. They must both go through the plane at the base of the ionosphere where $z = h_0$ is real. Here x takes the complex values $h_0 \tan\theta$ and D–$h_0 \tan\theta$ respectively; compare § 14.9 item (4). The ray path within the ionosphere is given by (12.47). It is a relation between the complex variables x, z and represents a Riemann surface. Within the ionosphere it has no solutions for which x and z are both real.

Some numerical results for complex rays in this and other models of the ionosphere were given by Budden and Terry (1971). They compared them with results obtained with real pseudo rays; see end of § 14.10.

A further application of complex rays can be illustrated by using the same model (12.46) of the ionosphere but now with $Z = 0$, so that α is real. This was used in § 12.7 to study real rays. Equation (12.48) gives the horizontal range D to a receiver on the ground, and it can be shown that the phase path is

$$P = 2h_0 \sec\theta + \frac{2f^2}{\alpha}(\sin^2\theta + \tfrac{1}{3}\cos^2\theta)\cos\theta. \tag{14.96}$$

Now (12.48) can be converted to a cubic equation for $\tan\theta$ with real coefficients. It always has one real solution and it may have three. This was illustrated in fig. 12.9. Consider a value of D that is small enough for the cubic to have only one real solution. Then the receiver is inside a skip distance. The other two solutions of the cubic give complex θ and correspond to complex rays. For these two θs the imaginary parts have opposite signs and it can be shown that the values of $\mathrm{Im}(P)$ also have opposite signs. On the complex ray for which $\mathrm{Im}(P)$ is negative, the signal is attenuated. It is present but with small amplitude. For the complex ray with positive $\mathrm{Im}(P)$ the signal would be enhanced, but this is not possible on energy grounds. This ray therefore cannot be present.

The skip distance is where a caustic surface crosses the ground. Near it the signal amplitude is given by an expression containing an Airy integral function; see (10.67) and § 10.22. On the dark side of the caustic, that is within the skip distance, the asymptotic approximation for the Airy integral function has only a subdominant term that decreases as we move further away from the caustic, and this is the signal from the complex ray with negative $\mathrm{Im}(P)$. The other complex ray with positive $\mathrm{Im}(P)$ would correspond to the absent dominant term. These conclusions can be confirmed, without recourse to ray theory, by using the method of the angular spectrum of plane waves, § 10.2. They were illustrated by Budden and Terry (1971). See also Connor and Felsen (1974).

14.12. Complex rays in anisotropic absorbing media

In an anisotropic medium with losses, the refractive index surfaces are complex surfaces in complex refractive index space. It was shown in § 5.3 by two different methods that the ray has the direction of the normal to the refractive index surface. In the second method, equations (5.29)–(5.31), the principle of stationary phase in an angular spectrum of plane waves was used. It can still be applied for complex $\boldsymbol{n}$, and the condition (5.31) still holds. Here ξ, η, ζ is a vector $\boldsymbol{g}$ parallel to the ray. The phase (5.30) now takes complex values. In general, for a given $\boldsymbol{n}$, the direction of $\boldsymbol{g}$ in (5.31) is complex. The formulae (5.33)–(5.35) for V, v, α are still true and all these quantities are now complex.

The equation of the refractive index surface is the dispersion relation. It may be written in the form (14.8) where (x, y, z) is a point in ordinary space and may be complex. Then the ray $\boldsymbol{g}$ is parallel to the normal, that is to the right-hand expression in (14.9), that is to $\partial D/\partial \boldsymbol{n}$. It is difficult to visualise a complex ray, normal to a complex refractive index surface, in the six-dimensional space of the real and imaginary parts of $\boldsymbol{n}$, and we have to be content with these formulae. On this basis the whole of the theory of the canonical equations of a ray, §§ 14.2–14.7, can be carried through for complex $\boldsymbol{n}$, x, y, z.

The first method used in § 5.3 cannot, however, immediately be taken over for media with losses. It used the time averaged Poynting vector $\mathbf{\Pi}_{av}$ which is necessarily a real vector in real space and cannot have a complex direction. We need to find a complex analogue of $\mathbf{\Pi}_{av}$ that has the direction of the complex ray. It will now be shown that the required vector is the bilinear concomitant vector $\boldsymbol{W}$. Its z component W in a stratified medium has already been used in §7.14, and defined by (7.89). The name comes from its use in the theory of differential equations. The following treatment is for monochromatic waves, that is for 'steady state' conditions.

Consider a fictitious adjoint medium identical with the original medium except that at each point the direction of the superimposed magnetic field is reversed. For an electron plasma this means that the direction cosines of $\boldsymbol{Y}$ are reversed in sign, so that any term containing an odd power of Y as a factor is reversed in sign. For the more general plasma it means that ε_1 and ε_2 are interchanged. Thus for an electron plasma (3.35) shows that the susceptibility matrix $\boldsymbol{M}$ and thence the dielectric constant $\boldsymbol{\varepsilon} = 1 + \boldsymbol{M}$ is transposed. Similarly, for the more general plasma, (3.53) or (3.54) or (3.55) shows that $\boldsymbol{\varepsilon}$ is transposed. Each of these forms is for a particular set of axes but the property still holds when $\boldsymbol{\varepsilon}$ is transformed to use any other set of Cartesian axes. Hence, in the subscript notation of (2.54), (5.27) its adjoint form is given by

$$\bar{\varepsilon}_{ij} = \varepsilon_{ji}. \tag{14.97}$$

In the adjoint medium there is a source of waves identical with the actual source in the original medium. The wave fields in the adjoint medium are the adjoint fields $\bar{\boldsymbol{E}}, \bar{\boldsymbol{\mathscr{H}}}$. An example for a stratified medium was given in § 7.14 item (4). They must satisfy the differential equations that are the adjoint of Maxwell's equations (2.44) for monochromatic waves. These are formed by reversing the sign of k and thence of $\omega = kc$; compare (7.95). In the constitutive relation, however, the sign of ω is not reversed. The dispersion relation at any point is the same in the original and the adjoint medium. Thus the refractive index surfaces and ray directions and ray paths are the same in the two media. For any point Q on a ray in the original medium, each component of the predominant field contains a factor

$$\exp\{\mathrm{i}(\omega t - kP)\} = \exp\left(\mathrm{i}\omega t - \mathrm{i}k\int \mathscr{M}\,\mathrm{d}s\right) \tag{14.98}$$

where P is the phase path, and $\mathscr{M} = n\cos\alpha$ is the ray refractive index; see (5.35), (14.42). Here ds is an element of the ray path, and the integral is evaluated along the ray up to the point Q. In a medium with losses, $\mathrm{Im}(\mathscr{M})$ and $\mathrm{Im}(P)$ are negative and in real space the wave is attenuated as it travels. For the corresponding fictitious ray in the adjoint medium the factor is $\exp\{\mathrm{i}(-\omega t + kP)\}$. Now P is the same in the two cases so that the wave is enhanced as it travels. This wave has no physical existence and is merely a mathematical adjunct. For a loss-free medium and real space, P is real, and the waves are neither attenuated nor enhanced, respectively.

Now, by analogy with (7.97), the bilinear concomitant vector is defined as follows:

$$\boldsymbol{W} = \boldsymbol{E} \wedge \bar{\boldsymbol{\mathscr{H}}} + \bar{\boldsymbol{E}} \wedge \boldsymbol{\mathscr{H}}. \tag{14.99}$$

Its use for radio propagation problems was suggested by Budden and Jull (1964). A full theory was given by Suchy and Altman (1975a).

In the theory of the dispersion relation for a loss-free medium, and in real space, i appears only where it multiplies Y, for example in (3.35) or (4.12). For the more general plasma i appears only where it multiplies $(\varepsilon_1 - \varepsilon_2)$, for example in (3.53) or (4.61). Thus reversal of the direction of the superimposed magnetic field has the same effect as reversing the sign of i. In this case, therefore, the adjoint fields $\bar{\boldsymbol{E}}, \bar{\boldsymbol{\mathscr{H}}}$ are proportional to the complex conjugates $\boldsymbol{E}^*$, $\boldsymbol{\mathscr{H}}^*$, and $\boldsymbol{W}$ is proportional to $\boldsymbol{\Pi}_{av}$ (2.63). But $\boldsymbol{E}^*$, $\boldsymbol{\mathscr{H}}^*$ are not analytic functions of x, y, z, n and so the process of analytic continuation cannot be applied to them. On the other hand $\bar{\boldsymbol{E}}, \bar{\boldsymbol{\mathscr{H}}}$ are analytic functions. They and $\boldsymbol{W}$ can be continued analytically into complex space, and used for media with losses.

The adjoint fields $\bar{\boldsymbol{E}}, \bar{\boldsymbol{\mathscr{H}}}$ have been defined as the fields in the adjoint medium from a source identical to the actual source, but this is not always the best choice. They must satisfy the adjoint of Maxwell's equations which are linear and homogeneous, so that $\bar{\boldsymbol{E}}, \bar{\boldsymbol{\mathscr{H}}}$ can be multiplied by any constant. The choice of this constant is a matter of convenience. This problem has been discussed for a stratified medium in §7.14(1), (7), (8), where the constant was chosen so that, for a real ray in a loss-free part of the medium, $\bar{\boldsymbol{E}}$, $\bar{\boldsymbol{\mathscr{H}}}$ are the same as $\boldsymbol{E}^*$, $\boldsymbol{\mathscr{H}}^*$. The same choice is permissible for a medium that is not plane stratified. But if this applies in one loss-free region, it does not apply for the same ray in a second loss-free region separated from the first by a region with losses, or by a surface where the refractive index shows a resonance. This is important in the theory of resonance tunnelling (Budden, 1979); see § 19.6. The choice of the constant is the problem of the normalisation of $\bar{\boldsymbol{E}}, \bar{\boldsymbol{\mathscr{H}}}$. It does not affect the theory in the rest of this section.

Now it can be shown that $\boldsymbol{W}$ is normal to the refractive index surface. The proof follows exactly the steps from (5.20) to (5.28) but with the following changes:

(a) $\boldsymbol{n}$ is not restricted to be real;

(b) the adjoints $\bar{\boldsymbol{E}}, \bar{\boldsymbol{\mathscr{H}}}$ are used instead of the complex conjugates $\boldsymbol{E}^*, \boldsymbol{\mathscr{H}}^*$;

(c) the adjoints of (5.20), (5.21) are used instead of their complex conjugates, for substituting in (5.24), (5.25);

(d) $\boldsymbol{\Pi}_{av}$ is replaced by $\boldsymbol{W}$.

(e) (14.97) is used in the stage (5.27).

Hence $\boldsymbol{W}$ is in the same direction as the ray. This applies for complex rays and in complex space.

These results have been established for monochromatic waves. Thus the only time dependence is through the factor $\exp(i\omega t)$ for $\boldsymbol{E}$, $\mathscr{H}$, and $\exp(-i\omega t)$ for their adjoints, and (14.99) shows that $\boldsymbol{W}$ is independent of time. From (14.99) together with Maxwell's equations (2.44) and their adjoints it can be shown that

$$\operatorname{div} \boldsymbol{W} = 0 \tag{14.100}$$

so that $\boldsymbol{W}$ is solenoidal and must be a flux of some indestructible quantity that exists in the medium with a density $\mathscr{Q}$ per unit volume. For a loss-free medium $\boldsymbol{W}$ is $4Z_0\boldsymbol{\Pi}_{av}$ so that, when steady state conditions do not hold, div $\boldsymbol{W} = -4Z_0\partial\mathscr{S}/\partial t$ where $\mathscr{S}$ is the total energy per unit volume; § 2.12. In a medium with losses $\mathscr{Q}$ is a complex generalisation of $4Z_0\mathscr{S}$.

The energy flux in a ray pencil in real space and in a loss-free medium was studied in § 10.17 and it was shown that the signal intensity is given by (10.47). This method can now be applied to a medium with losses and to use the flux of $\mathscr{Q}$ in a ray pencil of complex rays. The same expression (10.47) is obtained but it is now complex, so that it cannot represent the signal intensity, which is real. It can be shown, however, that (10.47) can still be used if the ray pencil enters a region of real space where the medium is loss-free. Then the signal intensity is found by taking the modulus of (10.47) and multiplying by $\exp\{2\,\mathrm{Im}(P)\}$ where P is the phase path. The proof is given by Budden and Terry (1971, Appendix A).

In the foregoing discussion an attempt has been made to explain the main properties of complex rays and the bilinear concomitant vector in anisotropic absorbing media. The theory has been carried considerably further in some recent papers. For example Suchy (1982) has studied the vector $\boldsymbol{W}$ where steady state conditions do not hold. Then $\mathscr{Q}$ varies with time t and (14.100) must be replaced by

$$\operatorname{div} \boldsymbol{W} = -\partial\mathscr{Q}/\partial t. \tag{14.101}$$

The theory has also been applied to media with spatial dispersion (Suchy, 1982).

14.13. Reciprocity and nonreciprocity with rays 1. The aerial systems

The reciprocity theorem of electromagnetism is not in general true for systems that include a gyrotropic medium, and so it does not apply to radio signals propagated through the ionosphere or magnetosphere. But in certain special cases there is reciprocity for a radio-communication link. The theory for the general case is discussed in § 18.12. In this and the two following sections it is examined when the approximations of ray theory are used, as in this chapter. The discussion is based on the more detailed treatment of Budden and Jull (1964).

When a signal travels along a ray, the path of the ray is unaltered if the direction of the wave normal is everywhere reversed. This was proved in § 10.5 for a stratified medium. In the more general case it follows from the properties of the refractive index surface, because the direction of the ray is everywhere perpendicular to the local refractive index surface. This has a centre of symmetry at the origin so that if the wave normal is reversed, the ray is then simply antiparallel to its original direction. The attenuation and phase change of a signal travelling along a ray are given by $\exp(-ik\int \mathscr{M}\,ds)$ where ds is an element of the ray path and $\mathscr{M} = \boldsymbol{n}\cos\alpha$ is the ray refractive index (14.42). This still applies when collisions are allowed for so that $\mathscr{M}$ is complex, and it applies for complex rays. But $\mathscr{M}$ is unchanged by a reversal of the ray direction. Thus the ray path, and the attenuation and phase change, are the same for a signal travelling either way along the path. There is an exception to this. It has been assumed that the rays can be described by the eikonal function (14.7), that is the 'phase memory' term, and that this depends on the refractive index vector $\boldsymbol{n}$. It is this term that has been used to find the ray direction, and the ray refractive index $\mathscr{M}$. Cases can occur where this is not accurate enough and another factor must be included in the eikonal. It is called 'additional memory' and is explained in § 16.9. It has the effect that the ray path from A to B is not the same as the path from B to A. It is a small effect especially at high frequencies, and in this chapter it is ignored.

We consider two aerials at fixed points A, B in free space though not necessarily on the ground; one or both could be on space vehicles. Each aerial is a linear electrical system that does not contain nonreciprocal circuit elements. When one aerial is used as a transmitter, a pencil of rays diverges from it. A small number, usually one or two, of these reaches the other aerial, and these are called the 'principal rays'. The divergence of the neighbouring rays affects the amplitude of the received signal, as discussed in §§ 10.17, 14.8. If the other aerial is now used as the transmitter, the principal rays follow the same paths as before, in the opposite direction, but their neighbouring rays follow different paths. The effect on the received signal amplitude of the divergence of the rays in a ray pencil is given, for a stratified medium, by (10.43)–(10.47). These contain integrals (10.43) along the ray path and it is easy to show that they are unchanged if the direction of the ray is reversed. For the more general case the signal amplitude is found by integrating the Buckley equations (14.71), (14.72) along the ray path, and again the result is unchanged if the direction is reversed.

When nonreciprocity occurs, therefore, it has nothing to do with the ray path, nor with the attenuation and phase shift, nor with the divergence of the rays in the pencils. It arises because of the interaction of the waves with the transmitting and receiving aerials, and depends on the polarisations of the waves in the signal, and of the waves that the aerials can radiate.

At any point on a ray within the plasma, the ray direction, the wave normal and the earth's magnetic field are coplanar, §§ 5.3, 5.4, and the plane containing them will be called the 'magnetic plane'. At each point on the ray, therefore, we define a right-handed Cartesian axis system such that any vector, for example the electric intensity $\boldsymbol{E}$, has components E_1, E_2, E_3 where E_3 has the direction of the wave normal for a signal going from A to B, and E_1 is in the magnetic plane. Thus the 1, 2, 3 axes are the same as the x, y, z axes of ch. 4, but not of later chapters. For a signal going from B to A the *same* axes are used so that E_3 is now *opposite* to the wave normal.

In the approximations of ray theory, the fields at a point on the ray are the same as those for a progressive plane wave, ordinary or extraordinary, in a homogeneous medium. For this wave, as in ch. 4, we use n for the refractive index, ρ for the polarisation E_2/E_1, and L for the ratio E_3/E_1 as given by (4.54). Subscripts O, E or superscripts (O), (E) will be attached later when we need to distinguish the ordinary and extraordinary waves. Then for a signal going from A to B, the fields in this axis system are

$$\boldsymbol{E} = (1, \rho, L)E_1, \quad \mathscr{H} = (-n\rho, n, 0)E_1. \tag{14.102}$$

For a ray going from B to A, the sign of n is reversed, but ρ and L are unchanged; see end of § 4.4.

For points outside the plasma the ray is straight. The ray and the wave normal coincide and give the direction of the 3-axis. When a ray emerges into free space, n_{O} and n_{E} become nearly equal since both are near to unity, and there is a cumulative coupling process that causes failure of the ray theory approximations. This is the process of 'limiting polarisation', §§ 17.10, 17.11. Suppose that the emerging wave is an ordinary wave within the plasma. After emerging it is no longer purely ordinary but attains a constant limiting polarisation $\breve{\rho}_{\mathrm{O}}$. Similarly an extraordinary wave attains a limiting polarisation $\breve{\rho}_{\mathrm{E}}$. For the ray through A these values are very close to the values $\rho_{\mathrm{O}}, \rho_{\mathrm{E}}$ for a point P on the ray just within the plasma. The exact position of P is unimportant because, when X is very small, ρ_{O} and ρ_{E} are insensitive to small changes of X. Hence to a good approximation we expect that

$$\breve{\rho}_{\mathrm{O}}\breve{\rho}_{\mathrm{E}} \approx 1 \tag{14.103}$$

from (4.29). On the part AP of the ray in free space, the direction of the 1 axis is parallel to the magnetic plane at P. This ensures that (14.103) is true at A. A similar argument applies for the ray through B.

In discussions of reciprocity we consider two experiments. In the first a voltage V is applied to the terminals of the aerial at A and the current i_1 at the short-circuited terminals of B is measured. In the second the same voltage V is applied to the terminals of the aerial at B and the current i_2 at the short-circuited terminals of the aerial at A is measured. If the electromagnetic fields are entirely within isotropic media, then $i_1 = i_2$, and this is the standard reciprocity theorem. It is discussed further in § 18.12 (see also, for example, Landau and Lifshitz, 1960; Ginzburg, 1970; Collin and Zucker, 1969). It is generally true for full wave solutions and is not restricted to ray theory. When anisotropic media are present the theorem is not generally true, but it may still be true in some special cases. There are other special cases where $i_1 = -i_2$ and this will be called 'antireciprocity'. Examples are given in §§14.15, 18.12.

To apply these ideas we next define certain constants of the aerials when they are used for conveying a signal along a particular path AB in either direction. First let the aerial at A be used as the transmitter, by applying to its terminals a voltage V at the signal frequency. Let E_1, E_2 be the resulting components of $\boldsymbol{E}$ at a point on the ray AB at distance D from A. This point must be in free space, but D must be large enough to ensure that the storage fields of the aerial are negligible compared with the radiation field. Then

$$E_1 = \gamma_A V \cos\psi_A, \quad E_2 = \gamma_A V \sin\psi_A \tag{14.104}$$

where γ_A, ψ_A are constants of the aerial. If it radiates linearly polarised waves, ψ_A is real and gives the plane of $\boldsymbol{E}$, and we shall then say that the aerial is linearly polarised in this plane. But in general the aerial radiates an elliptically polarised wave and ψ_A is complex. Similarly, if the aerial at B is used as the transmitter by applying the voltage V to its terminals, then the fields on the ray BA at distance D from B in free space are

$$E_1 = \gamma_B V \cos\psi_B, \quad E_2 = \gamma_B V \sin\psi_B. \tag{14.105}$$

If the aerial at A is used as a receiver, and the signal reaching it along the ray BA has electric field components E_1', E_2' at A, then the current at the short-circuited terminals at A is

$$i_2 = K\gamma_A(E_1' \cos\psi_A + E_2' \sin\psi_A). \tag{14.106}$$

If the aerial at B is used as a receiver, and if the electric field components there are E_1'', E_2'' then the current at the short-circuited terminals at B is

$$i_1 = K\gamma_B(E_1'' \cos\psi_B + E_2'' \sin\psi_B). \tag{14.107}$$

The constant K is a property of the aerials only, and not of the fields. It can be proved that the same K must be used in the two cases, by supposing that the path AB is entirely in free space and using the standard reciprocity theorem.

14.14. Reciprocity and nonreciprocity with rays 2. The electric and magnetic fields

Consider now one magnetoionic component, which will be taken to be the ordinary wave. The whole of the following discussion applies if the roles of the ordinary and extraordinary waves are reversed. For a point on a ray going from A to B, the field components are, from (14.102)

$$\boldsymbol{E} = (1, \rho_O, L_O)E_1^{(O)} \quad \mathscr{H}(-n_O\rho_O, n_O, 0)E_1^{(O)}. \tag{14.108}$$

The components of the corresponding adjoint field were defined in §14.12. They have the same ratios as the fields in a fictitious medium in which the direction of the earth's magnetic field is reversed, and the wave normal and ray directions are unchanged. Hence it can be shown from (4.54) and (4.20) that n_O, L_O are unchanged, and the sign of ρ_O is reversed (see problem 7.4). Then for the adjoint fields

$$\bar{\boldsymbol{E}} = (1, -\rho_O, L_O)\bar{E}_1^{(O)}, \quad \bar{\mathscr{H}} = (n_O\rho_O, n_O, 0)\bar{E}_1^{(O)}. \tag{14.109}$$

The bilinear concomitant vector (14.99) can now be found and is

$$\boldsymbol{W} = -2n_O(L_O, 0, \rho_O^2 - 1)E_1^{(O)}\bar{E}_1^{(O)}. \tag{14.110}$$

Its (complex) length is

$$W = -2n_O\{L_O^2 + (\rho_O^2 - 1)^2\}^{\frac{1}{2}}E_1^{(O)}\bar{E}_1^{(O)}. \tag{14.111}$$

It was shown in § 14.12 that $\boldsymbol{W}$ everywhere has the same direction, in general complex, as the ray.

Now (14.100) shows that div $\boldsymbol{W} = 0$, so that the flux of $\boldsymbol{W}$ across any closed surface is zero. In particular, if the surface is a very narrow tube whose sides are everywhere parallel to rays emanating from A, there is no flux of $\boldsymbol{W}$ across the sides so that the total flux for any cross section must be the same at all points of the tube. Such a tube will be called a 'ray tube'. Let a denote the area of any cross section normal to the axis of the tube. Then Wa is constant along a ray, whence it follows from (14.111) that

$$n_O a\{L_O^2 + (\rho_O^2 - 1)^2\}^{\frac{1}{2}}E_1^{(O)}\bar{E}_1^{(O)} = \text{constant} \tag{14.112}$$

where an unimportant factor -2 is omitted. Now, for any point on the ray, $E_1^{(O)}$ contains a factor $\exp\{i(\omega t - kP_A)\}$, from (14.98), where P_A is the phase path from A to the point considered. Similarly $\bar{E}_1^{(O)}$ contains a factor $\exp\{i(-\omega t + kP_A)\}$. These factors cancel in (14.112). Let

$$F_O = E_1^{(O)}n_O^{\frac{1}{2}}\{L_O^2 + (\rho_O^2 - 1)^2\}^{\frac{1}{4}} \tag{14.113}$$

and similarly with $\bar{F}_O$, $\bar{E}_1^{(O)}$ for the adjoints. Then $F_O\bar{F}_O$ is a complex generalisation of the signal intensity Π used in § 14.8. Now on the ray

$$F_O a^{\frac{1}{2}} = \text{constant} \times \exp(-ikP_A) \tag{14.114}$$

where the factor $e^{i\omega t}$ is omitted as usual.

Let a voltage V be applied to the terminals of the transmitting aerial at A. The fields E_1, E_2 at a distance D away in free space, on the ray, are given by (14.104) and must be resolved into components that will become ordinary and extraordinary after they enter the plasma. Then

$$E_1 = E_1^{(O)} + E_1^{(E)}, \quad E_2 = E_2^{(O)} + E_2^{(E)} = \rho_{OA}E_1^{(O)} + E_1^{(E)}/\rho_{OA} \tag{14.115}$$

where ρ_{OA} is the limiting value $\breve{\rho}_O$ of ρ_O on the ray through A, and (14.103) has been used. In the free space $L_O = 0$, $n_O = 1$. Hence at the same point, from (14.113) and (14.104)

$$\begin{aligned} F_O &= (\rho_{OA}E_2 - E_1)(\rho_{OA}^2 - 1)^{-\frac{1}{2}} \\ &= \gamma_A V(\rho_{OA}\sin\psi_A - \cos\psi_A)(\rho_{OA}^2 - 1)^{-\frac{1}{2}} \end{aligned} \tag{14.116}$$

which is the value at a distance D from A. The signal now travels along the ray AB through the anisotropic plasma and reaches B. To find F_O at B we must multiply by $J_O \exp\{-ik(P_{AB}-D)\}$ where P_{AB} is the phase path from A to B. J_O is determined solely by the geometry of the ray system and allows for the divergence of the rays from A. Hence at B, F_O is

$$F_{OB} = J_O \gamma_A V(\rho_{OA} \sin\psi_A - \cos\psi_A)(\rho_{OA}^2 - 1)^{-\frac{1}{2}} \exp\{-ik(P_{AB}-D)\} \qquad (14.117)$$

and from (14.113) the fields E_1, E_2 at B are

$$E_1'' = F_{OB}(\rho_{OB}^2 - 1)^{-\frac{1}{2}}, \quad E_2'' = \rho_{OB} F_{OB}(\rho_{OB}^2 - 1)^{-\frac{1}{2}}. \qquad (14.118)$$

These are now inserted in (14.107) to give the current i_1 at the short-circuited terminals of the aerial at B:

$$i_1 = M_O(\rho_{OA} \sin\psi_A - \cos\psi_A)(\cos\psi_B + \rho_{OB} \sin\psi_B) \qquad (14.119)$$

where

$$M_O = J_O K \gamma_A \gamma_B V(\rho_{OA}^2 - 1)^{-\frac{1}{2}}(\rho_{OB}^2 - 1)^{-\frac{1}{2}} \exp\{-ik(P_{AB}-D)\}. \qquad (14.120)$$

Next let the aerial at B be used as the transmitter. The divergence factor J_O in (14.117) is the same as before for the reasons given in § 14.13, and the exponential factor is also the same. The only difference is that the subscripts A and B are interchanged, so M_O is unchanged. The current at the short-circuited terminals of the aerial at A is

$$i_2 = M_O(\cos\psi_A + \rho_{OA} \sin\psi_A)(\rho_{OB} \sin\psi_B - \cos\psi_B). \qquad (14.121)$$

14.15. Reciprocity and nonreciprocity with rays 3. Applications

Suppose that there is a communication link between A and B with only one ordinary ray. Then the condition $i_1 = i_2$ for reciprocity gives, from (14.119), (14.121):

$$\rho_{OA} \sin\psi_A \cos\psi_B = \rho_{OB} \cos\psi_A \sin\psi_B. \qquad (14.122)$$

Similarly the condition $i_1 = -i_2$ for antireciprocity gives:

$$\cos\psi_A \cos\psi_B = \rho_{OA}\rho_{OB} \sin\psi_A \sin\psi_B. \qquad (14.123)$$

In many practical cases the limiting polarisations are very nearly circular. This is because ρ_{OA}, ρ_{OB} are insensitive to the angle Θ between the earth's magnetic field and the wave normal, except when it is close to $\pm\frac{1}{2}\pi$. If this transverse case is excluded, it is found that ρ_{OA}, ρ_{OB} are both close to $\pm$i. For ionospheric ray paths between ground stations that are fairly close together, Θ is acute for the ray near A and obtuse for the ray near B or vice versa. Then ρ_{OA}, ρ_{OB} are equal with opposite signs.

If $\rho_{OA} = -\rho_{OB}$, the reciprocity condition (14.122) is

$$\psi_A + \psi_B = 0. \qquad (14.124)$$

Thus there is reciprocity when both aerials are linearly polarised in the magnetic plane, and the condition still holds if the aerials are now rotated through equal angles but in opposite senses about the local direction of the ray AB.

If ρ_{OA} and ρ_{OB} satisfy the further condition $\rho_{\mathrm{OA}} = -\rho_{\mathrm{OB}} = \pm \mathrm{i}$, the antireciprocity condition (14.123) is

$$\psi_{\mathrm{A}} - \psi_{\mathrm{B}} = \tfrac{1}{2}\pi. \tag{14.125}$$

Thus there is antireciprocity when the aerials are linearly polarised, one in the magnetic plane and the other at right angles to it. The condition still holds if the aerials are rotated from these positions through equal angles and in the same sense about the local direction of the ray AB. These conclusions all apply equally well if the one ray is an extraordinary ray. For further discussion see Budden and Jull (1964).

It often happens that a signal goes over two different ray paths simultaneously, and the commonest case is when one ray is ordinary and the other extraordinary. For the second ray the subscript O in (14.119)–(14.121) must be replaced by E. The two rays may leave the aerials in slightly different directions but in practical cases we expect $\gamma_{\mathrm{A}}\,\psi_{\mathrm{A}}\,\gamma_{\mathrm{B}}\,\psi_{\mathrm{B}}$ to be the same for both rays. Then the contributions to i_1, i_2 from the two signals are

$$\left.\begin{aligned} i_1 &= M_{\mathrm{O}}(\rho_{\mathrm{OA}}\sin\psi_{\mathrm{A}} - \cos\psi_{\mathrm{A}})(\cos\psi_{\mathrm{B}} + \rho_{\mathrm{OB}}\sin\psi_{\mathrm{B}}) \\ &\quad + M_{\mathrm{E}}(\rho_{\mathrm{EA}}\sin\psi_{\mathrm{A}} - \cos\psi_{\mathrm{A}})(\cos\psi_{\mathrm{B}} + \rho_{\mathrm{EB}}\sin\psi_{\mathrm{B}}), \\ i_2 &= M_{\mathrm{O}}(\cos\psi_{\mathrm{A}} + \rho_{\mathrm{OA}}\sin\psi_{\mathrm{A}})(\rho_{\mathrm{OB}}\sin\psi_{\mathrm{B}} - \cos\psi_{\mathrm{B}}) \\ &\quad + M_{\mathrm{E}}(\cos\psi_{\mathrm{A}} + \rho_{\mathrm{EA}}\sin\psi_{\mathrm{A}})(\rho_{\mathrm{EB}}\sin\psi_{\mathrm{B}} - \cos\psi_{\mathrm{B}}). \end{aligned}\right\} \tag{14.126}$$

In rarer cases there could be three or more ray paths (see, for example, § 12.7 and fig. 12.9) and further terms could be added to deal with this.

These results display the following properties, which apply for any number of rays.

(a) There is reciprocity when $\psi_{\mathrm{A}} = \psi_{\mathrm{B}} = 0$, that is when both aerials are linearly polarised in the magnetic plane.

(b) There is reciprocity when $\psi_{\mathrm{A}} = \psi_{\mathrm{B}} = \tfrac{1}{2}\pi$, that is when both aerials are linearly polarised at right angles to the magnetic plane.

(c) There is antireciprocity when $\psi_{\mathrm{A}} = 0$, $\psi_{\mathrm{B}} = \pm\tfrac{1}{2}\pi$, or when $\psi_{\mathrm{B}} = 0$, $\psi_{\mathrm{A}} = \pm\tfrac{1}{2}\pi$, that is when both aerials are linearly polarised, one in the magnetic plane and the other at right angles to it.

These three examples are the results of a more general treatment of reciprocity and can be proved without recourse to ray theory; see §18.12.

An important case that often occurs in practice is for medium and high frequencies when there are two rays from A to B, one ordinary and the other extraordinary. The signal is assumed to be continuous, not pulsed or otherwise modulated. The two waves are of roughly equal amplitudes which means that the attenuations are small and about the same for the two paths. The phase paths, however, are different and change with time because of slow movements or changes of the ionosphere. Hence the difference of the phase paths changes with time so that the two contributions to the received signal have a variable phase difference, and alternately reinforce each other or interfere destructively. This gives a very regular variation of the signal called 'polarisation fading' because its occurrence depends on the different polarisations of the ordinary and extraordinary waves. To study it, assume that

$$\rho_{OA} = -\rho_{OB} = \mathrm{i}, \quad \rho_{EA} = -\rho_{EB} = -\mathrm{i} \tag{14.127}$$

which is typical for propagation between two ground stations at temperate latitudes in the same hemisphere. The divergence factors J_O, J_E are different but it would be expected that they are roughly equal in most practical cases. Then M_O, M_E are the same except for the exponential factor. Let

$$M_O = M_E \exp\{2\mathrm{i}\phi(t)\} \tag{14.128}$$

where $2\phi(t)/k$ is the difference between the two phase paths. Then (14.126) gives

$$\begin{aligned} i_1 &= -M_E[\exp\{\mathrm{i}(2\phi - \psi_A - \psi_B)\} + \exp\{\mathrm{i}(\psi_A + \psi_B)\}], \\ i_2 &= -M_E[\exp\{\mathrm{i}(2\phi + \psi_A + \psi_B)\} + \exp\{-\mathrm{i}(\psi_A + \psi_B)\}]. \end{aligned} \tag{14.129}$$

Let both aerials be linearly polarised so that ψ_A and ψ_B are real. Then

$$|i_1| = M|\cos(\psi_A + \psi_B - \phi)|, \quad |i_2| = M|\cos(\psi_A + \psi_B + \phi)| \tag{14.130}$$

where M is a real constant. The cosine factors give the polarisation fading. Now let

$$\psi_A + \psi_B = \tfrac{1}{2}r\pi \tag{14.131}$$

where r is any integer or zero. Then the fading for the direction A → B is exactly in step with that for B → A. This occurs for the two cases r even which gives reciprocity, cases (a) and (b) above, and r odd which gives antireciprocity, case (c). Alternatively let

$$\psi_A + \psi_B = (\tfrac{1}{2}r \pm \tfrac{1}{4})\pi. \tag{14.132}$$

Now the cosine factors in (14.130) are exactly out of step since one is a maximum when the other is zero. The result of using other values of ψ_A and ψ_B can be found from (14.130) in a similar way. These and other effects have been confirmed by observations; see Jull and Pettersen (1962, 1964).

These results have an alternative explanation. The two rays reaching a receiver have been assumed to be roughly circularly polarised with opposite senses and equal amplitudes. They therefore combine to give a linearly polarised wave whose plane undergoes Faraday rotation when the phase difference changes; see §13.7. The direction of this rotation is in opposite senses about the ray direction AB for signals going in the two directions. The fading occurs as the angle between the plane of polarisation of the wave and of the aerial varies. This angle can be expressed in terms of $\psi_A, \psi_B, \phi(t)$ and the above results then follow. See Budden and Jull (1964) for further details.

PROBLEMS 14

14.1. Show by a counter example that, for different rays through the same point, the directions of the effective stratification vector (defined near end of § 14.3) are not necessarily the same.

Hint: A possible counter example is as follows. Consider a collisionless electron plasma in which the electron concentration is independent of the coordinates, but the magnetic field varies in space so that $Y_x = Y_0 2^{-\frac{1}{2}} + \alpha z$, $Y_z = Y_0 2^{-\frac{1}{2}} + \alpha x, Y_y = 0$. The dispersion relation (4.65), (4.66) can be written

$$D \equiv (1 - X - Y^2)n^4 - 2(1 - X)^2 n^2 + (1 - X)^3 + Y^2\{(2 - X)n^2 - 1 + X\}$$
$$+ X(n^2 - 1)(\boldsymbol{Y}\cdot\boldsymbol{n})^2 = 0.$$

Consider two ordinary rays through the origin with wave normals parallel (a) to the x-axis, (b) to the z-axis. The angles between $\boldsymbol{Y}$ and the wave normals are the same so the two refractive indices n are the same. The two ray directions are in the n_x–n_z plane and make equal and opposite angles with $\boldsymbol{Y}$. Show that the angles between the effective stratification vector and the x axis are (a) $\arctan(n^2 - 1)(1 - X)/(n^2 - 1 + X)$, (b) $\arctan(n^2 - 1 + X)/\{(n^2 - 1)(1 - X)\}$ and so not in general the same.

14.2. It is required to find a ray that leaves a transmitter at the origin and goes to a receiver at the point $(x_0, 0, 0)$. For a particular ray at the transmitter the polar angles θ, ϕ of the wave normal are θ_i, ϕ_i and this ray goes through the point $(x_i, y_i, 0)$. This applies for three different rays, $i = 1$ to 3. By linear interpolation, that is by a two-dimensional form of the rule of false position (see (14.1)), show that estimates of the required values θ_0, ϕ_0, are

$$\theta_0 \approx \begin{vmatrix} \theta_1 & x_1 & y_1 & 1 \\ \theta_2 & x_2 & y_2 & 1 \\ \theta_3 & x_3 & y_3 & 1 \\ 0 & x_0 & 0 & 1 \end{vmatrix} \times \begin{vmatrix} x_1 & y_1 & 1 \\ x_2 & y_2 & 1 \\ x_3 & x_3 & 1 \end{vmatrix}^{-1}$$

and a similar expression with ϕs instead of θs.

14.3. The dispersion relation may be written $\omega(x, y, z; \kappa_x, \kappa_y, \kappa_z) = \Omega$ where $\boldsymbol{\kappa} = \omega\boldsymbol{n}/c$ and Ω is the angular frequency of the wave. The first three canonical ray tracing equations are $\mathrm{d}\boldsymbol{r}/\mathrm{d}\tau = A\partial\omega/\partial\boldsymbol{n}$. Let $\omega A\,\mathrm{d}\tau/c = \mathrm{d}t$. Use these equations to prove that the group velocity $\boldsymbol{\mathscr{U}} = \partial\omega/\partial\boldsymbol{\kappa} = \mathrm{d}\boldsymbol{r}/\mathrm{d}t$ and show that $1/A = \boldsymbol{n}\cdot\boldsymbol{\mathscr{U}}\omega/c$.

14.4. For a high frequency communication link between two points A and B there is only one ray, that goes through the ionosphere. Its limiting polarisations at A and B are ρ_A, ρ_B where $\rho_A = -\rho_B = \mathrm{i}$. The aerials at A and B are both designed to emit only this wave. Show that the signal emitted from A cannot be received at B, nor that emitted from B at A. The aerial at B is now changed so that it can receive the

signal from A. Show that now a signal from A can be received at B, but a signal from B cannot be received at A. Discuss how to design the aerials so that communication in both directions is possible, whatever may be the values of ρ_A, ρ_B.
(Note: to attain the required condition $E_2/E_1 = \mathrm{i}$ at A in (14.104) it is necessary that $\psi_A \rightarrow \mathrm{i}\infty$, but $\gamma_A \cos\psi_A$, $\gamma_A \sin\psi_A$ tend to bounded limits).

15

Full wave solutions for isotropic ionosphere

15.1. Introduction

This and the remaining chapters are mainly concerned with problems in which the approximations of ray theory cannot be made, so that a more detailed study of the solutions of the governing differential equations is needed. The solution of this type of problem is called a 'full wave solution'. The subject has been studied almost entirely for plane stratified media. When the medium is not plane stratified, for example when the earth's curvature is allowed for, a possible method is to use an 'earth flattening' transformation that converts the equations into those for an equivalent plane stratified system; see §§ 10.4, 18.8. See Westcott (1968, 1969) and Sharaf (1969) for examples of solutions that do not use such a transformation.

This chapter is written entirely in terms of a horizontally stratified ionosphere and for frequencies greater than about 1 kHz, so that the effect of positive ions in the plasma may be neglected. The earth's curvature also is neglected.

The need for full wave solutions arises when the wavelength in the medium is large so that the electric permittivity changes appreciably within one wavelength and the medium cannot be treated as slowly varying (§§ 7.6, 7.10). Thus it arises particularly at very low frequencies, say 1 to 500 kHz. But even at higher frequencies the wavelength in the medium is large where the refractive index n is small, and some full wave solutions are therefore often needed for high frequencies too.

In this chapter the earth's magnetic field is neglected. This means that the differential equations to be studied are of the second order, for example (7.6), (7.68). The standard differential equations of theoretical physics are nearly all of the second order. They have been very thoroughly investigated and the number of papers and books on them is extremely large. Many of the simpler problems of radio propagation can be reduced to a study of these standard forms. The particular standard differential equation to be used depends on two functions, the electron height distribution $N(z)$ and the collision frequency $\nu(z)$. In most of the examples studied here ν is independent of the height z. For an example with a variable ν see problem 15.1. Since $N(z)$ must conform to the requirements of the standard differential equations, the forms that it can take are restricted, but still a very wide range of functions $N(z)$ can be studied. Two important uses of these standard solutions are (1) the study of some physical phenomena, for example partial penetration and reflection, §§ 15.4, 15.10, 15.17, and the various forms of tunnelling, §§ 15.9, 17.6–17.9, 19.4, 19.6; (2) for testing whether computer programs intended for more general functions $N(z)$, $\nu(z)$, are free from errors.

The coordinate system used here is the same as in ch. 6 onwards, so that z is height. The wave normals are assumed to be in the plane $y = 0$ which is called the plane of incidence. This chapter deals with the problem where the incident wave enters the ionosphere from the free space below it. It is assumed to be a plane wave with its wave normal at an angle θ to the z axis, and $S = \sin\theta$, $C = \cos\theta$. The variable q used throughout this chapter is defined by (7.2).

The differential equations for the wave fields in a horizontally stratified ionosphere were derived in § 7.2 and it was shown that they separate into two independent sets, (7.3) leading to (7.6), and (7.4) leading to (7.68). The first of these is for horizontal polarisation and most of this chapter applies to it. The other is for vertical polarisation, equation (7.68), and this is more complicated because of the term in $\mathrm{d}\mathscr{H}_y/\mathrm{d}z$. It is dealt with in §§ 15.5–15.7 where it is shown that, in some conditions, results for horizontal polarisation can be applied also for vertical polarisation.

Many of the methods and results in this chapter have been given by Jacobsson (1965) who studied the reflection of light from isotropic stratified films.

15.2. Linear electron height distribution

The case where $N(z)$ within the ionosphere is a linear function of z has already been dealt with in ch. 8. The differential equation for horizontal polarisation was (8.5) and it was transformed to the standard equation (8.7) which is the Stokes equation. It uses the independent variable ζ given by (8.6). This method will now be applied to the model ionosphere described in §§ 12.3, 12.7 in which there is free space below the level $z = h_0$, and above this X, proportional to $N(z)$, is given by

$$X = p(z - h_0), \quad q^2 = C^2 - \frac{p}{1 - \mathrm{i}Z}(z - h_0) \tag{15.1}$$

where p is a real constant. It is assumed that the collision frequency and thence Z is a constant. Then in (8.2), (8.5), (8.6) the height z_0 where q is zero is

$$z_0 = h_0 + (1 - \mathrm{i}Z)C^2/p, \tag{15.2}$$

the constant a in (8.2) is

$$a = p/(1 - \mathrm{i}Z), \tag{15.3}$$

and (8.6) gives

$$\zeta = \left(\frac{k^2 p}{1 - \mathrm{i}Z}\right)^{\frac{1}{3}} \{z - h_0 - (1 - \mathrm{i}Z)C^2/p\}. \tag{15.4}$$

The new independent variable ζ is thus a new measure of height, in which the origin is taken where $q = 0$, that is at the height $z = z_0$ (15.2), which is complex if $Z \neq 0$. The scale is changed by the complex factor

$$g = \{k^2 p/(1 - \mathrm{i}Z)\}^{\frac{1}{3}} \tag{15.5}$$

and since this is a cube root it may be chosen in three different ways. In § 8.20, for the collisionless plasma $Z = 0$, it was chosen to be real and positive. We now choose the value that goes continuously into a real positive value when $Z \to 0$. Hence

$$\arg g = \tfrac{1}{3}\arctan Z. \tag{15.6}$$

The change of scale includes a rotation in the complex z plane through an angle arg g, which must be less than $\frac{1}{6}\pi$.

A solution of (8.7) must now be chosen that satisfies the physical conditions at great heights, that is when z is real, positive and large. Then $|\zeta|$ is large and $\arg\zeta \to \arg g$ so that $\arg\zeta$ is within the range $0 \leqslant \arg\zeta < \frac{1}{6}\pi$. The asymptotic approximation of the solution $\mathrm{Ai}(\zeta)$ is given by (8.52) which contains the factor $\exp(-\frac{2}{3}\zeta^{\frac{3}{2}})$ and $\zeta^{\frac{3}{2}}$ has a positive real part and a positive imaginary part. Hence this factor represents a wave whose amplitude decreases as $|\zeta|$ and $|z|$ increase, and whose phase is propagated upwards. Therefore $\mathrm{Ai}(\zeta)$ is an acceptable solution. The asymptotic approximation for $\mathrm{Bi}(\zeta)$ is given by (8.56) and includes a term with the factor $\exp(+\frac{2}{3}\zeta^{\frac{3}{2}})$. This represents a wave whose amplitude increases as $|\zeta|$ and $|z|$ increase, and whose phase is propagated downwards. It can be shown that the energy in this wave flows downwards, and it could not therefore be produced by a wave incident on the ionosphere from below. Hence the correct solution is proportional to $\mathrm{Ai}(\zeta)$. Notice that when collisions are neglected, $\arg g = 0$, the asymptotic approximation for $\mathrm{Ai}(\zeta)$ at great heights resembles an evanescent wave, § 2.14, since ζ is real and there is no variation of phase with height. When collisions are included there is some upward phase propagation since ζ is no longer real.

The horizontal component $\mathscr{H}_x$ of the magnetic field is found from the first equation (7.3), which comes from Maxwell's equations. Above the level $z = h_0$ the fields are therefore given by

$$E_y = K\,\mathrm{Ai}(\zeta), \quad \mathscr{H}_x = -\mathrm{i}\frac{gK}{k}\mathrm{Ai}'(\zeta) \tag{15.7}$$

where K is a constant. Below the level $z = h_0$ is free space so that $q = C$ is a constant. Then two independent solutions of (7.6) for E_y are $\exp\{\mp \mathrm{i}kC(z - h_0)\}$. The upper sign gives the obliquely upgoing wave, that is the incident wave, and the lower sign gives the downgoing, reflected wave. The solution of (7.3), (7.6) below the level $z = h_0$

is then

$$E_y = \exp\{-ikC(z-h_0)\} + R\exp\{ikC(z-h_0)\}$$
$$\mathscr{H}_x = -C\exp\{-ikC(z-h_0)\} + RC\exp\{ikC(z-h_0\} \quad (15.8)$$

where R is the reflection coefficient for $z = h_0$ as the reference level; § 11.2. Now the boundary conditions for the horizontal components E_y, $\mathscr{H}_x$ require that the values of each on the two sides of the boundary plane must be the same. Hence (15.7), (15.8) with $z = h_0$ give two equations from which K may be eliminated to give

$$R = \frac{Ck\,\mathrm{Ai}(\zeta_0) - ig\,\mathrm{Ai}'(\zeta_0)}{Ck\,\mathrm{Ai}(\zeta_0) + ig\,\mathrm{Ai}'(\zeta_0)} \quad (15.9)$$

where, from (15.4), (15.5):

$$\zeta_0 = -(Ck/g)^2. \quad (15.10)$$

If collisions are neglected, ζ_0 is real and the numerator and denominator of (15.9) are complex conjugates, so that $|R| = 1$. The reflection is therefore complete which was to be expected since all the energy in the incident wave should be reflected. The phase difference at $z = h_0$ between the incident and reflected waves is arg R.

The functions Ai and Ai′ in (15.9) may be replaced by their asymptotic approximations (8.53), (8.55), for $\frac{2}{3}\pi < \arg\zeta_0 \leqslant \pi$ from (15.10), if $|\zeta_0|$ is large enough, that is if the gradient p in (15.1) is small and C is not too small. This is equivalent to the condition that the W.K.B. solutions of (7.6) must be good approximations just above the boundary $z = h_0$, whence (8.29) gives as the condition

$$|\zeta_0| \gg 1. \quad (15.11)$$

Then (15.9) gives

$$R \approx \mathrm{i}\exp\{-\tfrac{4}{3}\mathrm{i}(-\zeta_0)^{\frac{3}{2}}\} = \mathrm{i}\exp\{-\tfrac{4}{3}\mathrm{i}(Ck/g)^3\}. \quad (15.12)$$

An expression for R, based on the W.K.B. solutions, was given at (7.152). For a reference level at $z = h_0$, the lower limit of the integral in it must be changed from 0 to h_0. Then on using q from (15.1) in the integral, the modified (7.152) gives the same expression (15.12).

The agreement between (15.12) and the exact value (15.9) is good even when (15.11) is not satisfied. For example if collisions are neglected, ζ_0 is real and negative, and both (15.12) and (15.9) give $|R| = 1$. Then arg R is greater for (15.12) when $-\zeta_0 > 0.5$. The difference is less than 2° when $-\zeta_0 \geqslant 2$; it has a maximum value of about 11.4° when $-\zeta_0 = 1$, and it is 6° when $-\zeta_0 = 0.63$ and $-25°$ when $-\zeta_0 = 0.2$.

15.3. Reflection at a discontinuity of gradient

Electromagnetic waves are reflected at the bounding surface between two media with different refractive indices, and §§ 11.8–11.12 were devoted to this topic. There is then a discontinuity in the refractive index. It is now of interest to ask whether reflection can occur at a surface where the refractive index is continuous, but the gradient of the refractive index is discontinuous.

A problem where there was such a discontinuity of gradient was studied in § 15.2. Below the level $z = h_0$ is free space with $N(z) = 0$, $\mathrm{d}N/\mathrm{d}z = 0$, and above it N is proportional to $z - h_0$ so that $N(z)$ is continuous but there is a discontinuity of $\mathrm{d}N/\mathrm{d}z$. The expression (15.9) was deduced for the reflection coefficient of the whole ionosphere. In §§ 7.18, 7.19 it was suggested that reflection of radio waves occurs near the level $z = z_0$ where $q = 0$, and in earlier chapters this level has been called a 'level of reflection'. If there is also some reflection from the discontinuity at $z = h_0$ then (15.9) must be the result of combining these two reflections. Hence we must now ask whether (15.9) includes any contribution from the level $z = h_0$.

The field immediately above $z = h_0$ is given by (15.7) with $\zeta = \zeta_0$. If $|\zeta_0|$ is large enough the functions Ai, Ai′ may be replaced by their asymptotic approximations (8.53), (8.55). Each of these contains two terms of which one is a downgoing wave. Some or all of this is transmitted through the discontinuity and gives a downgoing wave below it, but this is not part of the reflection at $z = h_0$ and must be excluded. We therefore need to consider a different field configuration in which there is only an upgoing wave just above the level $z = h_0$.

First suppose that the gradients of $N(z)$, $n(z)$, $q(z)$ just above $z = h_0$ are so small that the W.K.B. solutions may be used. For an upgoing wave the W.K.B. solution satisfies (7.62) with the $-$ sign, so that the wave admittance (11.36) is

$$A_{22} = -\mathscr{H}_x/E_y = q. \tag{15.13}$$

Immediately above the boundary, $q = C$ so that $A_{22} = C$ is exactly the same as for an upgoing progressive wave in the free space below the boundary. For these two upgoing waves, above and below the boundary, the boundary conditions are satisfied and so there is no reflected wave. We conclude that if the W.K.B. solutions are good approximations on both sides of a discontinuity of gradient of $N(z)$, the reflection coefficient is zero to the same degree of approximation.

Next suppose that $\mathrm{d}N/\mathrm{d}z$ above the discontinuity is so large that the W.K.B. solutions cannot be used there. This means that there is no field configuration that can be regarded as a progressive wave. It is therefore not possible to define a purely upgoing wave in the region above $z = h_0$. Thus the problem of finding the reflection coefficient has no meaning. The reflection coefficient of the whole ionosphere is still given by (15.9) and this is influenced by the presence of the discontinuity of gradient. For example it would be changed if the discontinuity of gradient was 'smoothed out' by replacing the sharp angle in the $N(z)$ curve near $z = h_0$ by an arc extending over a small range of z. But still the reflection coefficient (15.9) cannot be separated into a part from the discontinuity and a part from the rest of the ionosphere.

The reflection process associated with the zero of q at $z = z_0$ does not occur at this one point only but must be thought of as occurring in a region of the complex z plane surrounding it and extending out to where the W.K.B. solutions can take over. If there is a discontinuity within this region the reflection coefficient is modified, but the effects associated with the discontinuity and with the zero of q cannot be separated.

An example is given in fig. 15.6 where there is a discontinuity of $\mathrm{d}N/\mathrm{d}z$ at the base of the ionosphere. It shows evidence that the reflected wave contains two components that can interact and give an interference pattern when the frequency is varied.

15.4. Piecewise linear models

Two simple models of the ionosphere were of interest in the early days of the theory of radio reflections from the ionosphere, before computers were available. They are sketched in fig. 15.1(a, b). They were studied by Hartree (1929). The Airy integral functions were not then well known, and Hartree's solutions were given in terms of Bessel functions of order 1/3; see § 8.7. The reflection coefficients for these two models are still of some historical interest and their derivation is useful as an illustration of methods. A third model, fig. 15.1(c), has been studied by B. Jeffreys (1956) for the application to wave mechanics, but it could equally well be used as a model of the ionosphere.

We consider only horizontally polarised waves, and electron collisions are neglected. For the model of fig. 15.1(a) the base of the ionosphere is at $z = h_0$. In the height range $h_0 \leqslant z \leqslant h_1$, X, proportional to $N(z)$, and q^2 are given by (15.1) with $Z = 0$. For $z \geqslant h_1$, X and $N(z)$ are constant. Thus

$$q = \begin{cases} C^2 & \text{for } z \leqslant h_0, \\ C^2 - p(z - h_0) & \text{for } h_0 \leqslant z \leqslant h_1, \\ C^2 - p(h_1 - h_0) = q_1^2, \text{say,} & \text{for } h_1 \leqslant z. \end{cases} \tag{15.14}$$

The constant q_1 is chosen to be either negative imaginary, or positive.

Fig. 15.1. Piecewise linear models of the ionosphere. The abscissa is the electron concentration $N(z)$ and the ordinate is height z above the ground.

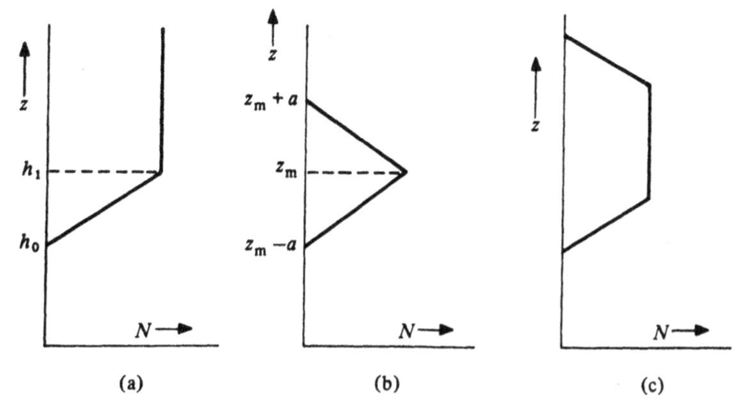

If $h_1 - h_0$ is zero there is then a sharp boundary between two homogeneous media. The use of a non-zero $h_1 - h_0$ has the effect of removing the sharpness of the boundary, and thereby modifying the reflection coefficient as given by the Fresnel formula (11.57). This is one reason why the study of this model is of interest.

For $z > h_1$ there can only be an upgoing wave, or an evanescent wave whose amplitude decreases with increasing z, so that here, from (11.36)

$$A_{22} = q_1. \tag{15.15}$$

This is independent of z and is the value at the boundary $z = h_1$.

For $h_0 \leqslant z \leqslant h_1$ the function $q^2(z)$ is linear, and E_y satisfies the Stokes equation (8.7) with ζ given by (15.4) with $Z = 0$. Then at the boundary $z = h_1$

$$\zeta = \zeta_1 = (k^2 p)^{\frac{1}{3}}(h_1 - h_0 - C^2/p) \tag{15.16}$$

and at the boundary $z = h_0$

$$\zeta = \zeta_0 = -(k^2 p)^{\frac{1}{3}} C^2/p. \tag{15.17}$$

Between these boundaries E_y is a linear combination of any two independent solutions $\mathscr{S}_A(\zeta)$, $\mathscr{S}_B(\zeta)$ of (8.7) and $\mathscr{H}_x$ is found from (7.3). Hence

$$E_y = \mathscr{S}_A(\zeta) + K\mathscr{S}_B(\zeta), \quad \mathscr{H}_x = -\mathrm{i}w\{\mathscr{S}'_A(\zeta) + K\mathscr{S}'_B(\zeta)\}, \tag{15.18}$$

where K is a constant and $w = (p/k)^{\frac{1}{3}}$. The wave admittance (11.36) is then

$$A_{22} = \mathrm{i}w\{\mathscr{S}'_A(\zeta) + K\mathscr{S}'_B(\zeta)\}/\{\mathscr{S}_A(\zeta) + K\mathscr{S}_B(\zeta)\}. \tag{15.19}$$

For $z = h_1$, $\zeta = \zeta_1$, this is the same as (15.15) so that

$$K = -\{\mathrm{i}q_1\mathscr{S}_A(\zeta_1) + w\mathscr{S}'_A(\zeta_1)\}/\{\mathrm{i}q_1\mathscr{S}_B(\zeta_1) + w\mathscr{S}'_B(\zeta_1)\}. \tag{15.20}$$

For $z = h_0$, $\zeta = \zeta_0$, (15.19) is the same as A_{22} in the free space below the boundary, so that the reflection coefficient R_{22} with reference level $z = h_0$ can be found from the third equation (11.50). Let

$$\left.\begin{matrix}\Delta_+ \\ \Delta_-\end{matrix}\right\} = \begin{vmatrix} \mathrm{i}q_1\mathscr{S}_A(\zeta_1) + w\mathscr{S}'_A(\zeta_1) & \mathrm{i}q_1\mathscr{S}_B(\zeta_1) + w\mathscr{S}'_B(\zeta_1) \\ \pm\mathrm{i}C\mathscr{S}_A(\zeta_0) + w\mathscr{S}'_A(\zeta_0) & \pm\mathrm{i}C\mathscr{S}_B(\zeta_0) + w\mathscr{S}'_B(\zeta_0) \end{vmatrix}. \tag{15.21}$$

Then it can be shown that

$$R_{22} = \Delta_+/\Delta_-. \tag{15.22}$$

The choice of the functions $\mathscr{S}_A$ and $\mathscr{S}_B$ is a matter of convenience in calculation. A fairly obvious choice would be $\mathscr{S}_A(\zeta) = \mathrm{Ai}(\zeta)$, $\mathscr{S}_B(\zeta) = \mathrm{Bi}(\zeta)$. Most large computers have library subroutines for calculating these functions and their derivatives.

Some special cases of (15.22) may be noted. Suppose that in the range $h_0 \leqslant z \leqslant h_1$ the W.K.B. approximations can be used in some part of the range.

(a) W.K.B. approximations good throughout the range

This can only happen if q is everywhere positive. Then the W.K.B. solutions are good on both sides of the discontinuities of gradient of q at $z = h_0$ and $z = h_1$. It was shown in § 15.3 that there is no appreciable reflection in these conditions. Hence the reflection coefficient of the ionosphere is negligibly small. It can be shown that if $\mathscr{S}_A$, $\mathscr{S}_B$ are the two W.K.B. solutions (8.27), (8.28), one column of Δ_+, (15.21) is a multiple of the other.

(b) W.K.B. approximations good at $z = h_1$ but q has a zero below this

In this case q_1 is negative imaginary so that the wave where $z > h_1$ is evanescent. Just below $z = h_1$ the W.K.B. solution is the corresponding evanescent approximation of the solution of the Stokes equation, that is the subdominant term. Hence the solution here is simply $\mathrm{Ai}(\zeta)$ which is the same as if the discontinuity at $z = h_1$ did not exist. The problem reduces to that solved in § 15.2, for $Z = 0$, and the reflection coefficient is (15.9). If the W.K.B. approximations are good also at $z = h_0$, the reflection coefficient can be further simplified and is given by (15.12).

(c) Sharp boundary limit

If $h_1 - h_0$ tends to zero it can be shown that (15.22) for R_{22} reduces to the Fresnel formula (11.57) for a sharp boundary.

Some further properties of the model (15.14), fig. 15.1(a), were given by Hartree (1929) and by Budden (1961a).

The model of fig. 15.1(b) is of interest because it simulates a rather crude form of symmetric ionospheric layer with free space below and above it. The plasma frequency at the maximum of this layer is the penetration frequency f_p. The theory is more complicated because there are now three discontinuities of $\mathrm{d}N/\mathrm{d}z$ where the boundary conditions must be imposed. It has been given in full by Hartree (1929) in terms of Bessel functions of order 1/3, and by B. Jeffreys (1942, 1956) in terms of Airy integral functions. Jeffreys also considered a symmetrical trapezoidal form, fig. 15.1(c), for the function corresponding to $N(z)$. This is still more complicated because it has four discontinuities of $\mathrm{d}N/\mathrm{d}z$. Her papers were concerned with wave mechanics and potential barriers but the physics is the same as for the radio problem.

The formula for R_{22} for the $N(z)$ of fig. 15.1(b) is too complicated to be worth repeating here but it is given, in the same notation as used here, by Budden (1961a). For normal incidence and for frequencies sufficiently less than f_p the reflection is almost total so that $|R_{22}| \approx 1$. For frequencies sufficiently greater than f_p the wave penetrates the layer and $|R_{22}| \approx 0$. For frequencies in a range near f_p there is partial penetration and partial reflection. Curves showing how $|R_{22}|$ depends on frequency were given by Budden (1961a). They are similar in general form to those of fig. 15.5 which are for a more realistic model of the ionospheric layer. When the thickness of the layer, $2a$ in fig. 15.1(b), is small, the transition of $|R_{22}|$ from 1 to 0 occurs over a large range of f/f_p. For a thicker layer with large $2a$ the transition is more abrupt. This same behaviour can be seen in fig. 15.5.

15.5. Vertical polarisation at oblique incidence 1. Introductory theory

The term 'vertical polarisation' was explained in §7.2. It refers to obliquely incident waves in which the magnetic field has only a horizontal component. The wave normal is in the x–z plane, so that $\mathscr{H}_y$ is the only non-zero component of $\mathscr{H}$. The electric field $\boldsymbol{E}$ has both a vertical component E_z and a horizontal component E_x.

The differential equations for vertical polarisation were formulated in §7.12. One form is (7.68) in which

$$q^2 = n^2 - S^2 \tag{15.23}$$

where n is the refractive index. It shows how $\mathscr{H}_y$ depends on the height z, and it has a term in $\mathrm{d}\mathscr{H}_y/\mathrm{d}z$. It may be reduced to its 'normal form', that is a form without a first derivative term, by a change of dependent variable. Let

$$V = \mathscr{H}_y/n \tag{15.24}$$

so that (7.68) gives

$$\frac{\mathrm{d}^2 V}{\mathrm{d}z^2} + k^2(M^2 - S^2)V = 0 \tag{15.25}$$

where

$$M^2 = n^2 + \frac{1}{2k^2n^2}\frac{\mathrm{d}^2(n^2)}{\mathrm{d}z^2} - \frac{3}{4k^2n^4}\left\{\frac{\mathrm{d}(n^2)}{\mathrm{d}z}\right\}^2. \tag{15.26}$$

Equation (15.25) should be compared with (7.6) for horizontal polarisation, in which (15.23) appears instead of $M^2 - S^2$, and E_y instead of V. Thus the refractive index n in (7.6) is now replaced by M which is therefore called the 'effective' refractive index in (15.25). Similarly let

$$Q^2 = M^2 - S^2. \tag{15.27}$$

Then Q is called the effective value of q.

On entering the ionosphere with z increasing, suppose that $N(z)$ increases monotonically from zero, and neglect the effect of collisions. Then at first n and Q are positive, but they decrease until Q is zero where n^2 is still positive. Just above this level Q^2 is negative and the waves are evanescent. When z and $N(z)$ increase still further, a level is reached where $n^2 = 0$ and Q^2 and M^2 are infinite. This is the new feature of the equations for vertical polarisation. It is this infinity of the effective refractive index, sometimes called a resonance, that we mainly wish to study. It is on the real z axis if $Z = 0$, and near the real axis if Z is small.

By making a suitable choice of the function $N(z)$ it is possible to arrange that the differential equation (15.25) can be transformed to a standard form whose solutions can be expressed in terms of known functions. See for example Wait (1962, ch. VI), Burman and Gould (1964), Heading (1969), Westcott (1962b, c, 1969a, 1970). For some of these choices of $N(z)$ the refractive index n is never zero at any point on or near the real z axis, so the effect of the infinity of Q is not clearly evident. A possible method that does not have this disadvantage was suggested by van Duin and Sluijter (1980), who proposed using a transformation that converts (15.25) into Heun's equation; see § 15.12.

Now suppose that $N(z)$ is linear and Z is constant, as in (15.1), and let

$$\xi = z - h_0 - (1 - \mathrm{i}Z)/p \tag{15.28}$$

so that

$$n^2 = 1 - \frac{p(z - h_0)}{1 - \mathrm{i}Z} = \frac{-p\xi}{1 - \mathrm{i}Z}, \quad Q^2 = q^2 - \frac{3}{4k^2\xi^2}. \tag{15.29}$$

Then Q^2 is infinite where $\xi = 0$ and this is a regular singular point of the differential equation (15.25). It can in some cases be close to the point where $Q^2 = 0$, and then the function $\{Q(z)\}^2$ is not even approximately linear there. The presence of the singularity means that, at the reflection level, (15.25) does not approximate to the Stokes equation, so that the formula (7.152) for the reflection coefficient cannot be used. The reflection process is more complicated for vertical than for horizontal polarisation.

Equation (15.26) is not the only possible form for the square of the 'effective' refractive index. By using other dependent variables, for example nE_x/q, a different form is obtained. But for vertical polarisation it is never possible to find a dependent variable that makes $M = n$, except in the special case $S = 0$, that is for vertical incidence.

Now let

$$\zeta = \left(\frac{k^2 p}{1 - \mathrm{i}Z}\right)^{\frac{1}{3}} \xi, \quad B = S^2 \left\{\frac{(1 - \mathrm{i}Z)k}{p}\right\}^{\frac{2}{3}}. \tag{15.30}$$

Then with ζ as independent variable, the differential equation (7.68) is

$$\frac{\mathrm{d}^2\mathscr{H}_y}{\mathrm{d}\zeta^2} - \frac{1}{\zeta}\frac{\mathrm{d}\mathscr{H}_y}{\mathrm{d}\zeta} - (\zeta + B)\mathscr{H}_y = 0. \tag{15.31}$$

This is not one of the standard differential equations of theoretical physics, but it must now be studied. In the special case of vertical incidence, $B = 0$, it becomes

$$\frac{\mathrm{d}^2\mathscr{H}_y}{\mathrm{d}\zeta^2} - \frac{1}{\zeta}\frac{\mathrm{d}\mathscr{H}_y}{\mathrm{d}\zeta} - \zeta\mathscr{H}_y = 0. \tag{15.32}$$

Now let $\mathscr{S}(\zeta)$ be any solution of the Stokes equation (8.7) so that $\mathscr{S}'' = \zeta\mathscr{S}$. Then it is easy to show that $\mathscr{H}_y = \mathscr{S}'(\zeta)$ is a solution of (15.32). For example one solution is $\mathrm{Ai}'(\zeta)$. Reflection at vertical incidence can equally well be studied for waves polarised with $\boldsymbol{E}$ parallel to the y axis. The solution is then as in § 15.2 where it was shown that the magnetic field is proportional to $\mathrm{Ai}'(\zeta)$, (15.7). Hence the solution $\mathrm{Ai}'(\zeta)$ of (15.32) satisfies the physical conditions for waves vertically incident from below.

It can be shown by the method of § 8.16 that the Stokes multiplier of (15.32) is $-\mathrm{i}$. For vertical polarisation the reflection coefficient of the ionosphere is R_{11} and (11.16) shows that it is the ratio of the magnetic fields of the downgoing and upgoing waves. Thus the formula for R_{11} analogous to (7.152) is

$$R_{11} = -\mathrm{i}\exp\left\{-2\mathrm{i}k\int_0^{z_0} n\,\mathrm{d}z\right\}. \tag{15.33}$$

If a reflection coefficient was defined as the ratio of the electric fields, a change of sign would be necessary; see end of § 11.8. Then (15.33) would give the same formula (7.152) as was derived for horizontal polarisation in §§ 7.19, 8.20.

15.6. Vertical polarisation 2. Fields near zero of refractive index

The study of the reflection of vertically polarised waves really requires a full treatment of the differential equation (15.31) similar to that given in ch. 8. for the Stokes equation. The following account is an outline only. It is based in part on a treatment given by Försterling and Wüster (1951). The equation (15.31) has also been studied by Denisov (1957), Hirsch and Shmoys (1965), and White and Chen (1974). The object of this section is to study the behaviour of the fields near the point $\zeta = 0$ where the refractive index n is zero. If $B \neq 0$, this is not the same as the 'reflection' point.

A solution of (15.31) can be found as a series in ascending powers of ζ, thus

$$v_1(\zeta) = \zeta^{\beta}(1 + a_1\zeta + a_2\zeta^2 + \cdots). \tag{15.34}$$

This is substituted in (15.31) and the coefficients of successive powers of ζ are equated to zero. This shows that if $B \neq 0$ the only possible value of β is 2, and the coefficients $a_1, a_2 \cdots$ can be found. Further $a_1 = 0$, so that

$$v_1(\zeta) = \zeta^2 + a_2\zeta^4 + a_3\zeta^5 + \cdots. \tag{15.35}$$

A second solution may be found (see, for example, Whittaker and Watson, 1927, § 10.32) as follows. Let

$$v_2(\zeta) = Kv_1(\zeta)\ln\zeta + 1 + b_1\zeta + b_2\zeta^2 + \cdots. \tag{15.36}$$

Substitute in (15.31) and equate coefficients of powers of ζ to zero. This gives

$$b_1 = 0, \quad K = \tfrac{1}{2}B, \quad b_3 = 1/3. \tag{15.37}$$

The value of b_2 is arbitrary and may be taken as zero. If it is not zero, the effect is simply to add a multiple of $v_1(\zeta)$ on to $v_2(\zeta)$. Then b_4, b_5 etc. may be found. For vertical incidence B is zero. Then K is zero and the logarithm term disappears from (15.36). The solutions $v_1(\zeta)$, $v_2(\zeta)$ are then simply the derivatives of the two series in (8.8).

The general solution of (15.31) is

$$\mathscr{H}_y = A_1 v_1(\zeta) + A_2 v_2(\zeta). \tag{15.38}$$

For vertical incidence the required solution is proportional to $\mathrm{Ai}'(\zeta)$ and (8.9) shows that

$$A_1 = \tfrac{1}{2} 3^{-\frac{2}{3}}/(-\tfrac{1}{3})!, \quad A_2 = -3^{-\frac{1}{3}}/(-\tfrac{2}{3})!, \quad \text{for } B = 0. \tag{15.39}$$

Hence A_2 is not zero when $B = 0$ and it cannot drop discontinuously to zero when $B \neq 0$. The solution (15.38) for oblique incidence must contain a multiple of $v_2(\zeta)$ including the logarithm term in (15.36)

The field components E_x and E_z may be found by applying Maxwell's equations (7.4) to (15.38). This shows that E_x contains a term $A_2 B \ln \zeta$ and E_z contains a term A_2/ζ. These are both infinite when $\zeta = 0$, that is when $n = 0$. This is usually above the reflection level where $q = 0$.

In the ionosphere the electrons always make some collisions so that Z is never exactly zero. Consequently we cannot have $n = 0$ at any real height z. For example (15.29) shows that z is complex when $n^2 = 0$. But if Z is small the condition $n^2 = 0$ may hold at a point very close to the real axis in the complex z plane. At real heights near this point the terms $\ln \zeta$ and $1/\zeta$ in E_x, E_z respectively may be very large. This can happen only for vertical polarisation at oblique incidence. It does not occur for horizontal polarisation. Computed curves showing how $\mathscr{H}_y$, E_x and E_z depend on height near $\zeta = 0$ were given by Booker, Fejer and Lee (1968), for vertical polarisation and small Z.

The large field components E_x, E_z would mean that the simple linear form (3.12) of the equation of motion of the electrons cannot be used. A more complicated non-linear equation would be needed, leading to a modification of the foregoing treatment to allow for non-linear effects in strong fields. The following results might be expected.

The field component E_z imparts vertical motions to the electrons. When E_z is large these motions are large, and, within one cycle of oscillation of the field, any one electron moves to different levels where E_z is different. Thus one electron encounters very different fields E_z within one cycle. The vertical force to which it is subjected does not, therefore, vary harmonically with time and the electron motion is not simple harmonic. The motion could be Fourier analysed into a series of frequencies equal to the wave frequency and its harmonics. The theory of this process has been developed by Feinstein (1950) and by Försterling and Wüster (1951) who showed that harmonics of the wave frequency can be generated in this way near a level where $n = 0$. This process would be most marked when Z is very small, for then the point where $n = 0$ is near the real axis in the complex z plane. Some energy must go into the harmonics generated, and this must come from the original wave which is therefore attenuated. The effect is thus similar to that of additional damping of the electrons' motions. Even if $Z = 0$, some energy would be removed from the wave as harmonics, and this should lead to a reflection coefficient with modulus less than unity. This is discussed in the following section. When the earth's magnetic field is allowed for the theory must be modified, but there is still a point in the complex z plane where the governing differential equations have a singularity analogous to the one at $n = 0$, in this section. This topic is discussed in § 19.5. Near this singularity the wave fields are very large and should still lead to the generation of harmonics.

The author does not know of any observations of harmonics of the wave frequency that could be attributed to a mechanism of this kind. This is in spite of the fact that in recent years many experiments have been done with radio waves of very large power, sufficient to cause heating and substantial artificial modification of the ionospheric plasma (see for example *Radio Science*, special issue Nov. 1974; Fejer, 1975, 1979).

15.7. Vertical polarisation 3. Reflection coefficient

In § 7.19 the phase integral form (7.152) was given for the reflection coefficient of the ionosphere. It was based on solutions of the differential equation (7.6) for horizontal polarisation, so that the reflection coefficient R in (7.152) is R_{22}, defined by (11.17). It was justified by an application of the principle of uniform approximation in Langer's (1937) form, given in § 8.20. It can be used if the function $q^2(z)$ in (7.6) has an isolated zero at the 'reflection level' where $q = 0$. This is the same as the requirement that $q^2(z)$ shall be sufficiently near linear at this level. An approximate upper limit to the curvature was given by the criterion (8.75). The first factor i in (7.152) came from the second term of (8.73), which was derived from the asymptotic approximation for $\mathrm{Ai}(\zeta)$. Thus this factor is the Stokes multiplier for the Stokes equation (8.7).

We must now enquire whether the same method can be used for vertical polarisation to give R_{11} (11.16). The differential equation (15.25) is of the same form as (7.6). It contains the function $Q^2(z)$ (15.27), instead of $q^2(z)$. Clearly the method would be expected to work if $Q^2(z)$ satisfies the criterion (8.75) at the 'reflection' level where $Q = 0$. If n^2 and Q^2 are given by (15.29) near the reflection level, it can be shown that this criterion leads to

$$S^3 \gg (\tfrac{3}{2})^{\frac{3}{4}} \left| \frac{p}{k(1 - \mathrm{i}Z)} \right| \quad \text{that is } |B| \gg 1.36 \tag{15.40}$$

where B is given by (15.30). This shows that for oblique incidence with large enough

S the separation of the zero of Q and the singular point where $n = 0$, $Q \to \infty$, in the complex z plane, is great enough for the zero of Q to be treated as 'isolated'. The reflection coefficient may then be written

$$R_{11} = \lambda(B) \exp\left(-2ik \int_0^{z_0} Q \, \mathrm{d}z \right) \tag{15.41}$$

where z_0 is the point where $Q = 0$. Here $\lambda(B)$ is the Stokes multiplier, and if (15.40) is satisfied it is approximately $+\mathrm{i}$. It was shown at (15.33) that, when S and B are zero, $\lambda(B) = -\mathrm{i}$ which is the Stokes multiplier for (15.32). For intermediate values of B, (15.41) might still be used provided that $\lambda(B)$ is given a suitable value.

The required $\lambda(B)$ is the Stokes multiplier for the differential equation (15.31). There is no closed expression for it, but it can be computed by the following method. The W.K.B. solutions of (15.31) can be found by the method of § 7.12. Equation (7.74) shows that they contain the factors $\exp\{\pm \int^{\zeta} (\zeta + B)^{\frac{1}{2}} \, \mathrm{d}\zeta\}$. They are to be used where $\zeta \gg B$ so the square root is expanded and only positive powers of ζ need be retained. Hence the W.K.B. solutions are

$$\mathscr{H}_y \sim \zeta^{-\frac{1}{4}} \exp\{\pm(\tfrac{2}{3}\zeta^{\frac{3}{2}} + B\zeta^{\frac{1}{2}})\}. \tag{15.42}$$

When $|\zeta|$ is very large the factor $\exp(\pm\frac{2}{3}\zeta^{\frac{3}{2}})$ predominates. It occurs also in the asymptotic approximations to the solutions of the Stokes equation. Hence the Stokes lines and anti-Stokes lines for (15.31) are the same as for the Stokes equation; see fig. 8.6.

At great heights $|\zeta|$ is large and (15.29), (15.30) show that $0 \leqslant \arg \zeta < \frac{1}{6}\pi$. Here the required solution contains only the subdominant term, representing either an upward travelling wave, or a field whose amplitude decreases as z increases. Well below the level of reflection $\frac{2}{3}\pi < \arg \zeta \leqslant \pi$. In going from great to small heights the Stokes line at $\arg \zeta = \frac{2}{3}\pi$ is crossed and there are two terms in the asymptotic approximation below the level of reflection. We now require a connection formula between the two asymptotic approximations. This is of the form (8.64) and may be written

$$\zeta^{-\frac{1}{4}}\{\exp(-\tfrac{2}{3}\zeta^{\frac{3}{2}} - B\zeta^{\frac{1}{2}}) + \lambda(B)\exp(\tfrac{2}{3}\zeta^{\frac{3}{2}} + B\zeta^{\frac{1}{2}})\} \leftrightarrow \zeta^{-\frac{1}{4}} \exp(-\tfrac{2}{3}\zeta^{\frac{3}{2}} - B\zeta^{\frac{1}{2}}) \tag{15.43}$$

where $\lambda(B)$ is the Stokes multiplier for (15.31). It is equal to $-\mathrm{i}$ when $B = 0$. For other values of B it must be computed. This was done by a numerical integration of (15.31). For details of the computing methods see Budden (1961a, § 16.16 and Appendix).

Some results are given in fig. 15.2. They show how $\lambda(B)$ goes from $-\mathrm{i}$ to $+\mathrm{i}$ when $|B|$ goes from 0 to large values. If electron collisions are neglected, B is real and the integral in (15.41) is also real so that the exponential has modulus unity. But fig. 15.2 shows that $|\lambda(B)|$ is in general less than unity and hence $|R_{11}| < 1$. The smallest value is about 0.72 and occurs when $B \approx 0.45$. This means that not all the incident energy is reflected from the ionosphere, even when there is no physical mechanism for converting the energy of the waves into heat. This might be because some energy is converted into waves of the harmonic frequencies, as mentioned at the end of § 15.6. A more likely explanation, however, is that this problem cannot be treated as a 'steady state' problem. The incident wave must first be switched on, and after that the energy accumulates near the level where $n = 0$. If this energy cannot be absorbed, it continues to accumulate near the level where $n = 0$, and equilibrium is never attained. See § 19.5 for further discussion.

In practice there must be some collisions so that the plasma is most strongly heated where the electric field is very large, § 15.6, that is near where $n^2 = 0$. The possibility of heating a plasma in this way was considered by Freidberg, Mitchell, Morse and Rudsinski (1972).

In the lower ionosphere the dependence of X on height z is roughly exponential so that the gradient $p = \mathrm{d}X/\mathrm{d}z$ (15.1), near where $X = 1$, is not strongly dependent

Fig. 15.2. The complex plane showing values of the Stokes multiplier λ for the differential equation (7.68). Numbers by the curves show the values of $|B|$ marked with bars, and of arg B in degrees marked with arrows. Only values of arg B from $-60°$ to $0°$ are of interest when finding reflection coefficients, but some results for arg $B = \pm 120°$ are included for mathematical interest.

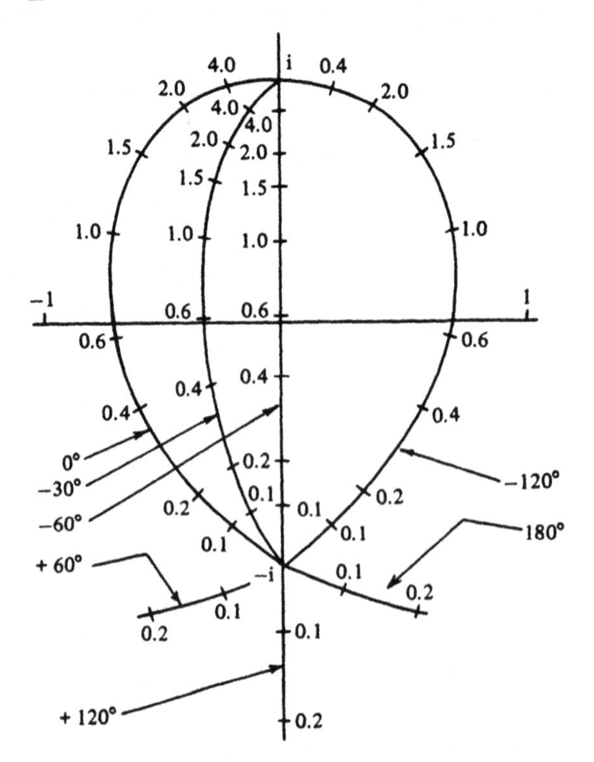

on frequency. If we take $p \approx 0.6\,\mathrm{km}^{-1}$ as typical for all frequencies, this gives the following values of B, from (15.30) with $Z = 0$:

Frequency	16 kHz	2 MHz	6 MHz
Approx. value of B	$1.6\,S^2$	$40\,S^2$	$80\,S^2$

The value $B = 0.45$ therefore corresponds to an angle of incidence of about 30° at 16 kHz, or 6° at 2 MHz.

The above theory is modified when the earth's magnetic field is allowed for; see §§ 19.4–19.6.

15.8. Exponential electron height distribution

Radio waves of very low frequency, less than about 500 kHz, are reflected low down in the ionosphere usually between 70 and 100 km. Here the electron concentration $N(z)$ increases with height z. If the E-layer is a Chapman layer, § 1.5, then in its lower part $N(z)$ is roughly exponential. The actual $N(z)$ is more complicated than this, see fig. 1.3, but still an exponential $N(z)$ is very useful for studying some general features of the reflection of low frequency radio waves. It also has the advantage that there are no discontinuities of gradient $\mathrm{d}N/\mathrm{d}z$ that can affect the solution. For horizontal polarisation and constant collision frequency the reflection coefficient R_{22} can be expressed in closed form (15.59). For vertical polarisation there is no corresponding simple solution. The exponential distribution has been studied by Westcott (1962b, d), Wait (1962) and others.

If N increases exponentially with height z, then X, proportional to N, is given by

$$X = \mathrm{e}^{\alpha z}. \tag{15.44}$$

It is convenient to choose the origin of z where $X = 1$. It is important to notice that this origin depends on frequency f, since X is inversely proportional to f^2 (3.6). It is assumed that Z is a constant. From (4.9), (7.2)

$$q^2 = C^2 - \frac{\mathrm{e}^{\alpha z}}{1 - \mathrm{i}Z}. \tag{15.45}$$

Then the differential equation (7.6) for E_y is

$$\frac{\mathrm{d}^2 E_y}{\mathrm{d}z^2} + k^2\left(C^2 - \frac{\mathrm{e}^{\alpha z}}{1 - \mathrm{i}Z}\right)E_y = 0. \tag{15.46}$$

Now let

$$\zeta = \frac{2\mathrm{i}k}{\alpha}(1 - \mathrm{i}Z)^{-\frac{1}{2}} \exp\left(\tfrac{1}{2}\alpha z\right) \tag{15.47}$$

where the factor $(1 - \mathrm{i}Z)^{-\frac{1}{2}}$ is chosen so that its argument is positive and less than $\frac{1}{2}\pi$, and let

$$\nu = 2\mathrm{i}kC/\alpha. \tag{15.48}$$

Throughout this section ν has this meaning and must not be confused with the ν used elsewhere for the electron collision frequency. Then (15.46) becomes

$$\frac{d^2E_y}{d\zeta^2}+\frac{1}{\zeta}\frac{dE_y}{d\zeta}+\left(1-\frac{\nu^2}{\zeta^2}\right)E_y=0 \tag{15.49}$$

which is Bessel's equation of order ν. A solution must now be found that represents an upgoing wave only, at great heights. When z is very large (15.45) gives

$$q\approx -\mathrm{i}(1-\mathrm{i}Z)^{-\frac{1}{2}}\exp(\tfrac{1}{2}\alpha z) \tag{15.50}$$

where the minus sign has been chosen to ensure that q has a positive real part and a negative imaginary part. Then

$$k\int_0^z q\,dz=-\zeta. \tag{15.51}$$

At a great height the upgoing W.K.B. solution of (15.46) contains the exponential of $-\mathrm{i}$ times (15.51), that is $e^{+\mathrm{i}\zeta}$. Similarly the downgoing W.K.B. solution contains the factor $e^{-\mathrm{i}\zeta}$ and this cannot appear in the required solution.

Two independent solutions of (15.49) are $H_\nu^{(1)}(\zeta)$ and $H_\nu^{(2)}(\zeta)$ where Watson's (1944) notation is used for Bessel functions of the third kind, sometimes called Hankel functions. The order ν of these functions here is purely imaginary (15.48). The appearance of a complex order is perhaps unusual in physical problems, but it does not affect the definition of the Hankel functions as given by Watson. Now (15.47) shows that when z is positive

$$\tfrac{1}{2}\pi\leqslant\arg\zeta<\pi \tag{15.52}$$

and for this range the asymptotic forms, based on formulae given by Watson (1944, p. 201) are

$$H_\nu^{(1)}(\zeta)\sim(2/\pi\zeta)^{\frac{1}{2}}\exp\{-\mathrm{i}(\tfrac{1}{2}\nu\pi+\tfrac{1}{4}\pi)\}e^{\mathrm{i}\zeta}, \tag{15.53}$$

$$H_\nu^{(2)}(\zeta)\sim(2/\pi\zeta)^{\frac{1}{2}}\exp\{\mathrm{i}(\tfrac{1}{2}\nu\pi+\tfrac{1}{4}\pi)\}(e^{-\mathrm{i}\zeta}+2\mathrm{i}\,e^{\mathrm{i}\zeta}\cos\nu\pi). \tag{15.54}$$

These apply for

$$-\tfrac{1}{2}\pi\leqslant\arg\zeta\leqslant\tfrac{3}{2}\pi. \tag{15.55}$$

Bessel's equation (15.49) has anti-Stokes lines where $\arg\zeta=0,\ \pm\pi$, and Stokes lines where $\arg\zeta=\pm\frac{1}{2}\pi$. Watson gives asymptotic approximations in the Poincaré sense, § 8.11, so that his version of (15.55) is $-\pi<\arg\zeta<2\pi$. The range of $\arg\zeta$ that he uses therefore extends nearly to anti-Stokes lines. It was explained in § 8.17 how this may lead to errors when $|\zeta|$ is not indefinitely large. The range given in (15.55) ends on Stokes lines. For (15.53) only the subdominant term is present when $0<\arg\zeta<\pi$ and this includes the range (15.52). For (15.54) both dominant and subdominant terms are present in this range. The reader may find it instructive to draw the Stokes diagram for Bessel's equation, and to prove that the Stokes multiplier is $2\mathrm{i}\cos\nu\pi$. Since Bessel functions are not in general single valued, the form of diagram used in fig. 8.6 is not suitable. An alternative form suggested by

Heading (1957) may be used in which $\arg\zeta$ is plotted as abscissa and the circle of fig. 8.6 is replaced by a horizontal line.

Hence the required solution is $H_\nu^{(1)}(\zeta)$ (15.53). This function must now be examined at very low levels, where it is to be separated into upgoing and downgoing waves. Bessel's equation (15.49) may be solved by assuming that there are solutions of the form $\zeta^\beta(a_0 + a_1\zeta + \cdots)$. This is substituted in (15.49) and coefficients of successive powers of ζ are equated to zero. In this way two solutions are found that are independent when ν is not an integer. In the present problem ν is never an integer. The two solutions are

$$J_\nu(\zeta) = (\tfrac{1}{2}\zeta)^\nu(\nu!)^{-1}(1 + a_2\zeta^2 + \cdots),$$
$$J_{-\nu}(\zeta) = (\tfrac{1}{2}\zeta)^{-\nu}\{(-\nu)!\}^{-1}(1 + b_2\zeta^2 + \cdots). \tag{15.56}$$

The coefficients a_2, b_2, etc. can be found but are of no interest in the present problem. The solution $H_\nu^{(1)}(\zeta)$ must be a linear combination of (15.56) and it is shown by Watson (1944, p. 74) that

$$H_\nu^{(1)}(\zeta) = \{J_{-\nu}(\zeta) - \mathrm{e}^{-\nu\pi\mathrm{i}}J_\nu(\zeta)\}/(\mathrm{i}\sin\nu\pi). \tag{15.57}$$

This provides the connection formula in the present problem.

At the ground z is very large and negative, so that $|\zeta|$, from (15.47), is very small. Hence only the first terms of the series (15.56) need be retained, and (15.57) shows that the solution near the ground is proportional to

$$\frac{(\frac{1}{2}\zeta)^{-\nu}}{(-\nu)!} - \mathrm{e}^{-\nu\pi\mathrm{i}}\frac{(\frac{1}{2}\zeta)^\nu}{\nu!} \tag{15.58}$$

since the denominator of (15.57) is a non-zero constant, and can therefore be omitted. The factor $\zeta^{-\nu}$ in the first term is, from (15.47), (15.48), proportional to $\mathrm{e}^{-\mathrm{i}kCz}$ which is the upgoing wave. Similarly the second term contains a factor $\mathrm{e}^{\mathrm{i}kCz}$ which is the downgoing wave. The ratio of the two terms is therefore the reflection coefficient

$$R_{22} = -\left(\frac{k}{\alpha}\right)^{2\nu}(1 - \mathrm{i}Z)^{-\nu}\frac{(-\nu)!}{\nu!}\exp(-2\mathrm{i}kCh_1) \tag{15.59}$$

where h_1 is the height above the ground where $X = 1$, $z = 0$. The reference level for R_{22} is at the ground. Since ν is purely imaginary all terms in (15.59) except $(1 - \mathrm{i}Z)^{-\nu}$ have modulus unity and hence

$$|R_{22}| = \exp\left\{-\frac{2kC}{\alpha}\arctan Z\right\}, \tag{15.60}$$

$$\arg R_{22} = \pi + \frac{4kC}{\alpha}\ln\left(\frac{k}{\alpha}\right) - \frac{kC}{\alpha}\ln(1 + Z^2) + 2\arg\{(-2\mathrm{i}kC/\alpha)!\} - 2kCh_1. \tag{15.61}$$

Table 15.1

kC/α	Frequency for normal incidence when $\alpha = 0.6\,\mathrm{km}^{-1}$	Phase difference (15.64) (degrees)
0	0	90
0.05	1.4 kHz	43
0.1	2.86 kHz	26
1.0	28.6 kHz	10
4.0	1.14 MHz	2.3

When electron collisions are neglected, $Z = 0$, (15.60) shows that $|R_{22}| = 1$. This was to be expected since no energy can be absorbed from the wave.

The approximate phase integral formula (7.152) may be used to find R_{22} for the exponential ionosphere (15.44). Let z_0 be the value of z where $q = 0$ in (15.45). Then the phase integral $\Phi = 2k\int_{-h_1}^{z_0} q\,\mathrm{d}z$ may be evaluated. The resulting expression contains $\exp(-\alpha h_1)$ but this is extremely small and may be neglected, so that

$$\Phi = 2kCh_1 + \frac{4kC}{\alpha}\{\ln(2C) - 1\} + \frac{kC}{\alpha}\ln(1 + Z^2) - \mathrm{i}\frac{2kC}{\alpha}\arctan Z. \qquad (15.62)$$

Then (7.152) gives $R_{22} = \mathrm{i}e^{-\mathrm{i}\Phi}$ whence $|R_{22}|$ is the same as (15.60) and

$$\arg R_{22} = \tfrac{1}{2}\pi - \frac{4kC}{\alpha}\{\ln(2C) - 1\} - \frac{kC}{\alpha}\ln(1 + Z^2) - 2kCh_1. \qquad (15.63)$$

The phase integral formula thus gives the correct $|R_{22}|$ for all values of α. The difference between the values (15.61), (15.63) for $\arg R_{22}$ is

$$\tfrac{1}{2}\pi + \frac{4kC}{\alpha}\left\{\ln\left(\frac{2kC}{\alpha}\right) - 1\right\} + 2\arg\left\{\left(\frac{-2\mathrm{i}kC}{\alpha}\right)!\right\}. \qquad (15.64)$$

For a 'slowly varying' ionosphere α is small and $2kC/\alpha$ is large. The first term of Stirling's formula may then be used for the factorial function, and this makes (15.64) zero. Some values of (15.64) for larger α are given in table 15.1. The value $\alpha = 0.6\,\mathrm{km}^{-1}$ used in the second column is roughly typical for the lower ionosphere. Hence in this example the phase integral formula (7.152) is remarkably accurate, since it always gives the correct value of $|R_{22}|$ and the error in phase never exceeds 90°.

15.9 Parabolic electron height distribution 1. Phase integrals

The parabolic model of the ionosphere (12.17) was discussed in §§ 12.4, 12.7 – 12.10 by the methods of ray theory. This model is important because, near the maximum of any ionospheric layer, $N(z)$ can be approximated by the square law function (12.17). It has the disadvantage that at the upper and lower boundaries, $z = z_m \pm a$ in (12.17), there is a discontinuity of gradient that does not occur in the actual ionosphere. The parabolic layer will now be studied with the main object of seeing what happens where ray theory fails. It is assumed that the electron collision frequency is independent of height so that Z is a constant. Only horizontal polarisation is discussed here, for then the differential equation (7.6) can be converted into a standard form. For vertical polarisation it is not possible to reduce the differential equation (7.68) to a simple standard form.

Consider a parabolic ionospheric layer with $N(z)$ given by (12.17). Let f_p be the penetration frequency for vertical incidence. Then with (12.18)

$$X = (1-\zeta^2)f_p^2/f^2, \quad q^2 = C^2 - \left(\frac{1-\zeta^2}{1-iZ}\right)\frac{f_p^2}{f^2} \tag{15.65}$$

where $C = \cos\theta$, and θ is the angle of incidence. The differential equation (7.6) to be studied is then

$$\frac{d^2E_y}{dz^2} + k^2\left\{C^2 - \left(\frac{1-\zeta^2}{1-iZ}\right)\frac{f_p^2}{f^2}\right\}E_y = 0. \tag{15.66}$$

In the rest of this section the subscript y will be omitted. Now let

$$\xi = \left(\frac{kf_p}{af}\right)^{\frac{1}{2}}(1-iZ)^{-\frac{1}{4}}(z-z_m), \quad A^2 = ka\left\{\frac{f_p}{f}(1-iZ)^{-\frac{1}{2}} - C^2\frac{f}{f_p}(1-iZ)^{\frac{1}{2}}\right\}, \tag{15.67}$$

where the fractional powers of $1-iZ$ are chosen so that they tend to $+1$ with argument zero, when $Z \to 0$. Then (15.66) is

$$\frac{d^2E}{d\xi^2} + (\xi^2 - A^2)E = 0. \tag{15.68}$$

If the functions X and q^2 in (15.65) are not exactly quadratic functions of ζ, the factor $1-\zeta^2$ then contains added small correction terms in higher powers of ζ. It is then possible to use a 'uniform approximation' analogous to (8.72), (8.68) so that the differential equation is still converted to the form (15.68). Further details are given in §17.3; see (17.12), (17.13).

Alternative forms of (15.68) are often used. Thus let

$$\xi = 2^{-\frac{1}{2}}e^{-\frac{1}{4}i\pi}u, \quad A^2 = -2i(n+\tfrac{1}{2}). \tag{15.69}$$

Then (15.68) becomes

$$\frac{d^2E}{du^2} + (n + \tfrac{1}{2} - \tfrac{1}{4}u^2)E = 0 \tag{15.70}$$

which is known as Weber's equation and is the form studied by Whittaker and Watson (1927). It was used for electromagnetic waves and a parabolic ionised layer by Rydbeck (1942a, 1943), Northover (1962), Heading (1962a). Its solutions are called parabolic cylinder functions. One of them, $D_n(u)$, corresponds to the solution $F(x;A)$ of (15.68) used below, § 15.10. Other forms of the differential equation are given by Miller (1955) and by Abramowitz and Stegun (1965). The form (15.68) was used by Heading (1962a, § A.4). It has the advantage that the independent variable is simply a scaled value of the height $z - z_m$ and if $Z = 0$ the scaling factor is real and positive. The two values $\xi = \pm A$ are where $q = 0$ and are thus the turning points of (15.68) or 'reflection points'.

Before studying (15.68) in detail it is useful to summarise some of the results of ray theory. First let $Z = 0$. The penetration frequency for obliquely incident waves is f_p/C. If $f < f_p/C$ the turning points are on the real z axis, at P and P′ in fig. 15.3(a), which is a sketch of the complex z plane. If the lower turning point P at $z = z_0 < z_m$ is sufficiently isolated, the reflection coefficient R, for waves incident from below, can be found from the phase integral formula (7.152). Let R have reference level at the lower boundary $z = z_m - a$. Then

$$R = \mathrm{i}\exp\left(-2\mathrm{i}k\int_{z_m - a}^{z_0} q\,\mathrm{d}z\right), \quad |R| = 1,$$

$$\arg R = \tfrac{1}{2}\pi - kaC\left\{1 - \frac{1}{2}\left(\frac{f_p}{Cf} - \frac{Cf}{f_p}\right)\ln\left(\frac{f_p + Cf}{f_p - Cf}\right)\right\}. \tag{15.71}$$

The integral here is of the same form as was used for (12.23). A criterion for P to be sufficiently isolated was given at (8.75), which can be shown to lead to

$$|A| \gg 0.84. \tag{15.72}$$

Now (15.67) shows that A is proportional to PP′, so that the criterion means that P and P′ must not be too close together, that is, f must not be too near the penetration frequency f_p/C. It also shows that A^2 is proportional to ka so the criterion is more easily satisfied for large ka, that is for thick layers. The phase integral formula (15.71) takes no account of the discontinuity of $\mathrm{d}N/\mathrm{d}z$ at the bottom of the layer. If the

Fig. 15.3. The complex z (height) plane showing the turning points PP′ or QQ′ of the differential equation (15.66). For (a) electron collisions are neglected, and for (b) they are included. PP′ are for a frequency less than the penetration frequency, and QQ′ are for a frequency greater than penetration.

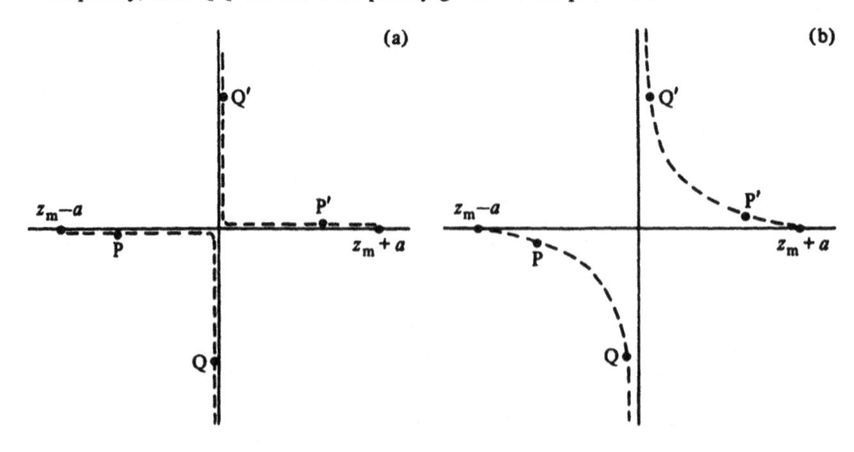

reflection here is small, it is necessary that the W.K.B. solution should be a good approximation just above the lower boundary (see § 15.3). A condition for this is that (7.59) should be satisfied where $\zeta = -1$, and this gives

$$kaC^3 f^2/f_p^2 \gg 1. \tag{15.73}$$

This is the criterion that the lower turning point P, that is the 'reflection' level, shall not be too close to the lower boundary. This too is most easily satisfied for large ka.

The range of z between P and P′ in fig. 15.3(a) is where q is purely imaginary, and the waves are evanescent. This is an exact analogy of a potential barrier in wave mechanics. Within the barrier the 'upgoing' wave decays as z increases, but if the barrier is not too wide the wave energy that reaches P′ may not be negligible. The waves can tunnel through the barrier and this is an example of the simplest form of tunnelling in which q^2 is zero at both ends of the barrier. These physical considerations suggest that if T is the transmission coefficient of the ionosphere then

$$|T| = \exp\left(-k\int_{(\mathrm{P})}^{(\mathrm{P}')} |q|\,\mathrm{d}z\right) = \exp\left\{-\frac{1}{2}\pi ak\left(\frac{f_p}{f} - \frac{C^2 f}{f_p}\right)\right\}. \tag{15.74}$$

Now suppose that $Z \neq 0$. The values z_0 of z at the turning points P, P′ are found from (15.65) and are now complex, as shown in fig. 15.3(b). But the phase integral formula (7.152) can still be used, as suggested in § 7.19. The integral is now a contour integral with a complex upper limit z_0. Equation (15.65) shows that in the first equation (15.71), in q, f_p is to be replaced by $f_p(1-\mathrm{i}Z)^{-\frac{1}{2}} = g$ say. Then

$$R = \mathrm{i}\exp\left[-\mathrm{i}kaC\left\{1 - \frac{1}{2}\left(\frac{g}{Cf} - \frac{Cf}{g}\right)\ln\left(\frac{g+Cf}{g-Cf}\right)\right\}\right]. \tag{15.75}$$

If $f \ll f_p/C$, the point P is near the real z axis. When f increases, P and P′ move to the left and right respectively along the dotted curves in fig. 15.3(b). Now PP′ is never zero but it may be small when f is near to f_p/C, and then (15.75) may be unreliable. But for still larger f, P and P′ move apart again to positions such as Q and Q′ in fig. 15.3(b). The turning point now at Q is again 'isolated' and so the approximation (15.75) can be good even when $f > f_p/C$. If now $Z = 0$ and $f > f_p/C$ the point P in fig. 15.3(a) has moved to the position Q. The formula (15.75) then gives

$$|R| = \exp\left\{-\frac{1}{2}\pi kzC\left(\frac{Cf}{f_p} - \frac{f_p}{Cf}\right)\right\}. \tag{15.76}$$

When $Z \neq 0$, q is never zero when z is real and for the upgoing wave Re(q) is positive, though it may be small within a barrier. The transmission coefficient is given by

$$T = \exp\left(-\mathrm{i}k\int_{z_m - a}^{z_m + a} q\,\mathrm{d}z\right) = \exp\left[-\mathrm{i}kaC\left\{1 - \frac{1}{2}\left(\frac{g}{Cf} - \frac{Cf}{g}\right)\ln\left(\frac{Cf+g}{Cf-g}\right)\right\}\right]. \tag{15.77}$$

Note the similarity of this and (15.75).

The approximate formulae (15.71)–(15.77) are based on the phase integral method, which is a generalisation of ray theory. They are compared below with results from full wave theory.

15.10. Parabolic electron height distribution 2. Full wave solutions

The differential equation (15.68) must now be studied. It has no singularities for bounded ξ. In this respect it resembles the Stokes differential equation (8.7). The only singularity is at infinity. In other words every point in the complex ξ plane, including the turning points $\xi = \pm A$, is an ordinary point. It means that two independent series solutions in ascending integer powers of ξ can be found and they are convergent for all ξ. Hence any solution of (15.68) is bounded and single valued for all ξ. It has no singularities except at infinity. Such a function is called an 'integral function', or 'entire function'. The form of (15.68) shows that for one of the series the powers must be all even, and for the other all odd. It proves more convenient to find series for the two functions $E\exp(\pm\frac{1}{2}i\xi^2)$. They are not needed here but are given by Heading (1962a, § A.4).

The asymptotic approximations for the solutions of (15.68) may be found by the W.K.B. formula (7.26) which gives

$$E \sim (\xi^2 - A^2)^{-\frac{1}{4}} \exp\left\{\pm i \int^{\xi} (\xi^2 - A^2)^{\frac{1}{2}} d\xi\right\}. \tag{15.78}$$

This is to be used for large ξ, and in particular for $|\xi| > |A|$, to give an asymptotic

Fig. 15.4. Stokes diagram for the function F(ξ; A). S denotes Stokes lines and A anti-Stokes lines. The branch I represents the incident wave, R the reflected wave and T the transmitted wave.

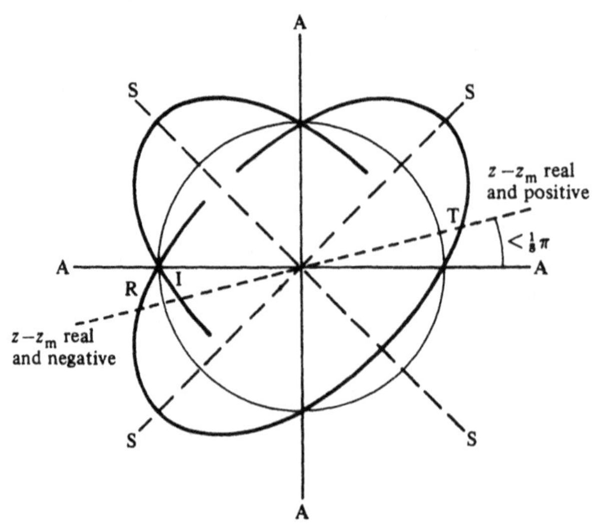

form multiplied by an asymptotic series in descending powers of ξ (see § 8.11). We first find the asymptotic form. The series is given later. The fractional powers in (15.78) are expanded by the binomial theorem. For the first factor only one term need be retained since later terms affect only the second and later terms of the asymptotic expansion. For the integrand, however, two terms must be retained since the second gives a logarithm on integration. The result is

$$E \sim \xi^{-\frac{1}{2}}\xi^{\frac{1}{2}\mathrm{i}A^2}\mathrm{e}^{-\frac{1}{4}\mathrm{i}\xi^2} \text{ or } \xi^{-\frac{1}{2}}\xi^{-\frac{1}{2}\mathrm{i}A^2}\mathrm{e}^{\frac{1}{4}\mathrm{i}\xi^2}. \tag{15.79}$$

When $|\xi|$ is very large, the behaviour of these functions is determined mainly by the last exponential. The exponent $\frac{1}{2}\mathrm{i}\xi^2$ is real when $\arg\xi = \pm\frac{1}{4}\pi, \pm\frac{3}{4}\pi$ and these are therefore the Stokes lines (see §8.13) for (15.68). Similarly $\frac{1}{2}\mathrm{i}\xi^2$ is purely imaginary when $\arg\xi = 0, \pm\frac{1}{2}\pi, \pi$ and these are the anti-Stokes lines; see fig. 15.4.

The fractional powers of ξ in (15.79) may be written $\exp(\pm\frac{1}{2}\mathrm{i}A^2\ln\xi)$. In general these functions have infinitely many values. We choose the values for which $-\frac{5}{4}\pi \leqslant \arg\xi < \frac{3}{4}\pi$. This is equivalent to the use of a branch cut in the complex ξ plane where $\arg\xi = \frac{3}{4}\pi$. Let the asymptotic series that must multiply the first of the functions (15.79) be

$$S(\xi) = 1 + C_1\xi^{-2} + C_2\xi^{-4} + \cdots. \tag{15.80}$$

The differential equation (15.68) shows that the powers must all be even. Substitute (15.80) in (15.68) and equate coefficients of successively decreasing powers of ξ. This gives

$$C_1 = \frac{\mathrm{i}}{16}(1-\mathrm{i}A^2)(3-\mathrm{i}A^2), \quad C_r = \frac{\mathrm{i}^r}{16^r r!}(1-\mathrm{i}A^2)\cdots(4r-1-\mathrm{i}A^2). \tag{15.81}$$

The series for the second function (15.79) is obtained from (15.81) by reversing the sign of i.

Suppose now that conditions are such that there is no appreciable reflection at the discontinuity of $\mathrm{d}N/\mathrm{d}z$ at the upper boundary. Equation (15.67) shows that when $z - z_{\mathrm{m}}$ is large and positive, $|\xi|$ is large and $0 \leqslant \arg\xi < \frac{1}{8}\pi$. Thus $\mathrm{Re}(\frac{1}{2}\xi^2)$ increases as z increases. This shows that the term with the factor $\exp(-\frac{1}{2}\mathrm{i}\xi^2)$ in (15.79) is the one that represents an upgoing wave. For radio waves incident from below, we seek a solution in which only this term is present when z is large and positive. This solution will be denoted by $\mathrm{F}(\xi; A)$. Fig. 15.4 shows the Stokes diagram for this function. It can be shown that

$$\mathrm{F}(\xi; A) = 2^{\frac{1}{4}(1-\mathrm{i}A^2)}\exp\{\tfrac{1}{8}\pi(\mathrm{i}+A^2)\}D_n(u) \tag{15.82}$$

where u and n are given by (15.69).

For large real z the term with the factor $\exp(-\frac{1}{2}\mathrm{i}\xi^2)$ is the dominant term if $Z \neq 0$, and $\mathrm{Re}(-\frac{1}{2}\mathrm{i}\xi^2)$ is then positive, so that the exponential increases with increasing z. This suggests that the wave increases in amplitude as it travels, which would be physically unacceptable. But the solution is to be used only within the ionosphere

$z \leqslant z_m + a$, and here the other factors in (15.79) must ensure that the amplitude does not increase with increasing z.

When $z - z_m$ is real, large and negative, $\mathrm{Re}(\frac{1}{2}\xi^2)$ decreases when z increases. Hence the term with the factor $\exp(-\frac{1}{2}\mathrm{i}\xi^2)$ is now the downgoing wave, that is the reflected wave. But here the second term of (15.79) is also present as shown by the Stokes diagram fig. 15.4, and this is the incident wave. The required asymptotic approximations are

$$\mathrm{F}(\xi; A) \sim \begin{cases} \xi^{-\frac{1}{2}}\xi^{\frac{1}{2}\mathrm{i}A^2}\mathrm{e}^{-\frac{1}{4}\mathrm{i}\xi^2} \quad \text{for} \quad -\dfrac{3\pi}{4} \leqslant \arg\xi \leqslant \dfrac{\pi}{4}, & (15.83) \\ \\ \xi^{-\frac{1}{2}}\xi^{\frac{1}{2}\mathrm{i}A^2}\mathrm{e}^{-\frac{1}{4}\mathrm{i}\xi^2} - \dfrac{\mathrm{i}(2\pi)^{\frac{1}{2}}}{\Gamma(\frac{1}{2} - \frac{1}{2}\mathrm{i}A^2)}\mathrm{e}^{\frac{3}{4}\pi A^2}2^{-\frac{1}{2}\mathrm{i}A^2}\xi^{-\frac{1}{2}}\xi^{-\frac{1}{2}\mathrm{i}A^2}\mathrm{e}^{\frac{1}{4}\mathrm{i}\xi^2} & \end{cases}$$

$$\text{for} \quad -\frac{5\pi}{4} \leqslant \arg\xi \leqslant -\frac{3\pi}{4}. \tag{15.84}$$

The last term can be derived from a knowledge of the Stokes multiplier for the Stokes line at $\arg\xi = -\frac{3}{4}\pi$. The Stokes multipliers for all four Stokes lines are given by Heading (1962a). Alternatively it can be found from (15.82) by using the asymptotic forms for $D_n(u)$ given by Whittaker and Watson (1935, p. 347).

In (15.83), (15.84) the ranges given for $\arg\xi$ end at Stokes lines; see § 8.17. In (15.83) the range could be given as $-\pi < \arg\xi < \pi/2$ ending just within anti-Stokes lines and this would be correct in the Poincaré sense. But then this range would include the value of $\arg\xi$ where $z - z_m$ is real and negative and it might be inferred that only the one term (15.83) is needed at both the top and bottom of the ionosphere. The use of the ranges given in (15.83), (15.84) ensures that at the bottom of the ionosphere the second term of (15.84) must be present.

When $z = z_m \pm a$ let $\xi = \xi_+$ and ξ_-, respectively, so that

$$\xi_- = \mathrm{e}^{-\mathrm{i}\pi}\xi_+. \tag{15.85}$$

In (15.84) the first term is the downgoing wave and the second is the upgoing wave, and their ratio is the reflection coefficient

$$R_{22} = \mathrm{i}(2\pi)^{-\frac{1}{2}}\Gamma(\tfrac{1}{2} - \tfrac{1}{2}\mathrm{i}A^2)\mathrm{e}^{\frac{1}{4}\pi A^2}2^{\frac{1}{2}\mathrm{i}A^2}\xi_+^{\mathrm{i}A^2}\mathrm{e}^{-\mathrm{i}\xi^2} \tag{15.86}$$

with reference level at the base of the ionosphere. Similarly (15.83) is the upgoing wave at the top of the ionosphere and its ratio to the second term of (15.84) is the transmission coefficient

$$T_{22} = R_{22}\exp\{-\tfrac{1}{2}\pi(A^2 + \mathrm{i})\} \tag{15.87}$$

where (15.85) has been used. For (15.86), (15.87) to be good approximations it is necessary that the condition (15.73) shall be satisfied. This means that the thickness of the parabolic layer must be several or many free space wavelengths.

When (15.73) is not adequately satisfied, it cannot be assumed that the solution of (15.68) just below the upper boundary is $\mathrm{F}(\xi; A)$ alone, and it would be necessary to add to it a multiple of a second solution. This subject was studied by Rydbeck (1943). The reflection coefficient would then be expressed in terms of these two functions and their derivatives at $\xi = \pm A$. It would be much more complicated than (15.86). For this general case it proves easier to find R_{22} and T_{22} by a numerical integration of (15.68), for example by the method of §§ 18.2, 18.6, 18.7. This was used for the results of figs. 15.5, 15.6.

If collisions are neglected, (15.67) shows that A^2 is real and ξ_+ is real and positive. Then the modulus of the gamma function in (15.86) can be found from the 'reflection formula' for gamma functions: $\Gamma(x)\Gamma(1-x) = \pi \operatorname{cosec}(\pi x)$ (Abramowitz and Stegun, 1965, 6.1.17), whence

$$|\Gamma(\tfrac{1}{2} - \tfrac{1}{2}\mathrm{i}A^2)|^2 = \Gamma(\tfrac{1}{2} - \tfrac{1}{2}\mathrm{i}A^2)\Gamma(\tfrac{1}{2} + \tfrac{1}{2}\mathrm{i}A^2) = \pi \operatorname{sech}(\tfrac{1}{2}\pi A^2) \tag{15.88}$$

and (15.86), (15.87) give

$$|R_{22}|^2 = (1 + \mathrm{e}^{-\pi A^2})^{-1}, \quad |T_{22}|^2 = (1 + \mathrm{e}^{\pi A^2})^{-1}, \quad |R_{22}|^2 + |T_{22}|^2 = 1. \tag{15.89}$$

This shows that no energy is lost from the waves. The ray theory formula (15.71) suggests that if $f < f_p/C$, that is $A^2 > 0$, the waves cannot penetrate the layer and $|R_{22}| = 1$. But (15.89) shows that $|T_{22}|$ is not zero and $|R_{22}| < 1$. Similarly if $f > f_p/C$, the ray theory result (15.77) with $Z = 0$ suggests that $|T_{22}| = 1$, but (15.89) shows that $|T_{22}| < 1$. For $f = f_p/C$ (15.89) shows that $|R_{22}|^2 = |T_{22}|^2 = \tfrac{1}{2}$. For a range of

Fig. 15.5. Dependence of $|R_{22}|^2$ and $|T_{22}|^2$ on frequency for a parabolic layer when collisions and the earth's magnetic field are neglected. In all these examples the vertical incidence penetration frequency f_p is 1 MHz, and the angle of incidence θ is 10°. The numbers by the curves are the half thickness a in km. The ordinate scale is linear.

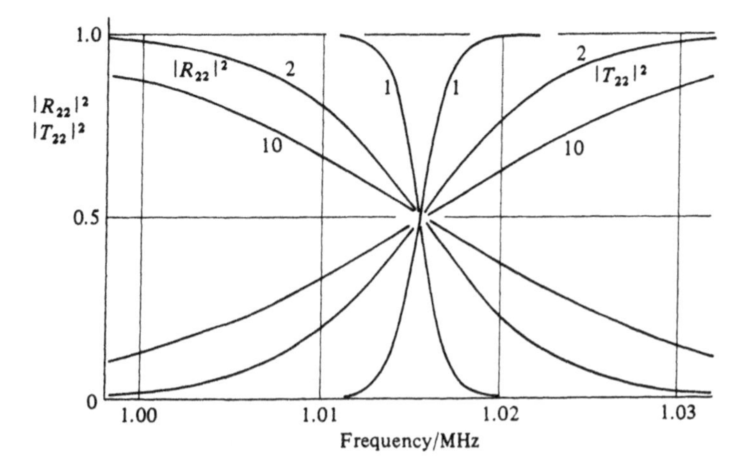

frequencies on either side of the penetration frequency f_p/C there is partial penetration and partial reflection. This range gets large when ka is made smaller. Fig. 15.5 shows how $|R_{22}|^2$ and $|T_{22}|^2$ depend on frequency for various values of the semi-thickness a, when collisions are neglected.

Fig. 15.6 shows further examples of the frequency dependence of $|R_{22}|$ and $|T_{22}|$. The continuous curves in both figs. 15.5, 15.6 were calculated by numerical integration of the differential equations with the method of §§ 18.2, 18.6, 18.7. The broken curves were calculated from the phase integral formulae (15.75), (15.77). In fig. 15.6(a) collisions are neglected. The values for $|R_{22}|$ from (15.89) are indistinguishable from the continuous curve where $f < 1.03$ MHz and indistinguishable from the broken curve when $f > 1.04$ MHz. The values for $|T_{22}|$ are indistinguishable from the continuous curve for the whole frequency range. In both figs. 15.6(a, b) the curves for $|R_{22}|$ show oscillations resembling an interference pattern, and suggesting that there are two contributions to the reflected wave, one from the discontinuity of dN/dz at the bottom of the layer and the other associated with the zero of q at the lower turning point.

15.11. Parabolic electron height distribution 3. Equivalent height of reflection

In § 12.4 ray theory was used to find the equivalent height of reflection $h'(f)$ (12.22), for radio waves vertically incident on a parabolic layer from below. This theory fails for frequencies near the penetration frequency and the full wave theory of § 15.10 will therefore now be used. Only vertical incidence, $C = 1$, is discussed, and electron collisions are neglected, so that $Z = 0$. The effect of discontinuities of dN/dz at the boundaries of the layer is ignored so that the reflection coefficient is given by (15.86) with $C = 1$, $Z = 0$. The theory taking account of collisions was given by Rydbeck (1942, 1943). Some results with collisions were given by Budden and Cooper (1962).

For a wave that travels over a given path the total change of phase is $2kP$ where P

Fig. 15.6. Two examples of the dependence of $|R_{22}|$ and $|T_{22}|$ on frequency f. The half thickness $a = 1$ km and θ, f_p are the same as in fig. 15.5. The continuous curves were calculated by numerical integration of the differential equations, and the broken curves from the phase integral formulae (15.75), (15.77). The ordinate scales are logarithmic. In (a) electron collisions are neglected, and in (b) $\nu = 5 \times 10^{-5}\,\mathrm{s}^{-1}$.

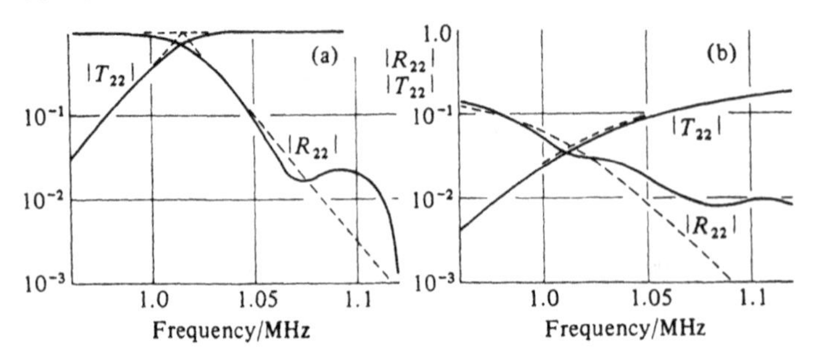

is the phase path, defined in § 10.16. It is related to the group path P' by (10.40). For vertical incidence $P' = 2h'(f)$ and $P = 2h(f)$ where $h(f)$ is called the phase height of reflection. In general h and h' may be complex. The amplitude of the wave is changed by a factor $\exp(-ikP)$, so that

$$R_{22} = \exp(-2ikh_L) \tag{15.90}$$

where h_L is the contribution to h from within the layer and R_{22} is the reflection coefficient (15.86), which is referred to the level $z = z_m - a$. Hence, for a transmitter and receiver at the ground

$$h(f) = z_m - a + \mathrm{i}(\ln R_{22})/2k \tag{15.91}$$

and on applying (10.40)

$$h'(f) = z_m - a + \tfrac{1}{2}\mathrm{i}\,\mathrm{d}(\ln R_{22})/\mathrm{d}k$$

$$= z_m - a - \mathrm{i}a\frac{f}{f_p}\frac{\mathrm{d}(\ln R_{22})}{\mathrm{d}(A^2)} \tag{15.92}$$

where (15.67) for A^2 has been used with $C = 1, Z = 0$. Let k_p be the value of k for the penetration frequency $f = f_p$. Then (15.67) shows that $\xi_+ = (ak_p)^{\frac{1}{2}}$ and (15.86) gives

$$\ln R_{22} = \ln\Gamma(\tfrac{1}{2} - \tfrac{1}{2}\mathrm{i}A^2) + \tfrac{1}{4}\pi A^2 + \mathrm{i}A^2(\tfrac{1}{2}\ln 2 + \ln\xi_+) + \text{constant}. \tag{15.93}$$

It was shown in § 11.16 that the time of travel of a wave packet is given approximately by $\mathrm{Re}\{h'(f)\}$ so only the imaginary part of (15.93) is needed in (15.92). The derivative of the logarithm of the gamma function is the psi function $\psi(x) = \mathrm{d}\{\ln\Gamma(x)\}/\mathrm{d}x$. Its properties are given by Abramowitz and Stegun (1965, § 6.3). From (15.93), (15.92)

$$h'(f) = z_m - a + \tfrac{1}{2}a\frac{f}{f_p}\{\ln(2ak_p) - \psi(\tfrac{1}{2} - \tfrac{1}{2}\mathrm{i}A^2) - \tfrac{1}{2}\mathrm{i}\pi\}. \tag{15.94}$$

Since A^2 is real, the real part of the psi function is symmetric about $A^2 = 0$, $f = f_p$, and $\mathrm{Re}\{h'(f)\}$ here has a maximum value

$$z_m - a + \tfrac{1}{2}a\{\ln(2ak_p) + \gamma + \ln 4\} = z_m + 0.3283a + \tfrac{1}{2}a\ln(ak_p) \tag{15.95}$$

where γ is Euler's constant and the formula $\psi(\tfrac{1}{2}) = -\gamma - \ln 4$ has been used.

For waves that travel right through the parabolic layer, the contribution to $h'(f)$ from the region between the ground and the top of the layer can be found as was done for ray theory at (12.24). The contribution is

$$h'(f) = z_m - a + \mathrm{i}\,\mathrm{d}(\ln T_{22})/\mathrm{d}k \tag{15.96}$$

where T_{22} is the transmission coefficient. Now (15.87) shows that the real part of the last term of (15.96) is simply twice the real part of the last term of (15.91). Hence the contribution from the layer to $\mathrm{Re}\{h'(f)\}$ is twice the last term of (15.94).

Fig. 15.7 shows how these values of $\mathrm{Re}\{h'(f)\}$ depend on frequency and the corresponding curves for the ray theory, (12.22), (12.24), are shown for comparison. The difference is small when $|A|^2 \gg 0$, but near the penetration frequency the ray theory curves approach infinity whereas the full wave theory shows that there is a bounded upper limit. The range of frequency where the difference is appreciable depends on the thickness $2a$ of the layer. It is greatest when a is small.

15.12. The differential equations of theoretical physics

In the study of the reflection and transmission of radio waves by a stratified isotropic ionosphere, the electron height distributions $N(z)$ discussed earlier in this chapter have been deliberately chosen so that the governing differential equation can be transformed to one of the standard differential equations of theoretical physics. The number of suitable standard equations is very large. Lists of suitable equations are given by Brekhovskikh (1960), Ginzburg (1970), Heading (1977a), and these authors give further references. The general problem of what functions $N(z)$ can be studied in this way has been examined by Heading (1965). Solutions for various functions $N(z)$, including many that are not described in this book, have been given by Försterling and Wüster (1950), Brekhovskikh (1960), Wait (1962, ch. III), Gould and Burman (1964), Ginzburg (1970, ch. IV). Solutions for a spherically stratified isotropic medium were studied by Westcott (1968), Sharaf (1969), and for cylindrically stratified media by Westcott (1969b). Discussion of all the possibilities is not the object of this book. The purpose here is to set out the basic ideas, so that the interested reader can then follow up the numerous references for further examples.

Fig. 15.7. The continuous curves show the contribution to the equivalent height of reflection from a parabolic layer according to full wave theory when collisions and the earth's magnetic field are neglected. In this example $k_p\, a = 12.5$. The lower curve is for reflection and the upper curve is the contribution for waves that have penetrated the layer. The broken curves show the corresponding results for ray theory.

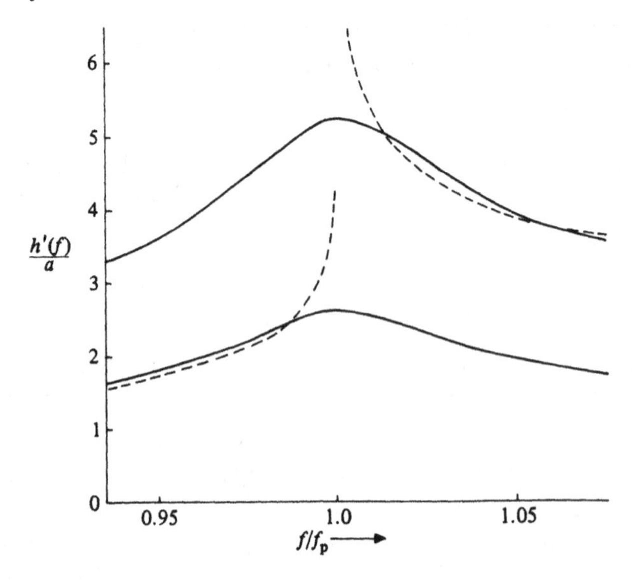

All these methods depend on the use of circuit relations. A given solution must first be separated into upgoing and downgoing component waves, both at a great height, above the reflecting levels, and at a low height that is usually in the free space below the ionosphere. The circuit relations are the connection formulae that relate the amplitudes of these component waves at the top and at the bottom (see § 8.19). In all the cases so far discussed the independent variable at great heights tends to infinity, and the differential equations studied, the Stokes equation, ch. 8, Bessel's equation, § 15.8, Weber's equation or its equivalent, §§ 15.9–15.11, all had an irregular singularity at infinity; see § 8.11. This led to the need for asymptotic approximations and the study of the Stokes phenomenon, when deriving the circuit relations.

There is one class of second order differential equations known as Fuchsian equations for which there are no irregular singularities. The most important is the hypergeometric equation, which has three regular singularities and no other singularity. Its solutions include many of the functions of theoretical physics as special cases or as limiting cases. It was used by Epstein (1930) and by Eckhart (1930) to find reflection and transmission coefficients. It can be used for a wide range of electron height distributions $N(z)$. An even wider range can be covered by an extension of this idea. The differential equation with four regular singularities is known as Heun's equation (Heun, 1889) and it has been used in some recent studies of radio propagation problems (van Duin and Sluijter, 1980).

The theory given by Epstein is summarised here as an example of this type of method. Some special cases of it are discussed in §§ 15.16–15.17.

15.13. The hypergeometric equation and its circuit relations

The theory summarised here is given in full by Whittaker and Watson (1927), Copson (1935), Olver (1974) and in other mathematical text books. A fuller version than the one that follows was given by Budden (1961a).

The hypergeometric equation may be written

$$\zeta(1-\zeta)\frac{\mathrm{d}^2u}{\mathrm{d}\zeta^2}+\{c-(a+b+1)\zeta\}\frac{\mathrm{d}u}{\mathrm{d}\zeta}-abu=0. \tag{15.97}$$

This is the form used by the authors cited above. Epstein (1930) used a different form with independent variable $u=-\zeta$, and the meanings of his symbols a, b, are different. The symbol c here is a new constant, not the speed of light. Equation (15.97) has regular singularities at $\zeta=0$, 1 and ∞. First, for the neighbourhood of $\zeta=0$ we seek a solution in ascending powers of ζ:

$$u=\zeta^{\beta}(1+a_1\zeta+a_2\zeta^2+\cdots). \tag{15.98}$$

Substitute in (15.97) and equate to zero the coefficients of successively increasing powers of ζ. The lowest power $\zeta^{\beta-1}$ gives

$$\beta = 0 \quad \text{or} \quad 1 - c. \tag{15.99}$$

The others give the coefficients $a_1, a_2, \ldots$. Hence for $\beta = 0$ a solution is

$$u \equiv F(a, b; c; \zeta) = 1 + \frac{ab}{c1!}\zeta + \frac{a(a+1)b(b+1)}{c(c+1)2!}\zeta^2 + \cdots$$
$$+ \frac{a(a+1)\ldots(a+r-1)b(b+1)\ldots(b+r-1)}{c(c+1)\cdots(c+r-1)r!}\zeta^r + \cdots. \tag{15.100}$$

This is called the hypergeometric series. Similarly for $\beta = 1 - c$ a solution is

$$u \equiv (-\zeta)^{1-c}F(a-c+1, b-c+1; 2-c; \zeta)$$
$$= (-\zeta)^{1-c}\left\{1 + \frac{(a-c+1)(b-c+1)}{(2-c)\,1!}\zeta\right.$$
$$\left. + \frac{(a-c+1)(a-c+2)(b-c+1)(b-c+2)}{(2-c)(3-c)2!}\zeta^2 + \cdots\right\}. \tag{15.101}$$

A factor $(-1)^{1-c}$ has here been introduced for later convenience. Both (15.100) and (15.101) can be shown to be convergent when $|\zeta| < 1$.

Alternatively we may take as a trial solution a series in descending powers of ζ. This is possible because the singularity of (15.97) at infinity is regular, which ensures that there must be at least one convergent series solution. Thus let

$$u = \zeta^{\beta}\left(1 + \frac{b_1}{\zeta} + \frac{b_2}{\zeta^2} + \cdots\right). \tag{15.102}$$

A similar treatment then gives

$$\beta = -a \quad \text{or} \quad -b \tag{15.103}$$

and thence two possible solutions are

$$u = (-\zeta)^{-a}F(a, a-c+1; a-b+1; \zeta^{-1}), \tag{15.104}$$

$$u = (-\zeta)^{-b}F(b, b-c+1; b-a+1; \zeta^{-1}). \tag{15.105}$$

Both are convergent if $|\zeta| > 1$. Again, factors $(-1)^{-a}, (-1)^{-b}$ have been introduced for later convenience. These two functions may have branch points at infinity, but they give solutions valid for any range 2π of $\arg\zeta$ and they do not display the Stokes phenomenon.

The solutions (15.100), (15.101), (15.104), (15.105) or their analytic continuations can have branch points at the singularities $\zeta = 0, 1, \infty$. A cut is therefore introduced in the complex ζ plane extending along the positive real axis. The solutions are then single valued provided that the cut is not crossed, and in future the values used will be those for which

$$|\arg(-\zeta)| < \pi. \tag{15.106}$$

The above method for deriving (15.101) might fail if c is an integer. Then the series would have to be replaced by a more complicated expression containing a logarithm. A similar failure could occur for (15.104) or (15.105) if a–b is an integer. But these integer values do not occur in the radio wave problem; see (15.118), (15.119) below.

The function (15.100) is defined by the series only in the range $|\zeta| < 1$, but the process of analytic continuation is now used to extend the definition outside this range. For $|\zeta| < 1$ it can be shown that

$$F(a,b;c;\zeta) = \frac{(c-1)!}{2\pi \mathrm{i}(a-1)!(b-1)!}\int_{-\mathrm{i}\infty}^{\mathrm{i}\infty} \frac{(a+s-2)!(b+s-2)!(-s)!}{(c+s-2)!}(-\zeta)^{s-1}\,\mathrm{d}s \tag{15.107}$$

where the contour runs to the right of the poles of $(a+s-2)!(b+s-2)!$ and to the left of the poles of $(-s)!$. Integrals of this kind were used by Barnes (1908). For a diagram showing the contour see Budden (1961a, fig. 17.6). For $|\zeta| < 1$ the contour can be closed by a large semi-circle to the right and the series (15.100) is the sum of the residues of the poles of $(-s)!$. For $|\zeta| > 1$ the integral still defines a function $F(a,b;c;\zeta)$ that is now called the hypergeometric function. The integral therefore provides the required analytic continuation. The contour can now be closed by a large semi-circle to the left, and thus the integral is expressed as the sum of two series from the residues of the poles of $(a+s-2)!$ and $(b+s-2)!$. The solution (15.100) is therefore equal to a linear combination of the two solutions (15.104), (15.105). Similarly the series (15.101) can be expressed as a Barnes integral, continued analytically into the region $|\zeta| > 1$, and then expressed as another linear combination of (15.104), (15.105). These two linear combinations are the required circuit relations. From them (15.104) can be expressed as a linear combination of (15.100), (15.101), and so also can (15.105). The results are

$$\frac{(-\zeta)^{-a}}{(a-b)!}F(a,1-c+a;1-b+a;\zeta^{-1}) = \frac{(-c)!}{(-b)!(a-c)!}F(a,b;c;\zeta) + \frac{(c-2)!(-\zeta)^{1-c}}{(c-b-1)!(a-1)!}F(a-c+1,b-c+1;2-c;\zeta), \tag{15.108}$$

$$\frac{(-\zeta)^{-b}}{(b-a)!}F(b,1-c+b;1-a+b;\zeta^{-1}) = \frac{(-c)!}{(-a)!(b-c)!}F(a,b;c;\zeta) + \frac{(c-2)!(-\zeta)^{1-c}}{(c-a-1)!(b-1)!}F(a-c+1,b-c+1;2-c;\zeta). \tag{15.109}$$

For the applications that follow, the hypergeometric functions have now fulfilled their purpose and need not be used again. Those on the right of (15.108), (15.109) will only be used when $|\zeta|$ is very small so that only the first terms (unity) of the series in (15.100), (15.101) are appreciable. Similarly those on the left will only be used for very large $|\zeta|$ and again only the first terms (unity) of the series are appreciable. Hence in the following sections, the hypergeometric functions will be set equal to unity.

15.14. Epstein distributions

In § 15.8 Bessel's equation (15.49) was studied. If it is compared with the hypergeometric equation (15.97) it is seen that both have a regular singularity at $\zeta = 0$. For Bessel's equation the substitution (15.47) was used so that ζ is proportional to $e^{\frac{1}{2}\alpha z}$ where z is height. Thus the singularity at $\zeta = 0$ is now where $z \to -\infty$ so that it is associated with the region of free space below the ionosphere. This suggests that a similar substitution should be used for the hypergeometric equation. We therefore now take

$$-\zeta = e^{\xi}, \quad \xi = (z/\sigma) + B. \tag{15.110}$$

This is equivalent to the substitution used by Epstein (1930). The constant σ determines the scale of the vertical structure of the model ionosphere. The constant B allows the origin of z to be chosen conveniently. In terms of ξ as independent variable, (15.97) becomes

$$(1 + e^{\xi})\frac{d^2u}{d\xi^2} + \{c - 1 + (a+b)e^{\xi}\}\frac{du}{d\xi} + abe^{\xi}u = 0. \tag{15.111}$$

This is now transformed to a version without a first derivative term by changing the dependent variable u to E thus

$$u = E\exp\{\tfrac{1}{2}(1-c)\xi\}(1 + e^{\xi})^{\frac{1}{2}(c-1-a-b)} \tag{15.112}$$

which gives

$$\frac{d^2E}{dz^2} + k^2q^2E = 0 \tag{15.113}$$

where

$$q^2 = \eta_1 + \frac{e^{\xi}}{(e^{\xi}+1)^2}\{(\eta_2 - \eta_1)(e^{\xi}+1) + \eta_3\} \tag{15.114}$$

and

$$\eta_1 = -\tfrac{1}{4}(c-1)^2/\sigma^2k^2, \tag{15.115}$$

$$\eta_2 = -\tfrac{1}{4}(a-b)^2/\sigma^2k^2, \tag{15.116}$$

$$\eta_3 = \tfrac{1}{4}(a+b-c+1)(a+b-c-1)/\sigma^2k^2. \tag{15.117}$$

Equation (15.113) is the same as (7.6) and E may be regarded as the electric field of a horizontally polarised wave obliquely incident on the ionosphere. Hence (15.114) determines what electron height distributions $N(z)$ can be studied with this theory. By suitably assigning the constants σ, B, η_1, η_2, η_3 a very wide range of models of the ionosphere can be investigated.

Below the ionosphere z is large and negative and (15.114) gives $q^2 \approx \eta_1$. But here $q = C$ so that $\eta_1 = C^2$ where $C = \cos\theta$ and θ is the angle of incidence. Then (15.115) gives

$$c - 1 = -2ik\sigma C. \tag{15.118}$$

The choice of sign here is arbitrary. The opposite sign would lead to the same final result. At a great height in the ionosphere z is large and positive and (15.114) gives $q^2 \approx \eta_2$. Hence q tends to a constant value, q_2 say, and (15.116) gives

$$a - b = -2ik\sigma q_2 \tag{15.119}$$

where again the choice of sign is arbitrary. At intermediate heights the dependence of q on z is determined by η_3 and (15.117) shows that

$$a + b - c = \pm(4k^2\sigma^2\eta_3 + 1)^{\frac{1}{2}} \tag{15.120}$$

where either sign may be chosen.

The ionospheric plasma is here assumed to be isotropic and (15.114) is equal to $n^2 - S^2 = C^2 - X/(1 - iZ)$. Suppose now that σ is real and positive and $B = -z_m/\sigma$ which is real. Then ξ is simply proportional to $(z - z_m)$, from (15.110). Suppose further that η_1, η_2, η_3 are all real. Then q^2 and n^2 are real when z is real, and $Z = 0$. Fig. 15.8 shows some typical curves of $n^2 = 1 - X$ vs ξ. In all the examples n^2 and thence X tend to constant values at small heights and at great heights. In between, the change may be monotonic or there may be one maximum, or one minimum. Two special cases are of particular interest. The first, discussed in § 15.16, is for $\eta_3 = 0$, so that n^2 makes a continuous monotonic change from unity below the ionosphere to a

Fig. 15.8. Typical Epstein distributions. All curves are for vertical incidence, $\theta = 0$. Curve (a) is a sech^2 distribution with $\eta_1 = \eta_2 = 1$, $\eta_3 = -3$. For curve (b) $\eta_1 = 1$, $\eta_2 = -0.4$, $\eta_3 = 0$. For curve (c) $\eta_1 = 1$, $\eta_2 = 0.4$, $\eta_3 = -3$.

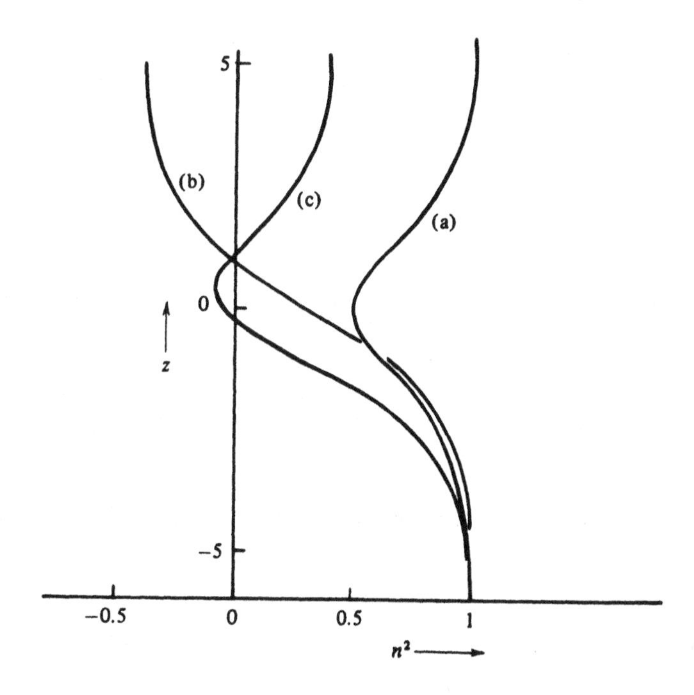

constant value at great heights, fig. 15.8(b), thus

$$n^2 = 1 + (\eta_2 - C^2)e^{\xi}/(1+e^{\xi}) = 1 + \tfrac{1}{2}(\eta_2 - C^2)\{1 + \tanh(\tfrac{1}{2}\xi)\}. \qquad (15.121)$$

The second case is when $\eta_1 = \eta_2 = C^2$ so that there is free space above and below the ionosphere, and

$$n^2 = 1 + \tfrac{1}{4}\eta_3 \operatorname{sech}^2(\tfrac{1}{2}\xi). \qquad (15.122)$$

This is the sech^2 distribution discussed in §§ 12.4, 15.17.

If B is complex a further group of distributions can be obtained from (15.114). Examples will be found in § 19.3 and in problem 15.1. Still other distributions can be obtained by letting some of the variables tend to infinity. For example let

$$B = \mathrm{i}\pi - \ln\{\eta_2(1 - \mathrm{i}Z)\}, \quad \eta_1 = C^2, \quad \sigma = 1/\alpha \qquad (15.123)$$

and then let $\eta_2 \to \infty$. Then (15.114) gives $q^2 = C^2 - e^{\alpha z}/(1 - \mathrm{i}Z)$ which is simply the exponential distribution (15.45). The reader should verify that the reflection coefficient for this case, (15.132) below, is the same as (15.59).

Epstein's transformation is about the simplest possible. When it is replaced by other transformations, a very wide range of height distributions $N(z)$ can be studied. The published papers on this topic are very numerous. Some of the most important are Rawer (1939), Burman and Gould (1954), Heading (1965).

15.15. Reflection and transmission coefficients for Epstein layers

In § 15.13 four independent solutions (15.100), (15.101), (15.104), (15.105) were found for the hypergeometric equation. The field variable E associated with each of these must now be examined. First, when z is large and negative, $|e^{\xi}| \ll 1$ and (15.112) becomes

$$E = u \exp\{\tfrac{1}{2}(c-1)\xi\}. \qquad (15.124)$$

In (15.100), (15.101) the hypergeometric functions are unity and when (15.110), (15.118) are used these solutions give

$$(15.100){:}\quad u = 1, \quad E = \exp\{\tfrac{1}{2}(c-1)\xi\} = \exp\{-\mathrm{i}kC(z + \sigma B)\}, \qquad (15.125)$$

$$(15.101){:}\quad u = (-\zeta)^{1-c}, \quad E = \exp\{\tfrac{1}{2}(1-c)\xi\} = \exp\{\mathrm{i}kC(z + \sigma B)\}. \qquad (15.126)$$

Hence (15.100) and (15.125) represent the upgoing or incident wave below the ionosphere and (15.101) and (15.126) represent the downgoing or reflected wave.

Next, when z is large and positive $|e^{z}| \gg 1$ and (15.112) becomes

$$E = u \exp\{\tfrac{1}{2}(a+b)\xi\}. \qquad (15.127)$$

In (15.104), (15.105) the hypergeometric functions are unity when z is large and positive and when (15.110), (15.119) are used they give

$$(15.104){:}\quad u = (-\zeta)^{-a}, \quad E = \exp\{\tfrac{1}{2}(b-a)\xi\} = \exp\{\mathrm{i}kq_2(z + \sigma B)\}, \qquad (15.128)$$

$$(15.105){:}\quad u = (-\zeta)^{-b}, \quad E = \exp\{\tfrac{1}{2}(a-b)\xi\} = \exp\{-\mathrm{i}kq_2(z + \sigma B)\}. \qquad (15.129)$$

Hence (15.105) and (15.129) represent the upgoing wave or transmitted wave above the ionosphere, and (15.104) and (15.128) represent a downgoing wave that cannot be present when the only source of energy is below the ionosphere.

To find the reflection and transmission coefficients we select a solution in which (15.104) that is (15.128) is absent. The required solution is (15.105) and this is a linear combination of the solutions (15.100), (15.101) given by the circuit relation (15.109). Hence the solution below the ionosphere is, from (15.109), (15.124),

$$E=\frac{(-c)!}{(-a)!(b-c)!}\exp\{-ikC(z+\sigma B)\}+\frac{(c-2)!}{(c-a-1)!(b-1)!}\exp\{ikC(z+\sigma B)\} \tag{15.130}$$

and above the ionosphere it is

$$E=\frac{1}{(b-a)!}\exp\{-ikq_2(z+\sigma B)\}. \tag{15.131}$$

The reflection coefficient R_{22} is the ratio of the second to the first term in (15.130) and for some reference level $z=z_1$ it is

$$R_{22}=\frac{(c-2)!(-a)!(b-c)!}{(c-a-1)!(b-1)!(-c)!}\exp\{2ikC(z_1+\sigma B)\}. \tag{15.132}$$

The transmission coefficient T_{22} is the ratio of (15.131) to the first term of (15.130). If the transmitted wave is observed at $z=z_2$ and the incident wave at $z=z_1$, then

$$T_{22}=\frac{(-a)!(b-c)!}{(b-a)!(-c)!}\exp[-ik\{q_2(z_2+\sigma B)-C(z_1+\sigma B)\}]. \tag{15.133}$$

15.16. Ionosphere with gradual boundary

As an illustration of the use of these formula consider the example of fig. 15.8 curve b, in which $\eta_3=0$ and n^2 is given by (15.121). Below the ionosphere where z is large and negative, $n^2=1$. At great heights where z is large and positive, n^2 is a constant equal to

$$n^2=\eta_2+S^2=1-X_2/(1-\mathrm{i}Z_2) \tag{15.134}$$

so that the medium is ultimately homogeneous and there is a continuous transition from this to free space as the height decreases. Let $B=-z_{\mathrm{m}}/\sigma$, $\eta_2=q_2^2$. Then the reflection coefficient (15.132) referred to the level $z_1=z_{\mathrm{m}}$ is

$$R_{22}=\frac{C-q_2}{C+q_2}\frac{(-2ik\sigma C)!}{(2ik\sigma C)!}\left[\frac{\{ik\sigma(q_2+C)\}!}{\{ik\sigma(q_2-C)\}!}\right]^2. \tag{15.135}$$

If the scale factor σ is very small, the transition from $q=C$ to $q=q_2$ occurs sharply where $z=0$. The factorials in (15.135) are then all unity and the reflection coefficient reduces to $(C-q_2)/(C+q_2)$ which is simply the Fresnel formula (11.57) for reflection at a sharp boundary.

The factor $(-2ik\sigma C)!/(2ik\sigma C)!$ in (15.135) is the ratio of two complex conjugates and has modulus unity. It therefore affects only the phase change at reflection. If electron collisons are neglected, n^2 is real and q_2 is either real and positive or purely imaginary. When it is purely imaginary the angle of incidence θ exceeds the critical angle. The last term of (15.135) is then the ratio of complex conjugates, and $|R_{22}| = 1$, so there is total reflection whether or not the boundary is sharp. When q_2 is real the moduli of the factorial functions may be found from the formula $x!(-x)! = \pi x/\sin(\pi x)$ whence it is easily shown that

$$|R_{22}| = \frac{\sinh\{\pi k\sigma(C - q_2)\}}{\sinh\{\pi k\sigma(C + q_2)\}}. \qquad (15.136)$$

When $\sigma \to 0$ this reduces to the Fresnel formula. When σ is large it gives

$$|R_{22}| = \exp(-2\pi k\sigma q_2) \qquad (15.137)$$

which tends to zero as $\sigma \to \infty$.

For large σ the medium varies very slowly with z, and the arguments of § 7.19 show that it ought then to be possible to find R_{22} from the phase integral formula (7.152). The upper limit z_0 of the integral in this formula is where $q = 0$, and (15.121) shows that then

$$e^{\xi} = -C^2/\eta_2, \text{ that is } z = z_m + \sigma\ln(-C^2/\eta_2) + 2\pi i\sigma r \qquad (15.138)$$

where r is a positive, negative or zero integer. Thus q has an infinite number of zeros and it must be decided which is to be used for z_0 in (7.152). Now (15.134) shows that $-C^2/\eta_2$ always has a negative imaginary part and we choose the logarithm in (15.138) so that its imaginary part is in the range $-\pi$ to 0.

Suppose that $Z_2 = 0$ and $\eta_2 = q_2^2$ is real and negative. Then q_2 is imaginary and $|R_{22}| = 1$ as shown above. In this case one of the zeros of q is on the real axis. This is the point where, in the language of ray theory, the wave would be reflected. It must be the point to be chosen for z_0, because it is the only one for which (7.152) gives $|R_{22}| = 1$. Thus in (15.138) we use $r = 0$. If now Z_2 takes a small positive value, $\ln(-C^2/\eta_2)$ acquires a negative imaginary part. The point z_0 moves away from the real z axis to the negative imaginary side. A similar situation occurred for the parabolic distribution; see § 15.9 and fig. 15.3.

Now let Z_2 and X_2 in (15.134) be changed continuously. The zero of q at $z = a_0$, with $r = 0$, changes position but it can never cross the real z axis and is always the nearest zero to the real z axis on the negative imaginary side. In particular if $Z_2 = 0$ and $\eta^2 = q_2^2$ is real and negative, this zero is where $\text{Im}(z_0) = -\pi\sigma$. It is the analogue of the point Q in fig. 15.3(a).

Evaluation of the phase integral for this problem is rather complicated. A method of doing it was given by Budden (1961a, § 20.5) for the special case $Z = 0$, $\theta = 0$, $C = 1$. If σ is large, the factorials in (15.135) can be given their asymptotic values from Stirling's formula. The values of R_{22} found from these two methods were the same. This confirms that the choice of the zero of q with $r = 0$ in (15.138) is correct.

The phase integral formula may therefore be used when the model ionosphere varies slowly enough, that is when σ is large. For smaller σ the more accurate formula (15.135) must be used, and when σ is infinitesimally small this reduces to the Fresnel formula (11.57) for the reflection coefficient of a sharp boundary. There is a continuous transition from reflection in a slowly varying medium at a zero of q, to reflection at a sharp boundary.

15.17. The 'sech^2' distribution

The distribution (15.122) is, for $B = -z_m/\sigma$,

$$n^2 = 1 + \tfrac{1}{4}\eta_3 \operatorname{sech}^2\{\tfrac{1}{2}(z - z_m)/\sigma\}. \tag{15.139}$$

There is free space both below and above the ionosphere where $|z - z_m|$ is large. The ionosphere is a symmetrical layer with its centre at $z = z_m$, where $n^2 = 1 + \tfrac{1}{4}\eta_3$. This distribution was used by Rawer (1939) to study partial penetration and reflection. It has also been studied by Westcott (1962c, d). It is in some respects better than the parabolic layer studied in §§ 15.9–15.11 since there is no discontinuity of dN/dz. Hence the 'sech^2' distribution can be used to study very thin layers and is not subject to the limitation (15.73).

Let

$$\eta_3 = -4X_m/(1 - iZ) \tag{15.140}$$

so that the collision frequency is independent of height and the electron concentration is given from

$$X = X_m \operatorname{sech}^2\{\tfrac{1}{2}(z - z_m)/\sigma\}. \tag{15.141}$$

From (15.118), (15.119) it follows that $a - b = c - 1 = -2ik\sigma C$. Let $4k^2\sigma^2\eta_3 + 1 = 4\gamma^2$ so that $a + b - c = 2\gamma$ from (15.120). Then the reflection coefficient (15.132) referred to the level $z_1 = z_m$ is

$$R_{22} = -\frac{(-2ik\sigma C)!\,(2ik\sigma C - \gamma - \frac{1}{2})!\,(2ik\sigma C + \gamma - \frac{1}{2})!}{(2ik\sigma C)!\,(-\gamma - \frac{1}{2})!\,(\gamma - \frac{1}{2})!} \tag{15.142}$$

and the transmission coefficient (15.133) when both incident and transmitted waves are referred to the same level, $z_1 = z_2$, is

$$T_{22} = 2ikC\frac{(2ik\sigma C - \gamma - \frac{1}{2})!\,(2ik\sigma C + \gamma - \frac{1}{2})!}{\{(2ik\sigma C)!\}^2}. \tag{15.143}$$

It can be verified that $R_{22} \to 0$ when $\sigma \to 0$, that is when the layer becomes indefinitely thin.

If $Z = 0$, η_3 is $-4k_p^2/k^2$ from (15.140) where k_p is the value of k for the penetration frequency f_p, and

$$4\gamma^2 = 1 - 16\sigma^2 k_p^2. \tag{15.144}$$

Thus γ is independent of frequency, and is either real or purely imaginary. In both cases the moduli of the factorial functions can be found from the formula $x!(-x)! = \pi x/\sin(\pi x)$, and (15.142), (15.143) give

$$|R_{22}|^2 = \frac{\cos 2\pi\gamma + 1}{\cos 2\pi\gamma + \cosh 4\pi k\sigma C}, \quad |T_{22}|^2 = \frac{\cosh 4\pi k\sigma C - 1}{\cos 2\pi\gamma + \cosh 4\pi k\sigma C}. \tag{15.145}$$

These formulae can be used to study the partial penetration and reflection near the penetration frequency of the layer.

It is interesting to compare the 'sech²' distribution with a parabolic distribution having the same penetration frequency and the same curvature at its maximum; see fig. 12.1. It can be shown that the parabola then has half thickness $a = 2\sigma$. For frequencies greater than about 2 MHz most ionospheric layers have $a/\lambda \geqslant 2$ so that in the cases of greatest interest $16\sigma^2 k_p^2 \gg 1$ and $i\gamma \approx 2\sigma k_p \gg 1$. Thus $\cos 2\pi\gamma \gg 1$ so that the 1 may be neglected in the first numerator of (15.145) and we may take $\cos 2\pi\gamma \approx \frac{1}{2}\exp(4\pi\sigma k_p)$. Then, for frequencies close to penetration, $k \approx k_p/C$, the first formula (15.145) reduces to (15.89) for the parabolic distribution.

When $Z \neq 0$ the formulae (15.142), (15.143) cannot be so easily simplified and the factorial functions with complex argument have to be computed. Rawer (1939) has computed $|R_{22}|^2$ and $|T_{22}|^2$ for various values of Z and curves are given in his paper.

Equations (15.145) show that $|T_{22}|/|R_{22}| = \sinh 2\pi k\sigma C/\cos \pi\gamma$. At high enough frequencies $\sinh 2\pi k\sigma C \approx \frac{1}{2}\exp(2\pi k\sigma C)$ so that

$$\ln|T_{22}/R_{22}| \approx 2\pi k\sigma C + \text{constant}. \tag{15.146}$$

If this is plotted against frequency, proportional to k, it gives a straight line whose slope gives σ, which is a measure of the thickness of the ionospheric layer. This method has been applied to observations at vertical incidence by Briggs (1951). Even if electron collisions are included, it may be assumed that $Z \ll 1$ at high frequencies so that γ is still nearly constant and the method can still be used.

The equivalent height of reflection for radio waves vertically incident on a 'sech²' layer can be found by the method used in §15.11 for the parabolic layer. Since the centre of the 'sech²' layer is at a height z_m above the ground

$$h'(f) = z_m + \tfrac{1}{2}i\,d(\ln R_{22})/dk \tag{15.147}$$

where R_{22} is given by (15.142) with $C = 1$; compare (15.92). Formulae derived from this, with collisions included, were given by Rawer (1939) who also gave curves of $h'(f)$ versus f for various values of Z and σ.

15.18. Other electron height distributions

Two other distributions $N(z)$ that come within the subject of this chapter should be briefly mentioned. The first is the sinusoidal distribution

$$X = \begin{cases} \frac{1}{2}X_m[1 + \cos\{\pi(z - z_m)/a\}] & \text{for } |z - z_m| \leqslant a, \\ 0 & \text{for } |z - z_m| \geqslant a. \end{cases} \tag{15.148}$$

It has no discontinuities of $\mathrm{d}N/\mathrm{d}z$ but there are discontinuities of curvature at the top and bottom of the layer. If Z is independent of height, the differential equation (7.6) for horizontal polarisation can be transformed into one of the standard forms of Mathieu's equation (see, for example, Whittaker and Watson, 1927, ch. XIX), by a change of the independent variable. But the reflection and transmission coefficients, R, T, for this problem can be found very easily by one of the numerical methods described in ch. 18. Some results for $Z = 0$ and vertical incidence were given by Budden (1961a, § 17.8), showing how $|R|$ depends on frequency for frequencies extending on both sides of penetration. They show the phenomenon of partial penetration and reflection very similar to that for the parabolic layer, fig. 15.5.

The second distribution $N(z)$ has the square law dependence

$$X = \begin{cases} \beta(z - h_0)^2 & \text{for } z \geqslant h_0, \\ 0 & \text{for } z \leqslant h_0. \end{cases} \tag{15.149}$$

It is of interest because there is a discontinuity of curvature $\mathrm{d}^2N/\mathrm{d}z^2$ at $z = h_0$. The equation (7.6) for horizontal polarisation can be transformed to Weber's equation (15.70). The reflection coefficient for this case was given by Hartree (1931a) and Rydbeck (1944). A closely related problem, with some allowance for the earth's magnetic field, was discussed by Wilkes (1940). Further details are given by Budden (1961a, § 17.7).

15.19. Collisions. Booker's theorem

The effect of electron collisions has been allowed for in several of the functions $n^2(z)$ studied in this chapter. The following discussion gives another way of dealing with collisions in certain types of isotropic ionosphere.

Suppose that for a loss-free isotropic ionosphere, in which $X(z)$ is a given analytic function of height z, the reflection coefficient R for some real reference level $z = z_1$ has been found for an angle of incidence θ. It is assumed that there is free space in a range of height below z_1. This will be called the 'original system'. The following results apply for any θ and for either horizontal or vertical polarisation. Define a new variable

$$\check{z} = z - \mathrm{i}b \tag{15.150}$$

where b is a real positive constant. Let $\check{z}$ now be treated as the height variable instead of z in a new model of the ionosphere so that real $\check{z}$ refers to real heights. This will be called the 'displaced system', because it is formed by a displacement b in the imaginary direction of the complex z plane. The real $\check{z}$ axis is where $\mathrm{Im}(z) = \mathrm{i}b$. Let $\check{R}$ be the reflection coefficient for reference level $\check{z} = z_1$ in the displaced system. Let F be the field component used to define R, so that F is $\mathscr{H}_y$ for vertical polarisation and E_y for horizontal polarisation. Then in the free space below the ionosphere the z

dependence of F is given by

$$\begin{aligned} F &= \exp\{-ik\cos\theta(z-z_1)\} + R\exp\{ik\cos\theta(z-z_1)\} \\ &= \exp(kb\cos\theta)[\exp\{-ik\cos\theta(\check{z}-z_1)\} \\ &\quad + R\exp(-2kb\cos\theta)\exp\{ik\cos\theta(\check{z}-z_1)\}], \end{aligned} \tag{15.151}$$

whence

$$\check{R} = R\exp(-2kb\cos\theta). \tag{15.152}$$

This new reflection coefficient $\check{R}$ is the same as if the reference level in the original system was changed to the complex value $z_1 + ib$. When $\check{z}$ is real $X(z)$ is complex and given by

$$X(\check{z}+ib) = C(\check{z}) + iD(\check{z}) = \mathscr{X}/(1+\mathscr{Z}^2) + i\mathscr{X}\mathscr{Z}/(1+\mathscr{Z}^2) \tag{15.153}$$

where $C, D, \mathscr{X}, \mathscr{Z}$ are functions of $\check{z}$ that are real when $\check{z}$ is real. The displaced model is a new model of the ionosphere in which X is $\mathscr{X}(\check{z})$ and Z is $\mathscr{Z}(\check{z})$, and the reflection coefficient is given by (15.152). This result was first given by Booker, Fejer and Lee (1968). Later writers have referred to it as 'Booker's theorem'. By choosing various values of b a whole range of new models can be derived from the original loss-free system.

For this method to work it is necessary first that $\mathscr{X}(\check{z})$ and $\mathscr{Z}(\check{z})$ shall be physically realisable functions, and second that at great heights in the displaced system the only wave present is an upgoing wave. As a simple example, given by Booker *et al.* (1968), suppose that for the original system $X(z) = e^{\alpha z}$ as in (15.44). Then (15.153) gives

$$\mathscr{X} = \exp(\alpha\check{z})\sec(\alpha b), \quad \mathscr{Z} = \tan(\alpha b) \tag{15.154}$$

so that $\mathscr{Z}$ is independent of height $\check{z}$. This is the same as in the example of § 15.8. It is easily checked that (15.152) is in agreement with (15.59), (15.60).

An essential requirement is that the original function $X(z)$ shall be analytic and shall not have any singularities in the region $0 < \mathrm{Im}(z) \leqslant b$. The theorem cannot be applied if $X(z)$ is only piecewise analytic. For example in the linear model of § 15.2, $X(z)$ has a discontinuity of gradient where $z = h_0$. When $X(z)$ and the solutions (15.8), (15.7) are continued analytically to use complex z, the two real regions $z < h_0$ and $z > h_0$ are continued separately to give two different complexions of the complex z plane. The two solutions (15.8), (15.7) were made to match at the real point $z = h_0$. They do not match at any other point in the two complexions of the z plane. For similar reasons the theorem cannot be applied to the parabolic model, §§ 15.9–15.11. It was shown by von Roos (1970) that if the earth's magnetic field is allowed for, the functions $\mathscr{X}(z)$, $\mathscr{Z}(z)$ that result from a displacement are not physically realisable.

The most useful application of the theorem is for an isotropic ionosphere where the original $X(z)$ is some combination of exponential functions. The Epstein functions, §15.14, are important examples. Heading (1972) has shown that the theorem can be generalised as follows. If the original real function $X(z)$ is multiplied by a complex constant A, it may still be possible to find the reflection coefficient R, even though this modification is physically unrealisable. But if a displacement b is now imposed as in (15.150), new functions $\mathscr{X}(z), \mathscr{Y}(z)$ can be found that are physically realisable. This extends the range of usefulness of the theorem. Heading has used this to derive three different examples of ionospheric models with collisions, using loss-free Epstein distributions as the original system.

PROBLEMS 15

15.1. Consider a model isotropic ionosphere in which the electron concentration is the same at all heights so that X is a constant, and the collision frequency is height dependent so that $Z = Z_g \exp(-z/\sigma)$. For z large and negative, Z is so large that $n^2 = 1 - X/(1 - iZ) \approx 1$ and the medium is like free space. At great heights Z is negligible and the medium is homogeneous with refractive index $n_2 = (1 - X)^{\frac{1}{2}}$. Show that the reflection coefficient R for vertical incidence can be found by using an Epstein distribution (15.110)–(15.117) with $\eta_1 = 1$, $\eta_2 = 1 - X$, $\eta_3 = 0$, $B = \frac{1}{2}i\pi - \ln Z_g$. Show that if n_2^2 is positive

$$|R| = \frac{\sinh\{\pi k\sigma |n_2 - 1|\}}{\sinh\{\pi k\sigma(n_2 + 1)\}} \exp(-\pi k\sigma)$$

and if n_2^2 is negative, $|R| = \exp(-\pi k\sigma)$. For an example of the use of a model of this type see Greifinger and Greifinger (1965).

For the solution see Budden (1961a, §17.17).

16

Coupled wave equations

16.1. Introduction

For wave propagation in a stratified medium, the idea of writing the governing differential equations as a set of coupled equations has already been used in §§ 7.8, 7.15, for the study of W.K.B. solutions. The term 'coupled equations' is usually given to a set of simultaneous ordinary linear differential equations with the following properties:

(1) There is one independent variable which in this book is the height z or a linear function of it.

(2) The number of equations is the same as the number of dependent variables.

(3) In each equation one dependent variable appears in derivatives up to a higher order than any other. The terms in this variable are called 'principal' terms and the remaining terms are called 'coupling' terms.

(4) The principal terms contain a different dependent variable in each equation, so that each dependent variable appears in the principal terms of one and only one equation.

It is often possible to choose the dependent variables so that the coupling terms are small over some range of z. Then the equations may be solved by successive approximations. As a first approximation the coupling terms are neglected, and the resulting equations can then be solved. The values thus obtained for the dependent variables are substituted in the coupling terms and the resulting equations are solved to give a better approximation. Some examples of this process are given in § 16.13. Clearly it cannot be used when any of the coupling terms becomes large.

In most of the coupled equations studied in this chapter the coupling terms are large near the turning points in the complex z plane, that is points where two roots q of the Booker quartic equation, ch. 6, are equal. Generally, coupled equations are not recommended for computing, because the transformations that lead to them introduce, at the turning points, singularities that are not originally present. See end of § 16.13 for further discussion of this. The importance of coupled equations is that they help in the study of the processes of reflection, and of coupling between ordinary and extraordinary waves, that is 'mode conversion'. It is therefore important to study coupled equations in the neighbourhood of a turning point, as is done in §§ 16.3, 16.4.

The basic equations for the general anisotropic stratified plasma are the set of four equations (7.80). These are already in the form of coupled equations as defined above. The first of them is

$$\mathrm{d}E_x/\mathrm{d}z + \mathrm{i}k\mathsf{T}_{11}E_x = \mathrm{i}k(\mathsf{T}_{12}E_y - \mathsf{T}_{14}\mathscr{H}_y) \tag{16.1}$$

since (7.81) shows that $\mathsf{T}_{13} = 0$. Here the terms in E_x on the left-hand side are the principal terms and the terms in $E_y, \mathscr{H}_y$ are the coupling terms. In these equations, however, the coupling terms are in general of comparable magnitude to the principal terms, even in a homogeneous medium, so that we could not use a method of successive approximation in which the coupling terms are at first neglected.

It has been assumed in earlier chapters, especially chs. 10, 13, that the ordinary and extraordinary waves can be considered to be propagated independently. This suggests that the dependent variables should be chosen so that one refers to the ordinary wave only, and the other to the extraordinary wave only. The coupling terms should then be very small if the assumption is justified. Equations of this kind were given first by Försterling (1942) for the case of vertical incidence; see § 16.11. One of Försterling's variables $\mathscr{F}_{\mathrm{O}}$ gave the field of the total ordinary wave including both upgoing and downgoing waves. The other $\mathscr{F}_{\mathrm{E}}$, similarly, gave the total extraordinary wave. The terms 'ordinary' and 'extraordinary' are used here only for convenience of description. They may be ambiguous, as explained in § 4.16. Thus Försterling did not separate the four characteristic waves completely. If the coupling terms in his equations are neglected, the resulting equations are two second order equations (16.92) governing the independent propagation of the ordinary and extraordinary waves.

The theory was extended to the general case by Clemmow and Heading (1954), who used four independent variables f_1 to f_4, (6.53), for the four characteristic waves; §§ 6.10, 7.15, 7.16. Their form of the coupled equations is (7.109). Where the coupling terms can be neglected, the resulting four equations (7.110) lead to the four W.K.B. solutions (7.112).

The theory of coupling and mode conversion in plasmas, presented from a somewhat different point of view from that used here, has been given by Fuchs, Ko and Bers (1981). They use a conformal mapping of the complex z plane into the complex plane of $\omega n(z)/c$ ($= k$ in their notation), and the mapping of the real z axis is particularly important. They introduce branch cuts in the z plane similar to those used here in §§ 16.5, 16.6. They give references to applications of the theory, including many from the literature of plasma physics.

16.2. First order coupled equations

The coupled equations studied in this chapter all apply for waves with the constant values $S_1 = S = \sin\theta$, $S_2 = 0$; § 6.2. Thus they apply, for example, when the incident wave below the ionosphere is an obliquely incident plane wave. The form of any set of coupled equations depends on how we define the partial waves that give the principal terms. In §§ 6.10, 7.14 the four partial waves were defined as follows. At any level z consider a fictitious homogeneous medium that has all the properties of the actual medium; see end of § 6.10. Then for each partial wave the ratios of the four field components E_x, $-E_y$, $\mathscr{H}_x$, $\mathscr{H}_y$ are to be the same as for one of the four possible progressive waves, § 6.2, with the given value of S in the fictitious medium. It was this form that led to the four coupled equations (7.109). In each equation the principal terms include a first order derivative of one of the field variables, and the coupling terms contain no derivatives of the field variables.

To study these first order coupled equations it is useful first to give explicit expressions for the elements Γ_{ij} of the coupling matrix $\mathbf{\Gamma} = -\mathbf{S}^{-1}\mathbf{S}'$ (7.108). The prime here means $k^{-1}\mathrm{d}/\mathrm{d}z$.

Now $\mathbf{S}^{-1}\mathbf{S}$ is a constant, unity, so that $\mathbf{S}^{-1}\mathbf{S}' + (\mathbf{S}^{-1})'\mathbf{S} = 0$. Hence

$$\mathbf{\Gamma} = \tfrac{1}{2}\{(\mathbf{S}^{-1})'\mathbf{S} - \mathbf{S}^{-1}\mathbf{S}'\}. \tag{16.2}$$

Substitution from (7.134), (7.135) now gives Γ_{ij}. The relation

$$[\{(A_iF_i)^{-\frac{1}{2}}\}'(A_jF_j)^{-\frac{1}{2}} - (A_iF_i)^{-\frac{1}{2}}\{(A_jF_j)^{-\frac{1}{2}}\}'] \times [(\bar{a}_5q_i + \bar{a}_6)(a_3q_j + a_4) + A_iA_j(q_i + q_j) + (\bar{a}_3q_i + \bar{a}_4)(a_5q_j + a_6)] = 0 \tag{16.3}$$

holds for all i, j since the first bracket is zero when $i = j$, and the second is zero when $i \neq j$ because $\mathbf{S}^{-1}\mathbf{S}$ is diagonal. The result is therefore

$$\begin{aligned}2\Gamma_{ij} = (A_iF_iA_jF_j)^{-\frac{1}{2}}\{&(\bar{a}_5q_i + \bar{a}_6)'(a_3q_j + a_4) - (\bar{a}_5q_i + \bar{a}_6)(a_3q_j + a_4)' \\ &+ (q_iA_i)'A_j - q_iA_iA_j' + q_jA_jA_i' - (q_jA_j)'A_i \\ &+ (\bar{a}_3q_i + \bar{a}_4)'(a_5q_j + a_6) - (\bar{a}_3q_i + \bar{a}_4)(a_5q_j + a_6)'\}.\end{aligned} \tag{16.4}$$

Finally, with the help of (7.131), a rearrangement gives

$$\begin{aligned}2\Gamma_{ii} = (A_iF_i)^{-1}\{&q_i^2(a_3\bar{a}_5' - a_3'\bar{a}_5 + a_5\bar{a}_3' - a_5'\bar{a}_3) + a_4\bar{a}_6' - a_4'\bar{a}_6 + a_6\bar{a}_4' - a_6'\bar{a}_4 \\ &+ q_i(a_3\bar{a}_6' - a_3'\bar{a}_6 + a_4\bar{a}_5' - a_4'\bar{a}_5 + a_5\bar{a}_4' - a_5'\bar{a}_4 + a_6\bar{a}_3' - a_6'\bar{a}_3)\},\end{aligned} \tag{16.5}$$

$$\begin{aligned}2\Gamma_{ij} = (A_iF_iA_jF_j)^{-\frac{1}{2}}\{&(q_i'q_j - q_j'q_i)(a_3\bar{a}_5 + a_5\bar{a}_3) \\ &+ q_iq_j(a_3\bar{a}_5' - a_3'\bar{a}_5 + a_5\bar{a}_3' - a_5'\bar{a}_3) + q_i(a_4\bar{a}_5' - a_4'\bar{a}_5 + a_6\bar{a}_3' - a_6'\bar{a}_3) \\ &+ q_j(a_3\bar{a}_6' - a_3'\bar{a}_6 + a_5\bar{a}_4' - a_5'\bar{a}_4) + (q_i + q_j)(A_i'A_j - A_iA_j') \\ &+ (q_i' - q_j')(a_4\bar{a}_5 + a_6\bar{a}_3 + A_iA_j) + a_4\bar{a}_6' - a_4'\bar{a}_6 + a_6\bar{a}_4' - a_6'\bar{a}_4\}.\end{aligned} \tag{16.6}$$

With these elements of $\mathbf{\Gamma}$ the coupled equations (7.109) are

$$f_i' + \mathrm{i}q_if_i = \Gamma_{i1}f_1 + \Gamma_{i2}f_2 + \Gamma_{i3}f_3 + \Gamma_{i4}f_4 \quad (i = 1, 2, 3, 4) \tag{16.7}$$

where, from (6.53) second equation, and (7.135)

$$f_i = (A_iF_i)^{-\frac{1}{2}}\{(\bar{a}_5q_i + \bar{a}_6)E_x - q_iA_iE_y + A_i\mathscr{H}_x + (\bar{a}_3q_i + \bar{a}_4)\mathscr{H}_y\}. \tag{16.8}$$

All the terms of the coupling coefficients Γ_{ij} (16.6) contain derivatives with respect to z. They are therefore small in a sufficiently slowly varying medium, provided that the denominators $(A_iF_iA_jF_j)^{-\frac{1}{4}}$ are not small. At points in the complex z plane where they can be neglected, the remaining terms in (7.109) are four independent differential equations whose solutions are the W.K.B. approximations, as defined and derived in § 7.15. These four solutions are associated with the four roots q_1 to q_4 of the Booker quartic, ch. 6, and will be referred to as wave 1 to wave 4 respectively.

As an illustration of the properties of the coupling coefficients consider the special case of Γ_{12} and Γ_{21} from (16.6). Their denominators $(A_1F_1A_2F_2)^{-\frac{1}{4}}$ contain a factor $(q_1-q_2)^{-1}$, so that they are both infinite at a point $z=z_p$ in the complex z plane where $q_1=q_2$. Such a point is called a coupling point. It may be a reflection point. In the general treatment of this section there is no distinction between reflection and coupling, and mathematically they are the same phenomena. Near z_p waves 1 and 2 are not independently propagated and are said to be 'strongly coupled'. But (16.6) also shows that $\Gamma_{13}, \Gamma_{14}, \Gamma_{31}, \Gamma_{41}, \Gamma_{23}, \Gamma_{24}, \Gamma_{32}, \Gamma_{42}$ all contain a factor $(q_1-q_2)^{-\frac{1}{2}}$ so that they too are infinite at z_p, even through the power is $-\frac{1}{2}$ giving an infinity of lower order than the power -1 in Γ_{12}, Γ_{21}. This might still suggest that there is strong coupling of wave 3 and wave 4 with waves 1 and 2. But this is not so. Provided that $q_1=q_2\neq q_3$ and $\neq q_4$, it can be shown that wave 3 and wave 4 are propagated independently of waves 1 and 2. The proof of this was first given by Booker (1936). Another proof follows from the theory in § 16.3, based on the original method of Heading (1961). An alternative proof for vertically incident waves is given by using second order coupled equations; see §§ 16.10–16.12.

The expression (16.6) for Γ_{ij} has a slightly different form if the matrix $\mathbf{S}$ in (6.53), (7.108) is normalised differently. For example if the normalisation factor $(A_jF_j)^{-\frac{1}{2}}$ is omitted from (7.134), the coefficient $(A_iF_iA_jF_j)^{-\frac{1}{2}}$ in (16.6) must be replaced by $(A_jF_j)^{-1}$. Then in the example just discussed $\Gamma_{13}, \Gamma_{14}, \Gamma_{23}, \Gamma_{24}$ are not now infinite at z_p where $q_1=q_2$, but $\Gamma_{31}, \Gamma_{41}, \Gamma_{32}, \Gamma_{42}$ now contain a factor $(q_1-q_2)^{-1}$ and so they are infinite to a higher order. The difficulty cannot be removed by a change in the normalisation of $\mathbf{S}$. Thus care is needed when using the first order coupled equations near a reflection or coupling point. Their main function is to illustrate physical principles, and they are not recommended for computing.

The expressions (16.5), (16.6) for the elements of $\mathbf{\Gamma}$ take simpler forms in some special cases.

(a) *Vertical incidence*

When $\theta=0, S=0, C=1$ the numbers a and $\bar{a}$ take the simple forms (7.137) and some of them are zero. Then (16.5) shows that $\Gamma_{ii}=0$ and (16.6) gives

$$2\Gamma_{ij}=(A_iF_iA_jF_j)^{-\frac{1}{2}}\{(A_i'A_j-A_iA_j')(q_i+q_j)+(A_iA_j-a_4^2)(q_i'-q_j')\}$$
$$=\left(\frac{A_i}{F_i}\frac{A_j}{F_j}\right)^{\frac{1}{2}}\left\{\left(\frac{A_i'}{A_i}-\frac{A_j'}{A_j}\right)(q_i+q_j)+\left(1-\frac{a_4^2}{A_iA_j}\right)(q_i'-q_j')\right\} \tag{16.9}$$

which is antisymmetric. The qs are now given by (7.139) and it can be shown from (7.140)–(7.144) that

$$\frac{A_1}{F_1} = -\frac{A_2}{F_2} = \frac{\rho_O^2}{2n_O(\rho_O^2 - 1)}, \quad -\frac{A_3}{F_3} = \frac{A_4}{F_4} = \frac{1}{2n_E(\rho_O^2 - 1)} \tag{16.10}$$

and from (7.143), (7.144)

$$\frac{A_1'}{A_1} = \frac{A_2'}{A_2} = \frac{a_4'}{a_4} + \frac{\rho_O'}{\rho_O}, \quad \frac{A_3'}{A_3} = \frac{A_4'}{A_4} = \frac{a_4'}{a_4} - \frac{\rho_O'}{\rho_O}. \tag{16.11}$$

For the square root in (16.9) a sign convention must be adopted and we take

$$(-n_O)^{-\frac{1}{2}} = \mathrm{i}n_O^{-\frac{1}{2}}, \; (-n_E)^{-\frac{1}{2}} = \mathrm{i}n_E^{-\frac{1}{2}}, \tag{16.12}$$

where $n_O^{-\frac{1}{2}}$, $n_E^{-\frac{1}{2}}$ have positive real parts. Consider now the case $i = 1$, $j = 3$. Then $1 - a_4^2/A_iA_j$ is zero from (7.143), (7.144), (4.29) and the remaining term in (16.9) gives

$$\Gamma_{13} = \tfrac{1}{2}\mathrm{i}\psi(n_O + n_E)(n_On_E)^{-\frac{1}{2}} \tag{16.13}$$

where

$$\psi = \rho_O'/(\rho_O^2 - 1) = \frac{1}{2k}\frac{\mathrm{d}}{\mathrm{d}z}\ln\left(\frac{\rho_O - 1}{\rho_O + 1}\right). \tag{16.14}$$

The other elements of $\boldsymbol{\Gamma}$ can similarly be found and when they are inserted in (7.109) they give the following four first order coupled equations

$$\left.\begin{aligned}
\mathsf{f}_1' + \mathrm{i}n_O\mathsf{f}_1 &= -(n_O'/2\mathrm{i}n_O)\mathsf{f}_2 + \tfrac{1}{2}\mathrm{i}\psi(n_O + n_E)(n_On_E)^{-\frac{1}{2}}\mathsf{f}_3 + \tfrac{1}{2}\psi(n_O - n_E)(n_On_E)^{-\frac{1}{2}}\mathsf{f}_4,\\
\mathsf{f}_2' - \mathrm{i}n_O\mathsf{f}_2 &= (n_O'/2\mathrm{i}n_O)\mathsf{f}_1 + \tfrac{1}{2}\psi(n_O - n_E)(n_On_E)^{-\frac{1}{2}}\mathsf{f}_3 - \tfrac{1}{2}\mathrm{i}\psi(n_O + n_E)(n_On_E)^{-\frac{1}{2}}\mathsf{f}_4,\\
\mathsf{f}_3' + \mathrm{i}n_E\mathsf{f}_3 &= -\tfrac{1}{2}\psi(n_O + n_E)(n_On_E)^{-\frac{1}{2}}\mathsf{f}_1 - \tfrac{1}{2}\psi(n_O - n_E)(n_On_E)^{-\frac{1}{2}}\mathsf{f}_2 + (n_E'/2\mathrm{i}n_E)\mathsf{f}_4,\\
\mathsf{f}_4' - \mathrm{i}n_E\mathsf{f}_4 &= -\tfrac{1}{2}\psi(n_O - n_E)(n_On_E)^{-\frac{1}{2}}\mathsf{f}_1 + \tfrac{1}{2}\mathrm{i}\psi(n_O + n_E)(n_On_E)^{-\frac{1}{2}}\mathsf{f}_2 - (n_E'/2\mathrm{i}n_E)\mathsf{f}_4,
\end{aligned}\right\} \tag{16.15}$$

where

$$\left.\begin{aligned}
\mathsf{f}_1 &= \{2n_O(\rho_O^2 - 1)\}^{-\frac{1}{2}}(n_OE_x - n_O\rho_OE_y + \rho_O\mathscr{H}_x + \mathscr{H}_y),\\
\mathsf{f}_2 &= -\mathrm{i}\{2n_O(\rho_O^2 - 1)\}^{-\frac{1}{2}}(n_OE_x - n_O\rho_OE_y - \rho_O\mathscr{H}_x - \mathscr{H}_y),\\
\mathsf{f}_3 &= -\mathrm{i}\{2n_E(\rho_O^2 - 1)\}^{-\frac{1}{2}}(-\rho_On_EE_x + n_EE_y - \mathscr{H}_x - \rho_O\mathscr{H}_y),\\
\mathsf{f}_4 &= \{2n_E(\rho_O^2 - 1)\}^{-\frac{1}{2}}(-\rho_On_EE_x + n_EE_y + \mathscr{H}_x + \rho_O\mathscr{H}_y).
\end{aligned}\right\} \tag{16.16}$$

Here f_1 to f_4 are associated with the four wave types upgoing ordinary, downgoing ordinary, upgoing extraordinary, downgoing extraordinary, respectively. This order is a matter of choice and is determined here by (7.139) for the four qs. Any other choice would be permissible and in § 18.10 a different order is used with $\mathsf{f}_2, \mathsf{f}_3$ interchanged. The parameter ψ in (16.14) is called the 'coupling parameter'. It was introduced by Försterling (1942) and plays an important part in all the theory of coupled equations. It may be regarded as a function of the state of the ionosphere at each level, and its properties are discussed in § 16.12. Since it is a derivative with respect to the height z, it is zero in a homogeneous medium.

The elements Γ_{12}, Γ_{21} on the right of (16.15) are large near levels where $n_O = 0$. They are associated with coupling between the upgoing and downgoing ordinary waves, that is with reflection of the ordinary wave. Similarly Γ_{34} and Γ_{43} are associated with reflection of the extraordinary wave. The elements Γ_{13} and Γ_{31} are associated with coupling between the upgoing ordinary and extraordinary waves, and similarly Γ_{24} and Γ_{42} are associated with the two downgoing waves. It has been suggested that these two kinds of coupling may be responsible for the 'Z-trace' sometimes observed in ionograms at high latitudes; see § 16.8. These terms become large when ψ is large, that is when $n_O \approx n_E$, for it is shown, (16.96), that ψ has a factor $(n_O - n_E)^{-2}$; see § 16.12. Thus Γ_{13}, Γ_{31}, Γ_{24} and Γ_{42} have infinities of this order.

The elements Γ_{14} and Γ_{41} would be associated with coupling between the upgoing ordinary wave and the downgoing extraordinary wave. This could be strong only near points where $n_O + n_E = 0$, and (16.96) indeed shows that ψ has a factor $(n_O + n_E)^{-2}$. But n_O and n_E are usually defined so that their real parts are not less than zero; see § 16.5 and (16.52). This condition, therefore, could never occur. A similar conclusion applies for Γ_{23}, Γ_{32}. It has been suggested that coupling of these types might be responsible for a reflection known as the 'coupling echo'. This subject, however, is more complicated, and is discussed in § 16.14.

When the earth's magnetic field is neglected so that $\psi = 0$, the equations (16.15) separate into two pairs, one pair containing only f_1 and f_2 and the other pair containing only f_3 and f_4. By the substitutions $\mathsf{f}_1 = n_O^{\frac{1}{2}} E_y^{(1)}$, $\mathsf{f}_2 = n_O^{\frac{1}{2}} E_y^{(2)}$ the first pair can be converted to the form (7.46), (7.47). Similarly the second pair can be converted to the same form.

(b) *Other special cases*

There are two other special cases where the first order coupled wave equations have a fairly simple form. The first is when the earth's magnetic field is vertical, see problem 16.2, and the second is when the earth's magnetic field is horizontal and in the plane of incidence, that is for north–south or south–north propagation at the magnetic equator. The equations are similar in general form to (16.15). They are not needed in this book, and they have not been extensively used. The algebraic details were given by Budden and Clemmow (1957) and by Budden (1961a, § 18.15).

16.3. Coupled equations near a coupling point

The derivation of the first order coupled equations in § 7.15 depended on the use of the transforming matrix $\mathbf{S}$ in (6.53) and of its inverse. Thus it is necessary that $\mathbf{S}$ shall be non-singular. At a coupling point, where two of the qs are equal, the two corresponding columns (7.118) or (7.134) of **S** are the same so **S** is there singular and $\mathbf{S}^{-1}$ does not exist. To derive a set of coupled equations that can be used throughout a domain of the z plane containing a coupling point, an alternative treatment must be used. The following treatment is based on the method of Heading (1961) who gave the general mathematical theory. It is here written in the language of radio propagation in stratified media (Budden, 1972). A treatment very similar to that of the present section was given by Ludwig (1970). Heading's method has also been used in the study of mode conversion, by Fuchs, Ko and Bers (1981).

Let z_p be a coupling point where $q_1 = q_2$ and consider a domain of the complex z plane containing z_p in which there is no other coupling point and in which the coefficients in the Booker quartic $F(q) = 0$, (6.15) or (7.128), are analytic functions of z. Then, where the roots q_i are distinct, each of them is analytic. A factor $(q - q_3)(q - q_4)$ can be removed from the quartic, leaving a quadratic for q whose coefficients are analytic in the domain. Each of its two roots, however, contains a square root proportional to $q_1 - q_2$, that is zero when $z = z_p$. Thus z_p is a branch point of q_1 and q_2. Let

$$\alpha = \tfrac{1}{2}(q_1 + q_2), \quad \beta = \tfrac{1}{2}(q_1 - q_2). \tag{16.17}$$

Then α is analytic but β has a branch point at z_p and β^2 is analytic there. These symbols must not be confused with the coefficients α, β in the Booker quartic (6.15), (6.16).

For the columns $\mathbf{s}_j$ of **S** we now choose the form (7.134) with the factor $(A_j F_j)^{-\frac{1}{2}}$ omitted. Thus in this section the method of normalising **S** is different from that used in §§7.14–7.16. Then on using (7.119) (16.17)

$$\tfrac{1}{2}(\mathbf{s}_1 - \mathbf{s}_2) = \beta \begin{pmatrix} a_3 \\ 2\alpha + a_1 \\ 3\alpha^2 + \beta^2 + 2\alpha a_1 + a_2 \\ a_5 \end{pmatrix} = \beta \mathbf{s}_\beta. \tag{16.18}$$

This is zero at $z = z_p$ solely because of the factor β. The other factor $\mathbf{s}_\beta$ is a column matrix whose elements are all analytic and in general not all zero in the domain. It is also convenient to define

$$\mathbf{s}_\alpha = \tfrac{1}{2}(\mathbf{s}_1 + \mathbf{s}_2) = \begin{pmatrix} a_3\alpha + a_4 \\ \alpha^2 + \beta^2 + a_1\alpha + a_2 \\ \alpha^3 + 3\alpha\beta^2 + a_1(\alpha^2 + \beta^2) + \alpha a_2 \\ a_5\alpha + a_6 \end{pmatrix} \tag{16.19}$$

whose elements are analytic and in general not all zero in the domain.

Now let **U** be a 4×4 matrix whose columns are $\mathbf{s}_\alpha, \mathbf{s}_\beta, \mathbf{s}_3, \mathbf{s}_4$. They are linearly independent in the domain so that **U** is analytic and non-singular, and $\mathbf{U}^{-1}$ exists. The matrix **U** will now be used instead of **S** to derive coupled equations. It can be shown from (16.17)–(16.19) and (7.85) that

$$\mathbf{TU} = \mathbf{UA} \quad \text{where} \quad \mathbf{A} = \begin{pmatrix} \alpha & 1 & 0 & 0 \\ \beta^2 & \alpha & 0 & 0 \\ 0 & 0 & q_3 & 0 \\ 0 & 0 & 0 & q_4 \end{pmatrix}. \tag{16.20}$$

Instead of $\mathbf{f}$ in (6.53), define four new independent variables c_i forming a column $\mathbf{c}$ where

$$\mathbf{e} = \mathbf{Uc}, \quad \mathbf{c} = \mathbf{U}^{-1}\mathbf{e} \tag{16.21}$$

and substitute for $\mathbf{e}$ in (7.80). Then $\mathbf{c}$ must satisfy

$$\mathbf{c}' + \mathrm{i}\mathbf{Ac} = \mathbf{\Lambda c} \tag{16.22}$$

where

$$\mathbf{\Lambda} = -\mathbf{U}^{-1}\mathbf{U}'. \tag{16.23}$$

The four equations (16.22) are the required new form of the first order coupled equations.

Any transforming matrix must satisfy two requirements throughout the chosen domain of the z plane: (a) it must be non-singular so that its inverse exists, (b) it must be analytic so that it can be differentiated. The new transforming matrix $\mathbf{U}$ satisfies both these requirements. Hence the new coupling matrix $\mathbf{\Lambda}$ is analytic and bounded throughout the domain.

In a sufficiently slowly varying medium the non-diagonal elements of $\mathbf{\Lambda}$ can be neglected to a first approximation. Then the waves associated with c_3, c_4, that is with q_3, q_4 are independently propagated and are the same as given by (7.117). Those associated with c_1, c_2, that is with q_1, q_2, however, remain strongly coupled because of the non-diagonal elements of $\mathbf{A}$. With this approximation (16.22) gives

$$\begin{pmatrix} c_1 \\ c_2 \end{pmatrix}' = -\mathrm{i}\begin{pmatrix} \alpha + \mathrm{i}\Lambda_{11} & 1 \\ \beta^2 & \alpha + \mathrm{i}\Lambda_{22} \end{pmatrix}\begin{pmatrix} c_1 \\ c_2 \end{pmatrix}. \tag{16.24}$$

Now $\Lambda_{11}, \Lambda_{22}$ contain derivatives with respect to z and are therefore small in a slowly varying medium. Suppose that they are small enough to be neglected. Let

$$\begin{pmatrix} c_1 \\ c_2 \end{pmatrix} = \begin{pmatrix} h_1 \\ h_2 \end{pmatrix} \exp\left(-\mathrm{i}k \int^z \alpha \mathrm{d}z\right). \tag{16.25}$$

Then (16.24) gives

$$\begin{pmatrix} h_1 \\ h_2 \end{pmatrix}' = -\mathrm{i}\begin{pmatrix} 0 & 1 \\ \beta^2 & 0 \end{pmatrix}\begin{pmatrix} h_1 \\ h_2 \end{pmatrix}. \tag{16.26}$$

If Λ_{11} and Λ_{22} may not be neglected but are equal, the α in (16.25) is replaced by $\alpha + \mathrm{i}\Lambda_{11}$. If they are unequal it is still possible to reduce the equations to the form

(16.26) by replacing (16.25) by a matrix transformation

$$\begin{pmatrix} c_1 \\ c_2 \end{pmatrix} = \boldsymbol{L} \begin{pmatrix} h_1 \\ h_2 \end{pmatrix} \tag{16.27}$$

and it can be shown that the 2×2 matrix $\boldsymbol{L}$ is analytic in the domain.

Now (16.26) gives

$$h_1'' + \beta^2 h_1 = 0, \quad h_2 = \mathrm{i}h_1' \tag{16.28}$$

and it is these equations that determine the behaviour of the solutions near an isolated coupling point, that is a zero of β. The first is of the same form as (7.6) with β instead of q, and we may therefore use Langer's (1937) form (8.72) of the uniform approximation. Let

$$\zeta = \left(\frac{3}{2} \mathrm{i}k \int_{z_p}^{z} \beta \mathrm{d}z \right)^{2/3} \tag{16.29}$$

(compare (8.68)). Then the solution analogous to (8.72) is

$$h_1 = \zeta^{\frac{1}{4}} \beta^{-\frac{1}{2}} \mathrm{Ai}(\zeta), \quad h_2 = -\zeta^{-\frac{1}{4}} \beta^{\frac{1}{2}} \mathrm{Ai}'(\zeta) + (\text{small terms})\mathrm{Ai}(\zeta) \tag{16.30}$$

where $\mathrm{Ai}'(\zeta)$ is $\mathrm{dAi}(\zeta)/\mathrm{d}\zeta$; this prime does *not* mean $k^{-1}\mathrm{d}/\mathrm{d}z$. The 'small terms' are zero if β^2 is proportional to $z - z_p$. Since ζ contains a cube root, it can take three different values and there are three different solutions (16.30) but only two are independent because of the linear relation (8.63) satisfied by the Airy integral functions. The physical conditions of the problem are used to choose the correct solution. For some examples see § 16.7.

If the chosen domain of the complex z plane extends out from z_p to where ζ is large enough, and if the uniform approximation (16.30) can still be used there, the Airy integral function may be set equal to a combination of its two asymptotic approximations (8.27), (8.28). These give for h_1 from (16.30) and (16.29) the two solutions

$$h_1 \propto \beta^{-\frac{1}{2}} \exp\left(\mp \mathrm{i}k \int_{z_p}^{z} \beta \mathrm{d}z \right) \tag{16.31}$$

and where these are used in (16.25) they give, with (16.17):

$$c_1 \propto (q_1 - q_2)^{-\frac{1}{2}} \exp\left(-\mathrm{i}k \int^{z} q_1 \,\mathrm{d}z \right), \quad c_2 \propto (q_2 - q_1)^{\frac{1}{2}} \exp\left(-\mathrm{i}k \int^{z} q_2 \,\mathrm{d}z \right) \tag{16.32}$$

and hence the field variables $\mathbf{e}$ can be found from (16.21). These two solutions give the two W.K.B. solutions (7.136) associated with the two roots q_1, q_2 of the Booker quartic. Thus the form (16.22) of the coupled equations leads to a solution (16.30) which is uniformly valid near the coupling point z_p and at points sufficiently remote from it that the W.K.B. solutions take over.

Some examples of the use of this method are given in the following §16.4. Equations (16.29)–(16.32) are used in the study of the phase integral method for coupling in §16.7.

The solution (16.30) can be used to supply a rough criterion to show when the W.K.B. solutions are good approximations. The method is the exact parallel of that used in §§7.10, 8.10. It was shown in §8.10 that the asymptotic approximations for the function Ai(ζ) can be used with an accuracy better than about 8 to 9% if $|\zeta| \geqslant 1$. Suppose that, near z_p, β^2 is a linear function of z so that $\beta^2 \approx K(z - z_p)$. Then (16.29) gives

$$|\zeta|^{\frac{3}{2}} \approx \left|\frac{k\beta^3}{K}\right| = \left|\frac{k\beta^3}{\mathrm{d}(\beta^2)/\mathrm{d}z}\right| = \tfrac{1}{2}k\left|\frac{\mathrm{d}}{\mathrm{d}z}\left(\frac{1}{\beta}\right)\right|^{-1} \tag{16.33}$$

and from (16.17) the criterion $|\zeta| \geqslant 1$ gives

$$\frac{1}{k}\left|\frac{\mathrm{d}}{\mathrm{d}z}\left(\frac{1}{q_1 - q_2}\right)\right| \leqslant \tfrac{1}{4}. \tag{16.34}$$

If this is used near a reflection point in an isotropic ionosphere, we have $-q_2 = q_1 = q$ and then (16.34) and (7.61) are identical. If (16.34) is used to test the W.K.B. solution $\mathbf{c}_1$ in (16.32), that is the one with q_1 in the exponent, it must be applied for all three pairs q_1, q_2 and q_1, q_3 and q_1, q_4.

16.4. Application to vertical incidence

The case of vertical incidence is important in radio probing of the ionosphere and is useful for illustrating the methods of §16.3. It is convenient to choose the x and y axes so that the earth's magnetic field is in the x–z plane. Some of the algebra for this case has been given in §§7.17, 16.2(a). The four roots q of the Booker quartic are now the refractive indices for the upgoing and downgoing ordinary and extraordinary waves. The polarisation ρ_O is the same for both upgoing and downgoing ordinary waves, as explained near the end of §4.4. Similarly, from (4.29) the polarisation of both upgoing and downgoing extraordinary waves is $\rho_E = 1/\rho_O$.

The theory of §16.3 dealt with the neighbourhood of only one isolated coupling point and ignored the others. It is possible, however, to deal with two coupling points, A, B, at the same time, provided that the two qs that are equal at A, say q_1, q_2, are different from the two that are equal at B, say q_3, q_4. Then for A the operations (16.18), (16.19) are applied to the columns $\mathbf{s}_1$, $\mathbf{s}_2$ of $\mathbf{S}$, and for B they are at the same time applied to the columns $\mathbf{s}_3$, $\mathbf{s}_4$. Two cases will now be considered.

(a) *Reflection points*

One form of coupling point is a reflection point A for the two ordinary waves, where $n_O = 0$. Another is a reflection point B for the two extraordinary waves, where $n_E = 0$.

We therefore take

$$q_1 = -q_2 = n_{\rm O}, \quad q_3 = -q_4 = n_{\rm E}. \tag{16.35}$$

The four columns of the matrix $\mathbf{S}$ may be chosen to be the four sets given in round brackets in (7.145)–(7.148). Thus

$$\mathbf{S} = \begin{pmatrix} 1 & 1 & \rho_{\rm O} & \rho_{\rm O} \\ -\rho_{\rm O} & -\rho_{\rm O} & -1 & -1 \\ -\rho_{\rm O} n_{\rm O} & \rho_{\rm O} n_{\rm O} & -n_{\rm E} & n_{\rm E} \\ n_{\rm O} & -n_{\rm O} & \rho_{\rm O} n_{\rm E} & -\rho_{\rm O} n_{\rm E} \end{pmatrix}. \tag{16.36}$$

For the ordinary waves (16.17) gives $\alpha = 0$, $\beta = n_{\rm O}$ and the first two columns $\mathbf{s}_\alpha$, $\mathbf{s}_\beta$ of $\mathbf{U}$ are $\frac{1}{2}(\mathbf{s}_1 + \mathbf{s}_2)$, $\frac{1}{2}(\mathbf{s}_1 - \mathbf{s}_2)/n_{\rm O}$. Similarly for the extraordinary waves, $\alpha = 0$, $\beta = n_{\rm E}$ and the last two columns of $\mathbf{U}$ are $\frac{1}{2}(\mathbf{s}_3 + \mathbf{s}_4)$, $\frac{1}{2}(\mathbf{s}_2 - \mathbf{s}_4)/n_{\rm E}$. Hence

$$\mathbf{U} = \begin{pmatrix} 1 & 0 & -\rho_{\rm O} & 0 \\ -\rho_{\rm O} & 0 & 1 & 0 \\ 0 & -\rho_{\rm O} & 0 & 1 \\ 0 & 1 & 0 & -\rho_{\rm O} \end{pmatrix}, \quad \mathbf{U}^{-1} = (1-\rho_{\rm O}^2)^{-1} \begin{pmatrix} 1 & \rho_{\rm O} & 0 & 0 \\ 0 & 0 & \rho_{\rm O} & 1 \\ \rho_{\rm O} & 1 & 0 & 0 \\ 0 & 0 & 1 & \rho_{\rm O} \end{pmatrix} \tag{16.37}$$

and from (16.23)

$$\Lambda = \frac{\rho_{\rm O}\rho_{\rm O}'}{\rho_{\rm O}^2 - 1}\mathbf{1} - \psi\mathbf{C} \tag{16.38}$$

where $\mathbf{1}$ is the unit 4×4 matrix, ψ is Försterling's (1942) coupling parameter (16.14), and

$$\mathbf{C} = \begin{pmatrix} 0 & 0 & 1 & 0, \\ 0 & 0 & 0 & 1 \\ 1 & 0 & 0 & 0 \\ 0 & 1 & 0 & 0 \end{pmatrix} \tag{16.39}$$

Further, from (16.20)

$$\mathbf{A} = \begin{pmatrix} 0 & 1 & 0 & 0 \\ n_{\rm O}^2 & 0 & 0 & 0 \\ 0 & 0 & 0 & 1 \\ 0 & 0 & n_{\rm E}^2 & 0 \end{pmatrix} \tag{16.40}$$

and (16.22) gives

$$\mathbf{c}' + \mathrm{i}\mathbf{A}\mathbf{c} = -\psi\mathbf{C}\mathbf{c} + \frac{\rho_{\rm O}\rho_{\rm O}'}{\rho_{\rm O}^2 - 1}\mathbf{c} \tag{16.41}$$

where $\mathbf{c}$ is given by (16.21). The last term can be removed by setting $\mathbf{c} = \mathbf{b}(\rho_{\rm O}^2 - 1)^{-\frac{1}{2}}$ (compare (16.25)). Then the equations (16.41) can be partitioned thus

$$\begin{pmatrix} \mathrm{b}_1 \\ \mathrm{b}_3 \end{pmatrix}' = -\mathrm{i}\begin{pmatrix} \mathrm{b}_2 \\ \mathrm{b}_4 \end{pmatrix} - \psi\begin{pmatrix} 0 & 1 \\ 1 & 0 \end{pmatrix}\begin{pmatrix} \mathrm{b}_1 \\ \mathrm{b}_3 \end{pmatrix},$$

$$\begin{pmatrix} \mathrm{b}_2 \\ \mathrm{b}_4 \end{pmatrix}' = -\mathrm{i}\begin{pmatrix} n_{\rm O}^2 & 0 \\ 0 & n_{\rm E}^2 \end{pmatrix}\begin{pmatrix} \mathrm{b}_1 \\ \mathrm{b}_3 \end{pmatrix} - \psi\begin{pmatrix} 0 & 1 \\ 1 & 0 \end{pmatrix}\begin{pmatrix} \mathrm{b}_2 \\ \mathrm{b}_4 \end{pmatrix} \tag{16.42}$$

Differentiate the first of these once, and use the second to eliminate b_2, b_4. This gives

$$\begin{pmatrix} b_1 \\ b_3 \end{pmatrix}'' + \begin{pmatrix} n_O^2 + \psi^2 & 0 \\ 0 & n_E^2 + \psi^2 \end{pmatrix} \begin{pmatrix} b_1 \\ b_3 \end{pmatrix} = -2\psi \begin{pmatrix} b_3 \\ b_1 \end{pmatrix}' - \psi' \begin{pmatrix} b_3 \\ b_1 \end{pmatrix}. \tag{16.43}$$

These are the second order coupled equations that Försterling (1942) derived by a different method. No approximations are used in their derivation. They are discussed in more detail in §§ 16.10–16.12. They or the first order equations (16.42) can be used near any reflection point where $n_O = 0$ or $n_E = 0$ because the transforming matrix **U** (16.37) was chosen to achieve this. They cannot be used near a coupling point where $n_O = n_E$, $\rho_O = -1$ because ψ and ψ' are infinite there.

(b) *Coupling points*

Coupling occurs between the two upgoing waves at points where $n_O = n_E$, and these are also coupling points for the two downgoing waves. These two coupling processes can be treated together. We again use (16.35) and by analogy from (16.17) we take

$$\alpha = \tfrac{1}{2}(n_O + n_E) = \tfrac{1}{2}(q_1 + q_3) = -\tfrac{1}{2}(q_2 + q_4),$$
$$\beta = \tfrac{1}{2}(n_O - n_E) = \tfrac{1}{2}(q_1 - q_3) = \tfrac{1}{2}(q_4 - q_2). \tag{16.44}$$

For the matrix **S**, instead of (16.36), we now use

$$\mathbf{S} = \begin{pmatrix} 1 & 1 & 1 & 1 \\ -\rho_O & -\rho_O & -\rho_E & -\rho_E \\ -\rho_O n_O & \rho_O n_O & -\rho_E n_E & \rho_E n_E \\ n_O & -n_O & n_E & -n_E \end{pmatrix}. \tag{16.45}$$

The columns of **U** are $\tfrac{1}{2}(\mathbf{s}_1 + \mathbf{s}_3)$, $(\mathbf{s}_1 - \mathbf{s}_3)/(n_O - n_E)$, $\tfrac{1}{2}(\mathbf{s}_2 + \mathbf{s}_4)$, $(\mathbf{s}_2 + \mathbf{s}_4)/(n_O - n_E)$ so that

$$\mathbf{U} = \begin{pmatrix} E & 0 \\ 0 & F \end{pmatrix} \begin{pmatrix} \mathbf{1} & \mathbf{1} \\ \mathbf{1} & -\mathbf{1} \end{pmatrix} \tag{16.46}$$

where

$$E = \begin{pmatrix} 1 & 0 \\ -\tfrac{1}{2}(\rho_O + \rho_E) & (\rho_E - \rho_O)/(n_O - n_E) \end{pmatrix},$$
$$F = \begin{pmatrix} -\tfrac{1}{2}(\rho_O n_O + \rho_E n_E) & (\rho_E n_E - \rho_O n_O)/(n_O - n_E) \\ \tfrac{1}{2}(n_O + n_E) & 1 \end{pmatrix}. \tag{16.47}$$

The matrix **A** from (16.20) is $\mathbf{U}^{-1}\mathbf{T}\mathbf{U}$ whence with (7.85) and (16.44)

$$\mathbf{A} = \begin{pmatrix} B & 0 \\ 0 & B \end{pmatrix}, \quad B = \begin{pmatrix} \alpha & 1 \\ \beta^2 & \alpha \end{pmatrix}. \tag{16.48}$$

The coupling matrix (16.23) is

$$\Lambda = \begin{pmatrix} X & W \\ W & X \end{pmatrix} \tag{16.49}$$

where

$$X = -\tfrac{1}{2}(F^{-1}F' + E^{-1}E'), \quad W = \tfrac{1}{2}(F^{-1}F' - E^{-1}E'). \tag{16.50}$$

Then the coupled equations (16.22) are

$$\begin{pmatrix} c_1 \\ c_2 \end{pmatrix}' = -\mathrm{i}\boldsymbol{B}\begin{pmatrix} c_1 \\ c_2 \end{pmatrix} + \boldsymbol{X}\begin{pmatrix} c_1 \\ c_2 \end{pmatrix} + \boldsymbol{W}\begin{pmatrix} c_3 \\ c_4 \end{pmatrix},$$
$$\begin{pmatrix} c_3 \\ c_4 \end{pmatrix}' = -\mathrm{i}\boldsymbol{B}\begin{pmatrix} c_3 \\ c_4 \end{pmatrix} + \boldsymbol{X}\begin{pmatrix} c_3 \\ c_4 \end{pmatrix} + \boldsymbol{W}\begin{pmatrix} c_1 \\ c_2 \end{pmatrix}. \quad (16.51)$$

where $\mathbf{c}$ is given by (16.21), (16.46). These exhibit the expected symmetry between the solution c_1, c_2 for upgoing waves and c_3, c_4 for downgoing waves. They can be used near a coupling point where $n_O = n_E$ because $\mathbf{U}$ in (16.46) was chosen to achieve this, and the coupling matrices $\boldsymbol{W}$, $\boldsymbol{X}$ (16.50) are small in a slowly varying medium. They cannot be used near a reflection point where $n_O = 0$ or where $n_E = 0$ because there the matrix $\boldsymbol{F}$ is singular and the coupling matrices $\boldsymbol{W}$, $\boldsymbol{X}$ are infinite.

16.5. Coupling and reflection points in the ionosphere

For the further study of coupled equations it is necessary to know the distribution of the coupling points in the complex z (height) plane. Reflection points are a special case of coupling points and the term 'coupling point' is therefore used to include them. They are now to be discussed for a stratified ionosphere in which only electrons are effective. A coupling point is where two roots of the Booker quartic are equal. It is therefore where the discriminant Δ, (6.39), (6.40), is zero. The equation $\Delta = 0$ was examined by Pitteway (1959) who showed that the solutions of interest are of eight types; see § 6.6. A summary of their main properties is given here. A full discussion is given by Pitteway (1959), Jones and Foley (1972) and Smith, M.S. (1974a). It is here assumed that Z is independent of height z, and the positions of the coupling points in the complex X plane are studied. This is equivalent to the use of a model ionosphere in which X increases linearly with height z.

Table 16.1

Location in X plane	Symbol	Description
U	O	Reflection. Ordinary wave
$U - Y$	E_-	Reflection. Extraordinary wave
$U + Y$	E_+	Reflection. Extraordinary wave (Z-mode)
$U + \mathrm{i}Z_t$	C_1	Coupling. Upgoing ordinary and extraordinary
$U - \mathrm{i}Z_t$	C_3	
$U + \mathrm{i}Z_t$	C_2	Coupling. Downgoing ordinary and extraordinary
$U - \mathrm{i}Z_t$	C_4	
X_∞	R	Reflection, coinciding with point of resonance W

First consider vertical incidence. The eight types of coupling point are then as given in § 4.10 and are as in table 16.1. The positions of these points in the complex X plane are shown in fig. 16.1. The first three are reflection points (4.74). The next four are where two values of n^2 are equal as given by (4.79). The last is where one value of n^2 is infinite, (4.76). The coupling point R and the resonance point W are really separate effects but they coincide for vertical incidence. For oblique incidence, when $l_x \neq 0$, they are at different positions.

The points C_1, C_2 coincide but they are associated with different coupling processes. Near C_1 the two upgoing waves are strongly coupled but they are propagated independently of the two downgoing waves, which are also strongly coupled near C_2. Similarly C_3 and C_4 coincide but are associated with independent coupling processes.

The various coupling points are branch points of the multiple valued functions $n(z)$ at vertical incidence, or $q(z)$ at oblique incidence. Here we continue to discuss vertical incidence. To aid the description it is useful to introduce branch cuts. The two values $\pm n_O(z)$ are equal (zero) at the reflection point O. This case is similar to the reflection point z_0 in an isotropic ionosphere, described in § 8.21. There a branch cut was used, shown as a wavy line in fig. 8.8. It runs from z_0 to infinity without crossing the real z axis. For a study of the reflection of the ordinary wave a similar branch cut running from 0 to ∞ may be used. Similarly the two values $\pm n_E(z)$ are equal (zero) at the two reflection points E_+ and E_- and branch cuts running from each of them to infinity, without crossing the real axis, may be used when studying the reflection of the extraordinary wave. The exact positions of these cuts are not important and their choice depends on the particular problem being solved. Since the refractive index of

Fig. 16.1. The complex X plane showing the positions of the reflection and coupling points. W is the point of resonance where one refractive index is infinite. (a) and (b) are for vertical incidence. The wavy lines indicate branch cuts. In (a) $Z < Z_t$ and conditions are greater than critical. In (b) $Z > Z_t$ and conditions are less than critical. In (c), $Z = 0$. The arrows show how the coupling points move when the angle of incidence is increased from zero. The black dots are their positions at vertical incidence.

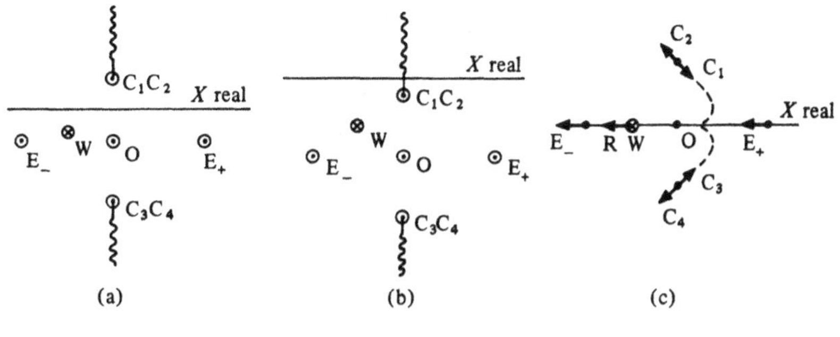

an upgoing wave is usually defined to be the value for which $\mathrm{Re}(n) > 0$, a convenient choice that is often used is to take as the branch cuts the lines whose equations are

$$\mathrm{Re}\{n_O(z)\} = 0, \quad \mathrm{Re}\{n_E(z)\} = 0. \tag{16.52}$$

The refractive indices n_O, n_E for the two upgoing waves are equal at C_1 and at C_3, and branch cuts running from these points to infinity are shown as wavy lines in fig. 16.1. Again the exact positions of these cuts are not important and their choice is a matter of convenience. In § 4.15 the terms ordinary and extraordinary were defined as given by $\mathrm{Re}(S_R)$ positive and negative respectively where S_R is given by (4.93). Thus a convenient choice for the branch cuts is the lines whose equations are

$$\mathrm{Re}(S_R) = 0 \tag{16.53}$$

so that S_R^2 in (4.93) is real and negative. These lines run from C_1 in the direction away from C_3, and from C_3 away from C_1, as in fig. 16.1.

In fig. 16.1a, $Z < Z_t$ and the points C_1, C_2 are on the positive imaginary side of the real X axis. This is the usual case for high frequencies, for then Z is small, and conditions are said to be 'greater than critical'. When $Z = Z_t$ the points C_1, C_2 are on the real X axis and this is called 'critical coupling'; see § 16.6. For $Z > Z_t$ as in fig. 16.1(b), all the coupling points, including C_1, C_2, are on the negative imaginary side of the real X axis and conditions are said to be 'less than critical'. Instead of the terms 'greater than...' and 'less than critical', some authors use the terms 'quasi-transverse' and 'quasi-longitudinal' respectively, and 'critical coupling' is called the 'QL–QT transition'. The terms 'quasi-longitudinal' and 'quasi-transverse' are also used to describe certain approximations in magnetoionic theory; see § 4.17. The two usages are related but not exactly the same. The concepts of greater than and less than critical do not involve approximations.

When X is a linear function of z, fig. 16.1 shows how the coupling points are arranged in the complex z (height) plane. When X is a more complicated function of z, the distribution of coupling points is less simple. For example if $X \propto e^{\alpha z}$, and if z_0 is the position of some coupling point, then the same kind of coupling point appears again at $z = z_0 + 2\pi i r/\alpha$ where r is any positive or negative integer. Thus every type of coupling point occurs an infinite number of times. Usually it is only those that are nearest to the real z axis that are important. Some examples of this case were given by Budden (1961a, figs. 20.6, 20.8). In practice Z also is a function of z and this affects the positions of the coupling points.

Next let the wave be obliquely incident. The behaviour of the coupling points now depends on many factors and is complicated. The description here is given for one case only, needed in later sections. For other cases see Pitteway (1959), Budden and Terry (1971), Jones and Foley (1972), Smith, M.S. (1974a). It is here assumed that the plane of incidence is the magnetic meridian plane so that $l_y = 0$, and that collisions are neglected, $Z = 0$, and that $Y < 1$ as in figs. 6.4–6.6. The angle of incidence $\theta =$

arcsin S is assumed to be real and increases from zero so that the wave normal of the incident wave moves nearer to the direction of the earth's magnetic field, that is of $\mathbf{Y}$. Some of the features of the process can be followed in figs. 10.4, 10.7. The reference line AB or MN is at first through the origin and moves to the right. See also figs. 6.4–6.7. For all values of S, the resonance point W remains fixed at $X = (1 - Y^2)/(1 - Y^2 l_z^2)$. When $S = 0$ the points E_+, E_-, O and R (same as W) are on the real X axis as shown in fig. 16.1(c). When S increases slightly, O does not move; the line A′B′, fig. 10.4, still intersects one of the refractive index surfaces for the ordinary wave where $X = 1$, and the wave is here reflected with a 'Spitze' in its ray, § 10.9. The point R moves to the left of W and can be seen in the insets of figs. 6.4, 6.5. The points E_+, E_- move to smaller values of X. The points C_1, C_2 now separate, and C_3, C_4 also separate, as shown by arrows in fig. 16.1(c). When S increases further, C_1 and C_3 move on to the real X axis and coincide there. At this stage three roots of the Booker quartic are equal. The value of S is between the values used in figs. 6.6(a, b), and is such that the upper continuous curve has a vertical tangent at its point of inflection. When S increases still further, C_1, C_3 separate again but remain on the real X axis, fig. 6.6(b). At this stage the line A′B′ in fig. 10.4 is still to the left of P and the reflection point O is still where $X = 1$. The ordinary wave is still reflected with a 'Spitze'. With S increasing, C_3 moves to the left and eventually coincides with O, at $X = 1$; fig. 6.6(c). This happens when $S = S_A = l_x\{Y/(Y+1)\}^{\frac{1}{2}}$, (6.47), fig. 6.3. The reference lines A′B′ in fig. 10.4 and MN in fig. 10.7 now go through the window point P. The coupling points O and C_3 here both apply to the same two roots of the Booker quartic. This is an example of a coalescence C2, studied later in §§ 17.4, 17.6. It is associated with the phenomenon of radio windows, §§ 17.6, 17.7. When S increases to a value greater than S_A, the point O moves to where $X < 1$ and there is no longer a Spitze in the ordinary ray. Now C_3 remains at $X = 1$. The point C_1 moves to smaller X and eventually C_1 and C_3 coincide at $X = 1$, fig. 6.6(e), when $S = \sin\Theta$, where Θ is the angle between $\mathbf{Y}$ and the vertical. Thereafter C_1 remains at $X = 1$, and C_3 moves down to where $X < 1$, fig. 6.6(f).

The sequence just described is modified when $l_y \neq 0$, when $Z \neq 0$ and when $Y > 1$. These cases are all discussed by Smith, M.S. (1974a).

Another example of coupling points has been mentioned in § 13.9. They are analogous to C_1, C_3 and denoted by $z = z_p$, z_q respectively. They are there used in the theory of ion cyclotron whistlers, in which the effects of heavy positive ions and collisions are allowed for.

16.6. Critical coupling

The term 'critical coupling' is customarily used only for vertically incident waves. It refers to one of the two coupling points C_1, C_2 described in § 16.5. C_1 is associated with the two upgoing waves and C_2 with the two downgoing waves, and for vertical incidence they are at the same position in the complex z plane. This position depends on frequency and on the functions $X(z)$, $Z(z)$. Critical coupling occurs when C_1 and C_2 are on the real z axis. The points C_1 and C_2 are where $n_O = n_E$ and the condition for this was discussed in §§ 4.4, 4.10. $X(z)$ and $Z(z)$ are both real, and if only electrons are allowed for, the condition is (4.26)–(4.28). For frequencies of a few hundred hertz or less, where heavy ions are important, critical coupling can also occur. It is associated with the theory of ion cyclotron whistlers and has been discussed in § 13.9. In this case critical coupling is also known as 'exact crossover', and the condition for it is (13.60)–(13.62). See also fig. 13.12(c), where z_p denotes the coupling points C_1 and C_2. In this section we deal only with the first case where only electrons are considered.

At critical coupling (4.28) shows that the collision frequency ν has the transition value $\omega_t = \frac{1}{2}\omega_H \sin^2 \Theta/|\cos \Theta|$. This is independent of frequency. The function $\nu(z)$ is believed to be a monotonically decreasing function so that it attains the value ω_t at one fixed height, to be called the 'transition height'. In temperate latitudes ω_t is about $9 \times 10^5\ \mathrm{s}^{-1}$ and the transition height is about 90 km. The second condition in (4.26) is that at the transition height $X = 1$, and this depends on frequency. Thus in general the relations (4.26)–(4.28) are not simultaneously true at any real height. There is, however, always one frequency, the plasma frequency at the transition height, that makes $X = 1$. It is called the 'critical coupling frequency', and denoted by f_c. It depends on the state of the ionosphere and therefore on time of day, season and geographical position. In temperate latitudes in the day time it is usually in the range 200 kHz to 1 MHz and is rarely less than 150 kHz. At night it is smaller, of the order of 20 to 50 kHz in temperate latitudes. In the tropics, however, the transition height is lower and f_c is smaller.

For frequencies greater than f_c, conditions are greater than critical. The coupling points C_1, C_2 are on the positive imaginary side of the real z axis, fig. 16.1(a). This is the usual condition when probing the ionosphere with frequencies greater than about 1 MHz. The branch cut in fig. 16.1(a) does not cross the real z axis and each wave remains either ordinary or extraordinary at all heights, without a change of name. The ordinary wave is reflected near where $X = 1$ and the extraordinary near where $X = 1 - Y$, and these two reflections are easily observed, and have different equivalent heights.

For frequencies less than f_c, conditions are less than critical, and this is the regime in temperate latitudes for frequencies less than about 50 kHz at night, and less than about 500 kHz to 1 MHz in the day time. The coupling points C_1, C_2 are now on the negative imaginary side of the real z axis, and the branch cut that ends at C_1, C_2 now crosses the real z axis, fig. 16.1(b). Consider, first, a wave that travels vertically upwards from the ground, whose polarisation after it enters the ionosphere is that of an ordinary wave. This wave has a reflection point where $X = 1 - \mathrm{i}Z$. But, at these low frequencies, $Z = \nu/2\pi f$ is large so that the real and imaginary parts of X at the reflection point are comparable. The effect of this is that the reflection coefficient is very small and this reflection is rarely observed. Because of the collisions, $\mu = \text{Re}(n_O)$ does not go near to zero at the real height where $X = 1$. This effect can be seen in figs. 4.7–4.9, 16.5, 16.6 which are for conditions less than critical. For other curves showing this see, for example, Lepechinsky (1956, fig. 3a), Ratcliffe (1959, figs. 7.4, 10.3), Budden (1961a, fig. 6.14), Rawer and Suchy (1967, fig. 31). At the level where $X = 1$ the condition $Z > Z_t$ now holds and the real z axis crosses the branch cut. The value of n_O at a point on one side of the cut is the same as n_E at the adjacent point on the other side. The field components are the same at these two adjacent points. Thus where $X > 1$ the wave is an extraordinary wave. This is simply a change of name, as explained in §4.16, and is not the result of any mode conversion process. The wave travels on up to near where $X = 1 + Y - \text{i}Z$ and here it is reflected. At this greater height Z, though still large, is less than $1 + Y$. This means that $n_E(z)$ does decrease towards zero near this level, and the resulting reflection of the extraordinary wave is great enough to be observed; see, for example, fig. 16.5–16.7. The downgoing extraordinary wave again passes the level where $X = 1$ and its name changes back to 'ordinary wave' below this. Thus the reflected wave observed at the ground has the polarisation of an ordinary wave, but is reflected as an extraordinary wave. This wave is called an 'initial ordinary' wave (Cooper, 1961). For very low frequencies it is the most commonly observed reflection.

Consider next a wave that travels vertically upwards from the ground and whose polarisation after it enters the ionosphere is that of an extraordinary wave. One reflection point for this is where $X = 1 - Y - \text{i}Z$ but since $Y > 1$ this point is not near the real z axis. Cooper (1961, 1964) has shown that it can give some reflection but this is too weak to be important. Now follow this wave upwards along the real z axis. When it goes above the level where $X = 1$ it is an ordinary wave because here $Z > Z_t$, and the branch cut is crossed. It is not therefore reflected near where $X = 1 + Y - \text{i}Z$ but travels on upwards into the magnetosphere as an ordinary wave, that is a whistler mode wave.

The upgoing extraordinary wave can also be followed along a path in the complex z plane that runs on the negative imaginary side of the coupling point C_1. This is a branch point of $n_O(z)$ and $n_E(z)$, and by going on the other side of it we ensure that the branch cut is not crossed and the wave does not change its name but remains an extraordinary wave where $\text{Re}(X) > 1$. Thus on the real axis where $X > 1$ there is an extraordinary wave, generated in the region near C_1 by a mode conversion process. The original unconverted wave is extraordinary where $X < 1$ and ordinary where $X > 1$. Paradoxically this initially extraordinary wave where $X < 1$ gives rise by mode conversion to an extraordinary wave where $X > 1$. An estimate of its amplitude can be made by the phase integral method, § 16.7, with the path running on the negative imaginary side of C_1. This extraordinary wave is now reflected where $X = 1 + Y - iZ$, and the reflected extraordinary wave travels downwards. If we follow it along the real z axis, its name changes to ordinary on going below where $X = 1$, and it reaches the ground as an ordinary wave. But it also gives rise, by mode conversion near the coupling point C_2, to a wave that is extraordinary where $X < 1$, and this too reaches the ground.

It is now interesting to enquire what would be observed if conditions changed continuously from greater than critical to less than critical. This might happen if the transmitter frequency was continuously decreased, or if the frequency was fixed and the electron concentration $N(z)$ was increasing. In describing the phenomenon we shall suppose that the ionosphere does not change and that the transmitter frequency is slowly decreased. It is assumed that $N(z)$ and therefore $X(z)$ are monotonically increasing functions of the height z. The critical coupling frequency f_c has some fixed value less that the electron gyro-frequency f_H. This discussion applies only for vertical incidence. The transmitted wave is assumed to be linearly polarised so that on entering the ionosphere it splits into ordinary and extraordinary waves with roughly equal amplitude.

For frequencies less than f_H the ordinary wave reflection from where $X = 1 - iZ$ is weak, as explained above. For all frequencies, therefore, the strongest reflection is of the extraordinary wave from where $X = 1 + Y - iZ$. For frequencies $f \gg f_c$ this returns to the ground as an extraordinary wave. On emerging from below the ionosphere its polarisation is the limiting polarisation ρ_E of the extraordinary wave. This is discussed in § 17.11 where it is shown that ρ_E is given by (4.21) for some limiting level in the base of the ionosphere where $X \approx 0$. Thus

$$\rho_E = i[Z_t + \{(1 - iZ)^2 + Z_t^2\}^{\frac{1}{2}}]/(1 - iZ) \tag{16.54}$$

where the real part of the square root is positive, and Z has its value at the limiting level.

For frequencies $f \ll f_c$ the reflected extraordinary wave is an ordinary wave when it goes below where $X = 1$ and so the strongest reflection reaching the ground has the limiting polarisation ρ_O of an ordinary wave, given by

$$\rho_O = i[Z_t - \{(1 - iZ)^2 + Z_t^2\}^{\frac{1}{2}}]/(1 - iZ). \tag{16.55}$$

Other reflected waves can be produced by mode conversion and can reach the ground but they are weak when f is not near f_c.

Thus there should be a change of polarisation from (16.54) to (16.55) as the frequency is decreased through f_c. There is a transition range of frequency near f_c where waves of both polarisations reach the ground and the resultant polarisation would have to be found by combining these waves. The extent of this transition range is not known with certainty. If it is small, then as the transmitter frequency is decreased the polarisation of the received wave would switch over quickly from

(16.54) to (16.55). These two polarisations are elliptical with opposite senses. The two polarisation ellipses are as shown in fig. 4.2(a).

The author has heard this effect described by colleagues but does not know of any published account of observations of it. It has been called 'flipover'. This name may have been used in describing experiments where the polarisation ellipse of the received wave was delineated on the screen of a cathode ray oscillograph. When conditions changed from greater to less than critical, the observed ellipse flipped over from (16.54) to (16.55).

For measurements with vertically incident radio pulses of frequency 150 kHz at night, changes of polarisation have been observed (Parkinson, 1955) but they are more complicated than could be explained by the 'flipover' described above. They are linked with the occurrence of a 'coupling echo', which is an extra reflection from the level where $X \approx 1$; see § 16.14.

The theory of the transition through critical coupling has been studied by Davids and Parkinson (1955), Altman, Cory and Fijalkow (1970), Altman and Fijalkow (1970), Altman and Postan (1971), Smith, M.S. (1973a). A related topic was discussed by Lepechinsky (1956), and by Landmark and Lied (1957). Although the subject is complicated, the results of all these authors seem to be compatible with the conclusion that a change of polarisation of the reflection from near $X \approx 1 + Y$ would be expected when conditions change from less than to greater than critical, but the change would not occur suddenly.

16.7. Phase integral method for coupling

The phase integral formula for finding ionospheric reflection coefficients was described in § 8.21. It was based on the use of an Airy integral function in the uniform approximation solution (8.72) of the differential equations. This solution was for an isotropic ionosphere and for horizontally polarised waves but it was explained near the end of § 8.21 that the method can also be applied for an anisotropic ionosphere, and for the reflection of the ordinary or the extraordinary wave. A similar solution for the neighbourhood of a coupling point has been given at (16.30) and this too contains an Airy integral function. It is very similar to (8.72) and can be used to derive a phase integral formula for coupling. To illustrate the method it is described here for vertical incidence and for upgoing waves. A similar treatment is possible for downgoing waves, and, more generally, for oblique incidence.

Consider first the case where $Z < Z_t$ so that conditions are 'greater than critical'; see § 16.5. The coupling point C_1 is then on the positive imaginary side of the real z axis, and is assumed to be so close to it that the W.K.B. approximations are not usable on the real z axis near C_1; see fig 16.2(a). This part of the real axis is called the 'coupling region'. It is assumed that at real heights below this the only wave present is an upgoing ordinary wave. It is required to study the waves at real heights above the coupling region and to find the amplitudes there of the ordinary and extraordinary waves. These upgoing waves are reflected at the appropriate levels in the ionosphere. The resulting downgoing waves may also be present near C_1 but they are assumed to be independently propagated and they can therefore be ignored.

Now take $q_1 = n_O, q_2 = n_E$ so that, from (16.17), $\beta = \frac{1}{2}(n_O - n_E)$ and from (16.29) the ζ used in Ai(ζ) (16.30) is

$$\zeta = \left\{\frac{3}{4}ik \int_{z_p}^{z} (n_O - n_E)dz\right\}^{2/3} \tag{16.56}$$

where z_p is the value of z at C_1. The Stokes lines are where $\arg\zeta = 0, \pm\frac{2}{3}\pi$ and the anti-Stokes lines are where $\arg\zeta = \pm\frac{1}{3}\pi, \pi$ so that their equations are

$$\text{Stokes:}\quad \mathrm{Re}\int_{z_p}^{z} (n_O - n_E)dz = 0; \quad \text{anti-Stokes:}\quad \mathrm{Im}\int_{z_p}^{z} (n_O - n_E)\,dz = 0. \tag{16.57}$$

These can be plotted in the complex z plane, and examples are shown in fig. 16.2(a) (compare fig. 8.9(a)). For the upgoing ordinary wave the variables $\mathrm{c}_1, \mathrm{c}_2$ of (16.21) contain the exponential factor $\exp(-ik\int_0^z n_O dz)$ and thence from (16.25) the variables h_1, h_2 contain the factor

$$\text{ordinary wave:}\quad \exp\left\{-\tfrac{1}{2}ik\int^{z}(n_O - n_E)\,dz\right\}. \tag{16.58}$$

Similarly for the extraordinary wave the variables h_1, h_2 contain an exponential factor (16.58) with $+\frac{1}{2}$ instead of $-\frac{1}{2}$. These are the exponential factors in the two terms used in the asymptotic approximation for Ai(ζ).

Fig. 16.2. (a) is the complex z (height) plane and shows the Stokes lines S and anti-Stokes lines A radiating from the coupling point C_1. In this example $X = e^{\alpha z}$ and the numbers by the axes are the values of αz. $Y = 4$, $\Theta = 30°$, $Z = 0.5$ and $Z_t = 0.557$, so that conditions are greater than critical. (b) is the Stokes diagram when the incident wave from below is an ordinary wave, and (c) is the same when it is an extraordinary wave.

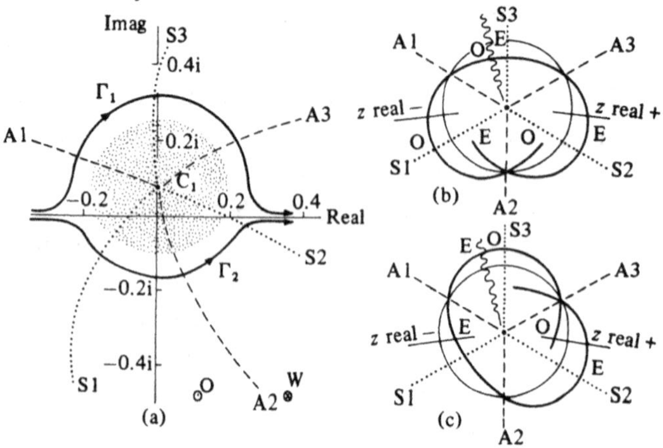

Suppose now that $Y > 1$, as in the example of fig. 16.2. On the real z axis where $X \ll 1$, the incident ordinary wave, containing (16.58), is the dominant term. This can be seen as follows. For $X \ll 1$ the example of fig. 4.6 shows that $\mathrm{Re}(n_E) > \mathrm{Re}(n_O)$. Here let z be given a positive imaginary increment so that the anti-Stokes line A1, fig. 16.2(a), is approached. The real part of the exponent in (16.58) then has a negative increment, so the modulus of the term decreases as A1 is approached. This shows that the term is dominant before A1 is crossed and subdominant after it is crossed. If $Y < 1$ a slightly different argument is needed but again it can be shown that the ordinary wave is the dominant term on the negative imaginary side of A1 (Budden 1961a, § 20.7). The function ζ, (16.56), contains a cube root and has three values. The value to be chosen is that for which $\mathrm{Ai}(\zeta)$ has only a dominant term when z is real and $X < 1$. The Stokes diagram for this $\mathrm{Ai}(\zeta)$ can now be drawn and is shown in fig. 16.2(b).

The W.K.B. solution below the coupling region may be written in the form (7.111), thus

$$\mathsf{f}_1 = \mathsf{K}_1 \exp\left\{ -ik \int_0^z n_O(z)\,\mathrm{d}z \right\} \tag{16.59}$$

where K_1 is a constant. Let this solution be followed as we move along a path, Γ_1 in fig. 16.2(a), that goes on the positive imaginary side of C_1 and is far enough from it for the W.K.B. solutions to be good approximations at all points on it. This solution is at first dominant. It becomes subdominant after A1 is crossed. It does not undergo the Stokes phenomenon when the Stokes line S3 is crossed because there is no dominant term present here. When the branch cut is crossed the name changes to extraordinary and n_O changes to n_E but (16.59) is continuous here. This solution, now an extraordinary wave, again becomes dominant after A3 is crossed and the real z axis is again reached. Thus this W.K.B. solution is continuous for the whole path Γ_1. The ratio of the values of (16.59) at two points z_a, z_b above and below the coupling region is

$$\mathsf{f}_1(z_a)/\mathsf{f}_1(z_b) = \exp\left\{ -ik \int_{z_b\,(\Gamma_1)}^{z_a} n(z)\,\mathrm{d}z \right\} \tag{16.60}$$

where n means n_O before the cut is crossed and n_E after it is crossed. The values of f_1 give the field components of the wave, from (6.53).

The formula (16.60) gives the amplitude of the extraordinary wave above the coupling region, produced by mode conversion from the incident ordinary wave below. It is an example of the phase integral formula for coupling, and was first given by Eckersley (1950).

The solution (16.59) can also be followed along a path, Γ_2 in fig. 16.2(a), that goes on the negative imaginary side of C_1. This does not cross the branch cut so the wave remains an ordinary wave. The Stokes diagram, fig. 16.2(b), shows that this term now becomes subdominant, and disappears because of the Stokes phenomenon, after the Stokes line S2 is crossed and before the real z axis is again reached where $X > 1$. This is above the level where $X = 1 - iZ$ associated with reflection of the ordinary wave. Here the wave is evanescent and therefore strongly attenuated. At greater heights its amplitude would be extremely small so that even if present it would be of little interest.

When the incident wave is an extraordinary wave, let its W.K.B. solution below the coupling region be

$$\mathsf{f}_2 = \mathsf{K}_2 \exp\left\{-ik\int_0^z n_{\mathrm{E}}(z)\,dz\right\}. \tag{16.61}$$

The phase integral method for coupling can be used in a similar way. The Stokes diagram is now as shown in fig. 16.2(c). If the path Γ_2 is used, the wave remains an extraordinary wave and does not undergo the Stokes phenomenon. Its amplitude $\mathsf{f}_2^{(\mathrm{E})}(z_a)$ above the coupling region is given by

$$\mathsf{f}_2^{(\mathrm{E})}(z_a)/\mathsf{f}_2(z_b) = \exp\left\{-ik\int_{z_b\,(\Gamma_2)}^{z_a} n_{\mathrm{E}}(z)dz\right\}. \tag{16.62}$$

If the path Γ_1 is used, again the solution does not undergo the Stokes phenomenon, but its name changes to ordinary after the branch cut is crossed. Its amplitude above the coupling region is given by

$$\mathsf{f}_2^{(\mathrm{O})}(z_a)/\mathsf{f}_2(z_b) = \exp\left\{-ik\int_{z_a\,(\Gamma_1)}^{z_b} n(z)\,dz\right\} \tag{16.63}$$

where n means n_{E} before the cut is crossed and n_{O} afterwards. This formula gives the amplitude of the ordinary wave produced by mode conversion from the incident extraordinary wave. It is likely to be small because the ordinary wave here is evanescent.

The method can similarly be used to deal with conditions that are less than critical. The coupling point C_1 in fig. 16.2(a) would then be on the negative imaginary side of the real axis. The details are not given here.

The path Γ_1 is a 'good path' because the W.K.B. solutions are good approximations at all points on it. Provided that such a path exists, it is not necessary to use it in (16.60) or (16.63). The contour can be distorted in any convenient way provided that no singularities of $n(z)$ are crossed. In particular it can be moved as close as desired to the point C_1. An example of this is given in the following section.

16.8. The Z-trace

When pulses of radio waves are reflected from the ionosphere at normal incidence and observed at the ground, using the ionosonde technique, §§ 1.7, 13.4, records of $h'(f)$ are obtained that normally consist of two curves, one for the ordinary and one for the extraordinary wave (figs. 13.2–13.4). Sometimes, especially at high latitudes, a third curve is observed in which $h'(f)$ for the F-layer has a greater value than for either of the other curves (Harang, 1936; Meek, 1948; Newstead, 1948; Toshniwal, 1935). The form of the $h'(f)$ record near the penetration frequency of the F-layer is then as sketched in fig. 16.3. The third echo is usually known as the Z-trace. Its polarisation shows that when it leaves the ionosphere it is an ordinary wave, but its equivalent height shows that it must have been reflected near the level where $X = 1 + Y$. In topside ionograms the Z-trace normally appears, fig. 13.6, and can be explained by simple ray theory, § 13.5, because the transmitter and receiver are immersed in the plasma. But this is not possible for ionograms observed at the ground since the transmitter and receiver are in free space.

If conditions were less than critical, § 16.5, the Z-trace would be observed at the ground with the polarisation of an ordinary wave, and the ordinary wave reflection from near where $X = 1$ would be either weak or absent. But this could only happen near a magnetic pole where $\sin\Theta$ and thence Z_t are small so that the transition height is above the maximum of the F-layer. For a detailed study of the theory of the ordinary wave reflection from near $X = 1$ in these conditions see Budden and Smith (1973). For ionograms observed at the ground in latitudes not near a pole, the transition height is well below the maximum of the F-layer and the critical coupling frequency is much less than the penetration frequency of the F-layer, so that conditions must be greater than critical. Then it is not possible to explain the Z-trace by simple ray theory.

It has been suggested (e.g. Rydbeck, 1950, 1951) that the effect is produced by a twofold mode conversion process, for the upgoing and downgoing waves, associated with the coupling points C_1 and C_2 respectively.

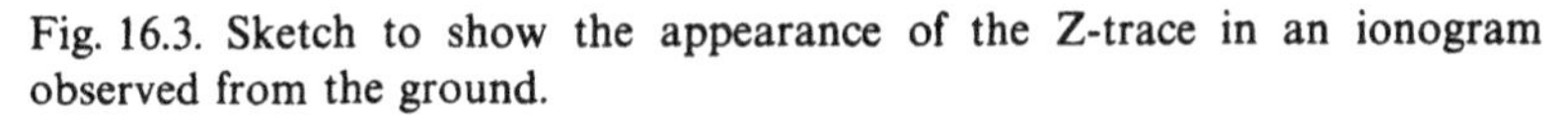

Fig. 16.3. Sketch to show the appearance of the Z-trace in an ionogram observed from the ground.

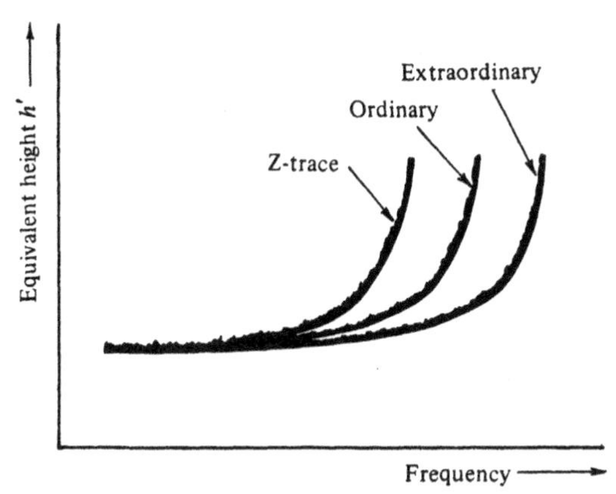

It is instructive to estimate the order of magnitude of the Z-trace reflection assuming that this is the correct explanation. We use the phase integral formula (16.60) for the amplitude of the extraordinary wave produced by mode conversion from the incident ordinary wave. The frequency is near the penetration frequency of the F-layer so that Z is small. For simplicity we take $Z=0$ so that on the real z axis n_O is real for $X<1$, and n_E is real for $X>1$. The contour Γ_1, fig. 16.2(a), can be distorted to coincide with the path AQC_1QB as sketched in fig. 16.4. The real part of the exponent in (16.60) is then determined entirely by the two sides of the line QC_1. Assume that the electron concentration varies linearly with height z so that $X=\alpha z$. Let $\mathscr{G}$ be the modulus of (16.60). Then

$$\mathscr{G}=\exp\left[\operatorname{Re}\left\{\frac{-ik}{\alpha}\int_{(Q)}^{(C_1)}(n_O-n_E)\,dX\right\}\right]. \tag{16.64}$$

At Q, $X=1$ and at C_1, $X=1+iZ_t$. At intermediate points let $X=1+i\zeta$. Then

$$\mathscr{G}=\exp\left\{\frac{k}{\alpha}\int_0^{Z_t}\operatorname{Re}(n_O-n_E)\,d\zeta\right\}. \tag{16.65}$$

When $\zeta=0$, $n_O-n_E=-1$ and when $\zeta=Z_t$, $n_O-n_E=0$. Rydbeck (1950, 1951) and Pfister (1953) have given curves that show that $\operatorname{Re}(n_O-n_E)$ varies smoothly and monotonically in the range $0\leqslant\zeta\leqslant Z_t$. Hence the integral in (16.65) is of order $-Z_t$, and approximately

$$\mathscr{G}\approx\exp(-kZ_t/\alpha)=\exp(-\omega_t/\alpha c). \tag{16.66}$$

For temperate latitudes take $\omega_t\approx 10^6\,s^{-1}$, $\alpha=0.1\,km^{-1}$. Then $\mathscr{G}\approx e^{-33}=4.7\times 10^{-15}$. To give the amplitude of the Z-trace this figure must be squared since the coupling process occurs twice. Hence the amplitude of the Z-reflection would be far too weak to be detected. But since it has often been observed even in low latitudes, for example in India (Toshniwal, 1935), there must be some other explanation. One possibility is that occasionally there can be a very steep gradient of electron

Fig. 16.4. The complex z plane, showing possible contours for use with the phase integral formula (16.56).

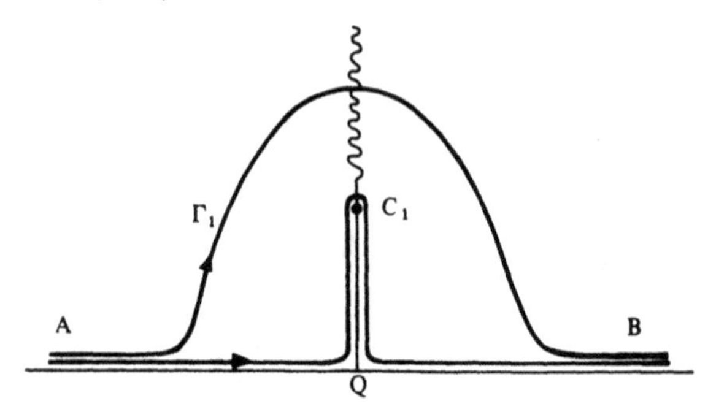

concentration near the coupling level, so that α has a much larger value than was assumed above.

A more likely explanation (Ellis, 1956) is that the Z-trace arises from waves that travel obliquely so that in the coupling region the wave normal is nearly parallel to the earth's magnetic field. Then the two coupling points C_1, C_3 are very near together and approach coalescence, fig 16.1(c). There is a radio window, the 'Ellis window', at the level where $X = 1$ and the upgoing ordinary wave goes through it and is an extraordinary wave on the upper side. The theory of this effect is given in §§ 17.6, 17.7. This extraordinary wave would not return to a receiver near the transmitter after reflection. But if there are irregularities of electron concentration near the level where $X = 1 + Y$, some energy can be scattered and return along the same path. If the refractive index within an irregularity differs by Δn_E from its value n_E in the surrounding plasma, the scattered amplitude is proportional to $\Delta n_E/n_E$ (Booker, 1959). This is largest near the level $X = 1 + Y$ where $|n_E|$ is smallest.

16.9. Additional memory

In this section the summation convention, §§ 2.11, 14.8, for repeated suffixes is not used.

The W.K.B. solution for the general anisotropic stratified medium was given at (7.112). Apart from the constant K_j it contained two factors $\mathbf{s}_j$ and $\exp(-ik\int_0^z q_j\,dz)$. The first is the jth column of $\mathbf{S}$. It is given, for example, by (7.134) and it is then a function of the composition of the medium at the height z only. It does not depend on how the medium varies with z. It will be called a 'local' property of the medium. The other factor has been called the 'phase memory' term; § 7.3. The exponent gives the change of complex phase of the wave in its passage to the point z. It depends on the whole of the medium that the wave has traversed, and is not a local property. In these solutions, therefore, it is important to distinguish between 'local' and 'memory' factors.

Because of the coupling process, one characteristic wave, with subscript i say, gives rise as it travels to some of the other characteristic waves. The amplitude of wave j thus generated in any element δz_1 at $z = z_1$ on the path of wave i is proportional to $\Gamma_{ij}\delta z_1 \exp(-ik\int_0^{z_1} q_i\,dz)$. This infinitesimal wave j then travels to another point z_2 with a phase memory factor $\exp\{-ik(\int_0^{z_1} q_i dz + \int_{z_1}^{z_2} q_j dz)\}$. The original wave i reaches z_2 with a phase memory factor $\exp(-ik\int_0^{z_2} q_i dz)$. Thus the complex phase difference at z_2 between the parent wave i and the infinitesimal wave j is

$$\phi_{ij} = k\int_{z_1}^{z_2} (q_i - q_j)\,dz. \tag{16.67}$$

Now at z_2 many of these infinitesimal waves j have to be added for all contributing

z_1. This could be done by an amplitude–phase diagram, which would give a spiral, and the resultant is small even when Γ_{ij} is moderately large, because of the partial destructive interference of the contributions with the variable phase (16.67).

Consider now the wave generated in δz_1 through the action of the element Γ_{jj} of the coupling matrix. It is of the same type as the parent wave j and travels with the same phase memory factor. At z_2 the phase difference between it and the parent wave is $\phi_{jj} = 0$ from (16.67). When the contributions are added, therefore, they all have the same phase and reinforce each other so that the resultant may be large even when Γ_{jj} is small. This shows that the diagonal elements Γ_{jj} have to be treated differently from the non-diagonal elements. It was suggested in §7.15 that $k\Gamma_{jj}$ should be included with $-ikq_j$ in the phase memory term as in (7.117). This suggestion must now be examined.

The coupling matrix $\boldsymbol{\Gamma} = -\mathbf{S}^{-1}\mathbf{S}'$ was given at (16.5), (16.6) and this form used the version (7.134) for $\mathbf{S}$. But a different $\mathbf{S}$ can be used by normalising the columns differently, as explained at the end of § 7.15; for an example see § 16.3. This means that a new version $\check{\mathbf{S}}$ of $\mathbf{S}$ can be found by multiplying by an arbitrary diagonal matrix $\mathbf{D}$ on the right, and the variable $\mathbf{f} = \mathbf{S}^{-1}\mathbf{e}$ in (16.53) must be replaced by $\check{\mathbf{f}}$ thus

$$\check{\mathbf{S}} = \mathbf{SD}, \quad \check{\mathbf{S}}^{-1} = \mathbf{D}^{-1}\mathbf{S}^{-1}, \tag{16.68}$$

and

$$\check{\mathbf{f}} = \check{\mathbf{S}}^{-1}\mathbf{e} = \mathbf{D}^{-1}\mathbf{S}^{-1}\mathbf{e}, \quad \mathbf{e} = \mathbf{SD}\check{\mathbf{f}}. \tag{16.69}$$

Now $\boldsymbol{\Gamma}$ is replaced by

$$\check{\boldsymbol{\Gamma}} = \mathbf{D}^{-1}\boldsymbol{\Gamma}\mathbf{D} - \mathbf{D}^{-1}\mathbf{D}'. \tag{16.70}$$

The diagonal elements of $\boldsymbol{\Gamma}$ and $\mathbf{D}^{-1}\boldsymbol{\Gamma}\mathbf{D}$ are the same, so the effect of using $\check{\boldsymbol{\Gamma}}$ instead of $\boldsymbol{\Gamma}$ is to replace Γ_{jj} by $\Gamma_{jj} - \mathsf{D}_{jj}^{-1}\mathsf{D}'_{jj}$. In cases where the Γ_{jj} are not zero, therefore, it is always possible to change the normalisation of $\mathbf{S}$ by using a $\mathbf{D}$ in (16.68) with

$$\mathsf{D}_{jj} = \exp\left(k\int^{z}\Gamma_{jj}\,\mathrm{d}z\right), \quad j = 1, 2, 3, 4, \tag{16.71}$$

to give a new coupling matrix $\check{\boldsymbol{\Gamma}}$ with $\check{\Gamma}_{jj} = 0$. Then the factor (16.71) does not appear in f_j. But it is not removed from the field components $\mathbf{e}$ because it has merely been transferred to the factor $\mathbf{D}$ in (16.69).

Every term of Γ_{jj} has a factor that is a derivative d/dz. It can happen that $\Gamma_{jj} = \mathrm{d}\mathsf{L}_j/\mathrm{d}z$ where L_j is a local function only. Then (16.71) is also a local function and can be combined with the local factor $\mathbf{s}_j$ of the W.K.B. solution, (7.112), and it does not appear in the memory term. But many cases occur where Γ_{jj} cannot be expressed in this way because $\Gamma_{jj}\mathrm{d}z$ is not a perfect differential. This was discussed by Smith, M.S. (1975) who gave examples for an electron plasma where Γ_{jj} depends on the two functions $X(z)$ and $Z(z)$. Then $\int^z \Gamma_{jj}\mathrm{d}z$ depends on how the medium varies in the range of z covered by the integral. In this respect it resembles $\int^z q\mathrm{d}z$. It has a memory content that is called 'additional memory'. Budden and Smith (1976) showed that there is additional memory for many types of wave propagation encountered in geophysics.

In ray tracing, the local factor of the W.K.B. solution is ignored. It is assumed that the ray paths are given solely by the eikonal function, § 14.2, and for stratified media this is the phase memory $k\int^z q\,\mathrm{d}z$. The local factor is slowly varying whereas the phase memory has a factor k that makes its effect large, especially at high frequencies. In some cases, at lower frequencies, the local factor $\mathbf{s}_j$ may contain a factor g_j that varies too rapidly to be ignored. It must then be allowed for in the eikonal by writing the phase memory as $k\int^z q\,\mathrm{d}z + \mathrm{i}\ln \mathrm{g}_j$. This is just the converse of the process for removing Γ_{jj} described above.

Smith, M.S (1976) studied the properties of the form of $\mathbf{\Gamma}$ given at (16.5), (16.6) and examined the effect on ray paths of incorporating Γ_{jj} into the eikonal. He showed that in general ray paths are not reversible. For example the path of an ordinary ray that goes from A to B via the ionosphere is not the same as the reverse of the path of an ordinary ray going from B to A. The effect is small but Smith gave examples at frequencies 100, 300 and 900 kHz where it could be important. The use of Γ_{jj} affects the calculation of lateral deviation §§ 10.13, 10.14 and of the bearing errors that result from it. In another paper Smith, M.S. (1975) gave a possible physical interpretation of Γ_{jj} and showed how its influence can be detected in the results of full wave calculations of the reflecting properties of the ionosphere.

The Γ_{jj} become important in ray theory at low frequencies where the factor k, in the conventional phase memory $k\int q\mathrm{d}z$, gets small. But this is also where the W.K.B. approximations used in ray theory are expected to be inaccurate. It may be asked whether these two effects are really separate. The reason for thinking that the effect of Γ_{jj} must be allowed for when the non-diagonal elements of $\mathbf{\Gamma}$ can still be neglected was explained at the beginning of this section. It is that the effect of Γ_{jj} is cumulative whereas the effect of the non-diagonal elements is smaller because of destructive interference. But this explanation has not been fully tested and there is a need for further study.

16.10. Second order coupled equations

Some forms of coupled equations have been used in which the principal terms contain derivatives up to the second order, and the coupling terms have derivatives up to the first order. An example has already been given at (16.43). Second order coupled equations of this kind have been used mainly for waves that are vertically incident on the ionosphere. A few special cases for oblique incidence have been given (Heading, 1953; Budden and Clemmow, 1957) but not extensively studied. The most important type for vertical incidence is the pair of equations, known as the Försterling equations (16.43) or (16.90), (16.91) and these are the main subject of study in

§§ 16.11–16.13. They were formulated (Försterling, 1942) long before the first order coupled equations were known.

For vertical incidence $S = 0$, $C = 1$, the differential equations (7.77), (7.78) take the simpler form

$$\frac{\mathrm{d}E_y}{\mathrm{d}z} = \mathrm{i}k\mathscr{H}_x, \quad \frac{\mathrm{d}E_x}{\mathrm{d}z} = -\mathrm{i}k\mathscr{H}_y, \quad \mathscr{H}_z = 0, \tag{16.72}$$

$$\frac{\mathrm{d}\mathscr{H}_y}{\mathrm{d}z} = -\frac{\mathrm{i}k}{\varepsilon_0}D_x, \quad \frac{\mathrm{d}\mathscr{H}_x}{\mathrm{d}z} = \frac{\mathrm{i}k}{\varepsilon_0}D_y, \quad D_z = 0. \tag{16.73}$$

If $\mathscr{H}_x$, $\mathscr{H}_y$ are eliminated:

$$\frac{\mathrm{d}^2E_x}{\mathrm{d}z^2} + \frac{k^2}{\varepsilon_0}D_x = 0, \quad \frac{\mathrm{d}^2E_y}{\mathrm{d}z^2} + \frac{k^2}{\varepsilon_0}D_y = 0. \tag{16.74}$$

Now D_x, D_y, D_z can be expressed in terms of E_x, E_y, E_z from (7.79) and (3.35). The last of (16.73) gives

$$M_{zx}E_x + M_{zy}E_y + (1 + M_{zz})E_z = 0 \tag{16.75}$$

whence E_z may be eliminated from the expressions for D_x, D_y. Then (16.74) become

$$\frac{1}{k^2}\frac{\mathrm{d}^2E_x}{\mathrm{d}z^2} + \left(1 + M_{xx} - \frac{M_{xz}M_{zx}}{1 + M_{zz}}\right)E_x + \left(M_{xy} - \frac{M_{xz}M_{zy}}{1 + M_{zz}}\right)E_y = 0 \tag{16.76}$$

$$\frac{1}{k^2}\frac{\mathrm{d}^2E_y}{\mathrm{d}z^2} + \left(1 + M_{yy} - \frac{M_{yz}M_{zy}}{1 + M_{zz}}\right)E_y + \left(M_{yx} - \frac{M_{yz}M_{zx}}{1 + M_{zz}}\right)E_x = 0. \tag{16.77}$$

Note that the four coefficients used here are the four elements $\mathbf{T}_{41}$, $-\mathbf{T}_{42}$. $\mathbf{T}_{32}$, $-\mathbf{T}_{31}$ of the matrix $\mathbf{T}$, (7.81). The x and y axes may now be chosen so that the vector $\mathbf{Y}$ is in the x–z plane, that is $l_y = 0$. Let n_{O}, n_{E} be the two refractive indices for waves travelling upwards in a fictitious homogeneous medium with the properties of the actual ionosphere at each level, and let ρ_{O}, ρ_{E} be the corresponding values of the wave polarisation E_y/E_x. Then the four quantities n_{O}, n_{E}, ρ_{O}, ρ_{E} are functions of z and are expressible in terms of the M_{ij}. They provide an alternative way of specifying the properties of the plasma at each level, and the coefficients in (16.76), (16.77) can be expressed in terms of them. It can be shown that the equations then become

$$\begin{aligned} &\frac{1}{k^2}\frac{\mathrm{d}^2E_x}{\mathrm{d}z^2} + \frac{\rho_{\mathrm{O}}n_{\mathrm{E}}^2 - \rho_{\mathrm{E}}n_{\mathrm{O}}^2}{\rho_{\mathrm{O}} - \rho_{\mathrm{E}}}E_x + \frac{n_{\mathrm{O}}^2 - n_{\mathrm{E}}^2}{\rho_{\mathrm{O}} - \rho_{\mathrm{E}}}E_y = 0, \\ &\frac{1}{k^2}\frac{\mathrm{d}^2E_y}{\mathrm{d}z^2} + \frac{\rho_{\mathrm{O}}n_{\mathrm{O}}^2 - \rho_{\mathrm{E}}n_{\mathrm{E}}^2}{\rho_{\mathrm{O}} - \rho_{\mathrm{E}}}E_y - \frac{n_{\mathrm{O}}^2 - n_{\mathrm{E}}^2}{\rho_{\mathrm{O}} - \rho_{\mathrm{E}}}E_x = 0. \end{aligned} \tag{16.78}$$

The proof is left as an exercise for the reader, who should also verify that the equations reduce to the correct form at the magnetic equator, and at the magnetic poles.

Equations (16.78) or equivalently (16.76), (16.77) are in a sense second order coupled equations. The first two terms of each equation are principal terms and the remaining term is the coupling term. But the coupling terms are not in general small and so this form is unsuitable for use with a method of successive approximation.

16.11. Försterling's coupled equations for vertical incidence

Försterling's equations (16.43) have already been derived in § 16.4 as a deduction from the first order coupled equations. They will now be derived by a shorter and more direct method that is similar to the original method (Försterling, 1942).

In the differential equations (16.74) the variables E_x, E_y, D_x, D_y refer to the total fields. Each of them is now to be expressed as the sum of the fields for the ordinary and extraordinary waves, thus

$$E_x = E_x^{(O)} + E_x^{(E)}, \quad E_y = E_y^{(O)} + E_y^{(E)} \tag{16.79}$$

and similarly for D_x. D_y. The four new variables on the right in (16.79) may be made to satisfy two further relations. We now choose the x and y axes so that $l_y = 0$. Then they may be made to satisfy

$$E_y^{(O)}/E_x^{(O)} = \rho_O, \quad E_y^{(E)}/E_x^{(E)} = \rho_E \tag{16.80}$$

where ρ_O, ρ_E are the two wave polarisations as given by magnetoionic theory, ch. 4. They apply for both upgoing and downgoing waves since the same system of axes is used for both; see end of § 4.4. Equations (16.80) are the required two further relations and they are used to *define* the ordinary and extraordinary waves in the variable medium. Now (4.29) gives $\rho_O\rho_E = 1$ so that $E_x^{(E)} = \rho_O E_y^{(E)}$ and hence (16.79) becomes

$$E_x = E_x^{(O)} + \rho_O E_y^{(E)}, \quad E_y = \rho_O E_x^{(O)} + E_y^{(E)}. \tag{16.81}$$

The electric displacement $\boldsymbol{D}$ is derived from the electric field by the constitutive relations, which are the same for a homogeneous and a variable medium. For the ordinary and extraordinary component waves, (4.42), (4.43) show that

$$D_x^{(O)} = \varepsilon_0 n_O^2 E_x^{(O)}, \quad D_y^{(O)} = \varepsilon_0 n_O^2 E_y^{(O)}, \quad D_x^{(E)} = \varepsilon_0 n_E^2 E_x^{(E)}, \quad D_y^{(E)} = \varepsilon_0 n_E^2 E_y^{(E)}. \tag{16.82}$$

The total $\boldsymbol{D}$ must be the sum of the contributions from the two component waves. Hence

$$\varepsilon_0^{-1} D_x = n_O^2 E_x^{(O)} + n_E^2 \rho_O E_y^{(E)}, \quad \varepsilon_0^{-1} D_y = n_O^2 \rho_O E_x^{(O)} + n_E^2 E_y^{(E)}. \tag{16.83}$$

Equations (16.81), (16.83) are now substituted in (16.74). A prime ′ is used to denote $k^{-1}\mathrm{d}/\mathrm{d}z$. Thus

$$E_x^{(O)\prime\prime} + \rho_O E_y^{(E)\prime\prime} + 2\rho_O' E_y^{(E)\prime} + \rho_O'' E_y^{(E)} + n_O^2 E_x^{(O)} + n_E^2 \rho_O E_y^{(E)} = 0 \tag{16.84}$$

$$\rho_O E_x^{(O)\prime\prime} + E_y^{(E)\prime\prime} + 2\rho_O' E_x^{(O)\prime} + \rho_O'' E_x^{(O)} + n_O^2 \rho_O E_x^{(O)} + n_E^2 E_y^{(E)} = 0. \tag{16.85}$$

Now (16.85) is multiplied by ρ_O and (16.84) is subtracted; similarly (16.84) is multiplied by ρ_O and (16.85) is subtracted:

$$(\rho_O^2 - 1)(E_x^{(O)''} + n_O^2 E_x^{(O)}) + \rho_O\rho_O'' E_x^{(O)} + 2\rho_O\rho_O' E_x^{(O)'} = 2\rho_O' E_y^{(E)'} + \rho_O'' E_y^{(E)} \quad (16.86)$$

$$(\rho_O^2 - 1)(E_y^{(E)''} + n_E^2 E_y^{(E)}) + \rho_O\rho_O'' E_y^{(E)} + 2\rho_O\rho_O' E_y^{(E)'} = 2\rho_O' E_x^{(O)'} + \rho_O'' E_x^{(O)}. \quad (16.87)$$

The equations are now in the required coupled form, but some further simplifications are possible. New dependent variables are chosen which remove the first order derivative from the principal terms. Thus let

$$E_x^{(O)} = \mathscr{F}_O(\rho_O^2 - 1)^{-\frac{1}{2}}, \quad E_y^{(E)} = \mathscr{F}_E(\rho_O^2 - 1)^{-\frac{1}{2}} \quad (16.88)$$

so that, from (16.81),

$$\mathscr{F}_O = (-E_x + \rho_O E_y)(\rho_O^2 - 1)^{-\frac{1}{2}}, \quad \mathscr{F}_E = (\rho_O E_x - E_y)(\rho_O^2 - 1)^{-\frac{1}{2}} \quad (16.89)$$

and use ψ from (16.14). Then

$$\mathscr{F}_O'' + (n_O^2 + \psi^2)\mathscr{F}_O = \psi'\mathscr{F}_E + 2\psi\mathscr{F}_E', \quad (16.90)$$

$$\mathscr{F}_E'' + (n_E^2 + \psi^2)\mathscr{F}_E = \psi'\mathscr{F}_O + 2\psi\mathscr{F}_O', \quad (16.91)$$

which are Försterling's coupled equations, arranged with the principal terms on the left and the coupling terms on the right. Note that in (16.43) $\mathsf{b}_1 = \mathscr{F}_O, \mathsf{b}_3 = -\mathscr{F}_E$.

For frequencies greater than about 1 MHz, the coupling parameter ψ (6.14) is negligibly small at real heights, in nearly all cases of practical importance. If it is neglected, the equations (16.90), (16.91) become

$$\mathscr{F}_O'' + n_O^2\mathscr{F}_O = 0, \quad \mathscr{F}_E'' + n_E^2\mathscr{F}_E = 0 \quad (16.92)$$

which shows that the ordinary and extraordinary waves are then propagated and reflected independently of each other. The equations can then be treated in exactly the same way as (7.6) was treated when discussing propagation in an isotropic ionosphere in chs. 7, 8, 12, 15. This provides the justification for studying the ordinary and extraordinary waves independently, for vertical incidence, as was done in ch. 13. Instead of the variable E_y used in (7.6), the field variables to be used are the Försterling variables (16.89). Then much of the theory of chs. 12, 15 can be applied also to an anisotropic ionosphere. In particular a phase integral formula for the reflection coefficient, analogous to (8.78), can be derived for each of the two waves.

The neglect of ψ in (16.90), (16.91) is not permissible, even at high frequencies, in the lowest part of the ionosphere where the waves enter it, or emerge into free space. It is here that the phenomenon of limiting polarisation occurs; § 17.10, 17.11. But this level is remote from the reflection levels, so that (16.92) can still be used as a basis for finding reflection coefficients and equivalent heights.

The properties of the coupling parameter ψ are discussed in the following section. Some applications of Försterling's equations are given in §§16.13–16.14.

16.12. Properties of the coupling parameter ψ

Equation (6.14) gives ψ and in view of the relation $\rho_O\rho_E = 1$ it can be written

$$\psi = \frac{\rho_O'}{\rho_O^2 - 1} = \frac{\rho_E'}{\rho_E^2 - 1} = \frac{1}{2k}\frac{d}{dz}\ln\left(\frac{\rho_O - 1}{\rho_O + 1}\right) = \frac{1}{4k}\frac{d}{dz}\ln\left(\frac{\rho_O + \rho_E - 2}{\rho_O + \rho_E + 2}\right). \quad (16.93)$$

Now (4.20) and (4.27) show that

$$\rho_O + \rho_E = \frac{iY\sin^2\Theta}{(U - X)\cos\Theta} = \frac{2iZ_t\,\mathrm{sgn}\,(\cos\Theta)}{U - X}. \quad (16.94)$$

For brevity it will be assumed that $\cos\Theta$ is positive as in the northern hemisphere so that $\mathrm{sgn}\,(\cos\Theta) = +1$. Then

$$\psi = \frac{-\frac{1}{2}iZ_t}{Z_t^2 + (U - X)^2}(X' + iZ') \quad (16.95)$$

Fig. 16.5. The function $|\psi|$, and the real parts μ and minus the imaginary parts χ, of the two refractive indices, plotted against height z for a frequency of 1 MHz in a typical ionospheric layer, chosen to simulate the E-layer in the daytime. In this example the ionosphere is a Chapman layer with its maximum at 115 km and having a penetration frequency of 4.4 MHz. The scale height H is 10 km. The collision frequency is given by $\nu = \nu_0 \exp(-z/H)$ and is equal to $10^6\,\mathrm{s}^{-1}$ at 90 km. $f_H = 1.12$ MHz, $\Theta = 23.27°$, $\omega_t = 5.98 \times 10^5\,\mathrm{s}^{-1}$.

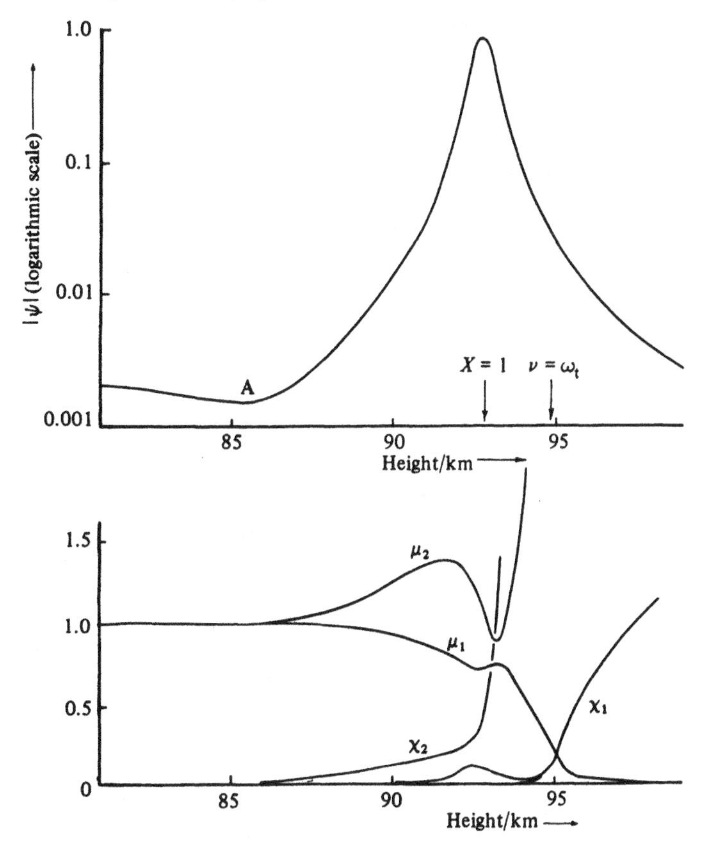

where the prime $'$ means $k^{-1}\mathrm{d}/\mathrm{d}z$. Thus ψ depends on the rates of variation with height z of X, proportional to electron concentration, and Z, proportional to electron collision frequency. In a homogeneous medium ψ is zero. It can also be shown from (4.67) that

$$\psi = \frac{-\tfrac{1}{2}\mathrm{i}X^2Y^3\sin^2\Theta\cos\Theta}{A^2(n_\mathrm{O}^2-n_\mathrm{E}^2)^2}(X'+\mathrm{i}Z') \tag{16.96}$$

where A is given by (4.66). The point where $A=0$ is not a singularity of ψ because there one of n_O^2, n_E^2 is infinite and the denominator is bounded and non-zero.

It is often convenient to treat X and Z as analytic complex functions of z, but real on the real z axis. The functions $X(z)$, $Z(z)$ must have singularities but it is assumed that these are not near the real z axis, and they are not considered further here. Then in the regions of the z plane that are of interest, X' and Z' are bounded. The only singularities of ψ are where the denominator is zero, that is where

$$X = 1-\mathrm{i}(Z\pm Z_\mathrm{t}) \tag{16.97}$$

and these are simple poles of ψ. They are the coincident coupling points C_1, C_2 for the upper sign, and C_3, C_4 for the lower sign; § 16.5.

Fig. 16.6. The same as fig. 16.5 but for waves of frequency 160 kHz.

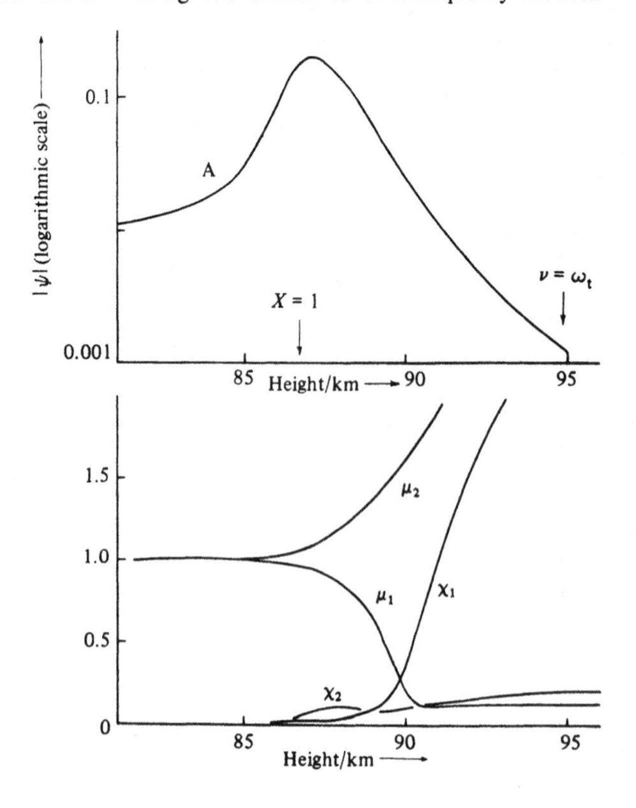

Below the ionosphere X is zero, but Z and Z' are not zero, and

$$\psi = \tfrac{1}{2}Z'Z_t/\{(1 - iZ)^2 + Z_t^2\}. \tag{16.98}$$

In this case there are no electrons and the medium behaves like free space. Yet (16.98) shows that the coupling parameter is not zero. This has important consequences in the theory of limiting polarisation and is discussed in § 17.10, 17.11.

Some typical curves showing how $|\psi|$ depends on height z are given in fig. 16.5–16.7. Here a model has been assumed for the ionosphere in which both X and Z vary with height z. Curves of the real parts and minus the imaginary parts, μ, χ respectively, of the two refractive indices are also shown for comparison. The transition height where $\nu = \omega_t$ and the level where $X = 1$ are indicated. All curves of $|\psi|$ show a bend marked A in the figures. At levels below this the electron concentration is negligible and the last factor of (16.95) is determined by Z'. Thus at these low levels it is the variation of collision frequency with height that ensures that ψ is not zero. At higher levels the electron concentration is appreciable and varies much more rapidly than the collision frequency. At levels above A, therefore, Z' is negligible compared with X'. The curves show a maximum of $|\psi|$ near the level

Fig. 16.7. Curves for a frequency of 160 kHz in an ionospheric layer chosen to simulate the lower part of the E-layer at night. The ionosphere is the same as in fig. 16.5 except that the penetration frequency is now 0.5 MHz.

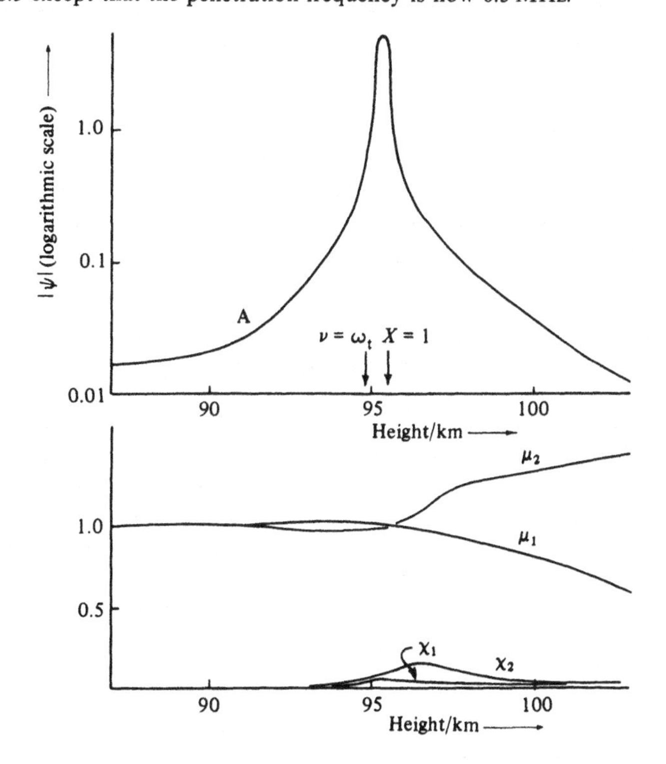

where $X = 1$. This is Re(X) for the poles (16.97) of ψ. The maximum is very marked when the level of $X = 1$ is near the transition height $\nu = \omega_t$, and less marked when these levels are well separated, fig. 16.6.

In figs. 16.5, 16.6 the height where $X = 1$ is below the transition height and conditions are greater than critical; see § 16.5. Thus at all heights μ_2, χ_2 refer to the ordinary wave and μ_1, χ_1 to the extraordinary wave. In fig. 16.7 the height where $X = 1$ is slightly above the transition height and conditions are less than critical. Hence μ_1, χ_1 refer to the ordinary wave below the level where $X = 1$, and to the extraordinary wave above this. Similarly μ_2, χ_2 refer to the extraordinary wave below, and to the ordinary wave, that is the whistler mode, above this level.

For frequencies greater than about 2 MHz, the level where $X = 1$ is nearly always well above the transition height and it is then permissible to neglect collisions, as was done in most of chs. 10, 12, 13. Then ψ is given by

$$\psi \approx \frac{-\frac{1}{2}\mathrm{i}Z_t X'}{Z_t^2 + (1-X)^2}. \tag{16.99}$$

If typical numerical values are inserted, it can be shown that at high frequencies $|\psi|$ remains small so that coupling effects are negligible. If ψ is neglected, Försterling's equations give the two separate equations (16.92) already discussed in § 16.11.

In this and the previous section the coupling, for vertical incidence, between the ordinary and extraordinary waves is expressed by the coupling parameter ψ and occurs because of the vertical gradients of X and Z in (16.95). It was assumed that the vector $\boldsymbol{Y}$ is in the x–z plane and that its magnitude Y and direction Θ are independent of the coordinates. Important instances can occur where these assumptions cannot be made. There can then be z dependence of Y, Θ and the azimuth angle Φ of the plane containing $\boldsymbol{Y}$ and the z axis, and these gradients can also give some coupling; see the comments at the end of § 18.8 on spatial variations of $\boldsymbol{Y}$. An extension of the theory to deal with this was given by Cohen (1960), for vertical incidence. He obtained coupled equations similar to (16.90), (16.91) but they contain two coupling parameters ψ_{O}, ψ_{E} instead of one, whose definitions are more complicated than (16.93). The ψs on the right of (16.90) must be replaced by ψ_{E} and those on the right of (16.91) by ψ_{O}. The ψ^2 on the left of both (16.90), (16.91) is replaced by $\psi_{\mathrm{O}}\psi_{\mathrm{E}}$. There are corresponding modifications in the definition (16.89) of the dependent variables. Cohen has discussed important applications of his equations to radio propagation in the sun's atmosphere and in the interplanetary plasma, as well as ionospheric problems. Both Cohen (1960) and Titheridge (1971c) have studied the effect of a varying Θ on Faraday rotation

16.13. The method of 'variation of parameters'

In a number of important papers (Gibbons and Nertney, 1951, 1952; Kelso, Nearhoof, Nertney and Waynick, 1951; Rydbeck, 1950; Davids and Parkinson, 1955)

a method of successive approximations has been applied to Försterling's coupled equations (16.90), (16.91). The method used is called 'variation of parameters'. It is a well known mathematical technique; see for example, Heading (1975c), Jeffreys and Jeffreys (1972). It is summarised here in the form needed for the Försterling equations.

Suppose that we seek a solution of the linear inhomogeneous equation

$$\mathrm{d}^2u/\mathrm{d}s^2 + uf(s) = g(s). \tag{16.100}$$

Consider the homogeneous equation

$$\mathrm{d}^2u/\mathrm{d}s^2 + uf(s) = 0 \tag{16.101}$$

and let $u_1(s)$, $u_2(s)$ be two independent solutions of it, whose Wronskian is

$$W(u_1, u_2) = u_1 \mathrm{d}u_2/\mathrm{d}s - u_2\, \mathrm{d}u_1/\mathrm{d}s. \tag{16.102}$$

We now seek a solution of (16.100) of the form

$$u = Au_1 + Bu_2 \tag{16.103}$$

where A and B are functions of s, and since there are two of them we can impose one relation between them. Now

$$\mathrm{d}u/\mathrm{d}s = u_1\, \mathrm{d}A/\mathrm{d}s + u_2\, \mathrm{d}B/\mathrm{d}s + A\, \mathrm{d}u_1/\mathrm{d}s + B\, \mathrm{d}u_2/\mathrm{d}s. \tag{16.104}$$

For the imposed relation take

$$u_1\, \mathrm{d}A/\mathrm{d}s + u_2\, \mathrm{d}B/\mathrm{d}s = 0. \tag{16.105}$$

Then

$$\frac{\mathrm{d}^2u}{\mathrm{d}s^2} = \frac{\mathrm{d}u_1}{\mathrm{d}s}\frac{\mathrm{d}A}{\mathrm{d}s} + \frac{\mathrm{d}u_2}{\mathrm{d}s}\frac{\mathrm{d}B}{\mathrm{d}s} + A\frac{\mathrm{d}^2u_1}{\mathrm{d}s^2} + B\frac{\mathrm{d}^2u_2}{\mathrm{d}s^2}. \tag{16.106}$$

Substitute this in (16.100) and use (16.101) with $u = u_1$, u_2. Then

$$\frac{\mathrm{d}u_1}{\mathrm{d}s}\frac{\mathrm{d}A}{\mathrm{d}s} + \frac{\mathrm{d}u_2}{\mathrm{d}s}\frac{\mathrm{d}B}{\mathrm{d}s} = g(s). \tag{16.107}$$

Now (16.105), (16.107) may be solved to give

$$\frac{\mathrm{d}A}{\mathrm{d}s} = -\frac{u_2 g}{W}, \quad \frac{\mathrm{d}B}{\mathrm{d}s} = \frac{u_1 g}{W} \tag{16.108}$$

and the solution (16.103) is

$$u = -u_1 \int_a^s \frac{u_2 g}{W}\mathrm{d}s - u_2 \int_s^b \frac{u_1 g}{W}\mathrm{d}s \tag{16.109}$$

where a and b are arbitrary constants to be determined from the physical conditions. In the cases of interest here the differential equations are in the normal form, that is, like (16.100) they have no first derivative term, so that the Wronskian W is independent of s and may be taken outside the integrals (16.109).

The application of this result to Försterling's equations (16.90), (16.91) will now be illustrated by considering a specific problem in which an ordinary wave is incident from below on the region around where $X = 1$ and where the coupling may be strong. The waves that go through this region may be reflected at a higher level, but the reflected waves will at first be ignored since the effect of coupling on waves coming down from above can be treated separately.

The coupling parameter ψ is assumed to be negligible except in the coupling region, and is there assumed to be so small that the right-hand sides of (16.90), (16.91) may be neglected in the zero order approximation. The factors ψ^2 on the left will be neglected completely; they could be allowed for by replacing n_O, n_E by $(n_O^2+\psi^2)^{\frac{1}{2}}$, $(n_E^2+\psi^2)^{\frac{1}{2}}$ respectively in the following argument. It is also assumed that conditions are greater than critical so that n_O, n_E can be used unambiguously. Finally it is assumed that n_O, n_E are not near zero in the coupling region (see figs. 16.5, 16.6) so that the reflection levels are not close to the coupling region and the W.K.B. solutions (16.111) below, can be used there. Take $s = kz$ so that in the zero order approximation the equations are

$$\mathrm{d}^2\mathscr{F}_O/\mathrm{d}s^2 + n_O^2\mathscr{F}_O = 0, \quad \mathrm{d}^2\mathscr{F}_E/\mathrm{d}s^2 + n_E^2\mathscr{F}_E = 0 \tag{16.110}$$

and their W.K.B. solutions are

$$\begin{aligned} \mathscr{F}_O^{(1)} &= n_O^{-\frac{1}{2}}\exp\left(-\mathrm{i}\int_0^s n_O\,\mathrm{d}s\right), \quad \mathscr{F}_O^{(2)} = n_O^{-\frac{1}{2}}\exp\left(\mathrm{i}\int_0^s n_O\,\mathrm{d}s\right), \\ \mathscr{F}_E^{(1)} &= n_E^{-\frac{1}{2}}\exp\left(-\mathrm{i}\int_0^s n_E\,\mathrm{d}s\right), \quad \mathscr{F}_E^{(2)} = n_E^{-\frac{1}{2}}\exp\left(\mathrm{i}\int_0^s n_E\,\mathrm{d}s\right). \end{aligned} \tag{16.111}$$

There is no upgoing extraordinary wave below the coupling region, and any downgoing waves above it are being ignored, so that at all levels the zero order approximation is

$$\mathscr{F}_O = \mathscr{F}_O^{(1)}, \quad \mathscr{F}_E = 0. \tag{16.112}$$

To obtain a more accurate solution, this result is substituted in the right-hand sides of (16.90), (16.91). This leaves (16.90) unaffected so that to the first order $\mathscr{F}_O$ is unchanged. Equation (16.91), however, becomes, if ψ^2 on the left is neglected:

$$\frac{\mathrm{d}^2\mathscr{F}_E}{\mathrm{d}s^2} + n_E^2\mathscr{F}_E = g(s) \text{ where } g(s) = \mathscr{F}_O^{(1)}\frac{\mathrm{d}\psi}{\mathrm{d}s} + 2\psi\frac{\mathrm{d}\mathscr{F}_O^{(1)}}{\mathrm{d}s}. \tag{16.113}$$

The formula (16.109) is now applied with $u_1 = \mathscr{F}_E^{(1)}$, $u_2 = \mathscr{F}_E^{(2)}$ so that $W = 2\mathrm{i}$. Thus to the first order

$$\mathscr{F}_E = \tfrac{1}{2}\mathrm{i}\mathscr{F}_E^{(1)}\int_a^s \mathscr{F}_E^{(2)}g(s)\,\mathrm{d}s + \tfrac{1}{2}\mathrm{i}\mathscr{F}_E^{(2)}\int_s^b \mathscr{F}_E^{(1)}g(s)\,\mathrm{d}s. \tag{16.114}$$

The limits a, b must now be found. When s is outside the coupling region the integrands in (16.114) are zero and the integrals are constant. When s is well below the coupling region, the first integral must be zero since $\mathscr{F}_E^{(1)}$ is an upgoing extraordinary wave which is absent there. Similarly the second integral must be zero when s is above the coupling region since $\mathscr{F}_E^{(2)}$ is a downgoing extraordinary wave. Thus a is any value of s well below the coupling region and b is any value well above it. Equation (16.114) shows that above the coupling region there is an upgoing extraordinary wave, the first term, and below it there is a downgoing extraordinary wave, the second term. These are generated by mode conversion, in the coupling region, from the incident upgoing ordinary wave. In fact (16.114) shows that the coupling region may be thought of as two distributed sources of these waves. The amplitudes of these sources at any level are the integrands.

For frequencies less than the electron gyro-frequency, the above method can also be used when the incident wave is an upgoing extraordinary wave, since this wave travels up through the coupling region to its reflection level near where $X = 1 + Y$. By mode conversion in the coupling region it gives rise to a downgoing ordinary wave below the coupling region, and an upgoing ordinary wave above it. Similarly the method can be used when the incident wave is a downgoing wave, ordinary or extraordinary, from above the coupling region.

The above theory is not only of great historical importance but also of physical interest, particularly in the interpretation of the integrands in (16.114) as distributed sources of waves produced by mode conversion. It was valuable as a method of computation in the days when the storage capacity of computers was less than it is today. But for the original basic differential equations (7.80), (7.81) the coupling and reflection points are ordinary points. Near them the field variables and the coefficients are bounded and continuous. They give no trouble when these basic equations are integrated numerically. Transformation of the equations to coupled form introduces, at the coupling points, singularities that are not present in the original equations. This and the need for repeatedly calculating n_O, n_E, ψ, or similar quantities, introduce complications when the Försterling equations or any other form of coupled equations are used in computing. The speed and large storage capacity of modern computers are now such that the original equations can be numerically integrated quickly and without approximations and this means that, for computing, the use of coupled equations and the method of variation of parameters are largely superseded.

16.14. The coupling echo

In measurements with radio pulses at vertical incidence with a frequency of 150 kHz, reflections have sometimes been observed that come from the level near where $X \approx 1$, as shown from the equivalent height of reflection. These are additional to the reflection that comes from the higher level where $X \approx 1 + Y$. They occur usually at night (Kelso *et al.*, 1951; Nertney, 1951, 1953; Lindquist, 1953; Parkinson, 1955).

At this frequency the values of n_O and n_E are close to unity when X is near to 1; for examples see figs. 16.6, 16.7 which are for 160 kHz. This is because of the large value of Z. The reflection point where $X = 1 - \mathrm{i}Z$ is a long way from the real z axis so that the zero of n_O is not apparent. But since there is a region of strong coupling near where $X \approx 1$, it was suggested that these reflections arise because of a coupling mechanism, and they are called 'coupling echoes'. Calculations using a method of successive approximation with variation of parameters, as in § 16.13 (Davids, 1952; Davids and Parkinson, 1955), have confirmed that reflections of this kind might be expected.

There is no coupling point associated with an upgoing ordinary and a downgoing extraordinary wave, nor with an upgoing extraordinary and downgoing ordinary wave. The coupling echo cannot be explained by a combination of separate simple coupling and reflection processes each associated with one coupling point. The regions where the W.K.B. solutions fail, surrounding the coupling points C_1 to C_4 and O, must overlap so that it is not possible to find a contour for use with the phase integral method. Smith, M.S. (1973a) has shown that for ionospheric models of a particular type, a coupling echo with detectable amplitude would not be expected when the function $N(z)$ is smoothly varying. Altman and Postan (1971) have made calculations for a range of models and frequencies. They show that coupling echoes can sometimes occur. The coupling processes that have to be invoked to explain them are more complicated than is generally assumed, and may be different for small and large values of the angle Θ between the earth's magnetic field and the vertical.

The possibility cannot be excluded that some 'coupling echoes' are in fact caused by scattering from irregularities of N in the D-region, giving partial reflections, as described in § 11.13, and indeed some authors have used the term 'backscatter' for the processes that generate these echoes.

Further study of the theory of the coupling echo and further experimental results would be of the greatest interest. In particular it would be interesting to know how the polarisation of the reflected wave changes when the polarisation of the incident wave is changed.

PROBLEMS 16

16.1. The coupling parameter ψ (16.14) has simple poles in the complex z (height) plane where $X = 1 - \mathrm{i}(Z \pm \mathrm{i}Z_t)$. Show that the residues are $\pm 1/(4k)$. Note that they are independent of $\mathrm{d}X/\mathrm{d}z$, $\mathrm{d}Z/\mathrm{d}z$.

16.2. For oblique incidence on the ionosphere when the earth's magnetic field is vertical, the Booker quartic is a quadratic equation for q^2, and its four solutions are $\pm q_{\mathrm{O}}, \pm q_{\mathrm{E}}$ for the ordinary and extraordinary waves respectively. The plane of incidence is the x–z plane with z vertical, and θ is the angle of incidence. For the ordinary wave the ratios E_y/E_x and $-\mathscr{H}_x/\mathscr{H}_y$ are not the same but their geometric mean is denoted by ρ_{O}, thus $\rho_{\mathrm{O}}^2 = -E_y\mathscr{H}_x/E_x\mathscr{H}_y$. In the same way ρ_{E} is defined for the extraordinary wave. Show that $\rho_{\mathrm{O}}\rho_{\mathrm{E}} = 1$. Let $\psi = \rho_{\mathrm{O}}'/(\rho_{\mathrm{O}}^2 - 1)$ where a prime $'$ means $k^{-1}\mathrm{d}/\mathrm{d}z$. Let $\eta^2 = (U - X)/(UC^2 - X)$ where $C = \cos\theta$. Show that the second order coupled equations may be written

$$h_O'' + \left\{ q_O^2 + \psi^2 + \frac{\eta''}{\eta(\rho_O^2 - 1)} \right\} h_O = \left\{ -\psi' + \frac{\eta''\rho_O}{\eta(\rho_O^2 - 1)} \right\} h_E - 2\psi h_E',$$

$$h_E'' + \left\{ q_E^2 + \psi^2 + \frac{\eta''}{\eta(\rho_E - 1)} \right\} h_E = \left\{ -\psi' + \frac{\eta''\rho_E}{\eta(\rho_E^2 - 1)} \right\} h_O - 2\psi h_O',$$

where

$$h_O = (\rho_O^2 - 1)^{-\frac{1}{2}}(\eta E_x - \rho_O E_y), \quad h_E = (1 - \rho_E^2)^{-\frac{1}{2}}(\eta E_x - \rho_E E_y).$$

(See Budden and Clemmow, 1957).

16.3. Sketch the form of the $h'(f)$ curves you would expect if the critical coupling frequency is greater than the electron gyro-frequency. How would the polarisation of the reflected components depend on frequency?

16.4. Show that if the square matrix $\mathbf{S}$ of (6.53), (7.85) is normalised so that the coupling matrix $\mathbf{\Gamma} = -\mathbf{S}^{-1}\,\mathrm{d}\mathbf{S}/\mathrm{d}s$ (7.108) has zero trace, then det ($\mathbf{S}$) is independent of height $z = s/k$.

Note. The trace of a square matrix such as $\mathbf{\Gamma}$ is the sum of its diagonal elements and is written Tr($\mathbf{\Gamma}$). The adjugate $\mathbf{A} = \mathbf{S}^{-1}\det(\mathbf{S})$ of $\mathbf{S}$ is the transpose of the matrix of its cofactors; see § 7.14(4). To solve the problem the result $(\mathrm{d}/\mathrm{d}z)\det(\mathbf{S}) = \mathrm{Tr}(\mathbf{A}\,\mathrm{d}\mathbf{S}/\mathrm{d}z)$ is needed, and the reader should prove this. The proof is not difficult but the author has not found it in any textbook. The solution of the problem is given by Inoue and Horowitz (1966a) and by Rawer and Suchy (1967, p. 154), for the special case where all the diagonal elements of $\mathbf{\Gamma}$ are zero.

16.5. For a radio wave vertically incident on a horizontally stratified ionosphere, the horizontal Cartesian components of the electric intensity are E_x, E_y. The earth's magnetic field is in the $x-z$ plane. Let the polarisation of an ordinary wave be $\rho_O = \mathrm{i}\tan\phi$. Show that the coupling parameter ψ (16.93) is $\phi' = k^{-1}\mathrm{d}\phi/\mathrm{d}z$. Show that if V is a two-dimensional vector with components $(E_x, -\mathrm{i}E_y)$, then if the x and y axes are rotated through a complex angle ϕ, the components of V in the new axes are $\pm(\mathrm{i}\mathscr{F}_O, \mathscr{F}_E)$ where $\mathscr{F}_O$, $\mathscr{F}_E$ are the Försterling variables (16.89). What is the significance of this transformation when V^2 is zero?
[This version of Försterling's theory was given by Saha, Banerjea and Guha (1951).]

16.6. The coupling matrix $\mathbf{\Gamma}$ is $-\mathbf{S}^{-1}\mathbf{S}'$ from (7.108). Prove that the non-diagonal elements of $\mathbf{\Gamma}$ are given by

$$\Gamma_{ij} = \frac{1}{(q_i - q_j)}(\mathbf{S}^{-1}\mathbf{T}'\mathbf{S})_{ij} \quad \text{(not summed; } i \neq j).$$

[See Arantes and Scarabucci (1975). This result is useful for computing because $\mathbf{T}'$ is easier to compute than $\mathbf{S}'$.]

17

Coalescence of coupling points

17.1. Introduction

The first order coupled wave equations (16.22) were derived and studied in § 16.3 for the neighbourhood of a coupling point $z = z_p$ where two roots q_1, q_2 of the Booker quartic equation are equal. If this coupling point is sufficiently isolated, a uniform approximation solution can be used in its neighbourhood and this was given by (16.29), (16.30). It was used in § 16.7 to study the phase integral formula for coupling. The approximations may fail, however, if there is another coupling point near to z_p, and then a more elaborate treatment is needed. When two coupling points coincide this is called 'coalescence'. This chapter is concerned with coupled wave equations when conditions are at or near coalescence.

A coupling point that does not coincide with any other will be called a 'single coupling point'. It is isolated only if it is far enough away from other coupling points and singularities for the uniform approximation solution (16.30) to apply with small error for values of $|\zeta|$ up to about unity. Thus an isolated coupling point is single, but the reverse is not necessarily true.

Various types of coalescence are possible. The two coupling points that coalesce may be associated with different pairs q_1, q_2 and q_3, q_4 of roots of the Booker quartic. This is not a true coalescence because the two pairs of waves are propagated independently of each other. The two coupling processes can then be treated at the same time as was done in the example of § 16.4, and the two coupling points still behave as though they are single. Alternatively the two pairs may have one member in common, for example q_1, q_2 and q_1, q_3. This type does occur, as was shown for example by Smith, M.S. (1974a) but he does not call it coalescence. (See Smith's figs. 1 and 6 where $S = S_a, S_c, S_d$ or S_f). It occurs when the curve of q versus X has a vertical tangent at a point that is also a point of inflection. There is an example in fig. 6.6(e). The theory has not been fully worked out and is not given in this book. Finally the two coupling points may both be associated with the same pair of roots, say q_1, q_2. This is the type that is now to be studied. It is shown that, even for this type alone, there are two different kinds of coalescence, to be called a 'coalescence of the first kind' C1, and a 'coalescence of the second kind' C2.

For C1 at exact coalescence, one incident wave gives rise to two other waves with roughly equal amplitudes. For example in fig. 15.5 it gives both a reflected and a transmitted wave. When the coupling points move apart one of these waves gets weaker and the other stronger, and when the points are well separated the incident wave gives only one other wave.

For C2 at exact coalescence, one incident wave travels right through the region near the coalescence just as it would if the coupling points were not there. There is then only one emergent wave. The fields of both the parent incident wave, before the coalescence is reached, and the product wave afterwards, are given by a W.K.B. solution that applies for the whole wave path going right through the coalescence. When the coupling points move apart a second wave is produced and the original product wave gets weaker. When the coupling points are well separated the original product wave has negligible amplitude, and the second wave now takes over as a new product wave.

The C1 is important in the theory of partial penetration and reflection, § 15.10, figs. 15.5, 15.6. The C2 is important in the theory of radio windows §§ 17.6–17.9, and of limiting polarisation §§ 17.10–17.11. For a version of the theory of coalescence more detailed than that gives here, see Budden and Smith (1974).

17.2. Further matrix theory

In this chapter some further properties of matrices are needed in continuation of § 7.14. They can be found in standard text books, for example, Aitken (1956), Heading (1958), Frazer, Duncan and Collar (1938), Graham (1979). We use the 4×4 square matrices $\mathbf{T}$ (7.81), and $\mathbf{S}$ whose columns $\mathbf{s}_i$ are the four eigen columns of $\mathbf{T}$ with eigen values q_i. Let

$$\mathbf{C}(q) = \mathbf{T} - q\mathbf{1} \tag{17.1}$$

where $\mathbf{1}$ is the unit 4×4 matrix. Then $\mathbf{C}$ is called the characteristic matrix of $\mathbf{T}$. The q s are given by the Booker quartic equation which may be written

$$F(q) \equiv \det \mathbf{C} = 0; \tag{17.2}$$

compare (7.84). The transpose of the matrix of the cofactors of the elements of $\mathbf{C}$ in (17.2) is denoted by adj $\mathbf{C}$. (Note that Budden and Smith (1974) use adj $\mathbf{C}$ to mean the untransposed matrix of the cofactors.) The elements of the column $\mathbf{s}_i$ may be chosen to be any linear combination of the columns of adj $\mathbf{C}$ when $q = q_i$, and it is now assumed that the same combination is used for all four $\mathbf{s}_i$s. At a coupling point, where two q_is are equal, say $q_1 = q_2$, the two values of any element of adj $\mathbf{C}$ are also equal so that $\mathbf{s}_1 = \mathbf{s}_2$. It might happen that, at a coupling point all cofactors of $\mathbf{C}$ are zero. Then $\mathbf{s}_1, \mathbf{s}_2$ must be found in some other way, and it can be shown that they can be chosen so that they are independent. Then it can be further shown that the coupling point cannot be single. It must be a coalescence. The proof is given by Budden and Smith (1974, appendix §(b)).

The notation (16.17) for α, β is now used. It was shown in § 16.3 that β^2 is analytic near a single coupling point z_p so it can be expanded in a series

$$\beta^2 = \tfrac{1}{4}(q_1 - q_2)^2 = -a(z - z_p) + b(z - z_p)^2 + \cdots. \tag{17.3}$$

In the special case $q_1 = -q_2$ this is the same as the series (8.3) for q^2. There is another coupling point $z = z_q$ where $q_1 = q_2$ and

$$z_q - z_p \approx a/b. \tag{17.4}$$

If $|z_q - z_p|$ is large enough, b/a must be small, and the two coupling points are then isolated. Coupled equations of the form (16.22) can then be used near each of them. But if z_p and z_q are close together, b/a is no longer small. Then one of two things can happen.

First it may happen that the matrix $\mathbf{U}$ in (16.20), (16.21) remains non-singular in the limit $z_q = z_p$ so that (16.21)–(16.30) can still be used. This would require that, in the limit, $\mathbf{s}_1$ and $\mathbf{s}_2$ are non-zero and equal. This always happens if at least one column of adj $\mathbf{C}$ is non-zero in the limit, because this column can be used for $\mathbf{s}_i$. Then we say that the coupling points approach a coalescence of the first kind, abbreviated by C1. Provided that only these two coupling points are near the domain considered, β^2 in (17.3) is with good approximation a quadratic function of z–z_p and equation (16.28) can be transformed to Weber's equation (15.70) or its equivalent (15.68) whose solutions are the parabolic cylinder functions or the function $\mathrm{F}(\zeta; A)$ (15.82). Further details are given in the following § 17.3.

Second it may happen that in the limit $z_p = z_q$ all elements of adj $\mathbf{C}$ are zero. Then the columns of adj $\mathbf{C}$ cannot be used to give the eigen columns $\mathbf{s}_i$, which must be constructed in some other way. It is then found that $\mathbf{s}_1$ and $\mathbf{s}_2$ are not parallel even though $q_1 = q_2$, and the coalescence is then said to be of the second kind, C2. For this case it can be shown that, when z_p and z_q are close together but not exactly coincident, there is a point close to them in the complex z plane where the first two columns $\mathbf{s}_\alpha, \mathbf{s}_\beta$ of $\mathbf{U}$ are parallel. Then $\mathbf{U}$ is singular. The transformation (16.21) cannot be used and a different one must be found. This is given in § 17.4. It introduces new variables w_1, w_2 that each satisfy a form of Weber's equation but it is different from the form used for C1.

For either C1 or C2, (17.4) shows that for exact coalescence $a = 0$, whence from (17.3) $\beta = \tfrac{1}{2}(q_1 - q_2)$ has a factor z–z_p, and is therefore analytic at and near the coalescent coupling points. Since $\alpha = \tfrac{1}{2}(q_1 + q_2)$ also is analytic, it follows that q_1 and q_2 are both analytic throughout a domain containing the coalescence.

17.3. Coalescence of the first kind, C1

At a coupling point two roots of the Booker quartic (17.2) are equal, and it is required that simultaneously

$$\det \mathbf{C} = 0 \quad \text{and} \quad \partial(\det \mathbf{C})/\partial q = 0. \tag{17.5}$$

Elimination of q from these gives

$$\Delta = 0 \tag{17.6}$$

where Δ is the discriminant (6.38)–(6.40) of the quartic and is a function of z. The solutions z of (17.6) are the coupling points in the complex z plane. To simplify the discussion it will now be assumed that electrons are the only species that appreciably affect the wave propagation. It will also be assumed that the collision frequency varies so slowly with z that $\partial Z/\partial z$ can be neglected. Then Δ depends on z only because of X for electrons, and (17.6) may be written

$$\Delta(X) = 0. \tag{17.7}$$

It is easy to extend the argument to deal with cases where these assumptions are not true.

At a coalescence, two solutions of (17.7) occur at the same z, which requires that simultaneously

$$\Delta(X) = 0 \quad \text{and} \quad \frac{\partial}{\partial z}\Delta(X) = \frac{\partial \Delta(X)}{\partial X}\frac{\partial X}{\partial z} = 0 \tag{17.8}$$

and the last of these can be satisfied either by $\partial\Delta/\partial X = 0$, or $\partial X/\partial z = 0$. The second occurs at a turning point of the function $X(z)$, such as the maximum of an ionospheric layer. The first is a property of $\Delta(X)$ and thence via (17.5) and (17.1) of the matrix $\mathbf{T}$, and does not depend on the ionospheric distribution functions.

Suppose now that (17.7) has a solution $X = X_c$, possibly complex, that is a coupling point where $q_1 = q_2$ and that is isolated in the complex X plane. Then the eigen columns $\mathbf{s}_1$, $\mathbf{s}_2$ can be chosen so that they are non-zero for all X near X_c and are identical when $X = X_c$. Suppose further that there is a value of X very close to X_c where $\partial X/\partial z = 0$, $z = z_m$. Then there are two values z_p, z_q of z where $X = X_c$, such that $z_p - z_m \approx z_m - z_q$. Thus conditions are near to coalescence of these two coupling points z_p, z_q in the z plane. Since the eigen columns $\mathbf{s}_1$, $\mathbf{s}_2$ are non-zero, the matrix $\mathbf{U}$ is non-singular in a domain containing z_p, z_q and the transformation (16.21) can be used. Conditions are near to C1.

Suppose that β^2 is exactly a quadratic function of z so that in (17.3) there are no terms after the term in $(z - z_p)^2$. Then

$$z_q = z_p + a/b, \quad \beta^2 = b(z - z_p)(z - z_q). \tag{17.9}$$

Let

$$\xi = (k^2 b)^{\frac{1}{4}}\{z - \tfrac{1}{2}(z_p + z_q)\}, \quad A = \tfrac{1}{2}(k^2 b)^{\frac{1}{4}}(z_q - z_p); \tag{17.10}$$

compare (15.67). Then the first equation (16.28) becomes

$$\mathrm{d}^2h_1/\mathrm{d}\xi^2 + (\xi^2 - A^2)h_1 = 0. \tag{17.11}$$

This is the same as (15.68) and can be used in a similar way. At exact coalescence C1, the coupling points z_p and z_q coincide and $A^2 = 0$.

Usually β^2 is not exactly a quadratic function of z. Suppose that it has zeros at z_p, z_q which are close together and that its other zeros are remote from them. Then near z_p, z_q the function β^2 must be approximately quadratic and it is still possible to reduce (16.28) to the form (17.11) by using a uniform approximation (see §§ 7.9, 8.20) first given by Miller and Good (1953). Instead of (17.10) let ξ and A be defined by

$$k\int_{z_\mathrm{p}}^{z} \beta\,\mathrm{d}z = \mathrm{i}\int_{-A}^{\xi} (A^2 - \xi^2)^{\frac{1}{2}}\,\mathrm{d}\xi, \tag{17.12}$$

$$k\int_{z_\mathrm{p}}^{z_\mathrm{q}} \beta\,\mathrm{d}z = \mathrm{i}\int_{-A}^{A} (A^2 - \xi^2)^{\frac{1}{2}}\,\mathrm{d}\xi = \mathrm{i}\pi A^2. \tag{17.13}$$

Here (17.12) maps the complex z plane into the ξ plane, and (17.13) ensures that it is a one-to-one mapping. It is easily checked that they reduce to (17.10) in the limit when β^2 is exactly quadratic, provided that the signs of the square roots are suitably chosen, and it is now assumed that this choice of sign is made. The zeros of β at z_p, z_q map into the points $\xi = \mp A$. With this transformation, (16.28) becomes

$$\mathrm{d}^2h_1/\mathrm{d}\xi^2 + (\xi^2 - A^2)h_1 = [\text{small term}]\,\mathrm{d}h_1/\mathrm{d}\xi. \tag{17.14}$$

The right-hand side is analogous to the second term in brackets in (7.52). It is zero if β^2 is exactly quadratic, and small enough to be neglected if β^2 is sufficiently near to quadratic. The main interest of (17.12)–(17.14) is that although it can be used for C1, it is not suitable for C2 and a different form (17.32), (17.33) must be used, for which A^2 is not zero at exact coalescence.

The most important example of a coalescence C1 is when the coupling point at $X = X_\mathrm{c}$ is a reflection point, for example when q_1 and q_2 apply to the upgoing and downgoing ordinary wave respectively. Since, in most cases of practical interest, X_c is isolated in the complex X plane, the other two waves associated with q_3 and q_4 are independently propagated and can be disregarded. The reflection process is described by the second order equation (16.28). This is of the same form as (7.6) used for an isotropic ionosphere. If β^2 in (16.28) is a quadratic function of z, the equation can be used in exactly the same way as (7.6) was used in §§ 15.9–15.11 to study propagation in an ionosphere with a parabolic electron height distribution $N(z)$. The theory of §§ 15.9–15.11 can therefore be used also for studying partial penetration and reflection of the ordinary wave, and of the extraordinary wave in an anisotropic ionosphere. The main difference is that in (15.68) the dependent variable is the horizontal electric field E of a wave obliquely incident on an isotropic ionosphere, whereas in (17.11) it is h_1 which is a linear combination of E_x, E_y, $\mathscr{H}_x$, $\mathscr{H}_y$ given by (16.21), (16.25).

Other kinds of C1 are possible. In the example just discussed, the C1 occurred because a reflection point X_c was near the maximum of an ionospheric layer where $\partial X/\partial z = 0$. Now X_c might be some other kind of coupling point, for example where the upgoing ordinary and extraordinary waves are critically coupled. When electrons only are allowed for, this usually occurs in the height range 90 to 98 km for vertical incidence and lower for oblique incidence, so that it is unlikely to be near a layer maximum, and therefore unlikely to lead to a C1. For frequencies small enough for ions to be important, however, the theory has not been so fully explored and it is possible that this might give C1s that are important.

In both the above examples (17.7) was satisfied because of the proximity of a zero of $\partial X/\partial z$ to a coupling point. It could alternatively be satisfied by a zero of $\partial\Delta/\partial X$, even in a medium where X varies monotonically with z. No cases of C1s of this type have yet been found but the possibility has not been excluded.

17.4. Coalescence of the second kind, C2

Consider now a domain of the complex z plane where all the ion and electron concentrations are monotonic functions of z. It is assumed that all concentrations and collision frequencies are slowly varying functions of z. Let z_p, z_q be two coupling points in the domain, with $q_1 = q_2$ at both. These points can be made to coalesce by changes in one or more of the parameters S ($= \sin\theta$), l_x, l_y, ω.

It often happens that at exact coalescence the columns $\mathbf{s}_1$ and $\mathbf{s}_2$ found from columns of adj $\mathbf{C}$ (see § 17.2) are identically zero. Then $\mathbf{s}_\alpha$ (16.19) is zero, the resulting form of $\mathbf{U}$ is singular, and the transformation (16.21) cannot be used. It is possible to find non-zero $\mathbf{s}_1$ and $\mathbf{s}_2$ by finding them at some point z near $z_p = z_q$ and then dividing them by some analytic factor that tends to zero at exact coalescence, for example $z - z_p$ or $q_1 - q_2$; see end of § 17.2. But then $\mathbf{s}_1$ and $\mathbf{s}_2$ tend to different limits when $z \to z_p$, and are not parallel. Consequently $\mathbf{S}$ is non-singular throughout a domain of the complex z plane containing $z_p = z_q$, and the transformation (6.53) can be used just as if the coupling points did not exist. But, at $z = z_p = z_q$, $\mathbf{s}_\beta$ (16.18) has some infinite elements because its denominator β is zero so that again the transformation (16.21) cannot be used.

Now let the two coupling points move apart. Then at each of them $\mathbf{S}$ is singular and the transformation (6.53) cannot be used. But $\mathbf{U}$ can now be defined as in § 16.3, and is non-singular at and near both z_p and z_q. Thus the transformation (16.21) can now be used throughout a domain containing both coupling points.

Neither $\mathbf{U}$ with (16.21) nor $\mathbf{S}$ with (6.53) can be used both at exact coalescence and at near coalescence. A new transforming matrix $\mathbf{V}$ is needed instead of $\mathbf{U}$ or $\mathbf{S}$. It must be uniformly analytic and non-singular throughout a domain containing z_p

and z_q, both at exact coalescence and at near coalescence. A method of finding it is now to be given in outline. Some further results from matrix theory are needed for this. Their derivation is too long to give here but they are given in full by Budden and Smith (1974, appendix).

The columns $\mathbf{s}_1, \mathbf{s}_2$ used in (16.19), (16.18) are any column of adj **C** or they may be a linear combination of several columns provided that the same combination is used for both. Then if $\mathbf{s}_1$ and $\mathbf{s}_2$ are non-zero at a coupling point where $q_1 = q_2$ they must be parallel there. If $\mathbf{s}_\alpha$ is zero at a coupling point, including a coalescence, both $\mathbf{s}_1$ and $\mathbf{s}_2$ must be zero. This suggests that a different combination of columns of adj **C** should be used, in the hope that it will give a non-zero $\mathbf{s}_\alpha$ at the coalescence. But this cannot be done if all the elements of adj **C**, that is all the cofactors of **C**, are zero. This would mean that, at the coalescence, **C** is of rank two or less. If **C** were of rank one it would follow from (17.2) that $\partial^2 F/\partial q^2 = 0$, as well as (17.5), and this would mean that three roots of the quartic (17.2) are equal. This case could occur but is excluded from the present discussion. A necessary condition for a C2, therefore, is that **C** shall be of rank two. It is proved by Budden and Smith (1974) that this condition is also sufficient.

Suppose, now, that a suitable linear combination of columns of adj **C** has been chosen for $\mathbf{s}_i$. Then all four elements are analytic functions of q and of z. At a single coupling point where $q_i = q_1 = q_2$ the function $q_i(z)$ is not analytic because it has a branch point there, but α and β^2 (16.17) are analytic. By the method used in § 16.3 it can then be shown that the only linear combinations of $\mathbf{s}_1, \mathbf{s}_2$ that are analytic at and near a coupling point are some combinations of $\mathbf{s}_\alpha$, $\mathbf{s}_\beta$. Thus any new transforming matrix **V**, replacing **U** in (16.21) must have columns that are all linear combinations of $\mathbf{s}_\alpha$, $\mathbf{s}_\beta$, $\mathbf{s}_3$, $\mathbf{s}_4$.

Since the $\mathbf{s}_\alpha$ found from columns of adj **C** is analytic, and zero where $z = z_p$, it must contain a factor that is some integral power of $z - z_p$. The power could be two or more if the C2 is at the maximum of an ionospheric layer. But in the cases studied here the power is unity and only this case will be considered. Hence the column

$$\mathbf{s}_\gamma = \mathbf{s}_\alpha/(z - z_p) \tag{17.15}$$

is analytic and non-zero in the domain containing z_p, z_q. At coalescence $z_p = z_q$, but $\mathbf{s}_\gamma$ is still non-zero and bounded. Both at and near coalescence $\mathbf{s}_\beta$ also is non-zero in the domain.

Now define a new transforming matrix **V** whose columns are $\mathbf{s}_\gamma$, $\mathbf{s}_\beta$, $\mathbf{s}_3$, $\mathbf{s}_4$. It is non-singular throughout the domain for conditions both at and near to coalescence. With this matrix the steps analogous to (16.20)–(16.26) are now followed through. It can be shown that

$$\mathbf{TV} = \mathbf{V}\mathfrak{B} \quad \text{where} \quad \mathfrak{B} = \begin{pmatrix} \alpha & z - z_p & 0 & 0 \\ \beta^2/(z - z_p) & \alpha & 0 & 0 \\ 0 & 0 & q_3 & 0 \\ 0 & 0 & 0 & q_4 \end{pmatrix} \tag{17.16}$$

(compare (16.20)). Let

$$\mathbf{e} = \mathbf{V}\mathbf{d}, \quad \mathbf{d} = \mathbf{V}^{-1}\mathbf{e}. \tag{17.17}$$

and substitute for $\mathbf{e}$ in (7.80). Then $\mathbf{d}$ must satisfy

$$\mathbf{d}' + \mathrm{i}\mathfrak{B}\mathbf{d} = \mathbf{\Psi}\mathbf{d} \tag{17.18}$$

where

$$\mathbf{\Psi} = -\mathbf{V}^{-1}\mathbf{V}'. \tag{17.19}$$

In a sufficiently slowly varying medium $\mathbf{V}'$ is small and $\mathbf{V}^{-1}$ is bounded so that the coupling matrix $\mathbf{\Psi}$ can be neglected to a first approximation (compare $\mathbf{\Gamma}$ in (7.109) and $\mathbf{\Lambda}$ in (16.22)). Then the waves associated with q_3, q_4 are independently propagated, so the two elements $\mathrm{d}_1, \mathrm{d}_2$ of $\mathbf{d}$ may be separated from the others to give

$$\begin{pmatrix} \mathrm{d}_1 \\ \mathrm{d}_2 \end{pmatrix}' = -\mathrm{i}k \begin{pmatrix} \alpha & z - z_\mathrm{p} \\ \beta^2/(z - z_\mathrm{p}) & \alpha \end{pmatrix} \begin{pmatrix} \mathrm{d}_1 \\ \mathrm{d}_2 \end{pmatrix} \tag{17.20}$$

(compare (16.24)). The further transformation

$$\begin{pmatrix} \mathrm{d}_1 \\ \mathrm{d}_2 \end{pmatrix} = \begin{pmatrix} g_1 \\ g_2 \end{pmatrix} \exp\left(-\mathrm{i}k \int^z \alpha \, \mathrm{d}z \right) \tag{17.21}$$

now leads to

$$\begin{pmatrix} g_1 \\ g_2 \end{pmatrix}' = -\mathrm{i}k \begin{pmatrix} 0 & z - z_\mathrm{p} \\ \beta^2/(z - z_\mathrm{p}) & 0 \end{pmatrix} \begin{pmatrix} g_1 \\ g_2 \end{pmatrix}. \tag{17.22}$$

This will now be studied first for the case where β^2 is exactly a quadratic function of z given by (17.9), so that (17.22) gives

$$\mathrm{d}g_1/\mathrm{d}z = -\mathrm{i}k(z - z_\mathrm{p})g_2, \quad \mathrm{d}g_2/\mathrm{d}z = -\mathrm{i}kb(z - z_\mathrm{q})g_1. \tag{17.23}$$

Now change the dependent variables to

$$w_1 = g_1 b^{\frac{1}{2}} + g_2, \quad w_2 = g_1 b^{\frac{1}{2}} - g_2 \tag{17.24}$$

and let

$$\xi = (k^2 b)^{\frac{1}{4}}\{z - \tfrac{1}{2}(z_\mathrm{p} + z_\mathrm{q})\}, \quad P = \tfrac{1}{2}(k^2 b)^{\frac{1}{4}}(z_\mathrm{q} - z_\mathrm{p}); \tag{17.25}$$

compare (17.10). Then the equations are

$$\mathrm{d}w_1/\mathrm{d}\xi = -\mathrm{i}(\xi w_1 - P w_2), \tag{17.26}$$

$$\mathrm{d}w_2/\mathrm{d}\xi = \mathrm{i}(\xi w_2 - P w_1). \tag{17.27}$$

If w_1 or w_2 is eliminated we obtain respectively

$$\mathrm{d}^2 w_1/\mathrm{d}\xi^2 + (\xi^2 - P^2 + \mathrm{i})w_1 = 0, \tag{17.28}$$

$$\mathrm{d}^2 w_2/\mathrm{d}\xi^2 + (\xi^2 - P^2 - \mathrm{i})w_2 = 0. \tag{17.29}$$

These are both of the form (15.68), (17.11) with $A^2 = P^2 \pm \mathrm{i}$. By a further change of independent variable they could be transformed to Weber's equation (15.70). At exact coalescence C2, $P = 0$, $A^2 = \pm\mathrm{i}$, whereas for exact coalescence C1, from (17.10), $A^2 = 0$. This is a very important difference.

When $P \neq 0$, let one of the solutions of (17.28) be selected. Then w_2 is found from

(17.26) and the recurrence relations for the solutions of these equations (Whittaker and Watson, 1927, p. 350, last formula in § 16.61) show that it also satisfies (17.29). Thence via (17.24), (17.21), (17.17) the components **e** of the electromagnetic field can be found. If $|\xi|$ is large so that z is far enough away from z_p, z_q, the appropriate W.K.B. solutions of (17.28), (17.29) can be used for w_1, w_2 and these in turn give the W.K.B. solutions for the characteristic waves. They display the Stokes phenomenon which means that the circuit relations for these functions show that there is mode coupling near z_p, z_q, between these characteristic waves.

When $P=0$ the behaviour is different. The equations (17.23)–(17.27) now have two independent exact solutions. The first is

$$w_1 = \exp(-\tfrac{1}{2}\mathrm{i}\xi^2), \quad w_2 = 0. \tag{17.30}$$

It is dominant on the Stokes lines where $\arg\xi = \tfrac{1}{4}\mathrm{i}\pi, \tfrac{5}{4}\mathrm{i}\pi$. Because it is exact there can be no subdominant term here and no Stokes phenomenon. The expressions given by Heading (1962a, p. 67) for the Stokes multipliers show that they are zero here. The Stokes multipliers for $\arg\xi = \tfrac{3}{4}\mathrm{i}\pi, -\tfrac{1}{4}\mathrm{i}\pi$ are not zero but here the solution (17.30) is subdominant. The Stokes diagram is shown in fig. 17.1(a). The thick curve has no breaks because the Stokes phenomenon does not occur at a C2. The second solution is

$$w_1 = 0, \quad w_2 = \exp(\tfrac{1}{2}\mathrm{i}\xi^2). \tag{17.31}$$

Its Stokes diagram fig. 17.2(b) has a continuous thick curve because, again, the Stokes phenomenon does not occur.

When β^2 is not exactly a quadratic function of z, a uniform approximation similar to (17.12), (17.13), but not quite the same, may be used. Instead of (17.25) let ξ and P

Fig. 17.1. Stokes diagrams for solutions of (17.26), (17.27), that is of (17.28), (17.29), when $P=0$. S denotes Stokes and A denotes anti-Stokes lines.

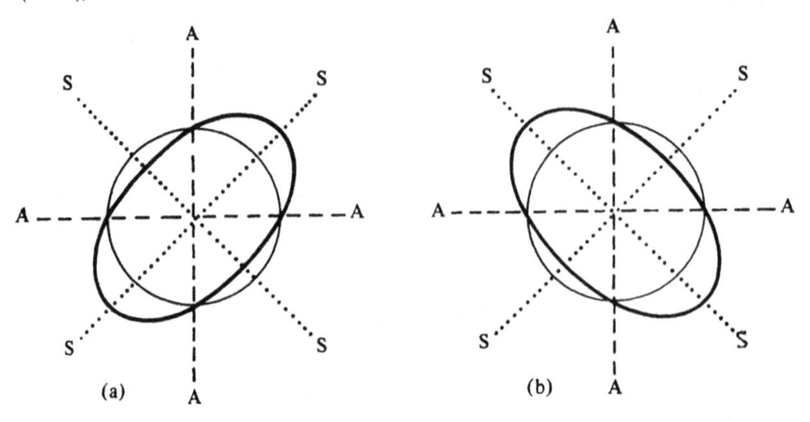

be defined by

$$k\int_{z_p}^{z}\beta\,dz = i\int_{-P}^{\xi}(P^2-\xi^2)^{\frac{1}{2}}\,d\xi, \tag{17.32}$$

$$k\int_{z_p}^{z_q}\beta\,dz = i\int_{-P}^{P}(P^2-\xi^2)^{\frac{1}{2}}\,d\xi = \tfrac{1}{2}i\pi P^2. \tag{17.33}$$

Then (17.24) and (17.26)–(17.29) must be modified by the inclusion of small terms. These are zero when β^2 is exactly quadratic. With a suitable choice of the signs of the square roots, (17.32), (17.33) are then the same as (17.25).

At exact coalescence C2, $P=0$ and (17.32) shows that

$$\exp(\pm\tfrac{1}{2}i\xi^2) = \exp\left(\pm ik\int_{z_p}^{z}\beta\,dz\right). \tag{17.34}$$

Let this now be used in the two solutions (17.30), (17.31). For one of them the factor $\exp(-ik\int^z\beta\,dz)$ appears in g_1, g_2 from (17.24), and the transformation (17.21) shows that d_1, d_2 and thence $\mathbf{e}$, from (17.17) contain the factor $\exp(-ik\int^z q_1\,dz)$, where (16.17) has been used. Similarly for the other solution the field components $\mathbf{e}$ contain the factor $\exp(-ik\int^z q_2\,dz)$. These are just the exponential factors that appear in the W.K.B. solutions for the two waves associated with q_1 and q_2 respectively.

At a C2, therefore, there is no coupling between these two waves. They are independently propagated and their fields are given by the W.K.B. solutions (7.112) for points z at and near $z_p = z_q$. This occurs because the Stokes phenomenon is absent. A physical explanation is that at a C2, if the eigen columns $\mathbf{s}_i$ are constructed by dividing some column of adj $\mathbf{C}$ by a suitable factor, then the columns $\mathbf{s}_1$, $\mathbf{s}_2$ are not parallel even though $q_1 = q_2$. This means that the two waves have different polarisations and wave impedances, just as they do at an ordinary point where the waves are distinct, and so they are independently propagated. It also means that at *exact* coalescence the matrix $\mathbf{S}$ in (7.85) is non-singular at all points in a domain containing $z_p = z_q$ so that the transformation (6.53) could be used. This behaviour should be contrasted with a C1 where the polarisations are the same at a coalescence, and the coupling is strong.

For conditions that are near but not exactly at C2, one pair of Stokes multipliers for each of the equations (17.28), (17.29) is still small but not exactly zero. Thus there is now a small amount of coupling. It gets stronger as the coupling points move further apart.

The best known example of coalescence C2 is the phenomenon of radio windows. This is described in §§ 17.6–17.9. Another example of a rather different kind is limiting polarisation, §§ 17.10, 17.11.

When conditions are near to coalescence C2, the physical description of what happens to the waves depends on $\arg(z_q - z_p)$. To illustrate this it is useful to discuss two extreme special cases. For the first $\arg(z_q - z_p) = \pm\frac{1}{2}\pi$, which arises in the study of ion cyclotron whistlers, § 17.5. For the second $\arg(z_q - z_p) = 0$ or π, which arises in the study of radio windows, § 17.6.

17.5. Ion cyclotron whistlers

The theory of ion cyclotron whistlers was given in § 13.9. It was studied for vertically incident waves and was associated with two coupling points z_p, z_q shown in fig. 13.12. Their position in the complex z plane is given by

$$\mathfrak{G}(z) = \pm \mathrm{i} \sin^2\Theta / \cos\Theta \tag{17.35}$$

where Θ is the angle between the earth's magnetic field and the vertical, and $\mathfrak{G}$ is given by (13.60). The two points coalesce only if $\Theta = 0$, that is at a magnetic pole of the earth. In practice when collisions are allowed for $\mathfrak{G}$ is complex when z is real, and the condition $\mathfrak{G} = 0$ requires a complex value of z. When $\Theta = 0$, therefore, the coalescence is not on the real z axis, and has only a small influence on the propagation. An upgoing electron whistler wave remains an electron whistler at all heights, and ion whistlers are not formed near the poles. When Θ differs from zero the two coupling points separate and for large enough Θ one of them, z_p in fig. 13.12(c), is on the real z axis. This is the condition of exact crossover, § 13.9, and is an example of critical coupling, § 16.6. For still larger Θ, $\mathrm{Im}(z_p)$ is positive, and an upgoing electron whistler now goes over continuously to an ion cyclotron whistler at points above $\mathrm{Re}(z_p)$. In the description given in § 13.9, the other coupling point z_q did not come into the discussion because it was assumed to be too far away from the real z axis.

If collisions were negligible the behaviour would be different, for then $\mathfrak{G}$ is real when z is real. The coalescence would be on the real z axis when $\Theta = 0$, and it would be necessary to consider both coupling points. For $\Theta \neq 0$, $\mathrm{Im}(z_p)$ is now always positive. Although the neglect of collisions is unrealistic, the description of what would happen to the waves is a useful illustration of a coalescence C2, and it is therefore given briefly here.

The dependence of the two values of n^2 on frequency was sketched in fig. 13.8(b). The continuous curves are for $\Theta = 0$, and the dotted curves for Θ non-zero but small. Similarly curves of n^2 versus height z can be plotted and are sketched in figs. 17.2. Both values of n^2 are real for all real z. For $\Theta = 0$, fig. 17.2(a), $n_{\mathrm{O}}^2 = \varepsilon_2$ and $n_{\mathrm{E}}^2 = \varepsilon_1$. The curves cross at a height $z = z_0$ and this is the point of coalescence. The ordinary wave is the electron whistler and has left-handed circular polarisation $\rho = \mathrm{i}$ at all heights. The other wave has right-handed circular polarisation $\rho = -\mathrm{i}$ at all heights. At exact coalescence there is no coupling between these waves and so no formation of an ion cyclotron whistler.

When Θ is small but non-zero the curves are as in fig. 17.2(b). The two coupling points have now separated slightly and are at complex conjugate values of z. There is no real value of z where the two values of n^2 are equal. This is to be contrasted, in §17.6, with fig. 17.3(b) where the slightly separated coupling points are both on the real z axis.

Suppose that an electron whistler wave travels up from below. Its squared refractive index is given by the part A of the upper curve in fig. 17.2(b), and its polarisation is very nearly left-handed circular $\rho \approx +\mathrm{i}$. It then encounters the coupling region near z_0. If the medium were extremely slowly varying, the squared refractive index would go continuously over to the part B of the curve and the polarisation would reverse its sense and go over to $\rho \approx -\mathrm{i}$. The electron whistler then changes its name to ion cyclotron whistler and there would be no mode conversion. But in practice this does not happen when Θ is small. In the coupling region the characteristic polarisation for the branch AB is changing very rapidly when z increases, even in a moderately slowly varying medium. Thus there is strong coupling with branch C, fig. 17.2(b), whose characteristic polarisation is nearly the same as for branch A. This topic is elucidated in more detail by Jones, D. (1970), and Wang (1971).

In this case therefore, mode conversion is the mechanism that makes an electron whistler remain an electron whistler. It is strong when Θ is small, and at coalescence when $\Theta = 0$ it dominates completely so that there is no ion cyclotron whistler. This will be called 'complete mode conversion'. For larger Θ the curves in fig. 17.2(b) are more widely separated, the polarisations change more slowly with changing z, and the coupling is much weaker. The electron whistler at great heights changes its name to ion cyclotron whistler and there is no mode conversion.

In this example, the two waves that are coupled at near coalescence are both vertically upgoing waves so that the phenomenon is coupling and not reflection.

Fig. 17.2. Dependence of n^2 on height z when collisions are neglected. In (a) the earth's magnetic field is vertical, $\Theta = 0$, and the frequency is the crossover frequency for the height z_0. At z_0 there is a C2. In (b) Θ is non-zero but small. Near z_0 conditions are near to, but not exactly at C2.

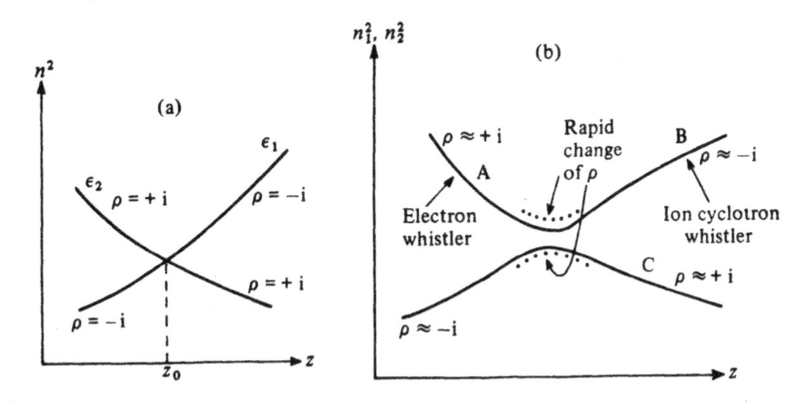

It should be contrasted with the example in the following section where the wave normals are oblique, and one wave is upgoing and the other downgoing.

It is paradoxical that complete mode conversion is invoked to explain why, at coalescence, an electron whistler remains an electron whistler at all heights, but this is a consequence of how 'mode conversion' is defined. For the two waves of fig. 17.2(b), the curves of n^2 versus z on the real z axis are continuous, and these waves are therefore taken to be the unconverted waves. This definition is retained in the limit of coalescence, fig. 17.2(a), where the two curves now touch and have a discontinuity of gradient. In the following section a case will be studied where the mode conversion at exact coalescence is a true and complete conversion from ordinary to extraordinary. When the coupling points are well separated the 'conversion' is a complete reflection of an upgoing ordinary wave to give a downgoing ordinary wave. For further details of the example in this section see Budden and Smith (1974) who give expressions for the columns $\mathbf{s}_\gamma$, $\mathbf{s}_\beta$ used in (17.16)–(17.19).

17.6. Radio windows 1. Coalescence

The most important practical application of the coalescence C2 is to the theory of radio windows and particularly the Ellis window; see end of § 16.8 and Ellis (1956). In this section a simple example is given to illustrate further the nature of a coalescence C2. The fuller theory is given in §§ 17.7–17.9.

Consider a stratified ionosphere and radio waves of frequency greater than f_H, so $Y < 1$. Thus only electrons need be considered and collisions will be neglected. The incident wave is an upgoing ordinary wave with its wave normal in the magnetic meridian plane, so $l_y = 0$. It is assumed that X is a monotonically increasing function of the height z. The angle of incidence is $\theta = \arcsin S$. We shall here study what happens when S is varied, with all other parameters held constant. Note that, in § 17.5, S was constant (zero) and it was the angle Θ between the earth's magnetic field and the vertical that was varied.

The coupling points in the complex X plane were studied for this case in § 16.5 and it was shown that when S is near to $S_A = l_x\{Y/(Y+1)\}^{\frac{1}{2}}$, (6.47), the two coupling points O and C_3 are on the real X axis. They are also on the real z axis at z_p, z_q respectively. They coalesce where $X = 1$ when $S \equiv S_A$. The formal proof that this is a coalescence C_2 was given by Budden and Smith (1974) who also discuss the expressions for the column matrices $\mathbf{s}_1, \mathbf{s}_2$ to be used in forming the transforming matrices $\mathbf{U}$, (16.20)–(16.23) and $\mathbf{V}$, (17.16)–(17.19).

Curves of q versus X for this case, with various values of S, were given in fig. 6.6, and fig. 6.6(c) applies for $S = S_A$. The relevant parts of these curves are sketched again in fig. 17.3 as curves of q versus z. Fig. 17.3(a) is for exact coalescence C2, that is $S = S_A$, and fig. 17.3(b) is for S slightly less than S_A. The coupling points z_p, z_q

are on the real z axis where the curves have vertical tangents in fig. 17.3(b). For $z < z_p$ the part of the curve marked q_2 is for the upgoing ordinary wave and the part q_1 is for the downgoing ordinary wave. Thus in fig. 17.3(b), $z = z_p$ is a reflection level for this wave. At each point on the curve the column matrix $\mathbf{s}$, meaning $\mathbf{s}_1$ or $\mathbf{s}_2$ (7.134) can be found, and it gives the polarisation and wave admittance of the wave field. Its elements are continuous as we proceed along the curve. Similarly for $z > z_q$ the part q_1 of the curve is for a downgoing extraordinary wave, that is a Z-mode wave, and the part q_2 is for an upgoing extraordinary wave. Again the column matrix $\mathbf{s}$ can be found for each point on this curve.

Between z_p and z_q is a region where q_1 and q_2 are complex conjugates. One of the waves decays and the other grows when z is increasing. This region, therefore, is like a potential barrier, and waves can tunnel through it. It is, however, different from the conventional simple barrier studied in § 15.9. When z_p and z_q are well separated, the barrier is wide and the tunnelling is weak. The two reflection processes at z_p and z_q then predominate. They are special cases of strong coupling between upgoing and downgoing waves. When $|z_q - z_p|$ is smaller the curves are as in fig. 17.3(b). There is some tunnelling because for the incident wave the column $\mathbf{s}$ changes rapidly as the part of the curve D near z_p is traversed. The value of $\mathbf{s}$ for curve D near z_p is closer to the value for curve F near z_q than it is to the value for curve E near z_p. Thus as the incident wave approaches z_p it is better matched in wave admittance and polarisation to the wave of branch F than it is to the reflected wave of branch E, and there is some mode conversion with tunnelling.

Fig. 17.3. Dependence of q on height z when collisions are neglected. In (a) $S = S_A$ and there is exact coalescence C2 for $z = z_p = z_q$. In (b) S is slightly less than S_A and conditions are near to, but not exactly at coalescence. Fig. (b) should be contrasted with fig. 17.2b.

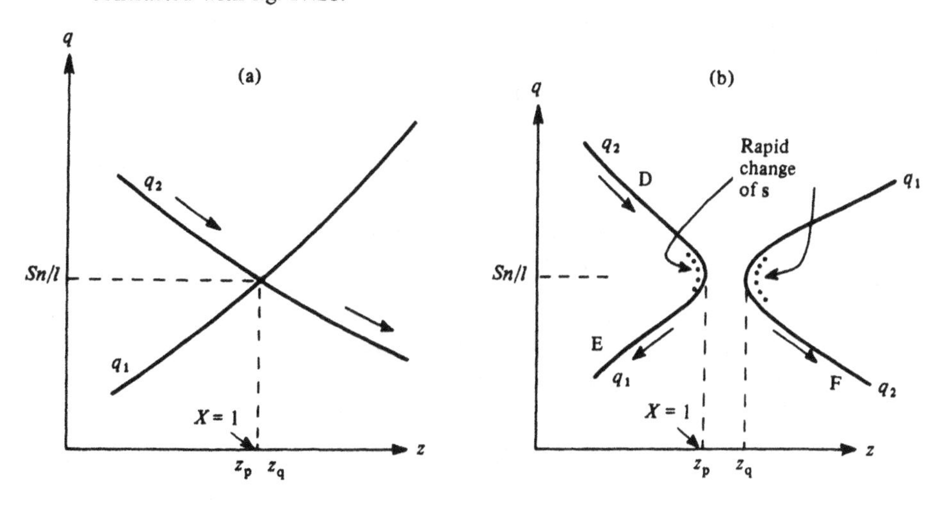

The incident wave is ordinary but the wave that emerges above z_q is extraordinary. When $z_q - z_p$ is zero, the barrier has closed to zero width. The matching of the wave admittance and polarisation, as given by **s**, for the upgoing ordinary wave and the upgoing extraordinary wave is now complete at $z = z_p = z_q$, and this **s** is very different from that for the downgoing waves. There is now no reflection of the ordinary wave. It continues on through the level $z = z_p = z_q$ and above this its name is changed to extraordinary. This wave is represented in fig. 17.3(a) by the continuous curve marked with arrows. Its fields are given by one of the W.K.B. solutions (7.112) at all heights including $z = z_p = z_q$ and its neighbourhood. It does not now undergo reflection, and therefore the downgoing wave is absent.

The amplitude of the transmitted wave above $z = z_q$ is therefore greatest at exact coalescence when $S = S_A$, $l_y = 0$, and this specifies a particular direction of the wave normal of the incident wave, to be called the 'central direction', or centre of the window. If now this wave normal direction is changed slightly so that it makes an angle η with the central direction, either S or l_y or both are changed. The coalescence is no longer exact and the transmitted wave is weaker. There is a range of angles η within a cone near the central direction such that the amplitude of the transmitted wave is greater than half its maximum value. This cone is called a 'radio window'. Note that it refers to a range of angles measured in free space. In practice for the ionosphere at frequencies greater than f_H, the cone is roughly circular with semi-vertical angle about 1.5°. The size of the cone depends on the gradient $dN(z)/dz$ of electron concentration at the level where $X = 1$, on the inclination arcsin S_A of the central direction to the vertical and on f and f_H. It is found to be almost independent of the collision frequency (neglected in this section). A method of calculating it and formulae (17.51), (17.52), are given in §17.7.

17.7. Radio windows 2. Formulae for the transparency

The meaning of the term 'radio window' was given in the preceding section, together with a description of the physical processes that the waves undergo near the level of the coupling points that cause it. The present section gives a more detailed theory and supplies formulae (17.51), (17.52) for calculating the transparency and angular width of a window. The following treatment is based on a paper by Budden (1980).

The idea of a 'window' for radio waves was first used by Ellis (1956, 1962) who applied it to explain the Z-trace; § 16.8. Several other applications have been suggested recently. Thus Jones, D. (1976a, 1977a, b) has suggested that they may play a part in one explanation of the terrestrial non-thermal continuum radiation (Gurnett and Shaw, 1973), and of the terrestrial kilometric radiation (Gurnett, 1975). These radiations are observed outside the earth and may possibly come from Cerenkov emission by energetic charged particles that enter the magnetosphere.

It has been suggested (Oya, 1974; Benson, 1975; Jones, D., 1976a, 1977a, b) that a similar mechanism might operate for the decametric radiation from Jupiter, and for the radio emission from pulsars. The properties of radio windows have been used in studies of radio emission from the solar corona and other astrophysical sources (see Melrose, 1974a, b, 1980, who gives further references). Radio windows are also of interest in the theory of thermo-nuclear plasmas (see, for example, Weitzner and Batchelor, 1979).

The theory is given here for a stratified cold plasma and for frequencies great enough to ensure that only electrons need be considered. Electron collisions are neglected, but their effect is mentioned at the end of this section. The transparency of a window depends on the direction of the wave normal of the incident wave. But any wave is refracted, so that the wave normal direction varies with level in the plasma. A reference level is needed that can be used to specify the wave normal direction unambiguously, and we choose for this a region of free space, as was done in § 17.6. This is a useful reference even if the waves never enter such a region. Various directions of the wave normal have to be considered and it is therefore best to assume that the earth's magnetic field is in the x–z plane, so that $l_y = 0$. Then S_1, S_2 are used for the x and y direction cosines of the wave normal of the incident wave in the reference free space; compare §§ 6.2–6.4. It is also assumed, at first, that the source is infinitely distant so that S_1, S_2 are necessarily real; see § 14.9, item (9). This applies for the astrophysical applications mentioned above. For sources on or near the earth it is necessary, in general, to use complex S_1, S_2 even when collisions are neglected. This is an application of complex rays (Budden and Terry, 1971, §§ 5, 6) and is mentioned in § 17.8.

Consider a wave that starts as an upgoing ordinary wave at some real level $z = z_a$ well below the coupling points z_p, z_q, and travels through the window to a real level $z = z_b$ well above the coupling points. Here it is an upgoing extraordinary wave. Suppose that there is a 'good path' Γ from z_a to z_b in the complex z plane, such that at all points on it the W.K.B. solution for this wave can be used. Then the ratio of the moduli of the variable f, (6.53) at z_b and z_a for this wave is, from (7.111)

$$\mathscr{W} = \exp\left\{\mathrm{Re}\left(-ik \int_{z_a\,(\Gamma)}^{z_b} q\mathrm{d}z \right)\right\}. \tag{17.36}$$

To show that there is such a good path, the Stokes diagram for the solutions must be studied. It can be shown that near z_a the W.K.B. term for the upgoing ordinary wave becomes more dominant when $\mathrm{Im}(z)$ increases; the reader is advised to check this. Hence for exact coalescence the Stokes diagram is as in fig. 17.1(b). If coalescence is not exact there is also a downgoing ordinary wave where $z \approx z_a$ and the Stokes diagram is as in fig. 17.4(b). In the region where $\mathrm{Im}(z)$ is positive it includes a continuous curve linking the incident ordinary wave and the resulting extraordinary wave. This shows that the W.K.B. solution is not affected by the Stokes phenomenon. There is a good path Γ as shown in fig. 17.4(a). This contour can now be distorted provided that no singularities of the integrand in (17.36) are crossed. Thus it can run along the real z axis, provided that it is indented near the branch points z_p and z_q of q, as in fig. 17.4(a). But now in the ranges z_a to z_p and z_q to z_b, q is purely real, since collisions are neglected. Hence (17.36) becomes

$$\mathscr{W} = \exp\left\{k\int_{z_p}^{z_q} \mathrm{Im}(q)\mathrm{d}z\right\}. \tag{17.37}$$

A simple formula for this, (17.51), (17.52) below, is now to be derived. At exact coalescence $z_p = z_q$ and $\mathscr{W} = 1$. For other conditions $\mathscr{W}$ is a measure of the transparency of the window.

The change of X in going from one edge of a window barrier to the other, that is from z_p to z_q in figs. 17.3(b) or 17.4(a), is of order 0.02 or less in practical cases; compare figs. 6.6(b, d). Thus the maximum thickness of a window that has appreciable transparency is about 1/50 of the scale height for the variation of the electron concentration $N(z)$. In the ionosphere this means that window barriers are unlikely to be thicker than about 1 km. In such small thicknesses it is useful to assume that $N(z)$ is linear and that the earth's magnetic field is constant. These assumptions are made here. In the magnetosphere the barriers are much thicker but there are many cases where the same assumption is useful. In other cases the barrier thickness can be several hundred km (Herring, 1980, 1984) and a separate theory is needed but not given here (see Weitzner and Batchelor, 1979).

Figs. 10.4, 10.7, 10.9 showed cross sections of the refractive index surfaces by the

Fig. 17.4. (a) is the complex z plane and shows possible contours for the integral in (17.36). (b) is the Stokes diagram when z_p and z_q are slightly separated, so that conditions are near but not exactly at coalescence C2. It should be compared with fig. 17.1.

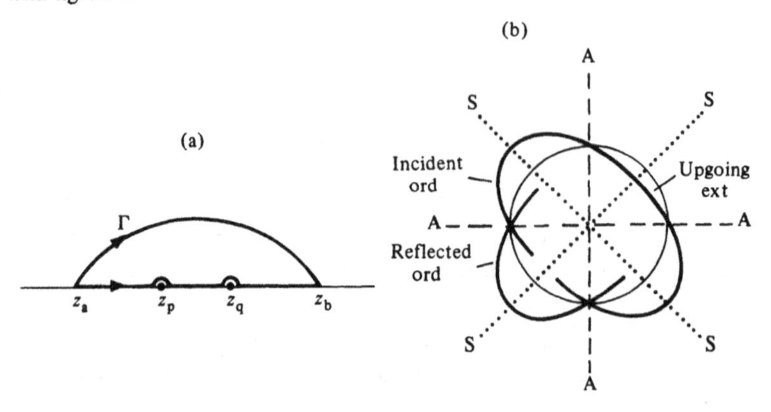

magnetic meridian plane when $Z=0$, and they were drawn with the direction of the earth's magnetic field at the correct angle to the vertical axis. Thus the direction cosines of the common axis of the refractive index surfaces are

$$l_x, \quad 0, \quad l_z, \tag{17.38}$$

and l_x, l_z are both assumed to be positive.

For any point on one of the curves the radius from the origin is the refractive index n, the abscissa is $n \sin\psi = \sin\theta = S_1$ and the ordinate is $n \cos\psi = q$; see § 10.8. For these figures $S_2 = 0$ because the wave normal is in the magnetic meridian plane. It was shown in ch. 5 that the surfaces for the ordinary and extraordinary waves cannot have any points in common when $X \neq 1$ and $X \neq 0$. When $Y < 1$ and $X \to 1$ the surface for the ordinary wave shrinks down to the line PQ forming part of the symmetry axis, and the surface for the extraordinary wave shrinks to the remaining parts of that axis together with the unit sphere; see § 5.2. Thus the two surfaces touch only when $X = 1$ and then only at the two window points P, Q. The form of these curves near the window point P is sketched in fig. 17.5. When $Y > 1$ and $X \to 1$ the surface for the ordinary wave shrinks to the line PQ as before, together with the parts of the symmetry axis extending to infinity beyond the two other window points P_2, Q_2; figs. 10.9, 17.7.

Fig. 17.5. Cross section by the magnetic meridian plane of parts of typical refractive index surfaces (not to scale) for $Y < 1$. Continuous curves are for the ordinary wave, and broken curves are for the extraordinary wave in the Z-mode. The numbers by the curves are the values of X.

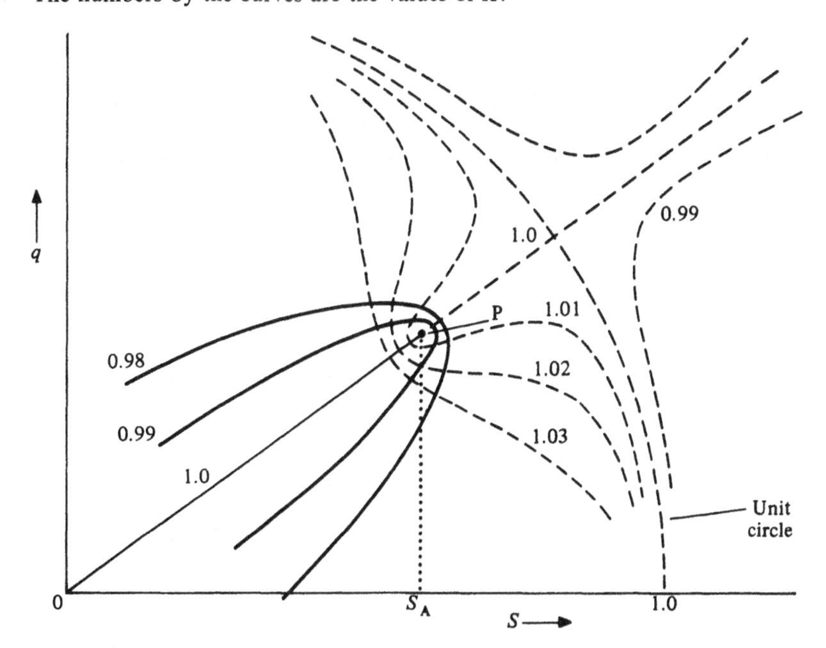

The surface for the extraordinary wave then shrinks to the segments PP_2 and QQ_2 of the symmetry axis, together with the unit sphere. Thus the surfaces touch only when $X = 1$ and then only at the four window points P, Q, P_2, Q_2. The form of the surfaces near P_2 is sketched in fig. 17.7. In this section the effect of the window near P is studied. Similar methods could be applied for Q. The points P_2, Q_2 are real only if $Y > 1$ and their effect is then sometimes called a 'second window'. The point P_2 is studied in § 17.9 and again similar methods could be applied for Q_2.

The refractive index surfaces are surfaces of revolution about the symmetry axis. Inspection of fig. 17.5 shows that near P they are approximately paraboloids. A quadratic approximation will now be derived that gives the equation of these paraboloids. An almost identical approximation has been used by Melrose (1980). Fig. 17.6 shows a cross section by any plane containing the symmetry axis of two typical refractive index surfaces near P. The line OA is the refractive index n and the angle AOP $= \Theta$ is the angle between the wave normal and the earth's magnetic field, the same as the Θ of chs. 4, 5. At P

$$n = n_1 = \{Y/(Y+1)\}^{\frac{1}{2}}, \quad \Theta = 0 \tag{17.39}$$

from § 5.2(7). Let

$$X = 1 + \sigma \tag{17.40}$$

where $|\sigma|$ is small. Choose new perpendicular axes u, v with the origin at P and with u along the symmetry axis. The form (4.65) of the dispersion relation is

$$An^4 - 2Bn^2 + C = 0 \tag{17.41}$$

Fig. 17.6. Cross section, by any plane containing the earth's magnetic field direction OP, of part of two refractive index surfaces near the window point P where they are, with good approximation, parabolae. The continuous curve is for $X < 1$ and the broken curve for $X > 1$.

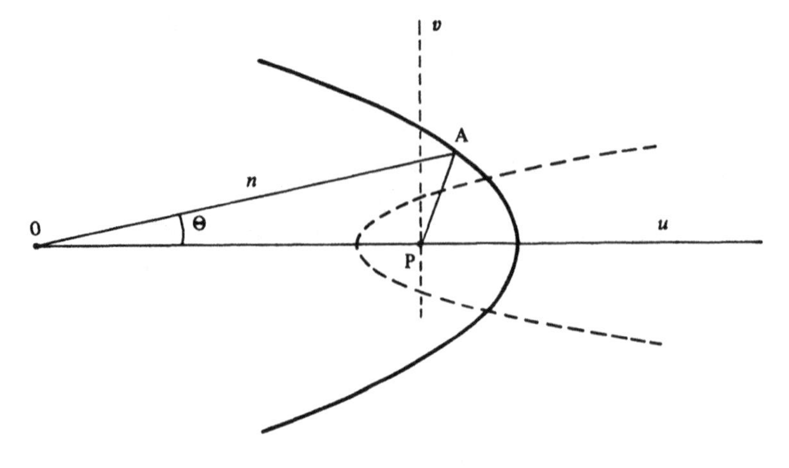

where, for $U = 1$ and with (17.40)

$$A = -\sigma - Y^2 + (1+\sigma)Y^2\cos^2\Theta, \quad B = \sigma^2 + \sigma Y^2 - \tfrac{1}{2}(1+\sigma)Y^2\sin^2\Theta,$$
$$C = Y^2\sigma - \sigma^3 \tag{17.42}$$

from (4.66). Now

$$n\cos\Theta = n_1 + u, \quad n\sin\Theta = v. \tag{17.43}$$

Eliminate n and Θ from (17.41)–(17.43) and neglect all products of the small quantities u, v, σ of degree 3 or greater. This gives

$$v^2 - 4\sigma u/n_1 - 2\sigma^2/Y = 0 \tag{17.44}$$

which is the required approximation. Further

$$\boldsymbol{n}\cdot\boldsymbol{Y} = nY\cos\Theta = Y(l_x S_1 + l_z q) \tag{17.45}$$

where (6.13) has been used, whence from (17.43)

$$u = l_x S_1 + l_z q - n_1, \quad v^2 = q^2 + S_1^2 + S_2^2 - (l_x S_1 + l_z q)^2 \tag{17.46}$$

and substitution in (17.44) gives

$$q^2 l_x^2 - q\{4\sigma l_z/n_1 + 2S_1 l_x l_z\} + S_1^2 l_z^2 + S_2^2 - 4\sigma S_1 l_x/n_1 - 2\sigma^2/Y + 4\sigma = 0. \tag{17.47}$$

This quadratic for q is the approximation, near the window, for the Booker quartic equation. Its solutions are

$$q = \frac{l_z}{l_x}\left(S_1 + \frac{2\sigma}{S_A}\right) \pm \frac{1}{l_x}\left\{\sigma^2\left(\frac{2}{Y} + \frac{4l_z^2}{S_A^2}\right) - 4\sigma\left(1 - \frac{S_1}{S_A}\right) - S_2^2\right\}^{\frac{1}{2}} \tag{17.48}$$

where (6.47) for S_A has been used. The square root in (17.48) has two zeros where $\sigma = \sigma_p, \sigma_q$ say. These correspond to the points in fig. 17.4(a) where $z = z_p, z_q$. For $\sigma_p < \sigma < \sigma_q$ the square root is imaginary and the waves are attenuated as they pass through the barrier. Outside this range q is real and the waves are unattenuated. The curves of q vs $\sigma = X - 1$ then have the shape shown in figs. 6.6(b, c, d) for $X \approx 1$.

The formulae (17.47), (17.48) are useful when arctan (l_x/l_z) exceeds about 10°. For smaller l_x they are not so reliable. Then both the earth's magnetic field and the wave normal are nearly vertical. This case has been treated by Budden and Smith (1973) but is not given in this book.

Suppose, now, that the electron concentration $N(z)$ varies linearly with z in the region of the window so that

$$\sigma = \frac{e^2 G z}{\varepsilon_0 m\omega^2} + \text{constant, where } G = \frac{\mathrm{d}N}{\mathrm{d}z}. \tag{17.49}$$

The attenuation of the waves by the barrier in nepers is, from (17.37)

$$-\ln\mathscr{W} = -\frac{\varepsilon_0 m\omega^2 k}{e^2|G|}\int_{\sigma_p}^{\sigma_q}\mathrm{Im}(q)\,\mathrm{d}\sigma. \tag{17.50}$$

Substitution of the square root in (17.48) for Im(q) gives finally

$$-\ln \mathscr{W} = 0.204\frac{f_H^{\frac{1}{3}} f^{\frac{5}{3}}}{|G|}\{J_1(S_1 - S_A)^2 + J_2 S_2^2\} \tag{17.51}$$

where f and f_H are in kHz, G is in m^{-4} and

$$J_1 = (Y+1)\{l_z^2(Y+\tfrac{1}{2})+\tfrac{1}{2}\}^{-\frac{3}{2}}, \quad J_2 = \{l_z^2(Y+\tfrac{1}{2})+\tfrac{1}{2}\}^{-\frac{1}{2}}. \tag{17.52}$$

to give the transparency of the window in decibels the numerical factor 0.204 in (17.51) should be replaced by 1.772.

These formulae show that the window is completely transparent when $S_1 = S_A$, $S_2 = 0$. The half angular width of the window may be defined as the values of $|S_1 - S_A|$ with $S_2 = 0$, or of S_2 with $S_1 = S_A$, that make the attenuation equal to 6dB, that is half power. These values are respectively, for $S_2 = 0$:

$$|S_1 - S_A|^2 = \frac{3.386|G|\{l_z^2(Y+\tfrac{1}{2})+\tfrac{1}{2}\}^{\frac{3}{2}}}{f_H^{\frac{1}{3}} f^{\frac{5}{3}}(Y+1)} \tag{17.53}$$

and for $S_1 = S_A$:

$$S_2^2 = \frac{3.386|G|\{l_z^2(Y+\tfrac{1}{2})+\tfrac{1}{2}\}^{\frac{1}{2}}}{f_H^{\frac{1}{3}} f^{\frac{5}{3}}}. \tag{17.54}$$

One of the solutions (17.48) is for waves travelling in the direction of X increasing and the other for the opposite direction. Since the imaginary parts are equal with opposite signs it follows that (17.50)–(17.54) apply for waves going through the window in either direction. Thus in the lower ionosphere they apply for a downgoing extraordinary wave in the Z-mode partially penetrating the window to give a downgoing ordinary wave below it.

In deriving (17.50)–(17.54) several approximations were used. The formulae were therefore tested (Budden, 1980) by comparing them with the results of calculations made by numerical integration of the governing differential equations for various values of S_1 and S_2. The agreement was found to be good. In Budden's formulae the factor $1/l_x$ in (17.48) was omitted in error. This is the same as cosec Θ in his notation – not the same Θ as used here. Thus in Budden's (1980) formulae (22), (23) a factor cosec Θ should be inserted, and in (24), (25) the sin Θ in the denominator should be deleted. In Budden's numerical example cosec Θ was 1.113. Insertion of this factor results in even better agreement with the numerically computed results.

When electron collisions were included in the numerical calculations it was found that the resulting additional attenuation was about the same for all values of S_1, S_2 near the window. Thus collisions did not alter the angular widths, and (17.53), (17.54) can still be used.

17.8. Radio windows 3. Complex rays

In §§ 17.6, 17.7 it was assumed that the source of the waves approaching a radio window is infinitely distant, so that the incident wave when in free space is a plane wave. The rays associated with it are the family of lines that are parallel in free space and whose x and y direction cosines S_1, S_2 are real and give the direction of the source. These various rays impinge on any plane $z=$ constant at various values of x and y, including complex values, as explained in § 14.9 items (9), (10). On entering the ionosphere the rays are refracted. If they enter an attenuating region, the refractive index n is complex and the rays are refracted at complex angles. Let a particular ray of the family pass through the point with coordinates x_1, y_1 before it enters the ionosphere. Suppose that this ray gets to a receiver with real coordinates x_2, y_2 in or beyond the attenuating region. Then x_1 or y_1 or both must be complex. A simple example of this was given in § 14.9 item (9).

The barrier of a radio window is like an attenuating medium in which the variable q, given by (17.48), is complex within the barrier. Thus for a ray that has real x, y before it enters the barrier, either x or y or both must be complex when z is real, after this ray leaves the barrier. For a ray that goes through the barrier and arrives at a receiver with real x, y, either x and y or both must be complex before this ray enters the barrier. Rays of this kind can be traced through the barrier and Budden (1980) gave some examples.

The use of complex rays for this problem is a form of the phase integral method; see § 17.7. It was shown that the results of this method are in agreement with the full numerical calculations, and therefore also with the approximate formulae (17.50)–(17.54) (Budden, 1980, § 8).

Suppose now that the source is not infinitely distant. For example it may be a point transmitter at the ground. Rays now leave it in all directions including complex directions. A ray that starts with real direction cosines and enters an attenuating part of the ionosphere is refracted to complex directions and subsequently has complex values of x or y or both when z is real. Such a ray cannot therefore reach a receiver with real x, y. The ray that gets to a receiver at a real point above the attenuating region must leave the source in a complex direction.

We now continue to assume that collisions are negligible so that the ionospheric plasma is loss-free. But in a window barrier q is complex and the waves are attenuated. The above conclusions for an attenuating region therefore apply also for a window barrier. The only exception is a ray that enters the window in its central direction. For this ray the barrier has zero thickness and there is no attenuation. A real ray with this direction can now go through real space from transmitter to receiver.

Any other ray that goes through the window must have a complex direction so that on it x or y or both are complex when z is real, except at the transmitter where it starts out and at the receiver where it arrives. This result was described by Budden and Terry (1971, § 6) who illustrated it with some numerical results.

For complex S_1, S_2 the formulae (17.36), and (17.39)–(17.49) can still be used, where Θ, n, u, v, are now complex. But (17.37) and (17.50)–(17.54) cannot be used and more complicated formulae are needed.

17.9. Radio windows 4. The second window

A second type of radio window associated with the window points P_2, Q_2, § 10.11, fig. 10.9, can occur if $Y > 1$ that is $f < f_H$. Fig. 17.7 shows part of the system of refractive index surfaces near a typical window point P_2. These can again be approximated by paraboloids of revolution as in § 17.7. It is only necessary to replace the value n_1 of n at P, (17.39) by its value at P_2 namely

$$n_2 = \{Y/(Y-1)\}^{\frac{1}{2}}, \quad \Theta = 0. \tag{17.55}$$

The final result is that (17.51) is still true, and (17.52) is replaced by

$$J_1 = (Y-1)\{l_z^2(Y-\tfrac{1}{2}) - \tfrac{1}{2}\}^{-\frac{3}{2}}, \quad J_2 = \{l_z^2(Y-\tfrac{1}{2}) - \tfrac{1}{2}\}^{-\frac{1}{2}}. \tag{17.56}$$

Fig. 17.7. The second window. Cross section by the magnetic meridian plane of parts of typical refractive index surfaces (not to scale) near the second window point P_2, for $Y > 1$. Continuous curves are for the ordinary wave in the whistler mode, and broken curves are for the extraordinary wave. The numbers by the curves are the values of X.

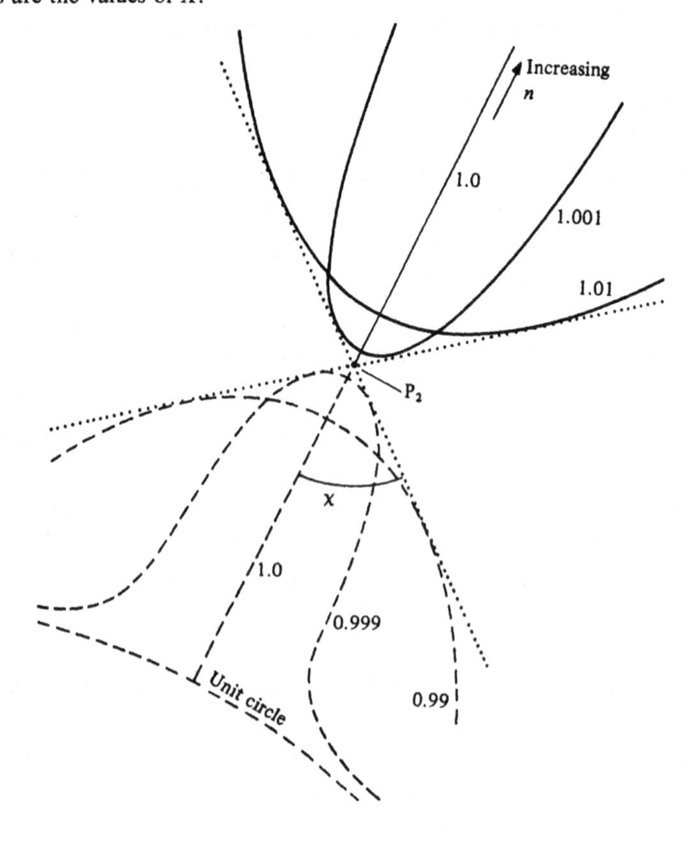

The refractive index surfaces near P_2 and Q_2 have a real envelope that is a cone with its apex at P_2 or Q_2. It can be shown (see problem 5.9) that at the apex the generators make with the axis an angle χ given by

$$\tan\chi = \{2(Y-1)\}^{\frac{1}{2}}. \tag{17.57}$$

In fig. 17.7 it is assumed that the medium is horizontally stratified. A vertical line may be drawn in this diagram, as for Poeverlein's construction § 10.8, whose abscissa is $S = \sin\theta$ where θ is the angle of incidence. Then, at any level, the point where it cuts a refractive index surface has ordinate q. When this line goes through P_2, the wave normal of the incident wave has the central direction of the window. If this line is to cut real refractive index surfaces when $X \neq 1$, it is clearly necessary that χ shall exceed the angle arctan (l_x/l_z) between the earth's magnetic field and the vertical, so that

$$l_x/l_z < \{2(Y-1)\}^{\frac{1}{2}}, \quad \text{that is} \quad Y > \tfrac{1}{2}(1 + l_z^{-2}). \tag{17.58}$$

This is also the condition that the expressions (17.56) shall be real. If this is not satisfied, no waves coming from where $X \neq 1$ can reach the level of the window. In the ionosphere this means that the second window can only be effective at high enough latitudes.

Jones, D. (1976b) has suggested that the second window might be a possible mechanism for explaining the propagation of auroral hiss.

17.10. Limiting polarisation 1. Statement of the problem

In the earlier chapters it has been assumed that when a characteristic wave, ordinary or extraordinary, travels through the ionosphere or magnetosphere its polarisation and wave admittance are given at most levels z by the W.K.B. solutions (7.112). There are regions of failure near coupling points in the complex z plane, and some of these have been studied in ch. 16 and the present chapter. There is another region where failure of a different kind occurs, that is where the wave emerges from the plasma into free space or enters the plasma from free space. This effect must now be examined.

The equation for finding coupling points is $\Delta = 0$, (6.38), (6.39), where Δ is the discriminant of the Booker quartic equation. It was mentioned, § 6.6 (Pitteway, 1959), that Δ has a factor X^4 so that there are four coupling points where $X = 0$. From the symmetry of the equations it is clear that two of these must apply for upgoing waves whose q values both tend to C when $X = 0$, and the other two for downgoing waves whose q values tend to $-C$. Thus in both cases there is a coalescence where $X \to 0$. For this reason the subject of limiting polarisation comes within the present chapter. The criteria in §§ 17.2, 17.4 show that it is a coalescence C2.

For many purposes it is useful to assume that in the lowest part of the ionosphere X is some exponential type of function such as $e^{-\alpha z}$, or the Chapman function

(1.9), (1.10). Then $X \to 0$ when $z \to -\infty$. The actual coalescence is therefore at infinity. There is a region where X is extremely small, that extends through a large range of height z. This idea of a large region where $q_1 - q_2$ is small, with the actual coalescence at infinity is very different from the coalescence at non-infinite $z_p = z_q$ discussed in earlier sections. The subject of limiting polarisation therefore needs a completely separate treatment.

The two coalescent coupling points cannot be made to separate by changing any of the parameters in the equations. We are therefore concerned only with exact coalescence. Then the matrix **S**, § 6.10, is non-singular and the transformation (6.53) can be used, as explained in § 17.4. There is no need for the matrix **V** nor for the transformation (17.17).

The subject is discussed here first for vertically incident waves. The extension for oblique incidence is mentioned at the end of the following section.

The polarisations ρ of the two characteristic waves are given by (4.21). When $X = 0$ this becomes

$$\rho = \frac{\frac{1}{2}\mathrm{i}Y\sin^2\Theta \pm \mathrm{i}\{\frac{1}{4}Y^2\sin^4\Theta + (1-\mathrm{i}Z)^2\cos^2\Theta\}^{\frac{1}{2}}}{(1-\mathrm{i}Z)\cos\Theta} \tag{17.59}$$

so that even when there are no electrons ρ has a value that depends on Z. In the free space below the ionosphere it is possible, even when electrons are absent, to speak of the electron collision frequency ν and therefore of Z. These would be roughly proportional to the pressure and would therefore change markedly within the height range of about 60 or 70 km below the ionosphere. Thus if a pure ordinary wave could travel downwards it would show a marked change of polarisation (17.59) in this range. But clearly the actual polarisation cannot change because the medium behaves just like free space. Here the assumption that a characteristic wave retains its correct polarisation must fail. It is shown in the following section that some of the other characteristic wave is generated by mode conversion, and the amount is just enough to keep the polarisation constant. This mode conversion is only effective where the electron concentration is small. Within the ionosphere where the electrons are numerous, the ordinary and extraordinary waves retain their correct polarisation (4.21), and are independently propagated, except in regions of strong coupling near coupling points, as discussed in ch. 16. This independence of the two characteristic waves was mentioned in § 16.11 and shown to be expected where the coupling parameter ψ in Försterling's equations (16.90), (16.91) can be neglected.

The following questions must therefore be examined:

(a) What conditions are necessary for a characteristic wave to retain its polarisation as given by (4.21)?

(b) When one characteristic wave travels down through the ionosphere, what is its limiting polarisation when it reaches the ground?

These problems have been discussed by Booker (1936, § 6), Eckersley and Millington (1939), Budden (1952), Hayes (1971).

This discussion is given in terms of an ordinary wave. An exactly similar argument would apply for an extraordinary wave. Consider therefore an ordinary wave travelling down through the lower ionosphere into the free space below. Here the coupling parameter ψ is small, and in a small height range the amount of extraordinary wave generated by mode conversion is also small. But the two refractive indices n_O, n_E are both unity, so that the ordinary wave and the small extraordinary wave travel with the same phase velocity. Thus as more of the extraordinary wave is generated, it adds to the amplitude of that already present, because the small contributions arriving from higher levels all have the same phase. In other words, although the mode conversion is small, it is cumulative, and in the following section it is shown that this keeps the polarisation constant. At higher levels where the electrons make $n_O \neq n_E$, the ordinary and extraordinary waves have different phase velocities, so that the small contributions to the extraordinary wave are not phase coherent, and interfere destructively. This process was mentioned in § 16.9. There is then no cumulative generation of the extraordinary wave and the polarisation of the resultant wave remains very close to that of an ordinary wave. It will be shown that the transition occurs near the level where $X \approx 0.0001 - 0.01$ in practical cases and it is here that the limiting polarisation is determined. This is called the 'limiting region'. It is quite different from the 'coupling region' discussed earlier. In the coupling region $X \approx 1$ and $|\psi|$ is large. In the limiting region X and $|\psi|$ are both very small.

The problem is discussed here for waves emerging into free space. Let $\rho_O(\text{out})$ be the limiting polarisation for a vertically emerging ordinary wave. A similar problem arises for waves entering the plasma. It is then required to find what polarisation $\rho_O(\text{in})$ a wave must have such that, when well within the plasma, it consists of an ordinary wave only. It was shown at the end of § 4.4 that if the same axis system is used the polarisations ρ_O are the same for upgoing and downgoing ordinary waves, and similarly the two polarisations ρ_E for the extraordinary waves are the same. Thence it follows that $\rho_O(\text{in}) = \rho_O(\text{out})$.

When the waves are oblique a similar conclusion holds. Let the wave normal of the emerging wave have direction cosines proportional to S_1, S_2, q_a where, in free space, $q_a = -C$. Consider now a wave that enters the plasma and let its wave normal have direction cosines proportional to $-S_1$, $-S_2$, q_b so that, in free space, $q_b = C$, and the wave normals of the two waves are antiparallel. The quartic (6.15), (6.16) shows that there is a solution with $q_b = -q_a$ at all levels, and a slight extension of

(7.136), together with the last equations in (7.77), (7.78) shows that, for this solution, the ratios of the components of $\boldsymbol{E}$, $\boldsymbol{\mathscr{H}}$ for the emergent wave are the same as those of $\boldsymbol{E}$, $-\boldsymbol{\mathscr{H}}$ for the entering wave. It then follows that the entering and emerging waves have the same polarisations in free space. Note that this conclusion applies only when the wave normals of the emerging and entering waves are antiparallel. It would not apply if the wave normal of the entering wave had direction cosines $+S_1$, $+S_2$, C.

17.11. Limiting polarisation 2. Theory

For vertically downgoing waves in the lower ionosphere the first order coupled equations (16.15) are used, where f_2, f_4 refer to the downgoing ordinary and extraordinary waves respectively. It might, alternatively, be possible to use the second pair of the equations (16.51), but the form (16.15) is simpler. These equations are first to be studied in free space, so that $n_{\mathrm{O}} = n_{\mathrm{E}} = 1$. Then the coefficients of $\mathsf{f}_1, \mathsf{f}_3$ in the second and fourth equations (16.15) are zero, which shows that the upgoing and downgoing waves are propagated independently. A prime $'$ denotes $k^{-1}\mathrm{d}/\mathrm{d}z = \mathrm{d}/\mathrm{d}s$ where $s = kz$. Then the second and fourth equations are

$$\mathsf{f}_2' - \mathrm{i}\mathsf{f}_2 = -\mathrm{i}\psi\mathsf{f}_4, \quad \mathsf{f}_4' - \mathrm{i}\mathsf{f}_4 = \mathrm{i}\psi\mathsf{f}_2. \tag{17.60}$$

Two independent exact solutions of these are

$$\mathsf{f}_2 = \mathsf{f}_{2a} = \mathrm{e}^{\mathrm{i}s}\exp\left(\int_p^s \psi\,\mathrm{d}s\right), \quad \mathsf{f}_4 = \mathrm{i}\mathsf{f}_2, \tag{17.61}$$

and

$$\mathsf{f}_2 = \mathsf{f}_{2b} = \mathrm{e}^{\mathrm{i}s}\exp\left(-\int_p^s \psi\,\mathrm{d}s\right), \quad \mathsf{f}_4 = -\mathrm{i}\mathsf{f}_2 \tag{17.62}$$

where the lower limit p is some fixed point in the complex s plane. From (16.93)

$$\exp\int_p^s \psi\,\mathrm{d}s = K\left(\frac{\rho_{\mathrm{O}} - 1}{\rho_{\mathrm{O}} + 1}\right)^{\frac{1}{2}} \tag{17.63}$$

where K is a constant. If now (17.61) with (17.63) is used with $\mathsf{f}_1 = \mathsf{f}_3 = 0$, in (16.16) to find the field components, it gives

$$E_x = E_y = -\mathrm{i}K2^{-\frac{1}{2}}\mathrm{e}^{\mathrm{i}kz}. \tag{17.64}$$

Similarly (17.62) gives

$$E_x = -E_y = \mathrm{i}K^{-1}2^{-\frac{1}{2}}\mathrm{e}^{\mathrm{i}kz}. \tag{17.65}$$

The factor $\mathrm{e}^{\mathrm{i}kz}$ shows that these waves are both travelling downwards with phase velocity c, and that their polarisations are linear and independent of the height z. Thus they behave just as though they are in free space. Any linear combination of (17.61), (17.62), that is of (17.64), (17.65), is a solution of (17.60) so the resulting wave may have any chosen polarisation $\rho = E_y/E_x$. Hence let

$$\mathsf{f}_2 = a\mathsf{f}_{2a} + b\mathsf{f}_{2b}, \quad \mathsf{f}_4 = \mathrm{i}(a\mathsf{f}_{2a} - b\mathsf{f}_{2b}). \tag{17.66}$$

Then

$$\rho = (a+b)/(a-b), \quad a/b = (\rho+1)/(\rho-1). \tag{17.67}$$

This completes the proof that, in the free space below the ionosphere, the mode conversion is just enough to keep the polarisation constant.

Now let the range of z be extended up to a level just above the limiting region where X is small but not zero. Here n_O, n_E differ from unity by small quantities of order X; compare (4.112). In the second equation (16.15) the coefficient of f_1 has a factor n'_O that is small because the medium is slowly varying. The coefficient of f_3 has factors $n_\mathrm{O} - n_\mathrm{E}$ and ψ that are both small. If any upgoing wave, represented by f_1 or f_3, is present, it can now generate a small amount of the downgoing wave f_2 by mode conversion, but the contributions to this from various levels are not phase coherent and therefore they interfere destructively. The same applies to the fourth equation (16.15). For this reason it is assumed that we can continue to ignore the effects of the upgoing waves and take $\mathsf{f}_1 = \mathsf{f}_3 = 0$. The coefficients of $\mathsf{f}_4, \mathsf{f}_2$ on the right of the second and fourth equations (16.15) respectively contain a factor $\frac{1}{2}(n_\mathrm{O} + n_\mathrm{E})(n_\mathrm{O} n_\mathrm{E})^{-\frac{1}{2}} = 1 + \mathrm{O}(X^2)$. Since this is multiplied by the small quantity ψ it is assumed that the small terms of order X^2 can be neglected. Then the equations become

$$\mathsf{f}'_2 - \mathrm{i}n_\mathrm{O}\mathsf{f}_2 = -\mathrm{i}\psi\mathsf{f}_4, \quad \mathsf{f}'_4 - \mathrm{i}n_\mathrm{E}\mathsf{f}_4 = \mathrm{i}\psi\mathsf{f}_2. \tag{17.68}$$

Now let

$$u = \mathsf{f}_4/\mathsf{f}_2 \quad \text{so that} \quad u' = \frac{\mathsf{f}'_4}{\mathsf{f}_2} - u^2\frac{\mathsf{f}'_2}{\mathsf{f}_4}. \tag{17.69}$$

Then u satisfies the differential equation

$$u' + \mathrm{i}u(n_\mathrm{O} - n_\mathrm{E}) - \mathrm{i}\psi(1+u^2) = 0 \tag{17.70}$$

which is a non-linear equation of Riccati type. If u can be found at some level $z = z_1$ below the ionosphere, the limiting polarisation ρ_1 can be found from (16.16) with $\mathsf{f}_1 = \mathsf{f}_3 = 0$, thus

$$\rho_1 = (\mathrm{i}\rho_\mathrm{O} + u)/(\mathrm{i} + \rho_\mathrm{O} u) \tag{17.71}$$

where ρ_O is the value of (17.59) for the ordinary wave where $z = z_1$. The wave above the limiting region is a pure ordinary wave so that there $\mathsf{f}_4 = 0$, $u = 0$. Equation (17.70) is suitable for numerical integration on a computer and has been used in this way by Barron (1960). The integration proceeds downwards starting just above the limiting region with $u = 0$. The same equation, with slightly different notation, has been used by Titheridge (1971b) in a study of the effect of a coupling region on Faraday rotation.

Now differentiate the first equation (17.68) once and use the result with (17.68) to eliminate f_4, f_4'. This gives

$$f_2'' - \{i(n_O + n_E) + \psi'/\psi\}f_2' + (in_O\psi'/\psi - in_O' - n_O n_E - \psi^2)f_2 = 0. \quad (17.72)$$

To remove the first derivative term take

$$v = f_2\psi^{-\frac{1}{2}}\exp\left\{-\tfrac{1}{2}i\int(n_O + n_E)ds\right\}. \quad (17.73)$$

Then

$$v'' + v\{\tfrac{1}{4}(n_O - n_E)^2 - \tfrac{1}{2}i(n_O' - n_E') + \tfrac{1}{2}i(n_O - n_E)\psi'/\psi - \psi^2 + \tfrac{1}{2}\psi''/2\psi - \tfrac{3}{4}(\psi'/\psi)^2\} = 0. \quad (17.74)$$

If the terms containing ψ are ignored, it can be confirmed that one solution is

$$v = \exp\left\{\tfrac{1}{2}i\int(n_O - n_E)ds\right\}, \quad f_2 \propto \exp\left(i\int n_O ds\right), \quad f_4 = 0 \quad (17.75)$$

so that the wave is a pure ordinary wave. Similarly if $n_O - n_E$ is ignored the solutions are (17.61), (17.62) and the wave travels as in free space. The limiting region is where the terms containing $n_O - n_E$ and those containing ψ are comparable, and it is here that the limiting polarisation is determined. There seems to be no simple way of solving this problem to give a concise expression for the limiting polarisation. To proceed further we shall therefore use a specific numerical example, typical of the lower ionosphere, to indicate the orders of magnitude of the various terms in the coefficient of v in (17.74).

Consider therefore a frequency $f \approx 2$ MHz so that $Y \approx \frac{1}{2}$ and take $\nu = 5 \times 10^5\,s^{-1}$, so that $Z \approx 0.04$. Let $\Theta \approx 30°$ as for temperate latitudes so that $Z_t \approx 0.07$ from (4.27). The value of ψ in the lower ionosphere is given by (16.98), where it is assumed that X is negligibly small and $X' \ll Z'$; compare figs. 16.5–16.7 at low levels. Then (16.98) gives $\psi \approx 0.04\,Z'$. The coefficient here is practically independent of Z and therefore of height z. Suppose that Z is proportional to $\exp(-z/H)$ where $H \approx 10$ km; compare § 1.6. Then $\psi \approx -0.04\,Z/kH \approx -4 \times 10^{-6}$, and ψ^2 is negligible compared with $\frac{1}{2}\psi''/\psi - \frac{3}{4}(\psi'/\psi)^2 = -\frac{1}{4}(kH)^{-2} \approx 1.4 \times 10^{-6}$. This would still be true for any other H because if H is changed, ψ^2, ψ''/ψ and $(\psi'/\psi)^2$ are all changed by the same factor.

When X is small the refractive indices are given to order X by (4.112), whence

$$n_O - n_E \approx \frac{XY(\frac{1}{4}Y^2\sin^4\Theta + U^2\cos^2\Theta)^{\frac{1}{2}}}{U(U^2 - Y^2)} \approx 0.6X. \quad (17.76)$$

Thus the orders of magnitude of the various terms in the coefficient of v in (17.74) are

$\frac{1}{4}(n_O - n_E)^2$	$\frac{1}{2}i(n_O' - n_E')$	$\frac{1}{2}i(n_O - n_E)\psi'/\psi$	$\frac{1}{2}\frac{\psi''}{\psi} - \frac{3}{4}\left(\frac{\psi'}{\psi}\right)^2$	ψ^2
$0.1X^2$	$0.3X'$	$7 \times 10^{-4}X$	1.4×10^{-6}	1.6×10^{-11}

If X in the lower ionosphere is proportional to $\exp(z/H_X)$, the second term is $0.3X/kH_X$ and is less than the third term if $H_X \gtrsim 10\,\text{km}$. Thus when X varies very slowly with height, the second term may be ignored. When the third term is comparable with the fourth, it exceeds the first term, and here $X \approx 0.002$. This gives an estimate of the level z where the limiting polarisation would be determined. If X follows the Chapman law (1.9), (1.10) with reasonable values of N_m and H, it can be shown that in the lowest part of the layer the second term is larger than the third. The second and fourth terms are now approximately equal where $X \approx 0.0002$.

It is thus clear that the location of the limiting region depends very much on how X varies with height in the lowest part of the ionosphere. But the above figures suggest that it is likely to be where X is between about 0.0001 and 0.01. This determines a range for the height z_l, whence the range of the value of Z can be found.

To find the limiting polarisation ρ_l the approximate values of X and Z must be inserted in (4.21). But X is so small that it can be neglected, and (17.59) can then be used. In this formula ρ is also insensitive to the exact value of Z since in most cases of practical interest $Z \ll 1$. Thus it is not necessary to know z_l and Z with great accuracy. In practice it is good enough if z_l can be estimated to within about $\pm 10\,\text{km}$. For the same reason it is not possible to use measurements of the limiting polarisation to give information about Z where $z = z_l$. The observed polarisations often vary with time (Landmark, 1955; Morgan and Johnson, 1955) probably because the ionosphere moves and is not exactly horizontally stratified.

For the study of limiting polarisation when the wave normal is oblique, the theory is more involved and the coupled equations are too complicated to be worth using. It is better to use numerical integration of the basic equations (7.80). This was done by Hayes (1971), who studied how the polarisation varied with height z for a wave that is initially a pure ordinary wave and comes obliquely down through the lower ionosphere. The results depend on the angle of incidence, the azimuth of the plane of incidence and on the function $N(z)$ in the lowest part of the ionosphere. A case was included where this $N(z)$ has a maximum and a minimum, as proposed by Deeks (1966a, b). The general conclusion was the same as above for vertical incidence, namely that the limiting polarisation is found from (17.59) for some z_l in a region of the lowest part of the ionosphere where X is of order 0.0001 to 0.01. Only a rough estimate is needed for the value of Z in this region.

18

Full wave methods for anisotropic stratified media

18.1. Introduction

This chapter deals with problems where the approximations of ray theory cannot be used and where it is necessary to take account of the anisotropy of the medium. It is therefore mainly concerned with low and very low frequencies, so that the change of the medium within one wavelength is large. The problem is discussed here for the ionosphere, which is assumed to be a plane stratified plasma in the earth's magnetic field. A radio wave of specified polarisation is incident obliquely and it is required to find the reflection coefficient as defined in §§ 11.2–11.6. In some problems it is also required to find the transmission coefficients. It is therefore necessary to find solutions of the governing differential equations, and the form used here is (7.80), (7.81). This type of problem has been very widely studied and there are many methods of tackling it and many different forms of the differential equations. These methods are too numerous to be discussed in detail, but the main features of some of them are given in §§ 18.3–18.5. The object in this chapter is to clarify the physical principles on which the various operations are based, and for this the equations (7.80), (7.81) are a useful starting point. In a few cases solutions can be expressed in terms of known functions, and some examples are given in §§ 19.2–19.6. Usually, however, it is necessary to use a numerical integration of the differential equations, and the present chapter is largely concerned with this.

The basic equations (7.80), (7.81), and most of the alternative forms described in §§ 18.4, 18.9, 18.10, apply for a single frequency and it is assumed that the incident wave is a plane wave. Thus the solutions cannot immediately give information about the propagation of modulated signals, and particularly of pulses, nor about the angular spectrum of waves from a point source; § 11.14. Information about pulse propagation can be obtained by using additional differential equations with dependent variables $\partial/\partial\omega$ of the field variables. These are integrated at the same time as the basic equations. For an example see § 18.10. Information about the angular dependence of the signals originating from a point source might similarly be obtained by using additional differential equations whose dependent variables are $\partial/\partial S_1$ and $\partial/\partial S_2$ of the field components, where S_1, S_2 are the x and y direction cosines of the wave normal of the incident wave in free space. But as far as the author knows this has never been done.

Throughout this chapter it is assumed that the wave normal of the incident wave is in the x–z plane so that $S_2 = 0$ and S is written for S_1. This has the advantage that for the matrix $\mathbf{T}$ (7.81) used in the basic equations (7.80), six of the sixteen elements are zero, which helps to shorten the computing time. The direction cosines of the vector Y are l_x, l_y, l_z. To study different magnetic azimuths of the plane of incidence l_x and l_y are changed. Some authors, for example Pitteway (1965), Pitteway and Jespersen (1966) have preferred to retain both S_1 and S_2 and to use a form of $\mathbf{T}$ with all sixteen elements non-zero.

The calculation of reflection and transmission coefficients is conveniently divided into three parts. They are described here for the form (7.80) of the equations in which the four dependent variables are the field components E_x, $-E_y$, $\mathscr{H}_x$, $\mathscr{H}_y$, that is the elements of the column matrix $\mathbf{e}$. But they apply also to other forms of the equations, that use other dependent variables; see §§ 18.9, 18.10. The three parts are:

(1) A solution is found that satisfies the physical conditions at a great height within the ionosphere. These conditions arise because the only influx of energy is in the incident wave below the ionosphere, and at a high enough level the solution must represent upgoing waves only. The differential equations are equivalent to a single equation of the fourth order, which has four independent solutions. At a high enough level where the W.K.B. solutions may be used, two of these are upgoing and two are downgoing waves. This is true when the angle of incidence θ is real, and was proved in § 6.4. For complex θ it is not necessarily true, and there may then be only one upgoing wave, or there may be three. These cases have not been fully explored. Complex θ is needed in problems of guided waves and in the study of the Goos–Hänchen effect, § 11.15. In this chapter it is assumed that there are two and only two upgoing waves at a great height, so that there are two solutions that satisfy the physical conditions. Methods of finding them are described in § 18.6.

The same methods can be used when the incident wave comes down from a great height. Then the starting solution is found at the bottom where it must consist of downgoing waves only, and the integration proceeds upwards.

(2) Starting with one of the solutions from (1), the step-by-step integration of the differential equations is performed, proceeding downwards until the free space below the ionosphere is reached. It is repeated with the second solution from (1). Some forms of the differential equations incorporate both solutions from (1) and then only one integration is needed; see § 18.9, 18.10. Some possible integration methods are described in § 18.2.

(3) The solution resulting from one integration gives values of the set E_x, $-E_y$, $\mathscr{H}_x$, $\mathscr{H}_y$ below the ionosphere. It is now separated into an upgoing and a downgoing wave, and each is in general elliptically polarised. The ratio of the amplitudes of these two waves would give a reflection coefficient, but it would apply only for an incident wave of a particular elliptical polarisation, of no special interest. The process is therefore repeated with the solution from the second integration, and the resulting upgoing and downgoing waves below the ionosphere in general have different polarisations from the first pair. Two linear combinations of the two solutions are now formed so that the incident wave of the combination has unit amplitude and is linearly polarised. For one combination it is in the plane of incidence, and for the other it is perpendicular to it. The associated reflected waves then give the two components R_{11}, R_{21} or R_{12}, R_{22} respectively of the reflection coefficient $\boldsymbol{R}$ (11.21). The details of these operations are given in § 18.7. If transmission coefficients are also needed, the same linear combinations are formed of the two solutions found in (1).

18.2. Integration methods

In stage (2) above, the differential equations are integrated over a range of height extending from the starting height $z = z_1$ used in stage (1) down to the finishing height, usually in the free space at the bottom of the ionosphere, used in stage (3). For doing this there are many different standard computer routines that use a step-by-step process. For the problems studied in this chapter the equations are written as r simultaneous first order differential equations. In (7.80) $r = 4$ and the dependent variables are the four elements of the column matrix $\mathbf{e}$, but it is sometimes necessary to use additional dependent variables. The equations may be written

$$\partial y_i/\partial z = F_i(y_1, y_2, \ldots, y_r, z), \quad i = 1, 2, \ldots, r. \tag{18.1}$$

The r dependent variables y_i are known at the starting height z_1. In any method the derivatives F_i must be calculated and the user must supply a computer subroutine for doing this. It is called the 'auxiliary subroutine'.

In one method, (18.1) is used first to calculate the derivatives where $z = z_1$ whence estimates of $y_1, \ldots y_r$ are made for a new height $z = z_2$. A first rough estimate would be $y_i(z_2) = y_i(z_1) + (z_2 - z_1)F_i$. This estimate is improved by recalculating the derivatives (18.1) at $z = z_2$ or at intermediate points. The final result is a corrected set of values of the y_i at the new height $z = z_2$. The interval $z_2 - z_1 = \delta z$ is called the 'step size'. One widely used method of this kind is the Runge–Kutta–Merson method. Details are given in books on computing methods, for example Hall and Watt (1976). The auxiliary subroutine is used four times for each step. Computer storage locations are needed for the r variables y_i and the r derivatives F_i. In addition storage locations for a further $4r$ variables are needed for numbers used during the calculation. For the applications of this chapter the dependent variables are complex so that each variable needs storage locations for two real numbers.

In another method the values of y_i and F_i are stored in the computer at four successive values of z with equal intervals δz, and the values after the next interval δz are computed. This is known as an 'Adams method'. It needs far more computer storage space for numbers, but the auxiliary subroutine is used only twice per step so this method is faster. For the first four steps, however, the four preceding $\acute{y}_i$ and F_i are not all available. The Runge–Kutta–Merson method can be used for the first four steps and the faster Adams process for the subsequent steps.

Usually the size of the step δz must be specified by the user. If it is too large accuracy is lost because of truncation errors, and if it is too small the computing time is wastefully long. Many computer library routines give error indications that show if the step is too large or too small. When first using a given set of differential equations it is advisable to make some trial integrations with various step sizes, so as to be sure that a suitable size has been found.

In most of the problems of this chapter the integration is downwards in the ionosphere so that the step δz is real and negative. In some applications, however, δz may be complex; see §§ 18.6, 18.13. For the basic equations (7.80) the dependent variables represent a wave motion, so that the wavelength λ in free space gives a guide to the step size. The wavelength in the ionosphere is of order $\lambda/|n|$ which may be small where the refractive index $|n|$ is large. For many purposes a step size $\delta z = -\frac{1}{50}\lambda$ has proved satisfactory in the author's experience. If, however, the electron concentration $N(z)$ varies very rapidly with height z, or if the path of integration passes close to a point of resonance where one refractive index n is very large (see § 18.13), it may be necessary to use a smaller step.

18.3. Alternative methods 1. Discrete strata

There are many ways of finding solutions of the basic equations (7.80), (7.81) or of equivalent equations as given in §§ 18.9, 18.10. Some of these methods are described briefly in this and the two following sections. The discussion of the main problem is resumed in § 18.6. The alternative methods are very numerous and no attempt has been made to include them all.

It has often proved convenient to imagine that the ionosphere is divided into discrete strata of thickness δz. By making δz small enough the stratified medium can be made to approximate as closely as desired to the actual ionosphere. In the commonest method of this kind the plasma in each stratum is assumed to be homogeneous. The method was used by Bremmer (1951) for an isotropic medium and was described in § 7.7 as one method for deriving and illustrating the W.K.B. solutions. It has been applied to the general anisotropic cold plasma by several authors, for example Johler and Harper (1962), Price (1964), Altman and Cory (1969a, b, 1970), Altman and Fijalkow (1970), Altman, Cory and Fijalkow (1970), Altman and Postan (1971), Nagano, Mambo and Hutatsuishi (1975), Nygrén (1981), Cheng and Fung (1977a, b), Fung and Cheng (1977).

It is assumed that the incident wave below the ionosphere is a plane wave with angle of incidence $\theta = \arcsin S$. Then in each homogeneous stratum there are in general four characteristic plane waves with refractive index vectors $(S, 0, q_i)$ where the q_i are the four roots of the Booker quartic equation, ch. 6, and are in general complex. The field components in each of the four waves are proportional to the elements E_x, $-E_y$, $\mathscr{H}_x$, $\mathscr{H}_y$ of the column matrix $\mathbf{s}_i$ as defined in § 7.14(1) and given by (7.118) or (7.134), and they contain a factor $\exp(-ikq_i z)$. These waves impinge on the boundaries between the strata and are partially reflected and transmitted. The reflection and transmission coefficients are found by imposing the boundary conditions for the electromagnetic fields. These conditions are that the column matrix $\mathbf{e}$, whose four elements are the horizontal components E_x, $-E_y$, $\mathscr{H}_x$, $\mathscr{H}_y$ of the total fields, is the same on the two sides of the boundary. The boundary conditions have been discussed by Johler and Walters (1960), Askne and Lisak (1976); see also §§ 11.8, 11.10.

Let $\mathbf{e}_r$ be the value of $\mathbf{e}$ at the bottom of the rth stratum. Then it is also the value of $\mathbf{e}$ at the top of the $(r-1)^{\text{th}}$ stratum next below. The amplitudes f_i of the four characteristic waves are then the column $\mathbf{f} = \mathbf{S}_{r-1}^{-1}\mathbf{e}_r$ from (6.53) where $\mathbf{S}_{r-1}$ is the matrix of the four columns $\mathbf{s}_i$, given in § 7.16. In the homogeneous stratum each of these amplitudes depends on height through a factor $\exp(-ikq_i z)$ so that at the bottom of the stratum

$$\mathbf{f} = \Delta_{r-1}\mathbf{S}_{r-1}^{-1}\mathbf{e}_r \tag{18.2}$$

where $\boldsymbol{\Delta}_{r-1}$ is a diagonal matrix with elements $\exp(-ikq_i\delta z)$ and the thickness δz of the stratum is negative. Then $\mathbf{e}$ at the bottom of the $(r-1)^{\text{th}}$ stratum is

$$\mathbf{e}_{r-1} = \check{\mathbf{P}}_{r-1}e_r, \quad \check{\mathbf{P}}_{r-1} = \mathbf{S}_{r-1}\Delta_{r-1}\mathbf{S}_{r-1}^{-1} \tag{18.3}$$

and $\mathbf{e}_{r-1}$ is also the $\mathbf{e}$ at the top of the $(r-2)^{\text{th}}$ stratum. In this way the fields in successive strata can be found, proceeding downwards. This form of the method was described by Nagano, Mambo and Hutatsuishi (1975). The topmost stratum is at a great height where it is known that there can be only upgoing waves, so that here only two of the four characteristic waves are present. Then by applying the matrix operations at successive boundaries, proceeding downwards, the amplitudes of the four waves in the free space below the ionosphere are found. These four waves are then combined to give the reflection coefficient matrix $\boldsymbol{R}$, (11.21).

This type of method has been widely and successfully used. A detailed account of one form of it was given by Johler and Harper (1962) who gave many results for various models of the ionosphere and included a study of the use of the Sen–Wyller formula, § 3.12, for dealing with collisions. The same type of method was used by Wait and Walters (1964) to study propagation from magnetic east to west and west to east at the magnetic equator. The waves then separate into two independent systems, see § 18.9, and they studied the case where $\boldsymbol{E}$ is in the plane of incidence so

$E_y = 0$, and $\mathscr{H}$ is perpendicular to the plane of incidence so $\mathscr{H}_x = 0$. In each stratum there are then only two characteristic waves, one upgoing and the other downgoing.

In any one stratum, an upgoing wave in one mode, say the ordinary wave, is partially reflected at the upper boundary to give downgoing waves of both modes, and these are again reflected at the bottom boundary to give upgoing waves of both modes. Thus the total upgoing ordinary wave is the resultant of an indefinite number of reflections, and similarly for the extraordinary wave and for the two downgoing waves. The full form of the theory takes account of this. But if the reflection coefficients are small enough all reflected waves after the first reflection are small and can be ignored in a first approximation. The reasons for this were discussed in § 7.7. When this is done the calculation can be simplified. A method of this kind was described by Altman and Cory (1969a, b, 1970) who showed that the resultant ionospheric reflection coefficients calculated in this way were in most cases within 10% of the results from integration of the differential equations without approximations.

Another interesting variant of this general method must be mentioned. In the basic equations (7.80) the matrix $\mathbf{T}$ is analytic and bounded at all points in the complex z plane except at points of resonance where $\varepsilon_{zz} = 0$, see § 18.13, and at singularities of $N(z)$ and at infinity. In practice, therefore, near any real point z_k, the matrix $\mathbf{T}$ can be expanded in a power series with matrix coefficients

$$\mathbf{T} = \mathbf{T}_0 + (z - z_k)\mathbf{T}_1 + (z - z_k)^2\mathbf{T}_2 + \cdots. \tag{18.4}$$

Similarly each of the four wave amplitudes can be expanded in a series of powers of $z - z_k$. Inoue and Horowitz (1966b) used strata that are not homogeneous, but with $\mathbf{T}$ given by the first two terms of (18.4). For the four solutions, the fields are matched at the boundaries as before. With this method it is found that the width δz of the strata can be about ten times the width needed with homogeneous strata, to obtain the same accuracy.

In this type of method it is necessary to solve the Booker quartic for each one of the strata, and then to sort the roots into those for upgoing and downgoing waves, in the same order for successive strata. This can use a lot of computer time and was a disadvantage with some of the older computers.

In the computer methods of integrating differential equations mentioned at the end of § 18.1, a finite step size δz is used so that they are in effect equivalent to dividing the media into discrete strata. But these strata are not homogeneous. The processes used are equivalent to the representation of $\mathbf{T}$ and of the solutions $\mathbf{e}$ by power series such as (18.4) that may go to fourth or higher powers. Moreover the computing of the elements of $\mathbf{T}$ by the auxiliary subroutine is very much simpler, and therefore less subject to human error, than calculating the coefficients (6.23) of the quartic, then solving the quartic and sorting the roots, then calculating the $\mathbf{s}_i$ (7.118), and then finally imposing the boundary conditions.

18.4. Alternative methods 2. Vacuum modes

The electromagnetic fields of a wave in a medium are in the vacuum that permeates the space between the particles. The fields act on each particle causing it to oscillate so that it emits its own field into the surrounding vacuum. The total field $\boldsymbol{E}, \boldsymbol{\mathscr{H}}$ at any point is the sum of the fields contributed by the incident wave and by the oscillating particles. It is this total field that acts on each particle. This idea was used by Darwin (1924) in a study of the refracting and dispersive properties of homogeneous matter. It was applied by Hartree (1929, 1931b) to a stratified magnetoplasma for incident plane waves at both normal and oblique incidence. For a plasma and for isotropic matter the oscillating particles are Hertzian dipole oscillators. Darwin (1924) also considered atoms or molecules for which higher order multipole oscillators have to be included.

To add the fields reaching any point, the contributions may first be found from the particles in an infinitesimally thin layer of thickness δz. For a homogeneous medium, or for normal incidence on a stratified medium this layer may be chosen to be perpendicular to the wave normal, so that the particles in it all oscillate in phase with each other. The sum of their fields can then be visualised as the resultant of an amplitude–phase diagram of the kind used in physical optics (Ratcliffe, 1959, figs. 3.1, 3.2). The sum of the contributions from the various layers δz is then expressed as an integral. For oblique incidence on a stratified variable medium, both stages of the summation are expressed as integrals. There are difficulties with the convergence of the integral over one layer δz, but these can be surmounted; see Darwin (1924, pp. 139, 144), Hartree (1929, p. 101). In this way the equations for finding the fields $\boldsymbol{E}, \boldsymbol{\mathscr{H}}$ are expressed as an integral equation, which can be thought of as an alternative to the basic differential equations (7.80) that are based on Maxwell's equations, and the constitutive relations. It can be shown that differentiation of the integral equation gives differential equations that agree with (7.80). The fields reaching any point are strongest from those particles that are nearest. There are many of these within a distance much less than one vacuum wavelength. The fields acting on any one particle therefore consist predominantly of the storage fields of the nearby particles. This is allowed for in the treatments of Darwin and Hartree.

When this method is used for studying the fields in a stratified medium and for finding the reflection coefficients, it shows how much of the resultant reflected wave is generated at each level. It thus shows which parts of the medium are the most important for the reflection process. It has been used in this way by Heading (1953, 1963, 1981a, b) and Westcott (1962a–d, 1964).

Poeverlein (1958a, b) has used the same idea in a slightly different way. The medium is imagined to be divided into plane layers perpendicular to the z axis and each layer is made very thin with vacuum on both sides of it. The waves propagate through the vacuum and interact with the successive thin layers. The electric fields cause currents to flow in each layer both parallel and perpendicular to its surfaces, and these can be found when the fields are known. The total fields on the two sides of a layer are related by boundary conditions at the surfaces. They consist of two wave systems, namely a 'partial wave' system and an 'additional' wave that goes right through the layer as if it were not there. Thus in this method the effect of the particles in a layer is already summed over the horizontal plane of the layer. The definition of the partial wave and the additional wave is a matter of choice, and Poeverlein (1958a) chooses them so that the partial wave is present only on the side from which the incident wave comes. Thus the partial wave has a component travelling towards the layer and another component travelling away. It cannot be said to be 'emitted' from the layer. But this choice is useful because the amplitudes of the upgoing and downgoing component waves in the vacuum at any level are very simply related to the reflection coefficient matrix variable $\boldsymbol{R}(z)$ defined at (18.45) in § 18.10.

The following method is another way of treating the waves as though they are propagated in a vacuum. The horizontal field components at any level are the four elements E_x, $-E_y$, $\mathscr{H}_x$, $\mathscr{H}_y$ of the column matrix $\mathbf{e}$. They can be resolved into four component waves in any chosen way and in § 6.10 they were resolved, by (6.53), into the fields of the four characteristic waves of magnetoionic theory, but now we use a different way. Imagine a thin layer of vacuum parallel to the plane $z = \text{constant}$ in the ionosphere. It can be supposed to be so thin that it has no appreciable effect on the fields. Within this vacuum layer the fields are resolved, as in § 11.7, into four linearly polarised waves, two upgoing, of which one has $\boldsymbol{E}$, and the other $\boldsymbol{\mathscr{H}}$, in the plane of incidence, and two downgoing with $\boldsymbol{E}$, $\boldsymbol{\mathscr{H}}$ respectively in the plane of incidence, that is the x–z plane. The elements of $\mathbf{e}$ for the first of these are proportional to C, 0, 0, 1, and similarly for the other three. Hence

$$\mathbf{e} = \mathfrak{f}_1\begin{pmatrix} C \\ 0 \\ 0 \\ 1 \end{pmatrix} + \mathfrak{f}_2\begin{pmatrix} 0 \\ -1 \\ -C \\ 0 \end{pmatrix} + \mathfrak{f}_3\begin{pmatrix} -C \\ 0 \\ 0 \\ 1 \end{pmatrix} + \mathfrak{f}_4\begin{pmatrix} 0 \\ -1 \\ C \\ 0 \end{pmatrix}. \tag{18.5}$$

Here $\mathfrak{f}_1$ to $\mathfrak{f}_4$ are the complex amplitudes of the four component waves, and they are now used as the four elements of a column matrix $\mathfrak{f}$. In §11.7 they were written a, b, c, d. Then (18.5) can be written in matrix form

$$\mathbf{e} = \mathbf{S}_v\mathfrak{f}, \quad \mathbf{S}_v = \begin{pmatrix} C & 0 & -C & 0 \\ 0 & -1 & 0 & -1 \\ 0 & -C & 0 & C \\ 1 & 0 & 1 & 0 \end{pmatrix}; \tag{18.6}$$

compare (11.44). This transformation is analogous to (6.53). The matrix $\mathbf{S}_v$ simply replaces the matrix $\mathbf{S}$ of (6.53), (7.85), (7.108). Now (18.6) can be substituted in the basic differential equation (7.80) to give the differential equation satisfied by $\mathfrak{f}$:

$$\mathfrak{f}' = -\mathrm{i}\mathbf{S}_v^{-1}\mathbf{T}\mathbf{S}_v\mathfrak{f} = -\tfrac{1}{2}\mathrm{i}\mathbf{W}_v\mathfrak{f} \text{ (say)}. \tag{18.7}$$

Expressions for $\mathbf{S}_v^{-1}$ and $\mathbf{W}_v$ are given later, (18.46), (18.47). Since $\mathbf{S}_v$ is independent of z, there is nothing in this equation corresponding to the coupling matrix $\boldsymbol{\Gamma}$ of (7.108). The amplitudes of the four 'vacuum modes' are the four elements of $\mathfrak{f}$. The coupling between them is given by the non-diagonal elements of $\mathbf{W}_v$. The equations (18.7) are now integrated, for example by the type of method described in § 18.1. This form of the equations wave suggested by Poeverlein (1967) and has been successfully used by Walsh (1967). The discussion of numerical swamping in Walsh's paper is particularly valuable; see § 18.11.

18.5. Alternative methods 3. The matrizant

If the matrices $\mathbf{e}(z)$ and $\mathbf{T}(z)$ in the basic equations (7.80) are replaced by scalars $\mathrm{e}(z)$ and $\mathrm{T}(z)$, the equation is

$$\mathrm{e}' = -ik\mathrm{T}\mathrm{e} \tag{18.8}$$

where, as usual, a prime $'$ means $k^{-1}\mathrm{d}/\mathrm{d}z$. This has the solution

$$\begin{aligned}\mathrm{e}(z) &= \mathrm{e}(z_1)\exp\left(-ik\int_{z_1}^{z}\mathrm{T}(\zeta).\mathrm{d}\zeta\right)\\ &= \mathrm{e}(z_1) - ik\mathrm{e}(z_1)\int_{z_1}^{z}\mathrm{T}(\zeta)\mathrm{d}\zeta + \frac{(ik)^2}{2!}\left(\int_{z_1}^{z}\mathrm{T}(\zeta)\mathrm{d}\zeta\right)^2 + \cdots.\end{aligned} \tag{18.9}$$

This would give $\mathrm{e}(z)$ at height z when $\mathrm{e}(z_1)$ at height $z = z_1$ is known, provided that the integral of T can be found. The terms in the expansion contain powers of this single integral of T. In fact e, T are really the matrices $\mathbf{e}$, $\mathbf{T}$, but a similar result can be derived. As a first approximation assume that $\mathbf{e}$ has the constant value $\mathbf{e}(z_1)$, and substitute this on the right of (7.80). Then integration gives the second approximation

$$\mathbf{e}(z) = \mathbf{e}(z_1) - ik\int_{z_1}^{z}\mathbf{T}(\zeta)\mathrm{d}\zeta\cdot\mathbf{e}(z_1). \tag{18.10}$$

If this is now substituted on the right of (7.80), a further integration gives

$$\mathbf{e}(z) = \left[1 - ik\int_{z_1}^{z}\mathbf{T}(\zeta)\mathrm{d}\zeta - k^2\int_{z_1}^{z}\mathbf{T}(\zeta_2)\left\{\int_{z_1}^{\zeta_2}\mathbf{T}(\zeta_1)\mathrm{d}\zeta_1\right\}\mathrm{d}\zeta_2\right]\mathbf{e}(z_1). \tag{18.11}$$

Note that in the last term the integral is a double integral and there is no factor 2! in the denominator. The process can now be continued indefinitely to give further terms of the series. The fourth term has a triple integral and subsequent terms have integrals of increasing multiplicity. The result is an infinite series analogous to (18.9). It can be shown to be absolutely convergent. Its sum gives

$$\mathbf{e}(z) = \mathbf{M}(z, z_1)\mathbf{e}(z_1) \tag{18.12}$$

and the 4×4 matrix $\mathbf{M}$ is called the 'matrizant'. It was used in the theory of ionospheric reflection by Volland (1962a, c) and Bossy (1971). If it is known for a sequence of height ranges $z_1 - z_2$, $z_2 - z_3$, etc. then clearly (18.12) can be applied to these ranges in succession to give

$$\mathbf{e}(z_r) = \mathbf{M}(z_r, z_{r-1}) \cdots \mathbf{M}(z_3, z_2)\mathbf{M}(z_2, z_1)\mathbf{e}(z_1). \tag{18.13}$$

One method of calculating the matrizants $\mathbf{M}$ is to use the series of multiple integrals whose first three terms are given by (18.11). If $|z_2 - z_1|$ is small enough it is found that only a small number of terms is needed. Further, for small $|z_2 - z_1|$, the matrix $\mathbf{T}$ can be expressed as a series of powers of $z - z_1$ and then $\mathbf{M}$ can also be expressed as a series of powers (Rawer and Suchy, 1967, p. 162). Bossy (1979) has found that if strata of thickness $z_2 - z_1 \approx 1$ km are used, and if $\mathbf{T}$ is expanded up to fifth powers of $z - z_1$, then $\mathbf{M}$ can be found with a relative precision better than 10^{-7}.

For any z_k, (18.12) shows that $\mathbf{M}(z_k, z_k)$ is the unit 4×4 matrix $\mathbf{1}$. Substitution of (18.12) in the basic equation (7.80) shows that

$$\frac{\mathrm{d}}{\mathrm{d}z}\mathbf{M}(z, z_1) = -ik\mathbf{T}\mathbf{M}(z, z_1). \tag{18.14}$$

Thus another way of finding $\mathbf{M}$ would be to start at $z = z_1$ with $\mathbf{M} = \mathbf{1}$ and integrate (18.14), for example by the methods described in § 18.1.

In its simplest form the matrizant method consists of calculating $\mathbf{M}$ for a series of strata and then using (18.13). It is therefore similar to the use of discrete strata as in § 18.3. Its advantage is that for the same amount of computation the accuracy is considerably greater. In (18.13) z_1 is at a great height and z_r is below the base of the ionosphere. For $\mathbf{e}(z_1)$ there are two different possible upgoing waves at the top and (18.13) must be applied to both. The results $\mathbf{e}(z_r)$ are then combined, as in § 18.7, to give the reflection coefficients. The form (18.13) does, however, have the disadvantage that, in many practical cases, one solution is a non-penetrating wave that is almost evanescent at great heights. With decreasing height its amplitude can increase so quickly that it gives rise to numerical swamping; see § 18.11. For this reason Bossy (1979) has used a different method for connecting the solutions at successive boundaries between strata. This will now be described in outline.

The matrix $\mathbf{e}$ of the field variables can be transformed by the relation

$$\mathfrak{f} = \mathbf{S}_v^{-1}\mathbf{e} \tag{18.15}$$

from (18.6). The four elements of $\mathfrak{f}$ are then the amplitudes of the four vacuum modes as described by (18.5). If this transformation is applied to (18.12) with $z = z_2$ it gives

$$\mathfrak{f}(z_2) = \mathbf{P}(z_2, z_1)\mathfrak{f}(z_1), \quad \mathbf{P} = \mathbf{S}_v^{-1}\mathbf{M}\mathbf{S}_v. \tag{18.16}$$

A matrix very similar to $\mathbf{P}$ was used by Lacoume (1967) (with different notation) who called it the 'propagator', but he applied it not to the vacuum modes but to the characteristic waves of magnetoionic theory. Thus he used $\mathbf{S}$ (6.53) instead of $\mathbf{S}_v$ and his $\mathbf{S}(z_2)$, $\mathbf{S}(z_1)$ were different, whereas in (18.16) the two $\mathbf{S}_v$s are the same. In (18.16) $\mathbf{P}$ relates the fields at the two boundaries $z = z_1, z_2$. Thus it plays the same part as the matrizant when $\mathfrak{f}$ is used instead of $\mathbf{e}$ to specify the wave fields. To find the fields at the boundaries of successive strata, the matrices $\mathbf{P}$ are multiplied in succession, just as the $\mathbf{M}$s are multiplied in (18.13). But by using $\mathbf{P}$ the trouble from numerical swamping is avoided.

The matrix $\mathbf{P}$ is related to the reflection and transmission coefficients of a stratum. This can be shown as follows (Volland, 1962a, b; Bossy, 1979). Let z_r, z_{r+1} be the boundaries of a stratum, with $z_r > z_{r+1}$. The elements $\mathfrak{f}_1$, $\mathfrak{f}_2$ of $\mathfrak{f}$ refer to upgoing waves and $\mathfrak{f}_3$, $\mathfrak{f}_4$ refer to downgoing waves, from (18.5). Let

$$\boldsymbol{u}_r = \begin{pmatrix} \mathfrak{f}_1(z_r) \\ \mathfrak{f}_2(z_r) \end{pmatrix}, \quad \boldsymbol{d}_r = \begin{pmatrix} \mathfrak{f}_3(z_r) \\ \mathfrak{f}_4(z_r) \end{pmatrix}. \tag{18.17}$$

The 4×4 matrix $\mathbf{P}(z_{r+1}, z_r)$ from (18.16) is partitioned into the four 2×2 matrices thus

$$\boldsymbol{P}_1 = \begin{pmatrix} \mathsf{P}_{11} & \mathsf{P}_{12} \\ \mathsf{P}_{21} & \mathsf{P}_{22} \end{pmatrix}, \quad \boldsymbol{P}_2 = \begin{pmatrix} \mathsf{P}_{13} & \mathsf{P}_{14} \\ \mathsf{P}_{23} & \mathsf{P}_{24} \end{pmatrix}, \quad \boldsymbol{P}_3 = \begin{pmatrix} \mathsf{P}_{31} & \mathsf{P}_{32} \\ \mathsf{P}_{41} & \mathsf{P}_{42} \end{pmatrix}, \quad \boldsymbol{P}_4 = \begin{pmatrix} \mathsf{P}_{33} & \mathsf{P}_{34} \\ \mathsf{P}_{43} & \mathsf{P}_{44} \end{pmatrix}. \tag{18.18}$$

Then (18.16) gives

$$\boldsymbol{u}_{r+1} = \boldsymbol{P}_1 \boldsymbol{u}_r + \boldsymbol{P}_2 \boldsymbol{d}_r, \quad \boldsymbol{d}_{r+1} = \boldsymbol{P}_3 \boldsymbol{u}_r + \boldsymbol{P}_4 \boldsymbol{d}_r \tag{18.19}$$

whence

$$\begin{pmatrix} \boldsymbol{d}_{r+1} \\ \boldsymbol{u}_r \end{pmatrix} = \begin{pmatrix} \boldsymbol{P}_3 \boldsymbol{P}_1^{-1} & \boldsymbol{P}_4 - \boldsymbol{P}_3 \boldsymbol{P}_1^{-1} \boldsymbol{P}_2 \\ \boldsymbol{P}_1^{-1} & -\boldsymbol{P}_1^{-1} \boldsymbol{P}_2 \end{pmatrix} \begin{pmatrix} \boldsymbol{u}_{r+1} \\ \boldsymbol{d}_r \end{pmatrix}. \tag{18.20}$$

The elements $\boldsymbol{u}_{r+1}$, $\boldsymbol{d}_r$ on the right represent waves going in to the stratum and $\boldsymbol{d}_{r+1}, \boldsymbol{u}_r$ on the left represent waves coming out. The 4×4 matrix on the right of (18.20) is called the S-matrix of the stratum (in French: matrice de diffusion). It is here denoted by $\mathscr{S}$. Suppose that there is no downgoing wave at the top z_r of the stratum. Then $\boldsymbol{d}_r = 0$ and (18.20) shows that the reflection coefficient matrix $\boldsymbol{R}_{r+1}$ at the bottom, defined by (11.20)–(11.24), is given by $\boldsymbol{P}_3 \boldsymbol{P}_1^{-1}$. Similarly the other 2×2 component matrices of $\mathscr{S}$ are reflection and transmission coefficient matrices of the stratum, thus

$$\mathscr{S} = \begin{pmatrix} \boldsymbol{R}_{r+1} & \boldsymbol{T}_{r,r+1} \\ \boldsymbol{T}_{r+1,r} & \boldsymbol{R}_r \end{pmatrix} \tag{18.21}$$

where the first subscript of a transmission coefficient matrix $\boldsymbol{T}$ denotes the boundary where the wave goes in to the stratum, and the second shows where it comes out.

18.6. Starting solutions at a great height

Integration of a wave equation such as (7.80) usually starts at a great height in the ionosphere and proceeds downwards. The set of four field variables $\mathbf{e}$ at the start must be suitably chosen. It must represent a wave that is travelling obliquely upwards and since there are two independent upgoing waves, two different choices are possible and two integrations are needed.

There are several different ways of calculating the starting values of $\mathbf{e}$. One way is to use the formula (7.118) for the eigen column $\mathbf{e} \propto \mathbf{s}_i$ at the starting level. It necessitates solving the Booker quartic (6.15) for the qs at the starting level, and then selecting the two that apply for upgoing waves. This is complicated, and requires computer storage space for the necessary algebra. With modern computers it presents no problem, but it was at one time a disadvantage when computers had less available storage for programs.

Two other methods of calculating the starting $\mathbf{e}$s are of some physical interest. The first was used by Pitteway (1965). Suppose that any arbitrary value of $\mathbf{e}$ is chosen. Then it must be expressible as the sum of multiples of the four eigen column $\mathbf{s}_i$ of $\mathbf{T}$ in (7.85), thus

$$\mathbf{e} = \sum_{i=1}^{4} a_i \mathbf{s}_i. \tag{18.22}$$

Multiply on the left by $\mathbf{T} + \Lambda\mathbf{1}$ where $\mathbf{1}$ is the unit 4×4 matrix, and Λ is a complex constant to be found as indicated below. Let this multiplication be repeated r times. The result is

$$\sum_{i=1}^{4} (q_i + \Lambda)^r a_i \mathbf{s}_i. \tag{18.23}$$

In the column (18.23), one of the eigen columns $\mathbf{s}_i$ has a coefficient very much larger than the other three. It is the one for which $|q_i + \Lambda|$ is greatest. If r is large enough this one eigen column predominates, and the others are negligible in comparison. This method is used as follows.

A value Λ is selected such that the largest $|q_i + \Lambda|$ occurs for the q_i corresponding to one of the two upgoing waves. Usually in practice one of these waves is almost evanescent so that its q_1 is almost negative imaginary. Then a positive imaginary value of Λ is suitable. The other wave is propagated with only small attenuation and its q_i is almost real and positive. Then a negative real value of Λ is suitable. It is not necessary to know the four q_is; see Pitteway (1965, appendix 1). An arbitrary column $\mathbf{e}$ is now chosen and multiplied by $\mathbf{T} + \Lambda\mathbf{1}$. It is then multiplied by some constant, which can be chosen, for example so that one of the elements of $\mathbf{e}$ is unity. This prevents the column (18.23) from becoming indefinitely large. If the new value of $\mathbf{e}$ differs from the old by more than a specified small amount, the multiplications are repeated. This iterative process is continued until two successive $\mathbf{e}$s are the same within the specified limit. The resulting $\mathbf{e}$ is then the required starting value.

For the second method, imagine a homogeneous medium which, for all heights z, has the same composition as the actual ionosphere at the starting level. In it the four characteristic waves are propagated independently. As an illustration, suppose first that the medium is loss-free. In practical cases for frequencies $f < f_H$ the value of q for one of the upgoing waves is then negative imaginary so that the wave amplitude decreases as z increases. This wave is an inhomogeneous wave; see § 2.15. If a value of the column matrix **e** is chosen arbitrarily it will include some of this wave; compare (18.22). Let this column be used as the starting value and let the equations (7.80) be integrated with a stepwise process for the fictitious homogeneous medium, with z decreasing, and with step size δz_h. Then the wave amplitude of the inhomogeneous wave increases, whereas the other three waves either decrease or remain constant in amplitude. Thus the inhomogeneous wave must ultimately predominate. During the integration **e** must, when necessary, be multiplied by a constant fraction to prevent its becoming too large. The absolute value of **e** is not important. It is only the ratios of its elements that are needed. The integration is continued until the ratios of corresponding elements of **e** after successive steps are all equal within specified limits. The resulting **e**, or any multiple of it, may then be used as a starting value for one of the main integrations in the actual ionosphere.

The other upgoing wave in a loss-free medium is a propagated wave in which all field quantities vary with height z through a factor $\exp(-ikqz)$ where q is real and positive. The column **e** for this wave can also be computed by a preliminary integration of (7.80) in a fictitious homogeneous medium. An imaginary step length $\delta z_h = il$ is used where l is real. Thus the amplitudes $\exp(-ikqz)$ increase at each step, whereas for the other three waves they either decrease or stay constant.

In the more general case the ionospheric medium has losses but the same method can be used. When the angle of incidence θ is real the two required starting solutions can be found by preliminary stepwise integrations with $\arg \delta z_h$ equal to π or $\frac{1}{2}\pi$ as in a loss-free medium. Occasionally it may be convenient to use other values of $\arg \delta z_h$. The derived starting **e** is an eigen column of **T**. This property is not affected by truncation errors, so the size $|\delta z_h|$ of the step can be made quite large to ensure rapid convergence. A suitable value is about one quarter of a vacuum wavelength.

This method of preliminary integration can be used with forms of the differential equations other than (7.80). It can be used, for example, with the non-linear forms (18.39), (18.49) of §§ 18.9, 18.10.

18.7. Finding the reflection coefficient

The two integrations with the two independent starting solutions of the preceding section give two sets of the four field variables **e** at some level $z = z_v$ below the ionosphere. These will be written $\mathbf{e}^{(a)}$, $\mathbf{e}^{(b)}$. From them the reflection coefficient matrix $\boldsymbol{R}$, § 11.5 is to be found, with reference level at $z = z_v$; see § 11.2. The free space below the ionosphere is like a vacuum and the fields $\mathbf{e}^{(a)}$ and $\mathbf{e}^{(b)}$ are here each resolved into the four linearly polarised vacuum modes. This is done by the method used in § 18.4, equations (18.5), (18.6), and in § 11.7, equations (11.44), (11.45). The column matrix $\mathfrak{f}$ then gives the amplitudes of these modes. For the two sets of fields, its values are, from (18.15):

$$\mathfrak{f}^{(a)} = \mathbf{S}_v^{-1}\mathbf{e}^{(a)}, \quad \mathfrak{f}^{(b)} = \mathbf{S}_v^{-1}\mathbf{e}^{(b)} \tag{18.24}$$

where $\mathbf{S}_v$ is given by (18.6). Now the two columns (18.24) are each partitioned as in (18.17) so that

$$\mathfrak{f}^{(a)} = \begin{pmatrix} \boldsymbol{u}^{(a)} \\ \boldsymbol{d}^{(a)} \end{pmatrix}, \quad \mathfrak{f}^{(b)} = \begin{pmatrix} \boldsymbol{u}^{(b)} \\ \boldsymbol{d}^{(b)} \end{pmatrix} \tag{18.25}$$

where the matrices $\boldsymbol{u}, \boldsymbol{d}$ are columns with two elements; compare (11.47). From the definition of the reflection coefficient $\boldsymbol{R}$ in §11.5 and (11.16), (11.17), or (11.46), it follows that, for either superscript (a) or (b), $\boldsymbol{d} = \boldsymbol{R}\boldsymbol{u}$. Hence let

$$\boldsymbol{U} = (\boldsymbol{u}^{(a)}, \boldsymbol{u}^{(b)}), \quad \boldsymbol{D} = (\boldsymbol{d}^{(a)}, \boldsymbol{d}^{(b)}). \tag{18.26}$$

Then

$$\boldsymbol{D} = \boldsymbol{R}\boldsymbol{U}. \tag{18.27}$$

Now the solutions $\mathbf{e}^{(a)}\mathbf{e}^{(b)}$ must be independent, that is one is not a multiple of the other. Then in general $\mathfrak{f}^{(a)}, \mathfrak{f}^{(b)}$ are independent, though there can be exceptions; see §18.11. When they are independent $\boldsymbol{U}$ is non-singular and (18.27) gives

$$\boldsymbol{R} = \boldsymbol{D}\boldsymbol{U}^{-1}. \tag{18.28}$$

If $\mathbf{e}^{(a)}, \mathbf{e}^{(b)}$ are replaced by any two independent linear combinations of $\mathbf{e}^{(a)}, \mathbf{e}^{(b)}$, then each pair $\boldsymbol{u}^{(a)}, \boldsymbol{u}^{(b)}$ and $\boldsymbol{d}^{(a)}, \boldsymbol{d}^{(b)}$ is replaced by the same linear combination. This is achieved by multiplying $\boldsymbol{U}, \boldsymbol{D}$ on the right by an arbitrary diagonal matrix. But in (18.28) this cancels out, so that the value of $\boldsymbol{R}$ is unaffected.

The elements of $\boldsymbol{R}$ in (18.28) could be written out in full in terms of the eight field quantities $E_x^{(a)}, E_y^{(a)}, \mathscr{H}_x^{(a)}, \mathscr{H}_y^{(a)}, E_x^{(b)}, E_y^{(b)}, \mathscr{H}_x^{(b)}, \mathscr{H}_y^{(b)}$, but the expressions are very complicated. Equation (18.28) shows, however, that the elements of $\boldsymbol{R}$ depend only on ratios such as $\mathscr{H}_i^{(s)}/E_j^{(s)}$, $i, j = x, y$, and not on the absolute values, and the superscript s = a or b, is the same for both; the reader should prove this. Thus $\mathbf{e}^{(a)}$ or $\mathbf{e}^{(b)}$ may each be multiplied by any constant without affecting the value of $\boldsymbol{R}$.

18.8. Allowance for the earth's curvature

In most work on full wave solutions for radio waves of low or very low frequency it is assumed that the earth is flat. This approximation is nearly always satisfactory when studying propagation to short distances with angles of incidence up to about 50° to 60°. But for angles of incidence near 90°, that is grazing incidence, the errors can be serious. This applies particularly in the study of guided waves, but in some cases it may apply for a single reflection. Some method of allowing for the earth's curvature is then needed.

If the ionosphere is assumed to be isotropic and spherically stratified, the wave equation can be formulated in spherical polar coordinates r, ϑ, φ with the origin at the earth's centre. The solution can then be expressed as the sum of terms each of which is the product of a spherical harmonic depending only on ϑ and φ and a radial part depending only on r. When the earth's magnetic field is allowed for however, so that the ionosphere is anisotropic, the problem is harder. There are some special cases where the solution can still be separated into radial and angular parts. This is possible, for example, for propagation from east to west or west to east at the magnetic equator (Wait and Walters, 1964); see § 18.3. In another case that has been studied (Wait, 1963, Krasnushkin, 1961 a, b) it was assumed that the earth's magnetic field is everywhere radial.

In the more general case, with a realistic magnetic field, these simplifications are not possible. An alternative approach is to use some 'earth flattening' approximation, and perhaps the best known, for isotropic media, is the method of the modified refractive index used by Booker and Walkinshaw (1946).

Consider a spherically stratified isotropic medium and assume that the approximations of ray theory can be used. Then for a wave at any radius r, the angle ψ that its wave normal makes with the radius is given by the Bouger law $rn \sin\psi = K$, (10.18), where K is a constant. Let the ray be in the plane $\varphi = 0$, and let $r, \vartheta, 0$ be the spherical polar coordinates of a point on it. Let r_0 denote some fixed radius in free space, for example, at the surface of the earth, and there let $n = 1$, $\psi = \arcsin S$. Then

$$S = \frac{r}{r_0} n \sin\psi. \tag{18.29}$$

To derive an earth flattening formula two approaches are now possible. First let

$$n_{\text{m}} = nr/r_0 \quad \text{so that} \quad n_{\text{m}} \sin\psi = S \tag{18.30}$$

and suppose that a Cartesian coordinate system is used with

$$x = r\vartheta, \quad z = r - r_0. \tag{18.31}$$

Then the last equation (18.30) is Snell's law (6.4) provided that the modified refractive index $n_{\text{m}}(z)$ is used. If the height z is small compared with the earth's radius r_0

$$n_{\text{m}} \approx n(1 + z/r_0). \tag{18.32}$$

This is in essence the method of Booker and Walkinshaw (1946) and has proved highly successful particularly for the study of guided waves.

Alternatively, for the second approach, let

$$S_{\text{m}}(z) = Sr_0/r \quad \text{so that} \quad n \sin\psi = S_{\text{m}}(z). \tag{18.33}$$

Again the last equation is the same as Snell's law. It now uses the unmodified refractive index n, but on the right-hand side S_{m} is dependent on height z, thus

$$S_{\text{m}} \approx S(1 - z/r_0). \tag{18.34}$$

With the approximations of ray theory, either (18.30) or (18.33) can be used without further approximations. For full wave solutions some further approximations are necessary. These are probably small in practical cases. The use of (18.32) for full wave solutions of guided wave problems has been fully discussed by Booker and Walkinshaw (1946).

When the medium is anisotropic, Bouger's law cannot be used. The refractive index n now depends on the angle ψ and has more than one value. Thus there is nothing directly analogous to the modified refractive index n_{m} (18.32). But S is the sine of the angle of incidence and is used for stratified anisotropic media as in earlier chapters. This suggests that (18.34) might be used as an earth flattening method for an anisotropic spherically stratified medium. The idea was proposed by Chatterjee (1953). It was used by Walker (1966) for studying guided whistler propagation in the upper ionosphere where the plasma has curved strata parallel to the lines of force of the earth's magnetic field. The suggestion is that S_{m}, (18.34), should be used instead of S wherever it appears in the coefficients (6.23) of the Booker quartic, and in the elements of $\mathbf{T}$ (7.81). The proof for the quartic in the special case $l_z = 0$ was given by Walker (1966). For the more general case the subject has not been fully studied. The method is difficult to test because there is no known exact solution with which the results of a test can be compared.

The assumption of a horizontally stratified medium implies also that the components of the vector $\boldsymbol{Y}$ are independent of x and y. The reflection of radio waves of very low frequency in the ionosphere occurs in a small range of height z in which the change of $\boldsymbol{Y}$ is very small, so that the approximation that $\boldsymbol{Y}$ is constant can safely be made. But for some problems this cannot be done. For most propagation problems in the earth–ionosphere wave guide, the direction of $\boldsymbol{Y}$ changes appreciably along the path. For the study of whistler propagation large ranges of height z are used and $\boldsymbol{Y}$ cannot be taken as constant. If the earth's magnetic field $\boldsymbol{B}$ is that of a magnetic dipole at the centre of the earth, then above any one place $\boldsymbol{B}$ and $\boldsymbol{Y}$ make a constant angle with the radius and they are proportional to $1/r^3$ where r is distance from the centre. Now $\boldsymbol{B}$ and $\boldsymbol{Y}$ must satisfy Maxwell's equations. If there are no currents in the upper atmosphere then curl $\boldsymbol{B} = 0$, curl $\boldsymbol{Y} = 0$. Thus if Y_x and Y_y vary with height z, it follows that Y_z must vary with x and y respectively. The properties of the medium are then no longer independent of x and y, so that it is no longer horizontally stratified. To avoid this difficulty it is sometimes implied that the components of $\boldsymbol{B}$ depend on z but are independent of x and y. This could only happen if there is a current density

$$\boldsymbol{j} = \mu_0^{-1} \operatorname{curl} \boldsymbol{B} = \mu_0^{-1}(-\partial B_y/\partial z, \partial B_x/\partial z, 0) \tag{18.35}$$

in the ionosphere.

Very few solutions of full wave radio reflection problems are known for the case where the spatial variation of $\boldsymbol{Y}$ is allowed for. This subject needs further study.

18.9. Admittance matrix as dependent variable

Various transformations of the basic differential equations (7.80) (7.81) have been used in earlier chapters, particularly the coupled equations in the forms (7.109), (16.22), (17.18) and the Försterling form (16.90), (16.91). These were introduced to elucidate some of the physical principles, but they are not recommended for computing.

Usually the electron concentration $N(z)$ and the collision frequency $\nu(z)$ are analytic functions of z, real when z is real. Then in the basic form (7.80) the elements of $\mathbf{T}$ are all bounded analytic functions of the height z at all points in the complex z plane except singularities of $N(z)$, $\nu(z)$ and at resonant points. The singularities of N and ν are nearly always remote from the real z axis or at infinity and give no trouble. The effect of resonant points is discussed in § 18.13. Thus, in particular, the coupling points are ordinary points of the differential equations (7.80). The matrix $\mathbf{T}$ and the solutions $\mathbf{e}$ are continuous and differentiable at and near coupling points. The path of integration can therefore go near or through any coupling point without invalidating the solution. In fact the part of the solution for the component waves that are coupled may vary more slowly near a coupling point than elsewhere because it behaves like an Airy integral function $\mathrm{Ai}(\zeta)$ near $\zeta = 0$; see fig. 8.5.

The transformations that give the coupled forms introduce singularities that are not present in the basic form (7.80). They use solutions of the Booker quartic, and any solution q has a branch point at a single coupling point associated with it; see § 16.3. Several of the elements of the coupling matrices $\mathbf{\Gamma}$ (7.108), or $\mathbf{\Lambda}$ (16.23), or $\mathbf{\Psi}$ (17.19) are infinite at the coupling points because these are singularities. Thus, for computing, the coupled equations have two disadvantages: (a) the path of integration must be carefully chosen so that it does not go near coupling points, and (b) the equations are much more complicated so that the computing time is much longer. If the amplitudes f_i (6.53) of the characteristic waves of magnetoionic theory are needed at any level z, it is best to integrate the basic equations (7.80). The integration can then be temporarily interrupted and the transformation $\mathbf{f} = \mathbf{S}^{-1}\mathbf{e}$ used, at each level where the f_i are needed.

There are, however, other forms of the equations that use transformations of a different kind and have some advantages. For finding reflection coefficients below the ionosphere the objective is to find the matrix $\boldsymbol{R}$ as given by (11.20)–(11.23) or equivalently by (18.28), from two solutions $\mathbf{e}^{(a)}$, $\mathbf{e}^{(b)}$ below the ionosphere. Instead of using two integrations to find these, as in § 18.7, we can formulate the differential equation satisfied by $\boldsymbol{R}(z)$ as given by (18.28) at all levels z, and then use only one integration of this new differential equation. The matrix $\boldsymbol{R}$ is related to the admittance matrix $\boldsymbol{A}$ (11.38)–(11.40) by the formulae (11.48)–(11.50). It is simpler to find the differential equation satisfied by $\boldsymbol{A}$ and then to find $\boldsymbol{R}$ afterwards.

As usual a prime $'$ is used to denote $k^{-1}\mathrm{d}/\mathrm{d}z$. Equation (11.40) is now differentiated once so that

$$A\begin{pmatrix}\mathscr{H}_y\\ E_y\end{pmatrix}=\begin{pmatrix}E_x\\ -\mathscr{H}_x\end{pmatrix},\quad A\begin{pmatrix}\mathscr{H}_y\\ E_y\end{pmatrix}'+A'\begin{pmatrix}\mathscr{H}_y\\ E_y\end{pmatrix}=\begin{pmatrix}E_x\\ -\mathscr{H}_x\end{pmatrix}'. \tag{18.36}$$

The basic equations (7.80), (7.81) are partitioned and rearranged thus:

$$\begin{aligned}\mathrm{i}\begin{pmatrix}E_x\\ -\mathscr{H}_x\end{pmatrix}'&=\begin{pmatrix}\mathsf{T}_{11} & 0\\ -\mathsf{T}_{31} & 0\end{pmatrix}\begin{pmatrix}E_x\\ -\mathscr{H}_x\end{pmatrix}+\begin{pmatrix}\mathsf{T}_{14} & -\mathsf{T}_{12}\\ -\mathsf{T}_{34} & \mathsf{T}_{32}\end{pmatrix}\begin{pmatrix}\mathscr{H}_y\\ E_y\end{pmatrix},\\ \mathrm{i}\begin{pmatrix}\mathscr{H}_y\\ E_y\end{pmatrix}'&=\begin{pmatrix}\mathsf{T}_{41} & 0\\ 0 & 1\end{pmatrix}\begin{pmatrix}E_x\\ -\mathscr{H}_x\end{pmatrix}+\begin{pmatrix}\mathsf{T}_{44} & -\mathsf{T}_{42}\\ 0 & 0\end{pmatrix}\begin{pmatrix}\mathscr{H}_y\\ E_y\end{pmatrix}.\end{aligned} \tag{18.37}$$

These are combined with (18.36) to give

$$\begin{aligned}\mathrm{i}A'\begin{pmatrix}\mathscr{H}_y\\ E_y\end{pmatrix}=&-A\begin{pmatrix}\mathsf{T}_{41} & 0\\ 0 & 1\end{pmatrix}A\begin{pmatrix}\mathscr{H}_y\\ E_y\end{pmatrix}+A\begin{pmatrix}-\mathsf{T}_{44} & \mathsf{T}_{42}\\ 0 & 0\end{pmatrix}\begin{pmatrix}\mathscr{H}_y\\ E_y\end{pmatrix}\\ &+\begin{pmatrix}\mathsf{T}_{11} & 0\\ -\mathsf{T}_{31} & 0\end{pmatrix}A\begin{pmatrix}\mathscr{H}_y\\ E_y\end{pmatrix}+\begin{pmatrix}\mathsf{T}_{14} & -\mathsf{T}_{12}\\ -\mathsf{T}_{34} & \mathsf{T}_{32}\end{pmatrix}\begin{pmatrix}\mathscr{H}_y\\ E_y\end{pmatrix}.\end{aligned} \tag{18.38}$$

The column with elements $\mathscr{H}_y$, E_y appears on the right of every term. There are two independent solutions $\mathbf{e}$ giving two independent columns $\mathscr{H}_y$, E_y and (18.38) must apply to both. For this it is necessary and sufficient that the equation got by omitting the column from (18.38) shall be satisfied. This gives the differential equation for A:

$$\mathrm{i}A'=-A\begin{pmatrix}\mathsf{T}_{41} & 0\\ 0 & 1\end{pmatrix}A+A\begin{pmatrix}-\mathsf{T}_{44} & \mathsf{T}_{42}\\ 0 & 0\end{pmatrix}+\begin{pmatrix}\mathsf{T}_{11} & 0\\ -\mathsf{T}_{31} & 0\end{pmatrix}A+\begin{pmatrix}\mathsf{T}_{14} & -\mathsf{T}_{12}\\ -\mathsf{T}_{34} & \mathsf{T}_{32}\end{pmatrix}. \tag{18.39}$$

As a partial check of this equation, suppose that the medium is isotropic. Then $Y=0$, the matrix M used in (7.81) is a scalar, $M=-X/U$, and only the trailing diagonal elements of T are non-zero. Then (18.39) gives

$$\mathrm{i}A'=-A\begin{pmatrix}\mathsf{T}_{41} & 0\\ 0 & 1\end{pmatrix}A+\begin{pmatrix}\mathsf{T}_{14} & 0\\ 0 & \mathsf{T}_{32}\end{pmatrix} \tag{18.40}$$

which shows that if A_{12} and A_{21} are zero at any level z, they must be zero at all levels. Then the equations for A_{11} and A_{22} are independent. The non-zero elements of T are

$$\mathsf{T}_{41}=1-X/U=n^2,\quad \mathsf{T}_{32}=C^2-X/U=q^2,\quad \mathsf{T}_{23}=1,\quad \mathsf{T}_{14}=\frac{C^2-X/U}{1-X/U}=\frac{q^2}{n^2} \tag{18.41}$$

and the equations are

$$\mathrm{i}A'_{11}=-n^2A_{11}^2+q^2/n^2,\quad \mathrm{i}A'_{22}=-A_{22}^2+q^2. \tag{18.42}$$

It can be verified that these are obtained directly from the definitions $A_{11}=E_x/\mathscr{H}_y$, $A_{22}=-\mathscr{H}_x/E_y$ (11.35), (11.36), by differentiating once and then using the forms (7.3), (7.4) of Maxwell's equations.

When equations of the type (18.39) are used for finding reflection coefficients, the calculation is divided into the three parts (1)–(3) as in § 18.1, but for any one set of conditions only one main integration (2) is needed instead of the two needed for (7.80). At a great height the starting value of $\boldsymbol{A}$, stage (1), may be found from (11.38), (11.39) by using the formula (7.118) for the ratios of the field components. Alternatively it may be found by a preliminary integration as described in § 18.6, with step size δz_{h} and starting with an arbitrarily chosen value of $\boldsymbol{A}$. The fields are to be the two W.K.B. solutions for upgoing waves, and hence $\boldsymbol{A}$ is to be a characteristic admittance matrix and therefore independent of z; see § 11.7 (2). Thus both sides of (18.39) are to be zero. For this preliminary integration, use of the full integration routines is wasteful of computer time. All that is necessary is to use the auxiliary subroutine to compute $\boldsymbol{A}'$, and repeat this with $\boldsymbol{A}$ replaced by $\boldsymbol{A} + k\boldsymbol{A}'\delta z_{\mathrm{h}}$ until the moduli of the elements of $\boldsymbol{A}'$ are all less than a specified small number. It is not permissible to multiply $\boldsymbol{A}$ by a constant during this process, as was done for (7.80), because (18.39) is a set of non-linear differential equations. A field with any two W.K.B. solutions gives $\boldsymbol{A}' = 0$. Care must be taken that the starting value found is for two upgoing waves, by suitable choice of $\arg \delta z_{\mathrm{h}}$. The value $\arg \delta z_{\mathrm{h}} \approx \frac{3}{4}\pi$ is usually satisfactory.

During the main integration, stage (2), $\boldsymbol{A}$ may represent fields in which both upgoing and downgoing waves are present. Its elements contain ratios of the fields and therefore have factors of order $\exp\{-ik\int(q_i - q_j)\mathrm{d}z\}$ where $\mathrm{Re}(q_i)$ and $\mathrm{Re}(q_j)$ may have opposite signs. This shows that the elements of $\boldsymbol{A}$ can vary with z at up to twice the rate of variation of the fields $\mathbf{e}$. This in turn means that the maximum permissible step size δz for the main integration is about half that for integration of (7.80). A step size of $\lambda/100$ is usually suitable, where λ is the vacuum wavelength. Thus although only one main integration is needed, it takes just as long as the two integrations needed for (7.80).

When $\boldsymbol{A}$ is found at the bottom of the ionosphere, it is to be used, stage (3), to find the reflection coefficient matrix $\boldsymbol{R}$. This is done very simply from the matrix operation (11.49).

Equations of the type (18.39) have been successfully used for calculating ionospheric reflection coefficients for radio waves of very low frequency, VLF, for example by Barron and Budden (1959), Barron (1961), who used a different form of $\boldsymbol{A}$ from that used here. One great advantage of this type of method is that it avoids troubles from numerical swamping; § 18.11. It cannot, however, be used for finding transmission coefficients. It can also have another disadvantage, especially at higher frequencies. This is best illustrated by considering the following simple special case.

Consider an isotropic ionosphere, so that $\boldsymbol{A}$ is diagonal and (18.39) separates into the two equations (18.42). We study the second of these, which applies for linearly polarised waves where E has only one non-zero component E_y. Suppose that there is a part of the ionosphere that is homogeneous so that q^2 is independent of z. Then the differential equation has the solution

$$A_{22} = -\mathrm{i}q\tan(kqz + \gamma) \tag{18.43}$$

where γ is a complex constant of integration. In general A_{22} represents two waves, one obliquely upgoing with $E_y = a\exp(-\mathrm{i}kqz)$ and the other downgoing with $E_y = b\exp(\mathrm{i}kqz)$. Then it can be shown that

$$b/a = \mathrm{e}^{2\mathrm{i}\gamma}. \tag{18.44}$$

If $|b/a| = 1$, γ is real and there are real heights z where (18.43) is infinite. Then the use of (18.42) must fail. This condition implies that there is a perfect standing wave, and failure occurs when the E_ys for the upgoing and downgoing component waves are in antiphase. Even when γ has a small imaginary part, there are real values of z where (18.43) and its derivative are very large, so that failure still occurs. This type of failure can occur in a variable medium and more generally in the anisotropic case (18.39).

For the application to VLF reflection, the values of Z used are large, of the order 30 for frequencies near 16 kHz, so that the amplitudes of the downgoing waves never come close to those of the upgoing waves, and there is no failure. But for higher frequencies where Z is smaller, the ionosphere can be strongly reflecting and (18.39) fails at points near the real z axis. It is then necessary to use either the basic equations (7.80), or other non-linear forms as described in the following section, or to use a path of integration displaced away from the real z axis; see end of § 18.13.

18.10. Other forms, and extensions of the differential equations

One way of avoiding the failure of (18.39) when standing waves are present is to use another independent variable instead of $\boldsymbol{A}$. One such variable is the reflection coefficient matrix $\boldsymbol{R}(z)$, as defined by (11.46)–(11.49). The differential equation satisfied by $\boldsymbol{R}(z)$ will now be derived.

In § 18.4 the transformation (18.5), (18.6) gave the amplitudes $\mathfrak{f}_1$ to $\mathfrak{f}_4$ of the four vacuum modes. These are the same as a, b, c, d of (11.46) (11.47), and (11.46) shows that

$$\begin{pmatrix} \mathfrak{f}_3 \\ \mathfrak{f}_4 \end{pmatrix} = \boldsymbol{R}\begin{pmatrix} \mathfrak{f}_1 \\ \mathfrak{f}_2 \end{pmatrix}. \tag{18.45}$$

Now the column $\mathfrak{f}$ satisfies (18.7), and (18.6) gives

$$\mathbf{S}_\mathrm{v}^{-1} = \tfrac{1}{2}\begin{pmatrix} 1/C & 0 & 0 & 1 \\ 0 & -1 & -1/C & 0 \\ -1/C & 0 & 0 & 1 \\ 0 & -1 & 1/C & 0 \end{pmatrix} \tag{18.46}$$

so that the 4×4 matrix on the right of (18.7) is

$$\mathbf{W}_v = \begin{pmatrix} T_{11}+T_{14}/C+CT_{41}+T_{44} & -T_{12}/C-T_{42} & -T_{11}+T_{14}/C-CT_{41}+T_{44} & -T_{12}/C-T_{42} \\ -T_{31}-T_{34}/C & C+T_{32}/C & T_{31}-T_{34}/C & -C+T_{32}/C \\ -T_{11}-T_{14}+CT_{41}+T_{44} & T_{12}/C-T_{42} & T_{11}-T_{14}/C-CT_{41}+T_{44} & T_{12}/C-T_{42} \\ T_{31}+T_{34}/C & C-T_{32}/C & -T_{31}+T_{34}/C & -C-T_{32}/C \end{pmatrix}. \tag{18.47}$$

It may be written in terms of its component 2×2 matrices thus

$$\mathbf{W}_v = \begin{pmatrix} w_{11} & w_{12} \\ w_{21} & w_{22} \end{pmatrix}. \tag{18.48}$$

Equation (18.7) is now partitioned into two equations for

$$\begin{pmatrix} f_1 \\ f_2 \end{pmatrix}' = \boldsymbol{u}' \quad \text{and} \quad \begin{pmatrix} f_3 \\ f_4 \end{pmatrix}' = \boldsymbol{d}'; \quad \text{compare (18.17).}$$

Then (18.45) is differentiated once and the values of $\boldsymbol{u}'$, $\boldsymbol{d}'$ are substituted. Finally $\boldsymbol{d}$ is set equal to $\boldsymbol{Ru}$ from (18.45). The result is an equation containing $\boldsymbol{R}$ and $\boldsymbol{R}'$ with $\boldsymbol{u}$ on the right of every term. It must be valid for either of the two independent solutions so the $\boldsymbol{u}$ may be omitted. This step is similar to the step from (18.38) to (18.39). Thus the equation satisfied by $\boldsymbol{R}$ is

$$2i\boldsymbol{R}' = w_{21} + w_{22}\boldsymbol{R} - \boldsymbol{R}w_{11} - \boldsymbol{R}w_{12}\boldsymbol{R}. \tag{18.49}$$

This equation was used by Budden (1955a, b) for studying ionospheric reflection at very low frequencies. Its use is very similar to the use of (18.39) with $\boldsymbol{A}$, as described in § 18.9. The preliminary integration used in stage (1), to find the starting value of $\boldsymbol{R}$ at the top, continues until $\boldsymbol{R} = 0$. Like $\boldsymbol{A}$, $\boldsymbol{R}$ contains ratios of the fields and may vary with z at up to twice the rate of variation of the fields $\mathbf{e}$, so that for the main integration, stage (2), a step size of about $\lambda/100$ is needed, as for $\boldsymbol{A}$. At the bottom of the ionosphere $\boldsymbol{R}$ is the required solution so there are no further operations for stage (3).

The original definition of $\boldsymbol{R}$ was formulated in ch. 11 leading to (11.46) that is (18.45), for the free space below the ionosphere. In (18.49) it is being used within the ionosphere, and is still defined by (18.45). It is the reflection coefficient for the vacuum modes (§ 18.4), but not for the upgoing and downgoing characteristic waves of magnetoionic theory. Suppose that the downward integration of (18.49) is stopped at some level before the bottom of the ionosphere is reached. This is equivalent to terminating the ionosphere by a sharp boundary plane at that level with free space below. The value of $\boldsymbol{R}$ that has been reached is the reflection coefficient at that boundary. This gives a useful physical interpretation of $\boldsymbol{R}$ within the ionosphere. It follows that no element of $\boldsymbol{R}$ can have modulus greater than unity at any real height, so that there is no failure of (18.49) even if the solution contains a perfect standing wave. Compare (18.43) and § 18.9 where it was shown that failure can occur if $\boldsymbol{A}$ is used as the dependent variable.

Suppose that the equations (18.49) are integrated for a homogeneous ionosphere, as for stage (1), and the integration is stopped when $\boldsymbol{R}'$ is zero. Then $\boldsymbol{R}$ is the reflection coefficient for a homogeneous medium with a sharp lower boundary. This method of finding it is equivalent to finding a solution of the matrix equation formed by setting the right hand side of (18.49) equal to zero, which can be done by a simple iteration method. It is not necessary to use accurate integration routines. This is an alternative to the method of § 11.10 that used solutions of the Booker quartic.

If the ionosphere is isotropic the only non-zero elements of $\mathbf{T}$ are the trailing diagonal elements given by (18.41). Then it can be shown that all matrices $\boldsymbol{w}_{ij}$ are diagonal, and (18.49) gives two separate equations:

$$2\mathrm{i}R'_{11} = Cn^2(1 - R_{11})^2 - \frac{q^2}{n^2C}(1 + R_{11})^2, \tag{18.50}$$

$$2\mathrm{i}R'_{22} = C(1 - R_{22})^2 - \frac{q^2}{C}(1 + R_{22})^2. \tag{18.51}$$

These two equations were given by Jacobsson (1965) in a study of the reflection of light from isotropic stratified films. See also Schelkunoff (1951).

One integration of (18.49) or (18.50) or (18.51) gives the reflection coefficient at one frequency only. Suppose that the reflection comes predominantly from one level in the ionosphere and it is required to find the equivalent path P' (10.40) at a given receiver. Then $\boldsymbol{R}$ alone does not contain the necessary information and we need to know also how it depends on frequency. To illustrate this consider a simple case of vertical incidence on an isotropic ionosphere with $\boldsymbol{E}$ having only one component E_x. Then the reflection coefficient is R_{11} and we now omit the subscripts. With $C = 1$, $q^2 = n^2$ (18.50) then gives

$$2\mathrm{i}R' = n^2(1 - R)^2 - (1 + R)^2. \tag{18.52}$$

The equivalent path P' is $2h'(f)$ where h' is the equivalent height of reflection. Below the ionosphere $-\arg R$ is the phase lag that results from the reflection. If the reflection process can be described in the language of ray theory, the phase height, § 12.2, is given by

$$h(f) = -\frac{c}{4\pi f}\arg R = \tfrac{1}{2}P(f) \tag{18.53}$$

whence from (10.40)

$$h'(f) = \partial(fh)/\partial f = -\frac{c}{4\pi}\frac{\partial}{\partial f}(\arg R) = -\frac{c}{4\pi}\mathrm{Im}\left(\frac{1}{R}\frac{\partial R}{\partial f}\right). \tag{18.54}$$

Thus if $\partial R/\partial f$ can be found, $h'(f)$ can be calculated. It is assumed that $\partial^2(fh)/\partial f^2$ and higher derivatives are small. If they are not small a radio pulse would become distorted or elongated, or there might be two reflected waves present from different levels. Then the concept of a single equivalent height ceases to be useful. Let $\partial R/\partial f = J$. Now (18.52) is to be differentiated partially with respect to f. The prime on the left means $k^{-1}\mathrm{d}/\mathrm{d}z$ and k is proportional to f. This must be allowed for, so that the result is

$$2\mathrm{i}J' = -2J\{n^2(1-R)+1+R\} + (1-R)^2\partial(n^2)/\partial f + \{n^2(1-R)^2 - (1+R)^2\}/f. \tag{18.55}$$

The whole integration process can now be applied to (18.52), (18.55) together. The method was used by Budden and Cooper (1962) to study $h'(f)$ for a Chapman layer and for a parabolic layer, including conditions where the reflection from the discontinuity of $\mathrm{d}N/\mathrm{d}z$ at the bottom could affect the results. They gave a somewhat simplified form of (18.52) (18.55) that was useful for reducing the computing time.

If the reflected wave contains two reflections from different levels, the problem is harder. In principle it could be solved by using (18.52), (18.55) together with a third differential equation for the height dependence of $\partial^2 R/\partial f^2$. This, however, is very complicated. The problem was studied by Smith, M. S. (1973b) who found that it is equivalent and easier to compute R at three different frequencies close together. Then the separate values of R and $h'(f)$ for the two reflections can be found. If the ionosphere is anisotropic, the matrix equation (18.49) must be used instead of (18.52) but the problem of separating the reflection into two components is now easier. For a single reflection of one magnetoionic component, the polarisation has its characteristic value for that component and is independent of the polarisation of the incident wave. Then $\det(\boldsymbol{R}) = 0$ and $\boldsymbol{R}$ is singular; see problem 11.1. Smith calls this a 'simple reflection'. If the reflected wave has two components with different polarisations, the $\boldsymbol{R}$ for the composite wave is not singular, but Smith showed that by calculating $\boldsymbol{R}$ at only two adjacent frequencies, it is possible to separate it into its two components and to find their polarisations and equivalent heights of reflection.

When full wave computer integrations of the differential equations are used at medium or high frequencies, the step size $\lambda/100 - \lambda/50$ is small and the number of steps needed can be extremely large with a consequent long computing time. But for large ranges of the height z the characteristic waves are not strongly coupled and the fields are given with very good accuracy by the W.K.B. solutions (7.112). The full wave equations are only needed for those parts of the path of integration where the coupling processes are appreciable, that is where the path runs near reflection or coupling points. Thus the path is divided into 'W.K.B. ranges' and 'full wave ranges'. When the equations being integrated are the basic equations (7.80), the field variables $\mathbf{e}$ are the dependent variables. At the bottom of a full wave range the transformation (6.53) is used to give $\mathbf{f}$ and this gives the amplitudes $\mathbf{K}_j$ in (7.112). Then $\mathbf{f}$ is used in the W.K.B. range that follows. The height dependence of its elements is given by the W.K.B. solutions (7.112), so that the four integrals $\int q_j \mathrm{d}z$ must be evaluated and this means finding and sorting the four roots q_j of the Booker quartic at a sequence of levels z, but this process is much faster than the full stepwise integration. At the bottom of the W.K.B. range the transformation (6.53) is used to give $\mathbf{e}$ and the full wave integration is resumed in the full wave range that follows.

When the equations being integrated are (18.39) or (18.49), the dependent variables, $\boldsymbol{A}$ or $\boldsymbol{R}$ respectively, depend only on ratios of the field components. But there is still a transformation that allows the W.K.B. solutions to be used in a W.K.B. range. Instead of using $\boldsymbol{R}$ defined in terms of vacuum modes, we define a matrix $\mathscr{R}$ that is the reflection coefficient for the characteristic modes. Thus suppose that the elements of $\mathbf{f}$ in (6.53) refer, in order, to the upgoing ordinary, upgoing extraordinary, downgoing ordinary and downgoing extraordinary waves. Then

$$\begin{pmatrix} \mathsf{f}_3 \\ \mathsf{f}_4 \end{pmatrix} = \mathscr{R} \begin{pmatrix} \mathsf{f}_1 \\ \mathsf{f}_2 \end{pmatrix}. \tag{18.56}$$

Note that (a) the order of the elements f_1 to f_4 is not the same as in (7.145)–(7.148) nor in (16.15), (16.16), (b) these $\mathbf{f}$s are given by (6.53) with the transforming matrix $\mathbf{S}$, and they are not the same as the vacuum mode $\mathfrak{f}$s of (18.45) that were given by (18.6) with transforming matrix $\mathbf{S}_v$. Some of the properties of $\mathscr{R}$ and the transformations that express it in terms of $\boldsymbol{A}$ and $\boldsymbol{R}$ were discussed by Smith (1974b).

Now let $\mathbf{f}_1$ to $\mathbf{f}_4$ be the W.K.B. solutions (7.111). Apply to (18.56) the operator $k^{-1}\mathrm{d}/\mathrm{d}z$, denoted by a prime $'$. Then it follows that

$$\mathrm{i}\mathscr{R}' = \begin{pmatrix} q_3 & 0 \\ 0 & q_4 \end{pmatrix} \mathscr{R} - \mathscr{R} \begin{pmatrix} q_1 & 0 \\ 0 & q_2 \end{pmatrix} \tag{18.57}$$

where a column $\begin{pmatrix} \mathbf{f}_1 \\ \mathbf{f}_2 \end{pmatrix}$ on the right has been omitted because (18.56) applies for two independent columns. Let

$$\mathsf{E}_j = \exp\left(-\mathrm{i}k \int_{z_1}^{z} q_j \mathrm{d}z \right). \tag{18.58}$$

These are just the factors used in the W.K.B. solutions. Then (18.56) gives

$$\mathscr{R}(z) = \begin{pmatrix} \mathsf{E}_3 & 0 \\ 0 & \mathsf{E}_4 \end{pmatrix} \mathscr{R}(z_1) \begin{pmatrix} \mathsf{E}_1^{-1} & 0 \\ 0 & \mathsf{E}_2^{-1} \end{pmatrix} \tag{18.59}$$

which can now be used to give $\mathscr{R}(z)$ at the bottom of a W.K.B. range.

The differential equation satisfied by $\mathscr{R}$ in the general case is complicated and rarely needed, but there are special cases where it is useful. For example, for vertical incidence when coupling between the ordinary and extraordinary waves is negligible, $\mathscr{R}$ is diagonal and its two elements then refer to the ordinary and extraordinary waves. This case, for the ordinary wave, was studied by Fejer and Vice (1959b) who used a variable R closely related to one diagonal element of $\mathscr{R}$. They were thus able to study absorption near the reflection level for waves of frequencies 1.83 and 2.63 MHz.

18.11. Numerical swamping

At a great height in the ionosphere where $X > 1$, in nearly all cases of practical interest, one of the two upgoing characteristic waves is a propagated wave and the other is an inhomogeneous wave whose amplitude decreases as the height increases. This inhomogeneous solution therefore increases in amplitude as z decreases, that is as the main integration proceeds downwards, whereas the propagated wave solution stays roughly constant. Even if the medium has losses, the inhomogeneous solution still increases far faster than the propagated solution. Moreover in an integration that starts at the top with a propagated solution, some of the inhomogeneous solution is generated by coupling in the medium as the integration proceeds, and once generated it grows rapidly so that it can get very large and the propagated solution is then negligible in comparison. This is numerical swamping. When it happens, both stepwise integrations of (7.80) lead to the same solution $\mathbf{e}$ at the bottom of the ionosphere and then it is not possible to find $\boldsymbol{R}$ because $\boldsymbol{U}$ in (18.28) is singular. This swamping can occur when the thickness of the ionosphere is many wavelengths. It did not occur in early calculations for 16 kHz (Budden, 1955a, b; Barron and Budden, 1959; Barron, 1961) when only reflection coefficients were calculated and the range of the integration was less than one vacuum wavelength. It can, however, occur (a) for higher frequencies, and (b) for problems where a large thickness of the ionosphere is studied, for example Pitteway's (1965) calculations of transmission coefficients.

There are two ways in which numerical swamping may be avoided. The first is to use one of the non-linear forms of the differential equation, (18.39) with dependent variable $\boldsymbol{A}$, or (18.49) with dependent variable $\boldsymbol{R}$, as described in the two preceding sections. These variables depend only on the ratios of the components of the electromagnetic field and are not affected by swamping. The same method can be used with a method of discrete strata; see for example Nygrén (1981).

The second way applies when the differential equations used are linear and homogeneous, for example (7.80) or (18.7). It uses the following property. When $\boldsymbol{R}$ is found at the bottom of the ionosphere from (18.28), two solutions $\mathbf{e}^{(a)}$, $\mathbf{e}^{(b)}$ are used. It was shown in § 18.7 that the value of $\boldsymbol{R}$ is unaffected if these two are replaced by any independent linear combinations of them, and this replacement can be made at any stage during the integration because the equations are linear and homogeneous. A starting solution that represents the inhomogeneous wave at the top is used first, and the resulting solution will be denoted by $\mathbf{e}^{(a)}$. During the downward stepwise integration a record is kept of $\mathbf{e}^{(a)}$ for a number of heights $z_A, z_B, \ldots$. If $\mathbf{e}^{(a)}$ gets large so that computer overflow is threatened, it can be multiplied by a fraction without affecting the results, because it is only the ratios of the elements of $\mathbf{e}$ that are finally used. When this integration is complete the second integration is started, with a solution that represents the propagated wave at the top, and this solution will be denoted by $\mathbf{e}^{(b)}$. If it gets too large this is because it has acquired a contribution from the inhomogeneous wave, that grows as z decreases. Now it is permissible to add to $\mathbf{e}^{(b)}$ any multiple of $\mathbf{e}^{(a)}$. Thus $\mathbf{e}^{(b)}$ is replaced by

$$\mathbf{e}^{(c)} = \mathbf{e}^{(b)} + \gamma \mathbf{e}^{(a)}. \tag{18.60}$$

The constant γ is to be chosen so that the elements of $\mathbf{e}^{(c)}$ are not large. The simplest choice, suggested by Walsh (1967), is to use the γ that makes one element of $\mathbf{e}^{(c)}$ zero. In this way most of the unwanted contribution from $\mathbf{e}^{(a)}$ is removed from $\mathbf{e}^{(b)}$. This process can be applied at each of the levels $z_A, z_B, \ldots$ where $\mathbf{e}^{(a)}$ has been recorded.

If the integration routine used is, for example, an Adams routine, § 18.2, that requires storage of $\mathbf{e}$ and its derivative for the four preceding steps, then it must be remembered that when $\mathbf{e}$ is changed in accordance with (18.60), or when $\mathbf{e}^{(a)}$ is multiplied by a fraction to avoid impending overflow, the stored numbers for the earlier steps are not now correct for the new $\mathbf{e}$. When the integration is resumed, therefore, it must be treated as though it is newly started.

An alternative choice for γ, used by Pitteway (1965), is to use the γ that makes

$$\mathbf{e}^{(c)\mathrm{T}} \mathbf{e}^{(a)*} = 0. \tag{18.61}$$

Then $\mathbf{e}^{(c)}$ and $\mathbf{e}^{(a)}$ are said to be 'Hermitian orthogonal'. For any column $\mathbf{e}$, the product $\mathbf{e}^{\mathrm{T}}\mathbf{e}^{*}$ is the sum of the squared moduli of the elements and cannot be zero unless all the elements are identically zero. Thus (18.61) shows that $\mathbf{e}^{(c)}$ can never be a multiple of $\mathbf{e}^{(a)}$, but must include an appreciable multiple of $\mathbf{e}^{(b)}$.

For a discussion of the avoidance of numerical swamping see Walsh (1967) who gives some instructive numerical examples.

For the solution $\mathbf{e}^{(a)}$ that consists of only the inhomogeneous wave at great heights, the downward integration gives a solution $\mathbf{e}^{(a0)}$, say, at the bottom of the ionosphere. Even if there were errors in the starting value used, this would not matter because as z decreases the true inhomogeneous wave component grows and the remaining part is then negligible in comparison. Thus the solution $\mathbf{e}^{(a0)}$ is known with precision. The complete solution $\mathbf{e}^{(a)} \to \mathbf{e}^{(a0)}$ is called by Pitteway (1965) the 'non-penetrating mode'. The other solution $\mathbf{e}^{(b)}$ is equal to $\mathbf{e}^{(b0)}$, say, at the bottom of the ionosphere, and at great heights it is the propagated wave. But we can add on to it a multiple of $\mathbf{e}^{(a)}$ for which the amplitude of the inhomogeneous wave at a great height is negligible compared with the propagated wave in $\mathbf{e}^{(b)}$, but for which the amplitudes of $\mathbf{e}^{(a0)}$ and $\mathbf{e}^{(b0)}$ are comparable. This new solution would also fulfil all the requirements for the original solution $\mathbf{e}^{(b)} \to \mathbf{e}^{(b0)}$. Thus the solution $\mathbf{e}^{(b)}$ is not known with precision. For a unique definition a further condition is necessary. Pitteway (1965) chooses $\mathbf{e}^{(b)}$ so that the ratio of the power flux in the propagated wave at a great height to that in the upgoing part of the wave $\mathbf{e}^{(b0)}$ is a maximum. This defines a unique $\mathbf{e}^{(b)}$ which he calls the 'penetrating mode'. At a great height the penetrating mode is the whistler wave, § 13.8. These definitions are important in the study of the propagation of whistlers through the ionosphere, both upwards and downwards. For details see Pitteway (1965), Pitteway and Jespersen (1966), Altman and Suchy (1980).

18.12. Reciprocity

Reciprocity in radio propagation problems has been discussed in §§ 14.13, 14.14, for conditions where the approximations of ray theory can be used. It must now be studied for the more general case. The reciprocity theorem was originally stated for electric circuits that contain only linear circuit elements, that is resistors, capacitors and inductors including mutual inductors. It used the voltages and currents at two pairs of terminals, at points A and B, and this form of it was given in § 14.13. Now it can be proved that it still applies if each terminal pair is connected to an aerial, so that the electromagnetic fields in the space containing the aerials are part of the system. The theorem is discussed here only for non-magnetic media. For it to be true it is necessary that the electromagnetic properties of the medium are linear and that the complex dielectric constant tensor $\boldsymbol{\varepsilon}$ is symmetric. Thus it does not apply for the plasmas of the ionosphere and magnetosphere, except when the earth's magnetic field can be neglected. The proof of the theorem is effected in two stages.

For the first stage let $\boldsymbol{E}^{(1)}, \mathscr{H}^{(1)}$ be the fields of some system of waves of fixed angular frequency, and let $\boldsymbol{E}^{(2)}$, $\mathscr{H}^{(2)}$ be those of a different system at the same frequency. Then at any point where Maxwell's equations and the constitutive relations are obeyed

$$\operatorname{div}(\boldsymbol{E}^{(1)} \wedge \mathscr{H}^{(2)} - \boldsymbol{E}^{(2)} \wedge \mathscr{H}^{(1)}) = 0. \tag{18.62}$$

To show this we use the vector property $\operatorname{div}(\boldsymbol{a} \wedge \boldsymbol{b}) = \boldsymbol{b}$ curl $\boldsymbol{a} - \boldsymbol{a}$ curl $\boldsymbol{b}$, and Maxwell's equations in the form (2.44). Then since $\partial/\partial t \equiv \mathrm{i}\omega$, the left-hand side of (18.62) is

$$\mathrm{i}Z_0(\boldsymbol{E}^{(2)}\cdot\boldsymbol{D}^{(1)} - \boldsymbol{E}^{(1)}\cdot\boldsymbol{D}^{(2)}) = \mathrm{i}c^{-1}\{\boldsymbol{E}^{(2)}\cdot(\boldsymbol{\varepsilon}\boldsymbol{E}^{(1)}) - \boldsymbol{E}^{(1)}\cdot(\boldsymbol{\varepsilon}\boldsymbol{E}^{(2)}). \tag{18.63}$$

Now write $\boldsymbol{E}$ and $\boldsymbol{D}$ in the suffix notation of (2.54), with the summation convention. Then (18.63) is

$$\mathrm{i}c^{-1}E_i^{(2)}E_j^{(1)}(\varepsilon_{ij} - \varepsilon_{ji}). \tag{18.64}$$

This is zero if, and in general only if $\boldsymbol{\varepsilon}$ is symmetric.

Next (18.62) is integrated through some volume V bounded by a surface Γ. This must not contain any sources because at all points within it the form (2.44) of Maxwell's equations must hold. If $\mathrm{d}\boldsymbol{\Gamma}$ is an element of the surface in the direction of the normal, the divergence theorem of vector analysis states that $\iiint \operatorname{div} \boldsymbol{a}\, \mathrm{d}V =$

$\iint_\Gamma \boldsymbol{a}\cdot\mathrm{d}\boldsymbol{\Gamma}$. Application of this to (18.62) gives

$$\iint_\Gamma (\boldsymbol{E}^{(1)}\wedge\boldsymbol{\mathscr{H}}^{(2)}-\boldsymbol{E}^{(2)}\wedge\boldsymbol{\mathscr{H}}^{(1)})\cdot\mathrm{d}\boldsymbol{\Gamma}=0. \tag{18.65}$$

Now Γ is chosen to consist of three parts. Two of them enclose the two terminal pairs of the aerials at A and B. The remaining one is a sphere of indefinitely large radius r in a vacuum, extremely distant from the sources. Here the fields must be outward travelling radiation fields each with amplitude proportional to $1/r$. Here also $\boldsymbol{E}^{(1)}$ is perpendicular to $\boldsymbol{\mathscr{H}}^{(1)}$ and similarly $\boldsymbol{E}^{(2)}$ to $\boldsymbol{\mathscr{H}}^{(2)}$ whence it can be shown that the integral (18.65) tends to zero as $r\to\infty$ (Clemmow, 1973, § 5, 4.6). Thus the sum of the remaining two contributions is zero. Finally, the fields $\boldsymbol{E}^{(1)}$, $\boldsymbol{\mathscr{H}}^{(1)}$ are chosen to be those that occur when a voltage is applied at the terminal pair at A, and $\boldsymbol{E}^{(2)}$, $\boldsymbol{\mathscr{H}}^{(2)}$ similarly when a voltage is applied at B. The result (18.65) is then sometimes called the 'Lorentz reciprocity theorem'.

For the second stage of the proof the two field systems must be linked to the voltages and currents at the terminal pairs at A and B. This needs care and is too long to give here. Some examples within the restrictions of ray theory were given in §§ 14.13, 14.14. The problem was discussed by Dällenbach (1942). A useful treatment of it is given by Collin and Zucker (1969, § 4.2). In this way the proof of the reciprocity theorem is completed when $\boldsymbol{\varepsilon}$ is everywhere symmetric.

If $\boldsymbol{\varepsilon}$ is not symmetric the reciprocity theorem is not in general true. An alternative theorem for magnetoplasmas can, however, be proved as follows. For the fields $\boldsymbol{E}^{(2)}$, $\boldsymbol{\mathscr{H}}^{(2)}$ imagine that at every point of the medium the direction of the superimposed magnetic field is reversed but no other change is made. Then for $\boldsymbol{E}^{(2)}$, $\boldsymbol{\mathscr{H}}^{(2)}$ the new dielectric constant $\bar{\boldsymbol{\varepsilon}}$ is the transpose of $\boldsymbol{\varepsilon}$; see end of § 3.10. The second term of (18.64) is now $\bar{\varepsilon}_{ji}-\varepsilon_{ij}$ and the expression is zero, so that (18.62), (18.65) are now true. We say that reciprocity holds provided that, for one of the two directions of propagation, the earth's magnetic field is imagined to be reversed.

In the full wave theory of this chapter it has been assumed that the incident, reflected and transmitted waves, where they are in free space, are plane waves, with their wave normals in the x–z plane. It is then more convenient and simpler to express the reciprocity relations in terms of the fields, or of the matrices $\boldsymbol{A}$ and $\boldsymbol{R}$. For both wave systems in (18.62) the fields are independent of y and the x dependence is only through factors $\exp(\mp ikSx)$ that cancel for the two waves. Thus the operator div is simply $\mathrm{d}/\mathrm{d}z$ of the z component, and (18.62) shows that

$$E_x^{(1)}\mathscr{H}_y^{(2)}-E_y^{(1)}\mathscr{H}_x^{(2)}-E_x^{(2)}\mathscr{H}_y^{(1)}+E_y^{(2)}\mathscr{H}_x^{(1)} \tag{18.66}$$

is independent of z. Note that this is very similar to the bilinear concomitant (7.97). Here, however, both field systems satisfy Maxwell's equations, unlike the adjoint fields in (7.97). For the ionospheric reflection problem we may imagine that at a great enough height there is free space. Both waves there are upgoing and it can be shown that (18.66) is zero. Hence it is zero also below the ionosphere. This condition can be written in terms of the admittance matrices $\boldsymbol{A}^{(1)}$, $\boldsymbol{A}^{(2)}$ for the two field systems, from (11.40) thus

$$(\mathscr{H}_y^{(2)}, E_y^{(2)})\left\{\begin{pmatrix}1 & 0\\ 0 & -1\end{pmatrix}\boldsymbol{A}^{(1)} - \boldsymbol{A}^{(2)\mathrm{T}}\begin{pmatrix}1 & 0\\ 0 & -1\end{pmatrix}\right\}\begin{pmatrix}\mathscr{H}_y^{(1)}\\ E_y^{(1)}\end{pmatrix} = 0. \tag{18.67}$$

This must be true for any arbitrary fields and so the expression within curly brackets is zero. Thence from (11.49) it can finally be shown that

$$\boldsymbol{R}^{(1)\mathrm{T}} = \begin{pmatrix}1 & 0\\ 0 & -1\end{pmatrix}\boldsymbol{R}^{(2)}\begin{pmatrix}1 & 0\\ 0 & -1\end{pmatrix} \tag{18.68}$$

which gives

$$R_{11}^{(1)} = R_{11}^{(2)},\quad R_{22}^{(1)} = R_{22}^{(2)},\quad R_{12}^{(1)} = -R_{21}^{(2)},\quad R_{21}^{(1)} = -R_{12}^{(2)}. \tag{18.69}$$

This result is mainly of theoretical interest. It holds only if the earth's magnetic field is imagined to be reversed for finding $\boldsymbol{R}^{(2)}$.

Many relations between the $\boldsymbol{A}$s or the $\boldsymbol{R}$s for the two fields can be found when the earth's magnetic field is not imagined to be reversed, and a few examples will now be given.

For the fields (1) let the plane of incidence, that is the x–z plane, make an angle ψ with the magnetic meridian plane measured east of north, and for (2) let the angle be $\pi - \psi$. Then for the two systems the direction cosine l_x of $\boldsymbol{Y}$ has opposite signs and there is no other difference. The effect of this on the elements of $\mathbf{T}$ (7.81) is

$$\mathrm{T}_{11}^{(1)} = -\mathrm{T}_{44}^{(2)},\quad \mathrm{T}_{12}^{(1)} = \mathrm{T}_{34}^{(2)},\quad \mathrm{T}_{31}^{(1)} = -\mathrm{T}_{42}^{(2)} \tag{18.70}$$

and a similar set with (1) and (2) interchanged. If these changes are made in the differential equation (18.39) for $\boldsymbol{A}$, and if the equation is then transposed, it shows that $\boldsymbol{A}^{(1)}$ and $\boldsymbol{A}^{(2)\mathrm{T}}$ obey the same differential equation, which is to be integrated proceeding downwards. At a very great height where there is free space, the starting value of $\boldsymbol{A}$ is diagonal and the same for both systems. Hence $\boldsymbol{A}^{(1)}$ and $\boldsymbol{A}^{(2)\mathrm{T}}$ are the same at the bottom of the ionosphere. Then from (11.49) it follows that

$$\boldsymbol{R}^{(1)} = \boldsymbol{R}^{(2)\mathrm{T}}. \tag{18.71}$$

The planes of incidence for the two fields here are not in general parallel so that this is not a reciprocity property, but results of this kind are customarily discussed under the same heading. Suppose, however, that $\psi = 0$. Then the two directions are from north to south and south to north. We can imagine that the aerials are electric dipoles at distant points A, B in the magnetic meridian plane. First let them both be vertical dipoles. Then the received signals for A → B and B → A are proportional to $R_{11}^{(1)}$, $R_{11}^{(2)}$ respectively and (18.71) shows that they are equal. Thus there is reciprocity for these aerials. Similarly there is reciprocity if both aerials are horizontal dipoles because $R_{22}^{(1)} = R_{22}^{(2)}$. Suppose now that there is a vertical dipole at A and a horizontal dipole at B. When A transmits, the received current at B is proportional to $R_{21}^{(1)}$ but when B transmits, the received current at A is proportional to $-R_{12}^{(2)}$. The minus sign is needed because of the sign convention used in the definition (11.16), (11.17) of $\boldsymbol{R}$.

The correct assignment of the signs for this problem is difficult. The following note may help. When a vertical dipole aerial transmits, then at a point many wavelengths away in free space, in the positive x direction, the field components E_z and $\mathscr{H}_y$ are in antiphase. When this same dipole receives a plane wave coming the opposite way from the same direction, the E_z and $\mathscr{H}_y$ in this wave are in phase. The $\mathscr{H}_y$s for the transmitted and received waves have opposite signs. The vertical dipole responds to E_z but it is $\mathscr{H}_y$ that is used in the definitions (11.16), (11.17) of the reflection coefficients. A horizontal dipole, parallel to the y axis, responds to E_y and there is no corresponding change of sign for transmission and reception.

Hence there is reciprocity for the amplitude in the two directions, but the phase shifts differ by 180°. The same result applies for a horizontal dipole aerial at A and a vertical dipole at B. This was called 'antireciprocity' in § 14.13, and some examples of it, with rays, were given in § 14.15.

These results have been proved for the reflection of a single plane wave but it can be shown (Budden, 1954) that they must also be true for a spherical wave from a small source, and when allowance is made for the earth's curvature. They are therefore applicable to radio transmission over any distance in the magnetic north–south direction.

Another relation for two wave systems was derived by Pitteway and Jespersen (1966). They used the property that (18.66) is independent of z and they applied it to two systems with azimuths ψ, $\pi - \psi$, as in the foregoing example. For system (1) the wave was a penetrating mode as defined at the end of § 18.11, giving an upgoing whistler wave at a great height. For system (2) the incident wave was a downgoing whistler wave at the great height. They showed that the transmission coefficients for these two systems are the same, and the polarisations of the incident wave in system (1) and of the downgoing wave below the ionosphere in system (2) are simply related.

For a stratified system a large number of relations between the reflection and transmission coefficients of two wave systems is possible, with incidence either from below or from above. This has been studied by Suchy and Altman (1975b), Altman and Suchy (1979a, b, 1980), Altman, Schatzberg and Suchy (1981), and in a series of papers by Heading (1971, 1973, 1975a, b, 1977c, 1980a). Heading finds that the structure of these relations is not clearly revealed if the equations are studied only for a three-dimensional space, and he has therefore generalised the equations so as to use an n-dimensional Cartesian axis system.

18.13. Resonance

Previous sections of this chapter have shown how the differential equations in the various forms, for example (7.80) or (18.39) or (18.49), can be integrated with respect to the height z. At the beginning of § 18.9 it was pointed out that nearly all points of the complex z plane are ordinary points of the basic differential equations (7.80), where the coefficients $\mathbf{T}$ are bounded and the dependent variables $\mathbf{e}$ all behave smoothly. Thus the path of integration may be chosen in any convenient way. Nearly always part of the real z axis is used. But the property fails where elements of $\mathbf{T}$ are infinite and (7.81), (7.82) show that this occurs where $1 + M_{zz} \equiv \varepsilon_{zz} = 0$, at $z = z_\infty$ say. This point is called a point of resonance. It is also the condition that the coefficient α (6.16) or (6.17) or (6.23) of the Booker quartic equation is zero so that one solution q is infinite. For a plasma with electrons only the condition is

$$\varepsilon_{zz} \equiv 1 - X(U^2 - l_z^2 Y^2)/\{U(U^2 - Y^2)\} = 0. \tag{18.72}$$

Usually in practice the zero of ε_{zz} is simple, so the point z_∞ is a simple pole of the non-zero elements of $\mathbf{T}$. It is also a singular point of the basic differential equations (7.80). It can be shown (Budden, 1979, §§2.4) that at least one of the four solutions behaves like $(z - z_\infty)^p$ where the power p is complex. Thus z_∞ is an essential singularity for this solution. A branch cut must be inserted running from z_∞ to infinity. It must not cross the real z axis because the physical solution must be continuous there.

There can be more than one point z_∞ where (18.72) is satisfied, but usually there is only one that needs attention when the integration is planned. If collisions are neglected, $U = 1$ and (18.72) is satisfied for some real value of X. Thus there can be a point z_∞ on the real z axis. It is at this point that the infinities occur of n^2 in figs. 4.3, 4.6 and of q in figs. 6.4–6.7. In the lower part of the ionosphere X is usually an increasing function of z. If a small amount of collision damping is introduced so that $U = 1 - \mathrm{i}Z$, then for (18.72) X must have a small negative imaginary part, and therefore $\mathrm{Im}(z_\infty)$ is negative. The branch cut must then run to infinity in the negative imaginary half of the complex z plane. In rarer cases, when $\mathrm{Im}(z_\infty)$ is positive, the branch cut must be in the positive imaginary half plane.

For calculations of ionospheric reflection coefficients at very low frequency, of order 10 to 20 kHz, Z is large, of order 30, and z_∞ is not near the real z axis. Then the resonance does not give any trouble. For higher frequencies, however, z_∞ may be too near the real z axis. Near z_∞ one value, say q_1, of q is very large, and one possible W.K.B. solution for $\mathbf{e}$ behaves like $\exp(-\mathrm{i}k \int q_1 \, \mathrm{d}z)$, so that it changes very rapidly with z. The usual step size δz of about 1/50 of a vacuum wavelength is therefore too large. There are two ways of dealing with this.

For the first way the real z axis is used as the path of integration and a smaller δz is used where z is near to $\mathrm{Re}(z_\infty)$. This part of the integration then takes longer. It is usually necessary to make several trials to find a suitable δz.

For the second way the path of integration is moved so that it does not go near to z_∞. The new path must not cross the branch cut so that if $\mathrm{Im}(z_\infty)$ is negative, the path is moved into the positive imaginary half of the z plane. This method was used by Smith, M.S. (1974). There are several ways of applying it. For example the path can be in four straight sections. Let a and b be real. Then the first section is the real z axis down to $z = \mathrm{Re}(z_\infty) + a$. The second is the straight line from $\mathrm{Re}(z_\infty) + a$ to $\mathrm{Re}(z_\infty) + ib$, and the third is from here to $\mathrm{Re}(z_\infty) - a$. The fourth is the real z axis below $\mathrm{Re}(z_\infty) - a$. An alternative way is to use a new variable of integration s say, that is real on the displaced part of the path. Then for example let

$$z = \mathrm{Re}(z_\infty) + s + ib(1 - s^2/a^2) \tag{18.73}$$

so that

$$\mathrm{d}z/\mathrm{d}s = 1 - 2ibs/a^2. \tag{18.74}$$

The real z axis is used for $z \geqslant \mathrm{Re}(z_\infty) + a$. Then (18.73) is used for $a \geqslant s \geqslant -a$, and finally the real z axis is again used for $z \leqslant \mathrm{Re}(z_\infty) - a$. The displaced path (18.73) is then a parabola in the complex z plane. For the middle stage here, the auxiliary subroutine must be modified. It must now include the calculation of z from (18.73), and when the derivative $\mathrm{d}v/\mathrm{d}z$ of any dependent variable v has been computed, it must be multiplied by (18.74) to convert it to $\mathrm{d}v/\mathrm{d}s$.

This technique of the displaced path can be used for the basic equations (7.80) or for the non-linear forms (18.39) or (18.49). In tests of the method Smith, M.S. (1974) found that the displaced path must not go too far from the real z axis because this may make the value of the reflection coefficient matrix $\boldsymbol{R}$ too small for accurate calculation.

The position of z_∞ is independent of the angle of incidence θ. Once a suitable integration path has been found, it can be used for a number of calculations with a sequence of values of θ.

A displaced path can also be used for avoiding failure of the form (18.39) of the equations, as illustrated by (18.43). Where this occurs, the variable $\boldsymbol{A}$ has simple poles. The integration path must not go too near these, but they are not branch points so the path may pass on either side. For an ionosphere with small losses these poles are near the real axis. It is then useful to use a line with fixed non-zero $\mathrm{Im}(z)$ for most of the integration path. Computing is then often quicker than with the more complicated equation (18.49) on the real axis.

PROBLEMS 18

18.1. A linearly polarised radio wave is vertically incident on the ionosphere from below. The earth's magnetic field is vertical and $Y = \frac{3}{8}$. The electron concentration increases linearly with height z so that $X = \alpha z$. Electron collisions are to be neglected. Prove that the reflected wave is always linearly polarised and show that its plane of polarisation is parallel to that of the incident wave when $\alpha = 1/r\lambda$, approximately, where r is an integer and λ is the vacuum wavelength. What happens if $Y > 1$?

18.2. In a full wave solution for finding reflection coefficients of the ionosphere at very low frequency, two downward integrations of the basic differential equations are done. The angle of incidence $\theta = \arccos C$ where C is positive. The first integration uses a starting solution for the non-penetrating mode so that it is almost evanescent at the top. At the bottom, in the free space below the ionosphere the field components of the upgoing part of this solution are found to be $E_x, E_y, \mathscr{H}_x, \mathscr{H}_y = C{:}\rho_1{:}-C\rho_1{:}1 = \boldsymbol{e}^{(\mathrm{a})\mathrm{T}}$. For the second integration the field components of the upgoing part of the solution at the bottom are $C{:}\rho_2{:}-C\rho_2{:}1 = \boldsymbol{e}^{(\mathrm{b})\mathrm{T}}$. To find the solution for the penetrating mode a combination $\gamma\boldsymbol{e}^{(\mathrm{a})\mathrm{T}} + \boldsymbol{e}^{(\mathrm{b})\mathrm{T}}$ is now to be found such that the vertical power flux in the upgoing part of the solution below the ionosphere is as small as possible. What value of the constant γ is needed for this? Explain what happens when (a) $\rho_1 = -\rho_2 = \mathrm{i}$, (b) $\rho_1 = \rho_2$, (c) $\rho_1 = 0$, ρ_2 bounded and $\neq 0$.

18.3. A and B are two points in a space where there are magnetoplasmas. Prove that for any aerial system at A it is possible to construct at least one aerial system at B so that there is reciprocity between the terminals of the two aerials.
Note: This theorem was enunciated and proved by Goubau (1942) and is sometimes called 'Goubau's reciprocity theorem'. For a proof that uses the notation of this book see Budden (1961a, § 23.3).

19

Applications of full wave methods

19.1. Introduction

This chapter is concerned with various applications of the full wave methods discussed in ch. 18. The object is to illustrate general principles but not to give details of the results. The number of possible applications is very large and only a selection can be given here. The topics can be divided into two groups: (a) problems where the solutions can be expressed in terms of known functions, and (b) problems where computer integration of the differential equations, or an equivalent method as in §§ 18.2–18.11, is used. In nearly all applications of group (a) it is necessary to make substantial approximations. The group (a) can be further subdivided into those cases where the fourth order governing differential equations are separated into two independent equations each of the second order, and those where the full fourth order system must be used. For the separated second order equations the theory is an extension of ch. 15 which applied for an isotropic medium, and many of its results can be used here.

In all the examples of this chapter the ionosphere is assumed to be horizontally stratified, with the z axis vertical. The incident wave is taken to be a plane wave with its wave normal in the x–z plane. It is assumed that the only effective charges in the plasma are the electrons.

In §§ 19.2, 19.3 it is assumed that the earth's magnetic field is vertical and it is convenient to give here the form of the basic differential equations (7.80) for this case. Thus $l_x = l_y = 0$, $l_z = 1$ for the northern hemisphere. Let

$$a = 1/(U^2 - Y^2). \tag{19.1}$$

Then the susceptibility matrix (3.35) is

$$\boldsymbol{M} = -X\begin{pmatrix} Ua & -\mathrm{i}Ya & 0 \\ \mathrm{i}Ya & Ua & 0 \\ 0 & 0 & 1/U \end{pmatrix}. \tag{19.2}$$

If this is now used in (7.81) for $\mathbf{T}$, the equations (7.80) are

$$\begin{aligned} E'_x &= \mathrm{i}\left(\frac{S^2}{1-X/U}-1\right)\mathscr{H}_y, \\ E'_y &= \mathrm{i}\mathscr{H}_x, \\ \mathscr{H}'_x &= XYaE_x + \mathrm{i}(C^2 - XUa)E_y, \\ \mathscr{H}'_y &= -(1-XUa)E_x + XYaE_y. \end{aligned} \tag{19.3}$$

19.2. Vertical incidence and vertical magnetic field

As a simple but important example where the basic equations are separable consider the case where both the vector $\boldsymbol{Y}$ and the wave normals are vertical. Then $S=0$, $C=1$ and when these are used in (19.3) they can be rearranged to give

$$(E_x+\mathrm{i}E_y)' = -(\mathscr{H}_x+\mathrm{i}\mathscr{H}_y), \quad (\mathscr{H}_x+\mathrm{i}\mathscr{H}_y)' = \{1-X/(U+Y)\}(E_x+\mathrm{i}E_y), \tag{19.4}$$

and

$$(E_x-\mathrm{i}E_y)' = (\mathscr{H}_x-\mathrm{i}\mathscr{H}_y), \quad (\mathscr{H}_x-\mathrm{i}\mathscr{H}_y)' = \{1-X/(U-Y)\}(E_x-\mathrm{i}E_y). \tag{19.5}$$

The equations have separated into these two independent pairs. By eliminating $\mathscr{H}_x \pm \mathrm{i}\mathscr{H}_y$ the equations are converted to the two independent second order equations

$$(E_x+\mathrm{i}E_y)'' + \{1-X/(U+Y)\}(E_x+\mathrm{i}E_y) = 0, \tag{19.6}$$

$$(E_x-\mathrm{i}E_y)'' + \{1-X/(U-Y)\}(E_x-\mathrm{i}E_y) = 0. \tag{19.7}$$

The expressions in brackets $\{\}$ are the two values of the squared refractive index n^2 as given by (4.95) for purely longitudinal propagation. Equations (19.6), (19.7) are a special case of Försterling's equations (16.90), (16.91). The coupling parameter ψ is zero in this case.

This separation is important for it shows that there are two waves that are propagated and reflected independently. For example suppose that a wave is present for which, at some level, $E_x - \mathrm{i}E_y$ and $\mathscr{H}_x - \mathrm{i}\mathscr{H}_y$ are both zero. Then (19.5) shows that they remain zero at all levels, so that the wave is circularly polarised and the rotation is clockwise to an observer looking upwards. Thus, where it is possible to speak of upgoing and downgoing waves, for example below the ionosphere, the upgoing wave has a right-handed sense and the downgoing wave a left-handed sense; see §§ 4.3, 11.6. The propagation is governed by (19.4) or (19.6) and they show that reflection would occur near where $X = U + Y$. Similarly if $E_x + \mathrm{i}E_y$ and $\mathscr{H}_x + \mathrm{i}\mathscr{H}_y$ are zero at some level, the wave is circularly polarised but now with the opposite sense. Its propagation is governed by (19.7) and reflection would occur near where $X = U - Y$. In practice the incident wave may be linearly polarised. It must then be resolved into two circularly polarised components whose reflection coefficients are found separately. The reflected waves are then recombined below the ionosphere. An example of this process follows.

For radio waves of very low frequency it is of interest to study the exponential form

$$N(z) \propto X(z) = e^{\alpha z} \tag{19.8}$$

for the electron height distribution low down in the ionosphere, for the reasons given in § 15.8. Thus the level $z = 0$ is where $X = 1$. As in that section we also let Z be independent of z. The equations (19.6), (19.7) can be considered together. Let

$$F_{\rm r} = E_x + {\rm i}E_y, \quad F_{\rm l} = E_x - {\rm i}E_y, \quad b_{\rm r} = 1/(U + Y), \quad b_{\rm l} = 1/(U - Y) \tag{19.9}$$

where the subscripts r and l indicate that the upgoing incident wave has right- or left-handed circular polarisation, respectively. Then both (19.6) and (19.7) may be written

$$d^2F/dz^2 + k^2(1 - be^{\alpha z})F = 0 \tag{19.10}$$

where F and b are to have the same subscript, r or l. This is of the same form as (15.46) and can be treated in the same way. Let

$$\nu = 2{\rm i}k/\alpha; \tag{19.11}$$

compare (15.48). This ν must not be confused with the ν used elsewhere for the electron collision frequency. The method of § 15.8 can now be applied to (19.10). The reflection coefficient is given by (15.59) with $C = 1$ and with $(1 - {\rm i}Z)^{-1}$ replaced by b, thus

$$R = -b^{\nu}\left(\frac{k}{\alpha}\right)^{2\nu}\frac{(-\nu)!}{\nu!}\exp(-2{\rm i}kh_1) \tag{19.12}$$

where h_1 is the distance from the reference level (§ 11.2) to the level where $X = 1$, $z = 0$. Since ν is purely imaginary, the two factorial functions are complex conjugates and their ratio has modulus unity. The factor $(k/\alpha)^{2\nu}$ also has modulus unity. Hence

$$|R| = |b^{2{\rm i}k/\alpha}| = \exp\left(-\frac{2k}{\alpha}\arg b\right). \tag{19.13}$$

Now (19.12) gives the elements (11.33) of the form $\boldsymbol{R}_0$ of the reflection coefficient matrix defined in 11.6. If the polarisation of the incident wave is right-handed circular, the reflected wave is left-handed circular and has no right-handed component. Hence $R_{\rm rr} = 0$, and similarly $R_{\rm ll} = 0$, so that

$$\boldsymbol{R}_0 = \begin{pmatrix} 0 & R_{\rm rl} \\ R_{\rm lr} & 0 \end{pmatrix}. \tag{19.14}$$

Now let b be $b_{\rm r}$. Then (19.12) gives $R_{\rm lr}$. Similarly if b is $b_{\rm l}$, (19.12) is $R_{\rm rl}$. Then (19.13) gives

$$|R_{1r}| = \exp\left(-\frac{2k}{\alpha}\arctan\frac{Z}{1+Y}\right), \quad |R_{r1}| = \exp\left(-\frac{2k}{\alpha}\arctan\frac{Z}{1-Y}\right). \tag{19.15}$$

If electron collisions are neglected, $Z=0$, then $|R_{lr}|=1$ so the wave is totally reflected. For this case the refractive index is zero where $X = 1 + Y$, and according to ray theory of ch. 13 there should be total reflection at this level in a slowly varying medium. We have now shown that for exponential $N(z)$ when collisions are neglected there is total reflection of this component even when the medium is not slowly varying. For frequencies greater than the gyro-frequency, $Y<1$, and (19.15) shows that if collisions are neglected $|R_{rl}| = 1$ so the other component also is totally reflected. Here the refractive index is zero where $X = 1 - Y$.

These formulae are more interesting, however, for low frequencies so that $Y>1$. Now (19.9) shows that $\arg b_l = \pi - \arctan\{Z/(Y-1)\}$ and the second equation (19.15) is

$$|R_{rl}| = \exp\left(-\frac{2\pi k}{\alpha} + \frac{2k}{\alpha}\arctan\frac{Z}{Y-1}\right) \tag{19.16}$$

where the arctan is in the range 0 to $\frac{1}{2}\pi$. If $Z=0$, $|R_{rl}| = \exp(-2\pi k/\alpha)$. In a slowly varying medium α is small so that the reflection coefficient is small. Now the condition $X = 1 - Y$ is not obeyed at any real height and according to the ray theory of ch. 13 there should be no reflection of this component. The full wave theory shows that for exponential $N(z)$ there is some reflection. It is small in a slowly varying medium but becomes larger when the medium varies more quickly.

If the incident wave is linearly polarised, we may take its plane of polarisation to be the x–z plane. Then the reflecting properties of the ionosphere are conveniently specified by the elements R_{11}, R_{21} of the matrix $\boldsymbol{R}$. This is found from the transformation (11.33) that uses the unitary matrix (11.30) and gives

$$\boldsymbol{R} = \boldsymbol{U}\boldsymbol{R}_0\boldsymbol{U}^{-1} = \tfrac{1}{2}\begin{pmatrix} -R_{rl} - R_{lr} & \mathrm{i}(R_{rl} - R_{lr}) \\ \mathrm{i}(R_{rl} - R_{lr}) & R_{rl} + R_{lr} \end{pmatrix} \tag{19.17}$$

whence, from (19.12):

$$\begin{aligned} R_{11} &= -R_{22} = \tfrac{1}{2}\left(\frac{k}{\alpha}\right)^{2\nu}\frac{(-\nu)!}{\nu!}(b_r^\nu + b_l^\nu), \\ R_{21} &= R_{12} = \tfrac{1}{2}\mathrm{i}\left(\frac{k}{\alpha}\right)^{2\nu}\frac{(-\nu)!}{\nu!}(b_r^\nu - b_l^\nu). \end{aligned} \tag{19.18}$$

These formulae are of interest in the special case of very low frequencies when $Y \gg 1$; for example at 16 kHz, $Y \approx 80$. If the 1 is neglected compared with Y in (19.9)

$$b_r = 1/(Y - \mathrm{i}Z), \quad b_l = -1/(Y + \mathrm{i}Z), \tag{19.19}$$

so that $|b_r| = |b_1|$. Then b_r^v, b_1^v have the same argument since v is purely imaginary, and b_r^v has the greater modulus. Then (19.18) shows that R_{11}, R_{21} are in quadrature, so that when the incident wave is linearly polarised, the polarisation ellipse of the reflected wave has its major axis in the plane of polarisation of the incident wave. It can be shown that in the northern hemisphere the rotation in the reflected wave has a left-handed sense. Further from (19.18)

$$|R_{11}| = \tfrac{1}{2}\left[\exp\left(-\frac{2k}{\alpha}\arctan\frac{Z}{Y}\right) + \exp\left\{-\frac{2k}{\alpha}\left(\pi - \arctan\frac{Z}{Y}\right)\right\}\right]$$
$$= \mathrm{e}^{-\pi k/\alpha}\cosh\left\{\frac{k}{\alpha}\left(\pi - 2\arctan\frac{Z}{Y}\right)\right\} \qquad (19.20)$$
$$|R_{21}| = \mathrm{e}^{-\pi k/\alpha}\sinh\left\{\frac{k}{\alpha}\left(\pi - 2\arctan\frac{Z}{Y}\right)\right\}.$$

These results were given by Stanley (1950) who derived them in a different way as a limiting case of the Epstein distribution, § 15.14. They are important because in observations with a frequency of 16 kHz at nearly vertical incidence (Bracewell *et al.*, 1951) it was found that the reflected wave was elliptically polarised with a left-handed sense and with the major axis of the ellipse parallel to the plane of polarisation of the incident wave.

The two differential equations (19.6), (19.7) are of the same form as (7.6) which was solved for various electron height distributions in ch. 15 where the earth's magnetic field was neglected. Thus all the distributions $N(z)$ used in ch. 15 can also be used for vertical incidence when the earth's magnetic field also is vertical. For example with the parabolic $N(z)$ used in § 15.9, the differential equation (15.66) is replaced by two equations obtained by putting F_r, F_l (19.9) for E_y, $1 \pm Y - \mathrm{i}Z$ for $1 - \mathrm{i}Z$, and $C^2 = 1$. The two reflection coefficients R_{lr} and R_{rl} are then found, by analogy with (15.86), and combined as in (19.17) to give the elements of $\boldsymbol{R}$. Some results for this case were given by Pfister (1949). Wilkes (1940) discussed the case where $N(z)$ is proportional to the square of the height; see § 15.18. He also discussed the case where N is independent of z, and $Z \propto 1/z$, which can be solved in terms of confluent hypergeometric functions. For a similar model see problem 15.1.

19.3. Oblique incidence and vertical magnetic field

When the waves are obliquely incident on the ionosphere, or when the earth's magnetic field is oblique, the equations (7.80) cannot in general be separated into two second order equations. They are equivalent to a single fourth order equation. There are very few cases where the properties of fourth order equations have been studied; most of the functions of theoretical physics are solutions of second order equations. Hence some approximations must be made before progress is possible.

A very important solution was given by Heading and Whipple (1952), for oblique incidence and vertical field, and another similar solution was given by Heading (1955) for vertical incidence when the earth's magnetic field is oblique and in the plane of incidence. The principles of the two methods are similar and only the first is described here in outline. Heading and Whipple made the following assumptions:

(1) $|1-\mathrm{i}Z| \ll Y$. Thus the electron collision frequency is very small compared with the gyro-frequency.

(2) X increases monotonically with the height z.

(3) At the level where $X \approx |1-\mathrm{i}Z|$, Z is constant and X is proportional to $\mathrm{e}^{\alpha z}$ where α is constant.

(4) At the level where $X \approx Y$, X is proportional to $\mathrm{e}^{\gamma z}$ where γ is another constant. Here Z is negligible because of (1).

The assumption (1) may not be true in the D-region, but still the method is useful as an illustration of some physical principles.

Let the ionosphere be divided into five regions as follows. The rather unusual numbering is used to agree with Heading and Whipple who speak only of regions I and II.

Region II(a) $X \gg Y$.
Region II $|1-\mathrm{i}Z| \ll X \approx Y$ (partially reflecting)
Region I(a) $|1-\mathrm{i}Z| \ll X \ll Y$
Region I $X \approx |1-\mathrm{i}Z|$ (partially reflecting)
Region 0 $X \ll |1-\mathrm{i}Z| \ll Y$ (like free space).

The lowest region, 0, is like free space because X is very small. In the highest region II(a), X is very large and both refractive indices are large, of order $X^{\frac{1}{2}}$. Here the W.K.B. solutions are good approximations. This is above all reflecting levels so the solution here must contain only upgoing waves. Some reflection occurs in regions I and II. In region I(a) there is no reflection but the waves have some interesting features.

In the three lowest regions the conditions $X \ll Y$ and $|U| \equiv |1-\mathrm{i}Z| \ll Y$ both hold. Then a (19.1) is negligible, and the equations (19.3) become

$$E_x' = -\mathrm{i}\frac{C^2 - X/U}{1 - X/U}\mathscr{H}_y, \quad \mathscr{H}_y' = -\mathrm{i}E_x, \tag{19.22}$$

$$E_y' = \mathrm{i}\mathscr{H}_x, \quad \mathscr{H}_x' = \mathrm{i}C^2 E_y. \tag{19.23}$$

They have separated into two independent pairs. The pair (19.23) refers to linearly polarised waves with the electric vector horizontal, and shows that these waves are propagated just as though regions 0, I, I(a) are free space.

The pair (19.22) refers to linearly polarised waves with the electric vector in the plane of incidence. It is shown below that these can undergo partial reflection in region I. In region I(a) $|X/U| \gg 1$ so the C^2 and the 1 in the first equation (19.22) are negligible, and the equations take the very simple form

$$E'_x = -\mathrm{i}\mathscr{H}_y, \quad \mathscr{H}'_y = -\mathrm{i}E_x. \tag{19.24}$$

In formulating the original equations (7.80) it was assumed, § 7.13, that all field quantities include a factor $\exp(-ikSx)$. If this is restored, one solution of (19.24) is

$$E_x = \mathscr{H}_y = \exp\{-\mathrm{i}k(Sx+z)\} \tag{19.25}$$

which represents a progressive wave travelling obliquely upwards with phase velocity $c(1+S^2)^{-\frac{1}{2}}$ so that the refractive index is $(1+S^2)^{\frac{1}{2}}$. Its wave normal is at an angle ψ to the vertical where $\tan\psi = S = \sin\theta$. The vertical component E_z can be found from (7.78) which gives $E_z = -S\mathscr{H}_y/(1-X/U)$, and this is negligible because $|X/U| \gg 1$. Hence the electric field $\boldsymbol{E}$ is horizontal and in the plane of incidence. There is a second solution of (19.24) namely

$$E_x = -\mathscr{H}_y = \exp\{-\mathrm{i}k(Sx-z)\} \tag{19.26}$$

that represents a wave travelling obliquely downwards.

The above results may be summarised in physical terms as follows. The assumption $Y \gg X$ means that the earth's magnetic field is extremely large so that electrons are prevented from moving across it. Hence the electrons move only vertically, and will not respond to forces tending to move them horizontally. This explains why a wave with its electric vector horizontal is unaffected by the electrons, so the three lower regions behave like free space for this wave. In region II and above, however, X is comparable with Y and the electrons are now so numerous that they have some effect on the wave.

For a wave with its electric vector in the plane of incidence there is a force eE_z tending to move each electron vertically. But in region I(a) where $X \gg |1-\mathrm{i}Z|$ the electrons are so numerous that the medium effectively has infinite conductivity in the z direction. Hence it cannot sustain a vertical component of $\boldsymbol{E}$ and the only electric field is horizontal, as in (19.25), (19.26). In region 0 there are so few electrons that the medium is like free space. The transition occurs in region I which is a region of finite conductivity, increasing as the height increases.

In a medium with indefinitely large Y, (3.50) shows that the principal axis elements of the electric permittivity tensor $\boldsymbol{\varepsilon}$ are $\varepsilon_1 = \varepsilon_2 = 1$, $\varepsilon_3 = 1 - X/U$. The form (3.53) of $\boldsymbol{\varepsilon}$ applies for the real axes x, y, z, and this is diagonal with elements $1, 1, 1 - X/U$. Thus for regions 0, I, I(a) the medium is uniaxial. This term is explained near the end of § 5.5. See also problem 4.7.

In region I it is convenient to treat $\mathscr{H}_y$ as the dependent variable. Let $z = z_1$ be the level where $X = |1-\mathrm{i}Z|$. Then assumption (3) gives $X = |1-\mathrm{i}Z|\exp\{\alpha(z-z_1)\}$. Further let $\chi = \arctan Z = -\arg U$. Then

$$X/U = \exp\{\alpha(z-z_1) + \mathrm{i}\chi\}. \tag{19.27}$$

and (19.22) gives

$$\frac{\mathrm{d}^2\mathscr{H}_y}{\mathrm{d}z^2}+k^2\left[1-\frac{S^2}{1-\exp\{\alpha(z-z_1)+\mathrm{i}\chi\}}\right]\mathscr{H}_y=0. \tag{19.28}$$

This is of the form considered by Epstein (1930) and discussed in §§ 15.13–15.17. The expression in brackets may be called the effective value of q^2 (15.114), with which it is identical provided that we take

$$\eta_1=C^2,\quad \eta_2=1,\quad \eta_3=0,\quad \sigma=1/\alpha,\quad B=-\alpha z_1+\mathrm{i}\chi-\mathrm{i}\pi. \tag{19.29}$$

Equations (15.115)–(15.117) can now be used to find a, b, c. Then (15.132), (15.133) are used to find the reflection and transmission coefficients for waves incident from below, and similar expressions (Heading and Whipple, 1952, (6.5), (6.6); or Budden, 1961a, (17.108), (17.109)) are used for waves incident from above.

In regions II and II(a) X is comparable with or greater than Y so that XUa is of order $1/Y$ and still negligible. There may be some level above region II where XUa is of order unity, but this requires that X is of order Y^2. Now $XYa\approx -X/Y$ and this is no longer negligible. Since X is now large the factor $S^2/(1-X/U)$ is negligible and equations (19.3) become

$$\begin{aligned} E_x'&=-\mathrm{i}\mathscr{H}_y,\quad \mathscr{H}_y'=-\mathrm{i}E_x-\frac{X}{Y}E_y,\\ E_y'&=\mathrm{i}\mathscr{H}_x,\quad \mathscr{H}_x'=\mathrm{i}C^2E_y-\frac{X}{Y}E_x. \end{aligned} \tag{19.30}$$

Elimination of $\mathscr{H}_x$, $\mathscr{H}_y$ now gives

$$E_x''+E_x=\mathrm{i}\frac{X}{Y}E_y,\quad E_y''+C^2E_y=-\mathrm{i}\frac{X}{Y}E_x. \tag{19.31}$$

These two equations are equivalent to a single fourth order equation. They were studied by Wilkes (1947) for the case where X is a linear function of z, and he showed that the solution can then be expressed as the sum of a number of contour integrals. They were solved by Heading and Whipple (1952) for the case where X is proportional to $\mathrm{e}^{\gamma z}$.

The method, in outline, was as follows. It should be compared with § 15.13. Let $z=z_2$ be the level where $X/Y=4\gamma^2/k^2$. Introduce the new independent variable $w=\exp\{2\gamma(z-z_2)\}$. It is analogous to the ζ in (15.110). Eliminate E_x from (19.31). This gives a fourth order equation for E_y with w as independent variable. It has a regular singularity at $w=0$; in this respect it resembles Bessel's equation. It is found that there are four independent series solutions in ascending powers of w. These are like hypergeometric series but more complicated, and they are called generalised hypergeometric series, (Heading, 1980b). Below region II, w is very small so that only the first term of each series need be retained. For each of the four solutions E_y there is a corresponding expression for E_x. Now these solutions can be identified with waves in region I(a). Two of them have $E_y = O(w) \to 0$ and E_x given by (19.25) (19.26). The other two have $E_x = O(w) \to 0$. They are solutions of (19.23) and are upgoing and downgoing waves travelling as though in free space. The possible ratios of the amplitudes of these four solutions must now be found. For this it is necessary to find circuit relations. In region II(a) the waves must be upgoing and there are two independent solutions with this property. Each of these can be expressed as a contour integral of Barnes type; see § 15.13 and (15.107) (Barnes, 1908). These two integrals are such that when w is large they can be evaluated by steepest descents (ch. 9) and this gives the two W.K.B. solutions for the upgoing waves. Then in region I(a) each Barnes integral is expressed as four series of the residues of poles of the integrand, and these are the generalised hypergeometric series already mentioned. This process gives the four elements of the reflection coefficient matrix of region II as measured below it in region I(a), with reference level $z = z_2$. They are very complicated expressions involving factorial functions containing γ, k and C.

The final step is to combine these reflection coefficients of region II with the reflection and transmission coefficients for region I. In region I(a) there are both upgoing and downgoing waves, whose relative phases must be allowed for. This depends on the height range $z_2 - z_1$, between the reference levels for regions II and I. The details and the final result are given by Heading and Whipple (1952, § 10). See also Budden (1961a, § 21.12).

Curves were given by Heading and Whipple to show how the overall reflection coefficients and the separate reflection coefficients for regions I and II depend on angle of incidence, for frequencies 16 kHz and 80 kHz. The formulae for the overall reflection coefficients were used by Budden (1955b) who showed that they agreed well with computer integrations of the differential equations. In the paper by Budden (1955b) there were errors in the signs of l_x and l_y so that results given for east to west propagation apply for west to east, and vice versa.

19.4. Resonance and barriers

One root q of the Booker quartic (6.15) is infinite when the coefficient α of q^4 is zero, and we need to study the effect of this infinity on wave propagation. At first collisions will be neglected, so that α is given by (6.24) and is zero for one real value of X satisfying (4.76) with $U = 1$. In this section the q that is infinite is denoted by q_1. It is assumed that the zero of α is simple and at some real height, since this is the most important case in practice. The origin $z = 0$ is chosen at this zero. For the lower side of an ionospheric layer X increases and α decreases when z increases, so that α is negative where $z > 0$. For no collisions the coefficients $\alpha, \beta, \ldots$ of the quartic are all real, and near the infinity

$$q_1 \approx -\beta/\alpha \tag{19.32}$$

so that q_1 is real on both sides of $z = 0$, when z is real. In this respect it differs from the infinity studied in §§ 15.5–15.7 where Q, the effective q, is purely imaginary when collisions are neglected. The behaviour of q near an infinity can be seen in figs. 6.4, 6.5, 6.7 and in figures given by Booker (1939), Smith, M.S. (1974a). Fig. 19.1(a) is a sketch of the typical behaviour. It would apply for the extraordinary wave in the Z-mode when $Y < 1$. It often happens that near the infinity there is a coupling point, $z = b < 0$ at B in fig. 19.1(a), where q_1 is equal to another root say q_2. An example is the reflection point R in fig. 6.5. Then for $z < b, q_1$ and q_2 are complex conjugates. There may also be another coupling point, $z = a < b$ at A in fig. 19.1(a), where the same two roots are again equal, so that they are again real when $z < a$; an example is the reflection point E_- of fig. 6.5. Then the range $a < z < b$ is similar to a conventional barrier because the waves here are inhomogeneous, but its properties are modified by the proximity of the resonance. The real values of q can be found by Poeverlein's construction, § 10.8, with refractive index surfaces for a sequence of values of z. Fig. 19.1(b) (compare fig. 10.7) is a sketch of the form these surfaces would take to give the two right-hand curves of fig. 19.1(a). The arrows indicate the directions of the ray. This shows that both the branches marked q_1 in fig. 19.1(a) refer to upgoing waves.

Consider now the special case of vertical incidence, so that the coefficient β of the quartic (6.15) is zero. Then, where $\alpha = 0$, two roots of the quartic are infinite. The reflection point R, fig. 6.5, has moved to coincidence with the infinity, so that $b = 0$, in fig. 19.1(a). This is therefore a degenerate case where a reflection point and a

Fig. 19.1. (a) is an example of how q may depend on $s = kz$ in and near a barrier when collisions are neglected. The chain curve shows $\text{Re}(q_1) = \text{Re}(q_2)$ in the barrier region where $(q_1 - q_2)^2$ is negative. (b) shows a family of refractive index surfaces for various values of the height z. The symbols +, 0, − by the curves refer to the values of z. The broken line is used for Poeverlein's construction.

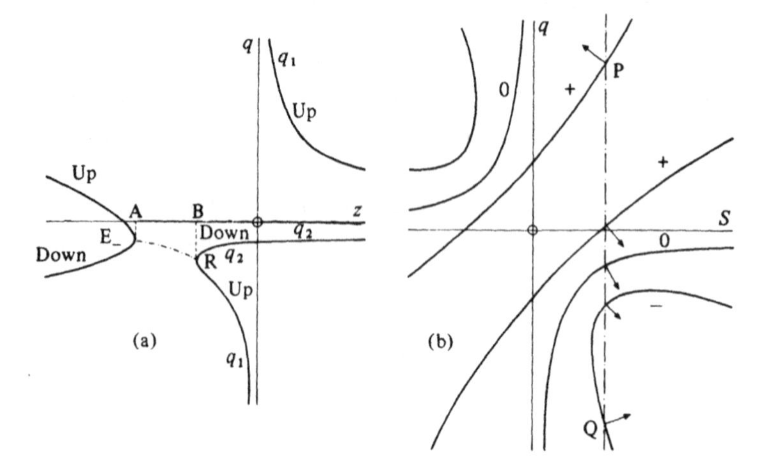

resonance point coincide. The qs are now the same as the refractive indices n, and the quartic is just the quadratic equation (4.65) for n^2. It is now convenient to plot n^2 against z, and the form of the curve is sketched in fig. 19.2(b). There is still a barrier with $n^2 < 0$, where $a \leqslant z \leqslant 0$. At its left-hand boundary $n^2 = 0$ as with a conventional barrier, but at the other boundary $n^2 \to \infty$. The case where $n^2 \to \infty$ at both boundaries of a barrier can occur and some aspects of it have been studied (Tang, Wong and Caron, 1975) but it is not considered here.

The objective now is to study the effect of the infinity of q_1 on the tunnelling through the barrier. A full theory that covers most of the important cases was given by Budden (1979). It is very long and only a few of the examples are given here. It is found that, by making approximations, the differential equations can be separated so that the tunnelling phenomenon is studied by using differential equations of only second order. Thus this theory is a continuation of the topics in ch. 15.

A point of resonance is a singular point of the governing differential equations. This was discussed in § 18.13 where it was explained that a branch cut must be inserted in the complex z plane running from the resonance point to infinity without crossing the real z axis. If there is a very small collision frequency, the resonance occurs where $\mathrm{Im}(X) < 0$. The case discussed here is for the lower ionosphere where X increases with increasing z as in fig. 19.1(a), so the resonance is where $\mathrm{Im}(z) < 0$. The branch cut runs to infinity in the negative imaginary half of the z plane. In the limit of no collisions this branch cut must be retained, and it now runs from $z = 0$ to $|z| \to \infty$ with $\mathrm{Im}(z) < 0$.

19.5. Isolated resonance

Suppose now that the incidence is oblique and that X increases very slowly with height z so that the reflection point at $z = b$ is at a great distance from the resonance. Then there is a large domain of the z plane, containing $z = 0$, where the solution q_1 of the quartic refers to a wave for which there are no reflection or coupling points. To study this wave we now examine the coupled equations (7.109) or (16.7). In the domain, (19.32) gives

$$q_1 \approx \Delta/s, \quad s = kz \tag{19.33}$$

where Δ is real and large. We take Δ to be positive as in fig. 19.1(a). If there are coupling points not involving q_1 in the domain, this can be easily dealt with. For simplicity it will be assumed that there are none. At the resonance some elements of $\mathbf{T}$ in (7.80), (7.81) have simple poles because of the factor ε_{zz}, proportional to α, in the denominator. But $\varepsilon_{zz}\mathbf{T}$ (7.82) is analytic and bounded at and near the resonance, and therefore its eigen values $\varepsilon_{zz}q$ and eigen columns are bounded and analytic. The eigen columns of $\varepsilon_{zz}\mathbf{T}$ are also eigen columns of $\mathbf{T}$. The matrix $\mathbf{S}$ formed from the four eigen columns is therefore bounded, non-singular and analytic. The coupling matrix

$\boldsymbol{\Gamma} = -\mathbf{S}^{-1}\mathbf{S}'$ (7.108) exists throughout the domain. It is very small because of the derivative $\mathbf{S}'$, since the medium is very slowly varying and there are no coupling points in the domain. If $\boldsymbol{\Gamma}$ is neglected, the first of the four coupled equations (7.109) is

$$\mathrm{f}_1' = -\mathrm{i}\mathrm{f}_1\Delta/s \tag{19.34}$$

with the solution

$$\mathrm{f}_1 = P\exp(-\mathrm{i}\Delta\ln s) \tag{19.35}$$

where P is a constant. The other three of the four coupled equations are associated with the remaining three values of q. These are bounded, and the solutions are not affected by the resonance. These three waves are propagated independently and we may assume that they are absent, so that $\mathrm{f}_2 = \mathrm{f}_3 = \mathrm{f}_4 = 0$.

Since Δ is positive, the phase propagation of the wave (19.35) is upwards when s is real and positive and downwards when s is real and negative. But in fig. 19.1(b) the corresponding points are P, Q respectively and the arrows show that the ray, that is the direction of group propagation, is obliquely upwards in both cases. They also show that the ray velocities have large horizontal components in opposite directions for positive and negative s.

Now the vertical component Π_z of the Poynting vector is to be found for real s above and below the resonance. For this we use some of the matrix theory of § 7.14. The following version is a summary. For the full theory see Budden (1979, §§ 3, 9).

The adjoint $\bar{\mathbf{e}}$ of the column $\mathbf{e}$ of the field variables was introduced in § 7.14(4). It satisfies the adjoint differential equation

$$\bar{\mathbf{e}}' = \mathrm{i}\bar{\mathbf{T}}\bar{\mathbf{e}} \tag{19.36}$$

from (7.95). Now from § 7.14(2) the eigen columns of the adjoint $\bar{\mathbf{T}}$ are the rows of $\mathbf{S}^{-1}$ with the order of its columns reversed, that is the rows of $\mathbf{S}^{-1}\mathbf{B}$ which are the same as the columns of $\mathbf{B}(\mathbf{S}^{\mathrm{T}})^{-1}$, where $\mathbf{B}$ is given by (7.92). The first equation (6.53) is $\mathbf{e} = \mathbf{S}\mathbf{f}$ and its adjoint is

$$\bar{\mathbf{e}} = \mathbf{B}(\mathbf{S}^{\mathrm{T}})^{-1}\bar{\mathbf{f}} \quad \text{whence} \quad \bar{\mathbf{e}}^{\mathrm{T}} = \bar{\mathbf{f}}^{\mathrm{T}}\mathbf{S}^{-1}\mathbf{B}. \tag{19.37}$$

The adjoint of the coupled equations may now be found. The coupling matrix $\bar{\boldsymbol{\Gamma}}$, like $\boldsymbol{\Gamma}$, is negligible. The adjoint of (19.34) is then, from (19.36)

$$\bar{\mathrm{f}}_1' = \mathrm{i}\bar{\mathrm{f}}_1\Delta/s \tag{19.38}$$

with the solution

$$\bar{\mathrm{f}}_1 = \bar{P}\exp(\mathrm{i}\Delta\ln s). \tag{19.39}$$

The bilinear concomitant W was given in 7.14(5), and if (19.37) is used, (7.97) shows that

$$W = \bar{\mathbf{e}}^{\mathrm{T}}\mathbf{B}\mathbf{e} = \bar{\mathbf{f}}^{\mathrm{T}}\mathbf{f} = \bar{\mathrm{f}}_1\mathrm{f}_1 = \bar{P}P. \tag{19.40}$$

The solution (19.39) contains the arbitrary constant $\bar{P}$. It was shown in § 7.14(8) that for any loss-free medium where s is real, $\bar{P}$ can be chosen so that $W = 4Z_0\Pi_z$. We apply this where s is real and positive, indicated by the + sign, so that

$$W = 4Z_0\Pi_z(+) \tag{19.41}$$

from (7.104). Now Π_z must also be real and positive so we take $\bar{P} = P^*$, $4Z_0\Pi_z = PP^*$, and then $\mathfrak{f}_1, \bar{\mathfrak{f}}_1$ are complex conjugates. This is true only when s is real and positive. If $\mathrm{Im}(s) \neq 0$, then $W \neq 4Z_0\Pi_z$ and $\bar{\mathfrak{f}}_1 \neq \mathfrak{f}_1^*$. Now from (7.87) W is independent of z, that is of s, and is therefore still given by (19.40) where s is real and negative. In going from positive to negative real s, the branch cut must not be crossed so the path is where $\mathrm{Im}(s) > 0$, and $\arg s$ increases from 0 to π. Then for s real and negative, from (19.35), (19.39)

$$\mathfrak{f}_1 = P|s|^{-i\Delta}e^{\pi\Delta}, \quad \bar{\mathfrak{f}}_1 = P^*|s|^{i\Delta}e^{-\pi\Delta}. \tag{19.42}$$

The product (19.40) of these is unchanged and W is the same as for positive s. But (19.42) are not now complex conjugates, and (19.41) is not now true.

To find $\Pi_z(-)$ where s is real and negative the product $\mathfrak{f}_1\bar{\mathfrak{f}}_1$ in (19.40) must be replaced by $\mathfrak{f}_1\mathfrak{f}_1^*$. This means that the arbitrary constant $\bar{P}$ in (19.39) must now have a new value $P^*e^{2\pi\Delta}$, and then

$$\Pi_z(-) = e^{2\pi\Delta}PP^* = e^{2\pi\Delta}\Pi_z(+). \tag{19.43}$$

The upward energy flux below the resonance level is greater than that above it by a factor $e^{2\pi\Delta}$. There must be a disappearance of energy at the resonance, in the loss-free case, if the solution (19.35), with a singularity at $s = 0$, is present. Another example of this is given in the following section.

The question as to what happens to the wave energy that disappears near a resonance is an intriguing physical problem. At least four different view points may be taken.

First it may be argued that the assumption of a loss-free medium is unrealistic. There is no medium, except for a vacuum, for which all forms of wave attenuation are absent. Therefore the refractive index n always has a non-zero imaginary part when $s = kz$ is real. In a plasma this could be caused by collision damping, Landau damping or other forms of damping. $\mathrm{Im}(n)$ may be negligibly small for most real values of z. But the pole at the resonance is not exactly on the real z axis. Near it both real and imaginary parts of n are very large, so that the energy absorption rate is large. The lost energy is simply converted to heat or other forms of energy in the region of the real z axis close to the pole. In this book non-linear effects have been ignored. The solutions for E_x, E_y, proportional to (19.35), are bounded near the resonance, so that for these components, the assumption of linearity may be justified for small wave amplitudes. But it can be shown from the last equation of (7.78) that E_z is not bounded at the resonance. Thus non-linear effects must occur. One result would be the generation of harmonics of the wave frequency. This provides another way by which energy could be lost.

Second, the illustrations in this book apply for a cold plasma whose dielectric constant is (3.55). If the plasma is warm its dielectric constant is not given exactly by (3.33). The result is that when ε_{zz} approaches zero, the solution q_1 gets very large, but it does not become infinite when $\varepsilon_{zz} = 0$. Instead, the characteristics of the wave go over continuously to those of a plasma wave. Curves showing how this happens have been given by Ginzburg (1970, figs. 12.2–12.4, 12.9–12.13). See also Golant and Piliya (1972). On this view the resonance at $z = 0$ does not exist. The waves have simply gone over to a mode of propagation governed by a modification of the basic equation, and they carry the 'lost' energy with them. Some authors call this 'mode conversion'. But the transition is continuous and it is only the name of the wave type that changes. If the medium is extremely slowly varying there is no coupling point near $z = 0$ and the transition is not a mode conversion. In an actual medium there might be some mode conversion near the transition, but this would be a separate phenomenon associated with a coupling point.

Third, for the theory of this chapter the incident wave is a plane wave and therefore extends indefinitely in the x and y directions. Any actual incident wave must be a laterally bounded beam. It was shown earlier in this section, from fig. 19.1(b), that the ray direction, that is the direction of energy propagation, gets more nearly horizontal as the resonance level is approached. Thus the upgoing energy is deviated sideways right out of the beam.

If any of these views is taken, the lost energy can be accounted for and there is nothing more to be said. But it is still of interest to pursue the enquiry for a fictitious medium which is a cold plasma and really has no damping mechanism at all. The following suggestion is offered. In any wave system some energy is stored in the medium. The time average of the stored energy per unit volume is constant in the steady state conditions assumed here, and is proportional to the dielectric constant and to the square of the electric field. Near a resonance at $z = 0$ the dielectric constant (3.55) is bounded, but E_z tends to infinity like $1/z$. Consequently if the stored energy is integrated through a volume near the resonance, the integral gets indefinitely large as the edge of the volume moves up to the resonance. The total energy stored in the part of the medium containing the resonance is infinite. But this state of affairs cannot be attained in a finite time. After the wave is first switched on, a state of physical equilibrium is never reached. The medium has an infinite capacity for storing energy. The energy that is apparently lost is simply going into storage in the medium.

19.6. Resonance tunnelling

The effect of a resonance on the penetration of a barrier is now to be studied. The theory is given here only for vertical incidence so that the reflection point R of fig.

19.1(a) coincides with the resonance, which is one boundary of the barrier. The full theory allowing for oblique incidence is given by Budden (1979). It is similar in form and leads to differential equations of the same type, but the algebra is more complicated. The theory is also of interest in the realm of thermo-nuclear plasmas and this aspect has been studied, for example, by White and Chen (1974) and Batchelor (1980).

When a vertically incident radio packet approaches a cut-off level where $n = 0$, and $X = X_0$, the vertical component $\mathscr{U}_z$ of the group velocity gets smaller. It is proportional to $(X_0 - X)^{\frac{1}{2}}$ from (5.78), (5.79). The wave packet slows down, reverses its direction and is reflected. Although $\mathscr{U}_z \to 0$ when $X \to X_0$, the group time of travel is bounded. It affects the calculation of equivalent height from the integral (13.8). When the wave packet approaches a resonance level where $n \to \infty$, $X = X_\infty$, again $\mathscr{U}_z$ gets smaller but it is now proportional to $(X - X_\infty)^{\frac{3}{2}}$ as shown by (5.82). The wave packet slows down, but because of the power 3/2, it would take an infinite time to reach the resonance level, and this crude ray type theory suggests that it cannot be reflected. It has sometimes been suggested that reflection from a resonance can occur, and the expression for $X = X_\infty$ has been called 'the fourth reflection condition' (the other three are $X = 1, 1 \pm Y$). The full wave theory that follows confirms that there is no reflection at a resonance. In those cases where reflection has been thought to occur at the level where $X = X_\infty$, it must have resulted from some other cause such as a nearby zero of n or a steep gradient of $N(z)$.

When an extraordinary wave travels vertically upwards into a collisionless ionosphere with $Y < 1$, it first encounters a cut-off level where $X = 1 - Y$ and here it is reflected. Above this is a barrier region where n_{E}^2 is negative. It extends up to where $X = X_\infty$ given by (4.76) with $U = 1$. Here n_{E}^2 is infinite, and above this it is positive and decreasing. This case will be discussed here and the subscript E will be omitted. It will be shown that for the upgoing wave there can be some penetration of the barrier, (19.63) below. Most of the energy is reflected at the cut-off level but some energy apparently disappears. For a wave coming down from above the resonance there is no reflection. Part of the wave energy penetrates the barrier and the rest apparently disappears; (19.72) below.

For vertical incidence Försterling's coupled equations (16.90), (16.91) can be used. The coupling parameter ψ depends on $\mathrm{d}N(z)/\mathrm{d}z$ and is small in a slowly varying medium, except possibly near the levels of the coupling points, that is $X \approx 1$, but this is not near the resonance. We are not concerned with coupling and therefore ψ will be neglected. The two Försterling equations are then independent, and (16.91) for the extraordinary wave will be written, without the subscripts, thus

$$\frac{\mathrm{d}^2\mathscr{F}}{\mathrm{d}s^2} + n^2\mathscr{F} = 0, \quad s = kz. \tag{19.44}$$

This is similar to (7.6) for an isotropic medium and may be discussed in a similar way.

Approximate solutions are the two W.K.B. solutions

$$\mathscr{F} = n^{-\frac{1}{2}} \exp\left\{ \mp \mathrm{i} \int^{s} n \, \mathrm{d}s \right\} \tag{19.45}$$

and the approximation is good provided that (7.59), with q replaced by n, is satisfied.

In a slowly varying ionosphere it may be assumed that, for a small range of s, X varies linearly with height. The origin of s is chosen at the resonance where $X = X_{\infty}$ so that

$$X - X_{\infty} = \alpha s \tag{19.46}$$

where α is a constant. Then it can be shown from (4.67) that, for the n^2 that has the infinity,

$$n^2 = \frac{A}{\alpha s} + B + \mathrm{O}(s) \tag{19.47}$$

where A and B are real positive constants expressible in terms of Y and Θ. If $|s|$ is small all terms except the first are negligible and substitution into (7.59) gives

$$|s| \gg 3\alpha/8A \tag{19.48}$$

as the condition for the validity of the W.K.B. solutions (19.45). The condition fails near the resonance at $s = 0$ and a full wave solution of the differential equation must be sought. The solutions for two important cases will now be given.

Assume first that α is so small that only the first term of (19.47) need be retained for values of $|s|$ up to that for which (19.48) becomes valid. Then the differential equation (19.44) is

$$\frac{\mathrm{d}^2\mathscr{F}}{\mathrm{d}s^2} + \frac{\beta}{s}\mathscr{F} = 0 \tag{19.49}$$

where β is a real positive constant. The squared refractive index is shown in fig. 19.2(a) (continuous curve). A solution is sought that gives the reflection coefficient for waves incident from the right. Thus when s is large and negative the solution must represent a wave travelling to the left or a field whose amplitude decreases as s gets more negative. If there is a small constant collision frequency, the infinity of n^2 occurs where $X = X_{\infty}$, and therefore also s, has a small negative imaginary part. The zero of s is then chosen to be where $X = \mathrm{Re}(X_{\infty})$, and (19.49) becomes

$$\frac{\mathrm{d}^2\mathscr{F}}{\mathrm{d}s^2} + \frac{\beta}{s + \mathrm{i}\gamma}\mathscr{F} = 0 \tag{19.50}$$

where γ is real and positive. The real part of n^2 is then as shown by the broken curve in fig. 19.2(a). Equations (19.49), (19.50) are of standard form (Watson, 1944, p. 97). A solution of (19.50) is

$$\mathscr{F} = (s + \mathrm{i}\gamma)^{\frac{1}{2}} \mathscr{C}\{2\beta^{\frac{1}{2}}(s + \mathrm{i}\gamma)^{\frac{1}{2}}\} \tag{19.51}$$

where $\mathscr{C}$ denotes a Bessel function. If, for real s, the sign of the square root is chosen so that

$$0 < \arg(s + i\gamma)^{\frac{1}{2}} < \pi \tag{19.52}$$

then the Bessel function that satisfies the physical conditions is $\mathscr{C} = H_1^{(1)}$, a Bessel function of the third kind, or Hankel function. When $|s|$ is large, the asymptotic formulae for this function give the following results (Watson, 1944, p. 197). For s real and negative

$$\mathscr{F} \sim (s + i\gamma)^{\frac{1}{4}} \exp[2\beta^{\frac{1}{2}}\{-|s|^{\frac{1}{2}} + \tfrac{1}{2}i\gamma|s|^{-\frac{1}{2}}\}] \tag{19.53}$$

and for s real and positive

$$\mathscr{F} \sim (s + i\gamma)^{\frac{1}{4}} \exp\{2\beta^{\frac{1}{2}}i|s|^{\frac{1}{2}}\}. \tag{19.54}$$

Here (19.54) represents a wave travelling to the left in fig. 19.2(a). This is the incident wave and may be identified with one of the W.K.B. solutions (19.45). There is no other term and therefore no reflected wave. The result applies in the limit $\gamma \to 0$ when the collision frequency is zero. The wave (19.53) is then evanescent and of negligible amplitude when s is large and negative. The energy in the incident wave has apparently disappeared. See end of § 19.5 for possible explanations.

To study the effect of the resonance on the penetration of the barrier, suppose now that B in (19.47) is comparable with $A/\alpha s$ when s has the smallest value that satisfies (19.48). Then the zero and infinity of n^2, (19.47), are so close together that we must seek a full wave solution that embraces both the zero and the infinity. If the electron collision frequency is zero, the differential equation (19.44) is now

$$\frac{d^2\mathscr{F}}{ds^2} + \left(\frac{\beta}{s} + \frac{\beta^2}{\eta^2}\right)\mathscr{F} = 0 \tag{19.55}$$

where β and η are constants. Fig. 19.2(b) shows how n^2 depends on s. It is zero where

Fig. 19.2. Height dependence of the squared refractive index n^2 near a resonance. In (a) the resonance is 'isolated'. The full curve is for $Z = 0$ (no collisions), and the broken curve is $\text{Re}(n^2)$ when Z is small and non-zero. In (b) $Z = 0$ and there is a zero of n^2 close to the resonance. In both figures the abscissa is $s = kz$ where z is height.

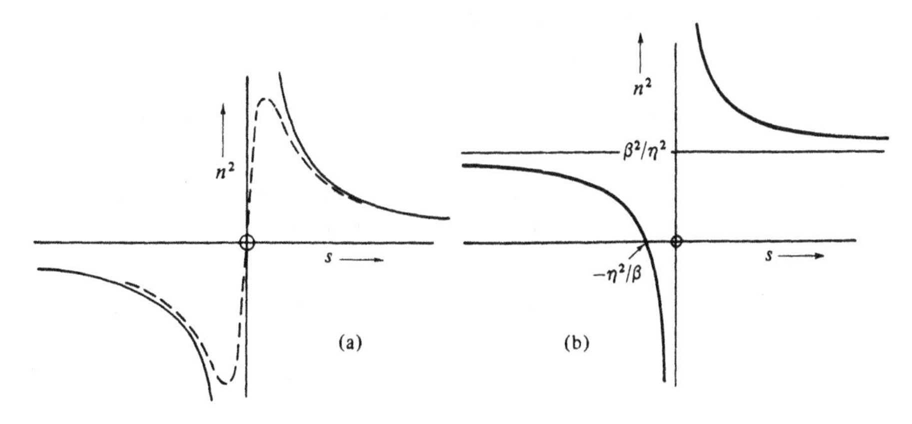

$s = -\eta^2/\beta$ and infinite where $s = 0$. If there is a small constant collision frequency, s in the second term of (19.55) is replaced by $s + i\gamma$ as in (19.50), where γ is real and positive. Then the pole and zero of n^2 both lie where Im(s) is small and negative. The pole is a singular point of the differential equation and there must therefore be a branch cut running from it to infinity in the negative imaginary half of the s plane. If $\gamma = 0$, this cut runs from $s = 0$.

Let

$$\zeta = 2i\beta s/\eta. \tag{19.56}$$

Then (19.55) becomes

$$\frac{d^2\mathscr{F}}{d\zeta^2} + \left(-\tfrac{1}{4} - \frac{\frac{1}{2}i\eta}{\zeta}\right)\mathscr{F} = 0 \tag{19.57}$$

which is the form given by Whittaker and Watson (1927, p. 337) for the confluent hypergeometric function..

Case (i). Wave incident from below

One solution of (19.57) is

$$\mathscr{F} = W_{k,m}(\zeta) \tag{19.58}$$

where $m = \pm\frac{1}{2}$, $k = -\frac{1}{2}i\eta$. When s is real, positive and large, $\arg s = 0$, $\arg\zeta = \frac{1}{2}\pi$ and (Whittaker and Watson, 1927, p. 343):

$$\mathscr{F} \sim e^{-\frac{1}{2}\zeta}\zeta^k = \exp\{-is\beta/\eta - \tfrac{1}{2}i\eta\ln s - \tfrac{1}{2}i\eta\ln(2i\beta/\eta)\}. \tag{19.59}$$

This represents an upgoing wave above the resonance level. The solution (19.58) may therefore be used to find the reflection and transmission coefficients for a wave incident from below. To do this it is necessary to find the asymptotic form of (19.58) when s is real and negative. Here $\arg s = \pi$, because in going from positive to negative s the branch cut, where Im(s) is negative, must not be crossed. Hence $\arg\zeta = \frac{3}{2}\pi$, and the asymptotic form for (19.58), given by Heading (1962b), is

$$W_{k,m}(\zeta) \sim \zeta^k \exp(-\tfrac{1}{2}\zeta) + \frac{2\pi i\exp(2\pi ik)}{(-\frac{1}{2}-m-k)!\,(-\frac{1}{2}+m-k)!}\zeta^{-k}\exp(\tfrac{1}{2}\zeta) \tag{19.60}$$

for $\pi \leqslant \arg\zeta \leqslant 2\pi$. For the special values $m = \pm\frac{1}{2}$, $k = -\frac{1}{2}\eta$ this becomes

$$\exp\{-is\beta/\eta - \tfrac{1}{2}i\eta\ln|s| + \tfrac{1}{2}\pi\eta - \tfrac{1}{2}i\eta\ln(2i\beta/\eta)\}$$
$$+\frac{2\pi i\exp(\pi\eta)}{(\frac{1}{2}i\eta)!\,(-1+\frac{1}{2}i\eta)!}\exp\{is\beta/\eta + \tfrac{1}{2}i\eta\ln|s| - \tfrac{1}{2}\pi\eta + \tfrac{1}{2}i\eta\ln(2i\beta/\eta)\}. \tag{19.61}$$

The first term represents an upgoing wave, the incident wave, and the second a downgoing wave, the reflected wave. The ratio of the moduli of these terms gives the modulus of the reflection coefficient R:

$$|R| = \frac{2\pi\exp(-\frac{1}{2}\pi\eta)}{|(\frac{1}{2}i\eta)!\,(-1+\frac{1}{2}i\eta)!|} = 1 - e^{-\pi\eta}. \tag{19.62}$$

Similarly the modulus of the transmission coefficient T is given, from (19.54), (19.61), by

$$|T| = e^{-\frac{1}{2}\pi\eta}. \tag{19.63}$$

Case (ii). Wave incident from above

Instead of using (19.56), let

$$\xi = -2i\beta s/\eta$$

so that (19.55) is

$$\frac{d^2\mathscr{F}}{d\xi^2} + \left(-\tfrac{1}{4} + \frac{\frac{1}{2}i\eta}{\xi}\right)\mathscr{F} = 0. \tag{19.64}$$

One solution is

$$\mathscr{F} = W_{k,m}(\xi) \tag{19.65}$$

where $m = \pm\frac{1}{2}$, $k = +\frac{1}{2}i\eta$. When s is real and positive $\arg\xi = -\frac{1}{2}\pi$ and for large $|s|$ (Heading, 1962b or Whittaker and Watson, 1927, p. 343):

$$\mathscr{F} \sim e^{-\frac{1}{2}\xi}\xi^k = \exp\{is\beta/\eta + \tfrac{1}{2}i\eta \ln s + \tfrac{1}{2}i\eta \ln(-2i\beta/\eta)\}. \tag{19.66}$$

This represents a downgoing wave from above, the incident wave. There is no upgoing wave and so the reflection coefficient is zero. When s is real and negative $\arg\xi = \frac{1}{2}\pi$. This is still within the range for which the expression (19.66) is valid, and for large $|s|$

$$\mathscr{F} \sim \exp\{is\beta/\eta + \tfrac{1}{2}in \ln|s| - \tfrac{1}{2}\pi\eta + \tfrac{1}{2}i\eta\ln(-2i\beta/\eta)\}. \tag{19.67}$$

This represents a downgoing wave at the bottom, the transmitted wave. The modulus of the transmission coefficient T is

$$|T| = e^{-\frac{1}{2}\pi\eta}. \tag{19.68}$$

The R and T used here are ratios of the values of $\mathscr{F}$ as given by (16.89). We are considering the extraordinary wave whose polarisation is $\rho_E = 1/\rho_O$, so that $E_x = \rho_O E_y$. For large $|s|$ the refractive index n is real and ρ_O is purely imaginary. Hence

$$\mathscr{F} = (\rho_O^2 - 1)^{\frac{1}{2}}E_y, \quad \mathscr{F}\mathscr{F}^* = (1 - \rho_O^2)E_yE_y^*. \tag{19.69}$$

Where the W.K.B. solutions are good approximations, (16.72) shows that $\mathscr{H}_x = -nE_y$, $\mathscr{H}_y = nE_x$. Hence from (2.64)

$$4Z_0\Pi_z = 2n(E_xE_x^* + E_yE_y^*) = 2n\mathscr{F}\mathscr{F}^*. \tag{19.70}$$

At very large s, for each W.K.B. solution, the vertical energy flux Π_z is independent of s. For $|s| \to \infty$ the two values of n are the same for positive and negative s. It follows that $|R|^2$ and $|T|^2$ are the ratios of the energy fluxes.

The results of this section may now be summarised as follows. For a wave incident from below

$$|R| = 1 - e^{-\pi\eta}, \quad |T| = e^{-\frac{1}{2}\pi\eta}, \quad |R^2| + |T^2| = 1 - e^{-\pi\eta} + e^{-2\pi\eta} < 1. \tag{19.71}$$

For a wave incident from above

$$|R|=0, \quad |T|=e^{-\frac{1}{2}\pi\eta}, \quad |R^2|+|T^2|=e^{-\pi\eta}<1. \tag{19.72}$$

In both cases the energy in the incident wave does not all reappear in the reflected and transmitted waves. Yet these results apply for the loss-free medium where there is no mechanism for absorbing energy. Possible explanations were discussed at the end of §19.5.

In this section it was assumed that near the resonance the height dependence of X was linear (19.46). An alternative method is to assume that X depends on an exponential function of height, as in (15.110), (15.114). Then instead of giving the confluent hypergeometric equation (19.57), (19.64), the equations are transformed to give the hypergeometric differential equation (15.97). Instead of a single pole and zero on the real axis, as in (19.55), there is an infinite number of each, on lines parallel to the imaginary s axis. This method was used by van Duin and Sluijter (1983). Their models had more assignable parameters than the one used here, so they were able to study more complicated cases. For example the squared refractive index n^2 could tend to different values, including negative values, above and below the resonance. One interesting result of this work is that for an incident wave approaching a resonance without first encountering a zero, the reflection coefficient does not now agree with (19.72) but has a small non-zero value.

19.7. Inversion of ionospheric reflection measurements

One of the main objects of using radio waves to probe the ionosphere from the ground is to find the height distribution $N(z)$ of electron concentration, and possibly also the electron collision frequency $\nu(z)$. At high frequencies, greater than about 1 MHz, this is done, for $N(z)$, by measuring the equivalent height function $h'(f)$ and using it to compute $N(z)$. Techniques for this were described in §§ 12.6, 13.6. This is an example of inversion of the data to give the required function. The method, however, gives very little information about the lowest parts of the ionosphere, that is the D-region. Here the refractive indices for the high frequency waves are close to unity and it has little effect on the function $h'(f)$. But radio waves in the very low frequency range, of order 10 to 100 kHz are reflected in the D-region and can be used to investigate its structure. Although radio pulses at frequencies as low as 50 kHz have been used (Watts 1957), the method is difficult and expensive because very large transmitting aerials are needed. The usual method is to measure the signals received from commercial transmitters. These emit morse or telex signals which can be received and measured just as if they were continuous unmodulated waves. They nearly always use vertical electric dipole aerials so that the waves are vertically polarised and the elements R_{11}, R_{21} of the reflection coefficient matrix $\boldsymbol{R}$ are measured. The dependence of their amplitude and phase on frequency, time of day and season, and distance from transmitter to receiver has been thoroughly studied (Bracewell *et al.*, 1951; Bracewell, 1952; Weekes and Stuart, 1952a; Bain *et al.*, 1952; Bracewell *et al.*, 1954; Straker, 1955; Belrose, 1968). By using a mobile receiver in a vehicle or aircraft, the dependence on angle of incidence has been measured (Hollingworth, 1926; Weekes, 1950; Bracewell *et al.*, 1951; Weekes and Stuart, 1952b). In some measurements, transmitting aerials that emit horizontally polarised waves have been constructed by laying aerial wires on poorly conducting ground such as desert soil, or ice, so that measurements of R_{12}, R_{22} were also possible (Macmillan, Rusch and Golden, 1960; Raghuram, Smith and Bell, 1974).

The effect on the calculated $\boldsymbol{R}$ of making changes in a given ionospheric model, that is in $N(z)$ and $\nu(z)$, was studied by Piggott, Pitteway and Thrane (1965). They were thus able to suggest modifications that might be made to improve the agreement with observations.

There is no direct method of using results of observations to calculate $N(z)$ and $\nu(z)$. One approach, used very successfully by Deeks (1966a, b) is a systematic application of trial and error. The data he used came from observations with many different frequencies and transmitter sites, and many receiving sites including mobile receivers. His method was to set up a computer program for calculating $\boldsymbol{R}$ for given functions $N(z)$, $\nu(z)$. Then by making a large number of trials with modifications made to different sections of the $N(z)$ curve, and with various functions $\nu(z)$, he was able to find those functions $N(z)$, $\nu(z)$ that best fit the observations. He concluded that if $N(z)$, at most heights z, is altered by 20%, the agreement with observations would be upset. Others have used the same method. Bain and May (1967) had more observations available, and suggested some modifications to Deeks' results. Thomas and Harrison (1970) applied the method for the D-region at night. The functions $N(z)$ and $\nu(z)$ deduced by Deeks' method have been widely accepted and used, and, as far as the author knows, no better results have since been published.

A general theory of the technique of inverting data to give the required results directly, without a trial and error process, has been given by Backus and Gilbert (1967, 1968, 1970). It was intended for application in the field of solid earth geophysics but the principle is equally applicable for the radio ionosphere problem at very low frequencies. The mathematical techniques that it uses are fairly advanced and include some functional analysis. A summary in the context of the radio observations was given by Mobbs (1978) in an unpublished thesis. The author is indebted to Dr A.J. Mobbs and Professor T.B. Jones of Leicester University for the opportunity of studying this work. The full general technique has not yet been brought to the stage where it can be applied more simply than the method of Deeks, and further work in this field would be of the greatest interest. Some simpler techniques have, however, been used, for example by Baybulatov and Krasnushkin (1966), Shellman (1970), Bailey and Jones (1974), Mambo, Nagano, Nakamura and Fukami (1983), Field, Warren and Warber (1983).

A very brief survey follows of the kinds of problem that have to be tackled in this study. We suppose that it is the electron height distribution $N(z)$ that is to be found, but the argument could be extended to include $\nu(z)$ also. Although $N(z)$ is a continuous function, it must be specified by a finite set of numbers. These might, for example, be tabulated values of N, or possibly $\ln N$, at equal intervals of height z, with the assumption that linear or higher order interpolation is used. When $N(z)$ is mentioned below it is this set of numbers that is referred to. They are to be found from a given set of data, such as $|R_{11}|$, $\arg R_{11}$, $|R_{21}|$, $\arg R_{21}$, etc. for various frequencies, angles of incidence, azimuths of propagation, etc. The following are some of the factors that have to be taken into account.

(a) All items of data are subject to error and those with least error must be given the greatest weight. The errors of different items may not be independent. In other words some items of data may be correlated and allowance can be made for this if the degree of correlation is known. For example, at 16 kHz, $|R_{11}|$ and $|R_{21}|$ may both vary from day to day but their ratio is found to be nearly constant.

(b) A given set of data is not enough to determine $N(z)$ uniquely. We try to find an $N(z)$ that minimises the difference between the calculated and observed values of the data.

(c) Some values of $N(z)$ may be 'insignificant' for some items of data. For example the values of N at a great height, z_2 say, cannot have any influence on $\boldsymbol{R}$ for low frequencies whose reflection level is well below z_2. The insignificant numbers in $N(z)$ are different for different items of data.

(d) The true function $N(z)$ may show some rapid variations. For example the work on partial reflections at high frequencies, § 11.13, shows that there are small scale irregularities in the D-region. For lower frequencies many of these variations of $N(z)$ would have a height scale very much less than one wavelength and could not appreciably influence the value of $\boldsymbol{R}$. The numbers that specify $N(z)$ give a running average over some height scale Δz. It is important to be able to find the greatest value Δz_0 of Δz that can be used without affecting the calculated $\boldsymbol{R}$. Then $\Delta z_0(z)$ is called the 'resolution' of the method, and it is usually dependent on z. If the tabulation interval for $N(z)$ is less than Δz_0, the table would contain redundant information that could not be found from the data.

It is item (b) above that is the main task of the inversion process. Baybulatov and Krasnushkin (1966) refer to it as the 'optimisation principle'. Shellman (1970) gives particular attention to the mapping of the standard errors of the items of data into error limits for the numbers that specify $N(z)$.

Suppose that a function $N(z)$ has been chosen and used to calculate the reflection coefficient $\boldsymbol{R}$. We now ask: how much change can be made in $N(z)$ at various heights z, without changing the elements of $\boldsymbol{R}$ by more than the limits of observational error? The answer to this was supplied by Bailey and Jones (1974). Their method illustrates well how the basic equations can be extended, and a summary of it is therefore given below.

The basic equations (7.80) are to be solved by a downward integration starting at the top $z = z_1$ and using two independent solutions $\mathbf{e}$ that represent only upgoing waves at the start. But the equation has four independent solutions. The other two can be chosen, for example, so that there are only downgoing waves at the top. The four independent columns $\mathbf{e}$ are used as the four columns of a square matrix $\mathbf{E}$ that satisfies (7.80) so that

$$\mathbf{E}' = -\mathrm{i}\mathbf{T}\mathbf{E} \tag{19.73}$$

where, as usual, a prime means $k^{-1}\,\mathrm{d}/\mathrm{d}z$. The columns that represent upgoing waves at the top are the first two columns of $\mathbf{E}$. Now for a given function $N(z)$ we wish to study how the reflection coefficient $\boldsymbol{R}$ is affected by changes of N. At the bottom, $z = z_0$, $\boldsymbol{R}$ is found, as explained in § 18.7, from the first two columns of $\mathbf{E}$. Thus we need to know how $\mathbf{E}(z_0)$ changes when $N(z)$ is changed. It is more convenient to use $\ln N$. Let the change of this be

$$\rho(z) = \delta(\ln N). \tag{19.74}$$

It is $\mathbf{T}$ that depends on $\ln N$ so let

$$\mathrm{d}\mathbf{T}/\mathrm{d}(\ln N) = \mathbf{J}(z). \tag{19.75}$$

Then if ρ is assumed to be very small, from (19.73):

$$\delta\mathbf{E}' = -\mathrm{i}(\mathbf{T}\delta\mathbf{E} + \mathbf{J}\mathbf{E}\rho). \tag{19.76}$$

Now

$$(\mathbf{E}^{-1}\delta\mathbf{E})' = \mathbf{E}^{-1}\delta\mathbf{E}' - \mathbf{E}^{-1}\mathbf{E}'\mathbf{E}^{-1}\delta\mathbf{E} = \mathbf{E}^{-1}(\delta\mathbf{E}' + \mathrm{i}\mathbf{T}\delta\mathbf{E}) = -\mathrm{i}\mathbf{E}^{-1}\mathbf{J}\mathbf{E}\rho \tag{19.77}$$

where (19.73), (19.76) have been used. At the top starting level $z = z_1$ it is assumed that $\rho(z_1) = 0$. Then the same starting value of $\mathbf{E}$ can be used for all functions $N(z)$ so that $\delta\mathbf{E}(z_1) = 0$. Now integrate with respect to z from z_0 to z_1. Then

$$\mathbf{E}^{-1}(z_0)\delta\mathbf{E}(z_0) = \mathrm{i}k\int_{z_0}^{z_1}\mathbf{E}^{-1}(\zeta)\mathbf{J}(\zeta)\mathbf{E}(\zeta)\rho(\zeta)\,\mathrm{d}\zeta. \tag{19.78}$$

This can be written

$$\delta\mathbf{E}(z_0) = \int_{z_0}^{z_1}\mathbf{K}_{\mathrm{E}}(\zeta)\rho(\zeta)\,\mathrm{d}\zeta, \text{ where } \mathbf{K}_{\mathrm{E}}(\zeta) = \mathrm{i}k\mathbf{E}(z_0)\mathbf{E}^{-1}(\zeta)\mathbf{J}(\zeta)\mathbf{E}(\zeta). \tag{19.79}$$

The four columns of $\mathbf{E}$ can be computed at the same time, so that $\mathbf{K}_{\mathrm{E}}(\zeta)$ can be found. A separate computation must be made for each different frequency, angle of incidence and azimuth of propagation used in the data.

When $\mathbf{E}(z_0)$ and $\mathbf{E}(z_0) + \delta\mathbf{E}(z_0)$ are known at the bottom of the ionosphere, they can be used to calculate the reflection coefficients $\boldsymbol{R}$ and $\boldsymbol{R} + \delta\boldsymbol{R}$ respectively, and thence the modulus and argument of any element. Some of these are the items used as data. Let γ_i be one such data item. Then if ρ is small $\delta\gamma_i$ is a linear combination of elements of the first two columns of $\delta\mathbf{E}(z_0)$ and

$$\delta\gamma_i = \int_{z_0}^{z_1} K_i(\zeta)\rho(\zeta)\,\mathrm{d}\zeta \tag{19.80}$$

where K_i is the same linear combination of the corresponding elements of $\mathbf{K}_\mathrm{E}$. The functions $K_i(\zeta)$ are called 'data kernels'. Bailey and Jones (1974) give an example where eight data items γ_i were used. They give curves of the function $K_i(\zeta)$ for these eight items, for typical functions $N(z)$. These curves are now to be used to estimate the error limits and resolution for the function $N(z)$. Clearly a large value of $K_i(\zeta)$ at height $z = \zeta$ indicates that only a small value of ρ at that height can be tolerated by the observations of data item γ_i. Similarly where any $K_i(\zeta)$ shows rapid oscillations, the height range for one oscillation leads to an estimate of the resolution Δz_0 needed for specifying $N(z)$. In a full treatment it is necessary to consider all the data items together and Bailey and Jones use a matrix method for doing this. One important result is that where the resolution range Δz_0 is small, the error limits of $N(z)$ are large, and vice versa. A very similar method was used by Mambo *et al.* (1983). They started with a test function $N(z)$, either found from rocket data or a simple exponential function. To this they applied the method by using radio reflection data for frequencies in the range 10 to 40 kHz, to give $\delta N(z)$, and thus they obtained an improved estimate of $N(z)$.

All these methods entail a very large amount of computing. They are of the greatest theoretical interest, but there is no indication yet that they are likely to replace the direct trial and error method of Deeks for elucidating the structure of the D-region.

19.8. Full wave solutions at higher frequencies

Although the main use of computed full wave solutions has been in the study of radio waves of very low frequency of order 10 to 200 kHz, there have been a few applications for higher frequencies. They were mainly investigations of reflection and transmission for very thin but intensely ionised layers, such as are believed to account for one form of sporadic E-layer, ($\mathrm{E_S}$), § 1.8. These have a thickness of order 1 km or less and a penetration frequency of a few MHz, and may give appreciable reflection for frequencies up to 7 to 10 MHz.

Measurements with rockets of the electron height distribution $N(z)$ in $\mathrm{E_S}$ layers have shown a great variety of shapes and thicknesses (Chessell 1971b). Some have very large $|\mathrm{d}N/\mathrm{d}z|$ at the top and bottom with almost constant N between, and others have much sharper maxima of $N(z)$ near the middle. Most workers on the theory have used model functions $N(z)$ with free space below and above. Thus Chessell (1971b) has used the functions of fig. 15.1(b,c), the parabolic model, fig. 12.3, and others. Miller and Smith (1977) have used $N(z)$ curves derived by fitting them to rocket data. All these authors have given many computed curves to show how the reflection and transmission coefficients depend on frequency. Nearly all the published results apply for vertically incident waves. This is the most important case because nearly all observations of E_S are made with ionosondes, § 1.7, at vertical incidence.

For a normal slowly varying ionosphere at frequencies of about 1 MHz or more we would expect to be able to use ray theory, and to find that the ordinary and extraordinary waves are propagated independently, and some examples were given in ch. 13. But in some parts of an E_S layer the gradient dN/dz must be very large. Here the various coupling points, § 16.5, can be close together and close to the real z axis. Thus there is strong coupling between the ordinary and extraordinary waves, and it is for this reason that full wave solutions are needed. An incident ordinary wave then gives rise to a large amount of the extraordinary wave component in both the reflected and transmitted waves, and similarly for an incident extraordinary wave. To study this it is convenient to calculate the reflection and transmission coefficients in a form that gives the ratios of the amplitudes of the ordinary and extraordinary waves. A form $\mathscr{R}$ of the reflection coefficient with this property was given by (18.56). Chessell (1971a, b) and Miller and Smith (1977) used a slightly different but essentially equivalent form, and they also used a transmission coefficient that relates the amplitudes of the ordinary and extraordinary waves.

The various coupling points can occur in the region of the complex z plane near where $N(z)$ is increasing in the lower half of the layer. The same coupling points appear again in a region near the upper half of the layer, where $N(z)$ is decreasing. Thus the coupling processes occur twice and there is a repeated exchange of energy between the ordinary and extraordinary waves. One effect of this is that for frequencies on either side of the penetration frequency, the range of frequency where partial penetration and reflection occurs is greater than it would be according to a simple ray theory without mode coupling. The occurrence of this coupling process was demonstrated by the authors cited above, and has been confirmed by Jalonen, Nygrén and Turunen (1982).

In this type of calculation there must be points of resonance which, for some frequencies, are on or near the real z axis. They may be allowed for by one of the methods of § 18.13. If collisions are neglected the resonance condition is (4.76) with $U = 1$, and this gives

$$f^4 - f^2(f_N^2 + f_H^2) + f_H^2 f_N^2 \cos^2\Theta = 0. \tag{19.81}$$

Hence at a point on the real z axis where the plasma frequency is f_N, a resonance will occur for two real frequencies f, one less than the gyro-frequency f_H and the other greater. Any given value of f_N, less than the value at the maximum of the layer, must occur twice, once in the lower half and once in the upper half of the layer. Thus for any E_S-layer with $N(z)$ going to zero at the boundaries, there must be two frequency ranges that give resonances on the real z axis. For any one frequency in these ranges there are two such resonances.

At a height near 115 km where E_S usually occurs, the electron collision frequency ν is of order $2 \times 10^4 s^{-1}$; see fig. 1.2. Thus at 1 MHz, $Z \approx 0.003$ and it is smaller still for higher frequencies. When collisions are allowed for the resonant points are no longer exactly on the real z axis. The path of integration can then be the real axis, but it must be remembered that near the resonance points one refractive index is very large so a very small step size must be used in the integration routine. An alternative method is to move the path of integration into the complex z plane away from the resonance points; see § 18.13. This method was used by Chessell (1971a). For the resonance in the lower half of the layer the resonance point is on the negative imaginary side of the real z axis, and there must be a branch cut running from it to infinity in the negative imaginary half plane. Similarly for the upper half of the layer, the resonance point and its branch cut are on the positive imaginary side. Thus the path of integration must be given a positive imaginary displacement in the lower half of the layer and a negative imaginary displacement in the upper half. Chessell (1971a) studied cases both with and without collisions. He confirmed that when collisions are neglected there can be an apparent disappearance of up to 65% of the incident energy. For a discussion of this see end of § 19.5.

For further discussion of the theory and observations of sporadic E-layers see Reddy and Rao (1968), Reddy (1968) and the references cited in §1.8.

Answers to problems

1.1. (a) 70 s. (b) 4 years.

2.2. (a) Circular. Area π. Normal 0, 0, 1.

(b) Ellipse. Area $\pi 2\sqrt{2}$, Axis ratio $2^{-\frac{1}{2}}$, Normal $\propto (1, 1, 0)$.

(c) Ellipse. Area 12π, Axis ratio $\frac{1}{2}$, Normal $\propto (-1, -1, 1)$.

(d) Linear. $|E| = \frac{1}{2}\sqrt{33}$. Direction $\propto (4, 4, 1)$.

(e) Circular. Area 2π, Normal 0, 0, 1.

(f) Ellipse. Area πab, Axis ratio a/b, Normal 0, 1, 0.

3.1. Ordered: 0.5 km s^{-1}. Thermal 150 km s^{-1}

3.8. (a) $\sim \omega_N^{-1}$. (b) 2×10^{15} s for temperature 800 K.

4.1. $\left\{\dfrac{1+|R^2|-|1+R^2|}{1+|R^2|+|1+R^2|}\right\}^{\frac{1}{2}}, \quad \frac{1}{2}|E_x^2|(1+|R^2|+|1+R^2|).$

4.2. $|(\rho - R)/(1 - R\rho)|$.

4.4. Smallest $0, 45°$. Greatest $(1-a)/(1+a)$, $90°$.

4.5. (a) Elliptical. Axis ratio $\frac{1}{3}$. (b) Partially circularly polarised. Energy flux ratio (circular component)/(unpolarised component) $= \frac{3}{2}$.

5.2. (a) $V = c$.

5.4. $Gn^2 + J = 0$. See (3.51) for meaning of G and J. If this is true for any Θ, it is true for all Θ.

5.5. $\tan\Theta = \{1 \pm (1 - 8\tan^2\beta)^{\frac{1}{2}}\}/2\tan\beta$. See fig. 5.19 for an example.

6.3. $$q^2 = C^2 - \frac{X(1-X)}{1 - X - \frac{1}{2}Y^2S^2 \pm Y\{(1-X)(C^2-X) + \frac{1}{4}Y^2S^4\}^{\frac{1}{2}}}.$$

7.1. $d^2y/dx^2 + y\omega^2 m/T = 0$.

$$y \sim m^{-\frac{1}{4}} \exp\left(\pm i\omega T^{-\frac{1}{2}} \int^x m^{\frac{1}{2}}\, dx\right).$$

Amplitude proportional to $m^{-\frac{1}{4}}$.

7.2. (M is the mean molecular mass, K is Boltzmann's constant and γ is the ratio of the specific heats)

$$\frac{d^2p}{dz^2} - \frac{T}{P}\frac{d}{dz}\left(\frac{P}{T}\right)\frac{dp}{dz} + \frac{\omega^2 M}{\gamma KT}p = 0,$$

$$\frac{d^2\zeta}{dz^2} + \frac{1}{P}\frac{dP}{dz}\frac{d\zeta}{dz} + \frac{\omega^2 M}{\gamma KT}\zeta = 0.$$

(a) $\zeta \sim T^{\frac{1}{4}} \exp\left\{\pm i\omega\left(\frac{M}{\gamma K}\right)^{\frac{1}{2}} \int^z T^{-\frac{1}{2}}\,dz\right\}$,

$$p \sim \mp i\omega\gamma P\left(\frac{M}{\gamma KT}\right)^{\frac{1}{2}}\zeta.$$

(b) $p \sim P^{\frac{1}{2}} \exp\left\{\pm i\omega\left(\frac{M}{\gamma KT}\right)^{\frac{1}{2}} z\right\}$,

$$\zeta \sim \frac{i}{\omega P}\left(\frac{KT}{\gamma M}\right)^{\frac{1}{2}} p,$$

7.3. $a \propto L^{\frac{1}{4}}$, $T \propto L^{\frac{1}{2}}$. $E \propto L - ^{\frac{1}{2}}$.

7.5. 0.25 km approx.

8.1. (a) $a_0 a_1$, (b) $a_0 \mathrm{Ai}'(0)$, (c) $-a_1 \mathrm{Ai}(0)$.

8.3. (a) $\pi\{\mathrm{Bi}(0)\mathrm{Ai}(z) - \mathrm{Ai}(0)\mathrm{Bi}(z)\}$.
(b) $\pi\{\mathrm{Bi}'(0)\,\mathrm{Ai}(z) - \mathrm{Ai}'(0)\,\mathrm{Bi}(z)\}$.
(c) $\{\mathrm{Bi}(-1)\,\mathrm{Ai}(z) - \mathrm{Ai}(-1)\,\mathrm{Bi}(z)\}/\{\mathrm{Bi}(-1)\,\mathrm{Ai}(0) - \mathrm{Ai}(-1)\,\mathrm{Bi}(0)\}$.
(d) No solution; homogeneous boundary conditions.
(e) $\mathrm{Ai}(z)/\mathrm{Ai}(0)$.
(f) As for (d)
(g) Any multiple of $\mathrm{Ai}(z)$.
(h) $\mathrm{Ai}(z)/\mathrm{Ai}'(-a)$.
(i) No solution; conditions incompatible

8.4. (a) a is smallest x that makes $\mathrm{Ai}(-x) = 0$. $y = \mathrm{Ai}(z-a)$. Homogeneous.
(b) a is any solution of $\mathrm{Ai}(-a)/\mathrm{Bi}(-a) = \mathrm{Ai}(1-a)/\mathrm{Bi}(1-a)$. $y = \mathrm{Bi}(-a)\mathrm{Ai}(z-a) - \mathrm{Ai}(-a)\mathrm{Bi}(z-a)$. Homogeneous.
(c) a can have any value provided $\mathrm{Ai}(-a) \neq 0$. $y = \mathrm{Ai}(z-a)/\mathrm{Ai}(-a)$. Inhomogeneous.
(d) a is solution of $\mathrm{Ai}'(-a) = 0$. $y = \mathrm{Ai}(z-a)$. Homogeneous.
(e) a is solution of $\mathrm{Ai}'(-a)/\mathrm{Bi}'(-a) = \mathrm{Ai}(1-a)/\mathrm{Bi}(1-a)$. $y = \mathrm{Bi}(1-a)\mathrm{Ai}(z-a) - \mathrm{Ai}(1-a)\mathrm{Bi}(z-a)$. Homogeneous
(f) a can have any value provided denominator of y is not zero. $y = [\mathrm{Bi}(1-a)\mathrm{Ai}(z-a) - \mathrm{Ai}(1-a)\mathrm{Bi}(z-a)]/$
$[\mathrm{Bi}(1-a)\mathrm{Ai}'(-a) - \mathrm{Ai}(1-a)\,\mathrm{Bi}'(-a)]$. Inhomogeneous.

8.5. $\dfrac{d^2u}{dz^2} - \dfrac{1}{z}\dfrac{dz}{dz} - zu = 0.$

8.6. (a) $y \sim z^{-\frac{1}{2}} \exp(\pm \frac{1}{4} i z^2)$.

(b) $\arg z = \pm\frac{1}{4}\pi,\ \pm\frac{3}{4}\pi$.

(c) $\arg z = 0,\ \pi,\ \pm\frac{1}{2}\pi$.

(d) $\mathrm{i}\sqrt{2}$ on all four Stokes lines.

9.1. $(2\pi/a)^{\frac{1}{2}}\left(1 - \dfrac{1}{8a} + \dfrac{9}{128a^2} - \cdots\right)$.

$(2\pi/a)^{\frac{1}{2}}\left(1 - \dfrac{5}{8a} + \dfrac{129}{128a^2} - \cdots\right)$.

In each case method (a) gives first term only.

9.2. $(\pi/a)^{\frac{1}{2}}\left(1 - \dfrac{1}{2a} + \dfrac{3}{4a^2} - \dfrac{3.5}{8a^3} + \cdots\right)$.

11.1. For reflected wave $E_y/\mathscr{H}_y = R_{21}/R_{11} = R_{22}/R_{12}$. Zero reflection when incident wave has $E_y/\mathscr{H}_y = R_{11}/R_{12} = R_{21}/R_{22}$.

11.2. $R_0 = \begin{pmatrix} 0 & -1 \\ -1 & 0 \end{pmatrix}$.

11.3. $(R_{11} - R_{22})^2 = -4R_{12}R_{21}$.

11.4. Incident wave $E_x = 12$, $E_y = 0$. Reflected wave circular $E_y = -\mathrm{i}E_x = 5$.

11.5. Electric fields of incident and reflected waves have same direction. Amplitude ratio $\frac{1}{3}$.

11.6. $R_{11} = (n + Z_0\tau/\rho - 1)/(n + Z_0\tau/\rho + 1)$ where n is refractive index, τ is thickness and ρ is resistivity. $\tau \approx 5 \times 10^{-10}$ m (about two atomic diameters).

11.7. $R_{11} = \dfrac{n^2\cos\theta - (n^2 - \sin^2\theta)^{\frac{1}{2}} - \gamma\sin\theta}{n^2\cos\theta + (n^2 - \sin^2\theta)^{\frac{1}{2}} + \gamma\sin\theta}$

where

$$n^2 = \frac{(U - X)^2 - Y^2}{U(U - X) - Y^2}, \quad \gamma = \frac{\mathrm{i}XY}{U(U - X) - Y^2}.$$

11.10. Angle shift $(z_0 - z_1)D/(D^2 + z_0^2)$, Range shift $2z_0(z_0 - z_1)(D^2 + Z_0^2)^{-\frac{1}{2}}$, for observer on ground at distance D from transmitter.

12.4. (a) $f_N^2 = \begin{cases} 0 & 0 \leqslant z < h_0 \\ f_1^2 + \alpha(z - h_0) & h_0 \leqslant z \end{cases}$

(b) $f_N^2 = \begin{cases} 0 & 0 \leqslant z \leqslant h_0 \\ \alpha(z - h_0) & h_0 \leqslant z \leqslant h_0 + f_1^2/\alpha \\ f_1^2 + \beta(z - h_0 - f_1^2/\alpha) & h_0 + f_1^2/\alpha \leqslant z \end{cases}$

14.4. The problem of the last sentence has no solution.

18.2. $\gamma = -(1 + \rho_2\rho_1^*)/(1 + \rho_1\rho_1^*)$.

(a) $\gamma = 0$. Solution (b) is the penetrating mode.

(b) $\gamma = -1$. In solution (b) the penetrating mode has been completely swamped. Its fields at the bottom cannot be found from the data.

(c) $\gamma = -1$. At the bottom the penetrating mode is linearly polarised with $E_y = \rho_2$, $\mathscr{H}_y = -C\rho_2$.

Bibliography

Abramovici, F. 1968. Diagnostic diagrams and transfer functions for oceanic wave guides. *Bull. seism. Soc. Am.* **58**, 427–56.

Abramowitz, M. and Stegun, I.A. 1965. *Handbook of mathematical functions.* New York: Dover.

Agy, V. and Davies, K. 1959. Ionospheric investigations using the sweep-frequency pulse technique at oblique incidence. *J. Res. Natn. Bur. Stand.* **63D**, 151–74.

Airy, G.B. 1838. On the intensity of light in the neighbourhood of a caustic. *Trans. Camb. Phil. Soc.* **6**, 379–402.

Airy, G.B. 1849. Supplement to a paper on the intensity of light in the neighbourhood of a caustic. *Trans. Camb. Phil. Soc.* **8**, 595–9.

Aitchison, G.J. 1957. Ionospheric demodulation of radio waves at vertical incidence. *Austral. J. Phys.* **10**, 204–7.

Aitken, A.C. 1956 (repr. 1964). *Determinants and matrices* (9th edn). Edinburgh: Oliver and Boyd.

Akhiezer, A.I., Lapshin, V.I. and Stepanov, K.N. 1976. Theory of damping of MHD waves in a high temperature plasma. *Zh. Eksp. Teor. Fiz.* **70**, 81–91. English Translation: *Sov. Phys. J.E.T.P.* **43**, 42–7.

Allis, W.P. 1956. Motions of ions and electrons. *Handbuch der Physik* **21**, 383–444.

Allis, W.P. 1959. Waves in a plasma. M.I.T. Research Laboratory Electronics Quarterly Progress Report 54.

Allis, W.P., Buchsbaum, S.J. and Bers, A. 1963. *Waves in anisotropic plasmas.* Cambridge, Mass: M.I.T. Press.

Al'pert, Ya.L. 1948. On the trajectories of rays in a magnetoactive ionised medium – the ionosphere (in Russian) *Izv. Akad. Nauk., S.S.S.R. Ser. Fiz.* **12**, 241–66.

Al'pert, Ya.L. 1980a. 40 years of whistlers. *J. Atmos. Terr. Phys.* **42**, 1–20.

Al'pert, Ya.L. 1980b. On the elements of the dielectric tensor, the refractive indices n_{12} and attenuation factors κ_{12} of a magnetically active collisional plasma. *J. Atmos. Terr. Phys.* **42**, 217–26.

Al'pert, Ya.L. 1983. *The near earth and interplanetary plasma,* Vols. 1 and 2. Cambridge University Press.

Al'pert, Ya.L., Budden, K.G., Moiseyev, B.S. and Stott, G.F. 1983. Electromagnetic radiation from a dipole source in a homogeneous magnetoplasma. *Phil. Trans. R. Soc. Lond.* **A. 309**, 503–57.

Altman, C. 1965. Use of the phase integral method to determine the reflection properties of the ionosphere. *J. Res. Natn. Bur. Stand.* **69D**, 511–19.

Altman, C. and Cory, H. 1969a. The simple thin film optical method in electromagnetic wave propagation. *Radio Sci.* **4**, 449–57.
Altman, C. and Cory, H. 1969b. The generalized thin film optical method in electromagnetic wave propagation. *Radio Sci.* **4**, 459–70.
Altman, C. and Cory, H. 1970. Multiple reflection processes in the D and E regions at low and very low frequencies. *J. Atmos. Terr. Phys.* **32**, 1439–55.
Altman, C., Cory, H. and Fijalkow, E. 1970. Coupling processes in the D and E regions at low and very low frequencies. I. Frequencies less than critical: a study of the daytime ionosphere. *J. Atmos. Terr. Phys.* **32**, 1457–73.
Altman, C. and Fijalkow, E. 1970. Coupling processes in the D and E regions at low and very low frequencies. II. Frequencies greater than critical: a study of the night time ionosphere. *J. Atmos. Terr. Phys.* **32**, 1475–88.
Altman, C. and Postan, A. 1971. Coupling processes in the D and E regions at low and medium frequencies. III. The 'coupling echo' and the transition through critical coupling. *J. Atmos. Terr. Phys.* **33**, 329–41.
Altman, C., Schatzberg, A. and Suchy, K. 1981. Symmetries and scattering relations in plane stratified anisotropic, gyrotropic and bianisotropic media. *Appl. Phys.* **B26**, 147–53.
Altman, C. and Suchy, K. 1979a. Eigenmode scattering relations for plane stratified gyrotropic media. *Appl. Phys.* **19**, 213–19.
Altman, C. and Suchy, K. 1979b. Generalisation of a scattering theorem for plane stratified gyrotropic media. *Appl. Phys.* **19**, 337–43.
Altman, C. and Suchy, K. 1980. Penetrating and non-penetrating mode propagation in the ionosphere in the light of eigenmode scattering relations. *J. Atmos. Terr. Phys.* **42**, 161–5.
Altman, C. *see* Fijalkow, Suchy.
Appleton, E.V. 1928. Some notes on wireless methods of investigating the electrical structure of the upper atmosphere I. *Proc. Phys. Soc.* **41**, 43–59.
Appleton, E.V. 1930. Some notes on wireless methods of investigating the electrical structure of the upper atmosphere II. *Proc. Phys. Soc.* **42**, 321–39.
Appleton, E.V. 1932. Wireless studies of the ionosphere. *J. Instn Elect. Engrs.* **71**, 642–50.
Appleton, E.V. 1935. A method of measuring the collision frequency of electrons in the ionosphere. *Nature, Lond.* **135**, 618–19.
Appleton, E.V. 1937. The Bakerian lecture. Regularities and irregularities in the ionosphere. *Proc. R. Soc. Lond.* **A. 162**, 451–79.
Appleton. E.V. and Beynon, W.J.G. 1940. The application of ionospheric data to radio communication problems. Part I. *Proc. Phys. Soc.* **52**, 518–33.
Appleton, E.V. and Beynon, W.J.G. 1947. The application of ionospheric data to radio communication problems. Part II. *Proc. Phys. Soc.* **59**, 58–76.
Appleton, E.V. and Beynon, W.J.G. 1955. An ionospheric attenuation equivalence theorem. *J. Atmos. Terr. Phys.* **6**, 141–8.
Appleton, E.V. and Builder, G. 1933. The ionosphere as a doubly refracting medium. *Proc. Phys. Soc.* **45**, 208–20.
Arantes, D.S. and Scarabucci, R.R. 1975. Full-wave analysis and coupling effects in a crossover region. *Radio Sci.* **10**, 801–11.
Arndt, D. *see* Dingle.
Askne, J. and Lisak, M. 1976. Wave propagation in an inhomogeneous multi-mode medium. *Radio Sci.* **11**, 969–76.
Aubry, M.P., Bitoun, J. and Graff, P. 1970. Propagation and group velocity in a warm magnetoplasma. *Radio Sci.* **5**, 635–45.
Aubry, M.P. *see* Bitoun.

Bachynski, M.P. *see* Shkarofsky.
Backus, G.E. and Gilbert, F. 1967. Numerical applications of a formalism for geophysical inverse problems. *Geophys. J. Roy. Astron. Soc.* **13**, 247–76.
Backus, G.E. and Gilbert, F. 1968. The resolving power of gross earth data. *Geophys. J. Roy. Astron. Soc.* **16**, 169–205.
Backus, G. and Gilbert, F. 1970. Uniqueness in the inversion of inaccurate gross earth data. *Phil. Trans. R. Soc. Lond.* A. **266**, 123–92.
Backus, G.E. *see* Gilbert.
Bailey, R.C. and Jones, T.B. 1974. The accuracy and resolution of model ionospheres derived from VLF propagation parameters. *J. Atmos. Terr. Phys.* **36**, 1059–69.
Bain, W.C., Bracewell, R.N., Straker, T.W. and Westcott, C.H. 1952. The ionospheric propagation of radio waves of frequency 16 kc/s over distances of about 540 km. *Proc. Inst. Elect. Engrs.* **99**, IV 250–9.
Bain, W.C. and May, B.R. 1967. D region electron-density distributions from propagation data. *Proc. Inst. Elect. Engrs.* **114**, II, 1593–7.
Banerjea, B.K. 1947. On the propagation of electromagnetic waves through the atmosphere. *Proc. R. Soc. Lond.* A. **190**, 67–81.
Banerjea, B.K. *see* Saha.
Baños, 1966. *Dipole radiation in the presence of a conducting half-space.* Oxford: Pergamon Press.
Barber, N.F. and Crombie, D.D. 1959. VLF reflections from the ionosphere in the presence of a transverse magnetic field. *J. Atmos. Terr. Phys.* **16**, 37–45.
Barnard, S. and Child, J.M. 1936. *Higher algebra.* London: Macmillan.
Barnes, E.W. 1908. A new development of the theory of hypergeometric functions. *Proc. Lond. math. Soc.* (2) **6**, 141–77.
Barrington, R.E. and Belrose, J.S. 1963. Preliminary results from the very low frequency receiver aboard Canada's Alouette satellite. *Nature, Lond.* **198**, 651–6.
Barrington, R.E., Belrose, J.S., and Mather, W.E. 1966. A helium whistler observed in the Canadian satellite Alouette II. *Nature, Lond.* **210**, 80–1.
Barrington, R.E. and Thrane, E. 1962. The determination of D-region electron densities from observations of cross-modulation. *J. Atmos. Terr. Phys.* **24**, 31–42.
Barrington, R.E. *see* Smith, R.L.
Barron, D.W. 1960. Numerical investigation of ionospheric problems. Ph. D. thesis, Cambridge (unpublished).
Barron, D.W. 1961. The numerical solution of differential equations governing the reflection of long radio waves from the ionosphere. IV. *Proc. R. Soc. Lond.* **A. 260**, 393–408.
Barron, D.W. and Budden, K.G. 1959. The numerical solution of differential equations governing reflection of long radio waves from the ionosphere III. *Proc. R. Soc. Lond.* A. **249**, 387–401.
Barrow, W.L. *see* Chu.
Batchelor, D.B. 1980. Budden tunnelling in parallel stratified plasmas. *Plasma Phys.* **22**, 41–55.
Batchelor, D.B. *see* Weitzner.
Bauer, P. 1975. Theory of waves incoherently scattered. *Phil. Trans. R. Soc. Lond.* A. **280**, 167–91.
Baybulatov, R.B. and Krasnushkin, P.Ye. 1966. Determination of the daytime electron concentration profile of the C and D layers of the ionosphere from measurements of the surface fields of ultralong radio waves and from the atmospheric pressure profile. *Geomagn. and Aeronomy* **6**, 819–27.
Becker, W. 1967. On the manual and digital computer methods used at Lindau for the

conversion of multifrequency ionograms to electron density-height profiles. *Radio Sci.* **2**, 1205–32.
Bell, T.F. *see* Raghuram.
Belrose, J.S. 1968. Low and very low frequency radio wave propagation. North Atlantic Treaty Organisation. AGARD lecture series XXIX: director Davies, K.
Belrose, J.S. 1970. Radio wave probing of the ionosphere by the partial reflection of radio waves (from heights below 100 km). *J. Atmos. Terr. Phys.* **32**, 567–96.
Belrose, J.S. and Burke, M.J. 1964. Study of the lower ionosphere using partial reflection. *J. Geophys. Res.* **69**, 2799–818.
Belrose, J.S., Burke, M.J., Coyne, T.N.R. and Reed, J.E. 1972. D-region measurements with the differential-absorption, differential phase partial-reflection experiments. *J. Geophys. Rev.* **77**, 4829–38.
Belrose, J.S. *see* Barrington, Smith, R.L.
Bennett, J.A. 1969. On the application of variation techniques to ray theory of radio propagation. *Radio Sci.* **4**, 667–78.
Bennett, J.A. 1974. Complex rays for radio waves in an absorbing ionosphere. *Proc. Instn. Elect. Electron. Engrs.* **62**, 1577–85.
Bennett, J.A. 1976. A new derivation of the W.K.B. solution for the coupled equations of radio waves in the ionosphere. *Proc. Camb. Phil. Soc.* **80**, 527–34.
Bennett, J.A. and Dyson, P.L. 1978. On the relation between phase path, group path and attenuation in a cold absorbing plasma. *J. Plasma Phys.* **19**, 325–48.
Bennett, J.A. *see* Dyson.
Benson, R.F. 1975. Source mechanism for terrestrial kilometric radiation. *Geophys. Res. Lett.* **2**, 52–5.
Bernstein, I.B. 1975. Geometric optics in space and time varying plasmas. *Phys. Fluids* **18**, 320–4.
Berreman, D.W. 1972. Optics in stratified and anisotropic media; 4×4 matrix formulation. *J. Opt. Soc. Amer.* **62**, 502–10.
Bers, A. *see* Allis, Fuchs.
Bertoni, H. *see* Ra.
Beynon, W.J.G. 1947. Oblique radio transmission in the ionosphere and the Lorentz polarization term. *Proc. Phys. Soc.* **59**, 97–107.
Beynon, W.J.G. 1974. Incoherent scatter sounding of the ionosphere. *Contemp. Phys.* **15**, 329–52.
Beynon, W.J.G. and Brown, G.M. (eds.) 1956. *Solar eclipses and the ionosphere.* Oxford: Pergamon Press.
Beynon, W.J.G. *see* Appleton.
Bibl, K. and Reinisch, B.W. 1978. The universal digital ionosonde. *Radio Sci.* **13**, 519–30.
Birkhoff, G. and Rota, G.C. 1962. *Ordinary differential equations.* Boston: Ginn and Co.
Bitoun, J., Graff, P. and Aubry, M.P. 1970. Ray tracing in warm magnetoplasma and applications to topside resonances. *Radio Sci.* **5**, 1341–9.
Bitoun, J. *see* Aubry, M.P.
Boardman, E.M. *see* Burton.
Bolt, B.A. *see* Bullen.
Booker, H.G. 1934. Some general properties of the formulae of the magnetoionic theory. *Proc. R. Soc. Lond.* A. **147**, 352–82.
Booker, H.G. 1935. The application of the magnetoionic theory to the ionosphere. *Proc. R. Soc. Lond.* A. **150**, 267–86.
Booker, H.G. 1936. Oblique propagation of electromagnetic waves in a slowly-varying non-isotropic medium. *Proc. R. Soc. Lond.* A. **155**, 235–57.
Booker, H. G. 1939. Propagation of wave packets incident obliquely on a stratified doubly refracting ionosphere. *Phil. Trans. R. Soc. Lond.* A. **237**, 411–51.

Booker, H.G. 1947. The elements of wave propagation using the impedance concept. *J. Inst. Elect. Engrs.* **94**, **III**, 171–98.
Booker, H. G. 1949. Application of the magnetoionic theory to radio waves incident obliquely upon a horizontally stratified ionosphere. *J. Geophys. Res.* **54**, 243–74.
Booker, H.G. 1959. Radio scattering in the lower ionosphere. *J. Geophys. Res.* **64**, 2164–77.
Booker, H.G. 1975. Electromagnetic and hydromagnetic waves in the ionosphere and magnetosphere. *Phil. Trans. R. Soc. Lond.* A. **280**, 57–93.
Booker, H.G. 1977. Fitting of multi-region ionospheric profiles of electron density by a single analytic function of height. *J. Atmos. Terr. Phys.* **39**, 619–23.
Booker, H.G. 1984. *Cold plasma waves.* The Hague: Martinus Nijhoff.
Booker, H.G. and Clemmow, P.C. 1950. The concept of an angular spectrum of plane waves and its relation to that of polar diagram and aperture distribution. *Proc. Instn. Elect. Engrs.* **97**, III, 11–17.
Booker, H.G., Fejer, J.A. and Lee, K.F. 1968. A theorem concerning reflection from a plane stratified medium. *Radio Sci.* **3**, 207–12.
Booker, H.G. and Gordon, W.E. 1950. A theory of radio scattering in the troposphere. *Proc. Instn. Radio Engrs.* **38**, 401–12.
Booker, H.G. and Seaton, S.L. 1940. Relation between actual and virtual ionospheric height. *Phys. Rev.* **57**, 87–94.
Booker, H.G. and Smith, E.K. 1970. A comparative study of ionospheric measurement techniques. *J. Atmos. Terr. Phys.* **32**, 467–97.
Booker, H.G. and Vats, H.O. 1985, Angular approximations for waves in a cold magnetoplasma. (in preparation)
Booker, H.G. and Walkinshaw, W. 1946. The mode theory of tropospheric refraction and its relation to wave-guides and diffraction. Report: Meteorological Factors in Radio Wave Propagation. pp. 80–127. London: Physical Society.
Born, M. and Wolf, E. 1970. *Principles of optics* (4th edn). Oxford: Pergamon Press.
Bossy, L. 1971. La détermination précise du matrizant pour la résolution des équations de Maxwell dans une ionosphère stratifiée. *Annales de la Société scientifique de Bruxelles.* **85**, I, 29–38.
Bossy, L. 1979. Wave propagation in stratified anisotropic media. *J. Geophys.* **46**, 1–14.
Bowles, K.L. 1958. Observation of vertical incidence scatter from ionosphere at 41 Mc/sec. *Phys. Rev. Lett.* **1**, 454–5.
Bowles, K.L. 1961. Incoherent scattering by free electrons as a technique for studying the ionosphere and exosphere: some observations and theoretical considerations. *J. Res.Natn. Bur. Stand.* **65D**, 1–14.
Boyd, R.L.F. 1968. Langmuir probes on spacecraft. In *Plasma diagnostics*, Ch. 12, pp. 732–76, ed. W. Lochte-Holtgreven. Amsterdam: North Holland.
Bracewell, R.N. 1952. The ionospheric propagation of radio waves of frequency 16 kc/s over distances of about 200 km. *Proc. Instn. Elect. Engrs.* **99**, IV, 217–28.
Bracewell, R.N., Budden, K.G., Ratcliffe, J.A., Straker, T.W. and Weekes, K. 1951. The ionospheric propagation of long and very long radio waves over distances less than 1000 km. *Proc. Instn Elect. Engrs.* **98**, III, 221–36.
Bracewell, R.N., Harwood, J. and Straker, T.W. 1954. The ionospheric propagation of radio waves of frequency 30–65 kc/s over short distances. *Proc. Instn Elect. Engrs.* **101**, IV, 154–62.
Bracewell, R.N. *see* Bain.
Bradbury, N.E. 1938. Ionisation, negative ion formation and recombination in the ionosphere. *Terr. Magn. Atmos. Elect.* **43**, 55–66.

Bradley, P.A. and Dudeney, J.R. 1973. A simple model of the vertical distribution of electron concentration in the ionosphere. *J. Atmos. Terr. Phys.* **35**, 2131–46.

Breit, G. and Tuve, M.A. 1926. A test of the existence of the conducting layer. *Phys. Rev.* **28**, 554–75.

Brekhovskikh, L.M. 1960. *Waves in layered media.* New York: Academic Press.

Bremmer, H. 1949. The propagation of electromagnetic waves through a stratified medium, and its W.K.B. approximation for oblique incidence. *Physica* **15**, 593–608.

Bremmer, H. 1951. The W.K.B. approximation as the first term of a geometric-optical series. *Commun. Pure Appl. Math.* **4**, 105–15.

Brice, N.M. *see* Gurnett, Smith, R.L.

Briggs, B.H. 1951. The determination of the collision frequency of electrons in the ionosphere from observations of the reflection coefficient of the abnormal E-layer. *J. Atmos. Terr. Phys.* **1**, 345–8.

Briggs, B.H. and Spencer, M. 1954. Horizontal movements in the ionosphere. A survey of experimental results. *Rep. Progr. Phys.* **17**, 245–80.

Brillouin, L. 1926. Remarques sur la méchanique ondulatoire. *J. Phys. Radium.* **7**, 353–68.

Brillouin, L. 1936. Propagation d'ondes électromagnétiques dans un tuyau. *Rev. Gen. d'Elec.* **40**, 227–39.

Brown, G.M. *see* Beynon.

Brown, J.N. *see* Watts.

Buchsbaum, S.J. *see* Allis.

Buckley, R. 1982. On the calculation of intensity in dispersive inhomogeneous media in the ray approximation. *Proc. R. Soc. Lond.* A. **380**, 201–9.

Budden, K.G. 1952. The theory of the limiting polarisation of radio waves reflected from the ionosphere. *Proc. R. Soc. Lond.* A. **215**, 215–33.

Budden, K.G. 1954. A reciprocity theorem on the propagation of radio waves via the ionosphere. *Proc. Camb. Phil. Soc.* **50**, 604–13.

Budden, K.G. 1955a. The numerical solution of differential equations governing reflection of long radio waves from the ionosphere. *Proc. R. Soc. Lond.* A. **227**, 516–37.

Budden, K.G. 1955b. The numerical solution of differential equations governing reflection of long radio waves from the ionosphere – II. *Phil. Trans. R. Soc. Lond.* A. **248**, 45–72.

Budden, K.G. 1961a (repr. 1966). *Radio waves in the ionosphere.* Cambridge University Press.

Budden, K.G. 1961b. *The wave-guide mode theory of wave propagation.* London: Logos Press.

Budden, K.G. 1961c. The edge focusing of a radio signal received from outside the earth by an aerial within the ionosphere. *Proc. R. Soc. Lond.* A. **263**, 552–66.

Budden, K.G. 1964. *Lectures on magnetoionic theory.* London: Blackie.

Budden, K.G. 1965. Effect of electron collisions on the formulas of magnetoionic theory. *Radio Sci. – J. Res. Natn. Bur. Stand.* **69D**, 191–211.

Budden, K.G. 1972. The theory of coupling of characteristic radio waves in the ionosphere. *J. Atmos. Terr. Phys.* **34**, 1909–21.

Budden, K.G. 1976. Radio caustics and cusps in the ionosphere. *Proc. R. Soc. Lond.* A. **350**, 143–64.

Budden, K.G. 1979. Resonance tunnelling of waves in a stratified cold plasma. *Phil. Trans. R. Soc. Lond.* A. **290**, 405–33.

Budden, K.G. 1980. The theory of radio windows in the ionosphere and magnetosphere. *J. Atmos. Terr. Phys.* **42**, 287–98.

Budden, K.G. 1983a. Approximations in magnetoionic theory. *J. Atmos. Terr. Phys.* **45**, 213–18.

Budden, K.G. 1983b. Heating processes in the ionosphere. *Plasma Phys.* **25**, 113–28.

Budden, K.G. and Clemmow, P.C. 1957. Coupled forms of the differential equations governing radio propagation in the ionosphere. II. *Proc. Camb. Phil. Soc.* **53**, 669–82.

Budden, K.G. and Cooper, E.A. 1962. Ionospheric equivalent heights of reflection calculated by a full wave method and by the phase integral method. *J. Atmos. Terr. Phys.* **24**, 609–18.

Budden, K.G. and Daniell, G.J. 1965. Rays in magnetoionic theory. *J. Atmos. Terr. Phys.* **27**, 395–415.

Budden, K.G. and Hugill, J. 1964. The theory of the cosmic background radio signal received within the ionosphere. *Proc. R. Soc. Lond.* A. **277**, 365–84.

Budden, K.G. and Jull, G.W. 1964. Reciprocity and nonreciprocity with magnetoionic rays. *Can. J. Phys.* **42**, 113–30.

Budden, K.G., Ratcliffe, J.A. and Wilkes, M.V. 1939. Further investigations of very long waves reflected from the ionosphere. *Proc. R. Soc. Lond.* A. **171**, 188–214.

Budden, K.G. and Smith, M.S. 1973. Theory of the polarisation of the ordinary wave reflected from the ionosphere in the limit of vertical incidence and vertical magnetic field. *J. Atmos. Terr. Phys.* **35**, 1909–26.

Budden, K.G. and Smith, M.S. 1974. The coalescence of coupling points in the theory of radio waves in the ionosphere. *Proc. R. Soc. Lond.* A. **341**, 1–30.

Budden, K.G. and Smith, M.S. 1976. Phase memory and additional memory in W.K.B. solutions for wave propagation in stratified media. *Proc. R. Soc. Lond.* A. **350**, 27–46.

Budden, K.G. and Stott, G.F. 1980. Rays in magnetoionic theory II. *J. Atmos. Terr. Phys.* **42**, 791–800.

Budden, K.G. and Terry, P.D. 1971. Radio ray tracing in complex space. *Proc. R. Soc. Lond.* A. **321**, 275–301.

Budden, K.G. *see* Al'pert, Barron, Bracewell.

Builder, G. *see* Appleton.

Bullen, K.E. and Bolt, B.A. 1985. *Introduction to the theory of seismology* (4th edn), Cambridge University Press.

Burke, M.J. *see* Belrose.

Burman, R. and Gould, R.N. 1954. The reflection of waves in a generalised Epstein profile. *Can. J. Phys.* **43**, 921–34.

Burman, R. and Gould, R.N. 1964. On certain exact wave functions for electromagnetic fields in planar stratified media. *J. Atmos. Terr. Phys.* **26**, 1127–9.

Burman, R. *see* Gould.

Burnside, W.S. and Panton, A.W. 1912. *The theory of equations* Vol. 1 (7th edn). London: Longmans Green. (Repr. 1960. New York: Dover.)

Burton, E.T. and Boardman, E.M. 1933. Audio-frequency atmospherics. *Proc. Instn. Radio Engrs.* **21**, 1476–94.

Businger, J.A. *see* Fleagle.

Calvert, W. *see* Fejer.

Capon, I.N. 1961. The application of ray tracing methods to radio signals from satellites. *Proc. Phys. Soc. Lond.* **77**, 337–45.

Capon, I.N. 1962. The application of ray tracing methods to calculations of skip distances for radio communications. *Proc. Phys. Soc. Lond.* **79**, 808–15.

Caron, P.R. *see* Tang.

Carpenter, D.L. 1963. Whistler evidence of a 'knee' in the magnetospheric ionization density profile. *J. Geophys. Res.* **68**, 1675–82.

Carpenter, D.L. 1968. Recent research on the magnetospheric plasmapause. *Radio Sci.* **3**, 719–25.

Carpenter, D.L. 1983. Some aspects of plasmapause probing by whistlers. *Radio Sci.* **18**, 917–25.

Carpenter, D.L., Dunckel, N. and Walkup, J.F. 1964. A new phenomenon: whistlers trapped below the protonosphere. *J. Geophys. Res.* **69**, 5009–17.

Carpenter, D.L. and Park, C.G. 1973. On what ionospheric workers should know about the plasmapause–plasmasphere. *Rev. Geophys. Space Phys.* **11**, 133–54.

Chapman, S. 1931a. The absorption and dissociative or ionizing effect of monochromatic radiation in an atmosphere on a rotating earth. *Proc. Phys. Soc.* **43**, 26–45.

Chapman, S. 1931b. The absorption and dissociative or ionizing effect of monochromatic radiation in an atmosphere on a rotating earth. Part II. Grazing incidence. *Proc. Phys. Soc.* **43**, 483–501.

Chapman, S. 1939. The atmospheric height distribution of band absorbed solar radiation. *Proc. Phys. Soc.* **51**, 93–109.

Chatterjee, B. 1952. Effect of magnetic field in oblique propagation over equatorial region. *Ind. J. Phys.* **26**, 297–312.

Chatterjee, B. 1953. Oblique propagation of radio waves over a curved earth. *Ind. J. Phys.* **27**, 257–68.

Chen, F.F. *see* White.

Cheng, F.T. and Fung, P.C.W. 1977a. New approach of studying electromagnetic mode coupling in inhomogeneous magnetized plasma. I. Normal incidence. *Astrophys. Space Sci.* **49**, 367–88.

Cheng, F.T. and Fung, P.C.W. 1977b. The equation of polarisation transfer in an inhomogeneous magnetized plasma. *Astrophys. Space Sci.* **49**, 427–42.

Cheng, F.T. *see* Fung.

Chessell, C.I. 1971a. The numerical calculation of reflection and transmission coefficients for thin highly ionized layers including the effect of the earth's magnetic field. *J. Atmos. Terr. Phys.* **33**, 1515–32.

Chessell, C.I. 1971b. Results of numerical calculation of reflection and transmission coefficients for thin highly ionized layers, and their application to sporadic E reflections. *J. Atmos. Terr. Phys.* **33**, 1803–22.

Chester, C., Friedman, B. and Ursell, F. 1957. An extension of the method of steepest descents. *Proc. Camb. Phil. Soc.* **53**, 599–611.

Child, J.M. *see* Barnard.

rectangular cross section. *Proc. Inst. Radio Engrs.* **26**, 1520–55.

CIRA 1972. Cospar International Reference Atmosphere 1972. Compiled by Cospar working Group 4. Akademie-Verlag. Berlin.

Clemmow, P.C. 1966. *The plane wave spectrum representation of electromagnetic fields.* Oxford: Pergamon Press.

Clemmow, P.C. 1973. *An introduction to electromagnetic theory.* Cambridge University Press.

Clemmow, P.C. and Dougherty, J.P. 1969. *Electrodynamics of particles and plasmas.* Reading, Mass.: Addison-Wesley.

Clemmow, P.C. and Heading, J. 1954. Coupled forms of the differential equations governing radio propagation in the ionosphere. *Proc. Camb. Phil. Soc.* **50**, 319–33.

Clemmow, P.C. and Mullaly, R.F. 1955. The dependence of the refractive index in magnetoionic theory on the direction of the wave normal. *The physics of the ionosphere*, pp. 340–50. London: The Physical Society.

Clemmow, P.C. *see* Booker, Budden.

Cohen, M.H. 1960. Magnetoionic mode coupling at high frequencies. *Astrophys. J.* **131**, 664–80.

Collar, A.R. *see* Frazer.

Collin, R.E. and Zucker, F.J. 1969. *Antenna theory* (Part I). New York: McGraw-Hill.
Computation Laboratory, Cambridge, Mass. 1945. *Tables of modified Hankel functions of order one third, and of their derivatives.* Harvard University Press.
Connor, K.A. and Felsen, L.B. 1974. Complex space-time rays and their application to pulse propagation in lossy media. *Proc. Instn. Electr. Electron. Engrs.* **62**, 1586–98.
Cooper, E.A. 1961. The properties of low frequency radio waves reflected from the ionosphere, calculated by the phase-integral method. *J. Atmos. Terr. Phys.* **22**, 122–41.
Cooper, E.A. 1964. Phase integral calculations of ionospheric reflections including a variable collision frequency. *J. Atmos. Terr. Phys.* **26**, 995–1005.
Cooper, E.A. *see* Budden.
Copson, E.T. 1935 (repr. 1976). *Theory of functions of a complex variable.* Oxford University Press.
Cory, H. *see* Altman, Fijalkow.
Coyne, T.N.R. *see* Belrose.
Crampin, S. 1970. The dispersion of surface waves in multilayered anisotropic media. *Geophys. J. Roy. Astr. Soc.* **21**, 387–402.
Crombie, D.D. *see* Barber.
Crompton, R.W., Huxley, L.G.H. and Sutton, D.J. 1953. Experimental studies of the motions of slow electrons in air with applications to the ionosphere. *Proc. R. Soc. Lond.* A. **218**, 507–19.
Crompton, R.W. *see* Huxley.
Dällenbach, W. 1942. Der Reziprozitätssatz des elektromagnetischen Feldes. *Arch. Für Elektrotechnik.* **36**, 153–65.
Daniell, G.J. 1966. The effect of the curvature of the earth on radioastronomical observations with satellites. *J. Atmos. Terr. Phys.* **28**, 1–8.
Daniell, G.J. *see* Budden.
Da Rosa, A.V. *see* Garriott.
Darwin, C.G. 1924. The optical constants of matter. *Trans. Camb. Phil. Soc.* **23**, 137–67.
Darwin, C.G. 1934. The refractive index of an ionised medium. *Proc. R. Soc. Lond.* A. **146**, 17–46.
Darwin, C.G. 1943. The refractive index of an ionised medium II. *Proc. R. Soc. Lond.* A. **182**, 152–66.
Davids, N. 1952. Theoretical group heights of reflection of 150 kc/s radio waves vertically incident on the ionosphere. *J. Atmos. Terr. Phys.* **2**, 324–36.
Davids, N. 1953. Optic axes and critical coupling in the ionosphere. *J. Geophys. Res.* **58**, 311–21.
Davids, N. and Parkinson, R.W. 1955. Wave solutions for critical and near critical coupling conditions in the ionosphere. *J. Atmos. Terr. Phys.* **7**, 173–202.
Davies, K. 1969. *Ionospheric radio waves.* Waltham, Mass.: Blaisdell.
Davies, K. 1981. Review of recent progress in ionospheric predictions. *Radio Sci.* **16**, 1407–30.
Davies, K., Watts, J.M. and Zacharisen, D.H. 1962. A study of F2 layer effects as observed with a Doppler technique. *J. Geophys. Res.* **67**, 601–9.
Davies, K. *see* Agy.
Debye, P. and Hückel, E. 1923. Zur Theorie der Elektrolyte. *Phys. Zeits.* **24**, 185–206 and 305–25.
Deeks, D.G. 1966a. D-region electron distributions in middle latitudes deduced from the reflection of long radio waves. *Proc. R. Soc. Lond.* A. **291**, 413–37.
Deeks, D.G. 1966b. Generalised full wave theory for energy dependent collision frequencies. *J. Atmos. Terr. Phys.* **28**, 839–46.
de Groot, W. 1930. Some remarks on the analogy of certain cases of propagation of

electromagnetic waves and the motion of a particle in a potential field. *Phil. Mag.* **10**, 521–40.

Denisov, N.G. 1957. On a singularity of the field of an electromagnetic wave propagated in an inhomogeneous plasma. *Soviet Phys. JETP.* **4**, 544. Russian original 1956: *Zh. Eksp. Terr. Fiz.* **31**, 609–19.

Deschamps, G.A. 1972. Ray techniques in electromagnetics. *Proc. Instn. Electr. Electron. Engrs.* **60**, 1022–35.

Deschamps, G.A. *see* Ziolkowski.

de Witt, R.N. *see* Warren.

Dingle, R.B. 1973. *Asymptotic expansions, their derivation and interpretation.* London: Academic Press.

Dingle, R.B., Arndt, D. and Roy, S.K. 1956a. The integrals $\mathscr{A}_p(x) = (p!)^{-1}\int_0^\infty \varepsilon^p(\varepsilon + x)^{-1}e^{-\varepsilon}d\varepsilon$ and $\mathscr{B}_p(x) = (p!)^{-1}\int_0^\infty \varepsilon^p(\varepsilon + x)^{-2}e^{-\varepsilon}d\varepsilon$ and their tabulation. *Appl. Sci. Res.* **6B**, 144–54.

Dingle, R.B., Arndt, D. and Roy, S.K. 1956b. The integrals $\mathscr{C}_p(x) = (p!)^{-1}\int_0^\infty \varepsilon^p(\varepsilon^2 + x^2)^{-1}e^{-\varepsilon}d\varepsilon$ and $\mathscr{D}_p(x) = (p!)^{-1}\int_0^\infty \varepsilon^p(\varepsilon^2 + x^2)^{-2}e^{-\varepsilon}d\varepsilon$ and their tabulation. *Appl. Sci. Res.* **6B**, 155–64.

Dougherty, J.P. and Farley, D.T. 1960. A theory of incoherent scatter of radio waves by a plasma. *Proc. R. Soc. Lond.* A. **259**, 79–99.

Dougherty, J.P. and Monaghan, J.J. 1965. Theory of resonances observed in ionograms taken by sounders above the ionosphere. *Proc. R. Soc. Lond.* A. **289**, 214–34.

Dougherty, J.P. *see* Clemmow.

Downing, A.M. 1979. Digitizing the conventional ionosonde. *Radio Sci.* **14**, 863–73.

Dudeney, J.R. 1978. An improved model of the variation of electron concentration with height in the ionosphere. *J. Atmos. Terr. Phys.* **40**, 195–203.

Dudeney, J.R. 1983. The accuracy of simple methods for determining the height of the maximum electron concentration of the F2-layer from scaled ionospheric characteristics. *J. Atmos. Terr. Phys.* **45**, 629–40.

Dudeney, J.R. *see* Bradley.

Duncan, W.J. *see* Frazer.

Dunckel, N. *see* Carpenter.

Dyson, P.L. and Bennett, J.A. 1979. General formulae for absorption of radio waves in the ionosphere. *J. Atmos. Terr. Phys.* **41**, 367–77.

Dyson, P.L. *see* Bennett.

Eckersley, T.L. 1931. On the connection between the ray theory of electric waves and dynamics. *Proc. R. Soc. Lond.* A. **132**, 83–98.

Eckersley, T.L. 1932a. Radio transmission problems treated by phase integral methods. *Proc. R. Soc. Lond.* A. **136**, 499–527.

Eckersley, T.L. 1932b. Long wave transmission, treated by phase integral methods. *Proc. R. Soc. Lond.* A. **137**, 158–73.

Eckersley, T.L. 1935. Musical atmospherics. *Nature. Lond.* **135**, 104–5.

Eckersley, T.L. 1950. Coupling of the ordinary and extraordinary rays in the ionosphere. *Proc. Phys. Soc.* B. **63**, 49–58.

Eckersley, T.L. and Millington, G. 1939. The limiting polarisation of medium waves reflected from the ionosphere. *Proc. Phys. Soc.* **51**, 110–28.

Eckhart, C. 1930. The penetration of a potential barrier by electrons. *Phys. Rev.* **35**, 1303–9.

Einziger, P.D. and Felsen, L.B. 1982. Evanescent waves and complex rays. *I.E.E.E. Trans. Ant. and Prop.* **AP-30**, 594.

Ellis, G.R.A. 1956. The Z propagation hole in the ionosphere. *J. Atmos. Terr. Phys.* **8**, 43–54.

Ellis, G.R.A. 1962. Use of Z-mode propagation for observing cosmic radio noise from earth satellites. *Nature. Lond.* **193**, 862–3.

Epstein, P.S. 1930. Reflection of waves in an inhomogeneous absorbing medium. *Proc. Nat. Acad. Sci., Wash.* **16**, 627–37.

Evans, J.V. 1969. Theory and practice of ionospheric study by Thomson scatter radar. *Proc. Instn. Elect. Electron. Engrs.* **57**, 496–530.

Evgrafov, M.A. and Fedoryuk, M.V. 1966. Asymptotic behaviour as $\lambda \to \infty$ of the solution of the equation $w''(z) - p(z, \lambda)w(z) = 0$ in the complex z-plane. *Uspekhi Matem. Nauk.* **21**, 3. English Translation: *Russ. Math. Survs.* **21**, 1–48.

Farley, D.T. *see* Dougherty.

Farmer, F.T. and Ratcliffe, J.A. 1935. Measurements of absorption of wireless waves in the ionosphere. *Proc. R. Soc. Lond.* A. **151**, 370–83.

Farmer, F.T. and Ratcliffe, J.A. 1936. Wireless waves reflected from the ionosphere at oblique incidence. *Proc. Phys. Soc.* **48**, 839–49.

Farmer, F.T. *see* Appleton.

Fedoryuk, M.V. *see* Evgrafov.

Feinstein, J. 1950. Higher order approximations in ionospheric wave propagation. *J. Geophys. Res.* **55**, 161–70.

Fejer, J.A. 1955. The interaction of pulsed radio waves in the ionosphere. *J. Atmos. Terr. Phys.* **7**, 322–32.

Fejer, J.A. 1960. Scattering of radio waves by an ionized gas in thermal equilibrium. *Can. J. Phys.* **38**, 1114–33.

Fejer, J.A. 1961. Scattering of radio waves by an ionized gas in thermal equilibrium in the presence of a uniform magnetic field. *Can. J. Phys.* **39**, 716–40.

Fejer, J.A. 1975. Alteration of the ionosphere by man-made waves. *Phil. Trans. R. Soc. Lond.* A. **280**, 151–65.

Fejer, J.A. 1979. Ionospheric modification and parametric instabilities. *Rev. Geophys. Space Phys.* **17**, 135–53.

Fejer, J.A. and Calvert, W. 1964. Resonance effects of electrostatic oscillations in the ionosphere. *J. Geophys. Res.* **69**, 5049–62.

Fejer, J.A. and Vice, R.W. 1959a. An investigation of the ionospheric D-region. *J. Atmos. Terr. Phys.* **16**, 291–306.

Fejer, J.A. and Vice, R.W. 1959b. The use of full-wave solutions in the interpretation of ionospheric absorption measurements. *J. Atmos. Terr. Phys.* **16**, 307–17.

Fejer, J.A. *see* Booker.

Felsen, L.B. 1964. Propagation and diffraction in uniaxially anisotropic regions. Parts 1 and 2. *Proc. Instn. Elect. Engrs.* **111**, 445–64.

Felsen, L.B. 1967. Lateral waves. In: *Electromagnetic wave theory*, ed. Brown, J., Part 1, pp. 11–44. Oxford: Pergamon Press.

Felsen, L.B. *see* Connor, Einziger, Ra.

Ferraro, A.J. *see* Weisbrod.

Ferraro, V.C.A. and Plumpton, C. 1966. *An introduction to magneto-fluid mechanics* (2nd edn). Oxford: Clarendon Press.

Feshbach, H. *see* Morse.

Field, E.C., Warren, R.E. and Warber, C.R. 1983. Calculation of ionospheric conductivity profiles by inverting VLF/LF reflection data. I. Isotropic propagation. *Radio Sci.* **18**, 452–60.

Fijalkow, E., Altman, C. and Cory, H. 1973. A full-wave study of ion-cyclotron whistlers in the ionosphere. *J. Atmos. Terr. Phys.* **35**, 317–37.

Fijalkov, E. *see* Altman.

Fleagle, R.G. and Businger, J.A. 1980. *An introduction to atmospheric physics.* Academic Press: New York.

Foley, G. *see* Jones, T.B.

Försterling, K. 1942. Über die Ausbreitung elektromagnetischer Wellen in einem magnetisierten Medium bei senkrechter Incidenz. *Hochfreq. Elek.* **59**, 10–22.

Försterling, K. and Lassen, H. 1933. Kurzwellenausbreitung in der Atmosphäre. *Hochfreq. Tech. Elektroakust.* **42**, 158–78.

Försterling, K. and Wüster, H.O. 1950. Über die Reflektion in einem inhomogenen Medium. *Ann. Phys. Lpz.* (6) **8**, 129–33.

Försterling, K. and Wüster, H.O. 1951. Über die Entstehung von Oberwellen in der Ionosphäre. *J. Atmos. Terr. Phys.* **2**, 22–31.

Frazer, R.A., Duncan, W.J. and Collar, A.R. 1938 (repr. 1963). *Elementary matrices.* Cambridge University Press.

Freidberg, J.P., Mitchell, R.W., Morse, R.L. and Rudsinski, L.I. 1972. Resonant absorption of laser light by plasma targets. *Phys. Rev. Lett.* **28**, 795–9.

Friedman, B. *see* Chester.

Fuchs, V., Ko. K. and Bers, A. 1981. Theory of mode conversion in weakly inhomogeneous plasma. *Phys. Fluids.* **24**, 1251–61.

Fukami, T. *see* Mambo.

Fung, P.C.W. and Cheng, F.T. 1977. New approach to studying electromagnetic mode coupling in inhomogeneous magnetized plasma. II. Oblique incidence. *Astrophys. Space Sci.* **50**, 361–81.

Fung, P.C.W. *see* Cheng.

Furry, W.H. 1947. Two notes on phase integral methods. *Phys. Rev.* **71**, 360–71.

Gans, R. 1915. Fortpflanzung des Lichts durch ein inhomogenes Medium. *Ann. Phys. Lpz.* **47**, 709–36.

Gardner, F.F. and Pawsey, J.L. 1953. Study of the ionospheric D-region using partial reflections. *J. Atmos. Terr. Phys.* **3**, 321–44.

Garrett, A.J.M. 1982a. A study of the mass ratio dependence of the mixed species collision integral for isotropic velocity distribution functions. *J. Plasma Phys.* **27**, 135–48.

Garrett, A.J.M. 1982b. A study of the mass ratio dependence of the mixed species collision integral. *J. Plasma Phys.* **28**, 233–54.

Garrett, A.J.M. 1983. Thermal relaxation and entropy for charged particles in a heat bath with fields. *J. Phys. A: Math. Gen.* **16**, 1505–16.

Garrett, A.J.M. 1985. Thermal relaxation of light dilute particles in a heat bath. To be published in *Physics Reports.*

Garriott, O.K., Da Rosa, A.V. and Ross, W.J. 1970. Electron content obtained from Faraday rotation and phase path length variations. *J. Atmos. Terr. Phys.* **32**, 705–27.

Garriott, O.K. *see* Rishbeth.

Gendrin, R. 1960. Guidage des sifflements radioélectroniques par le champ magnétique terrestre. *C.r. hebd., Acad. Sci.* Paris. **251**, 1085–7.

Gendrin, R. 1961. Le guidage des whistlers par le champ magnétique. *Plan. Space Sci.* **5**, 274–82.

Gibbons, J.J. and Nertney, R.J. 1951. A method for obtaining the solutions of ionospherically reflected long waves. *J. Geophys. Res.* **56**, 355–71.

Gibbons, J.J. and Nertney, R.J. 1952. Wave solutions, including coupling, of ionospherically reflected long radio waves for a particular E-region model. *J. Geophys. Res.* **57**, 323–38.

Gibbons, J.J. and Rao, R. 1957. Calculation of group indices and group heights at low frequencies. *J. Atmos. Terr. Phys.* **11**, 151–62.

Gibson, G.A. 1929. *An elementary treatise on the calculus.* London: Macmillan.

Gilbert, F. and Backus, G.E. 1966. Propagator matrices in elastic wave and vibration problems. *Geophysics.* **31**, 326–32.

Gilbert, F. *see* Backus.
Gillmor, C.S. 1976. The history of the term 'Ionosphere'. *Nature. Lond.* **262**, 347–8.
Gillmor, C.S. 1978. Early history of upper atmospheric physics research in Antarctica. In: *Upper atmosphere research in Antarctica*, ed. Lanzerotti, L.J. and Park, C.G. Vol. 29, pp. 236–62. Washington, D.C.: Amer. Geophys. Union.
Gillmor, C.S. 1981. Threshold to space: early studies of the ionosphere. In: *Space science comes of age*, ed. Hanle, P.A. and Chamberlain, V. del, pp. 101–14. Washington, D.C.: Smithsonian Instn. Press.
Gillmor, C.S. 1982. Wilhelm Altar, Edward Appleton and the magnetoionic theory. *Proc. Amer. Philos. Soc.* **126**, 395–440.
Ginzburg, V.L. 1970. *The propagation of electromagnetic waves in plasmas.* Oxford: Pergamon Press.
Giraud, A. and Petit, M. 1978. *Ionospheric techniques and phenomena.* Dordrecht: Reidel.
Golant, V.E. and Piliya, A.D. 1972. Linear transformation and absorption of waves in a plasma. *Soviet Phys. Uspekhi* **14**, 413–37. Russian original 1971: *Usp. Fiz. Nauk.* **104**, 413–57.
Golden, R.M. *see* Macmillan.
Goldstein, S. 1928. The influence of the earth's magnetic field on electric transmission in the upper atmosphere. *Proc. R. Soc. Lond.* A. **121**, 260–85.
Good, R.H. *see* Miller, S.C.
Goos, F. and Hänchen, H. 1947. Ein neuer und fundamentaler Versuch zur Totalreflexion. *Ann. Phys. Lpz.* (6) **1**, 333–46.
Gordon, W.E. 1958. Incoherent scattering of radio waves by free electrons with applications to space exploration by radar. *Proc. Instn. Radio Engrs.* **46**, 1824–9.
Gordon, W.E. *see* Booker.
Goubau, G. 1942. Reziprozität der Wellenausbreitung durch magnetisch doppelbrechende Medien. *Hochfreq. Elek.* **60**, 155–60.
Gould, R.N. and Burman, R. 1964. Some electromagnetic wave functions for propagation in stratified media. *J. Atmos. Terr. Phys.* **26**, 335–40.
Gould, R.N. *see* Burman.
Graff, P. *see* Aubry, Bitoun.
Graham, A. 1979. *Matrix theory and applications for engineers and mathematicians.* Chichester: Horwood.
Green, G. 1837. On the motion of waves in a variable canal of small depth and width. *Trans. Camb. Phil. Soc.* **6**, 457–62.
Greifinger, C. and Greifinger, P. 1965. Transmission of micropulsations through the lower ionosphere. *J. Geogphys. Res.* **70**, 2217–31.
Guha, U.C. *see* Saha.
Gurevich, A.V. 1978. *Non-linear phenomena in the ionosphere.* New York: Springer-Verlag.
Gurnett, D.A. 1975. The earth as a radio source; the nonthermal continuum. *J. Geophys. Res.* **80**, 2751–63.
Gurnett, D.A. and Shaw, R.R. 1973. Electromagnetic radiation trapped in the magnetosphere above the plasma frequency. *J. Geophys. Res.* **78**, 8136–49.
Gurnett, D.A., Shawhan, S.D., Brice, N.M. and Smith, R.L. 1965. Ion cyclotron whistlers. *J. Geophys. Res.* **70**, 1665–88.
Gurnett, D.A. *see* Rodriguez, Smith, R.L.
Hall, G. and Watt, J.M. (eds.). 1976. *Modern numerical methods for ordinary differential equations.* Oxford University Press.
Hamilton, W.R. 1931. Mathematical papers Vol. I. *Geometrical optics.* Cambridge University Press.

Hänchen, H. *see* Goos.

Hara, E.H. 1963. Approximations to the semiconductor integrals $\mathscr{C}_p(x)$ and $\mathscr{D}_p(x)$ for use with the generalized Appleton–Hartree magnetoionic formulas. *J. Geophys. Res.* **68**, 4388–9.

Harang, L. 1936. Vertical movements of the air in the upper atmosphere. *Terr. Magn. Atmos. Elect.* **41**, 143–60.

Hargreaves, J.K. 1979. *The upper atmosphere and solar–terrestrial relations.* New York: Van Nostrand Reinhold.

Harper, J.D. *see* Johler.

Harrison, M.D. *see* Thomas.

Hartree, D.R. 1929. The propagation of electromagnetic waves in a stratified medium. *Proc. Camb. Phil. Soc.* **25**, 97–120.

Hartree, D.R. 1931a. Optical and equivalent paths in a stratified medium, treated from a wave standpoint. *Proc. R. Soc. Lond.* A. **131**, 428–50.

Hartree, D.R. 1931b. The propagation of electromagnetic waves in a refracting medium in a magnetic field. *Proc. Camb. Phil. Soc.* **27**, 143–62.

Hartree, D.R. 1958. *Numerical analysis* (2nd edn.), Oxford: Clarendon Press.

Hartree, D.R., Michel, J.G.L. and Nicolson, P. 1946. Practical methods for the solution of the equations of tropospheric refraction. In: *Meteorological factors in radio wave propagation*, pp. 127–68. London: Physical Society.

Harvey, C.C. 1968. Low frequency waves above the ionosphere. Ph.D. thesis. Cambridge (unpublished).

Harwood, J. *see* Bracewell.

Haselgrove, C.B. and Haselgrove, J. 1960. Twisted ray paths in the ionosphere. *Proc. Phys. Soc.* **75**, 357–63.

Haselgrove, J. 1954. Ray theory and a new method for ray tracing. *The physics of the ionosphere*, pp. 355–64. London: Physical Society.

Haselgrove, J. 1957. Oblique ray paths in the ionosphere. *Proc. Phys. Soc.* **70B**, 653–62.

Haselgrove, J. 1963. The Hamilton ray path equations. *J. Atmos. Terr. Phys.* **25**, 397–9.

Haselgrove, J. *see* Thomas.

Hashimoto, K., Kimura, I. and Kumagai, H. 1977. Estimation of electron temperature by VLF waves propagating in directions near the resonance cone. *Plan. Space. Sci.* **25**, 871–7.

Hayes, M.G.W. 1971. Theory of the limiting polarization of radio waves emerging obliquely from the ionosphere. *Proc. R. Soc. Lond.* A. **324**, 369–90.

Heading, J. 1953. Theoretical ionospheric radio propagation. Ph.D. thesis. Cambridge (unpublished).

Heading, J. 1955. The reflection of vertically incident long radio waves from the ionosphere when the earth's magnetic field is oblique. *Proc. R. Soc. Lond.* A. **231**, 414–35.

Heading, J. 1957. The Stokes phenomenon and certain n^{th} order differential equations. *Proc. Camb. Phil. Soc.* **53**, 399–418 and 419–41.

Heading, J. 1958. *Matrix theory for physicists.* London: Longman.

Heading, J. 1961. The non-singular embedding of transition processes within a more general framework of coupled variables. *J. Res. Natn. Bur. Stand.* **65D**, 595–616.

Heading, J. 1962a. *An introduction to phase integral methods.* London: Methuen.

Heading, J. 1962b. The Stokes phenomenon and the Whittaker function. *J. Lond. Math. Soc.* **37**, 195–208.

Heading, J. 1962c. Phase-integral methods. *Q. J. Mech. Appl. Maths.* **15**, 215–44.

Heading, J. 1963. Composition of reflection and transmission coefficients. *J. Res. Natn. Bur. Stand.* **67D**, 65–77.

Heading, J. 1965. Refractive index profiles based on the hypergeometric and the confluent hypergeometric equation. *Proc. Camb. Phil. Soc.* **61**, 897–913.

Heading, J. 1969. Polarisation of obliquely reflected waves from an isotropic plane-stratified plasma. *Radio Sci.* **4**, 441–7.

Heading, J. 1970a. The equality of the moduli of certain ratios occurring in the connexion formulae of solutions of some transcendental differential equations. *Proc. Camb. Phil. Soc.* **67**, 347–61.

Heading, J. 1970b. *Mathematical methods in science and engineering.* London: Arnold.

Heading, J. 1971. Identities between reflection and transmission coefficients and electric field components for certain anisotropic modes of oblique propagation. *J. Plasma Phys.* **6**, 257–70.

Heading, J. 1972b. Generalisation of Booker's theorem. *Q. J. Mech. Appl. Maths.* **25**, 207–24.

Heading, J. 1973. Generalised reciprocity relations and conditions relating to solutions of *n* second-order partial differential equations. *Q. J. Mech. Appl. Maths.* **26**, 513–28.

Heading, J. 1975a. A new and extended concept for reflection and transmission coefficients for waves propagating in isotropic and anisotropic plasma. *J. Plasma Phys.* **13**, 147–65.

Heading, J. 1975b. A six-fold generalisation of ionospheric reciprocity. *Q. J. Mech. Appl. Maths.* **28**, 185–206.

Heading, J. 1975c. *Ordinary differential equations, theory and practice.* London: Elek Science.

Heading, J. 1976. Phase-integral methods. *Encyclopaedic dictionary of mathematics for engineers and applied scientists,* pp. 535–40. Ed. I.N. Sneddon. Oxford: Pergamon Press.

Heading, J. 1977a. Deductions from some artificial refractive index profiles based on the elementary functions. *Proc. Camb. Phil. Soc.* **81**, 121–32.

Heading, J. 1977b. Global phase-integral methods. *Q. J. Mech. Appl. Maths.* **30**, 281–302.

Heading, J. 1977c. Properties of two-way transmission in generalized wave propagation problems. *Proc. Camb. Phil. Soc.* **82**, 485–8.

Heading, J. 1979. Generalized approximate methods for transmission through a barrier governed by a differential equation of order 2*n*. *Proc. Camb. Phil. Soc.* **85**, 361–77.

Heading, J. 1980a. A unified approach to some integral formulae for reflection processes in an isotropic plasma. *J. Plasma Phys.* **24**, 519–32.

Heading, J. 1980b. Properties of some generalised hypergeometric functions governing coupling at transition points. *J. Instn. Maths. Appl.* **25**, 311–24.

Heading, J. 1981a. Theorem relating to the development of a reflection coefficient in terms of a small parameter. *J. Phys. A: Math. Gen.* **14**, 357–67.

Heading, J. 1981b. The equality of multiple integrals in the series development of a reflection coefficient for waves governed by a general linear differential equation. *J. Phys. A: Math. Gen.* **14**, 2881–95.

Heading, J. and Whipple, R.T.P. 1952. The oblique reflection of long wireless waves from the ionosphere at places where the earth's magnetic field is regarded as vertical. *Phil. Trans. R. Soc. Lond.* A. **244**, 469–503.

Heading, J. *see* Clemmow.

Heisler, L.H. *see* Munro.

Helliwell, R.A. 1965. *Whistlers and related ionospheric phenomena.* Stanford University Press.

Herlofson, N. 1951. Plasma resonance in ionospheric irregularities. *Arkiv för Fysik.* **3**, 247–97.

Herring, R.N. 1980. Tracing of Z-mode rays in the magnetosphere. *J. Atmos. Terr. Phys.* **42**, 885–97.
Herring, R.N. 1984. Theories of magnetospheric radio emissions. Ph.D. thesis. Cambridge (unpublished).
Heun, K. 1889. Zur Theorie der Riemann'schen Functionen zweiter Ordnung mit vier Verzweigungspunkten. *Math. Ann.* **33**, 161–79.
Hibberd, F.H. 1957. Self-distortion of radio waves in the ionosphere, near the gyro frequency. *J. Atmos. Terr. Phys.* **11**, 102–10.
Hines, C.O. 1957. Heavy ion effects in audio-frequency radio propagation. *J. Atmos. Terr. Phys.* **11**, 36–42.
Hirsch, P. and Shmoys, J. 1965. E-mode propagation in a plane-stratified plasma. *J. Res. Natn. Bur. Stand.* **D69**, 521–7.
Hollingworth, J. 1926. The propagation of radio waves. *J. Instn. Elect. Engrs.* **64**, 579–89.
Holt, O. 1963. Some experimental studies of the ionospheric D-region at high latitudes. Norwegian Defence Res. Estab. NRDE Rept. 46.
Horowitz, S. *see* Inoue.
Huang, X. and Reinisch, B.W. 1982. Automatic calculation of electron density profiles from digital ionograms 2. True height inversion of topside ionograms with profile fitting method. *Radio Sci.* **17**, 837–44.
Huang, X. *see* Reinisch.
Hückel, E. *see* Debye.
Hugill, J. *see* Budden.
Hutatsuishi, G. *see* Nagano.
Huxley, L.G.H. 1959. A discussion of the motion in nitrogen of free electrons with small energies with reference to the ionosphere. *J. Atmos. Terr. Phys.* **16**, 46–58.
Huxley, L.G.H. and Crompton, R.W. 1974. *The diffusion and drift of electrons in gases.* New York: John Wiley.
Huxley, L.G.H. and Ratcliffe, J.A. 1949. A survey of ionospheric cross-modulation. *Proc. Inst. Elect. Engrs.* **96**, III, 433–40.
Huxley, L.G.H. *see* Crompton.
Ince, E.L. 1927. *Ordinary differential equations.* London: Longmans Green.
Inoue, Y. and Horowitz, S. 1966a. Magnetoionic coupling in an inhomogeneous anisotropic medium. *Radio Sci.* **1**, 427–40.
Inoue, Y. and Horowitz, S. 1966b. Numerical solution of full wave equation with mode coupling. *Radio Sci.* **1**, 957–70.
Institute of Electrical and Electronics Engineers, Inc. 1969. I.E.E.E. standard definitions of terms for radio wave propagation. *I.E.E.E. Trans. Antennas and Propagation* **AP-17**, 271–5.
Inston, H.M. and Jeffs, R.M. 1968. Ground illumination by an h.f. transmitter through a twilight ionosphere. *Proc. Instn. Elect. Engrs.* **115**, 1089–96.
International Astronomical Union. 1973. Proceedings of 15th. General Assembly, Sydney 1973. Report of Commission 40, item 8. *Trans. Internatn. Astron. Union* **15B**, 166.
International Reference Atmosphere. *see* C.I.R.A. 1972.
Jackson, J.E. 1969. The reduction of topside ionograms to electron density profiles. *Proc. Inst. Elect. Electron. Engrs.* **57**, 960–76.
Jackson, J.E., Schmerling, E.R. and Whitteker, J.H. 1980. Mini-review on topside sounding. *I.E.E.E. Trans. Antennas and Propagation.* **AP-28**, 284–8.
Jacobsson, R. 1965. Light reflection from films of continuously varying refractive index. In: *Progress in optics.* Vol 5. 247–86. Ed. Wolf, E. Amsterdam: North Holland.
Jalonen, L., Nygrén, T. and Turunen, T. 1982. On the partial transparency of sporadic

E-layers at high latitudes. *J. Atmos. Terr. Phys.* **44**, 731–5.
Jeffreys, B. 1942. Note on the transparency of a potential barrier. (Research note). *Proc. Camb. Phil. Soc.* **38**, 401–5.
Jeffreys, B. 1956. The use of Airy functions in a potential barrier problem. *Proc. Camb. Phil. Soc.* **52**, 273–9.
Jeffreys, H. 1924. On certain approximate solutions of linear differential equations of the second order. *Proc. Lond. Math. Soc.* **23**, 428–36.
Jeffreys, H. 1931. *Cartesian tensors.* Cambridge University Press.
Jeffreys, H. 1962. *Asymptotic expansions.* Oxford: Clarendon Press.
Jeffreys, H. 1976. *The earth* (6th edn). Cambridge University Press.
Jeffreys, H. and Jeffreys, B.S. 1972. *Methods of mathematical physics* (3rd edn). Cambridge University Press.
Jeffs, R.M. *see* Inston.
Jenkins, F.A. and White, H.E. 1976. *Fundamentals of optics* (4th edn) Tokyo: McGraw-Hill Kogakusha Ltd.
Jespersen, J.L. *see* Pitteway.
Johler, J.R. and Harper, J.D. 1962. Reflection and transmission of radio waves at a continuously stratified plasma with arbitrary magnetic induction. *J. Res. Natn. Bur. Stand.* **66D**, 81–99.
Johler, J.R. and Walters, L.C. 1960. On the theory of reflection of low and very low radiofrequency waves from the ionosphere. *J. Res. Natn. Bur. Stand.* **64D**, 269–85.
Johnson, W.C. *see* Morgan.
Johnston, T.W. *see* Shkarofsky.
Jones, D. 1969. The effect of the latitudinal variation of the terrestrial magnetic field strength on ion cyclotron whistlers. *J. Atmos. Terr. Phys.* **31**, 971–81.
Jones, D. 1970. *The theory of the effect of collisions on ion cyclotron whistlers.* Plasma waves in space and in the laboratory. Vol 2, pp. 471–85. Ed. Thomas, J.O. and Landmark, B.J. Edinburgh University Press.
Jones, D. 1972. Refractive index and attenuation surfaces in the vicinity of the cross-over level. *Plan. Space Sci.* **20**, 1173–84.
Jones, D. 1976a. Source of terrestrial non-thermal radiation. *Nature, Lond.* **260**, 686–9.
Jones, D. 1976b. The second Z-propagation window. *Nature, Lond.* **262**, 674–5.
Jones, D. 1977a. Jovian and terrestrial 'harmonic' radiations. *Geophys. Res. Lett.* **4**, 121–4.
Jones, D. 1977b. Mode coupling of Z mode waves as a source of terrestrial kilometric and jovian decametric radiations. *Astron. Astrophys.* **55**, 245–52.
Jones, R.M. 1970. Ray theory for lossy media. *Radio Sci.* **5**, 793–801.
Jones, T.B. 1964. Ionospheric disturbances observed by means of standard frequency transmissions at oblique and vertical incidence. *J. Atmos. Terr. Phys.* **26**, 1294–8.
Jones, T.B. and Foley, G. 1972. Reflection conditions of a radio wave propagated obliquely through a horizontally stratified ionosphere. *J. Atmos. Terr. Phys.* **34**, 837–44.
Jones, T.B. and Spracklen, C.T. 1974. Ionospheric effects of the Flixborough explosion. *Nature. Lond.* **250**, 719–20.
Jones, T.B. and Wand, I.C. 1965. Doppler studies of complex reflections produced by travelling ionospheric disturbances. *J. Atmos. Terr. Phys.* **27**, 1111–13.
Jones, T.B. *see* Bailey.
Journal of Atmospheric and Terrestrial Physics. 1962. Special volume on 'Radio wave absorption in the ionosphere'. **23**, 1–379.
Jull, G.W. and Pettersen, G.W.E. 1962. Presentation at Fall U.R.S.I. meeting, Ottawa

(unpublished). Some results are quoted by Davies, K. (1969, p. 383).
Jull, G.W. and Pettersen, G.W.E. 1964. Origin of non-reciprocity on high frequency ionospheric paths. *Nature. Lond.* **201**, 483–4.
Jull, G.W. *see* Budden.
Kane, J.A. 1962. Re-evaluation of ionospheric electron densities and collision frequencies derived from rocket measurements of refractive index and attenuation. *J. Atmos. Terr. Phys.* **23**, 338–47.
Katsufrakis, J. *see* Smith, R.L.
Kelso, J.M. 1964. *Radio ray propagation in the ionosphere*, New York: McGraw-Hill.
Kelso, J.M., Nearhoof, H.J., Nertney, R.J. and Waynick, A.H. 1951. The polarisation of vertically incident long radio waves. *Ann. de. Géophys.* **7**, 215–44.
Kennel, C.F., Lanzerotti, L.J. and Parker, E.N. (eds). 1979. *Solar system plasma physics. Vol. II. Magnetospheres.* Amsterdam: North Holland.
Kennett, B.L.N. 1983. *Seismic wave propagation in stratified media.* Cambridge University Press.
Kimura, I. 1966. Effects of ions on whistler-mode ray tracing. *Radio Sci.* **1**, 269–83.
Kimura, I. *see* Hashimoto.
King, G.A.M. 1957. Relation between virtual and actual heights in the ionosphere. *J. Atmos. Terr. Phys.* **11**, 209–22.
King, J.W. 1959. Ionospheric self-demodulation and self-distortion of radio waves. *J. Atmos. Terr. Phys.* **14**, 41–9.
Knecht, R.W. *see* Wright.
Ko, K. *see* Fuchs.
Kramers, H.A. 1926. Wellenmechanik und halbzahlige Quantisierung. *Z. Physik.* **39**, 828–40.
Krasnushkin, P.E. 1961a. The boundary value problem of the propagation of electromagnetic waves in a spherically stratified anisotropic medium. *Doklady Akad. Nauk. S.S.S.R.* **138**, 813–16. English translation: *Sov. Phys. Doklady* **6**, 466–69.
Krasnushkin, P.E. 1961b. The propagation of long and ultra-long radio waves about the earth in the lower (C, D and E) layers of the ionosphere in the light of information theory. *Doklady Akad. Nauk. S.S.S.R.* **139**, 67–70. English Translation: *Sov. Phys. Doklady* **6**, 576–9.
Krasnushkin, P.E. *see* Baybulatov.
Kraus, J.D. 1966. *Radio astronomy.* New York: McGraw-Hill.
Kravtsov, Yu.A. 1964a. A modification of the geometrical optics method (in Russian). *Radiofizika.* **7**, 664–73.
Kravtsov, Yu.A. 1964b. Asymptotic solutions of Maxwell's equations near a caustic (in Russian). *Radiofizika.* **7**, 1049–56.
Kumagai, H. *see* Hashimoto.
Lacoume, J.L. 1967. Etude de la propagation des ondes électromagnétiques dans un magnétoplasma stratifié. *Ann. de Géophys.* **23**, 477–93.
Landau, L.D. and Lifshitz, E.M. 1960. *Electrodynamics of continuous media.* Oxford; Pergamon Press. Translated from the Russian by J.B. Sykes and J.S. Bell.
Landmark, B. 1955. A study of the limiting polarisation of high frequency waves reflected vertically from the ionosphere. Forsv. Forskn. Inst. Årb. (Oslo, Norway). Report No. 4.
Landmark, B. and Lied, R. 1957. Notes on a QL–QT transition level in the ionosphere. *J. Atmos. Terr. Phys.* **10**, 114–16.
Landmark, B. and Lied, F. 1962. Observations of the D-region from a study of ionospheric cross modulation. *J. Atmos. Terr. Phys.* **23**, 92–100.

Lange-Hesse, G. 1952. Vergleich der Doppelbrechung im Kristall und in der Ionosphäre. *Arch. Elektr. Übertr.* **6**, 149–58.

Langer, R.E. 1934. The asymptotic solutions of ordinary linear differential equations of the second order, with special reference to the Stokes phenomenon. *Bull. Amer. math. Soc.* **40**, 545–82.

Langer, R.E. 1937. On the connection formulas and the solutions of the wave equation. *Phys. Rev.* **51**, 669–76.

Lanzerotti, L.J. *see* Kennel.

Lapshin, V.I. *see* Akhiezer.

Lassen, H. 1927. Über den Einfluss des Erdmagnetfeldes auf die Fortpflanzung der elektrischen Wellen der drahtlosen Telegraphie in der Atmosphäre. *Elektrische Nachrichten-Technik*, **4**, 324–34.

Lassen, H. *see* Försterling.

Lawrence, R.S. *see* Little.

Lee, H.S. *see* Weisbrod.

Lee, K.F. *see* Booker.

Leinbach, H. *see* Little.

Lepechinsky, D. 1956. On the existence of a 'QL–QT' 'transition-level' in the ionosphere and its experimental evidence and effect. *J. Atmos. Terr. Phys.* **8**, 297–304.

Lied, R. *see* Landmark.

Lifshitz, E.M. *see* Landau.

Lindquist, R. 1953. An interpretation of vertical incidence equivalent height versus time recordings on 150 kc/s. *J. Atmos. Terr. Phys.* **4**, 10–27.

Lindsay, R.N. *see* Walker.

Liouville, J. 1837. Sur le développement des fonctions ou parties de fonctions en séries... *J. Math. Pures Appl.* [1], **2**, 16–35.

Lipson, S.G. and Lipson, H. 1969. *Optical physics.* Cambridge University Press.

Lisak, M. *see* Askne.

Little, C.G. and Lawrence, R.S. 1960. The use of polarisation fading of satellite signals to study electron content and irregularities in the ionosphere. *J. Res. Natn. Bur. Stand.* **64D**, 335–46.

Little, C.G. and Leinbach, H. 1959. The riometer–a device for the continuous measurement of ionospheric absorption. *Proc. Instn Radio Engrs. N.Y.* **47**, 315–19.

Liu, C.H. *see* Yeh.

Lobb, R.J. and Titheridge, J.E. 1977. The valley problem in bottomside ionogram analysis. *J. Atmos. Terr. Phys.* **39**, 35–42.

Lobb, R.J. *see* Titheridge.

Lockwood, G.E.K. 1963. Plasma and cyclotron spike phenomena observed in topside ionograms. *Can. J. Phys.* **41**, 190–4.

Lorentz, H.A. 1909. *Theory of electrons.* Leipzig: B.G. Teubner. (2nd ed. 1915 reprinted 1952 by Dover Publications, New York.)

Lotsch, H.K.V. 1970. Beam displacement at total reflection. The Goos–Hänchen effect. *Optik.* **32**, 116–37, 189–204, 299–319, 553–69.

Ludwig, D. 1966. Uniform asymptotic expansions at a caustic. *Comm. Pure Appl. Math.* **19**, 215–50.

Ludwig, D. 1970. Modified W.K.B. method for coupled ionospheric equations. *J. Atmos. Terr. Phys.* **32**, 991–8.

Macmillan, R.S., Rusch, W.V.T. and Golden, R.M. 1960. A very low frequency antenna for investigating the ionosphere with horizontally polarised waves. *J. Res. Natn. Bur. Stand.* **14**, 19–26.

Mambo, M., Nagano, I., Nakamura, K. and Fukami, T. 1983. A method of estimating the electron density profile of the D layer from a knowledge of VLF reflection coefficients. *Radio Sci.* **18**, 119–27.

Mambo, M. *see* Nagano.

Manning, L.A. 1947. The determination of ionospheric electron distribution. *Proc. Instn. Radio Engrs.* **35**, 1203–8.

Manning, L.A. 1949. The reliability of ionospheric height determinations. *Proc. Instn. Radio Engrs.* **37**, 599–603.

Margenau, H. 1946. Conduction and dispersion of ionized gases at high frequencies. *Phys. Rev.* **69**, 508–13.

Martyn, D.F. 1935. The propagation of medium radio waves in the ionosphere. *Proc. Phys. Soc.* **47**, 323–39.

Maslin, N.M. 1974. Theory of energy flux and polarisation changes of a radio wave with two magnetoionic components undergoing self demodulation in the ionosphere. *Proc. R. Soc. Lond.* A. **341**, 361–81.

Maslin, N.M. 1975a. Theory of the modifications imposed on the ionospheric plasma by a powerful radio wave reflected in the D or E region. *Proc. R. Soc. Lond.* A. **343**, 109–31.

Maslin, N.M. 1975b. Theory of heating in the lower ionosphere by obliquely incident waves from a powerful point transmitter. *Proc. R. Soc. Lond.* A. **346**, 37–57.

Maslin, N.M. 1976a. Caustics and cusps in an isotropic ionosphere. *J. Atmos. Terr. Phys.* **38**, 239–50.

Maslin, N.M. 1976b. Fields near a caustic and cusp in an isotropic ionosphere. *J. Atmos. Terr. Phys.* **38**, 1251–63.

Maslin, N.M. 1976c. The effect of a time varying collision frequency on a radio wave obliquely incident on the lower ionosphere. *Proc. R. Soc. Lond.* A. **348**, 245–63.

Maslin, N.M. 1976d. Theory of the modulation imposed on an obliquely incident radio wave reflected in a disturbed region of the lower ionosphere. *Proc. R. Soc. Lond.* A. **349**, 555–70.

Maslin, N.M. 1976e. Estimating ionospheric cross-modulation. *Proc. R. Soc. Lond.* A. **351**, 277–93.

Mather, W.E. *see* Barrington.

Mathews, J. and Walker, R.L. 1970. *Mathematical methods of physics* (2nd edn). New York: Benjamin.

Matsushita, S. *see* Smith, E.K.

May, B.R. *see* Bain.

Meek, J.H. 1948. Triple splitting of ionospheric rays. *Nature, Lond.* **161**, 597.

Melrose, D.B. 1974a. Mode coupling in the solar corona. I. Coupling near the plasma level. *Austral. J. Phys.* **27**, 31–42.

Melrose, D.B. 1974b. Mode coupling in the solar corona. II. Oblique incidence. *Austral. J. Phys.* **27**, 43–52.

Melrose, D.B. 1980. Mode coupling in the solar corona. VI. Direct conversion of the Langmuir waves into O-mode waves. *Austral. J. Phys.* **33**, 121–37.

Michel, J.G.L. *see* Hartree.

Miller, J.C.P. 1946. The Airy integral. *Brit. Ass. Math. Tables*, part vol. B. Cambridge University Press.

Miller, J.C.P. 1955. *Tables of Weber parabolic cylinder functions.* London: H.M.S.O.

Miller, K.L. and Smith, L.G. 1977. Reflection of radio waves by sporadic E layers. *J. Atmos. Terr. Phys.* **39**, 899–911.

Miller, S.C. and Good, R.H. 1953. A W.K.B.-type approximation to the Schrödinger equation. *Phys. Rev.* **91**, 174–9.
Millington, G. 1938a. Attenuation and group retardation in the ionosphere. *Proc. Phys. Soc.* **50**, 561–80.
Millington, G. 1938b. The relation between ionospheric transmission phenomena at oblique incidence and those at vertical incidence. *Proc. Phys. Soc.* **50**, 801–25.
Millington, G. 1949. Deviation at vertical incidence in the ionosphere. *Nature, Lond.* **163**, 213.
Millington, G. 1951. The effect of the earth's magnetic field on short wave communication by the ionosphere. *Proc. Instn Elect. Engrs.* **98**, IV. 1–14.
Millington, G. 1954. Ray path characteristics in the ionosphere. *Proc. Instn. Elect. Engrs.* **101**, IV, 235–49.
Millington, G. *see* Eckersley.
Mitchell, R.W. *see* Freidberg.
Moiseyev, B.S. *see* Al'pert.
Monaghan, J.J. *see* Dougherty.
Morgan, M.G. and Johnson, W.C. 1955. The observed polarisation of high-frequency sky-wave signals at vertical incidence. The Physics of the Ionosphere, pp. 74–7. London: Physical Society.
Morse, P.M. and Feshbach, H. 1953. *Methods of theoretical physics.* New York: McGraw-Hill.
Morse, R.L. *see* Freidberg.
Mott, N.F. 1952 (repr. 1962). *Elements of wave mechanics.* Cambridge University Press.
Mullaly, R.F. *see* Clemmow.
Munro, G.H. 1953. Reflexions from irregularities in the ionosphere. *Proc. R. Soc. Lond.* A. **219**, 447–63.
Munro, G.H. and Heisler, L.H. 1956a. Cusp type anomalies in variable frequency ionospheric records. *Austral. J. Phys.* **9**, 343–58.
Munro. G.H. and Heisler, L.H. 1956b. Divergence of radio rays in the ionosphere. *Austral. J. Phys.* **9**, 359–72.
Murata, H. 1974. Wave motions in the atmosphere and related ionospheric phenomena. *Space Sci. Rev.* **16**, 461–525.
Musgrave, M.J.P. 1970. *Crystal acoustics.* San Francisco: Holden-Day.
Nagano, I., Mambo, M. and Hutatsuishi, G. 1975. Numerical calculation of electromagnetic waves in an anisotropic multilayered medium. *Radio Sci.* **10**, 611–18.
Nagano, I. *see* Mambo.
Nakamura, K. *see* Mambo.
Namba, S. *see* Yokoyama.
National Bureau of Standards (U.S.A.) 1948. Ionospheric radio propagation. Circular 462.
Nearhoof, H.J. *see* Kelso.
Nertney, R.J. 1951. Comments concerning the paper 'Fine structure of the lower ionosphere' by Helliwell, Mallinckrodt and Kruse. *J. Geophys. Res.* **56**, 449–51.
Nertney, R.J. 1953. The lower E and D region of the ionosphere as deduced from long wave radio measurements. *J. Atmos. Terr. Phys.* **3**, 92–107.
Nertney, R.J. *see* Gibbons, Kelso.
Newstead, G. 1948. Triple magnetoionic splitting of rays reflected from the F2 region. *Nature, Lond.* **161**, 312.
Nicolet, M.J. 1953. The collision frequency of electrons in the ionosphere. *J. Atmos. Terr.*

Phys. **3**, 200–211.
Nicolet, M.J. 1959. Collision frequency of electrons in the terrestrial atmosphere. *Phys. Fluids.* **2**, 95–9.
Nicolson, P. *see* Hartree.
Northover, F.H. 1962. Reflection of electromagnetic waves from thin ionized gaseous layers. *J. Res Natn. Bur. Stand.* **66D**, 73–80.
Nuttall, J. 1965a. Singularities of the Green's function for a collisionless magnetoplasma. *Phys. Fluids.* **8**, 286–96.
Nuttall, J. 1965b. Theory of collective spikes observed by the Alouette topside sounder. *J. Geophys. Res.* **70**, 1119–25.
Nygrén, T. 1981. A simple method for obtaining reflection and transmission coefficients and fields for an electromagnetic wave in a horizontally stratified ionosphere. *Planet. Space Sci.* **29**, 521–8.
Nygrén, T. *see* Jalonen.
Olver, F.W.J. 1974. *Asymptotics and special functions.* New York: Academic Press.
Ott, H. 1942. Reflexion und Brechung von Kugelwellen; Effekte 2. Ordnung. *Ann. Phys. Lpz.* **41**, 443–66.
Oya, H. 1974. Origin of Jovian decameter wave emissions–conversion from the electron cyclotron plasma wave to the ordinary mode electromagnetic wave. *Planet. Space Sci.* **22**, 687–708.
Pack, J.L. *see* Phelps.
Panton, A.W. *see* Burnside.
Park, C.G. *see* Carpenter.
Parker, E.N. *see* Kennel.
Parkinson, R.W. 1955. The night time lower ionosphere as deduced from a theoretical and experimental investigation of coupling phenomena at 150 kc/s. *J. Atmos. Terr. Phys.* **7**, 203–34.
Parkinson, R.W. *see* Davids.
Paul, A.K. *see* Suchy.
Pawsey, J.L. *see* Gardner.
Pearcey, T. 1946. The structure of an electromagnetic field in the neighbourhood of a cusp of a caustic. *Phil. Mag.* **37**, 311–17.
Pedersen, P.O. 1927. *The propagation of radio waves.* Copenhagen: Danmarks Naturvidenskabelige Samfund.
Petit, M. *see* Giraud.
Pettersen, G.W.E. *see* Jull.
Pfister, W. 1949. Effect of the D ionospheric layer on very low frequency radio waves. *J. Geophys. Res.* **54**, 315–37.
Pfister, W. 1953. Magnetoionic multiple splitting determined with the method of phase integration. *J. Geophys. Res.* **58**, 29–40.
Phelps, A.V. 1960. Propagation constants for electromagnetic waves in weakly ionised dry air. *J. Appl. Phys.* **31**, 1723–9.
Phelps, A.V. and Pack, J.L. 1959. Electron collision frequencies in nitrogen and in the lower ionosphere. *Phys. Rev. Lett.* **3**, 340–2.
Pickle, C.B. 1951. Atlas of ionosphere records with interpretations and instructions on scaling. National Bureau of Standards (U.S.A). Report 1106.
Piggott, W.R., Pitteway, M.L.V., and Thrane, E.V. 1965. The numerical calculation of wave-fields, reflection coefficients and polarisations for long radio waves in the lower ionosphere. II. *Phil. Trans. R. Soc. Lond.* A. **257**, 243–71.

Piggott, W.R. and Rawer, K. 1972, 1978. *U.R.S.I. handbook of ionogram interpretation and reduction* (2nd edn). Report UAG-23, (1972). Revision of chapters 1–4. Report UAG-23A (1978). World Data Centre A for Solar Terrestrial Physics, N.O.A.A., Boulder. Col. U.S.A.

Piggott, W.R. and Thrane, E.V. 1966. The electron densities in the E and D regions above Kjeller. *J. Atmos. Terr. Phys.* **28**, 467–79.

Piliya, A.D. *see* Golant.

Pitteway, M.L.V. 1958. The reflexion of radio waves from a stratified ionosphere modified by weak irregularities. *Proc. R. Soc. Lond.* A **246**. 556–69.

Pitteway, M.L.V. 1959. Reflexion levels and coupling regions in a horizontally stratified ionosphere. *Phil Trans. R. Soc. Lond.* A **252**, 53–68.

Pitteway, M.L.V. 1960. The reflexion of radio waves from a stratified ionosphere modified by weak irregularities. II. *Proc. R. Soc. Lond.* A **254**, 86–100.

Pitteway, M.L.V. 1965. The numerical calculation of wave-fields, reflection coefficients and polarisations for long radio waves in the lower ionosphere I. *Phil. Trans. R. Soc. Lond.* A **257**, 219–41.

Pitteway, M.L.V. and Jespersen, J.L. 1966. A numerical study of the excitation, internal reflection and limiting polarisation of whistler waves in the lower ionosphere. *J. Atmos. Terr. Phys.* **28**, 17–43.

Pitteway, M.L.V. *see* Piggott, Wright.

Plumpton, C. *see* Ferraro.

Poeverlein, H. 1948. Strahlwege von Radiowellen in der Ionosphaäre. S.B. Bayer. Akad. Wiss., Math–nat Klasse: 175–201.

Poeverlein, H. 1949. Strahlwege von Radiowellen in der Ionosphäre. II. Theoretische Grundlagen. *Z. Angew. Phys.* **1**, 517–25.

Poeverlein, H. 1950. Strahlwege von Radiowellen in der Ionosphäre. III. Bilder theoretisch ermittelter Strahlwege. *Z. Angew. Phys.* **2**, 152–60.

Poeverlein, H. 1958a. Low-frequency reflection in the ionosphere. I. *J. Atmos. Terr. Phys.* **12**, 126–39.

Poeverlein, H. 1958b. Low frequency reflection in the ionosphere. II. *J. Atmos. Terr. Phys.* **12**, 236–47.

Poeverlein, H. 1967. Ionospheric wave theory using coupled vacuum modes. *Radio Sci.* **2**, 905–11.

Postan, A. *see* Altman.

Poston, T. and Stewart, I.N. 1978. *Catastrophe theory and its applications.* London: Pitman.

Price, G.H. 1964. Propagation of electromagnetic waves through a continuously varying stratified anisotropic medium. *Radio Sci.* **68D**, 407–18.

Proc. Inst. Elect. Electronic Engrs. 1969. Special issue on topside sounding and the ionosphere. **57**, June 1969.

Ra, J.W., Bertoni, H. and Felsen, L.B. 1973. Reflection and transmission of beams at a dielectric interface. *S.I.A.M. J. Appl. Math.* **24**, 396–412.

Radio Science. 1967. Analysis of ionograms for electron density profiles. Special issue. **2**, No. 10. October 1967.

Radio Science. 1968. Special issue on ray tracing. **3**, 1–119.

Radio Science. 1974. Special issue: ionospheric modification by high power transmitters. **9**, 881–1090.

Raghuram, R., Smith, R.L. and Bell, T.F. 1974. VLF antarctic antenna: impedance and efficiency. *I.E.E.E. Trans. Antenn. Prop.* **AP-22**, 334–8.

Rao, M.M. *see* Reddy.

Rao, R. *see* Gibbons.

Ratcliffe, J.A. 1939. The effect of the Lorentz polarisation term in ionospheric calculations. *Proc. Phys. Soc.* **51**, 747–56.
Ratcliffe, J.A. 1951. A quick method for analysing ionospheric records. *J. Geophys. Res.* **56**, 463–85.
Ratcliffe, J.A. 1954. The physics of the ionosphere (Kelvin lecture). *Proc. Instn. Elect. Engrs.* **101**, I, 339–46.
Ratcliffe, J.A. 1956. Some aspects of diffraction theory and their application to the ionosphere. *Rep. Prog. Phys.* **19**, 188–267.
Ratcliffe, J.A. 1959. *The magnetoionic theory and its applications to the ionosphere.* Cambridge University Press (repr. 1962).
Ratcliffe, J.A. 1970. *Sun, earth and radio.* London: Weidenfeld and Nicolson (World University Library).
Ratcliffe, J.A. 1972. *An introduction to the ionosphere and magnetosphere.* Cambridge University Press.
Ratcliffe, J.A. and Shaw, I.J. 1948. A study of the interaction of radio waves. *Proc. R. Soc. Lond.* A **193**, 311–43.
Ratcliffe, J.A., *see* Appleton, Bracewell, Budden, Farmer, Huxley.
Rawer, K. 1939. Elektrische Wellen in einem geschichteten Medium. Zur Frage der partiellen Reflexion and zur Berechnung der scheinbaren Höhe von Ionosphärenschichten. *Ann. Phys. Lpz.* (5). **35**, 385–416.
Rawer, K. 1948. Optique géométrique de l'ionosphère. *Rev. Scient. Paris.* **86**, 585–600.
Rawer, K. 1952. *The ionosphere.* London: Crosby Lockwood.
Rawer, K. 1962. Propagation problems with space radio communications *J. Res. Natn. Bur. Stand.* **66D**, 375–95.
Rawer, K. 1981. International reference ionosphere–I.R.I. 1979. (Report UAG-82, ed. Lincoln, J.V. and Conkright, R.O). Boulder, Colorado: World data centre A for solar-terrestrial physics.
Rawer, K. 1984. Note concerning the international reference ionosphere. *J. Atmos. Terr. Phys.* **46**, 91.
Rawer, K. and Suchy, K. 1967. Radio observations of the ionosphere. *Handb. der Phys.* **49/2**, 1–546.
Rawer, K. and Suchy, K. 1976. Remarks concerning the dispersion equation of electromagnetic waves in a magnetised cold plasma. *J. Atmos. Terr. Phys.* **38**, 395–8.
Rawer, K. *see* Piggott, Suchy.
Rayleigh, (J.W. Strutt), Baron 1912. On the propagation of waves through a stratified medium, with special reference to the question of reflection. *Proc. R. Soc. Lond.* A. **86**, 207–26. Scientific papers, VI. 71–90.
Reddy, C.A. 1968. Physical significance of the E_S parameters f_bE_S, fE_S, f_OE_S. 2. Causes of partial reflections from E_S. *J. Geophys. Res.* **73**, 5627–47.
Reddy, C.A. and Rao, M.M. 1968. On the physical significance of the E_S parameters f_bE_S, fE_S. and f_OE_S. *J. Geophys. Res.* **73**, 215–24.
Reed, J.E. *see* Belrose.
Reinisch, B.W. and Huang, X. 1982. Automatic calculation of electron density profiles from digital ionograms. 1. Automatic O and X trace identification for topside ionograms. *Radio Sci.* **17**, 421–34.
Reinisch, B.W. and Huang, X. 1983. Automatic calculation of electron density profiles from digital ionograms. 3. Processing of bottomside ionograms. *Radio Sci.* **18**, 477–92.
Reinisch, B.W. *see* Bibl, Huang.
Rishbeth, H. and Garriott, O.K. 1969. *Introduction to ionospheric physics.* New York: Academic Press.
Robbins, A. *see* Thomas.

Rodriguez, P. and Gurnett, D.A. 1971. An experimental study of very low frequency mode coupling and polarisation reversal. *J. Geophys. Res.* **76**, 960–71.
Rönnmark, K. 1984. Ray tracing in dissipative media. *Annales Geophysica.* **2**, 57–60.
Ross, W.J. *see* Garriott.
Rota, G.C. *see* Birkhoff.
Roy, S.K. *see* Dingle.
Royal Society. 1975. *Quantities, units and symbols* (2nd edn). London: Royal Society.
Rudsinski, L.I. *see* Freidberg.
Rusch, W.V.T. *see* Macmillan.
Rydbeck, O.E.H. 1942a. The reflection of electromagnetic waves from a friction-free ionized layer. *J. Appl. Phys.* **13**, 577–81.
Rydbeck, O.E.H. 1942b. A theoretical survey of the possibilities of determining the distribution of the free electrons in the upper atmosphere. Trans. Chalmers Univ. of Technology, Gothenburg, Sweden. Nr. 3.
Rydbeck, O.E.H. 1943. The reflection of electromagnetic waves from a parabolic ionised layer. *Phil. Mag.* **34**, 342–8.
Rydbeck, O.E.H. 1944. On the propagation of radio waves. Trans. Chalmers Univ. of Technology, Gothenburg, Sweden. Nr. 34.
Rydbeck, O.E.H. 1950. Magnetoionic triple splitting of ionospheric waves. *J. Appl. Phys.* **21**, 1205–14.
Rydbeck, O.E.H. 1951. The theory of magnetoionic triple splitting. *Commun. Pure Appl. Math.* **4**, 129–60.
Saha, M.N., Banerjea, B.K. and Guha, U.C. 1951. Vertical propagation of electromagnetic waves in the ionosphere. *Proc. Nat. Inst. Sci. India.* **17**, 205–26.
Scarabucci, R.R. *see* Arantes.
Schatzberg, A. *see* Altman.
Schelkunoff, S.A. 1938. The impedance concept and its applications to problems of reflection, refraction, shielding and power absorption. *Bell Syst. Tech. J.* **17**, 17–48.
Schelkunoff, S.A. 1951. Remarks concerning wave propagation in stratified media. *Comm. Pure Appl. Math.* **4**, 117–28.
Schmerling, E.R. *see* Jackson.
Seaton, S.L. *see* Booker.
Sen, H.K. 1967. *The generalized magnetoionic theory.* Bedford, Mass.: Air Force Cambridge Research Laboratories.
Sen, H.K. and Wyller, A.A. 1960. On the generalisation of the Appleton–Hartree magnetoionic formulas. *J. Geophys. Res.* **65**, 3931–50.
Sharaf, A.L. 1969. Exact solutions for fields of electric type in spherically stratified isotropic media. *Proc. Camb. Phil. Soc.* **66**, 119–27.
Shaw, I.J. *see* Ratcliffe.
Shaw, R.R. *see* Gurnett.
Shawhan, S.D. *see* Gurnett, Smith, R.L.
Shellman, C.H. 1970. Electron density distributions in the lower ionosphere with associated error limits derived from VLF and LF sounder data. *Radio Sci.* **5**, 1127–35.
Shinn, D.H. 1955. Tables of group refractive index for the ordinary ray in the ionosphere. *The Physics of the Ionosphere*, pp. 402–6. London: Physical Society.
Shinn, D.H. and Whale, H.A. 1952. Group velocities and group heights from the magnetoionic theory. *J. Atmos. Terr. Phys.* **2**, 85–105.
Shmoys, J. *see* Hirsch.
Shkarofsky, I.P. Johnston, T.W. and Bachynski, M.P. 1966. *The particle kinetics of plasmas.* Reading, Mass.: Addison-Wesley.

Sluijter, F.W. *see* van Duin.

Smith, E.K. and Matsushita, S. (eds.) 1962. *Ionospheric sporadic E*. Oxford: Pergamon Press.

Smith, E.K. *see* Booker.

Smith, L.G. *see* Miller.

Smith, M.S. 1973a. The theory of the reflection of low frequency radio waves in the ionosphere near critical coupling conditions. *J. Atmos. Terr. Phys.* **35**, 51–62.

Smith, M.S. 1973b. Frequency analysis of calculated ionospheric reflection coefficients. *J. Atmos. Terr. Phys.* **35**, 1363–75.

Smith, M.S. 1974a. Coupling points of the Booker quartic equation for radio wave propagation in the ionosphere. *Proc. R. Soc. Lond.* A **336**, 229–50.

Smith, M.S. 1974b. A reflection coefficient matrix for radio waves within the ionosphere. *J. Atmos. Terr. Phys.* **36**, 1165–72.

Smith, M.S. 1975. Phase memory in W.K.B. and phase integral solutions of ionospheric propagation problems. *Proc. R. Soc. Lond.* A. **346**, 59–79.

Smith, M.S. 1976. Non-reversibility for radio rays in the ionosphere. *J. Atmos. Terr. Phys.* **38**, 37–44.

Smith, M.S. *see* Budden.

Smith, N. 1937. Extension of normal incidence ionosphere measurements to oblique incidence radio transmission. *J. Res. Natn. Bur. Stand.* **19**, 89–94.

Smith, N. 1938. Application of vertical incidence ionosphere measurements to oblique incidence radio transmission. *J. Res. Natn. Bur. Stand.* **20**, 683–705.

Smith, N. 1939. The relation of radio sky-wave transmission to ionosphere measurements. *Proc. Instn Radio Engrs.* **27**, 332–47.

Smith, N. 1941. Oblique incidence radio transmission and the Lorentz polarisation term. *J. Res. Natn. Bur. Stand.* **26**, 105–16.

Smith, R.L. 1964. An explanation of subprotonic whistlers. *J. Geophys. Res.* **69**, 5019–21.

Smith, R.L. and Brice, N.M. 1964. Propagation in multicomponent plasmas. *J. Geophys. Res.* **69**, 5029–40.

Smith, R.L., Brice, N.M., Katsufrakis, J., Gurnett, D.A., Shawhan, S.D., Belrose, J.S. and Barrington, R.E. 1964. An ion gyrofrequency phenomenon observed in satellites. *Nature, Lond.* **204**, 274–5.

Smith, R.L. *see* Gurnett, Raghuram.

Snyder, M.A. 1966. *Chebyshev methods in numerical approximation*. Englewood Cliffs, N.J.: Prentice-Hall.

Sommerfeld, A. 1909. Über die Ausbreitung der Wellen in der drahtlosen Telegraphie. *Ann. Phys. Lpz.* **28**, 665–736.

Sommerfeld, A. 1949. *Partial differential equations in physics*. New York: Academic Press.

Sommerfeld, A. 1954. *Optics*. New York: Academic Press.

Spencer, M. *see* Briggs.

Spracklen, C.T. *see* Jones, T.B.

Stanley, J.P. 1950. The absorption of long and very long waves in the ionosphere. *J. Atmos. Terr. Phys.* **1**, 65–72.

Stegun, I.A. *see* Abramowitz.

Stepanov, K.N. *see* Akhiezer.

Stevens, E.E. 1961. The complexity of high latitude ionograms. D.R.T.E. report No. 1064. Ottawa: Dept. of National Defence.

Stewart, I.N. *see* Poston.

Stix, T.H. 1962. *Theory of plasma waves*. New York: McGraw-Hill.

Stokes, G.G. 1858. On the discontinuity of arbitrary constants which appear in divergent developments. *Trans. Camb. Phil. Soc.* **10**, 105–128. also in *Math. and Phys. Papers* **4**, 77–109, 1904.

Stoneley, R. 1955. The propagation of surface elastic waves in a cubic crystal. *Proc. R. Soc. Lond.* A **232**, 447–58.

Stoneley, R. 1963. The propagation of surface waves in an elastic medium with orthorhombic symmetry. *Geophys. J. Roy. Astron. Soc.* **8**, 176–86.

Storey, L.R.O. 1953. An investigation of whistling atmospherics. *Phil Trans. R. Soc. Lond.* A **246**, 113–41.

Stott, G.F. 1981. Aspects of dipole radiation in a magnetoplasma. Ph.D. thesis. Cambridge (unpublished).

Stott, G.F. 1983. Refractive index surfaces. *J. Atmos. Terr. Phys.* **45**, 219–29.

Stott, G.F. *see* Al'pert, Budden.

Straker, T.W. 1955. The ionospheric propagation of radio waves of frequency 16 kc/s over short distances. *Proc. Instn. Elect. Engrs.* **102C**, 122–33.

Straker, T.W. *see* Bain, Bracewell.

Stratton, J.A. 1941. *Electromagnetic theory.* New York: McGraw-Hill.

Stuart, R.D. *see* Weekes.

Sturrock, P.A. 1965. Dipole resonances in a homogeneous plasma in a magnetic field. *Phys. Fluids* **8**, 88–96.

Suchy, K. 1972a. The velocity of a wave packet in an anisotropic absorbing medium. *J. Plasma Phys.* **8**, 33–51.

Suchy, K. 1972b. Ray tracing in an anisotropic absorbing medium. *J. Plasma Phys.* **8**, 53–65.

Suchy, K. 1974a. The propagation of wave packets in inhomogeneous anisotropic media with moderate absorption. *Proc. Instn Electr. Electron. Engrs.* **62**, 1571–7.

Suchy, K. 1974b. Definition and use of collision frequencies. Methods of measurements and results of lower ionosphere structure (COSPAR symposium), pp. 23–33, ed. Rawer, K. Berlin: Akademie-Verlag.

Suchy, K. 1981. Real Hamilton equations of geometric optics for media with moderate absorption. *Radio Sci.* **16**, 1179–82.

Suchy, K. 1982. Geometric optics for warm and hot plasmas with absorption. *Kleinheubacher Berichte* **25**, 245–68.

Suchy, K. and Altman, C. 1975a. The Maxwell field, its adjoint field and the 'conjugate' field in anisotropic absorbing media. *J. Plasma Phys.* **13**, 299–316.

Suchy, K. and Altman, C. 1975b. Reflexion and transmission theorems for characteristic waves in stratified anisotropic absorbing media. *J. Plasma Phys.* **13**, 437–49.

Suchy, K. and Paul, A.K. 1965. Hamilton's equations of geometric optics. University of Maryland, the Institute for Fluid Dynamics and Applied Mathematics. Technical Note BN-424.

Suchy, K. and Rawer, K. 1971. The definition of collision frequencies and their relation to the electron conductivity of the ionosphere. *J. Atmos. Terr. Phys.* **33**, 1853–68.

Suchy, K. *see* Altman, Rawer.

Sutton, D.J. *see* Crompton.

Tang, T., Wong, K.C. and Caron, P.R. 1975. Reflection and transmission of extraordinary waves in an inhomogeneous magnetized plasma. I. Infinite plasma. *I.E.E.E. Trans. Plasma Sci.* PS-**3**, 76–81.

Tellegen, B.D.H. 1933. Interaction between radio waves. *Nature, Lond.* **131**, 840.

Terry, P.D. 1971. Complex ray theory for ion cyclotron whistlers. *Nature phys. Sci. Lond.* **229**, 200–202.

Terry, P.D. 1978. A complex ray tracing study of ion cyclotron whistlers in the ionosphere. *Proc. R. Soc. Lond.* A. **363**, 425–43.

Terry, P.D. *see* Budden.

Thomas, J.O., Haselgrove, J. and Robbins, A. 1958. The electron distribution in the ionosphere over Slough. I. Quiet days. *J. Atmos. Terr. Phys.* **12**, 46–56.

Thomas, J.O. and Robbins, A. 1958. The electron distribution in the ionosphere over Slough. II. Disturbed days. *J. Atmos. Terr. Phys.* **13**, 131–9.

Thomas, J.O. and Vickers, M.D. 1959. *The conversion of ionospheric virtual height–frequency curves to electron density–height profiles.* Dept. of Scientific and Industrial Research, Radio Research Special Report No. 28. London: H.M.S.O.

Thomas, L. and Harrison, M.D. 1970. The electron density distributions in the D-region during the night and pre-sunrise period. *J. Atmos. Terr. Phys.* **32**, 1–14.

Thrane, E.V. *see* Barrington, Piggott.

Titheridge, J.E. 1959. The use of the extraordinary ray in the analysis of ionospheric records. *J. Atmos. Terr. Phys.* **17**, 110–25.

Titheridge, J.E. 1961a. A new method for the analysis of ionospheric $h'(f)$ records. *J. Atmos. Terr. Phys.* **21**, 1–12.

Titheridge, J.E. 1961b. The effect of collisions on the propagation of radio waves in the ionosphere. *J. Atmos. Terr. Phys.* **22**, 200–217.

Titheridge, J.E. 1964. The refraction of satellite signals. I. Theoretical calculations. *J. Atmos. Terr. Phys.* **26**, 159–75. II. Experimental results. *J. Atmos. Terr. Phys.* **26**, 177–91.

Titheridge, J.E. 1966. Continuous records of the total electron content of the ionosphere. *J. Atmos. Terr. Phys.* **28**, 1135–50.

Titheridge, J.E. 1967a. The overlapping polynomial analysis of ionograms. *Radio Sci.* **2**, 1169–75.

Titheridge, J.E. 1967b. Direct manual calculation of ionospheric parameters using a single polynomial analysis. *Radio Sci.* **2**, 1237–53.

Titheridge, J.E. 1968. The characteristics of large ionospheric irregularities. *J. Atmos. Terr. Phys.* **30**, 73–84.

Titheridge, J.E. 1969. Single polynomial analysis of ionograms. *Radio Sci.* **4**, 41–51.

Titheridge, J.E. 1971a. On the ambiguity in Faraday rotation measurements of the electron content of the ionosphere. *J. Atmos. Terr. Phys.* **33**, 1115–17.

Titheridge, J.E. 1971b. Magnetoionic mode coupling of HF radio waves. *J. Atmos. Terr. Phys.* **33**, 1533–40.

Titheridge, J.E. 1971c. Mode coupling of HF waves in the ionosphere. *J. Atmos. Terr. Phys.* **33**, 1541–57.

Titheridge, J.E. 1975. The relative accuracy of ionogram analysis techniques. *Radio Sci.* **10**, 589–99.

Titheridge, J.E. 1979. Increased accuracy with simple methods of ionogram analysis. *J. Atmos. Terr. Phys.* **41**, 243–50.

Titheridge, J.E. and Lobb, R.J. 1977. A least square polynomial analysis and its application to topside ionograms. *Radio Sci.* **12**, 451–9.

Titheridge, J.E. *see* Lobb.

Toshniwal, G.R. 1935. Threefold magnetoionic splitting of radio echoes reflected from the ionosphere. *Nature, Lond.* **135**, 471–2.

Turunen, T. *see* Jalonen.

Tuve, M.A. *see* Breit.

Ursell, F. 1980. Integrals with a large parameter: a double complex integral with four nearly coincident saddle-points. *Math. Proc. Camb. Phil. Soc.* **87**, 249.

Ursell, F. *see* Chester.

Uscinski, B.J. 1977. *The elements of wave propagation in random media.* New York: McGraw-Hill.

van Duin, C.A. and Sluijter, F.W. 1980. Refractive index profiles with a local resonance based on Heun's equation. *Radio Sci.* **15**, 95–103.

van Duin, C.A. and Sluijter, F.W. 1983. Reflection and transmission coefficients for local resonances in a cold magnetoplasma. *Radio Sci.* **18**, 337–43.

Vats, H.O. *see* Booker.

Vice, R.W. *see* Fejer.

Vickers, M.D. *see* Thomas.

Volland, H. 1962a. The propagation of plane electromagnetic waves in a horizontally stratified ionosphere. *J. Atmos. Terr. Phys.* **24**, 853–6.

Volland, H. 1962b. Die Streumatrix der Ionosphäre. *Arch. Elektr. Übertr.* **16**, 328–34.

Volland, H. 1962c. Kopplung und Reflexion von elektromagnetischen Wellen in der Ionosphäre. *Arch. Electr. Übertr.* **16**, 515–24.

Von Roos, O. 1970. A note on the applicability of the Booker theorem to magnetoactive plasmas. *Radio Sci.* **5**, 693–5.

Wait, J.R. 1962. *Electromagnetic waves in stratified media.* Oxford: Pergamon Press.

Wait, J.R. 1963. Concerning solutions of the V.L.F. mode problem for an anisotropic curved ionosphere. *J. Res Natn. Bur. Stand.* **67D**, 297–302.

Wait, J.R. and Walters, L.C. 1964. Reflection of electromagnetic waves from a lossy magnetoplasma. *Radio Sci.* **68D**, 95–101.

Walker, A.D.M. 1966. The theory of guiding of radio waves in the exosphere. I. Guiding of whistlers. *J. Atmos. Terr. Phys.* **28**, 807–22.

Walker, A.D.M. 1968a. Ray tracing in the ionosphere at VLF-I. *J. Atmos. Terr. Phys.* **30**, 403–9.

Walker, A.D.M. 1968b. Ray tracing in the ionosphere at VLF-II. *J. Atmos. Terr. Phys.* **30**, 411–21.

Walker, A.D.M. 1976. The theory of whistler propagation. *Rev. Geophys. Space Phys.* **14**, 629–38.

Walker, A.D.M. 1977a. The phase velocity, ray velocity and group velocity surfaces for a magnetoionic medium. *J. Plasma Phys.* **17**, 467–86.

Walker, A.D.M. 1977b. The ray velocity surface and the C.M.A. diagram. *J. Plasma Phys.* **18**, 339–46.

Walker, A.D.M. and Lindsay, R.N. 1975. Excitation of the earth–ionosphere waveguide by downgoing whistlers III. *J. Atmos. Terr. Phys.* **37**, 1599–1600.

Walker, R.L. *see* Mathews.

Walkinshaw, W. *see* Booker.

Walkup, J.F. *see* Carpenter.

Walsh, E.J. 1967. Full wave solutions in terms of coupled vacuum modes. *Radio Sci.* **2**, 913–25.

Walters, L.C. *see* Johler.

Wand, I.C. *see* Jones, T.B.

Wang, T.I. 1971. Intermode coupling at ion whistler frequencies in a stratified collisionless ionosphere. *J. Geophys. Res.* **76**, 947–59.

Warber, C.R. *see* Field, Warren.

Warren, R.E., de Witt, R.N. and Warber, C.R. 1982. A numerican method for extending ray trace calculations of radio fields into strong focusing regions. *Radio Sci.* **17**, 514–20.

Warren, R.E. *see* Field.

Watson, G.N. 1944 (repr. 1966). *Theory of Bessel functions* (2nd edn). Cambridge University Press.

Watson, G.N. *see* Whittaker.

Watt, J.M. *see* Hall.
Watts, J.M. 1957. The interpretation of night-time low frequency ionograms. *J. Geophys. Res.* **63**, 717–26.
Watts, J.M. and Brown, J.N. 1954. Some results of sweep-frequency investigation in the low frequency band. *J. Geophys. Res.* **50**, 71–86.
Watts. J.M. *see* Davies.
Waynick, A.H. *see* Kelso.
Weekes, K. 1950. The ground interface pattern of very low frequency radio waves. *Proc. Instn Elect. Engrs.* **97**, III, 100–107.
Weekes, K. and Stuart, R.D. 1952a. The ionospheric propagation of radio waves with frequencies near 100 kc/s over short distances. *Proc. Instn elect. Engrs.* **99**, IV, 29–37.
Weekes, K. and Stuart, R.D. 1952b. The ionospheric propagation of radio waves with frequencies near 100 kc/s over distances up to 1000 km. *Proc. Instn. electr. Engrs.* **99**, IV, 38–46.
Weekes, K. *see* Bracewell.
Weisbrod, S., Ferraro, A.J. and Lee, H.S. 1964. Investigation of phase interaction as a means of studying the lower ionosphere. *J. Geophys. Res.* **69**, 2337–48.
Weitzner, H. and Batchelor, D.B. 1979. Conversion between cold plasma modes in an inhomogeneous plasma. *Phys. Fluids* **22**, 1355–8.
Wentzel, G. 1926. Eine Verallgemeinerung der Quantenbedingungen für die Zwecke der Wellenmechanik. *Z. Physik.* **38**, 518–29.
Westcott, B.S. 1962a. Ionospheric reflection processes for long radio waves.–I. *J. Atmos. Terr. Phys.* **24**, 385–99.
Westcott, B.S. 1962b. Ionospheric reflection processes for long radio waves. –II. *J. Atmos. Terr. Phys.* **24**, 619–31.
Westcott, B.S. 1962c. Ionospheric reflection processes for long radio waves. –III. *J. Atmos. Terr. Phys.* **24**, 701–13.
Westcott, B.S. 1962d. Ionospheric reflection processes for long radio waves. –IV, *J. Atmos. Terr. Phys.* **24**, 921–36.
Westcott, B.S. 1964. Ionospheric reflection processes for long radio waves. –V. *J. Atmos. Terr. Phys.* **26**, 341–50.
Westcott, B.S. 1968. Exact solutions for electromagnetic wave propagation in spherically stratified isotropic media. *Proc. Camb. Phil. Soc.* **64**, 227–35.
Westcott, B.S. 1969a. Exact solutions for vertically polarized electromagnetic waves in horizontally stratified isotropic media. *Proc. Camb. Phil. Soc.* **66**, 675–84.
Westcott, B.S. 1969b. Exact solutions for transverse electromagnetic wave propagation in a cylindrically stratified axially magnetized medium. *Proc. Camb. Phil. Soc.* **66**, 129–43.
Westcott, B.S. 1970. Oblique reflection of electromagnetic waves from isotropic media with limiting Epstein-type stratifications. *Proc. Camb. Phil. Soc.* **68**, 765–71.
Westcott, C.H. *see* Bain.
Westfold, K.C. 1949. The wave equations for electromagnetic radiation in an ionised medium in a magnetic field. *Austral. J. Sci. Res.* **2**, 168–83.
Whale, H.A. *see* Shinn.
Whipple, R.T.P. *see* Heading
White, H.E. *see* Jenkins
White, R.B. and Chen, F.F. 1974. Amplification and absorption of electromagnetic waves in overdense plasmas. *Plasma Phys.* **16**, 565–87.
Whittaker, E.T. and Watson, G.N. 1927 (repr. 1935). *Modern analysis* (4th edn). Cambridge University Press.
Whitteker, J.H. *see* Jackson.

Wieder, B. 1955. Some results of a sweep-frequency propagation experiment over a 1100 km. east–west path. *J. Geophys. Res.* **60**, 395–409.
Wilkes, M.V. 1940. The theory of reflection of very long wireless waves from the ionosphere. *Proc. R. Soc. Lond.* A **175**, 143–63.
Wilkes, M.V. 1947. The oblique reflection of very long wireless waves from the ionosphere. *Proc. R. Soc. Lond.* A **189**, 130–47.
Wilkes, M.V. *see* Budden.
Willmore, A.P. 1970. Electron and ion temperatures in the ionosphere. *Space Sci. Rev.* **11**, 607–70.
Wolf, E. *see* Born.
Wong, K.C. *see* Tang.
Woodward, P.M. and Woodward, A.M. 1946. Four-figure tables of the Airy function in the complex plane. *Phil. Mag.* **37**, 236–61.
Wright, J.W. and Knecht, R.W. 1957. Atlas of ionograms. Boulder, Col.: National Bureau of Standards, Central Radio Propagation Laboratory.
Wright, J.W. and Pitteway, M.L.V. 1979. Real-time data acquisition and interpretation capabilities of the Dynasonde. 1. Data acquisition and real time display. *Radio Sci.* **14**, 815–25. 2. Determination of magnetoionic mode and echo location using a small spaced array. *Radio Sci.* **14**, 827–35.
Wüster, H.O. *see* Försterling.
Wyller, A.A. *see* Sen.
Yabroff, I.W. 1957. Reflection at a sharply bounded ionosphere. *Proc. Instn Radio Engrs.* **45**, 750–3.
Yabroff, I. 1961. Computation of whistler ray paths. *J. Res. Natn. Bur. Stand.* **65D**, 485–505.
Yeh, K.C. and Liu, C.H. 1982. Radio wave scintillations in the ionosphere. *Proc. Instn Electr. Electron. Engrs.* **70**, 324–60.
Yokoyama, E. and Namba, S. 1932. Theory of the propagation of low frequency waves. *Rep. Radio. Res. Japan.* **2**, 131–55.
Zacharisen, D.H. *see* Davies.
Ziolkowski, R.W. and Deschamps, G.A. 1984. Asymptotic evaluation of high frequency fields near a caustic: An introduction to Maslov's method. *Radio Sci* **19**, 1001–25.
Zucker, F.J. *see* Collin.

Index of definitions of the more important symbols

Symbol	Definition	Page
A	coefficient of n^4 in dispersion relation for cold magnetoplasma	78
	(also used with other meanings)	
A_j $(j = 1, 2, 3, 4)$	quantities used in S_{ij}	190
	(also used with other meanings)	
$\boldsymbol{A}$	2×2 admittance matrix	302
A_{ij} $(i, j = 1, 2)$	elements of $\boldsymbol{A}$	303
$\boldsymbol{A}$, A_{ij}	used with different meanings in §13.6.	371
a	half thickness of parabolic layer	332, 456
	(also used with other meanings)	
$a_1, \ldots, a_6$, $\bar{a}_1, \ldots, \bar{a}_6$	quantities used in elements of $\mathbf{S}$ and $\mathbf{S}^{-1}$	190, 191
$\boldsymbol{B}$	magnetic induction of earth's magnetic field	23, 45
	(used with different meanings in §§1.1, 13.6)	
B	coefficient of $-2n^2$ in dispersion relation for cold magnetoplasma	78
	(also used with other meanings)	
$\mathbf{B}$	4×4 matrix whose trailing diagonal elements are unity	184
$\boldsymbol{b}$	(real) magnetic induction	25
C	$\cos\theta$, cosine of angle of incidence	142
	coefficient of n^0 in dispersion relation for cold magnetoplasma	78
$\mathscr{C}$	cylinder function; solution of Bessel's equation	598
$\mathscr{C}_p(S)$	$\mathscr{C}$ integrals	59
c	speed of light in a vacuum	22
	(also used with other meanings)	

Symbol	Definition	Page
D	dispersion of a whistler	378
	function used in $a_1, \ldots, a_6$	190
	function used in dispersion relation	403
	horizontal range	342
	(also used with other meanings)	
D_G	horizontal range measured over earth's curved surface	349
D	4×4 diagonal matrix	506
$\boldsymbol{D}$	complex electric displacement in wave with harmonic time dependence	26, 29
	(also used as 2×2 matrix)	
D_x, D_y, D_z	components of electric displacement $\boldsymbol{D}$	26, 28
$\mathscr{D}$	Schwarzian derivative	176
$\boldsymbol{d}$	(real) electric displacement	25
	(also used with other meanings)	
$\boldsymbol{E}$	complex electric intensity in wave with harmonic time dependence	26
E_x, E_y, E_z	components of $\boldsymbol{E}$ (also used with superscripts)	26
E	magnitude of $\boldsymbol{E}$	26
	short for E_y	457
E_1, E_2, E_3	contravariant components of $\boldsymbol{E}$ in complex principal axes	51
	(used with other meanings in §§14.13–14.14).	
$\mathscr{E}$	eikonal function	402
e	exponential	
e	charge on electron (a negative number)	39, 46
$\boldsymbol{e}$	(real) electric intensity	23
e	column matrix with elements $E_x, -E_y, \mathscr{H}_x, \mathscr{H}_y$	162, 182
$F(q)$	left side of Booker quartic equation	144
F	function used in equation of ray surface	410
	(also used to denote other functions)	
F_1, F_2, F_3, F_4	$F_1 = (q_1 - q_2)(q_1 - q_3)(q_1 - q_4)$ etc.	191
F_H	proton gyro-frequency	383
$\mathscr{F}_O, \mathscr{F}_E, \mathscr{F}$	field variables in Försterling's equations	510, 597
f	frequency	39
	(also used with other meanings)	
f_c	critical coupling frequency	496
f_{cr}	crossover frequency	383
f_{crs}	f_{cr} at satellite	384
f_H, f_{He}	gyro-frequency for electrons	46, 54
f_{Hi}	gyro-frequency for ions of species i	55

f_{L}	lower hybrid frequency	57
f_N, f_{Ne}	plasma frequency for electrons	39, 55
f_{Ni}	plasma frequency for ions of species i	55
f_{p}	penetration frequency (also used with superscripts)	332, 357
f_{R}	value of f_N where reflection occurs	360
f_{s}	with various superscripts: transition frequencies of the plasma near a satellite	366
f_{U}	$(f_N^2+f_H^2)^{\frac{1}{2}}$; upper hybrid resonance frequency	54
$\mathbf{f}$	column matrix with four elements	163, 188
$\mathrm{f}_1, \mathrm{f}_2, \mathrm{f}_3, \mathrm{f}_4$	elements of $\mathbf{f}$; amplitudes of the four characteristic waves	163, 188
G	$\frac{1}{2}(\varepsilon_1+\varepsilon_2)-\varepsilon_3$	52
	function used in dispersion relation	408
	(also used with other meanings)	
$\mathfrak{G}$	$\varepsilon_3(\varepsilon_1-\varepsilon_2)/\{\varepsilon_1\varepsilon_2-\frac{1}{2}\varepsilon_3(\varepsilon_1+\varepsilon_2)\}$	383
H	scale height of atmosphere	7
	rate of heating of unit volume of plasma	396
$\boldsymbol{H}$	complex magnetic intensity in wave with harmonic time dependence	26
H_x, H_y, H_z	components of $\boldsymbol{H}$	28
$\boldsymbol{\mathscr{H}}$	$Z_0\boldsymbol{H}$, alternative measure of magnetic intensity	31
$\mathscr{H}_x, \mathscr{H}_y, \mathscr{H}_z$	components of $\boldsymbol{\mathscr{H}}$	31
$h, h(f)$	phase height	329
h_0	height of base of ionosphere	331
$h', h'(f)$	equivalent (or group) height of reflection	329, 359
	also used with subscripts, O for ordinary or X for extraordinary	362
$\boldsymbol{h}$	(real) magnetic intensity	23
i	$\surd(-1)$	
i	integer label to indicate species of positive ion	55
$\boldsymbol{i}, \boldsymbol{j}, \boldsymbol{k}$	unit vectors parallel to x, y, z axes respectively	28
J	$\frac{1}{2}\varepsilon_3(\varepsilon_1+\varepsilon_2)-\varepsilon_1\varepsilon_2$	52
	(also used with other meanings)	
J_{ij} $(i,j = x, y, z)$	3×3 tensor	415
$\boldsymbol{j}$	(real) current density	24
K	Boltzmann's constant	7, 45
	(also used with other meanings)	
K_{ij} $(i,j = x, y, z)$	3×3 tensor	415

k	$\omega/c = 2\pi f/c$. Propagation constant in a vacuum	29
k_0	predominant value of k in wave packet	258
k_p	value of k at the penetration frequency	465
l_D	Debye length	45
l_x, l_y, l_z	direction cosines of $\boldsymbol{Y}$, antiparallel to the earth's magnetic field	144
	(also used for components of other unit vectors $\boldsymbol{l}$)	
$\boldsymbol{l}$	unit vector in direction of wave normal	36
M	effective refractive index	446
$\mathbf{M}$	4×4 matrizant	558
$\boldsymbol{M}$	3×3 susceptibility matrix of magnetoplasma with elements $M_{ij}(i, j = x, y, z)$	50
$\mathscr{M}$	$n \cos \alpha$; ray refractive index	111
m	mass of electron	39
m_i	mass of ion of species i	55
N, N_e	concentration of electrons	23, 55
N_m	N at maximum of ionospheric layer	9
N_i	concentration of ions of species i	55
	(N alone or with various subscripts is also used with other meanings)	
$\mathscr{N}_0$	power per unit solid angle in ray pencil	280
n	refractive index (in general complex)	29, 66
n_O, n_E	n for ordinary and extraordinary wave, respectively	80, 81
$\boldsymbol{n}$	refractive index vector (in general complex)	78, 103
n_i	Cartesian components of $\boldsymbol{n}$	
	$(i = x, y, z)$	78
	$(i = \xi, \eta, \zeta)$	103
	(n is also used with other subscripts, and with superscripts, for the refractive index in specified conditions)	
n'	$\partial(fn)/\partial f$; group refractive index (in general complex)	131
$\mathrm{O}(x)$	quantity such that $\lim_{x \to 0} \{\mathrm{O}(x)/x\}$ is bounded	
$\boldsymbol{P}$	complex electric polarisation in wave with harmonic time dependence	26
P_x, P_y, P_z	Cartesian components of $\boldsymbol{P}$	46
P_1, P_2, P_3	contravariant components of $\boldsymbol{P}$ in complex principal axes	52
P	phase path (usually for oblique incidence)	278
P'	equivalent path (usually for oblique incidence)	278
$\boldsymbol{p}$	(real) electric polarisation	24
p	gradient of X in linear model of ionosphere	262, 439

Symbol	Definition	Page
$\mathbf{Q}$	4 × 4 diagonal matrix whose 4 elements are solutions q of Booker quartic	183
Q	average thermal energy per electron in plasma	396
	effective value of q	446
q	solution of Booker quartic equation	143
q_i	($i = 1, 2, 3, 4$) the four values of q	183
$q_0, q_1, \ldots, q_m, \ldots$	values of q in successive strata	172
	(q alone, or with subscripts, is also used with other meanings)	
R	reflection coefficient	194
$R_0, R_1 \ldots, R_D$	values of R in specified conditions	296, 297
$\boldsymbol{R}$	2 × 2 reflection coefficient matrix (also used with subscripts 0, c)	299
	(also used with other meanings)	
$\mathscr{R}$	2 × 2 reflection matrix for ordinary and extraordinary component waves	573
R_{ij}	($i, j = 1, 2$) elements of $\boldsymbol{R}$ for linearly polarised fields	298
	($i, j = \mathrm{l, r}$) elements of $\boldsymbol{R}_0$ for circularly polarised fields	301
	($i, j = \mathrm{O, E}$) elements of $\boldsymbol{R}_c$ for ordinary and extraordinary waves	310
$\boldsymbol{r}$	average displacement of an electron	39
r	radius from centre of earth	260
	(also used to denote an integer, and with other meanings)	
S	$\sin\theta$, sine of angle of incidence	146, 342
S_A, S_B, S_C	transition values of $\sin\theta$	156
S_R	square root in Appleton–Lassen formula	88
S_1, S_2	x and y direction cosines of wave normal of incident wave in free space	142
S_{10}, S_{20}	predominant values of S_1, S_2 in wave packet	257, 258
$\mathbf{S}$	4 × 4 matrix of eigen columns of $\mathbf{T}$	162, 183
S_{ij}	($i, j = 1, 2, 3, 4$) elements of $\mathbf{S}$	190, 192
$\mathscr{S}$	energy per unit volume in a plasma	33
$\mathscr{S}_E, \mathscr{S}_M$	electric and magnetic parts, respectively, of $\mathscr{S}$	32
$\mathscr{S}_A(\zeta), \mathscr{S}_B(\zeta)$	solutions of Stokes equation	444
s	distance along ray path	409
	kz; height measured in units of $\lambda/2\pi$	516
	(also used with other meanings)	

Symbol	Meaning	Page
s_i	($i = 1, 2, 3, 4$) eigen column of T	183
T	temperature	7, 58
	time of travel of wave along a ray	378
	transmission coefficient	297
	(also used with other meanings)	
$T_0, T_1, \ldots, T_m, \ldots$	values of transmission coefficient T in discrete strata	172, 173
$\boldsymbol{T}$	2×2 transmission coefficient matrix	299
$T_{ij}\,(i,j = 1,2)$	elements of $\boldsymbol{T}$	299
T	4×4 matrix in basic equations (7.80)	182
$\mathsf{T}_{ij}(i,j = 1, 2, 3, 4)$	elements of T	182
T	$\boldsymbol{\Gamma}^{\mathrm{T}}$ is transpose of any matrix $\boldsymbol{\Gamma}$	183
t	time	25
	(also used with other meanings)	
$\boldsymbol{U}$	2×2 unitary matrix	300
	(also used with other meanings)	
U	4×4 transforming matrix	486
U	$1 - \mathrm{i}Z$ for electrons	44
U_e, U_i	U for electrons and for ions of species i respectively	55
$\boldsymbol{\mathscr{U}}$	3×3 unitary matrix	51
	group velocity vector	130
$\mathscr{U}$	magnitude of group velocity	130
$\mathscr{U}_z$	component of group velocity parallel to wave normal	129, 130
u	$\frac{1}{2}mv^2/KT$ for electrons	58
	(also used with other meanings)	
$\boldsymbol{V}$	ordered part of electron's velocity	10
	ray velocity vector	111
$V_\xi, V_\eta, V_\zeta; V_x, V_y, V_z$	components of ray velocity $\boldsymbol{V}$	112, 410
V	ray velocity $c/\mathscr{M}$; magnitude of $\boldsymbol{V}$	111
	volume	33
	(also used with other meanings)	
V	4×4 transforming matrix	526
v	speed of an electron	11, 57
	wave velocity	111
	(also used with other meanings)	
W	bilinear concomitant (stratified medium)	184, 185
	(also used with other meanings)	

$\boldsymbol{W}$	bilinear concomitant vector	427
X	$Ne^2/(\varepsilon_0 m\omega^2)$	39
X_e, X_i	X for electrons and ions of species i respectively	54, 55
X_∞	value of X where one value of n or q is infinite	79, 149
x	Cartesian coordinate	28, 29, 141
$\boldsymbol{Y}$	$e\boldsymbol{B}/(m\omega)$ (antiparallel to earth's magnetic induction $\boldsymbol{B}$)	49
Y	magnitude of $\boldsymbol{Y}$	49
$\boldsymbol{Y}_e$, $\boldsymbol{Y}_i$, Y_e, Y_i	$\boldsymbol{Y}$, Y for electrons and ions of species i respectively	54, 55
y	Cartesian coordinate	28, 29, 141
Z	ν/ω for electrons	44
Z_e, Z_i	Z for electrons and for ions of species i respectively	55
Z_t	transition value of Z	71
Z_0	$(\mu_0/\varepsilon_0)^{\frac{1}{2}}$; characteristic impedance of a vacuum	31
z	Cartesian coordinate (height in stratified medium)	28, 29, 141
z_0	level of reflection; (complex) value of height z where $n=0$ or $q=0$; 'true' height of reflection	195, 262, 329
z_p, z_q	(complex) values of height z at coupling points	387, 520–2
z_m	height of maximum N in ionospheric layer	9, 332
α	coefficient of q^4 in Booker quartic	145
	coefficient of z in exponent, for exponential height distribution	331
	angle between wave normal and ray	110
	$\frac{1}{2}(q_1+q_2)$	486
	(also used with other meanings)	
β	coefficient of q^3 in Booker quartic	145
	angle between ray and $\boldsymbol{Y}$	110
	$\frac{1}{2}(q_1-q_2)$	486
	(also used with other meanings)	
$\boldsymbol{\Gamma}$	4×4 coupling matrix	188
Γ_{ij}	$(i,j=1,2,3,4)$ elements of $\boldsymbol{\Gamma}$	482
γ	coefficient of q^2 in Booker quartic	145
Δ	discriminant of Booker quartic equation	152
	(also used with other meanings)	
δ	coefficient of q in Booker quartic	145
	(also used to indicate arbitrarily small quantity)	
$\boldsymbol{\varepsilon}$	3×3 electric relative permittivity tensor of magnetoplasma	33
ε_{ij}	$(i,j=x,y,z)$ elements of $\boldsymbol{\varepsilon}$	33

ε_{ijk}	$(i, j = x, y, z)$ isotropic tensor or rank three	415
$\varepsilon_1, \varepsilon_2, \varepsilon_3$	diagonal elements of $\boldsymbol{\varepsilon}$ in complex principal axes	52
$\varepsilon_a, \varepsilon_b, \varepsilon_c$	real elements of diagonalised $\boldsymbol{\varepsilon}$ for biaxial crystal	113
ε_0	permittivity of a vacuum	22
ε	relative electric permittivity; n^2 for isotropic plasma	29
	coefficient of q^0 in Booker quartic	145
ζ	Cartesian coordinate, parallel to $\boldsymbol{Y}$	103
	scaled value of height z (the method of scaling depends on the problem)	9, 200 332, 440, 457
	(also used with other meanings)	
η	Cartesian coordinate, perpendicular to wave normal and to earth's magnetic field	103
η_1, η_2, η_3	parameters in Epstein theory	470
Θ	angle between wave normal and the vector $\boldsymbol{Y}$	68
Θ_r	value of Θ where one refractive index is infinite (resonance)	105
Θ_s	value of Θ at Storey cone	106
Θ_t	transition value of Θ	90
	(Θ alone or with subscripts is also used with other meanings)	
θ	angle of incidence; angle between wave normal and vertical in reference free space	142, 342
	(also used with subscripts to indicate specified conditions, and with other meanings)	
$\boldsymbol{\kappa}$	$k\boldsymbol{n}$, $\omega\boldsymbol{n}/c$ wave propagation vector, magnitude κ	130
κ_i	$(i = x, y, z$ or $\xi, \eta, \zeta)$ components of $\boldsymbol{\kappa}$	130
λ	wavelength in free space c/f	289
λ_e	mean free path of electron	11, 58
$\lambda_1, \lambda_2, \lambda_3$	Stokes multipliers	212
μ	real part of refractive index	34
μ_0	magnetic permeability of a vacuum	22
ν, $\nu(v)$	average collision frequency for electron with speed v	10
ν_{eff}	effective collision frequency for electrons	10
ν	short for ν_{eff}	43
ν_{av}	average collision frequency for electrons	58
ξ	Cartesian coordinate (coplanar with wave normal and earth's magnetic field)	103
	scaled value of height z	447

$\mathbf{\Pi}$	real Poynting vector	34
$\mathbf{\Pi}_{av}$	time average of Poynting vector	34
Π	magnitude of $\mathbf{\Pi}_{av}$	415
Π_i	$(i = x, y, z)$ components of $\mathbf{\Pi}_{av}$	415
ρ	wave polarisation	68
ρ, ρ_O, ρ_F	electric charge density in plasma	48
ρ_O, ρ_E	polarisation ρ of ordinary and extraordinary waves respectively	71
ρ, ρ_0	density of air	7
σ	scaling factor in Epstein distributions (also used with other meanings)	470
τ	variable that increases as a ray is traversed	403
Φ	azimuth angle of wave normal	103
ϕ	azimuth angle of spherical polar coordinates	408
	generalised phase of wave in isotropic ionosphere (also used with other meanings)	170
φ	(complex) phase of wave (also used with other meanings)	109
χ	minus imaginary part of refractive index n	34
	zenith angle of sun's radiation (also used with other meanings)	7
ψ	coupling parameter	484, 510
	angle between wave normal and earth's radius	260
	angle between wave normal and vertical (also used with other meanings)	264
Ω	angular modulation frequency	396
ω	$2\pi f$; angular frequency	26
$\omega_c, \omega_{c1}, \omega_{c2}$	angular cut-off frequencies where one value of refractive index is zero	118, 126
ω_H	$2\pi f_H$; angular gyro-frequency for electrons	46
ω_N	$2\pi f_N$; angular plasma frequency	39
ω_t	transition value of collision frequency	71
ω_U	$2\pi f_U$; angular upper hybrid resonance frequency	124
ω_∞	angular frequency where one refractive index is infinite	127
ω	with one or two subscripts: transition values of ω or ω_N	124

Subject and name index

A page number in bold type refers to a definition or description, and (p) denotes a problem.

Abel's integral 339–42
Abramovici, F. 6, 612
Abramowitz, M. 205, 233, 236, 249, 360, 457, 463, 465, 612
absorbing medium 277f, 391–3, 413, 417ff
absorption of energy in radio wave 13, 20, 391, 412f, 595; *see also* attenuation
absorption of sun's radiation 7
accumulation of energy 452, 596
acoustic waves *see* sound waves
acoustic gravity waves 6, 20
Adams method of integration 553, 575
additional memory 189, 261, 429, **505**ff
adjoint differential equation **185**, 426, 594
adjoint fields 185, 189, 426f, 431f
adjoint matrix 185
adjoint medium 426
adjoint of polarisation 196(p)
adjugate of matrix 185
admittance matrix **302**ff, 307, 566–9
admittance, wave **301**ff, 442, 444
aerials 429–35, 436f(p), 578, 582(p), 602, 603
Agy, V. 347, 612
Airy, G.B. 204, 223, 612
Airy integral functions, Ai, Bi, 177, 197(ch. 8), 201, **204**, 214f, 227f(p), 240, 251, 282, 285, 288f, 345, 348, 425, 440–5, 488, 499ff, 566
Airy's integral 204, 227, 248
Aitchison, G.J. 394, 612
Aitken, A.C. 521, 612
Akhiezer, A.I. 62, 612
Allis, W.P. 43, 60, 116, 612
Al'pert, Ya.L. xv, xvi, 19, 57, 62, 84, 89, 104, 106, 108, 109, 112, 115, 251, 376, 407, 612
Altman, C. 185, 362, 390, 427, 499, 518, 553, 555, 576, 579, 612, 613, 622, 638
ambiguity in refractive index 267, 269
ambiguity in terms ordinary and extraordinary **88**, 89–94, 154
Ampère's circuital theorem 28
amplitude and imaginary part of phase 279, 331, 352, 425
amplitude–phase diagram 173, 223, 235, 506, 556
amplitude, planes of constant 36, 37(p)
analogues of progressive waves 178
analytic continuation 35, 330, 361, 427, 469
analytic property of qs 486
angular spectrum of plane waves 109, 130, 142, 254ff, 282, 313ff, 320, 379, 418, 420, 425
angular width of window 540
anisotropic ionosphere
  differential equations 181ff
  W.K.B. solutions 187ff, 192
anisotropic medium 4, 6
anisotropic plasma 68ff, 308–12, 356(ch. 13), 480(ch. 16), 520(ch. 17), 550(ch. 18)
antennas *see* aerials
antireciprocity **431**ff, 579
anti-Stokes line 200, **210**ff, 220, 228(p), 244, 246, 451, 454, 461, 500
aperture, diffraction by 322
apparent bearing 277
apparent loss of energy 595–7, 599, 608
Appleton, E.V. 9, 39, 44, 46, 75, 328, 331, 335, 349, 353, 355, 358, 613
Appleton–Hartree formula xv, 75
Appleton laboratories xv
Appleton–Lassen formula 59, 60, **75**, 86, 88, 102(p), 106, 121, 407
  for low frequency 98
  for high frequency or small N 98, 99
approximations for refractive index and polarisation 94ff
Arantes, D.S. 391, 519, 613
Arecibo 47
arbitrary constants (Stokes phenomenon) 209
Argand diagram 235
Arndt, D. 59, 61, 621
Askne, J. 554, 613

astigmatic pencil 317
asymptotes 242ff
asymptotic approximations 169, 441f, 460ff, 488
range of validity 213
asymptotic expansions, series 59, 61, 95, 207f, 461
asymptotic form **209**, 243, 454, 461
atmospherics 48, 376
attachment coefficient **8**, 15
attenuation 10, 255, 312, 323–5, 352f, 391–3, 412f, 450
from Booker quartic 277f, 293(p)
see also absorption
Aubry, M.P. 131, 407, 613, 615
audible frequencies, audio frequency amplifier 376
auroral hiss 543
auxiliary subroutine 552, 555, 581
average collision frequency 11, 58, 59
average electromagnetic fields 23–5, 44f
average force on electron 40, 41
axis ratio **73**ff, 99(p), 100(p)

Bachynski, M.P. 5, 11, 42, 636
backscatter 518
Backus, G.E. 6, 17, 603, 614, 623
Bailey, R.C. 604, 606, 614
Bain, W.C. 603, 614
Banerjea, B.K. 50, 519, 614, 636
Baños, A. 3, 229, 239, 241, 614
Barber, N.F. 304, 326, 614
Barnard, S. 152, 614
Barnes, E.W. 469, 591, 614
barrier 459, 533f, 536, 541, **591**ff
Barrington, R.E. 379, 380, 394, 614, 637
Barron, D.W. 302, 308, 547, 568, 574, 614
Barrow, W.L. 222, 614
base of ionospheric layer 362f, 370, 398(p)
basic equations 22(ch. 2)
Batchelor, D.B. 535, 536, 597, 614, 641
Bauer, P. 14, 614
Baybulatov, R.B. 604, 614
bearing error 277, 507
Becker, W. 371, 614
Bell, T.F. 603, 634
Belrose, J.S. 312, 379, 380, 603, 614, 615, 637
Bennett, J.A. 184, 391, 392, 407, 412, 417, 615, 621
Benson, R.F. 535, 615
Bernstein, I.B. 414, 615
Berreman, D.W. 182, 615
Bers, A. 481, 486, 612, 623
Bertoni, H. 317, 634
Bessel functions 204f, 443, 445, 454f, 598f
Bessel's equation 205, 454, 467, 470, 590
Beynon, W.J.G. 14, 20, 42, 335, 349, 353, 355, 613, 615
Bi *see* Airy integral function
biaxial 115
Bibl, K. 13, 615
bilinear concomitant **184**–7, 189, 196(p), 577, 594
bilinear concomitant vector 186, **426**ff, 432
binomial theorem 171, 181
Birkoff, G. 184, 615
Bitoun, J. 131, 407, 613
Boardman, Z.M. 376, 618
Bolt, B.A. 6, 618
Boltzmann distribution function 398
Boltzmann–Vlasov equations 4, 42
Booker, H.G. xvi, 1, 14, 20, 57, 89, 94, 97–9, 124, 137, 139, 142, 145, 149, 151, 152, 154, 155, 160, 259–61, 278, 279, 301, 312, 349, 351, 449, 478, 505, 545, 564, 565, 592, 615
Booker quartic (equation) 89, **141**(ch. 6), **144**, 163(p), 185, 191, 221, 256, 259, 273, 301, 307, 483, 486, 488, 492, 554, 555, 561, 571, 573, 580,
at magnetic equator 151
at magnetic pole 163(p)
attenuation from 277f
derivation 144ff
determinant form 183, 521
discriminant 152f, 160, 492, 523
double root, equal roots 148, 150, **152**f, 157f, 160, 163(p), 192, 262f, 265, 483, 485ff, 492ff, 520(ch. 17)
east-west and west-east 151, 153ff
effect of collisions 160ff
north-south and south-north 155ff
properties 146ff
special cases 151ff
three equal roots 150, 159, 160, 495, 526
Booker ray tracing equations 257, 259
Booker's theorem 477–9
Born, M. 114, 115, 616
Bossy, L. 558–60, 616
Bouger's law 260, 291, 292(p), 349, 564f
boundary conditions 142, 172, 303, 305, 441f, 554
at infinity 217, 440
boundary frequency 383, 385, 387
bound electrons 373
Bowles, K.L. 14, 616
Boyd, R.L.F. 13, 616
Bracewell, R.N. 298, 587, 603, 614, 616
Bradbury, N.E. 15, 616
Bradley, P.A. 349, 351, 617
branch cut 89, 212, 219, 239, 468, 481, **493**f, 496, 501, 580, 593, 595, 600, 608
branch point 239, 486, 493, 566; *see also* coupling point, turning point
Breit, G. 13, 346, 617
Breit and Tuve's theorem 346, 350f, 356
Brekhovskikh, L.M. 1, 172, 288, 466, 617
Bremmer, H. 172, 304, 617
Brewster angle 306
Brice, N.M. 44, 57, 380, 382, 383, 624, 637
Briggs, B.H. 20, 476, 617
Brillouin, L. 169, 180, 222, 617
broadening of pulse, wave packet 137, 414
Brown, G.M. 20, 615
Brown, J.N. 365, 641
Buchsbaum, S.J. 612

Buckley, R. 404, 414, 416f, 429, 617
Budden, K.G. many references in text, 612, 614, 616–18;
Builder, G. 358, 613
Bullen, K.E. 6, 618
Burke, M.J. 312, 615
Burman, R. 447, 466, 472, 618, 624
Burnside, W.S. 152, 618
Burton, E.T. 376, 618
Businger, J.A. 622

calculus of variations 410
Calvert, W. 368, 622
canonical equations for a ray 403–9, **404**, 436(p)
Haselgrove form 407ff
vector form 405
Capon, I.N. 351, 407, 618
Carlini, F. 170
Caron, P.R. 593, 638
Carpenter, D.L. 18, 376, 379, 618, 619
carrier frequency, wave 321, 356
Cartesian coordinates 28, 103, 141, 402–8, 412, 414, 429f
catastrophes 288
cathode ray oscillograph 320, 499
Cauchy relations 229ff
caustic
field near 281–92
intensity of light near 223ff
caustic surface 222, 279, **281**ff, 343, 345, 416, 425
cavity definitions of fields 23, 25, 28, 40
CCIR *see* Comité Consultatif International des Radiocommunications
centre of window **534**, 541
Čerenkov emission 534
change of adjective, description, name 92, 93, 128, 139, 155, 382, 387, 496f, 501f, 531, 534, 596
change of reference level 296, 313–315
Chapman layer 7ff, 196(p), 332–5, 453, 511, 543, 572
Chapman law 8, **9**, 15, 549
Chapman, S. 7, 619
characteristic admittance **302**
characteristic admittance matrix **303**, 307, 568
characteristic impedance of free space 31
characteristic matrix 521
characteristic waves **162**, 165, 188, 557, 559, 572, *see also* ordinary, extraordinary
Chatterjea, B. 152, 155, 160, 565, 614
Chebyshev polynomials 372
Chen, F.F. 448, 597, 641
Cheng, F.T. 553, 619, 623
Chessell, C.I. 606–8, 619
Chester, C. 251, 286, 619
Child, J.M. 152, 614
chorus 4
Chu, L.J. 222, 619
cigar-shaped surface 268
CIRA 619 *see* International Reference Atmosphere
circuit elements (electric) 429, 576
circuit relations (connection formulae) 180, **216**, 467, 473, 591
circular polarisation **69**, 82, 309, 372ff, 435, 530, 584
circularly polarised components, resolution into 300, 312, 372f, 585
Clemmow, P.C. 5, 11, 43, 45, 47, 49, 57, 106, 107, 109, 116, 119, 142, 182, 184, 188, 282, 481, 485, 507, 519, 577, 618, 619
C.M.A. type diagrams 107, **116**ff, 124, 134, 135
coalescence and limiting polarisation 521, 543f
and windows 505, 521, 532ff
at infinity 544
C1 521, **522**, 523–5
C2 388, 495, 521, **522**, 525–34, 543f
of coupling points 520(ch. 17)
Cohen, M.H. 514, 619
coincident coupling points 155, 489, 491, **493**f, 520(ch. 17)
col 231
cold plasma **4**, 63(p), 66
Collar, A.R. 521, 623
Collin, R.E. 430, 577, 620
collision frequency 10ff, 42ff, 57–62
height dependence 12, 479(p)
measurement 312, 331, 602
transition **71**ff, 86ff
velocity dependence 5, 11, **57**ff
collisionless plasma **44**, 53, 68, 75, 104, 112, 117, 129, 134, 327(p), 328ff, 345ff, 372ff, 376ff, 530
collisions 5, **42**ff, 57–62
and Booker's theorem 477–479
and equivalent height, phase height 330f, 360–362
and equivalent path 331
and group refractive index 137, 138
and Martyn's theorem 352f
effect on Booker quartic 160ff
effect on refractive index 86, 89
electron–electron, electron–ion 61f
in wave interaction 394–8
Comité Consultatif International des Radiocommunications (CCIR) 349, 352
Committee on Space Research (COSPAR) 14, 619
comparison medium 175f, 216
compensating singularities 205
complete mode conversion 531f
complex angle shift 317–319
complex angle of incidence 149, 420, 428
complex axes 51
complex Brewster angle 306
complex conjugate and adjoint 187, 427
complex conjugate roots of Booker quartic 148, 187, 592
complex curvature of wave front 317
complex direction of electric field 27
complex equivalent height 330f, 361
complex equivalent path 279, 324f, 331

complex functions 229ff
complex group refractive index, velocity 325, 361
complex height 68, 79, 88, 152, 330
of reflection **195**, 330, 361f, 440, 459, 517f
complex mass and collisions 44
complex numbers and harmonic time variation 26, 32
complex path of integration 569, 580, 608
complex phase 168, 170, 193, 195, 316, 421, 505
complex phase path 286, 324f, 331, 421
complex point source 418
complex Poynting vector 34
complex principal axes 50ff
complex range shift 317–19
complex rays 417–28, 429, 432, 535, 540f
anisotropic medium 388, 391, 425–8
isotropic medium 282, 286, 424f
complex reference level 478
complex refractive index 6, **34**, 35, 68, 89ff, 137, 417ff
complex space coordinates 35, 317, 417ff
complex step size 420, 553, 562, 568
complex time 324, 418
complex values
of q 147ff, 160ff, 171, 541, 554
of X and Z 79, 88, 152, 512
see also complex height, complex z plane
complex variables 229ff
complex vectors 25, **26**ff, 37(p), 38
complex wave normal 36, 37
complex X plane 492ff
complex z plane 152, 195, 218–21, 361, 387, 477, 494, 499–502, 512, 580f, 608
Computation Laboratory, Cambridge, Mass. 205, 620
computing xiii, 10, 223, 352, 400f, 418, 439, 451, 480, 517, 519(p), 551ff, 555, 561, 566, 572, 583, 603–6
Ai and Bi 201, 205, 444
concentration of electrons 7, 23f, 38f
condenser 32
confluent hypergeometric function 587, 600f
conformal mapping 481, 524
connection formula **216**, 451, 455, 467
Connor, K.A. 418, 425, 620
constancy of energy flow 168ff
constant electron concentration 479(p)
constitutive relations 38(ch. 3), 162, 509
continuous waves 62(p), 602
contour integral 195, 202, 208, 218ff, 229, 291, 361, 459, 469, 500–2; *see also* phase integral
contour map 230ff, 234, 237f
contravariant 51, 63(p), 64(p)
convex lens 223f
convolution theorem 322f
Cooper, E.A. 361, 464, 497, 572, 618, 620
coordinate system 28, 66, 103, 141
Copson, E.T. 467, 620
Cornu spiral 223, 235f, 248
Cory, H. 390, 499, 553, 555, 613, 622
COSPAR *see* Committee on Space Research
Coulomb force 42, 44
coupled equations 175, 188, **480**(ch. 16)
first order **482**–92
matrix form 188, 487, 491, 527
near coupling point 485–92, 522
second order 481, 491, 507ff, 519(p)
vertical incidence 483ff, 489ff
coupling 83, 94, 152, 505
between upgoing and downgoing waves 174ff, 485
coupling echo 485, 499, **517**f
coupling matrix 188, **482**f, 487, 491f, 519(p), 527, 593f
coupling parameter ψ484, 490, **511**ff, 518(p), 544
coupling, phase integral method 387f, 489, **499**–502
coupling point
isolated 488
turning point 80, 94, 152, **153**, 154, 157, 160, 161, 222, 387, 458, 480, 483, **485**ff, 491, 492ff, 496ff, 503, 512, 520ff, 566, 607
coupling region **499**, 516, 545
coupling terms **480**, 482
neglect of 188f, 480
covariant 51, 64(p)
Coyne, T.N.R. 312, 615
Crampin, S. 6, 620
critical angle 306, 474
critical coupling 388, **494**, 495–499, 530
critical coupling frequency **496**, 498, 519(p)
critical height *see* transition height
Crombie, D.D. 304, 326, 614,
Crompton, R. W. 10, 58, 620, 627
cross modulation *see* wave interaction
crossover 105, 383ff
crossover frequency **57**, 63(p), 381, 383, 531
crossover level 382, 384, 386
crystal 6, 51, 104
crystal optics 113ff
cubic equation 149, 425
cubic lattice 40
cumulative coupling 430, 507, 545
cumulative phase change 168, 195, 220
curl 28, 64
current density $\boldsymbol{j}$, $\boldsymbol{J}$ 24, 26
current in ionosphere 565
curvature of
earth 260, 346, 349, 351, 563ff, 579
$h'(f)$ curve 322f
$N(z)$ 216–18, 477
$N(z)$ at maximum 9, 15, 332, 334, 476
$q^2(z)$ curve 199
ray 404
ray surface, wave front 139, 142
refractive index surface 95, 106, 107
cusp 283, 287f
in ray path 267
in ray surface 113, 136, 139(p)
cut *see* branch cut
cut-off 6, **79**, 105, 112, 124, 126, 139(p), 140(p), 597

frequency **54**, 68, 125, 126, 135
cyclotron frequency 4, **46**
cylindrical lens 223
cylindrical structures xiv, 20
cylindrical waves 224

D-layer 9, 15ff
D-region 19, 364, 393, 394, 518, 588, 602, 603, 606
Dällenbach, W. 577, 620
damping of electron motion 10–12, 42–4, 57ff
Daniell, G. J. 111, 139, 291, 618, 620
dark side of caustic 282, 287, 425
Da Rosa, A.V. 375, 623
Darwin, C.G. 41, 101, 556, 620
Davids, N. 52, 104, 499, 514, 518, 620
Davies, K. xiii, 10, 13, 347, 349, 365, 368, 612, 620
Debye length 24, **44**ff, 64(p)
Debye, P. 45, 620
Deeks, D.G. 10, 17, 60, 319, 549, 603, 620
de Groot, W. 339, 620
Denisov, N.G. 448, 621
Deschamps, G.A. 281, 317, 418, 621, 642
destructive interference 506, 545, 547
determinant 152, 416
determinant form of curl 28
deviative absorption 393
de Witt, R.N. 287, 640
diagonal elements of coupling matrix 506, 519(p)
diagonal matrix 51, 52, 183, 188, 506, 554
dielectric constant, tensor; 113 *see also* permittivity tensor
differential equations
coupled form 188
for $\boldsymbol{A}$ 567
for anisotropic ionosphere 181ff, **182**, 550(ch. 18)
for isotropic ionosphere 166f
for limiting polarisation 546–548
for $\boldsymbol{R}$ 569, **570**–2
for $\mathscr{R}$ 573
for ray path 257, 259f, **404**, 408f
for vertical incidence 484, 490f, 508, **509**, 546ff, 571f, 584f, 597, 599
for 'vertical' polarisation 180
fourth order 182, 188
of Ricatti type 547
of theoretical physics 207, 447, **466**
diffraction 225, 235, 322
diffusion of electrons 8
digital ionosonde 13
Dingle, R.B. 59, 61, 208, 621
dipole, magnetic field of earth 2, 17, 565
dipole moment per unit volume 24, 40
directed beam 134, 258
direction cosines of $\boldsymbol{Y}$ **46**, 53, 68, 141, 551
direction
of energy flow 34, 37(p), 76f, 101(p), **108**f, 148
of ray **108–10**, 134, 140(p),254f, 257, 592
of wave normal 104, 110, **111**, 134, 139–40(p), 391, 408

disappearance of energy 595–597, 599, 608
discontinuity
of dN/dz 335, 441f, 458, 464, 478, 572
of the constants 209
discrete strata 142, 172, 254, 553–5
discriminant of Booker quartic **152**f, 160, 523, 543
dispersive medium 5
dispersion of whistler 378
dispersion relation 403, 406, 408, 425
cold plasma xv, 64, 74ff, **77**ff, **78**, 101(p), 103, 144, 146
crystal 114
displacement, electric $\boldsymbol{d}$, $\boldsymbol{D}$ **25**, 26, 66
displacement current 20(p), 25, 28
distortion of pulse 137
disturbances, ionosphere and magnetosphere 19ff
disturbing wave 393
divergence theorem 34, 576
divergent series 207f
dominant term **210**, 213, 425, 454
Doppler shift
of ionospheric reflection 13
on ray 407
double image point 319
double refraction 4
double root 148, 150, **152**f, 157f, 160, 163(p), 192, 262f, 265, 483, 485ff, 492ff, 520(ch. 17)
double saddle point 232, 252, 253(p), 256, 285, 313, 421
double steepest descents **251**ff, 256, 283, 285, 313, 316, 421
doubly refracting medium 4, 142
Dougherty, J.P. 5, 11, 14, 43, 45, 49, 368, 619, 621
downgoing wave 148, 263
Downing, A.M. 13. 621
ducts in magnetosphere 376, 379
Dudeney, J.R. 351, 617, 621
Duncan, W.J. 521, 623
Dunckel, N. 379, 619
duration of encounter 64(p)
dynamics 221
dynasonde 13
Dyson, P.L. 391, 392, 615, 621

E-layer 2, 9, 14, 15, 335f
E-region 9, 393, 394
earth flattening approximation 260f, 438, 564f
earth
atmosphere 1
curvature 3, 141, 260, 346, 349, 351, 563ff, 579
magnetic field: and electron, ion, motion 4, 45, 46; form of 17, 565; horizontal 308, 485; spatial variation 436(p), 565; vertical 309, 485, 564, 518(p), 583–91
east-west propagation 153ff, 163(p), 326(p), 554, 564
Eckersley law 379
Eckersley, T.L. 42, 221, 222, 376, 378, 379, 501, 545, 621

Eckhart, C. 467, 621
eclipses 20
edge focusing 290
effective collision frequency 11, 42, **43**, 45, 57, 60, 62, 65(p)
effective refractive index 373, 446f
effective stratification vector **405**, 436(p)
effective value of q 221, 446, 592
eigen columns, rows, of matrix 183, 186, 521, 561, 593
eigen values of matrix 51, 183, 521, 593
eigen vector of matrix 51
eikonal equation 402
eikonal function 170, **402**f, 429, 507
Einziger, P.D. 621
elastic waves 6
electric displacement $\boldsymbol{d}$, $\boldsymbol{D}$ **25**, 26, 66
electric field intensity $\boldsymbol{e}$, $\boldsymbol{E}$ **23**, 26
electric permittivity
  of plasma 33, 50, 52–57, 146
  of vacuum ε_0 22
electric polarisation $\boldsymbol{p}$, $\boldsymbol{P}$ **24**, 26, 39
electric vector
  horizontal 166ff, 194f, 199ff, 439ff
  in plane of incidence 180f, 301, 446–453
electromagnetic fields of characteristic waves 162f
electron concentration 23, 24
  calculation from $h'(f)$ 15, 337–342, 354(p), 369–372
  height dependence xiv, 7–10, 14–16
  measurement methods 13, 14, 313, 602ff
electron–electron collisions 61, 62
electron–ion collisions 61, 62
electron plasma **46**, 52, 53, 104, 112
electron production 7
electrons only plasma **46**, 52, 53, 55, 492, 535, 583ff
electron temperature 80
  measurement methods 14
electron velocities **4**f, 40, 45, 58, 394–7
electron viscosity 62
electron waves 6
electrons
  prevented from moving 100(p), 589
  rate of production 7
ELF *see* extremely low frequency
ellipse and complex vector 27, 36, 37(p)
ellipse in plane of propagation 36, 37(p)
elliptical polarisation 30, 31, 36, 67, **69**, 99(p), 374
Ellis, G.R.A. 505, 532, 534, 621, 622
Ellis window 505, 532
energy balance, electrons 394–7
energy
  in ray pencil 280, 414ff, 428
  stored 63(p), 428, 596
energy flow, flux 36, 279
energy input to plasma 26, 32, 62(p)
energy loss at reflection 450, 452
entire function **460**
envelope
  of pulse 321, 329
  of rays 281
  of refractive index surfaces 140(p), 543
Epstein distribution, layer 470–6, 478f, 479(p), 587
Epstein, P.S. 467, 470, 590, 622
equal refractive indices 80ff
equal roots of Booker quartic *see* double root
equation of motion of electron 39, 45, 449
equator (magnetic) 293(p), 326(p), 485, 508, 554
equivalent frequency at vertical incidence 351
equivalent height, $h'(f)$ **13**, **329**, 333, 350, 353–4(p), 476, 571, 602
  bounded limit, ext. wave 359
  calculation of 359–62, 398(p)
  effect of lower edge of ionosphere 362f, 370, 398(p)
  linear and exponential $N(z)$ 353–4(p)
  parabolic $N(z)$ 332–6, 464ff
  sech^2 $N(z)$ 335
equivalent path **258–9**, 278, 321, 329, 345–8, 412f, 465, 571
error
  in Abel method 340–342
  in data 604
  in W.K.B. approx. for Ai 206
error accumulation 406
error function **233**ff, 325
error integral 233, 248
error upper bound 177
E_s *see* sporadic E
essential singularity 480
Euler's constant 465
Euler's equations 410f
Euler's theorem 408, 410
evanescent wave 5, **35**, 68, 77, 83, 85, 179, 366, 440, 446, 459, 502, 559, 599
Evans, J.V. 14, 622
Evgrafov, M.A. 222, 622
exact coalescence 521ff, 532, 544
exact crossover 383, 385, 388, 399(p), 496, 530
exponential $N(z)$ 10, 331, 353–4(p), 453–6, 472, 478, 585ff
exterior caustic **282**, 287, 290
extraordinary (ray, wave) 81, 85, **88**, 154, 307–12, 430, 481f, 480(ch. 16)
  C.M.A. type diagram 117
  coupling 83, 480(ch. 16)
  cut-off 126, 597
  equivalent height 337, 359–68
  group refractive index 125ff
  penetration frequency 357ff
  polarisation 71, 73, 193, 372ff, 489ff
  ray direction 106, 134
  ray paths 263f, 268–77, 407, 434f
  ray surface 120–123
  reflection 492ff, 573
  refractive index 96, 160, 193
  refractive index surface 120–3, 147
  resonance 127, 592, 597
  slow wave 84

transition 92ff
see also Z-mode
extremely low frequency (ELF) 97, 378, 399(p)

F-layer 2, 9, 14, 15, 328, 335f, 366ff, 372, 503
F1-layer 15, 337, 365
F2-layer 15, 42, 337, 365
F-region 393
factorial function 253(p), 455f, 474, 476, 585, 591
fadeout 19
Faraday rotation 13, 99, 372–6, **373**, 399(p), 435, 514, 547
Faraday's law of electromagnetic induction 28
Farley, D.T. 14, 621, 622
Farmer, F.T. 41, 331, 622
Fedoryuk, M.V. 222, 622
Feinstein, J. 449, 622
Fejer, J.A. 14, 47, 312, 368, 394, 397, 449, 450, 478, 573, 616, 622
Felsen, L.B. 57, 239, 317, 418, 425, 620, 621, 622, 634
Fermat's principle 224, 293(p), **409**–12
Ferraro, A.J. 394, 641
Ferraro, V.C.A. 2, 622
Feshbach, H. 228, 632
fictitious homogeneous medium 163, 482
Field, E.C. 604, 622
fields of characteristic waves 162
Fijalkow, E. 390, 499, 553, 613, 622
first order coupled equations 482ff, 487
Fleagle, R.G. 622
flipover 499
Flixborough explosion 20
flow of energy 33, 168
flow of energy in progressive wave 76ff
flux of particles 169
focus 280f, 290
Fokker–Planck equation 42
Foley, G. 153, 492, 494, 628
forecasting of MUF 349–352
forecasting of $N(z)$ 349
formation of layers 7ff, 20
Försterling, K. 280, 448, 449, 466, 481, 484, 490, 491, 508, 509, 623
Försterling's coupled equations 481, 491, 507ff, **509**f, 514–17, 544, 584, 597
Försterling variables **510**, 519(p)
four characteristic waves 142f, **162**, 165, 188, 481f
four parts of $h'(f)$ curve 364
four reflection, transmission, coefficients 298f
Fourier analysis 49, 320–325, 449
Fourier integral 49, 50, 109, 128, 130, 320
Fourier transform 321ff, 329
fourth order differential equation 4, 87, 165, 188, 551, 587–91
fourth reflection condition 597
fractional hop whistler 377, **380**
Frazer, R.A. 521, 623
free space
below ionosphere 141, 513, 562f, 588ff
reference 535
wave in 31, 544
freezing in of magnetic field 2, 17
Freidberg, J.P. 452, 623
Fresnel diffraction pattern 322
Fresnel integral 233, **236**, 248
Fresnel reflection and transmission coefficients 306, 309, 444f, 473ff
frictionless slope 339
Friedman, B. 251, 286, 619
Fuchsian equations 467
Fuchs, V. 481, 486, 623
Fukami, T. 604, 631
full wave solutions
anisotropic ionosphere 388–391, 550(ch. 18), 583(ch. 19), 606
isotropic ionosphere 438(ch. 15)
full wave theory 5, 16, **194**, 335, 507, 606
fully ionised plasma 1, 2, 17
functional analysis 603
Fung, P.C.W. 553, 619, 623
Furry, W.H. 212f, 623

gamma function 463; *see also* factorial function
Gans, R. 169, 623
Gardner, F.F. 312, 623
Garrett, A.J.M. 398, 623
Garriott, O.K. 4, 7, 8, 15, 20, 375, 623, 635
Gauss's theorem 28
Gaussian curvature 106
Gaussian function 414
Gaussian quadrature 360
Gaussian units 22, 23, 31
Gendrin, R. 98, 623
generalisation of Snell's law 405
generalised hypergeometric series 590f
generalised magnetoionic theory 59
geometrical image point 314–19, 327(p)
geometrical optics 168, 194, 419; *see also* ray theory
geophysics 6, 507, 603
Gibbons, J.J. 137, 514, 623
Gibson, G.A. 408, 623
Gilbert, F. 6, 17, 603, 614, 623, 624
Gillmor, C.S. xv, 624
Ginzburg, V.L. xiii, xv, 45, 58, 62, 80, 89, 430, 466, 596, 624
Giraud, A. 624
Golant, V.E. 80, 596, 624
Golden, R.M. 603, 631
Goldstein, S. 75, 624
good path 502, 535f
Good, R.H. 524, 632
Goos–Hänchen shifts 315ff, **317**, 327(p), 551
Goos, F. 317, 624
Gordon, W.E. 1, 13, 616, 624
Goubau, G. 582, 624
Goubau's reciprocity theorem 582(p)
Gould, R.N. 447, 466, 472, 618, 624
gradual boundary of ionosphere 473f
Graff, P. 131, 407, 613, 615
Graham, A. 51, 521, 624
gravity 1, 2

great heights, conditions at 217, 221, 389, 440, 444, 451, 454, 461, 471–3, 478, 551, **560**ff, 568, 574
greater than critical 93, **494**, 496–499, 503, 514, 516
Green, G. 170, 176, 624
Greifinger, C., Greifinger, P. 479, 624
ground wave 3
group path **278**; *see also* equivalent path
group refractive index **131**, 137, 140(p), 346, 360, 412
  complex 137, 361
group retardation in lower layer 336, 363ff
group velocity 103(ch. 5), **128**ff, 135, 140(p), 255, 346, 379, 412, 436(p)
group velocity surface 137, 139(p)
Guha, U.C. 519, 636
guided waves 3, 6, 149, 222, 288, 297, 551, 563f
guiding of whistlers 376, 379, 391
Gurevich, A.V. 47, 395, 624
Gurnett, D.A. 380, 382, 383, 390, 534, 624, 636, 637
gyro-frequency 4, **46**
  measurement 275, 358

half thickness (semi-thickness) **332**, 463f, 476
Hall, G. 552, 624
Hamilton, W.R. 404f, 408, 624
Hamilton's canonical equations 404f
Hänchen, H. 317, 624
Hankel function 252(p), 454, 599
Hara, E.H. 59, 625
Harang, L. 503, 625
Hargreaves, J.K. 7, 625
harmonic waves 23, **25**ff
harmonics 449, 450, 596
Harper, J.D. 60, 553, 554, 628
Harrison, M.D. 603, 639
Hartree, D.R. 75, 101, 288, 401, 443, 445, 477, 556, 625
Harvey, C.C. 414, 625
Harwood, J. 603, 616
Haselgrove form of ray equations 407ff, 414
Haselgrove, C.B. 407, 408, 625
Haselgrove, J. 370, 407, 408, 412, 625, 639
Hashimoto, K. 376, 625
Hayes, M.G.W. 545, 549, 625
Heading, J. xv, 170, 176, 180, 182, 216, 222, 223, 447, 455, 457, 458, 460, 462, 466, 472, 478, 479, 481, 483, 486, 507, 515, 521, 528, 556, 579, 587, 588, 590, 591, 600, 601, 619, 625, 626
heating 33, 37(p), 391, 394–8, 452, 595
  of ionosphere 47, 450
heavy ions xiv, 380
height as complex variable 68, 79, 88, 152, 330, 387, 477, 494, 499–502
height, *see* equivalent height, phase height, true height, Cartesian coordinates
Heisler, L.H. 7, 632
helium ions 63(p), 379, 381–3
helium whistler 380, 385
Helliwell, R.A. 376, 626
Herlofson, N. 63, 626
Hermitian matrix **33**, 50
Hermitian orthogonal, 575
Herring, R.N. 351, 401, 536, 627
Hertzian dipole 282, 313, 556
Heun, K. 467, 627
Heun's equation 447, 467
Hibberd, F.H. 394, 627
high frequencies 222
Hines, C.O. 42, 380, 627
Hirsch, P. 448, 627
Hollingworth interference pattern 319
Hollingworth, J. 319, 603, 627
Holt, O. 312, 627
homogeneous boundary conditions 228(p)
homogeneous differential equations 26, 515
homogeneous function 408
homogeneous medium, plasma 38, 66
horizontal dipole aerials 578f
horizontal polarisation 166ff, 439ff, 477f, 588f
  differential equation 166, 199f, 453, 457ff, 470ff
  reflection 194f
horizontal range 342–351
horizontal ray 261f, 285f
horizontal variations in ionosphere 20, 380, 400
horizontally stratified 1, 2
Horowitz, S. 519, 555, 627
Huang, X. 13, 372, 627, 635
Hückel, E. 45, 620
Hugill, J. 106, 618
Hutatsuishi, G. 553, 554, 632
Huxley, L.G.H. 10, 58, 394, 396, 620, 627
Huyghens's principle 223ff
hybrid frequencies **57**, 368
hyperbola 242
hyperboloid surfaces 116
hypergeometric equation 467–9, 602
hypergeometric function 469, 472; *see also* confluent hypergeometric function
hypergeometric series 468, 590

illuminated side of caustic 224, 282, 286
image of transmitter 315–19, 327(p)
imaginary refractive index 35, 36, 37(p)
imaginary step size 562
impedance, wave 301ff
impulse 376
Ince, E.L. 184, 627
incoherent scatter xiv, 5, 13, **14**
independence of characteristic waves 510
independent second order equations 166, 584, 588
induction, electric *see* electric displacement
infinite equivalent height 333, 334, 337, 358, 362–5, 465f
infinite fields, vertical polarisation 449
infinite group refractive index 360
infinite refractive index 79, 80ff, 86, 95, 96, 98, 105, 366, 446, 580
infinite root of Booker quartic 149, 154, 155, 580, 591ff

infinitely distant source 281, 421f
infinitely high mountain 230, 235, 237
infinitely massive ions 53, 55
infinities of ε_1 ε_2 54, 56
infinity
  in q curve 150, 591ff
  point at 207
inflection point
  in refractive index surface 106, 116, 148, 160
  in ray surface 113
inhomogeneous differential equation 515
inhomogeneous medium, plasma 38
inhomogeneous waves **36**, 37(p), 100(p), 282, 306, 562, 574
initial ordinary, extraordinary wave 497
Inoue, Y. 519, 555, 627
Institute of Electrical and Electronics Engineers 69, 627
Inston, H.M. 280, 627
integral equation 338, 369, 556
integral function 460
integrals of Barnes type 469, 591
integrating factor 184
integration
  by steepest descents 229(ch. 9), **238–40**, 241ff
  by stationary phase **247**f, 254, 283, 285
integration methods, numerical 552f
interchange *see* change
interference 364, 394
interference pattern 288f, 319, 348, 464
interior caustic 282, 287
internal conical refraction 114
International Astronomical Union 69, 627
International Reference Atmosphere (CIRA) 14, 619
International Reference Ionosphere (IRI) **14**, 16, 349
International Telecommunications Union (ITU) 352
International Union of Radio Science (URSI) 14
interplanetary plasma 21
interpolation 361, 401
inverse square law 28
inversion
  of ionospheric data 17, 602–606
  of matrix 49, 52
ion-acoustic wave 5
ion cyclotron whistlers 93, 377, **380**–91, 495f, 530ff
ion gyro-frequency **55**, 57
ion–ion hybrid frequency 57
ionograms **362**ff, 503
ionosonde **12**ff, 80, 138, 346, 356, 362, 391, 397, 607
  at oblique incidence 347f
ionosphere 1(ch. 1), **14**ff
  artificial modification 47, 450
  observations of 12-14, 602f
ionosphere storm 19
ion plasmas xiv, 3ff, 54ff
ion plasma frequency 55
ions xiv, 54ff, 182
  neglect of 4, 39, 46, 98
ions and electrons, force between 12, 40
irregular singularity 200, 207ff, 467
irregularities of electron concentration, refractive index xiv, 288, 311, 312, 348, 365, 368, 505, 518, 604
irrotational vector field 402
isolated coupling point 485ff, 489, 520
isolated reflection (turning) point 459
isolated resonance 593–6, 598f
isolated saddle point 249f, 284
isolated zero of n or q 198ff, 451
isotropic ionosphere 328(ch. 12), 394, 438(ch. 15)
isotropic plasma 4, 67ff, 409
isotropic stratified medium 143, 165ff, 328(ch. 12)
isotropic tensor 414–417
iterative process 561
ITU *see* International Telecommunications Union

Jackson, J.E. 368, 370, 627
Jacobsson, R. 439, 571, 627
Jalonen, L. 607, 627
Jeffreys, B. 169, 170, 185, 208, 215, 228, 443 445, 515, 628
Jeffreys, H. 6, 33, 169, 170, 185, 208, 215, 216, 228, 414, 416, 515, 628
Jeffs, R.M. 280, 627
Jenkins, F.A. 173, 373, 628
Jespersen, J.L. 551, 576, 579, 634
Johler, J.R. 60, 97, 308, 553, 554, 628
Johnson, W.C. 549, 632
Johnston, T.W. 5, 11, 42, 636
Jones, D. xv, 108, 273, 382, 383, 385, 531, 534, 535, 543, 628
Jones, R.M. 417, 628
Jones, T.B. xv, 13, 20, 153, 492, 494, 603–6, 614, 628
Journal of Atmospheric and Terrestrial Physics 10, 393, 628
Jull, G.W. 417, 427, 428, 434, 435, 618, 628, 629
Jupiter 535

Kane, J.A. 10, 629
Katsufrakis, J. 380, 637
Kelso, J.M. 73, 89, 280, 514, 517, 629
Kennel, C.F. 19, 629
Kennett, B.L.N. 6, 629
Kimura, I. 376, 380, 407, 625, 629
kinetic energy of electrons 63–4(p), 394–397
kinetic theory 4, 43, 57
  wave interaction 397f
King, G.A.M. 370, 629
King, J.W. 394, 629
Knecht, R.W. 365, 642
Ko, K. 481, 486, 623
Kramers, H.A. 169, 180, 629
Krasnushkin, P.E. 564, 604, 614, 629
Kraus, J.D. 375, 629
Kravtsov, Yu.A. 281, 629
Kumagai, H. 376, 625

laboratory measurements of ν 58, 60
Lacoume, J.L. 184, 559, 629
lamination method for $N(z)$ 369
Landau, L.D. 399, 430, 629
Landau damping 595
Landmark, B. 394, 499, 549, 629
Lange-Hesse, G. 52, 630
Langer, R.E. 216, 218, 450, 488, 630
Langmuir probes 13
Langmuir wave 5, 49
Lanzerotti, L.J. 19, 629
Laplace integral 202
Laplace transform 338
Lapshin, V.I. 62, 612
Lassen, H. 75, 280, 623, 630
lateral deviation 131, 260, **273–7**, 315, 366, 375, 391, 401, 507, 596
  at vertical incidence 274
  east-west and west-east 275
  general azimuth 276
Lawrence, R.S. 407, 630
ledge 336f, 362, 365, 368
Lee, H.S. 394, 641
Lee, K.F. 449, 478, 616
left-handed polarisation 69
Leinbach, H. 13, 630
Lenz's law 2
Lepechinsky, D. 497, 499, 630
less than critical 93, **494**, 496–9, 502f, 514
level line **230**ff, 247
level of reflection 262–273, 442
L.G. approximations 170; *see also* W.K.B.
Lied, R. 394, 499, 629
Lifshitz, E.M. 399, 430, 629
lightning flash 47, 376, 378f
light waves 373, 571
limiting points of spiral 236
limiting polarisation 70, 153, 192, 256, 430, 498, 510, 513, 521, 529, **543–9**
limiting region **545**, 547ff
Lindquist, R. 517, 630
Lindsay, R.N. 182, 640
linear and homogeneous equations xiii, 26, 574
linear differential equations 26, 480, 515
linear electric circuit, system 429, 576
linear height distribution of electrons 10, 197ff, 331, 342f, 347, 353(p), 424f, 439–41, 443–5, 478
linear interpolation 351, 401, 436(p)
linear wave polarisation 29, 34, 36, 67, 69, 73, 114, 498, 546, 581(p)
linearly polarised aerials **431**, 434, 435
line of steepest descent, ascent **230**ff, 237, 241ff
line of steepest descents 230ff, 237
Liouville, J. 170, 176, 630
Liouville method 175ff
Lipson, S.G. and Lipson, H. 114, 173, 630
Lisak, M. 554, 613
Little, C.G. 13, 407, 630
Liu, C.H. 20, 642
Lobb, R.J. 371, 630, 639
local property 505ff
Lockwood, G.E.K. 368, 630
logarithm of frequency 363
logarithmic term in E_x 449
long whistler 377
longitudinal component of electric field, and polarisation 75ff, 101(p)
longitudinal propagation **76**, 82, 85, 89, 124, 132, 140(p), 267, 269, 392, 584ff
Lorentz, H.A. 5, 43, 82, 630
Lorentz polarisation term **40**ff, 378
Lorentz reciprocity theorem 577
Lorentz treatment of collisions 5, **10**ff, 43
loss-free medium **35**, 147, 168
Lotsch, H.K.V. 317, 630
low frequency 84, 127, 138, 496ff, 550(ch. 18), 586f *see also* very low frequency
low frequency ionograms 365
lower hybrid frequency **57**, 98
lower triangular matrix 370
lowest ionosphere 58
Ludwig. D. 281, 486, 630
Luxembourg effect *see* wave interaction
Lyman $-\alpha$, $-\beta$ 9

Macmillan, R.S. 603, 631
magnetic energy 32, 63(p)
magnetic equator 293(p), 326(p), 485, 508, 554
magnetic field
  of earth: and motions of electrons, ions 4, 45, 46
  of wave 46–8
magnetic field intensity
  $\boldsymbol{h}$, $\boldsymbol{H}$ **23**, 26, 46f
  $\mathscr{H}$ **31**, 66
magnetic induction $\boldsymbol{b}$, $\boldsymbol{B}$ 25
magnetic meridian, propagation in 263, 266, 268–73, 407, 494f, 537
magnetic permeability 326(p)
magnetic plane **429**, 433
magnetic pole 22, 24
  of earth 163(p), 293(p), 309, 503, 508
magnetic storm 19
magnetic stress 2
magnetohydrodynamics 97
magnetoionic splitting **256**, 263f, 356, 365, 498
magnetoionic theory xiv, 66(ch. 4), 103(ch. 5)
magnetopause 17
magnetoplasma 4, 66
magnetosheath 17
magnetosphere xiii, 2, **17**ff, 400
magnetotail xiii, 3, 18
Mambo, M. 553, 554, 604, 606, 631, 632
Manning, L.A. 338, 631
Margenau, H. 58, 60, 631
Martyn, D.F. 346, 353, 631
Martyn's theorem
  for attenuation **352**f, 356
  for equivalent paths **346**, 350f, 356
Maslin, N. M. 281, 283, 284, 288, 292, 343, 394, 631
mass of electron 39
mathematics xiv, 165

Mather, W.E. 380, 614
Mathews, J. 51, 631
Mathieu's equation 477
matrix equations 6, 46, 163, 308, 310, 369ff
 for inverting integral equation 369ff
matrix forms
 of differential equations 182
 of Maxwell's equations 145
matrix inversion 370–372
matrix theory 183ff, 521ff, 526
matrix susceptibility 49
matrizant 558ff
Matsushita, S. 15, 365, 637
maximum
 in h′(*f*) curve 337, 364f
 of electron concentration 221, 523
maximum usable frequency, MUF, **289**, 328, 348–52, 354(p)
Maxwellian velocity distribution 11, 58, 396, 398
Maxwell's equations 23, 25, **27**ff, **31**, 48, 108, 113, 145, 162, 166, 174, 181, 326(p), 402, 412, 428, 440, 449, 556, 565, 576
 in principal axes 64(p), 101(p)
May, B.R. 603, 614
mean free path of electron 11, 58, 60, 397
mechanical model 339
Meek, J.H. 503, 631
Melrose, D.B. 535, 538, 631
Michel, J.G.L. 288, 625
Miller, J.C.P. 205, 457, 631
Miller, K.L. 607, 632
Miller, S.C. 524, 632
Millington, G. 138, 155, 160, 275–7, 292, 346, 349, 353, 360, 545, 621, 632
Mitchell, R.W. 452, 623
Mobbs, A.J. 603
mode condition (guided waves) 222
mode conversion 92, 387–90, 481, 497f, 501, 503ff, 517, 531, **532**f, 544, 547, 596
modified refractive index 261, 564
modulation of disturbing and wanted wave 394, 396
Moiseyev, B.S. xvi, 62, 89, 104, 106, 109, 112, 115, 251, 612
molecules 1, 40
Monaghan, J.J. 49, 368, 621
monoenergetic 59
Morgan, M.G. 549, 632
Morse, P.M. 229, 632
Morse, R.L. 452, 623
Mott, N.F. 169, 632
MUF *see* maximum usable frequency
Mullaly, R.F. 106, 107, 116, 119, 619
multiple reflections 142, 173, 555
Munro, G.H. 7, 632
Murata, H. 7, 632
Musgrave, M.J.P. 6, 632

Nagano, I. 553, 554, 604, 606, 631, 632
Nakamura, K. 604, 606, 631
Namba, S. 304, 642
National Bureau of Standards 351, 632
narrow beam 134, 257f
Nearhoof, H.J. 73, 514, 629
negative ions, massive, 4
Nertney, R.J. 73, 514, 517, 623, 629, 632
neutrality of plasma 4, **48**, 49
neutral particles in ionosphere 2, 42
Newbern Smith see Smith, N.
Newstead, G. 503, 632
Nicolet, M.J. 12, 633
Nicolson, P. 288, 625
non-deviative absorption 393
non-linear differential equation(s) 170, 188, 449, 568
non-linear effects in ionosphere 47, 48, 395, 449, 595f
non-penetrating mode 559, **575**, 582(p)
nonreciprocity 326(p), 429
non-singular matrix 485, 486, 544, 563
normal
 to ray surface 111, 112, 115
 to refractive index surface 108ff, 114, 115, 404, 425, 427
normal form of differential equation 446, 515
normal incidence
 reflection at sharp boundary 306, 308–12
 see also vertical incidence
normalisation 51, 184, **186**f, 190f, 427, 483, 506, 519(p)
north-south propagation 150, 155ff, 163(p), 485, 579
Northover, F.H. 457, 633
nose frequency 136, **379**
nose whistler 377, **379**
numerical integration of differential equations 388–91, 463f, 477, 517, 540, 547, 549, 550(ch. 18)
numerical swamping 558, 559, 568, **574**
Nuttall, J. 368, 633
Nygren, T. 553, 574, 607, 627, 633

oblate refractive index surface 157
oblique axes 51
oblique incidence 494, 505
 horizontal polarisation 199ff, 494, 505
 rays at 342–53
 vertical magnetic field 587–91
 vertical polarisation 446–52
observations of ionosphere 12ff
ocean as inverted ionosphere 6
Olver, F.W.J. 170, 177, 205, 206, 208, 210, 467, 633
one magnetoionic component 309–11, 431–3
optical activity 113, 398(p)
optical waves 182
optic axes 104, 114
optic ray axes 115, 119, 139(p)
optimisation principle 604
ordered part of electron velocity 11, 62(p), 395
ordinary point of differential equation 206f, 460, 517, 566, 580
ordinary (ray, wave) 81, 85, **88**, 154, 307–12, 430ff, 481ff, 480(ch. 16)

C.M.A. type diagram 116
coupling 83, 480(ch. 16)
cut-off 126
equivalent height 359–68
group refractive index 125ff
penetration frequency 357ff
polarisation 71, 73, 193, 372ff, 489ff
ray direction 106, 134
ray paths 263, 266ff, 274–7, 407, 434f
ray surface 117, 119, 147
refractive index 160, 193, 376
refractive index surface 117, 119
resonance 127
transition 92ff
see also whistler mode
Ott, H. 239, 633
oxygen ion 379, 382, 383
Oya, H. 535, 633

Pack, J.L. 10, 58, 633
Panton, A.W. 152, 618
parabolic cylinder function 214, 457, 461f
parabolic height distribution of electrons 10, 221, 332–4, 343f, 347–50, 361, 456–66, 476, 478, 524, 572, 587, 607
parabolic ray path 262, 342
paraboloidal refractive index surface 538, 542
parallel plate condenser 32
parallel sided slab of plasma 63(p), 326f(p)
parametric instabilities 47
Park, C.G. 18, 619
Parker, E.N. 19, 629
Parkinson, R.W. 499, 514, 517, 518, 620, 633
partial penetration and reflection 15, 194, 439, 445, 463f, 475f, 477, 521, 524, 607
partial reflection method 10, 13, 311, **312**, 518
partial standing wave 30
partial fields, waves 188, **482**, 557
partitioning of matrix 304, 310, 490, 560, 563, 567, 570
passive medium **35**, 86
path
of integration **230**, 242ff
of wave packet *see* ray, ray path
Paul, A.K. 404, 638
Pawsey, J.L. 312, 623
Pearcey, T. 288, 633
Pedersen, P.O. 344, 633
Pedersen ray 344, 348
pencil of rays 224, 258, 279ff, **414**–17
pendulum 196(p)
penetrating mode **576**, 582(p)
penetration 334, 424
penetration frequency 14, 15, 194, 199, **332**, 357f, 445, 457ff
perfect differential 184, 506
permeability 326(p)
permittivity
electric 33
of vacuum ε_0 22
permittivity matrix, tensor 50, 52–7, 146
Petit, M. 624
Pettersen, G.W.E. 435, 628, 629
Pfister, W. 504, 587, 633
phase 26, 109, 168, 170, 175
planes of constant 36
see also complex phase
phase advance on reflection 195, **218**, 221f, 262, 285
phase change on reflection 316, 474
phase height **329**, 333, 353–354(p), 465, 571
phase integral formula
for coupling 501
for reflection **220**, 456, 458, 464, 474, 510
phase integral method **218**ff, 541
for coupling 223, 387f, 489, **499**–502
for reflection 195, **218**ff, 456–9
phase memory **167**ff, 189, 193, 220, 256, 402f, 429, 505f
phase path **278**, 286, 320, 329, 421, 424, 426, 432, 465
phase velocity 129; *see also* wave velocity
Phelps, A.V. 10, 58, 633
physical optics 173, 223ff, 556
Pickle, C.B. 365, 633
piecewise linear models 443–6
Piggott, W.R. 361, 365, 603, 633, 634
Piliya, A.D. 80, 596, 624
Pitteway, M.L.V. 13, 63, 146, 152, 161, 319, 492, 543, 551, 561, 574, 575, 576, 579, 603, 633, 634
plane polarised *see* linear wave polarisation
planes of incidence 295, **439**
of constant amplitude **36**, 37(p)
of constant phase 36
plane wave 3, 29, 30, 66, 165
planetary waves 6, 20
plasma frequency 3, 39, 48
plasma instability 35, 47
plasma oscillations 3, 48ff, 63(p), 83
plasma physics xiii, 4, 481
plasma wave 5, 49, 80, 596
plasmapause 18
plasmas 3ff, 35, 47
plasmasphere 18
plate-like cavity 25
Platteville, Col. 47
Plumpton, C. 2, 622
pocket calculator 55, 307, 347
Poeverlein, H. 264, 267, 270, 556–8, 634
Poeverlein's construction **264**ff, 274, 543, 592
Poincaré criterion, definition **208**, 213f, 250, 454, 462
point at infinity 207f
point of inflection
in *q* curve 520
in refractive index surface 106, 160
in ray surface 113
point source 111, 280f, 313, 316, 327(p), 414ff, 418ff, 541
polar cap absorption (PCA) 19
polar coordinates 290, 408f, 563f
polarisability of molecules 1, 40
polarisation

electric *see* electric polarisation
of whistler 377, 381ff, 386, 391
on ray 261, 430
see also limiting polarisation
polarisation ellipse 37(p), 72, 75, 587
major axes 74
polarisation equation **70**ff
polarisation fading 435
polarisation, reversal of sense 382–387
polarisation term 40
polarisation, wave **68**ff, 99–100(p), 399(p), 509, 529, 533
pole
near saddle point 238f, 315
of ψ 512, 514, 518(p)
of refractive index *see* infinite refractive index, infinite root, resonance
polynomial method for finding $N(z)$ 370f
positive ions 54ff, 379
Postan, A. 288, 518, 553, 613
Poston, T. 288, 499, 634
potential function 230, 402
potential barrier 459, 533
power flux in ray pencil 279–81, 415
power input to plasma **32**, 33, 62, 63
Poynting's theorem 34
Poynting vector 26, **33**ff, 76, 100(p), 108, 111, 168, 178f, 185–7, 254, 292(p), 401, 413, 415, 426, 594
precursor 323
prediction of MUF 348–352
predominant frequency 128
predominant values 110, 254, 258
predominant wave 257, 314, 316, 320
predominant wave normal 110, 391
pressure of radiation 47
Price, G.H. 553, 634
principal axis of magnetoplasma **50**ff
principal axis coordinates 63(p), 64(p)
principal axis elements **53**ff
principal rays 429
principal terms of coupled equations **480**, 482
principal values of integral 342
probes 13
probing frequency 356ff
probing the ionosphere 80, 328ff
Proc. Inst. Elect. Electronic Engrs. 368, 634
progressive wave 29, 30, 35, 37(p), 66, 162, 165f, 169, 170, **174**, 176, 183, 186, 227(p), 402, 412, 442, 482
projected path 268, 275–7
prolate refractive index surface 157
propagation constant, k **29**, 130, 481
propagator 559
protonosphere 380
protons 4, 19, 379, 381
proton whistler 377, **380**
pseudo rays 422–424
psi (ψ) coupling parameter 484, 490, **511**f, 518(p), 544
psi function 465
pulse 128ff, 258, 328ff, 354(p), 394, 550, 571; *see also* wave packet
pulse shape 320–5

q(Booker's variable) 142ff, **143**, 166ff, 197ff, 263, 301, 407, 439, 483, 486, 554, 580
as eigen value 183, 521, 561
equal values 262, 265, 521
zeros of 262
quadratic approximation for refractive index surface 538
quadratic equation
for n^2 **77**f, 103, 593
for polarisation **70**, 71–4
for q^2 151, 518(p)
for qs 486, 539
quantum mechanics *see* wave mechanics
quartic equation 419; *see also* Booker quartic
quasi-longitudinal (QL, QL2) 95ff, 494
quasi-transverse (QT, QT2) 95ff, 494

Ra, J.W. 317, 634
radar 13
radiation field 431
radiation pressure 47
radio astronomy 290, 292, 328, 375, 535
radio fadeout 19
radio windows 83, 85, 150, **268**, 495, 505, 521, 529, **532–43**
Radio Science 47, 407, 450, 634
Radio waves in the ionosphere (RWI) xiii, xiv, many references in text, 617
Raghuram, R. 603, 634
random thermal velocities 4, 39, 62
rank
of matrix 526
of tensor 414
Rao, M.M. 608, 635
Rao, R. 137, 623
Ratcliffe, J.A. xvi, 9, 19, 20, 41, 43, 68, 89, 298, 331, 339, 349, 365, 394, 397, 497, 556, 587, 603, 616, 618, 622, 627, 635
rationalised units 22
Rawer, K. xiii, 11, 13, 14, 15, 43, 69, 75, 89, 280, 290, 365, 368, 393, 472, 475, 476, 497, 519, 559, 634, 635, 638
ray 103(ch. 5), 108ff, 110, 129, 131, 148, 254ff, 257ff, 594
entering ionosphere from free space 407
ray path 255–277, **257**, 263
horizontal 262
in magnetic meridian plane 263f
equations 257, 259
not reversible 507
ray pencil 224, 255, 413, 418, 429
signal intensity in **414**–417
ray refractive index **111**, 374, 410f, 426, 429
ray surface 108ff, **111**ff, 115, 116–124, 136, 139(p), 376
ray theory 168, 193f, 222, 382, 428–435
anisotropic ionosphere 356(ch. 13)
isotropic ionosphere 328(ch. 12)
ray tracing 222, 380
general 351, 400(ch. 14)

stratified medium 195, 254(ch. 10)
through reflection level 262
uses 255, 401
ray tube 432
ray vector 111, 112
ray velocity **111**, 115, 129, 140(p), 403f, 409f
Rayleigh, Baron 169, 170, 635
rays, horizontal, field near 285f
real pseudo rays 422
reciprocal property of surfaces 112
reciprocity 261, 278, **576ff**
with full wave solutions 576–97
with rays 428–35
reciprocity theorem 428, **430ff**, 576
recombination coefficient, electrons 8
rectangular pulse 322–325
Reddy, C.A. 608, 635
Reed, J.E. 312, 615
reference free space 534, 535
reference level
for reflection coefficients **296f**, 313ff, 455, 458 462, 477f
for transmission coefficients 297f
reference line 264ff, 495
reflection 152, 178, 485
at discontinuity of gradient 441–443
at sharp boundary, isotropic medium 172, 178, 304ff
at sharp boundary, anisotropic medium 307–312
at vertical (or normal) incidence 308–312, 489ff
of pulse, wave packet 26ff
reflection coefficient(s) 172, 174, 194f, 218, 221, 295(ch. 11), 441, 455, 462, 472ff, 552, 579, 600–2
reflection coefficient matrix
$\boldsymbol{R}$ **299ff**, 303f, 307–9, 325(p) 557, 562f, 566, 568–72
$\boldsymbol{R}_0$ 301, 309, 325(p), 585
$\boldsymbol{R}_c$, $\mathscr{R}$ 309–311, 573, 607
reflection level 149, 256, 280, 282, 357, 442, 447, 450, 459
reflection point (turning point) 79, 152, **153**, 458, 480, 483, **492ff**, 496ff, 592
reflection process 166, 169, 175, 178, 197, 220f, 442f, 447
refraction of satellite signals 13
refractive index **29**, 66(ch. 4), 74ff, 103(ch. 5)
as vector 78, 103, 108, 109, 113, 146, 254, 403f, 408, 429
dependence on frequency 124ff
dependence on electron concentration 80ff
effect of collisions 86ff, 89ff
equal values 79
infinity of 79
isotropic plasma 67ff
of air 1
space 104, 264, 403f, 408
surface **104ff**, 110, 114, 116–24, 131, 136, 139(p), 146, 147, 264, 404, 416, 425, 428, 538, 592
zeros of 79, 357
see also group refractive index
region I, I(a), II 588
regular singularity 206f, 447, 467
Reinisch, B.W. 13, 372, 615, 627, 635
relative velocity of electrons and ions 62
remainder in series 208
removal of electrons 8, 15
repeated suffixes 33, 188, **414**, 505
resonance 5, 6, **79**, 112, 116, 127, 132, 133, 135, 427, 446, 579–81, **591**–602
in topside ionograms 368
resonance cone 105, 106, **113**, 117–119, 135
resonance frequency 135, 140(p)
resonance point 160, 493, 495, 553, 593, 607f
resonance tunnelling 427, 596–602
retarding force on electron 10, 43
reversal of sense (wave polarisation) 381–90
reversed resonance cone 106, **113**, 117, 118, 120, 121, 135, 157
reversed skip 343
reversed Storey cone 106, 117, 118, 122, 124, 134, 136
reversibility of ray path 261, 428
reversion of series 249
Riemann surface 420, 424
right-handed polarisation 69
riometer 13
Rishbeth, H. 4, 7, 8, 15, 20, 623, 635
Robbins, A. 370, 635, 639
rockets xiii, 10, 13, 379, 606f
Rodriguez, P. 390, 636
Rönnmark, K. 407, 636
Ross, W.J. 375, 623
Rota, G.C. 184, 615
rotation of axes 71, 519(p)
rounding errors 405
Roy, S.K. 59, 61, 621
Royal Society 22, 636
Rudsinski, L.I. 452, 623
rule of false position 401, 436(p)
Runge–Kutta–Merson method 552f
Rusch, W.V.T. 603, 631
Rutherford Appleton Laboratory xv, 349
Rydbeck, O.E.H. 304, 322, 457, 463, 464, 477, 503, 504, 514, 636

S-matrix 560
saddle point **231**, 237, 242ff
Saha, M.N. 519, 636
satellites 13, 356, 375, 379, 380
scale height **7**, 11, 536
Scarabucci, R.R. 391, 519, 613
scattering from irregularities 20, 312, 505, 518
Schatzberg, A. 579, 613
Schelkunoff, S.A. 301, 571, 636
Schmerling, E.R. 368,627
Schrödinger equation 167, 169
Schwarzian derivative **176**, 216
scintillation 1, 20
screening of ions in plasma 45
sea surface 325(p)

Seaton, S.L. 349, 616
$sech^2$ distribution 332–5, 475f,
second order coupled equations 481, 491, 519(p)
second order differential equations xiii, 165, 170, 438(ch. 15), 584ff, 593ff
second window 273, 538, **542**f
seismic waves, seismology 6
semi-thickness *see* half thickness
Sen, H.K. 5, 58, 59, 60, 636
Sen–Wyller formula **57**ff, 60, 145, 182, 312, 554
separation
  constants 144
  into upgoing and downgoing 174f
  of differential equations 439, 583f, 588
  of variables 143
series solution of differential equation 201, 207f, 448
shape of pulse 320–5
Sharaf, A.L. 438, 466, 636
sharp boundary, reflection at 178, 296, 304–11, 473–5, 570
sharply bounded anisotropic medium 307–12
sharply bounded isotropic medium 304ff
sharply bounded model ionosphere 304, 319
Shaw, I.J. 397, 635
Shaw, R.R. 534, 624
Shawhan S.D. 380, 382, 624, 637
Shellman, C.H. 604, 636
Shinn, D.H. 131, 360, 398, 636
Shkarofsky, I.P. 5, 11, 42, 636
Shmoys, J. 448, 627
shock front (earth's bow) 17
shooting method 401, 418
short wave fadeout 19
short whistler 377
S.I. units 22, 301
side-band frequencies 323
signal intensity 414–17, 432
silvered glass 326(p)
simple reflection 572
single coupling point 520, 566
singular matrix 485f, 522, 525, 572, 574
singular point, singularity 200, 223, 230, 447, 450, 517, 593
singularities 200, 205, 480, 493ff, 566
sinusoidal layer, distribution of $N(z)$ 476
six-dimensional space 108, 403, 425
sixth degree equation 106, 111
skip distance 280, **288**f, 343, 345, 348, 425
slab model of ionosphere 326f(p)
slit, diffraction by 322
slowly varying function 109, 240
slowly varying medium 38, 165(ch. 7), 170, 172, **178**, 456
Sluijter, F.W. 447, 467, 602, 640
Smith, E.K. 14, 15, 61, 365, 637
Smith, L.G. 607, 632
Smith, M.S. 60, 153, 155, 189, 261, 492, 494, 495, 499, 503, 506, 507, 518, 520, 521, 526, 532, 539, 572, 580, 581, 592, 618, 637
Smith, N. 41, 349–51, 637
Smith, R.L. 44, 57, 379, 380, 382, 383, 603, 614, 624, 634, 637
smoothing out in a plasma 24, 25, 40, 44, 48
Snell's law 142, 254, 260, 264, 305f, 346, 403, 405, 418, 421f, 564f,
Snyder, M.A. 372, 637
solar corona 535
solar flare 19
solar wind 3, **17**, 19
Sommerfeld, A. 3, 170, 239, 253, 402, 637
Sommerfeld contour integral 252(p)
sound waves 6, 167, 196(p), 414
source
  at infinity 421
  dimensions of 255
  distributed 517
south-north propagation 150, 155ff, 163(p), 485, 579
space charge (in plasma) 3
space vehicle xiii, 134, 156, 266, 429
spatial dispersion **38**, 113, 398(p)
spectral analyser, spectrogram 133, 377ff, 380
Spencer, M. 20, 617
spherical distribution of charge density 41
spherical polar coordinates 290, 408f, 563f
spherical refractive index surface 104, 105
spherical stratification 260, 290, 292(p), 564f
spherical wave 3, 141, 313ff
spikes 368
spiral 225–7, 506; *see also* Cornu spiral
spitze 150, **266**ff, 495
splitting *see* magnetoionic splitting
sporadic E, E_s 15, 20, 354(p), 365, 380, 606–8
Spracklen, C.T. 20, 628
square law N(z) 477, 587
square root in Appleton–Lassen formula 88, 153, 494
standing wave 30, 227(p), 569f
Stanley, J.P. 587, 637
starting solutions 551, 560ff, 568, 574
stationary phase 109, 129, 130, 137, 236, 252, 254, 257f, 330
  method of **247**, 283, 285
stationary time *see* Fermat's principle
steepest descent, ascent, lines of 230ff
steepest descents 110, 214, 229(ch. 9), **238**ff, 418, 591
  double **251**ff, 283, 285, 421
  higher order approximations **249**ff
Stegun, I.A. 205, 233, 236, 249, 360, 457, 463, 465, 612
step size 552f, 568, 570, 580, 608
step-by-step process for integration 401, 551
Stepanov, K.N. 62, 612
Stevens, E.E. 365, 637
Stewart, I.N. 288, 634
Stirling's formula 253(p), 456, 474
Stix, T.H. 5, 55, 637
Stokes: name used for two distinct purposes 200
Stokes, G.G. 200, 210, 211, 637
Stokes constant *see* Stokes multiplier
Stokes diagram 200, **211**, 220, 454, 460f, 535f
Stokes (differential) equation 178, 200–4, 212,

214f, 227f(p), 241–7, 439–41, 443–5, 447, 450, 467
Stokes line 200, 206, **210**ff, 214, 220, 228(p), 242, 245, 451, 454, 461, **500**
Stokes multiplier **212**ff, 216, 221, 228(p), 246, 448, 450ff, 454, 462, 528
Stokes phenomenon 200, **209**, 220, 241, 246, 467, 501, 528
Stoneley, R. 6, 638
storage fields 431, 556
stored energy
in plasma 32, 63, 596
in wave 179
Storey cone 106, 116–19, 122, 124, 134, 136, 139(p)
Storey, L.R.O. 106, 376, 638
Stott, G.F. xvi, 56, 57, 62, 89, 104, 106, 109, 111, 112, 115, 135–7, 140, 251, 419, 612, 618, 638
Straker, T.W. 298, 587, 603, 614, 616, 638
strata 142, 143, 146, 162, 172ff, 301
stratified medium 141(ch. 6), 143, 165(ch. 7), 216f
Stratton, J.A. 32, 34, 137, 638
string 195(p)
strong coupling 483
structure of ionosphere **14**ff
Strutt, J.W. *see* Rayleigh, Baron
Stuart, R.D. 603, 641
Sturrock, P.A. 368, 638
subdominant term **210**, 213, 425, 454
subprotonic whistler 156, 266, **379**
subscript notation 33, 51, 108, **414**, 426, 576
successive approximations 171, 175, 480f, 509, 515ff, 558
Suchy, K. xiii, xv, 11, 13, 15, 43, 69, 75, 89, 185, 325, 365, 368, 393, 404, 407, 412, 414, 427, 428, 497, 519, 559, 576, 579, 613, 635, 638
sudden commencement 20
sudden ionospheric disturbance (SID) 19
suffix notation *see* subscript notation
summation convention 33, 51, 188, **414**, 505, 576
sunspot number 16, 19
sun
atmosphere 514
radiation 7
zenith angle χ 9
surface integral 34, 576f
surface of revolution 104, 112
surface waves 3, 6
susceptibility matrix **49**ff, 146, 182, 426
Sutton, D.J. 58, 620
swamping, numerical 558–60, 568, **574**f
symmetric tensor 113, 576
symmetrical ionosphere, distribution *see* parabolic distribution, sech^2, sinusoidal layer

tables of Airy integral functions 205
of $N(z)$ 361
tail of magnetosphere 18
tail of pulse 323
Tang, T. 593, 638
Taylor series 95, 231, 239, 247, 321, 555
Tellegen, B.D.H. 393, 638
temperate latitudes 504
temperature
in ionosphere 4, 14
of electrons 394
tensors 414–17
terrestrial radiations 534
Terry, P.D. 252, 391, 417, 418, 422, 424, 425, 428, 494, 535, 541, 618, 638
thermal motions 394, 395
thermo-nuclear plasmas 535, 597
thin layer 354(p), 475, 606
Thomas, J.O. 370, 639
Thomas, L. 603, 639
Thomson scatter *see* incoherent scatter
Thrane, E.V. 361, 394, 603, 614, 633, 634, 639
three equal roots of Booker quartic 150, 159, 160, 495, 526
tides 20
tilt angle 73ff
tilted ionosphere 365
time average
of energy flow 36, 37, 76
in harmonic fields 32, 34
time of travel
of pulse or wave packet 255, 361; *see also* equivalent height, equivalent path
of wave crest or wave front 111, 403, 409–12; *see also* phase height, phase path, ray velocity
Titheridge, J.E. 13, 138, 370, 371, 375, 514, 547, 630, 639
top of trajectory 342, 344
topside ionogram 367, 369, 371f
topside of ionosphere xiii, 13, 365ff
topside sounding 13, 356f, **365**ff, 503
topside Z-trace 367f
Toshniwal, G.R. 503, 504, 639
total electron content 375
total internal reflection 307, 317
total reflection 586
trace of matrix 519(p)
trailing diagonal of determinant, matrix 184, 186
trains of whistlers 378
transforming matrix 485, 487, 526, 532, 557f, 573
transition collision frequency 71ff, 86ff, 496
transition cone 104, 115
transition direction of wave normal 90ff
transition height 496, 513f
transition through critical coupling 498f
transition values of $S = \sin\theta$ **156**, 158–60, 532
transmission coefficient 295(ch. 11), 421, 459, 462, 472f, 568, 579, 600–2, 607
transmission coefficient matrix **299**ff, 303, 310
transparency of radio window **535–40**, 542f
transport area of cross section 60, 64(p)
transpose of matrix 183
transverse propagation 73, **76**, 80, 81, 84, 89, 124, 132, 392; *see also* east-west propagation

travelling ionospheric disturbance, TID 7
trial and error 603
triangulated path 346
troposphere 175
true height of reflection **329**, 353–54(p)
truncation errors 553
tunnelling 439, 459, 533, 593, 596–602
turning point of ray (also called 'coupling point', 'reflection point') **152**, 458, 480, **492**ff
turning points of Ai, Bi, 205f
Turunen, T. 607, 627
Tuve, M.A. 13, 346, 617
two separate parabolic layers 336, 363

ultra-violet radiation 1, 9, 19
uniaxal medium **57**, 100(p), 115, 116, 589
uniform approximation 216ff, **218**, 282, 286, 330, 450, 457, 488, 499ff, 520, 524, 528f
unitary matrix 51, 300, 586
unitary transformation 301
units **22**, 301
upgoing wave 148, 263, 561, 592
upper bound of error 177
upper hybrid frequency, resonance **54**, 57, 81, 124, 126, 127, 139(p), 368
Ursell, F. 251, 286, 619, 639
Uscinski, B.J. 639

vacuum modes 556f, 562, 570
validity of W.K.B. solutions 176ff, 181
valleys 232, 234, 237, 247
  in $N(z)$ 371
van Duin, C.A. 447, 467, 602, 640
variation of parameters 515ff
Vats, H.O. 97, 98, 616
vector diagram *see* amplitude–phase diagram
vector, refractive index as 78, 103, 108f, 113, 146, 254, 402ff, 408
velocity dependence of collision frequency 5, 11, 43, 57ff
vertical component of group velocity **131**, 391
vertical dipole aerials 578f, 602
vertical incidence 131, 192ff, 200, 273f, 329ff, 356–72, 391–3, 449, 483ff, 489ff, 493, 499ff, 507ff, 544ff, 584–7, 592, 607; *see also* normal incidence
vertical magnetic field 309, 485, 518(p), 583–91
vertical motions of electrons 440, 589
vertical polarisation 166, 180f, 200, 221, 301, 446–53
vertical propagation in ionosphere 73, 80, 86, 303, 329ff
vertical tangents to q curves 150, 157, 158, 160, 495, 520, 533
vertical wave normal 379, 383
very low frequency (VLF) 13, 16, 60, 194, 223, 304, 317, 424, 438, 453, 496ff, 550(ch. 18), 568, 570, 580, 585, 602ff; *see also* low frequency
Vice, R.W. 312, 573, 622
Vickers, M.D. 370, 639
Volland, H. 558, 560, 640
von Roos, O. 478, 640

Wait, J.R. 175, 222, 447, 453, 554, 564, 640
Walker, A.D.M. 107, 111, 124, 137, 139, 156, 182, 376, 380, 565, 640
Walker, R.L. 51, 631
Walkinshaw, W. 1, 261, 564, 565, 616
Walkup, J.F. 379, 619
Walsh, E.J. 558, 575, 640
Walters, L.C. 97, 308, 554, 564, 628, 640
Wand, I.C. 12, 638
Wang, T.I. 390, 531, 640
wanted wave 394
Warber, C.R. 287, 604, 622, 640
warm plasma xiv, 80, 99, 596
Warren, R.E. 287, 604, 622, 640
water vapour 1, 175
Watson, G.N. 205, 338, 342, 448, 454, 455, 457, 462, 467, 477, 528, 598–601, 640, 641
Watson's lemma 250
Watt, J.M. 552, 624
Watts, J.M. 13, 365, 602, 620, 641
wave admittance 301ff, 442, 444, 533
wave admittance matrix 302ff
wave crest, trough 403, 409
wave front from point source 111, 410
wave guide xiv, 1, 68, 129, 222, 297, 376, 565
wave impedance 30, 529; *see also* wave admittance
wave interaction xiv, 5, 10, 11, 13, 58, 393–8
wave mechanics 6, 129, 167, 169, 221, 443, 445, 459
wave normal 110
wave packet 129, 130, 137, 221, **258**, 278, 320–5, 328ff, 354(p), 391, 412, 418
wave velocity 34, 67, **111**
wave length
  in free space 289, 553
  in medium 330, 553
Waynick, A.H. 73, 514, 629
Weber's equation 228(p), **457**, 467, 477, 522, 527
Weekes, K. 298, 587, 603, 616, 641
Weisbrod, S. 394, 641
Weitzner, H. 535, 536, 641
Wentzel, G. 169, 180, 641
west-east propagation 153ff, 163(p), 326(p), 554, 564
Westcott, B.S. 438, 447, 453, 466, 475, 556, 641
Westcott, C.H. 603, 614
Westfold, K.C. 52, 641
Whale, H.A. 131, 360, 636
Whipple, R.T.P. 587, 588, 590, 591, 626
whistler 4, 42, 106, 119, 134–6, **376**–80, 565, 576, 579
whistler mode 85, **86**, 106, 140(p), 158, 160, 273, 367, **376**, 497, 514, 542
whistler polarisation 377
whistler ray paths 376
White, H.E. 173, 373, 628
White, R.B. 448, 597, 640
Whittaker, E.T. 338, 342, 448, 457, 462, 467, 477, 528, 600, 601, 641

Whitteker, J.H. 368, 627
Wieder, B. 347, 641
Wilkes, M.V. 9, 477, 587, 590, 618, 642
Willmore, A.P. 13, 642
Wilts, J.R. xv
window frequency 54, 56, 105, 136
window point 83, 85, 105, 119, 122, 125, 134, 150, 155, 156, 268ff, 293(p), 495, 537f
windows, radio 83, 85, 150, **268**, 495, 505, 521, 529, **532–43**
wind shear 15
W.K.B. method 170f
W.K.B. ranges 572
W.K.B. solutions xiii, 165(ch. 7), 169ff, 219, 243, 256, 388, 402, 441f, 481, 488f, 516, 528f, 543, 572, 588, 598
  anisotropic ionosphere 187ff, 192, 483
  at great height 551
  conditions of validity 176ff, **177**, 181, 251, 489
  derivation 168f, **170**f, 172–177, 180f, **187**ff
  failure of 174, 192, 507
  for parabolic $N(z)$ 460
  history 170
  of Stokes equation 206, 209, 441
  properties 178ff
  source not at infinity 286
  vertical incidence 192ff, 451
  vertical polarisation 180f
Wolf, E. 114, 115, 616
Wong, K.C. 593, 638
Woodward, P.M., and Woodward, A.M. 205, 642
Wright, J.W. 13, 365, 642
Wronskian 227f(p), 515
Wüster, H.O. 448, 449, 466, 623, 642
Wyller, A.A. 5, 58–60, 636

x-rays 1

Yabroff, I.W. 308, 376, 642
Yeh, K.C. 20, 642
Yokoyama, E. 304, 642

Z-mode **84**, 157f, 163(p), 268, 357, 533, 540, 592
Z-trace 357ff, 362, 367f, 485, **503**–5
Zacharisen, D.H. 13, 620, 642
zenith angle of sun χ 7
zero order approximation 516
zero
  of Ai and Bi 205f
  of ε_1 ε_2 ε_3 54, 56
  of q 149, 155, 198
  of refractive index 79, 357
Ziolkowski, R.W. 281, 642
Zucker, F.J. 430, 577, 642

in the United States
okmasters